MW01599163

Drosophila Cells in Culture

Front cover illustration: Dividing *Drosophila Kc* cell (anaphase). Confocal microscopy after double staining. The centrosomes at the two poles of the spindle appear in yellow-green (detected with antibody against γ-tubulin). The two separating sets of chromosomes are stained in red (propidium iodide). Our *Kc* line has remained primarily diploid. (1 cm = 1.7 μm). Courtesy of Dr. Alain Debec (URA II35, CNRS-Université Paris VI).

Copyright © 1997 by ACADEMIC PRESS

Academic Press
a division of Harcourt Brace & Company
15 East 26th Street, 15th Floor, New York, New York 10010, USA
http://www.apnet.com

Academic Press Limited
24-28 Oval Road, London NW1 7DX, UK
http://www.hbuk.co.uk/ap/

Library of Congress Cataloging-in-Publication Data

Echalier, Guy.
    Drosophila cells in culture / by Guy Echalier.
        p.   cm.
    Includes bibliographical references and index.
    ISBN 0-12-229460-2 (alk. paper)
    1. Drosophila--Cytology--Technique.   I. Title.
QL537.D76E23   1997
571.6'15774--dc21   96-47587
                                        CIP

PRINTED IN THE UNITED STATES OF AMERICA
97 98 99 00 01 02 BB 9 8 7 6 5 4 3 2 1

Drosophila Cells in Culture

Guy Echalier
Groupe de Génétique Cellulaire et Moleculaire
Unité Associée 1135
Centre National de la Recherche Scientifique
Université Pierre et Marie Curie
Paris, France

Academic Press
San Diego New York Boston London Sydney Tokyo Toronto

A Catherine,
pour m'avoir "supporté"—au sens français
comme au sens anglo-saxon du terme—tout au
long de cette histoire, depuis les premières mises a
pondre de Mouches, après minuit dans la salle de
bains familiale, jusqu'aux ultimes "ruminations"
de ce manuel . . .
Avec tout mon amour,
Guy

Contents

2 Primary Cell Cultures of *Drosophila* Cells

3 *Drosophila* Continuous Cell Lines

5 Biology and Biochemistry of Cultured Cell Lines: 1. Nucleic Acids

6 Biology and Biochemistry of Cultured Cell Lines: 2. Proteins

8 Experimental Models of Gene Regulation: 2. Cell Responses to Hormone

9 Gene Transfer into Cultured *Drosophila* Cells

10 Transposons

11 *Drosophila* Viruses and Other Infections of Cultured Cells

Introduction

Cellulairement
P. Verlaine

The French adverb "cellulairement" ("cellularly" in English), which provides the epigraph to this Introduction, is a deliberate misappropriation of the title of a famous collection of poems written by one of our great poets, Paul Verlaine, when in jail. It applies exactly to the content of this book: that is to say, the cellular approach that has been used in the diverse and successful "assaults" made on *Drosophila* in the past few decades in relation to major biological problems.

THE INEVITABLE "COMEBACK"
OF *DROSOPHILA*

Around 1910 the small "vinegar fly" was adopted by T. H. Morgan and his group of early geneticists because of its ease of handling (small size, cheap food), its rapid generation time (12 days), and the simplicity of its chromosome complement. For almost a century, generations of investigators throughout the world accumulated vast volumes of information on the biology, primarily the genetics, of this organism, which has no rival among the higher eukaryotes.

When molecular biology began to emerge, the "superstar" *Drosophila melanogaster* went into eclipse. First a "reductionist" approach had to be developed on the genomes of bacteria and viruses in order to unravel the fundamental relation existing between nucleic acids and protein synthesis, and also certain basic mechanisms of gene regulation. However, during the 1970s, a number of technical advances, such as the use of "restriction enzymes," enabled molecular biologists to cope with the much more complex genomes of higher organisms. One of the main areas of contemporary research in biology has become the analysis of the intricate processes whereby "differentiated" tissue cells control the initiation of their specialized program of gene expression and its maintenance throughout development.

In this pluridisciplinary "attack," *Drosophila* has recovered all its assets, the first of which is the relative simplicity of its karyotype with its four easily identifiable chromosome pairs. The unusual smallness of this genome favors biochemical approaches: 0.18 pg of nuclear DNA (per haploid genome) of a *Drosophila* cell is approximately halfway between the genetic materials of bacteria and mammalian cells. Broadly speaking, its DNA content is only about 50-fold higher than that of *Escherichia coli* and 30 times smaller than that of mammals (6 to 7 pg in humans).

The unhoped for rediscovery of giant polytene chromosomes* in its larval salivary glands, in just the animal on which the chromosome theory of heredity had been built, permitted the exceptional precision of their cytogenetic mapping. Today, with the *in situ* hybridization method, the accurate chromosome site of any given DNA sequence can be routinely defined. Moreover, "puffing" is a visible manifestation of gene activation.

The main advantage of *Drosophila* still lies in the extreme sophistication of its genetics and the availability of numerous mutations (spontaneous or induced) affecting all aspects of its development and physiology. In addition, current use, as a vector, of the transposable element P with its well-controlled mobility (see Chapter 11) has greatly facilitated transgenesis.

---

* Very few organisms possess such polytenic chromosomes (see Ashburner, *1970*). On this short list is also a plant, the pea, Mendel's experimental material! When, in the 1930s, polytene chromosomes were "rediscovered" in Diptera such as *Drosophila,* N. W. Timoféeff-Ressovsky, a renowned Russian geneticist, used to say, as a joke, that the presence of giant chromosomes in genetic's animal of choice was, according to him, "the cytological proof of the existence of God"! (this was attributed to this author by the late Professor E. Hadorn from Zurich, a great pioneer in the developmental biology of *Drosophila*).

Last, it should not be forgotten that the many different developmental stages of this higher insect, which are due to its total metamorphosis, offer a wide variety of valuable experimental situations. Two quite separate organisms—the semiaquatic maggot and the winged fly—are successively constructed, by two quite distinct cell populations, from the information contained in this one tiny genome.

Such is the rich context in which *Drosophila* cell cultures were developed and are now extensively used. They can greatly facilitate biochemical investigations since they provide unlimited amounts of homogeneous cellular material. Moreover, especially with the DNA-mediated method of gene transfer, they give rise to multifarious bioassays as will be attested to throughout this book.

BRIEF HISTORICAL SURVEY OF *DROSOPHILA* CELL CULTURE

A few years after the epochal *in vitro* culture of the amphibian embryonic neural tube by Harrison (*1907*), whereby he demonstrated the growth of axons from the somas of neuroblasts, thus providing direct proof of the structural unity of the neuron, Goldschmidt (*1915*) reported silkworm spermatogenesis in similar "hanging drop" cultures. This latter work may be considered as the birth of invertebrate tissue culture.

Of the many pioneers of the following decades, I will quote only two, because their work, in spite of the rudimentary techniques involved and limited cell survival, influenced the major aims of insect cell culture. W. Trager, as early as *1935,* studied the development of the grasserie virus in small explants from silkworm ovarian tissue. Carlson, in *1946,* observed *in vitro* multiplication of neuroblasts from the embryonic neural cord of a grasshopper [a material which in developmental biology is still used to characterize the recognition/adhesion molecules that mediate the progression of axon fascicles (see Chapter 6, Section I,B)].

A decisive breakthrough in cell culture techniques was made when, during the 1950s, cell monolayers were grown in liquid culture media. Once more, insect cell cultures took about 10 years to catch up with mammalian cell technology, which is not surprising since the number of laboratories involved in the field was much smaller.

Moreover, it is important to emphasize that no significant success was achieved until the specific biochemical features of insect body fluids were taken into consideration. Thus, a detailed analysis of *Bombyx* larval hemolymph resulted in the designing by the Wyatts of an original culture medium in which ovarian sheath cells were able to migrate and multiply (S. S. Wyatt, *1956;* Wyatt and Wyatt, *1976*). Finally, a few slight modifications in this "recipe" allowed T. Grace, in 1962, to grow *in vitro* the first permanent insect cell lines from ovaries of a large lepidopteran, *Antheraea.*

As for *Drosophila* tissues, early attempts at culture were focused on imaginal disks and their capacity for morphogenesis *in vitro*. Among precursory work, the numerous short notes by Kuroda (see the Bibliography), the limited successes of Cunningham (1961) and Hanly (1961), and the much more substantial papers of I. Schneider (1963, 1964, 1966) should be remembered.

The first *bona fide** primary cell cultures from dissociated *Drosophila* embryos were obtained by Echalier *et al.* (1965) in Paris, then by the Russians Gvozdev and Kakpakov (1968a,b). The establishment of continuous cell lines was reported in the spring of 1969 by Echalier and Ohanessian (1969, 1970) and, only a few months later, by Kakpakov *et al.* (1969; see also Gvozdev and Kakpakov, 1970; see Fig. I.1).

From then on a large number of interesting new *Drosophila melanogaster* cell lines were grown, especially by I. Schneider (1972) in Washington, Barigozzi's laboratory (Mosna and Dolfini, 1972a,b) in Milan, and Professor Sang and collaborators in Great Britain. It is difficult to draw up a complete catalogue of available *Drosophila* cell lines (see a tentative list in Chapter 3). Most of them are derived from dissociated wild-type or mutant embryos. Yet, Gateff *et al.* (1980) succeeded in establishing two lines from mutant tumoral hemocytes, and, more recently, a series of lines was obtained from imaginal disk cells (Ui *et al.*, 1987; Currie *et al.*, 1988) and from larval neural material (Ui *et al.*, 1994). Even haploid cell lines are available (Debec, 1982). In addition, several cell lines could be grown from other *Drosophila* species. The particular merits of all such experimental material will be discussed in this book.

RELIABILITY OF *DROSOPHILA* CULTURED CELLS AS EXPERIMENTAL MATERIAL

When it was shown, at first by A. M. Courgeon (1972) in our laboratory, that some *Drosophila* cell lines, such as *Kc* cells, are sensitive "tar-

---

* During the late 1960s, the issue was confused by a series of papers in which the claim was made that long-term cultures of *Drosophila* embryonic cells were obtained with disconcerting ease. The whole story sank into deserved oblivion, and therefore no further mention will be made of it in this book, although these papers are all listed in the Bibliography.

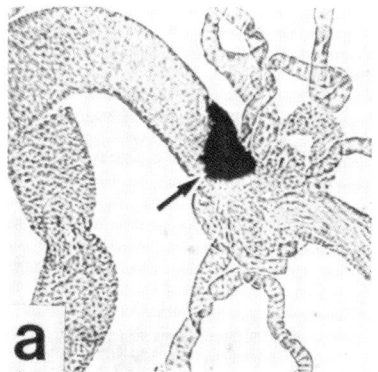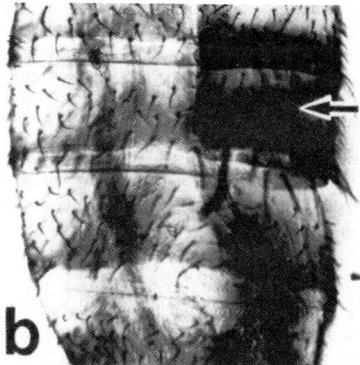

FIGURE I.1   Chimeric *Drosophila* larva and fly with tissue contribution from nuclei of *in vitro* cultured cells after transplantation into *y sn3 mal* cleavage embryos. (a) Larval posterior midgut with a cluster of about 30 cells (arrow) clonally derived from a nucleus of Schneider's cell line *S1*. The nuclear transplant area is histologically visualized by the presence of aldehyde oxidase in these cells, whereas the surrounding tissue of the host does not express this enzyme due to the mutant gene *maroon-like (mal)*. (b) Abdomen of a female fly whose 2nd and 3rd tergites (arrow) partly originated from a nucleus of Echalier and Ohanessian *Kc* cell line. The nuclear transplant area shows distinct deviations in size and pattern from normal cuticular pigmentation and bristle formation, cryptically expressing the *Minute* phenotype of the haplo-IV genotype of the *Kc* clone 52-84 [from Illmensee K. (1978) *In "Genetic Mosaics and Cell Differentiation"* (W. Gehring, ed.), Springer-Verlag. Reproduced with the kind authorization of Dr. K. Illmensee and Springer-Verlag].

gets" for ecdysteroid hormones, many traditional insect endocrinologists questioned the physiological significance of these *in vitro* hormone–cell interactions. The strict specificity of such hormonal stimulations was, however, soon demonstrated. Not only did cultured cells respond exclusively to steroid molecules that are active hormones in various *in vivo* assays, but also the threshold hormone concentration was even lower than that measured in the body fluid of the insect, and cell responses proved satisfactorily dose-dependent. Moreover, hormone activity was shown to be mediated by typical cell surface-specific receptors.

So, the convenient simplicity of this cell culture system, which in easily controlled media avoids the many unknown parameters of the organism context, rapidly compelled full recognition and today is considered a priceless tool for the analysis, at the molecular level, of the mechanisms of action of ecdysteroid hormones (see Chapter 8).

A more fundamental criticism was directed at possible modifications of the genetic material of cultured cells during their countless multiplication cycles (about one cycle per day in standard conditions of culture). It is true that the control of unicellular life is much less exacting than the construction and maintenance of a higher organism, and, in addition, the "normalizing" screen of meiotic processes is lacking here. Hence, despite the fact that a paradiploid karyotype is retained, at least by a few cell lines (see Chapter 3), punctual mutations may have accumulated throughout the years, altering the original *Drosophila* genome.

Fortunately, a decisive response to such arguments was provided by the spectacular experimental results of Illmensee (1976, 1978). He succeeded in transplanting isolated nuclei from *Drosophila* permanent line cells (*S2* and *Kc*) into fertilized eggs. They were incorporated into the cellularizing blastoderm and could be later identified with the use of proper genetic markers (always available with *Drosophila*!) in many different tissues of resulting larvae or adults (see the illustration). This means that several years after their establishment *in vitro,* the genetic material of these *Drosophila* cultured cells had conserved their ability to drive efficiently a number of differentiation programs! Somewhat less significant was the injection, with simple proliferation throughout the process of metamorphosis, of whole *Kc* cells into *Drosophila* instar larvae; at least, they were not encapsulated as exogenous material by the host's immune defenses (Echalier and Proust, 1973).

In addition, every time that, at the molecular level, the structure or chromosomal environment of a given gene, as cloned from cultured cells, was examined, no obvious modification, compared with the fly genome, was observed. For instance, Couderc *et al.* (1983) established that, as is the rule in *Drosophila,* six actin-coding sequences were present in the genome of *Kc0* cells. Moreover, between cultured cell DNA and Canton S embryo DNA they found no differences in the size of *Eco*RI genomic fragments containing the actin genes, which suggests that no important rearrangements, in this respect, had occurred, under culture conditions, in this *Kc* subline (even after some 15 years of growth *in vitro*!).

Nevertheless, as will be emphasized throughout this book, recently cloned sublines are preferable, and their karyotypes must be carefully monitored. Moreover, in order to ensure satisfactory homogeneity of cell responses in the course of any long series of experiments, it is always more advisable to thaw, at regular intervals, frozen samples from the same clonal subline.

In conclusion, the validity of *Drosophila* cultured cells as an experimental model no longer needs to be proved, even though the limitations of any experimental material should, of course, never be underestimated. It especially pleases me, as the father of the *Kc* cell line, that, during the many comparisons which have been made, and should continue to be made, with tissue cells of the whole organism, cultured cells have never, so far, been found to lie!

CONTENT

This monograph summarizes 30 years of experience in the handling of *in vitro* cultured *Drosophila* cells. Beyond being a critical discussion of all the technical aspects of *Drosophila* cell culture, the book is primarily a tentative survey of the multifarious experimental work that has been carried out on this attractive cellular material in many laboratories throughout the world for a wide range of purposes.

This impressive volume of research and its attendant results—which has completed so successfully, in many respects, the huge amount of work done on the whole fly in the dynamic fields of both genomic regulation and developmental genetics—have not always received the attention justly deserved from the scientific community, because many papers have been published in specialized journals for tissue culturists or in symposium reports, and thus remain difficult to obtain. However, our Bibliography, intentionally restricted to research performed on *in vitro* cultures of *Drosophila* cells or tissues,* comprises more than 1500 titles. The important series of papers published in the last few years demonstrate that the field is still very much alive.

The essential aim of this book is to convince, if still necessary, today's scientists that insect cell culture, especially the use of standard *Drosophila* cell lines, has become part and parcel of the wide range of efficient tools available to molecular biologists.

Chapter 1 assembles all available data on the *milieu intérieur* of *Drosophila*, and a comparative description of published culture media clearly

---

* We must point out that our analytical survey has been restricted to cell cultures (*sensu stricto*), which thus excludes tissue or organ cultures because they require quite distinct techniques and are habitually used for somewhat different purposes. Two separate reviews written by I. Schneider and collaborators in the "bible" of drosophilists, i.e., *The Genetics and Biology of Drosophila* (M. Ashburner and T. R. F. Wright, eds., 1978, Vol. 2a, Academic Press), document this point very clearly. Nevertheless, it seemed useful to also include tissue or organ cultures in the Bibliography of this book so that studies devoted to the *in vitro* maintenance or true growth of any *Drosophila* material whatsoever would all be assembled in one list. Any other references, concerning other experimental materials or techniques, have been singled out in the text by setting their date in *italic type,* and are listed at the end of each relevant chapter.

points out that the most successful of them took into account the main biochemical properties of the insect. Recipes for their preparation can be found in the Appendixes (this is the rule for all practical information throughout the book). Thereafter a critical analysis of each component of a medium should help in the subsequent design of a "new generation" of culture media which would permit the complex functioning of *in vitro*-kept differentiated cells.

The two next chapters (2 and 3) deal with the setting up of primary cultures and the establishment *in vitro* of permanent cell lines, respectively. Most work refers to *Drosophila* embryonic material, although, more recently, promising results have been obtained from imaginal disks, larval hemocytes, and neural ganglia. **Chapter 3** comprises a tentatively exhaustive catalog of all cell lines described to date with, of course, emphasis on those most widely used or of special interest (haploid cell lines, for instance).

Cell culture has not played in the karyological study of *Drosophila* the crucial role that it has for mammals. But the simplicity of the chromosome complement of *Drosophila* has made it possible, as surveyed in **Chapter 4,** to analyze on a wide scale various chromosome rearrangements and the evolution of the karyotype, over the years, among huge populations of dividing cells. Moreover, cytological observations of accidental spreading of heterochromatic segments can be linked to the fundamental problem of the functional organization of genetic material in eukaryotes. In this research field, which is currently especially active and in which *Drosophila* plays yet again a dominant role, cell cultures should be very helpful for a biochemical approach (see, for instance, the chromatin structure of heat shock loci as analyzed in Chapter 7).

Generally speaking, molecular biologists have already made good use of the abundant and particularly homogeneous cell material supplied by cultured cell lines, with an extreme diversity of aim and purpose. Therefore, the most significant results have been arbitrarily arranged into two successive chapters: **Chapter 5** which deals with nucleic acid research in which DNA repair mechanisms and the constitution of transcription complexes have been, during the last few years, the most actively studied problems, and **Chapter 6** which, under the undefined title of "proteins," includes a variety of approaches, among which particularly interesting developments have concerned cell adhesion molecules and insect nonspecific humoral defenses.

The central problem posed in contemporary biology is, indisputably, the analysis of the molecular mechanisms that control gene expression of higher eukaryotes during growth and cell differentiation. In this respect, *Drosophila* cell lines have provided two fruitful experimental models (which have been shown to be more complementary than competitive), namely, cell responses to heat shock or other stresses (studied in **Chapter 7**) and cell responses to steroid hormones (**Chapter 8**). As a matter of fact, the famous heat-shock system, though present in all living cells, was first discovered in *Drosophila,* and all major breakthroughs have been made using this organism and, especially, cultured cells. Similarly, the specific responses (i.e., morphological changes and enzymatic inductions) elicited in the cells of our standard *Kc* line by ecdysteroid treatment made it possible for several active laboratories to carry out a thorough analysis, which completed in a very timely fashion the *in vivo* physiological approach. See, for instance, the decisive functional testing of the hormone receptors after their gene cloning by cell transfection.

DNA-mediated gene transfer is now an irreplaceable method for either the molecular "dissection" of the control sequences of a cloned gene or the analysis *in vivo* of the functional domains and interactions of any gene product. This is probably today the main reason for the use of cultured *Drosophila* cell lines. This is the subject of **Chapter 9.** All technical aspects, from the wide variety of possible transfection methods to the comparative efficiency of available promoters, are discussed at length; emphasis is on "conditional promoters" because *Drosophila* cell lines stably transformed with this type of construct should provide an interesting alternative in biotechnology (for biotechnological production of substances of interest). Furthermore, the gene transfer method in *Drosophila* cultured cells is currently used in developmental biology to (1) determine by cotransfection of several genes their rank in the "cascade" of developmental genes that occurs during early embryogenesis (according to the strategy devised by Hogness' laboratory in Stanford) and (2) test the adhesive properties of putative cell adhesion molecules (in a cell rotation–aggregation assay adapted for transfected *Drosophila* cells by Goodman's group in Berkeley).

The study of mobile genetic elements is another important research theme in which *Drosophila* asserts itself as a material of choice (as is developed in **Chapter 10**). In fact, after the discovery of transposons in maize and bacteria, *Drosophila* was the first higher animal in which they could be characterized, and cell cultures played a dominant role in their detection and analysis. In pioneering molecular approaches to the expression of a eukaryotic genome carried out using *Drosophila* cell lines it was shown that the most abundant mRNAs corresponded to various repeated, dispersed, and mobile genomic sequences. It is noteworthy that contrary to previous assumptions the amplification in the number of copies of certain families of retrotransposons, which characterizes most *Drosophila* established cell lines, is not due to a continuous process of replication and transposition throughout years of culture, but is the result of a spectacular "burst" of transposition which takes place during the first weeks of the establishment of the cells *in vitro*.

Finally, **Chapter 11** covers the active field of insect virology. If the *Drosophila* fly is not a natural vector in the biological transmission of mammalian viral diseases, it may, nevertheless, be very useful as an experimental model. This is due once again to the sophistication of its genetics and to its complex but well-understood relations with its endogenous "hereditary" *sigma* virus (closely related to the important group of vesicular stomatitis viruses). Several viral variants, as well as host fly mutants, permissive or not to specific viral strains, have been characterized; and cell cultures have allowed careful analyses of the establishment of a "carrier state" in infested cell lines with very low production of virions, which is an important process in the natural transmission of vertebrate viral diseases by other vector insects. Furthermore, for the attention of *Drosophila* cell users, it seemed useful to give a complete description of the variety of endogenous or exogenous viruses that can, incidentally or experimentally, infest cell cultures (especially because most descriptive papers are dispersed as short notes throughout a number of heterogenous scientific journals). As a by-product of these virological studies, particularly efficient viral promoters (such as the black beetle virus promoter) can be engineered for a very high expression of transferred gene constructs.

Our laboratory has made some contribution to almost every chapter of the book. However, this monograph attempts to be a summary of the experimental results obtained by everyone involved in the *in vitro* culture of *Drosophila* cells. For about 30 years, we have remained in friendly contact or in fair competition with all leading groups in the field, and most of them have kindly communicated to us their experimental documents and laboratory recipes. I hope they will forgive me for having so amply borrowed their data, interpretations, or suggestions, and, occasionally, some of their own formulations. It was always "for a just cause," that is, to facilitate and promote the use of a mutual and so highly attractive cell material.

I am sincerely convinced that this subject concerns a wide spectrum of biologists: not only people who are involved in fundamental research in cell biology and developmental biology, but also those working in applied entomology (interested in virology and pest control, or arthropod-borne diseases); moreover, the specific features of *Drosophila* cell lines should qualify them for use in biotechnology (see Chapter 9).

References

Ashburner, M. (1970). Function and structure of polytene chromosomes during Insect development. *Adv. Insect Physiol.* **7,** 1–95.

Carlson, J. G. (1946). Protoplasmic viscosity changes in different regions of the grasshopper neuroblast during mitosis. *Biol. Bull.* (Woods Hole) **90,** 109.

Frew, J. G. H. (1928). A technique for the cultivation of Insect tissues. *J. Exp. Biol.* **6,** 1–11.

Goldschmidt, R. (1915). Some experiments on spermatogenesis *in vitro. Proc. Natl. Acad. Sci. U.S.A.* **1,** 220.

Grace, T. D. C. (1962). Establishment of four strains of cells from insect tissues grown *in vitro. Nature* **195,** 788–789.

Harrison, R. G. (1907). Observations on the living developing nerve fibre. *Proc. Soc. Exp. Biol.* (NY) **4,** 140.

Trager, W. (1935). Cultivation of the virus of grasserie in silkworm tissue cultures. *J. Exptl. Med.* **61,** 501.

Wyatt, G. R., and Wyatt, S. S. (1976). The development of an insect tissue culture medium. *In "Insect Tissue Culture, Applications in Medicine, Biology and Agriculture"* (E. Kurstak and K. Maramorosch, eds.), pp. 249–255. Academic Press, NY.

Wyatt, S. S. (1956). Culture *in vitro* of tissue from the Silkworm *Bombyx mori. J. Gen. Physiol.* **39,** 841–852.

Acknowledgments

I want particularly to acknowledge the friendly help of faithful colleagues, who agreed to read and criticize most of the chapters, contributing their priceless experience in the field. Keeping to strict alphabetical order, my warmest thanks go to Professor Claudio Barrigozzi (University of Milano, Italy), Dr. Lucy Cherbas (Indiana University, Bloomington, Indiana), Dr. Silvana Dolfini-Faccio (University of Milano, Italy), Dr. Andreas Dübendorfer (University of Zurich, Switzerland), Professor James Sang (Sussex University, Brighton, UK), Dr. Imogene Schneider (Walter Reed Army Research Center, Washington, D.C.), and Professor Robert Tanguay (University of Laval, Quebec, Canada). Many other investigators have generously provided unpublished information or technical protocols and authorized the reproduction in this book of many documents or figures.

I am indeed grateful for the constant support, for three decades now, of all my co-workers in Paris, with special mention of Dr. Annie Ohanessian and Mrs. Danielle Rouillé, without whom the *Drosophila Kc* cell line would probably not have been born some 30 years ago!

The rich and well-kept library of the Institut Jacques Monod (French CNRS) in Paris has been invaluable in my quest for information. I want to thank its two librarians, Mrs. Kropfinger and Mrs. Peutat, for their kind reception, especially Mrs. Kropfinger for having helped me eradicate many of the barbarisms and grammatical errors from my English text.

Last, I need to acknowledge that, if this book has become what I hope it is, a helpful manual for experimentalists, it is largely due to Shirley Light of Academic Press and to the professional pressure she has continuously exerted on me.

1

Composition of the Body Fluid of *Drosophila* and the Design of Culture Media for *Drosophila* Cells

When an artificial culture medium for growing cells outside the organism is being designed, it seems logical to try to copy, as closely as possible, the natural biological fluid in which cells are immersed *in situ*. Experience has proved, however, that such an "imitative approach" (Waymouth, *1965*)* for *in vitro* cultures should always be tempered with a large dose of empiricism. Thus, the composition of successful culture media is always the result of a compromise (Paul, *1973*).

---

* Throughout this volume reference citations in which the year is set in italic type can be found in the reference list at the end of the chapter. All other references are located in the Bibliography at the end of the volume.

I. MAIN CHARACTERISTICS OF THE "MILIEU INTERIEUR" OF *Drosophila*

Only comparative physiologists fully realize how markedly the composition of the "milieu intérieur" (Cl. Bernard) of most other zoological classes—and particularly of insects*—differs from that of blood plasmas of higher vertebrates. Yet, it must be remembered that the chemical features common to mammalian sera have inspired all the classical "physiological saline solutions" used for handling tissues or cells, such as the Ringer–Locke solution and its numerous derivatives. Therefore, none of them, nor culture media derived from them, could be ideally suited for invertebrate tissues.

Among insects the great diversity of the distinct orders should not be ignored, since there are considerable differences in the composition of their body fluids from one family to another and even among the species of the same genus. This might partly depend on specific biological habits and feeding, but seems to be mainly a question of taxonomy and the revelation of evolutionary tendencies (see review by Florkin and Jeuniaux, 1964). Moreover, within the same species, certain biochemical characteristics may vary greatly from one developmental stage to another. For the dipteran *Drosophila*, for instance, it is easy to understand that its larval instars (maggots living in a semiliquid paste) are completely different organisms, in their structure and mode of life, from the aerial flies that will arise during their metamorphosis.

These facts should all be considered when one is elaborating an adequate tissue culture medium.

Because of its importance in genetics and developmental biology, it is not surprising that we have at our disposal quite a considerable mass of information about the "milieu intérieur" of *Drosophila melanogaster*. For practical reasons—since it is much more difficult to draw off enough fluid from adult flies—most of these analytical data refer to the hemolymph of the late 3rd larval instar.**

---

* Let us clarify a meaningful point of terminology: because of their rudimentary and "open" circulatory system, insects possess only one extracellular fluid which circulates throughout the body and bathes all cells, that is, which combines the functions of the blood plasma and the interstitial lymph of vertebrates; so, it is more accurate to call it hemolymph.

** Most bibliographical references date from the beginning of the 1960s, because this was a time when the analytical chemistry and physiology of insects were actively studied.

TABLE 1.I   Osmotic Pressure of Hemolymph of 3rd Instar Larva of
Drosophila melanogaster

Wild-type strains	Method of measure	Osmolarity (mOsm)[a]	Equivalent NaCl solution (g/liter)	Refs.
Berlin	Microcryoscopy ($\Delta =$ 0.70° ± 0.01°C)	378.4 ± 5.4	11	Zwicky (*1954*)
	Thermoelectric		10.5	Begg (*1955*)
Oregon K	Microcryoscopy	342 ± 2	10	Croghan and Lockwood (*1960*)
Canton S	Thermoelectric measure of vapor pressure	366	10.5	Begg and Cruickshank (*1963*)

[a] Osmolarity expressed as mmol dissolved ions per liter.

The following section and accompanying tables will summarize all published data relevant to the design of *Drosophila* cell culture media.

A. Osmotic Pressure

The osmotic pressure of the hemolymph of *Drosophila melanogaster* 3rd larval instar is significantly higher than the 300 mOsm* figure that characterizes uniformly higher vertebrate plasmas. It was measured to be close to 360 mOsm (see Table 1.I), which corresponds to a sodium chloride solution of 10.5 g/liter (instead of the classical 8.5 g/liter NaCl solution that is assumed to be isotonic for mammalian cells). This value might, however, vary slightly according to the environmental or physiological conditions encountered (Croghan and Lockwood, *1960*).**

In contrast with the situation in vertebrates where Cl^- and Na^+ ions are largely predominant, ionized salts of insect body fluids account for a relatively small proportion of the total osmotic pressure. The main osmotic effectors are instead organic molecules and especially amino

* The effect of a solute on osmotic pressure is proportional to the number of particles in solution, i.e., to the number of moles when the compound is not ionizable (e.g., glucose) and to the number of ions when the substance is an electrolyte (e.g., NaCl). The convenient concept of "Osmoles" (Osm) was derived from this fact, and thus directly states the number of osmotically effective particles per liter.

** Singleton and Woodruff (1994) recently measured the osmolarity of *Drosophila* adult female hemolymph using a freezing point depression technique: 251 ± 9 mOsm.

TABLE 1.II   pH of Hemolymph of *Drosophila melanogaster*

Developmental stage	Measurement method	Mean values	Refs.
Larvae	Colorimetric indicators	7.1	Boche and Buck (*1938*)
	Microelectrodes		
	glass electrode	7.1	
	quinhydrone	7.12	
Mature larvae	Microelectrodes		Boche and Buck (*1942*)
	glass–quinhydrone	7.1	
	quinhydrone	7.12	
Last instar larvae	Colorimetric method (bromothymol blue)	6.6–6.7	Begg and Cruickshank (*1963*)

acids. For instance, in *Drosophila* larval hemolymph, amino acids (and other ninhydrin-positive substances) might be responsible for one-third (Zwicky, *1954*), or even 40% (Begg and Cruickshank, *1963*) of the total osmolarity.

B. pH

In spite of notable differences, insect body fluids tend to be slightly acidic. The hydrogen ion concentration in freshly drawn hemolymph from *Drosophila melanogaster* 3rd instar larvae (see Table 1.II) was estimated, using indicator dye bromothymol blue, to lie *between 6.6 and 6.7 pH units*. The higher figures reported by Boche and Buck (*1938, 1942*) might be explained by a loss of CO_2 during measurement with microelectrodes; whereas, in the colorimetric method, the capillary tubes used to collect the hemolymph can be immediately sealed.

According to various observations from other insects, the buffering capacity of the hemolymph might be the sum of several overlapping systems of varied importance: bicarbonates, inorganic and organic phosphates, but also, and perhaps predominantly, amino acids (with their carboxylic and amino groups*) and proteins.

---

* Yet, several authors pointed out that dissociation constants of amino acids are so different from the normal pH of the hemolymph that they are unlikely to have any appreciable buffering effect.

C. Inorganic Ions

1. CATIONIC PATTERNS

For cells to be maintained in good functioning condition, the extracellular fluid has to comply with definite relative proportions of sodium, potassium, calcium, and magnesium.

In this respect, the hemolymph of most insect orders contradicts the too hasty generalization according to which "animal cells would not tolerate any appreciable departure from the normal sea-water-like composition so far as Na^+, K^+, and Ca^{2+} are concerned" (Baldwin, *1967*). This is particularly obvious in advanced orders of insects in which the cationic patterns of the body fluid vary extensively. It is not, as was first assumed, a question of diet. The differences seem rather to depend on the phylogeny. Moreover, one report by Croghan and Lockwood (*1960*) on *Drosophila* larvae fed a diet high in concentrations of NaCl or KCl indicated that well-developed mechanisms of regulation control the ionic composition of the hemolymph.

If sodium remains the predominant cation in the hemolymph of *Drosophila* (see Table 1.III), potassium is present in almost equal amounts: thus, the sodium/potassium proportion, as expressed as an ionic ratio (Na^+/K^+), is only 1.4 (as opposed to its 37.4 value in human plasma).

The concentrations of the two divalent cations Ca^{2+} ($8mM$) and mainly Mg^{2+} ($21mM$) are strikingly high; so much so that it has often been suggested (based mainly on indirect evidence, however) that a fraction of both elements is bound to proteins or other organic complexes. Besides, such a Ca^{2+}/Mg^{2+} ratio of less than unity, while remarkable, is common to the majority of insects.

All these noticeable deviations from the standard cationic pattern observed in higher vertebrates, in human serum for instance, are clearly shown in Table 1.III.

2. INORGANIC ANIONS

A salient feature of insect hemolymph is the reduced titer of the chloride anion, which contrasts with its predominance in the body fluid of all other animals. In *Drosophila* larval hemolymph, chloride ions amount to $42mM$ and do not represent more than 10% of the total osmolarity. Thus, Cl^- contributes only modestly to the cation/anion balance.

Nothing is known about sulfate anions in *Drosophila*, and the only figure for inorganic phosphates was given by Begg and Cruickshank (*1963*): 270 mg/liter of larval hemolymph [i.e., 8.4 milliequivalents

TABLE 1.III Inorganic Ions of *Drosophila melanogaster* Larval and Adult Hemolymphs[a]

| Ions | 3rd Instar larval hemolymph | | | | | | Adult fly hemolymph | | Egg perivitelline fluid Van der Meer and Jaffe (1983) | Human serum Evans (1952) |
	Gloor and Chen (1950)	Zwicky (1954)	Croghan and Lockwood (1960)	Begg and Cruickshank (1963)	Larrivee (1979)	Stewart et al. (1994)	Larrivee (1979)	Van der Meer and Jaffe (1983)		
Na^+			52 ± 1	56.5	63	77 ± 6	87 ± 2	106 ± 7	98 ± 6	143.5
K^+			36 ± 1	40.2	55	40 ± 2	24 ± 1	25 ± 2	84 ± 7	3.8
Na^+/K^+			1.44	1.4	1.15	1.57	3.6	4.24	1.17	37.6
Ca^{2+}				8		1.5 ± 0.7	10.6 ± 0.5	7.2 ± 0.8	5 ± 0.3	2.5
Mg^{2+}				20.8			26 ± 2	14.4 ± 0.8	16.5 ± 1.5	1
Cl^-	41	38 ± 0.3	30 ± 1	42.2	33			58 ± 3	62 ± 3	101
PO_4^{3-}								39 ± 5	41 ± 2	3
SO_4^{2-}				8.4				7.3 ± 0.4	30 ± 1	

[a] Comparison with egg perivitelline fluid and human serum (mmol/liter).

(mEq)/liter]. According to several analyses carried out on other insect species, orthophosphates probably constitute only a small fraction of the total acid-soluble phosphates. The principal organic phosphates identified in *Drosophila* larval hemolymph were phosphoethanolamine, glycerophosphoethanolamine, phosphoserine, and tyrosine phosphate (Mitchell *et al.*, *1960*; Mitchell and Simmons, *1962*; Chen and Hanimann, *1965*; Mitchell and Lunan, *1964*; Burnet and Sang, *1968*).

The large anion deficit (some 100 mEq), which is apparent if one considers only choride and phosphate concentrations, is compensated for with various organic acids (see below).

D. Amino Acid Pool

Insects differ strikingly from all other zoological classes in the biochemical composition of their body fluids, which contain an exceptionally high concentration of free amino acids. For instance, whereas the total amount of amino acids in mammalian plasma does not exceed 0.5 g/liter, their level in the hemolymph of *Drosophila* late 3rd instar larvae reaches 15 g/liter, that is a 30-fold higher value: 112.5 mmol/liter according to Zwicky (*1954*) and 137mmol/liter, as reported by Begg and Cruickshank (*1963*).

In addition to its important contribution to the osmotic pressure and a more debatable participation in buffering (see above), this abundant amino acid pool serves primarily as an active reservoir for protein synthesis. Therefore, all of the 20 or so amino acids that are the customary building blocks of proteins have been identified in insect hemolymph, and it is not surprising that their individual concentrations vary greatly, not only from one species to another, but also from one developmental or physiological stage to another. Likewise, several lethal mutant larvae of *Drosophila*, studied mainly in Hadorn's laboratory in Zurich, accumulate amino acid and peptides abnormally, presumably as a consequence of pathological perturbations in their protein metabolism. Moreover, Burnet and Sang (*1968*) showed that the composition of this amino acid pool can be easily altered by changes in diet.

All published data on amino acid levels of wild-type *Drosophila* larvae are summarized in Table 1.IV and, for a better comparison, each figure has been converted into millimoles per liter of hemolymph or per kilogram of fresh body extract. This extensive compilation fully illustrates (beyond obvious and pronounced discrepancies partly due to technical progress made since the first pioneering analyses with paper electrophoresis) the large variations of concentration of any given amino acid in relation to the physiological age of the larvae (see particularly the figures from Chen and Hanimann, 1965) and probably to feeding conditions. Although it seems difficult to make general conclusions from such a complex survey, one may, however, point out the quantitative importance, in larval hemolymph, of a few amino acids and particularly of the following: (1) glutamic acid, and to a lesser degree aspartic acid, and their respective amides (especially glutamine); this is a common feature of many endopterygote insects and probably in keeping with the crucial role played by these substances in transamination pathways; (2) basic amino acids, like arginine (it must be remembered that arginine phosphate is the "phosphagen" of insects), lysine (an essential component of histones) and histidine; and (3) proline (an important constituent of cuticular proteins) and alanine.

On the other hand, it may be noted that tyrosine and tryptophan, whose derivatives will become essential at metamorphosis for the formation of the exoskeleton and pigments, remain in moderate amounts during larval stages.

Several uncommon amino acids, such as β-alanine, taurine, ornithine, α- and γ-aminobutyric acids, which are not found in protein molecules have been identified in *Drosophila* and in other insects. They may fulfill specific functions in intermediary metabolisms, or simply reflect uptake from special diets.

E. Organic Acids

Organic acids of the Krebs cycle, namely pyruvate, succinate, malate, and mainly citrate, are present in unusual and, sometimes, extraordinarily high concentrations in many insect hemolymphs, so much so that this unexplained situation should be considered to be another biochemical characteristic of the class. In some species, organic acids account for half of the total anion titer.

In *Drosophila*, the only available figures have been provided by Levenbook and Hollis (*1961*) and concern the citrate contents of fully grown larvae (Oregon R strain), i.e., 4.3 to 5 mmol/kg wet weight, and of adult flies, i.e., only 1.1 to 1.2 mmol/kg.

F. Sugars

In insects, as in most animals, glucose is the principal source of energy. The respiratory quotient in flying *Drosophila* equals 1 (Chadwick, *1947*) and, when fed to exhausted flies, glucose was the most efficient sugar for

TABLE 1.IV Identification and Quantitation of Amino Acids in Hemolymph or Total Extracts from *Drosophila melanogaster* Larvae (Wild Type)[a]

Amino acid	Auclair and Dubreuil (1953) 3rd instar larvae	Stumm-Zollinger (1954) 3rd instar larvae (ltr/+)	Chen and Hadorn (1954) "Sevelen" wild stock 72 h	96 h	Faulhaber (1959) 3rd instar larvae	Benz (1955) Mature 3rd instar larvae (96 h)	Crone-Gloor (1959) Hatching embryos "Sevelen"	Begg and Cruickshank (1963) 3rd Instar larvae "Canton-S"	Shinoda (1964) Last larval instar	Chen and Hanimann (1965) Various larval stages 1st day	2nd day	3rd day	4th day	Szabo and coll. (1967) 3rd instar larvae (120 b) "Oregon-R"	Burnet and Sang (1968) 2nd larval instar "Edinburgh"	Widmer (1973) Mature larvae "Sevelen"	Rapport and Yang (1981) Old larvae (85 b) "Oregon-R"
α-Alanine	2.7	7.8	20.8	5.8	+++	++++	++		27.5	13	17.9	2.9	6	+++	13.5	49	+++ 11%
β-Alanine		+	+	+	(+)	++	+	4.7	5.3	0.7	0.9	1.5	0.9		0.7		++ 5%
γ-Amino-butyric acid			+	+	+		+			1.3	0.5	0.3	0.04		0.5		
Arginine	30.6	+	+	?	+	++	+	10.4	11.8	2	1.7	1.5	0.9	+++[k]	0.8		+
Asparagine	4									+[d]	+[d]	+[d]	+[d]	+	+[b]		++++ +[b] 25%
Aspartic acid	6	+	+	+	+	+	+	3	3.2	0.8	0.7	0.1	0.1	+++	0.9	27.5	+
Cysteine/Cystine									3.3						0.1[g]		
Glutamic acid	2.7	+	+	+	+	+++	++	7.6	5.8	10	4.5	5.5	3.2	++	3	36.9	+
Glutamine	10.9	15	23.8	12	+++	+	+++	12.9	25	+[d]	+[d]	+[d]	+[d]	++	12[b]	8.9	++++ +[b] 25%
Glycine	1.9	2[b]	6.3	3.3	+	++	+		23.7	4.7	4	5.5	1.6	++[b]	2.8	28.9	+
Histidine		+	6.6	2.6	+	+	+	(+)	13.5	2.5	2.4	2.5	0.9	+++[k]	2		
Leucine/Isoleucine	5.4	2.4	3.3	2.4	+			2	1.1	2[e] / 1.45	1.3[e] / 0.9	2[e] / 1.3	0.5[e] / 03	+	1.1[e] / 0.8	5.4	+

10

Lysine	6.8	+	6[c]	21[c]	13[c]	++	++	+	+	9.7		6.7	3.2	3.8	6.4	0.3	+++[k]	3.4	34.7	+
Methionine				+	+							0.9	0.6	0.4	0.5	0.16		0.3	3.1	
Ornithine													0.5	0.4	0.5	0.04		0.35		
Phenylalanine					++							2.9	1	0.8	0.7	0.1	+	0.5	1.9	+
Proline	17.4	29.6	10	15.5	+	+						24.5	2.3	2	4.2	2.4	+	3.6		+++ 12%
Hydroxyproline	1.5																			
Serine	2.2	2[b]	7.6	2	+	++	+			3.9		7.7	8.5[d]	7.5[d]	11[d]	4.2[d]	++[b]	2.2	11.9	++
Phosphoserine													0.35	0.17	0.2	0.14		0.1[g]		
Taurine				+[l]	+	++							1	1	0.98	0.8		0.3		
Threonine	0.8	2.6	5.7	3	+	+	+			2.7		3.6	2.9	3.1	5.3	0.9	+	2.2	10.1	+
Tryptophan												(+)					+[f]	(+)		+
Tyrosine	7.3	6.9	2.4	2.8	+++	+++	+			3.7		0.5	1	1.2	1.3	0.9	++[f]	1.8	7.9	++ 7%
Tyrosine O-phosphate			+	+	+	+												2		
Valine	2.5	2.3	3.4	2.6	+	++				2.2		1.3	2.5	1.8	2.6	0.6	+	1.6	9.1	+

a The symbol + corresponds to the presence of a given amino acid. Concentrations are expressed in mmol/liter for the hemolymph and mM/kg (fresh weight) for whole body extracts.
b Glycine and serine.
c Lysine and arginine.
d Serine/asparagine/glutamine.
e Upper number = leucine.
f In pupating larvae.
g Phosphoserine/cysteine.
h Glutamine/asparagine.
i Phosphoserine/cysteine.
j Tyrosine phosphate/glycerophosphate/ethanolamine.
k Arginine/histidine/lysine.
l Lysine/ornithine.

11

rapidly restoring the capacity for continuous flight (Wigglesworth, *1949*). In the majority of insects, however, the actual glucose concentration in the body fluid is low: for instance, its titer in *Drosophila* larval hemolymph does not exceed 2.5 mmol/liter (Zwicky, *1954*).*

This paradox was cleared up when Wyatt and Kalf (*1957*) demonstrated that the major sugar characteristic of insect body fluids is, in fact, the nonreducing disaccharide α,α-trehalose and that the molecule (formed by the linkage of two of its glucose moieties) can be rapidly cleaved by a specific enzyme (trehalase) also present in the hemolymph. Trehalose was identified in *Drosophila* by the same authors, but not titrated.

II. DESCRIPTION AND COMPARISON OF THE MAIN CULTURE MEDIA AVAILABLE

A series of different liquid media was designed for culturing *Drosophila* cells *in vitro*. None is completely satisfactory (see Section IV), even though most of them made it possible to establish continuous cell lines, which is the only criterion for their relative adequacy. Furthermore, the wide diversity of their compositions implies the great adaptability of cells in culture (see Section IV).

Currently, the most widely used media are Echalier–Ohanessian's *D22* medium, Schneider's medium, and Shields and Sang's medium. All three are today commercially available:

> *D22* medium: powder medium (Sigma, St. Louis, MO) or liquid medium (Sigma)
> Schneider's medium: powder medium (Sigma) or liquid medium (Sigma; GIBCO, Grand Island, NY)
> Shields and Sang's medium: (Sigma)

The following discussion is limited to media and physiological solutions of general use for *Drosophila* cell culture. Other special formulas, devised for organ or tissue cultures, are intentionally omitted because they correspond to different approaches and aims that fall outside the scope of this book (see Introduction).

---

* Glucose represents only a small part (10%) of the "reducing substances" in *Drosophila* hemolymph; most of the reducing substances probably correspond to the complex of phenolic compounds implicated in melanosis and cuticle hardening.

A. *D22* Medium and General Principles for Formulation of a Culture Medium for *Drosophila* Cells

The formula for the *D22* medium is given in Table 1.V and its preparation can be found in Appendix 1.A. Initially called *D20* (Echalier and Ohanessian, 1970), it became *D22* after slight modification (Echalier, 1976). This medium is indeed a precise illustration of the general trend that consists in imitating, as closely as possible, the principal characteristics of the body fluid of *Drosophila* 3rd instar larvae. The letter *D* stands for *Drosophila,* while the number *22* reveals the many empirical improvements that were necessary to achieve the successful growth of the first established *Drosophila* cell lines (among them, the well-known *Kc* line; see Chapter 3).

TABLE 1.V Echalier and Ohanessian *D22* Medium[a]

Components	g/liter	mM
Potassium glutamate monohydrate	5.0	24.6
Sodium glutamate monohydrate	8.0	42.8
Glycine	5.0	66.7
$MgCl_2 \cdot 6H_2O$	0.9	4.4
$MgSO_4 \cdot 7H_2O$	3.36	13.7
$NaH_2PO_4 \cdot 2H_2O$	0.43	2.75
$CaCl_2$	0.8	7.2
Sodium acetate trihydrate	0.023	0.17
Succinic acid	0.055	0.47
Malic acid	0.6	4.5
D(+)-Glucose	1.8	10.0
Lactalbumin hydrolyzate	13.6	
Yeastolate (Difco)	1.36	

Vitamins B[b] 2 ml of 500× stock solution[c]
Adjust to pH 6.6–6.7 with 1 N KOH
Osmotic pressure, 360 mOsm

[a] From Echalier (1976). This is the final version of the original *D20* (Echalier and Ohanessian, 1970). See preparation in Appendix 1.A.
[b] From Grace (1962).
[c] See Appendix 1.B.

Consistent with the physiological norms of *Drosophila* hemolymph, the osmotic pressure of *D22* is close to 360 mOsm, which is equivalent to a 10.5 g/liter solution of NaCl and corresponds to a freezing point depression $\Delta = -0.66°C$.

Likewise, the pH is adjusted to a final value of 6.7. Because it was relatively difficult to control the partial pressure of CO_2 in the gaseous phase of the small vessels that were employed for primary cell cultures (see Chapter 2), the use of bicarbonate for buffering the medium was soon abandoned.

The principal difference of our *D22* formula, in contrast to all other culture media, except for the last version of Shields and Sang's medium that was based on the same principle, lies in the special attention given to the minor role played by chloride anions in the body fluid of most insects. Accordingly, these anions were replaced in the culture medium, at least to a large extent, by inorganic anions, namely glutamates. This anion substitution was directly inspired by a physiological solution devised by Shaw (*1956*) for prolonged maintenance *in vitro* of grasshopper neuroblasts.

As for the other anions, the titer of phosphates conforms to the analytical data of Begg and Cruickshank (*1963*), whereas, lacking information about sulfates in *Drosophila,* we arbitrarily took the figure that S. Wyatt (*1956*) recommended in medium for silkworm cells. Moreover, the noted abundance in insects of organic acids from the Krebs cycle led us to introduce some of them at concentrations close to those also suggested by S. Wyatt and later adopted by Schneider (1964). Moreover we added some sodium acetate, because it plays a well-known role in intermediary metabolism.

On the other hand, as can be seen in Table 1.XI, *D22* medium satisfactorily respects the specific cationic patterns of *Drosophila* hemolymph: the Na^+/K^+ ratio, which is reasonably close to its physiological value, and the high levels of bivalent cations Ca^{2+} and Mg^{2+} are of particular note.

A high content of free amino acids is globally supplied, in addition to the substantial amounts of glutamic acid and glycine in the basic solution, by some 14 g/liter of a lactalbumin enzymatic hydrolyzate. We would like to point out that lactalbumin is a nutritionally important protein from milk, and several of its commercially available hydrolyzates have been successfully used in different culture media, perhaps because their amino acid residues are in well-balanced proportions. In any case, it gave better results at the time than any tested mixture of individual amino acids, although the success of better defined media (see especially Shield and Sang's medium, Section II.C) has since contradicted this assertion.*

Sugars are represented by glucose only (1.8 g/liter, i.e., a value not far from the analytical figures of Zwicky, *1954*).

As for vitamins, studies carried out on axenically cultured *Drosophila* flies established their requirements, especially for all B vitamins (see Section IV.F). This vitamin need is probably fully met by the 1.36 g/liter of total yeast extract [Yeastolate (Difco)] introduced into the medium. Although it was perhaps unnecessary, we considered it safer to add the complex vitamin mix used in Grace's medium (*1962*) for Lepidoptera cells (see Appendix 1.B).

This yeast extract also covers all other possible and unknown nutritional needs (nucleotides, trace metals, lipids).

According to the general rule for cell cultures, D22 must also be routinely supplemented with 5–10% (v/v) fetal bovine serum (see Section IV.I.1).

B. Schneider's Culture Medium

Like all successful culture media, Schneider's medium has undergone—since its original formulation for the culture of *Drosophila* imaginal discs—a series of modifications (Schneider, 1964, 1966, 1972). The last and currently used version is that published by Schneider and Blumenthal (1978) in their chapter review in Volume 2a of "The Genetics and Biology of *Drosophila*" (Ashburner and Wright, eds., Academic Press). Its formula can be found in Table 1.VI.

It seems useful to point out that the composition of the "revised Schneider's medium" prepared by the Grand Island Biological Company (GIBCO) and which was the result of a personal communication from Imogene Schneider (1985, unpublished) is strictly identical to the Schneider and Blumenthal (1978) reference version, except that it does not contain any bacteriological peptone. On the other hand, and surprisingly, the so-called "Schneider's Insect medium," more recently available from Sigma, appears to be an unexplained compromise between the 1978 formula (in the composition of its saline solution and its double content of Yeastolate) and the 1964 recipe (amino acids and organic acids

---

* Nevertheless, as reported by Nagasawa *et al.* (*1988*), commercially available lactalbumin hydrolyzate possibly contains several unidentified water-soluble and heat-stable factors that stimulated the growth of an embryonic cell line from the flesh fly *Sarcophaga*. (See also page 48.)

TABLE 1.VI Schneider's Medium[a]

Components	mg/liter	mM
Salts		
CaCl$_2$	600	5.4
KCl	1600	21.5
MgSO$_4$ · 7H$_2$O	3700	15.0
KH$_2$PO$_4$	450	3.3
Na$_2$HPO$_4$	700	4.9
NaCl	2100	35.9
NaHCO$_3$	400	4.7
Sugars		
Glucose	2000	11.0
Trehalose	2000	5.5
Supplements		
TC Yeastolate	2000	
Bacteriological peptone[b]	5000	
Organic acids		
α-Ketoglutaric acid	200	1.4
Fumaric acid	100	0.8
Malic acid	100	0.75
Succinic acid	100	0.8
Amino acids		
β-Alanine	500	5.6
L-Arginine	400	2.3
L-Aspartic acid	400	3.0
L-Cysteine	60	0.5
L-Cystine	100	0.4
L-Glutamic acid	800	5.4
L-Glutamine	1800	12.3
Glycine	250	3.3
L-Histidine	400	2.6
L-Isoleucine	150	1.1
L-Leucine	150	1.1
L-Lysine hydrochloride	1650	9.0
L-Methionine	800	5.4
L-Phenylalanine	150	0.9
L-Proline	1700	14.8
L-Serine	250	2.4
L-Threonine	350	2.9
L-Tryptophan	100	0.5
L-Tyrosine	500	2.7
L-Valine	300	2.5

Final pH 6.7–6.8. Osmotic pressure: $\Delta = -0.67°C$

[a] From Schneider and Blumenthal (1978).
[b] The Schneider's medium commercially available (GIBCO, Grand Island, NY) is identical, except that it does not contain any bacteriological peptone.

being maintained at their original levels); it does not contain any Bacto-peptone either.

Let us now point out the main features of Schneider's medium:

The basic saline solution is, traditionally, dominated by chloride salts, which led to certain departures from the analytical data for hemolymph (see Table 1.XI). In particular, the Cl⁻ anion titer is theoretically too high and the Na⁺/K⁺ ratio (after addition of 15% fetal bovine serum) is over 3, instead of 1.4.

The most important variations among the successive "avatars" of Schneider's medium lay in its amino acid supply. Lactalbumin hydrolyzate (10 g/liter) was replaced by a defined mixture of amino acids (total amount: 12 g/liter). There remains, however, some doubt as to the necessity, at least in primary cultures, for supplementing it with bacteriological peptone (5 g/liter in the reference formulation of 1978).

On the other hand, a solution of individual B vitamins, probably inspired by *NCTC109*, a defined culture medium for mammalian cells, was not alone a substitute for yeast extracts, and its extra supply, as in the 1972 version, seems unnecessary, especially when the Yeastolate content was doubled. Eighteen percent inactivated fetal bovine serum has to be routinely added.

PUBLISHED MODIFICATIONS

A *"reduced Schneider's medium,"* better defined and also more economical, was proposed by Sederoff and Clynes (1974). They observed a normal growth of Schneider's *S2* cells in suspension culture, after having eliminated Bacto-peptone, the four organic acids, trehalose (whereas, *per contra,* the glucose content was doubled) and *NCTC109* vitamins. Moreover, because it is difficult to dissolve cystine and tyrosine, they left the former out (although they retained cysteine) and slightly reduced the amount of tyrosine (420 mg/liter instead of 500). A 5 per 100 serum supplement was found to be sufficient.

When studying myocyte and neuroblast differentiation in primary cultures (see Chapter 2), Seecof and Donady (1973) obtained better results with a slightly *"modified Schneider's medium."* By this, they meant that modifications were made in the formula published in the GIBCO catalog and which does not contain any peptone.* Small amounts of glutathione

---

* Therefore, it is erroneous that, in his "Drosophila Laboratory Manual," Ashburner (1989) confuses this Seecof and Donady's version with the "revised Schneider's medium" available from GIBCO.

(6 mg/liter) and asparagine (30 mg/liter) were added; and, while NaHCO₃ was omitted, phosphate levels were changed (680 mg of monopotassium phosphate and 430 of disodium phosphate per liter), which permitted a better control of the pH to near optimal values (pH 6.8–6.9).

C. Shields and Sang's Medium M3

After hesitating in their choice between various chloride mixtures (Shields and Sang, 1970; Shields *et al.,* 1975), Shields and Sang (1977) finally adopted the principle on which *D22* medium is based, that is to eliminate most of the chloride salts and to introduce Na^+ and K^+ ions essentially as glutamate salts. The composition of this so-called "improved medium" is given in Table 1.VII; it was, later, designated "*M3*" (Cross and Sang, 1978; Dübendorfer and Eichenberg-Glinz, 1980; Sang, 1981) and has been recently made available commercially by Sigma. A version is proposed for protein labeling, that is without yeast extract, leucine, and methionine.

The improvements, in comparison with *D22,* lie in the following:

Defined concentrations of individual amino acids advantageously replace the miscellaneous lactalbumin hydrolyzate, although some imprecision remains, since the medium cannot dispense with yeast extracts. The amino acid composition was based on the analyses by Chen and Hanimann (*1965*) for the free amino acid content of the whole body extract from 1st instar larvae (see Table 1.IV).

Organic acids (except oxalacetic acid, see below) could be eliminated. The extra B vitamin solution was also omitted, since Yeastolate was retained.

An organic buffer (Bis–Tris) is used and, as had been shown in the case for mammalian cell cultures, some oxalacetic acid must be added to satisfy the cells' need for carbon dioxide (see Section IV.B). Shields and Sang consider, however, that some bicarbonate remains useful, and, to minimize its breakdown under conditions that are too acid, they recommend introducing it at the end of the preparation, after preliminary adjustment of the pH.

M3 medium is, of course, routinely supplemented with 10% fetal bovine serum.

Variants of Shields and Sang's Medium

M3(BF) *(Cross and Sang, 1978):* The omission of potassium bicarbonate and a concomitant raising of the level of potassium glutamate (from 38.8 to 43.7 mM in order to keep the same K^+ titer) give more stable conditions for culture in small drops.

M3X medium (Sinclair et al., *1983, 1986)*is used for xanthine–guanine phosphoribosyltransferase (*gpt*) selection (see Chapters 5 and 7).

M3(NS), i.e., "nonserum" (Shields and Sang, personal communication; Edwards et al., *1978; Milner, 1985):*The same formula is simply called MM3, that is, "modified M3", in Currie *et al.* (1988). The use of Shield and Sang's medium with a very low (2%) or nil serum supplement makes compensation for the different ions necessary: the sodium glutamate concentration is increased to 46.5 mM (instead of 38.6). Potassium glutamate and potassium bicarbonate being, conversely, omitted, K^+ ions are reintroduced as KCl (39.9 mM). Moreover, let us point out that some cholesterol (30 mg/liter), although absent not only from the standard *M3* medium (because the usual 10% serum supplement provides a sufficient amount),[*] but also from the formulation given by Milner in DIS (1985), appears in the *M3(NS)* composition published in Ashburner's *Drosophila Laboratory Manual* (1989).

In a *minimal medium* devised for the establishment of pyrimidine auxotrophic cell lines (see Chapters 3 and 5), Regenass and Bernhard (1979) omitted yeast extract, but added vitamins, namely (in μg/liter): pyridoxine 20; biotin 1; thiamin 3; nicotinic acid 280; riboflavin 20; folic acid 34; Ca^{2+}-pantothenate 0.014; vitamin B_{12} 0.005.

D. Gvozdev and Kakpakov's Culture Medium: C-15 or S-15

This culture medium was designated "*C-15*" in the original Russian papers (Gvozdev and Kakpakov, 1968; Kakpakov *et al.,* 1969), most probably from the word "Cpeдa" which means medium. This medium

[*] See Section IV.G.

TABLE 1.VII Shields and Sang's *M3* Medium[a]

Components	g/liter	mM
Salts		
$MgSO_4 \cdot 7H_2O$	4.4	17.9
$CaCl_2 \cdot 6H_2O$	1.5	6.8
$Na_2PO_4 \cdot 2H_2O$	0.88	5.6
$KHCO_3$	0.5	5.0
Monopotassium glutamate monohydrate	7.88	38.8
Monosodium glutamate	6.53	38.6
Amino acids		
L-α-Alanine	1.5	16.8
L-β-Alanine	0.25	2.8
L-Arginine	0.5	2.9
L-Aspartic acid	0.3	2.25
L-Asparagine	0.3	2.27
L-Cysteine hydrochloride	0.2	1.27
L-Glutamine	0.6	4.1
Glycine	0.5	6.6
L-Histidine	0.55	3.5
L-Isoleucine	0.25	2.0
L-Leucine	0.4	3.0
L-Lysine hydrochloride	0.85	4.6
L-Methionine	0.25	1.6
L-Phenylalanine	0.25	1.5
L-Proline	0.4	3.5
L-Serine	0.35	3.3
L-Threonine	0.5	4.2
L-Tryptophan	0.1	0.5
L-Tyrosine	0.25	1.4
L-Valine	0.4	3.6
Glutathione	0.005	
Sugars		
Glucose	10.0	555.0
Vitamins, etc.		
TC Yeastolate (Difco)	1.0	
Choline hydrochloride	0.05	
Oxalacetic acid	0.25	1.9
Buffer		
Bis–Tris (Sigma)	1.05	5.0

Adjust to pH 6.8 with 1% NaOH. Add 10% fetal bovine serum. Osmotic pressure: 333 mOsm.

(continues)

TABLE 1.VII (*continued*)

---

[a] From Dis (1977).

[b] Preparation Procedure:
Dissolve in about 950 ml doubly glass-distilled water, adjust pH to 6.6 with
1% NaOH, and then add the $KHCO_3$ to bring to the final value of 6.8 (loss
of CO_2 due to breakdown of the bicarbonate under acid conditions is thus
minimized. Make up the volume to 100 ml and sterilize by Millipore filtration.
Add fetal bovine serum to 10% (v/v) (i.e., one part added to nine parts of
medium). *Note* [in Ashburner (1989)]: The amino acids are very close to their
saturating concentrations. Raising the pH to 6.8 may cause precipitation (this
apparently varies between lots of amino acids). Some cell lines may increase
the pH of the medium and cause precipitation. If this happens, routinely either
drop the pH slightly or decrease amino acid concentration slightly. L. and P.
Cherbas find that *Kc* cells do not grow well in M3 without supplementation
with Bacto-peptone (Difco) to a final concentration of 2.5 g/liter.

is most often called *C-15* in English publications, but, because the Cyrillic
character "C" corresponds to our letter "S", I prefer to name it *S-15*.

The principal merit of *S-15* is, of course, that it supported the growth
of the second series of permanent *Drosophila* cell lines ever to be established *in vitro* (Kakpakov *et al.,* 1969; see Introduction and Chapter 3).

Its composition is traditional (see Table 1.VIII), being essentially a
mixture of various chlorides, buffered with bicarbonate and Tris, and
supplemented with sugars, lactalbumin hydrolyzate, and yeast extracts.
Moreover, a very long, and probably useless (see Section IV.F), list of
individual vitamins is added. It comprises not only the B vitamin solution
from Grace's medium, but also several additional ones (such as vitamins
A and B_{12}).

Note that medium *S-15* only roughly emulates the main characteristics
of *Drosophila* body fluid (see Table 1.XI): its osmotic pressure (corresponding to a freezing point depression of $-0.82°C$) is close to 440 mOsm
(instead of 360 mOsm for larval hemolymph); its pH is 7.2 (instead of
6.6–6.8); as for ionic patterns, its Na^+ concentration is too high and K^+
titer too low, from a theoretical point of view. The amount of Cl^- ions
approximates vertebrate values: 120 mM (instead of 35 mM in *Drosophila* hemolymph).

S-15 medium has to be supplemented, as usual, with fetal bovine serum
(15%), but also, at least for primary cultures, with 10% of a *Drosophila*
pupal extract (see Section IV.J.2 and Appendix 1.H.2), which introduces
a further complexity.

TABLE 1.VIII   Gvozdev and Kakpakov's *C-15* Medium[a]

Components	g/liter	mM
Salts		
$NaH_2PO_4 \cdot 2H_2O$	0.5	3.2
NaCl	4.0	68.4
KCl	1.56	20.9
$CaCl_2$	0.5	4.5
$MgCl_2 \cdot 6H_2O$	2.5	12.3
$NaHCO_3$	0.35	4.2
Organic acids		
Sodium acetate	0.025	0.3
Malic acid	0.67	5.0
Succinic acid	0.06	0.5
Sugars		
Sucrose	5.0	13.9
Glucose	5.0	27.8
Amino acids		
Lactalbumin hydrolyzate	17.5	
L-Tryptophan	0.1	
L-Cysteine	0.025	
Glutathione	0.005	
Vitamins		
Yeast extracts	1.5	
Nicotinamide dinucleotide	0.005	
Ascorbic acid	0.1	
	mg/liter	
Vitamin B_{12}	0.02	
Thiamin	0.02	
Vitamin A	0.02	
Calcium pantothenate	0.02	
Niacinamide	0.1	
Pyridoxine	0.02	
p-Aminobenzoic acid	0.02	
Folic acid	0.02	
Inositol	0.02	
Choline chloride	0.02	
Biotin	0.02	
Riboflavin	0.02	
Buffer	*g/liter*	
Tris(hydroxymethylaminomethane)	3.0	
Indicator		
Phenol red	0.01	

Adjust to pH 7.2; osmotic pressure Δ: $-0.82°C$, i.e. *440 mOsm.*
+ Fetal bovine serum 15% (+ pupa extracts 10%)

[a] *S-15* medium. From Gvozdev and Kakpakov (1968).

It seems likely, nevertheless, that it is for mainly nonscientific reasons that the use of the medium *S-15* and the cell lines that were established in it is not widespread on the Western side of the former Iron Curtain.

SEMISYNTHETIC DERIVATIVE(S)

Russian investigators have striven to simplify and optimize their original medium. Among the many formulas that were succinctly quoted in later papers, it seems that only the so-called *C-46 medium* (Kakpakov, 1989 and personal communication to the author, 1993) is worth retaining. Its composition is given in Appendix 1.C.

Although lactalbumin hydrolyzate is replaced by defined concentrations of individual amino acids, this medium is only semisynthetic since yeast extract is kept and, in most cases, a FBS supplement is required. Organic acids and a few dispensable vitamins have been omitted in a formula which remains, nevertheless, very close to the original one and, thus, incurs the same theoretical critiques. In particular, its osmotic pressure is even higher (freezing depression point $-0.87°C$, instead of $0.82°C$).

Nevertheless, *C-46* is currently, in Russia, the most widely used medium for growing insect cells. It is even available commercially from the Vavilov Institute of General Genetics (Prof. Kakpakov) in Moscow.

E. Robb's Medium for Maintenance of Imaginal Discs: *R-14*

The most interesting fact about this medium, or more precisely of the buffered saline on which it is based, is that it was empirically devised by exploring the ranges of pH, osmolarity, and ionic concentrations that ensured high levels of incorporation of tritiated uridine into RNA and of tritiated amino acids into proteins, when whole imaginal discs were maintained *in vitro* for several hours (up to 2–3 days) (Robb, 1969).

It is particularly noteworthy, however, that for most tested parameters the optimal conditions, thus defined by a pragmatic approach, approximate the measured values of *Drosophila* larval hemolymph. This is true for osmolarity (optimal range from 285 to 345 mOsm), pH (between pH 6.75 and 7.35), sodium and potassium concentrations (around the physiological figures of 40 mmol/liter K^+ and 56 mmol/liter Na^+) as well as for the inorganic phosphates (between 1.8 and 3.6 mmol). On the other hand, the proper concentration range of divalent cations, Ca^{2+} and Mg^{2+}, was found to lie well below the measured hemolymph levels (0.3 to 1.5 mmol/liter for Ca^{2+}, instead of 8mmol/liter; 0.5 to 3.5 mmol/liter of

Mg^{2+}, instead of 21 mmol/liter), even though their optimal ratio ($Mg^{2+}/Ca^{2+} = 2.5$) remained close to *Drosophila* norms (see Section IV). The formula for medium *R-14* integrates all these experimental data (see Table 1.IX).

On the other hand, it must be emphasized that the many other components of this particularly complex medium, i.e., amino acids, sugars, vitamins, trace metals, and some rather "esoteric" ingredients, have not been submitted to the same systematic evaluation. Their addition and suggested concentrations were derived, theoretically, from the composition of synthetic media which were, at that time, devised for mammalian cells. There is, however, one exception, which is the high organic acid content, based on Wyatt's observations of insect hemolymphs (see Section I.E).

In spite of its very complex formulation, which is out of proportion to its limited capacities, Robb's medium (or simplified versions, for instance in Fristrom *et al.*, 1973) remains quite popular among insect physiologists for short-period maintenance *in vitro* of *Drosophila* imaginal discs or tissues.

In any case, medium *R-14* cannot be considered a reliable cell culture medium. Robb himself noticed that even a small perforation of the peripheral membrane of the discs resulted in a rapid decrease in uridine or amino acid incorporation.

F. Wyss's ZW Medium

This medium represents a serious attempt to devise a chemically defined medium for *Drosophila* cells. Its composition was indeed the result of Wyss's sustained investigations (in Zurich, which explains the letter "Z" in its denomination) of the various requirements of certain sublines from *Kc* and *Ca* cells (i.e., two cell strains originally established in *D22* medium). *ZW* may not, however, be considered a completely synthetic medium because, if it is to support the growth of primary cultures, it must be supplemented with undefined fly extracts (Wyss, 1982a; see Section IV.J).

It seems more fruitful to discuss elsewhere (see Section IV) the important conclusions reached by Wyss, in comparison to other available data. Only the main stages of the long elaboration of this semisynthetic medium (each one being marked by an intermediate formula) will be briefly reviewed here.

At first, Wyss and Bachmann (1976) endeavored to replace the lactalbumin hydrolyzate present in *D22* medium with a defined mixture of

TABLE 1.IX   Robb's *R-14* Medium[a]

Components	mM
Salts	
NaCl	25
NaOH	25
KCl	38
$CaCl_2 \cdot 2H_2O$	1.2
Na_2HPO_4	2.0
KH_2PO_4	0.37
$MgCl_2 \cdot 6H_2O$	1.2
$MgSO_4 \cdot 7H_2O$	1.2
$FeSO_4 \cdot 7H_2O$	0.0024
$ZnSO_4 \cdot 7H_2O$	0.0024
$CuSO_4 \cdot 5H_2O$	0.00008
Organic acids	
Fumaric acid (Na salt)	1.5
α-Ketoglutaric acid (Na salt)	1.9
Malic acid	4.3
Pyruvic acid (Na salt)	0.8
Succinic acid (Na salt)	4.5
Vitamins	
Biotin	2.5×10^{-5}
Choline choride	0.10
Folic acid	0.002
myo-Inositol	0.78
Nicotinamide	0.0066
Calcium pantothenate	0.0033
p-Aminobenzoic acid	0.006
Pyridoxal hydrochloride	0.0046
Thiamin hydrochloride	0.0026
Riboflavin	2×10^{-4}
Vitamin B_{12}	8×10^{-4}
Amino acids	
α-Alanine	0.16
β-Alanine	0.1
L-Arginine hydrochloride	2.0
L-Asparagine	0.2
L-Aspartic acid	0.16
L-Cystine hydrochloride	0.26
L-Glutamine	4.0
Glycine	0.16
L-Histidine hydrochloride	0.4
L-Isoleucine	0.64
L-Leucine	0.64

(*continues*)

TABLE 1.IX (*continued*)

Components	mM
L-Lysine hydrochloride	0.8
L-Methionine	0.2
L-Phenylalanine	0.4
L-Proline	0.32
I-Serine	0.16
L-Threonine	0.74
L-Tryptophan	0.08
L-Tyrosine	0.32
L-Valine	0.64
Sugars	
Sucrose	88.0
Glucose	10.0
Other substances	
Putrescine dihydrochloride	8×10^{-4}
Linoleic acid	2.4×10^{-4}
Lipoic acid	0.008
Thymidine	0.0024
Cholesterol (saturated at 23°C, i.e., about 2 mg/liter)	

Adjust to pH 7.1 with 0.25 N NaOH; osmotic
pressure: 275 mOsm/liter.

---

[a] For maintenance of imaginal discs (Robb, 1969).

amino acids (formula called ZA). Moreover, the elimination of the exceed-ingly high content of glycine and glutamic acid and the subsequent lower-ing of osmolarity (to 250 mOsm) were shown to be without effect on cell proliferation (formula "ZD").

In a second step, the possibility of replacing yeast extract with its nondialyzable high molecular fraction—while the medium is only com-pleted with trace metals, 4 vitamins, and a minimal addition of vertebrate serum (formula named ZH 1%)—permitted the authors to evaluate the influence on cell multiplication of, first, the addition of various purines and pyrimidines (Wyss, 1977) and then of graded concentrations of B vitamins (Wyss, 1979).

Finally, a partially purified "growth factor" (?) (see Section IV.J.2) extracted from whole flies could, alone, supplement a chemically defined medium (formula "ZO") and support the long-term growth of a Kc subline. Wyss (1982a,b,c) was thus in a situation whereby he could reevaluate the possible dispensability of each component of the culture medium, or, conversely, its optimal concentration. These thorough studies led to the formulation of a minimal medium "ZR," then to a better balanced "ZW" medium (see Table 1.X).

The ZW formula, provided that a suitable fly extract, together with some insulin and ecdysterone, was added, allowed the proliferation of primary cultures from embryonic cells, as well as from dissociated imaginal disc cells (Wyss, 1982a). Unfortunately, long-term cultures were apparently not obtained. Moreover, other established cell lines seemed to be more exacting than the *Kc* subline used for such analyses, and therefore proliferated poorly in *ZW* medium.

Even though the use of this ZW formula has been rather limited, Wyss's careful studies (see also below) remain a model for optimizing future new media (see Section V). However, it would probably be more advisable to test medium competence on primary culture cells.

III. BUFFERED SALINE SOLUTIONS FOR *Drosophila* TISSUES

An impressive number of saline solutions have been proposed for handling various *Drosophila* organs or tissues *in vitro*. Their compilation fills seven pages in the appendixes of Ashburner's "Drosophila Laboratory Manual" (1989). Almost every investigator has devised his/her own recipe and has retained it by simple routine.

This is clearly illustrated by the widespread use for several decades of the so-called "Insect Ringer",* first used by Ephrussi and Beadle (1936) in their pioneering experiments of disc transplantation in *Drosophila*. Let me point out, incidentally, that, later, both authors denied any paternity with regard to its formulation (Ephrussi, personal communication in the 1960s)! This saline is indeed a rudimentary mix of three chloride salts in which the vertebrate-like Na^+/K^+ ratio (= 27) is in complete contradiction to the specific situation of most insect body fluids; its osmolarity, too, corresponds to vertebrate standards (about 300 mOsm). The inadequate composition of this popular saline, as well as that of many others, proves either the resistance, at least during a short exposure, of *Drosophila* tissues, or the relative impermeability of the limiting membranes of imaginal discs (see Section V.A).

---

* NaCl 7.5, KCl 0.35, and $CaCl_2$ 0.21 g/liter.

TABLE 1.X Wyss's ZW Medium[a]

Components	mg/liter	mM
Salts		
NaCl	3200	54.7
KCl	2000	26.8
$CaCl_2 \cdot 2H_2O$	147	1.0
$MgSO_4 \cdot 7H_2O$	1230	5.0
$NaH_2PO_4 \cdot H_2O$	420	3.0
$CuSO_4$	0.001	
$FeSO_4 \cdot 7H_2O$	1.0	
$MnSO_4 \cdot 7H_2O$	1.0	
$ZnSO_4 \cdot 7H_2O$	1.0	
Organic acids		
Malic acid	670	
Oxalacetic acid	250	
Sodium acetate trihydrate	25	
Sodium pyruvate	110	
Succinic acid	60	
Sugar		
Glucose	2000	11.0
Vitamins		
Biotin	0.05	
Calcium pantothenate	5	
Choline hydrochloride	50	
Folic acid	0.05	
myo-Inositol	50	
Niacinamide	0.5	
Pyridoxal hydrochloride	5	
Riboflavin	0.01	
Thiamin hydrochloride	5	
Vitamin B_{12}	0.05	
Other substances		
DL-Carnitine hydrochloride	1000	
1-Thioglycerol	5.4	
Glutathione (reduced)	50	
Amino acids		
L-Alanine	45	0.5
β-Alanine	45	0.5
L-Arginine	174	1.0
L-Asparagine	150	0.1
L-Aspartic acid	133	1.0
L-Cysteine hydrochloride	315	2.0
L-Glutamine	680	4.65
L-Glutamic acid	294	2.0

(*continues*)

TABLE 1.X (*continued*)

Components	mg/liter	mM
Glycine	750	10.0
L-Histidine	620	4.0
L-Hydroxyproline	65	0.5
L-Isoleucine	131	1.0
L-Leucine	131	1.0
L-Lysine hydrochloride	182	1.0
L-Methionine	149	1.0
L-Ornithine hydrochloride	84	0.5
L-Phenylalanine	165	1.0
L-Proline	115	1.0
L-Serine	525	5.0
L-Threonine	119	1.0
L-Tryptophan	102	0.5
L-Tyrosine	90	0.5
L-Valine	117	1.0
Nucleosides		
Inosine	2.7	
Thymidine	2.4	
Uridine	2.4	
Lipids		
Cholesterol	0.01	
Linoleic acid	0.01	

Adjust to pH 6.6 with 1 N NaOH; osmolarity, 250 mOsm.

[a] From Wyss (1982a).

Table 1.XII shows a comparison of ionic concentrations, osmolarity, and pH of most published formulas, with regard to the physiological values of *Drosophila* larval hemolymph and the composition of a typical buffered saline for vertebrate cells. Despite the striking variability of their relative ionic proportions, most of those salines consist essentially of a mixture of three or four chlorides, generally buffered with bicarbonate, phosphates, or some organic buffer (Tris, HEPES). Sugars are often added, generally glucose, as a source of energy, but also sucrose, mainly, in this latter case, for raising the osmolarity to an acceptable level. This parameter, nevertheless, remains very often below the physiological value.

With each author extolling, on pragmatic grounds, the virtues of his/her own recipe, it is difficult to make a purely objective choice. However, to help beginners, we selected three formulas that seem to be the most reasonable, at least on theoretical grounds:

Robb (1969) "*Drosophila* Phosphate-Buffered Saline" (DPBS) (see Table 1.XII and Appendix 1.D). The experimental criteria used to optimize this saline for the handling of imaginal discs *in vitro* have already been mentioned (see Section II.E).

Chan and Gehring (1971) "Balanced Saline" (see Table 1.XII and Appendix 1.E). This saline allows the dissociation of healthy blastoderm cells, as proved by their behavior after transplantation *in vivo*.

Stewart *et al.* (1994) "Minimal Hemolymph-Like Solution" (see Table 1.XII and Appendix 1.F): This *HL3* formula and other more complex derivatives are based on new ion measurements of the major cations of larval hemolymph (see Table 1.III) obtained with ion-selective electrodes. It could maintain neuromuscular preparations from *Drosophila* 3rd larval instar in better condition (long survival, constant level of the membrane potential) than other standard physiological solutions and, even, Schneider's medium.

See also, for adult tissues, the so-called ionically matched adult *Drosophila* saline (IMAD) [Singleton and Woodruff (1994), Appendix 1-G].

IV. GENERAL DISCUSSION

At this stage of our comparative survey, a few general comments should be made. Even though the available culture media have made it possible to maintain, with a high rate of multiplication and for more than 25 years, a series of independent *Drosophila* embryonic cell lines, one must admit that none of them can be considered to be fully satisfactory. This is clearly demonstrated (1) by the difficulty of growing isolated cells in any of them, even from long-established cell lines (see cloning procedures in Chapter 3), and (2) by the still highly variable growth of primary cultures (see Chapter 2). No currently used *Drosophila* cell culture medium has proved to be really superior to the others. It is undeniable that the early commercialization of Schneider's medium, even more than its intrinsic qualities, promoted its widespread use. Each of these media supported the growth *in vitro* of several cell lines, which is the only conclusive criterion of their approximate adequacy.

Most of these cell lines were usually kept in their original medium but, occasionally, they could be "adapted," often without too much difficulty, to other formulas. For instance, in many laboratories, the *S2* line of Schneider is now routinely grown in Shields and Sang's *M3* medium (Cherbas, personal communication; Benyajati and Dray, 1984; Courey and Tijan, 1988; van der Straten *et al.*, 1989; Dorsett *et al.*, 1989). This *S2* line also multiplied in Kakpakov's *C-46* medium, and a subline, called *E*, grew in Echalier's medium (Sederoff and Clynes, 1974).

Conversely, *Kc* cells (Davies *et al.*, 1986), one Kakpakov *et al.*'s diploid line (Metakovsky and Gvosdev, 1978), and cells from *GM1* and *GM2* lines could be subcultured in Schneider's medium (Nakajima and Miyake, 1976). Likewise, *Kc* cells could be transferred to *M3(BF)* medium (Yamaguchi *et al.*, 1991).

More surprisingly, a subline derived from Schneider's *S2* line even became able, after gradual replacement, to grow in slightly modified versions of Eagle's medium, i.e., a typical medium for vertebrate cells, providing the pH remained below 7 (Lengyel *et al.*, 1975). Cherbas *et al.* (1994) recently reported that *HyQ-CCM3*, a serum-free medium marketed by Hyclone, Ltd., Logan, Utah, for use with lepidopteran *Spodoptera* cell lines (and whose composition is, unfortunately, a proprietary secret), is an excellent medium for *Drosophila Kc167* cells. *S2, S3,* and *GM-3* cells could also be readily adapted to it, whereas D cells required a supplement of 2% FBS.

It is obvious that such "adaptations" correspond to some selection of nutritional variants from huge *in vitro* cell populations and that the "plasticity," with regard to the different media in which they can grow, of newly explanted tissue cells must differ drastically. This selection process, from established cell lines, has been well documented in vertebrate cell cultures, and people have denounced at length a sort of "connivance" between investigators and permanent cell strains which, because of graded adaptations, may give an oversimplified impression of the true nutritional needs of many types of animal cells.

As was pointed out in the Introduction to this book (see also Echalier, 1980), the present ambitions of *in vitro* culture of *Drosophila* cells have significantly changed. We must now learn how to maintain *in vitro* various types of differentiated *Drosophila* cells, so that they can continue to enact, at least to some extent, their specialized programs. For this purpose, it should be clear that each type of culture, and that means not only every type of tissue but also every type of problem under study, possibly demands a proper medium.

The "optimization" of culture media requires iterative testing of each constituent in order to determine its preferential concentrations. This must be done in close relationship with all other ingredients, because, in such multivariant systems, all parameters are interconnected. Thus, a component may appear to be "essential" in one context and nonessential in another.

Moreover, the true indispensability of any given substance has to be proved by careful quantitative analyses which would establish its actual utilization by cultured cells or a possible beneficial effect on the metabolism.

Unhappily, this type of time-consuming and thankless research is, today, somewhat old fashioned, and there have been only a few reports about the metabolism and the specific needs of *Drosophila* cells in culture. There is one notable exception, however, namely the aforementioned, careful investigations of Wyss, although they were primarily focused on the needs of one single *Kc* subline.

In regards to further developments in culture of *Drosophila* cells or tissues, it seems worthwhile to confront, for each physical parameter and each category of nutrients, the scarce analytical data available with (1) the empirical and fragmentary observations made on permanent cell lines and (2) the large bulk of information collected concerning the nutritional needs of *Drosophila* larvae or flies (see a critical review by Sang, 1978). It is clear, however, that differentiated cells cultured *in vitro* may require nutrients that are not required by the entire animal, because, in the whole organism, these might be provided by the synthetic activities of other tissues, such as fat body.

A. Osmolarity

At the beginning of this general evaluation, we should recall that most of the culture media currently used for *Drosophila* cells were devised at a time when the only available analytical data about *Drosophila* extracellular fluids concerned the hemolymph of 3rd instar larvae.

Thus, the osmotic pressure was generally adjusted to the physiological value of 360 mOsm (which corresponds to a freezing point depression $\Delta = -0.66°C$, i.e., to a 10.5 g/liter NaCl solution). This is true for Schneider's medium and for our own *D22* medium. In Shields and Sang's *M3*, however, osmolarity is slightly lower (330 mOsm), whereas in Gvozdev and Kakpakov *S-15* it rises to 440 mOsm. Surprisingly, many buffered salines recommended for the *in vitro* handling of imaginal discs or tissues also deviate significantly from the mean values of larval hemolymph; for instance, Robb's DPBS as well as Chan and Gehring's saline does not exceed 310 to 315 mOsm.

In fact, Croghan and Lockwood (*1960*) showed that, even in the whole organism, the osmotic pressure and composition of larval body fluid may appreciably vary according to environmental modifications in spite of evident regulatory mechanisms. For example, the hemolymph of wild larvae, after "adaptation" to a standard diet supplemented with extra 7% NaCl, reached almost 400 mOsm.

There are very few reports about the behavior of cultured *Drosophila* cells or tissues, with respect to graded variations in the osmotic pressure.

Robb (1960) was not able to observe any differences in the incorporation rates of labeled uridine or amino acids into imaginal discs over a relatively large range of osmolarity (between 285 and 345 mOsm); it must be remembered, however, that his assays were limited to a few hours and that it was imperative that the external membrane of the discs remain intact.

The observations by Wyss are much more significant. During tenacious efforts to design a chemically defined medium (see Section II.F), he replaced lactalbumin hydrolyzate from *D22* medium with a mixture of amino acids, which resulted in an important decrease in the osmotic pressure. Then, by adding various osmotic effectors, Wyss and Bachmann (1976) were able to demonstrate that permanent lines *Kc* and *Ca* could proliferate at osmolarities ranging from 225 to 400 mOsm and that this range is dependent on the osmotic agent used. To maintain suitable ionic ratios and, at the same time, avoid a possibly detrimental osmotic effector (for instance, the very large amounts of sucrose that are frequently added, for this osmotic purpose, to insect culture media), Wyss preferred to retain a low 250 mOsm value for all his later formulas. A systematic reevaluation of each parameter in the final and more elaborate medium (called *ZW* medium) gave him the opportunity to confirm that a simple increase in osmolarity (using additional sucrose or chlorides, by steps of 25 mOsm, from 250 up to 375 mOsm) did not improve the cloning efficiency of a *Kc* subline, nor the proliferation of primary cultures (Wyss, 1982a,c).

B. pH and Buffers

With the exception of Gvozdev and Kakpakov's medium (pH 7.2) and Robb's *R-14* medium (pH 7.1), most culture media for *Drosophila* cells conform to the slightly acidic pH value (6.6 to 6.9) that characterizes *Drosophila* larval hemolymph.

The traditional buffering with sodium or potassium bicarbonates, in imitation of the naturally occurring system in vertebrate blood plasma, implies a delicate equilibrium with a gaseous phase comprising a few percent CO_2 in air; and it is not very easy to ensure its control in the tiny culture vessels and small volumes of liquid in which insect primary cultures are usually set up.

Therefore, Echalier and Ohanessian (1969, 1970) preferred to rely merely on the presumed, although controversial (see Section I.B) buffering capacities of the large amount of amino acids contained in their medium. Likewise, Seecof and Donady (1973) eliminated bicarbonate from Schneider's medium for primary cultures of embryonic cells. When trying to define an optimal pH for neuron and myocyte differentiation, they varied the pH between 6.3 and 7.1, and the best cultures were obtained at pH 6.8 to 6.9 (see Chapter 2).

Occasionally, a phosphate buffer (i.e., a proper mixture of disodic and monopotassic phosphates) was used, for instance in Robb's *R-14* medium or DPBS. It must be pointed out, however, that the phosphate concentrations required for a truly effective pH stabilization seem to be poorly tolerated by cultured cells (Echalier, unpublished observations, late 1960s).

If Shield and Sang's medium still contains 0.5 g/liter bicarbonate [which, according to Cross and Sang (1978), should be omitted for microdrop cultures (see Chapter 2)], the pH of this *M3* medium is essentially regulated by an organic buffer, namely 5 mM Bis–Tris. As has been well documented in mammalian cells, when such organic buffers are substituted for bicarbonate, it is advisable to add some dicarboxylic acid to satisfy the CO_2 requirement of the cells; therefore, 2 mM oxalacetic acid were introduced into *M3* medium.

Similarly, another organic buffer (Tricine) enters into the composition of Chan and Gehring's (*1971*) balanced saline; but its pK (8.15) is too far from the optimal pH for *Drosophila* cells. Among the series of organic buffers already tested in physiological studies, PIPES (whose pK is close to 6.8) is probably the most satisfactory for *Drosophila* cell culture media (Echalier, unpublished observations, 1980).

For an efficient monitoring of pH throughout the evolution of cell cultures, a colored pH indicator, which, as in vertebrate cell cultures, has been almost exclusively phenol red, is routinely added to culture media, at concentrations of 1 to 10 mg/liter (which were shown to be nontoxic for vertebrate cells). Because the pK of phenol red (7.9) is rather high and it is not very sensitive to pH changes around 6.7, Sederoff and Clynes (1974) recommended the use of a more suitable indicator, bromothymol blue (pK at 7.3), which is yellow at acid pH, green at neutral pH, and blue at alkaline pH (the suggested concentration is 10 mg/liter).

C. Major Ions

The appreciable differences that can be observed between various successful culture media, with respect to their ionic patterns (see Table 1.XI), prove that *Drosophila* cells can tolerate a large range of variation.

Nevertheless, it seems particularly significant that when Robb (1969) determined—using a rather empirical approach—the best medium parameters for the incorporation of labeled metabolites into cultured imaginal discs, he found the optimal ranges of sodium and potassium concentrations to be centered on measured 3rd instar larval hemolymph values (that is, 40 mM potassium and 56 mM sodium). On the other hand, the *in vivo* figures for calcium and magnesium concentrations were detrimental and had to be considerably reduced, although in the same proportion: down to 1 mM Ca^{2+} (instead of 8 mM) and 2.5 mM Mg^{2+} (instead of 21 mM).

As has already been pointed out in Section I.C.1., the exceedingly high concentration of both divalent cations, which is a common feature of the majority of insect hemolymphs and contrasts with the vertebrate situation, raises a true problem of physiology, so that it was often assumed that these two ions might be partly bound to proteins or organic complexes. The introduction of such large amounts into culture media, especially of calcium, demands some precaution in the preparation to avoid noxious precipitates. Therefore, in many media and buffered salines (see Tables 1.XI and 1.XII), the concentrations of calcium and magnesium were reduced, sometimes drastically, with no apparent problem for the cells. For instance, Wyss's ZW medium, which is able to support the growth of not only established cell lines but also primary cultures, contains only 1mmol Ca^{2+} and 5 mmol Mg^{2+}, i.e., two figures that are close to the optimal values determined by Robb.

From their serial assays for constructing a better defined culture medium, Wyss and Bachmann (1976) concluded that the ratio of Na^+ to K^+, whose physiological value approaches 1.5, can be both increased and decreased considerably, without any damage. Likewise, the chloride ion concentration could be modified between 4 and 80 mmol, without much effect on cell proliferation of *Kc* and *Ca* sublines. It must be pointed out (Sang, 1981), however, that all these experiments were carried out on well-established cell lines that were probably exceptionally robust and

TABLE 1.XI Comparison Between Ionic Concentrations of Culture Media[a]

Ions	Larval hemolymph	Echalier and Ohanessian D22 + 10% FCS	Schneider's medium + 18% FCS	Shields and Sang M3 + 10% FCS	Gvozdev and Kakpakov S-15 + 15% FCS	Robb R14	Wyss ZW	Hanks' balanced salt solution (BSS) for vertebrate cells
Na^+	55 (52–56)	48	67	59	86	67	58	141
K^+	38 (36–40)	22.5	21	40	18	38	27	6
Na^+/K^+	1.4 (1.3–1.5)	~2	>3	~1.5	4.7	1.8	2.18	>24
Ca^{2+}	8	7	5	6	4	1.2	1	1.3
Mg^{2+}	21	16	12.5	16	10.6	2.4	5	0.8
Cl^-	35 (30–42)	26	82	22	120	71	87	145
PO_4^{3-}	<3	2.5	7	5	2.8	2.4	3	0.8
SO_4^{2-}	?	12.5	12	16	0.05	1.2	5	0.8

[a] Currently used culture media for *Drosophila* cells, hemolymph of 3rd instar larvae) and a classical saline solution for vertebrate cells. Data in mmol/liter. Serum supplementation is taken into consideration.

TABLE 1.XII Comparison of Ionic Concentrations, Osmolarity, and pH of Published Saline Solutions for *Drosophila* Tissues[a]

Components	*Drosophila* larval hemolymph (3rd instar)	Ephrussi and Beadle (1936) insect ringer	Becker (1959) ringer	Robb (1969) DPBS	Chan and Gehring (1971) balanced saline	Seecof (1971) Phosphate-buffered saline (PBS)	Cohen and Gotchell (1971) Medium G	Garcia-Bellido and Nothiger (1976)	Jan and Jan (1976) Solution K	Budnik et al. (1986) Mg-rich Drosophila Ringer	Jacobs Lorena (in Ashburner 1989)	Stewart et al. (1994) HL3	Dulbecco and Vogt (1954) PBS for vertebrate cells
Na⁺	55 (50–56)	128	113.5	56	55	118	50	139	57	111	88	80	153
K⁺	38 (36–40)	4.7	1.9	40.5	40	41	40	7.6	40	5.5	64	5	4
Na⁺/K⁺	1.4 (1.25–1.55)	27	60	1.38	1.37	2.9	1.25	18	1.42	20	1.37	16	38
Ca²⁺	8	1.9	1	1	5	2.25	3	1.8	8	1.8	1	1.5	0.9
Mg²⁺	21			2.5	15	6.4	10	0.9	21	20	4	20	1
Cl⁻	35 (30–42)	136.5	115	96.5	105	150	56	138	154	158	144	118	143
PO₄³⁻	<3		0.07	2.4	15								
SO₄²⁻	?			1.2	20	10	10	8.5		1	5		9.5
Glucose				10								Trehalose (5)	
Sucrose				100	50		162		96			Sucrose (115)	
Other substances			Bicarbonate (2.4 mM)		Tricine (10 mM) BSA (1 g/ liter)		Disodium glycero-phosphate (25)		HEPES (5)			Sodium bicarbonate (10) HEPES (5)	
pH	6.7			6.8	6.95	6.75	6.78	7	7.1		6.8	7.2	7
Calculated osmolarity (mOsm)	360	271	231.5	310	315	327	356	296	381	297	306	343	302

[a] Data in mmol/liter. See also IMADS (Singleton and Woodruff, 1994), Appendix 1.G.

"adaptable" (see above). In contrast, Echalier and Ohanessian (1970), then Shields and Sang (1977) were of the opinion that chloride levels that are too high are detrimental to primary cultures. They, therefore, preferred to replace most of this anion with glutamates in their respective culture media.

Besides these major ions, trace metals (Cu, Fe, Zn, Mn) are necessary, at least for long-term cultivation of *Drosophila* cells, as they are for vertebrate cells.

These four minor cations, together with ten other elements, could be measured in adult flies by neutron inactivation (Christie *et al.*, 1983); their figures were the following [in μmole per gram of dry(?) weight]: Cu 0.1, Fe 0.79, Zn 0.59, and Mn 0.064.

Although they might be present as contaminants in most commercially available components and are probably supplied in sufficient amounts with the serum supplement (Glassman *et al.*, 1980), it might be advisable to introduce deliberately very low concentrations of their respective sulfates into any synthetic medium, as in Robb's *R-14* and in Wyss's media. This latter author claimed that he could not obtain any clonal growth after omission of these minor salts.

Notes: 1. For the possible future of *Drosophila* cell culture, it seems of utmost interest that the development of micromethods has made it possible today to measure the ionic concentrations of various *Drosophila* extracellular fluids, as distinct from larval hemolymph; and their ionic patterns may differ significantly from what have been considered up until now to be the physiological norms of *Drosophila*. Thus, Larrivee (1979) reported data concerning the major ions of adult fly hemolymph. Even more interestingly, the perivitelline fluid, that is, the thin layer of liquid that is found under the vitelline membrane of eggs and in direct contact with blastoderm cells, could be sucked off and analyzed with an electron microprobe (Van der Meer and Jaffe, 1983). It can be seen, from Table 1.III, that this perivitelline fluid has almost the same elemental composition as adult hemolymph, with the exception of potassium and sulfur which are, here, three to four times more concentrated. It is especially noteworthy that most ion concentrations significantly differ between the two polar pockets formed by this fluid, which suggests the occurrence of important variations in local extracellular environments (see Section V.1).

2. Because use of *Drosophila hydei* cell lines (see Chapter 3) is common, it should be noted that the hemolymph composition of this other species of *Drosophila* has been published (Hevert, 1974). In its larval hemolymph(?), ionic concentrations were found to be, respectively,

56 mM Na$^+$, 31 mM K$^+$, 9 mM Ca^{2+}, 36 mM Cl$^-$, with an osmotic pressure of 299 mOsm/liter. Remember, however, that *D. hydei,* as well as *D. virilis* and *D. simulans* cell lines have been grown, so far, in the same media as *D. melanogaster* cells (see Chapter 3).

D. Amino Acids

The 10 to 12 amino acids that are "essential" for the rat and other higher vertebrates, that is, which must be fed to these animals because they cannot be synthesized by them, were generally found essential in the diet of insects also. In addition, a few other amino acids, although not absolutely necessary, may be beneficial to growth. Exhaustive nutritional studies on *Drosophila* larvae or adults (reviewed by Sang, *1978*) led to the development of synthetic foods in which the initially used casein was replaced by a defined mixture of amino acids. For instance, 14 amino acids were introduced into Geer's synthetic medium (*1965*), but only 11 of them into its minimal version devised by Hunt (*1979*). It is noteworthy that, in several members of Diptera, an unusual nutritional need was reported for the simplest amino acid, glycine. In addition, insects can show certain patterns of utilization of D-amino acids that are quite different from those of other animals. In some cases, these unnatural forms could partly replace corresponding L-isomers.

Conversely, Widmer (*1973*) investigated the capacity for amino acid synthesis by *Drosophila* larvae of different genotypes after feeding them a semidefined diet with added [^{14}C]glucose. In the wild type, the following amino acids were found to be labeled: alanine, aspartic acid, glutamic acid, glutamine, glycine, phenylalanine, proline, serine, threonine, and tyrosine.

Because each tissue has its own specific metabolism, it would not be surprising if each cell strain, *in vitro*, has, qualitatively as well as quantitatively, its own distinctive requirements. There is, nevertheless, a strikingly consistent basic pattern of some 10 to 13 amino acids that may be considered to be "essential" for animal cells, namely, arginine, cysteine, glutamine, histidine, isoleucine and leucine, lysine, methionine, phenylalanine, threonine, tryptophan, tyrosine, and valine. It is however highly advisable, especially when attempting to isolate new cell strains, and in order to anticipate any possible additional need, to include in culture media a full complement of the 20 or more amino acids that are the common building blocks of most proteins.

In culture media for *Drosophila* cells, the exceptionally high amino acid content generally observed in insect body fluids was taken into account so that amino acids were generously supplied, either (1) globally, as hydrolyzates of a nutritionally significant protein: for instance, *D22* medium contains 13.6 g/liter of lactalbumin enzymatic hydrolyzate ($+134$ mM of glycine and glutamates), or (2) as a defined mix of amino acids: in Schneider's medium there are 79 mM from 20 different amino acids, while Shields and Sang's *M3* medium contains 77.5 mM of glutamates $+ 72$ mM of 21 other amino acids (see Table 1.VII). The respective proportions of individual amino acids were usually based on available analyses of free amino acids from *Drosophila* larvae (see Section I.D). Moreover, it should be remembered that in all these media, yeast extract and, to a lesser degree, the serum supplement furnish additional amino acids. Such high levels of amino acids in all insect culture media have to be compared with the much lower amino acid content of classical media for vertebrate cells. For example, in Eagle's minimum essential medium (MEM) or in Dulbecco's modified Eagle's medium (MEM), a common list of 20 amino acids totals only 6 and 14 mM per liter, respectively.

Systematic research on the quantitative amino acid requirements of established cell lines of *Drosophila* was conducted by Wyss in *Kc* and *Ca* sublines. In *D22* medium, in which these cells had been initially grown, the lactalbumin hydrolyzate could be completely replaced by a mixture of 19 amino acids, each at a 1 mM concentration. Then, one by one, each was removed or reduced to 0.1 mM. The omission of any one of eight amino acids—namely, arginine, asparagine, cysteine, histidine, methionine, proline, serine, and threonine—reduced cell proliferation to less than half that of control cultures and, therefore, they could be considered to be essential. Moreover, it was shown that cysteine, glutamine, glycine, and serine were relatively close to their limiting concentrations (Wyss and Bachmann, 1976). In the provisional medium formulas (named *ZA* and *ZD*), which derived from these investigations, it should not be forgotten that there was an additional, and undefined, supply of amino acids provided by yeast extract, the serum supplement and the 47 mM of glutamic acid that were constitutive of the original *D22* medium.

A more thorough reinvestigation became possible as soon as a better defined medium had been designed, which could, with the single addition of a relatively purified so-called "growth factor" extracted from *Drosophila* flies (and not containing any amino acid), support long-term proliferation of a *Kc* subline (Wyss, 1982a–c). The test chosen for the systematic assaying of all components of the medium was the cloning efficiency, in soft agar (see Chapter 3), of this *KcHP* subline. Among the 23 most common amino acids that were studied over a large range of concentrations, six of them (β-alanine, alanine, aspartic acid, glutamic acid, hydroxyproline, and ornithine) could be eliminated, whereas the omission of any one of the 17 others completely prevented the formation of cell colonies. It is interesting that asparagine is required, even in the presence of aspartic acid and glutamine, and the same is true for glutamine in the presence of glutamic acid and asparagine. Such a precise delineation of the concentrations of each ingredient still compatible with cloning led to the formulation of (1) a minimal medium (*ZR*) in which nonessential amino acids were omitted and other amino acids adjusted to the lowest levels necessary, and (2) an optimized medium (*ZW*, see Section II.F) in which, to prevent any deficiency, a full complement of the usual 23 amino acids was included, most of them at concentrations of 0.5 to 1 m*M*, but five (namely, glycine, serine, glutamine, histidine, cysteine) in deliberately higher amounts (see detailed composition in Table 1.X). On the whole, this medium comprises 42 m*M* of amino acids, which remains a threefold higher concentration than that of standard Dulbecco's MEM for vertebrate cells.

E. Sugars

If trehalose is usually the dominant carbohydrate in insect hemolymphs, and this also seems to be true in *Drosophila*, this diholoside has to be cleaved, *in vivo*, into its two glucose moieties before it can be used by cells. Therefore, it was logical to supply cultured cells directly with glucose, as the principal source of carbon and energy. Most culture media contain a reasonable level of it: from 1.8 to 2 g/liter in *D22* and Schneider's medium, to 5 g/liter in *S-15* and up to 10 g/liter in Shields and Sang's *M3*.

Clements and Grace (*1967*) confirmed that the metabolization of trehalose by Lepidoptera cells from Grace's line was not superior to that of glucose, so that its use in culture media would be unnecessarily expensive. They also showed that sucrose, in spite of a concentration some fortyfold higher than that of glucose in Grace's medium, was not used by cells during the first days of a subculture and that its later utilization coincided with a fall in the glucose level.

The "historical" success of Grace's medium, in which carbohydrates amount to some 28 g/liter, proved, in any case, that insect cells can tolerate exceedingly high concentrations of sugars. Thereafter, relatively large quantities of sucrose were added to several other media or physiological solutions, as a mere osmotic effector, to reach what was considered the proper osmolarity; but such excess is perhaps not quite as innocuous as generally thought (Wyss, 1982c).

F. Vitamins

Ever since pioneering studies began on the nutrition of *Drosophila,* it has been known that this organism requires the water-soluble B vitamins, but not vitamin C and the fat-soluble vitamins A, D, and E (Bacot and Harden, *1922;* see the extensive bibliography in Sang's review, *1978*). As soon as they could be biochemically characterized, the major vitamins of this B complex (thiamin, niacinamide, riboflavin, pyridoxine, pantothenate, biotin, folic acid and, more controversially, cyanocobalamin) were all found to be essential for *Drosophila,* as for insects in general. They are supplied, together with other important nutrients, by yeast,* which is considered to be the natural food of this fly. All the individual B vitamins (with the exception of B_{12}, which, however, may be present as a contaminant) are, therefore, indispensable constituents of any synthetic diet for *Drosophila* (Sang's "C medium", *1956*) and its numerous derivatives). They are usually included in more than sufficient amounts to allow for loss during autoclaving and storage, and it has since been demonstrated that, at least for most of them, considerable excesses can be fed without detriment.

As for *in vitro* cultured *Drosophila* cells, it may, therefore, be safely assumed that they need, at least, the major B vitamins, like most animal cell types. In currently used culture media for *Drosophila,* this requirement is met with a significant amount of yeast extract (Yeastolate: 1 to 2 g/liter). However, most investigators prefer, for safety's sake, to add a supplementary mixture of individual vitamins.

This extra supply of B vitamins seems, however, to be needless when yeast extract is present. Sederoff and Clynes (1974) could eliminate it from earlier versions of Schneider's medium, and it no longer appears in the present formulation of this medium. Similarly, in Shields and Sang's medium *M3,* yeast extract satisfies all vitamin requirements (except for choline, but see Section IV.G.3). Therefore, it is very likely that the long list of vitamins still included in our own *D22* medium could be left out, although we have no experimental confirmation of this, especially so

---

* "Yeast is, however, very deficient in folic acid, and larvae, rather surprisingly, can synthesize it" (Sang, personal communication, 1993).

far as the possible maintenance of our serum-less growing *Kc0* subline is concerned.

Conversely, the same mix of individual vitamins, copied from Grace's formula (*1962*) and present in many subsequent insect culture media, was unable, on its own, to substitute for Yeastolate. Beyond the fact that yeast extract may provide several other essential nutrients or growth factors (see below), Wyss remarked that such commonly used vitamin concentrations were much lower than those for vertebrate cells. In fact, a comparison between Grace's medium and the standard Eagle's MEM medium for vertebrate cells shows that Grace's figures for vitamins are about fiftyfold below the vertebrate media values (see Table 1.XIII). The use of a semisynthetic medium (in which Yeastolate was replaced by its high molecular fraction) enabled Wyss (1979) to test the influence of

TABLE 1.XIII   Vitamin Contents of Culture Media[a]

| | *Medium* | | | | |
Vitamins	*Grace's*[b] (1962)	*Schneider's* (1964)	*Wyss's* ZW (1982)	*McQuilkin, Evans, and Earle's* NCTC 109 (1957)	*Eagle's* MEM[c] (1959)
Thiamin hydrochloride	0.02	0.02	5.0	0.025	1.0
Riboflavin	0.02	0.02	0.01	0.025	0.1
Calcium pantothenate	0.02	0.02	5.0	0.025	1.0
Pyridoxine hydrochloride	0.02	0.02	5.0	0.062	1.0
p-Aminobenzoic acid	0.02	0.01		0.125	
Folic acid	0.02	0.02	0.05	0.02	1.0
Niacin or	0.02	0.02		0.062	
Niacinamide			0.5	0.062	1.0
iso-Inositol	0.02	0.02	50.0	0.125	2.0
Biotin	0.01	0.01	0.05	0.025	
Choline chloride	0.2	0.05	50.0	1.25	1.0
Vitamin B$_{12}$			0.05		

[a] Data in mg/liter.

[b] Vitamin values of Grace's medium for lepidopteran cells were adopted in Echalier and Ohanessian *D22* [+ 1.36 g/liter Yeastolate (Difco)] and, partially, in Gvozdev *et al.* *C-15* medium.

[c] In Dulbecco's MEM; additions included: biotin, 0.1; *p*-aminobenzoic acid, 1; and vitamin B$_{12}$, 0.1.

each B vitamin. Even at concentrations as high as 10 mg/liter, he did not observe any detrimental effect (except for riboflavin) on the cloning efficiency of a *Kc* subline. Qualitatively, the only B vitamin that improved clonal growth, in a dose-dependent manner and up to a maximum of 73% with 20 μg/ml, was pyridoxal (but, curiously, neither pyridoxine nor pyridoxamine was active, which contrasts with their habitual convertibility and suggests the possible intervention of some contaminant in the pyridoxal preparation used*).

After the formulation of an almost completely defined medium, with only the addition of a so-called "growth factor" isolated from fly extracts, Wyss (1982c) could reinvestigate more carefully these vitamin requirements, by varying their concentration over quite a large range. His tested *Kc* subline required calcium pantothenate (or coenzyme A), niacinamide (or nicotinic acid, or NAD), and riboflavin (or FAD) for good cloning efficiency. Surprisingly, the other B vitamins (biotin, folic acid, *myo*-inositol, pyridoxal, thiamin) did not seem to be strictly essential, but this might simply have been due to the short duration of this type of assay. In their presence, however, a strong stimulatory effect was recorded. The detailed results of such thorough analyses and other experiments performed on mass cultures led Wyss (1982a) to raise considerably the concentrations of most B vitamins in the formula for his semisynthetic ZW medium. They are up to 5 times higher, and for inositol 25 times higher, than their respective figures in Eagle's medium for vertebrate cells (see Table 1.XIII).

It must be emphasized that deficiencies in vitamins do not generally produce any immediate effects. Deleterious effects on cells may take several weeks (or even months) to become apparent, which makes the determination of their true requirements tedious and delicate. It is probably why, for instance, the real status of vitamin B_{12} (cyanocobalamin) for insect cells remains equivocal. It was, however, reported (Landureau and Steinbuch, *1969*; Landureau, *1970*) that vitamin B_{12} seems to be essential for cultured cockroach cells and that the level of requirement depends on the presence of folic acid and on the availability, in chemically defined media, of certain metabolites (nucleotides, methionine) in the synthesis of which cyanocobalamin is probably involved (Becker *1976*; Becker and Landureau, *1981*). This is, again, a perfect illustration of the complex interrelations that occur between the various ingredients of a culture medium and of the difficulty of developing a balanced formula.

---

* "This pyridoxal story could not be repeated" (Sang, personal communication, 1993).

G. Lipids, Sterols, Inositol, and Choline

1. FATTY ACIDS

Fatty acids, the main components of all lipids, may be conveniently classified into two categories: (1) saturated and monounsaturated acids that can be synthesized by most animals, including insects, and (2) polyunsaturated acids, such as linoleic acid (18 carbon-chain with two double bonds) and linolenic acid, which are generally essential and must be supplied in the diet or, in tissue culture, by the medium.

In insects, however, the lipid metabolism of Diptera seems to be rather peculiar. *Drosophila melanogaster* may be physiologically independent of linoleic and linolenic acids, since it could be reared aseptically, for at least four generations, on a fat-free synthetic medium. These two polyunsaturated acids, however, could not be synthesized from [^{14}C]acetate (Keith, 1967). Moreover, it was shown that dietary linoleate affected the normal fatty acid composition of the larvae (Sang, 1978).

With this being so, the question as to whether polyunsaturated acids should be added to defined and serum-free media for *Drosophila* cells remains unanswered. Whatever the case, the omission of a lipid stock (containing linoleic acid, but also lecithins, cholesterol, ethanolamine, and phosphorylcholine) from an earlier version of the synthetic medium tested by Wyss (1982c) decreased the cloning efficiency and size of developing *Kc* cell colonies.

2. STEROLS

Sterols cannot be synthesized by insects and are essential in the diet for full development. Cooke and Sang (1970) determined that normal growth of *Drosophila melanogaster* larvae can only be supported by sterols that display (1) an intact planar ring structure, (2) a side chain similar to that of cholesterol, and (3) a 3β-OH grouping. Moreover, phytosterols were preferable to cholesterol.

It is commonly assumed that, besides their role as precursors in the biosynthesis of ecdysteroid hormones (see Chapter 8), sterols or related isopentenoid molecules are required for building up the cell membranes of eukaryotes.

In most tissue culture media, sterol needs are met with yeast extract, but mainly with the serum supplement. Cholesterol levels in mammalian sera range from 167 to 540 μg/ml, with the addition of 473 to 937 μg/ml cholesteryl esters (quoted in Brooks *et al.*, 1980), so that the usual supplement of 10% (v/v) serum results in a final concentration, in culture media, of 20 to $55\mu g$ cholesterol per ml.

On the other hand, in completely defined media, the addition of some cholesterol might be necessary, for instance, in the form of a saturated ethanolic solution. More precisely, Brooks *et al.* (*1980*) suggested that cholesterol be boiled in ethanol, then precipitated with distilled water, because cells might ingest particulate cholesterol by endocytosis. Surprisingly, Tween 80, which is currently used for emulsifying cholesterol during the preparation of media for vertebrate cells, was found to be toxic for cockroach cells.

Silberkang *et al.* (1983; see also abstract in Havel *et al.*, 1980) pointed out that *Kc0* cells (that is, a *Kc* subline adapted to growth without serum) proliferate, although they cannot synthesize any sterol, in a medium containing only 0.01 to 0.03 μg/ml ergosterol (brought with the yeast extract). After addition of serum or cholesterol, the cell growth rate was unchanged and, although cell membranes accumulated cholesterol, their fatty acid composition was not markedly altered. These results seriously challenge the current opinion whereby cultured cells are supposed to have a strict sterol requirement and that these latter are crucially involved in the structure of plasma and intracellular membranes. See also a series of papers on sterol metabolism in *Kc* cells from the same laboratory (Brown *et al.*, 1983; Watson *et al.*, 1985; Havel *et al.*, 1986; Gonzalez-Pacanowska *et al.*, 1988).

3. CHOLINE AND INOSITOL

The function in fat physiology of choline and inositol differs essentially from that of B vitamins—among which they were originally included because of their common occurrence in the growth-promoting yeast and egg lecithins, which were widely used in early nutritional studies. Whereas true vitamins act as coenzymes in basic metabolic transformations, these two factors have a more structural function (as suggested by their much higher dietary requirements). Choline is a component of lecithins (phosphatidylcholines), a major class of phospholipids in all insects. Phoshatidylinositols seem to be quantitatively minor, albeit important. In their above mentioned paper, Silberkang *et al.* (1983) presented details of the phospholipid composition of *Kc* and Schneider's cells (see Table 1.XIV). So, in defined culture media, choline and inositol should be supplied in significantly higher quantities than true B vitamins (see Table 1.XIII).

Conversely, it seems needless to introduce carnitine, a compound related to choline, even though it was shown to replace partially this latter

TABLE 1.XIV   Phospholipid Composition of *Drosophila Kc* and *S2* Cells[a]

| Cells | Phospholipid[b] | | | |
	PC	PS	PE	PG + DPG
			%	
Kc				
Fetal calf serum	9.7	12.6	72.0	4.9
Cab-O-Sil	16.9	15.1	62.0	6.0
S2				
Fetal calf serum	28.7	26.9	41.3	3.2
Cab-O-Sil	34.2	25.2	38.2	2.4

[a] From Silberkang, M., Havel, C., Friend, D. S., McCarthy, B. J., and Watson, J. A. (1983). Isoprene synthesis in isolated embryonic *Drosophila* cells. I. Sterol-deficient eukaryotic cells. *J. Biol. Chem.* **258**, 8509. Kc and S2 cells were grown in *D-20* medium and Schneider's medium, respectively, supplemented with 10% fetal bovine serum (treated or not treated with Cab-O-Sil for removing cholesterol). 30% confluent cultures were labeled with sodium [^{32}P]phosphate for 64 to 72 h, then the cells were processed for phospholipid analysis by TLC.
[b] PC, Phosphatidylcholine; PS, phosphatidylserine; PE, phosphatidylethanolamine; PG + DPG, phosphatidylglycerol plus diphosphatidylglycerol.

metabolite in nutritional studies and was incorporated into unnatural "pseudolecithins" (Geer *et al., 1971*).

H. Other Ingredients

1. NUCLEIC ACIDS

The majority of cells are capable of synthesizing *de novo* the various pyrimidic and puric constituents of their nucleic acids. However, when nucleotides, or their amino bases, are available, cells usually spare their energy and incorporate them directly through reasonably efficient "salvage" pathways (see Chapter 5). Therefore, in animal diets as in tissue culture media, the addition of nucleotides is found to be nonessential, although frequently growth stimulating.

As for *Drosophila*, nutritional studies have shown (see review by Sang, 1978) that, when larvae can grow on sterile media lacking nucleic acids, the addition of RNA, ribonucleosides, or ribonucleotides—but not DNA—to the food can considerably speed up their development rate. These nutrients are, in nature, widely supplied with the yeast diet. As emphasized by Geer *et al.* (*1971*), "the extent to which each strain requires or can utilize a specific nucleotide is a unique characteristic of that strain."

This assertion, of course, holds true for cultured cell strains as well. Marunouchi and Miyake (1978) established that the growth enhancing effect of the Yeastolate present in Echalier's *D22* medium can be replaced by adenine or inosine, without any appreciable change in the generation time of *GM1* and *GM2* cell lines. Their optimal concentration was between 0.1 and 1 m*M*. Fourteen other nucleosides, nucleotides, or constitutive bases were found to be ineffectual. Moreover, the extremely low cloning efficiency observed even with complete *D22* medium could be raised to 10% in the yeast-free medium supplemented with 0.1 m*M* inosine.

Similarly, Wyss (1982c) noted that no clonal colonies of a *Kc* subline were formed when the three nucleosides (inosine, thymidine, uridine) normally contained in his defined medium (see Table 1.X) were omitted. In previous work, the same author (Wyss, 1977, 1979, 1981) had explored the purine and pyrimidine salvage pathways of several other cell lines, but his results will be more fully discussed, together with those of a thorough investigation conducted by Becker (from 1974a–b to 1980), in Chapter 5 which is devoted to nucleic acid metabolism.

In connection with the above, it might be of some interest that Yamane and Murakami (*1973*) were able to identify as 6,8-dihydropurine a crystalline substance, previously isolated from Bacto-peptone (Difco), and which was responsible for the major part of the growth-enhancing activity of this commonly used ingredient of culture media. This compound (a more water-soluble isomer of xanthine), at an optimal concentration of 0.1 to 0.3 mg/liter, increased the yield from certain vertebrate cell lines. (See footnote page 15.) There is no direct evidence, however, that this substance has any stimulating effect on insect cells.

2. TRICARBOXYLIC ACIDS

The unusual levels, observed in most insect hemolymphs, of di- and tricarboxylic acids related to the citric acid cycle led the pioneers of insect tissue culture (Wyatt, *1956*; Grace, *1962*) to include four of these acids (namely, succinic, fumaric, malic, and α-ketoglutaric acids) in culture media, and they seemed, indeed, to stimulate growth of Lepidoptera ovarian cultures.

Therefore, a similar combination of organic acids was retained in the formulas of most media devised for *Drosophila* cells. Shields and Sang's *M3* medium contains, however, only 2 m*M* oxalacetic acid, whose main function is probably to supply CO_2 groups, in correlation with the use of an organic buffer instead of bicarbonate (see Section IV.B).

Wyss (1982c), with an experimental series of single or multiple omissions from his defined medium, demonstrated the importance of five organic acids, and especially of malic acid, on the cloning efficiency of a *Kc* subline; but, again, these results are perhaps specific to the cell strain studied and, moreover, may depend on the relative proportions of all other ingredients present in the medium.

In contrast, Schneider (*1969*) noticed, not without some surprise, that the elimination of all organic acids from the Wyatt–Grace medium, in which cells were grown, stimulated rather than retarded growth of the mosquito (*Anopheles stephensi*) cell line.

In any case, it is interesting to recall that pyruvate, oxalacetate, and α-ketoglutarate are present at concentrations close to 0.1 to 0.2 mmol/liter in the mammalian serum that is routinely added to most culture media.

3. GLUTATHIONE

The tripeptide glutathione (γ-L-glutamyl-L-cysteinylglycine) was shown to improve mammalian cell survival in several culture systems. Because of its SH groups that can be oxidized to give S–S linkage, this molecule can act specifically as an oxidation buffer. In fact, the redox potential is a parameter that has been curiously ignored in most culture media, even though it might reveal itself to be of the utmost importance, at least for certain types of cultured cells (Landureau and Toulmond, *1980*; see also Section V).

Wyss (1982c) confirmed the positive effect of glutathione, with an optimal concentration of 50 mg/liter in his synthetic media, on clonal proliferation of several continuous lines, as well as on primary cultures of *Drosophila* embryonic cells.

I. Serum and Hormones

1. SERUM SUPPLEMENTS

Except for a few experimental series in which, for special purposes, cells have to be grown in a chemically defined medium, it has been a constant rule, in vertebrate tissue cultures, to enrich the medium with about 5 to 20% concentration of a mammalian serum. This is generally bovine serum and preferentially fetal bovine or newborn calf serum; human, horse, or porcine sera may also be used, according to the cell strain. The addition of such a serum appears to be, in most cases and even for established cell lines, absolutely essential for cell proliferation; its omission results in a rapid arrest of nucleic acid synthesis and cell multiplication.

After manifold biochemical investigations, the object of which was to identify putative serum mitogenic factors, and after much controversy as to the respective responsibilities of either the macromolecular components (such as albumin and fetuin) or of small molecules adsorbed to them, the current research on growth factors has finally confirmed the existence, in vertebrate sera, of several specific mitogens. The most important ones seem to be the platelet-derived-growth factor (PDGF, released from platelets during blood coagulation), the insulin-like factors (IGFs) and the so-called Serum Factor.

Therefore, multiple substitutes for serum are now commercially available, which correspond essentially to various "cocktails" of growth factors, plus some minor nutrients and vitamins (for taking into account the other ancillary nutritional functions of serum).

By analogy, the first culture media for insect tissues, for instance Grace's medium for Lepidoptera ovarian sheath cells, were supplemented with 5% (v/v) heat-treated homologous hemolymph.

As for cell cultures from *Drosophila*, collecting sufficient amounts of its own hemolymph was not feasible. Moreover, one disturbing problem encountered with insect hemolymph is its rapid melanization that makes the use of potentially noxious tyrosinase inhibitors (such as phenylthiourea) necessary. After a few disappointing attempts with Lepidoptera hemolymphs (Schneider, 1964; Echalier, unpublished data, 1964; Kuroda, 1973), cultivators had successful recourse to heterologous mammalian sera (Echalier *et al.*, 1965; Echalier and Ohanessian, 1969, 1970; Gvozdev and Kakpakov, 1968, 1969; Schneider, 1964, 1972). Since then, bovine serum, and, more particularly, fetal bovine serum, has been routinely added to culture media, as in vertebrate cell cultures, and at similar concentrations (5 to 20% in volume).

Many batches of mammalian sera are toxic for cultured cells. *Drosophila* cells seem to be especially sensitive, so they may be affected by a stock of serum that has been tested and found to be nontoxic for vertebrate cells. As is well known, this cytotoxicity can be eliminated by simple heating at 56–60°C, in a water bath, for 1/2 hour.

Moreover, different serum batches vary widely in their growth-promoting capacities. Thus, the selection of a proper serum, based on the observation of growth rate, cloning efficiency or modification of cell morphology for every cell type studied, is the decisive stage in successful cultures. Commercial guarantees should not be trusted and one should test the choice oneself.

When a suitable batch has been found, it is preferable to use it without heat decomplementation (Echalier, unpublished observations, 1965). Likewise, Shields *et al.* (1975) stated that a noninactivated serum gives best results in embryonic primary cultures, since a slight lytic activity might stimulate separation and multiplication of some cell types. They advised (Shields and Sang, 1977), however, to keep the complete serum-supplemented medium in the refrigerator (5°C) for 1 to 2 weeks before use (but not for a longer period), in order to temper its initial lytic properties.

The optimal supplement with serum, as already mentioned, corresponds to final concentrations of 10–20%. This level can usually, at least for the simple maintenance of well-established cell lines, be lowered to 1–2%, without any dramatic decrease in their multiplication rates.

We have even selected, from our *Kc* line, a subline which proliferates in medium *D22* without any serum. It was called *Kc0* (for 0 serum), and has been extensively used, since its growth is very economical when large numbers of cells are needed for biochemical investigation. Unfortunately, the karyotype of this strain was found to be rather unstable (see Chapter 4), although no true correlation between its aneuploidy and the absence of serum was established. It is obvious that such serum-independent growth must be due to the "autocrine" capacities of this cell line. The factor(s) responsible have not yet been identified, even though a TGFβ-like activity was characterized in *Kc0* cell extracts (Benzakour *et al.*, 1990; Benzakour, 1992).

Mosna (1972, 1973) similarly reported the gradual elimination of fetal bovine serum, by successive steps of 2%, from the *D22* medium in which the *GM1, 2,* and *3* cell lines were grown. After 2 months with no serum, the only abnormality in *GM2* line was a twofold decrease of the multiplication rhythm. It is not clear, unfortunately, how long the experience could be maintained.

Because fetal bovine serum is rather expensive, various substitutive sera have been tested, with success, at least for maintaining continuous cell lines. For instance, horse serum or porcine serum supported normal proliferation of *Kc* and *Ca* cell lines (Wyss and Bachmann, 1976). Surprisingly, their optimal concentration was found to be only 1%, when, for maximal growth, FBS was needed at concentrations above 2%; when both were used at higher concentrations, a marked reduction in cell yield was observed.

See, at the beginning of Section IV, the successful adaptation of *Kc167* cells and other *Drosophila* cell lines to a serum-free synthetic medium,

HyQ-CCM3, devised for lepidopteran cells and marketed by Hyclone, Ltd. (Cherbas *et al.,* 1994).

2. INSULIN

The discovery of the activity of hormones on *in vitro* cultured tissue goes back many years: Gey and Thalhimer (*1924*) reported a marked increase in the area of outgrowth, in primary explant cultures growing in a medium to which the then recently characterized insulin had been added. This growth-promoting activity, optimal in concentrations varying from 10^{-2} to 1 unit (U)/ml, was, later, amply documented in both vertebrate cell and organ cultures. Although it is associated with the stimulation of glucose uptake and its utilization, and with other anabolic effects that are considered to be typical of this pancreatic hormone, such a generally beneficial action of insulin on cell growth and differentiation remained poorly understood until the discovery of a new family of growth factors whose structure is closely related to insulin. These IGFs, circulating in large amounts in vertebrate plasmas, are the physiological mediators of hypophyseal growth hormone; they are also known as "nonsuppressible insulin" (NSILA) because their insulin-like activity, as assayed on different classical tests, cannot be neutralized by insulin antibody.

Seecof and Dewurst (1974) were the first to report that, in *Drosophila* embryonic cell cultures too, differentiation of certain cell types can be enhanced by the addition of bovine insulin. For instance, the number of myotubes was increased in proportion to insulin concentration (up to a maximum of 3.7 mU/ml). Moreover, after 24 hr *in vitro,* in the presence of a slightly higher insulin concentration (11.1 mU/ml), a 50% increase in the total protein content per culture was observed. Initiation of continuous cell cultures could also be facilitated. Such beneficial effects could be prevented by the removal of all available insulin from the medium with an excess of antibody.*

To explain the intriguing activity of mammalian insulin in insect cells, the same investigators looked for a possible insulin-like substance in the body fluid of *Drosophila.* With radioimmunoassay, they did indeed detect, and measure, some "immuno-reactive insulin" at a concentration equivalent to 410 μU/ml larval hemolymph. The presence of an insulin-like activity in protein extracts from *Drosophila* was rapidly confirmed by Meneses and Ortiz (*1975*), and similar observations were made in

---

* It is noteworthy, however, that Sang and Dübendorfer (Sang, personal communication, 1993) were not able to obtain better cell differentiation in primary cultures, by adding insulin, when cells were grown in better media.

many insects and, in particular, other Diptera. A high-affinity insulin receptor was isolated from adult *Drosophila* (Petruzzelli *et al., 1985a,b*) and, independently, a cDNA was cloned which encodes for a polypeptide similar to the human insulin receptor (Nishida *et al., 1986*). For further information, see Chapter 6.

Wyss (1981a, 1982, 1988) carefully reviewed the beneficial properties of heterologous insulin on *Drosophila* cell cultures. In the restrictive conditions of his defined culture medium ZW (see Section II.F), he confirmed that addition of as little as 1 ng/ml of bovine insulin (with increasing effect up to 10 μg/ml) enables up to 70% of the embryonic cells to survive *in vitro* and undergo spontaneous differentiation (as opposed to less than 1% in nonsupplemented medium). This effect was abolished by anti-insulin antibody. When fly extract and ecdysterone were combined with insulin, extensive differentiation and proliferation, not only of embryonic cells but also of dissociated imaginal disc cells, could be obtained (Wyss, 1982c). As for continuous cell lines, and always in stringent conditions of synthetic media, insulin seems to stimulate their growth rates, although with significant differences from one cell strain to another.

Interestingly, Wyss showed (1981a, 1988) that human IGF-1 had no effect up to 1 μg/ml. As he pointed out, testing of genuine Diptera insulin on culture systems is absolutely necessary. This is also true for all the other growth factor-like substances that have been recently characterized in *Drosophila* (see Chapter 6).

In any case, mammalian insulin [at a concentration of approximately 0.125 International Unit (IU) per ml] has today become a routinely used additive in *Drosophila* culture media (Wyss, 1988; Currie *et al., 1988*; Peel and Milner, 1990).

3. ECDYSTEROIDS AND OTHER INSECT HORMONES

See Chapter 8 which deals with the effects of ecdysteroids and juvenile hormone on *Drosophila* established cell lines.

J. Tissue Extracts

It seems logical to try to correct any possible deficiency in artificial culture media with various extracts, either from eggs or from adult tissues of *Drosophila*. Recent achievements favor this methodological approach.

1. EGG EXTRACTS

Recalling the frequent use of chick embryo extract in early vertebrate tissue cultures, Dewurst and Sang (1977) tried to substitute a saline extract from *Drosophila* eggs for the usual serum supplement. Although 50% of *D1* line cells had died by Day 1 in *M3* medium containing 1% egg extract, the others recovered rapidly and, at the stationary phase (on Day 8) had even increased in number to above the control rate. With an optimized concentration (2% egg extract) and the addition of bovine albumin, to make the protein concentration equal to that of a 10% serum-supplemented medium, *M3* medium supported a normal multiplication of *D1* and several other continuous lines. Preliminary attempts in primary cultures also seemed encouraging.

2. PUPA AND FLY EXTRACTS

The success of *in vivo* cultures of fragmented imaginal discs or even embryos, transplanted into the body cavity of female flies, as developed by Hadorn and his Zürich laboratory (Hadorn and Garcia-Bellido, *1964*; see also Chan and Gehring, *1971*), strongly suggests the presence in the adult organism of some unknown metabolite(s) or, perhaps, more specific growth factor(s) and their beneficial effect on cell multiplication. This specificity is supported by the fact that the proliferation rate is much higher in females than in males and that host flies have to be fed a protein diet, and not only sugars.

In relation to this (see Chapter 2, Section V.A), imaginal disc cells, from fragmented embryos, could generate, *in vitro*, complex adult structures when they were exposed to ecdysteroids, providing they had been implanted into intact animals 4 days before being cultured (Dübendorfer *et al.*, 1974, 1975).

In their pioneering works, Kakpakov *et al.* (1969) pointed out the positive effect of an extract from *Drosophila* pupae, at a concentration of 10%, on the prolonged survival and proliferation of primary cultures and, thereby, on the establishment of continuous cell lines. These permanent lines acquired rapidly, however, the capacity for growth in a medium without pupal extract, providing it was supplemented with calf serum or bovine crystalline albumin (see also Braude-Zolotarjova *et al.*, 1986). The unknown growth-promoting factors of such extracts were not removed by dialysis, but seemed to be considerably inhibited by heating (see Appendix 1.H).

During the search for a defined culture medium (see above), Wyss (1982a) reported some positive results, in primary cultures, when he replaced fetal bovine serum with a specific fraction (called *FX*) from a fly extract. Optimal conditions were reached with 1 to 5% concentration of this fraction and some insulin. He even obtained a limited proliferation of dissociated imaginal disc cells when insulin and a low concentration of ecdysterone were also added.

A highly cationic low molecular mass (less than 400 Da) factor, so far unidentified and called *CalGF*, could be isolated from acid ethanol extracts of adult *Drosophila* flies (Wyss, 1982c). Although it could support, in a nanomolar concentration, the growth of a *Kc* subline in a synthetic medium, it was found to be inadequate for other cell lines and, in primary cultures, did not account for the entire growth-promoting activity of total fly extract (Wyss, 1988). Let us point out that this component is not strictly speaking a growth factor, according to current terminology, because analysis of both hydrolyzed and unhydrolyzed preparations did not reveal the presence of any amino acids.

The real achievement represented by the establishment of continuous lines from imaginal disc cells by Currie and colleagues (1988) (see Chapter 3) seems to have been facilitated by using water extracts from adult flies (essentially females). In *MM3* medium supplemented with 50 μl/ml of fly extract (see Preparation procedure in Appendix 1.H), and also with 2% fetal bovine serum, 1 ng/ml 20-hydroxyecdysone, and 0.125 IU insulin, they obtained the proliferation of primary cultures, then the development of permanent cell lines. It is fair, however, to point out that the independent (and parallel) success of Miyake's laboratory (Ui *et al.*, 1987, 1988) did not require such extracts, but only a previous "conditioning" of serum-supplemented Cross and Sang's *M3(BF)* medium.

K. Antibiotics

Unless cell cultures are used for metabolic studies or any special purpose, it is highly advisable, in order to protect them from contamination, to routinely add low concentrations of wide-spectrum antibiotics to the culture media, for instance: penicillin 50–200 IU/ml and streptomycin 50–100 μg/ml. (See also Chapter 11, Section IV.A.2, for the eradication of mycoplasmas)

V. PROSPECTS

The "imitative approach", that is, the formulation of culture media according to the composition of *Drosophila* larval hemolymph, has demonstrated its possibilities, with the development of some 40 available cell lines, but also its limits. To culture differentiated cell types *in vitro*, two important aspects of the functional organization of the insects should probably be given greater consideration, as was pointed out several years ago by Echalier (1980).

1. Hemolymph, which carries all the metabolites and wastes for the entire organism, reflects in its composition, at a given time, the global activities and reorganization, so extensive in holometabolous Diptera, of all tissues. In fact, it probably does not correspond to the optimal microenvironment of any one of them in particular. Indirect but meaningful evidence leads to the supposition that the "open" circulatory system of insects imposes, as a counterpart, an unusually high degree of tissue "regionalism." For example, it should be recalled that cockroach neurons from a denuded ventral nerve cord were no longer able to function correctly in a physiological solution that copied the unorthodox Na^+/K^+ ratio and the unusually high Ca^{2+} and Mg^{2+} concentrations which characterize the hemolymph of this insect. For normal action potentials to be recovered, it was necessary to provide neurons with, for instance, a Na^+ concentration of some 180 m M/liter, instead of 15 m M/liter in the hemolymph (Treherne, *1974*). Such observations prove without a doubt that the necessity for an efficient "blood–brain barrier" is even greater in insects than in mammals.*

Similarly, the occurrence of a physiological "blood–testis barrier" was demonstrated in each testis follicle of the orthopteran *Locusta,* with the use of electron-opaque tracers. The apical compartment housing spermatogonia was readily penetrated by the tracers, whereas the basal compartment, which contains cysts of differentiating germ cells, appeared tightly closed (Szollosi and Marcaillou, *1977*; see also review by Abraham, *1991*).

Therefore, to return to the formulation of proper culture media for differentiated cells, it is likely that each *Drosophila* tissue requires its own special medium or, at least, the addition of special metabolites, in order to accomplish its complete functional program *in vitro*. In the case of insects, this assertion might go further than the simple truism which claims that specialized cells have specific needs.

It is clearly for practical reasons, primarily the availability of analytical data, that the hemolymph of the full grown *Drosophila* larva was chosen as a model and imitated in most "classical" culture media. Yet, the 3rd larval instar is an especially ambiguous stage, since the important turnover(s) of metamorphosis then commences. Today, the development of micromethods should provide us with further information on *Drosoph-*

---

* See, in Chapter 6, gliotactin and formation of blood–nerve barrier in *Drosophila*.

ila tissue interstitial fluids. For instance, the figures published by Van der Meer and Jaffe (*1983*) concerning the ionic composition of embryo perivitelline fluid (see Table 1.III) deviate somewhat from what was considered to be the norms for *Drosophila*. Even more meaningfully, they observed certain differences between the two polar pockets of this fluid. In very preliminary experiments, we tried likewise, with micropuncture and subtitration methods, to study the actual environment of imaginal disc cells; no definite data are available, but we found the relative concentrations of main ions rather peculiar (Echalier, unpublished data, 1979). Let us recall (Section II.E) that Robb's experiments proved the low permeability of the basal membranes in imaginal discs.

2. Another striking characteristic of insects is the fact that the hemolymph, contrary to vertebrate blood with its respiratory pigments, is not involved in the conveyance of oxygen, since a special ramifying network of tracheoles distributes the latter to the core of tissues and even, sometimes, of individual and particularly active cells. So, *in vivo*, oxygen availability is an essential component of the cell microenvironment, and we should probably take this into account for culturing specialized tissues.

The proper oxygenation of cells grown *in vitro* depends, of course, on the culture methods used (monolayer, suspension cultures; see following chapters), but the composition of the media might be of the utmost importance. Yet, their redox potential is a parameter that has been curiously ignored in most culture systems.

In connection with the above, Landureau and Toulmond (*1980*) reported that hemocyte lines from the cockroach, but not other cell lines tested, could be grown under drastic hypoxic and even anoxic conditions, and thus apparently rely on anaerobic metabolism. Further successes in the culture of various insect or mammal cell types may have been due mainly to careful monitoring of the reducing components of synthetic media (Landureau and Lenoir-Rousseaux, *1988*; Landureau, personal communication).

Furthermore, the response capacities of several sensitive *Drosophila* cell lines to ecdysteroid hormones were tested under various conditions of oxygenation, with the underlying hypothesis that only aerobic pathways support the high energetic cost of the complex programs of differentiation (Ropp *et al.*, 1986; see Chapter 8, Section II.E).

In conclusion, even though the present formulas of culture media have engendered many valuable achievements in *Drosophila* cell culture, significant improvements could obviously be made, if sufficient time and trouble were devoted to do so.

APPENDIX 1

1.A. Preparation of *D22* Medium*

According to our standard procedure for 1 liter, prepare separately the following solutions

1. Potassium glutamate monohydrate   5 g
   Sodium glutamate monohydrate   8 g
   Glycine   5 g
   Dissolve in 100 ml distilled (or deionized) water
2. $MgCl_2 \cdot 6H_2O$ 0.9 g
   $MgSO_4 \cdot 7H_2O$ 3.36 g
   $NaHPO_4 \cdot 2H_2O$ 0.43 g
   Dissolve in 100 ml distilled water
3. $CaCl_2$ in 50 ml distilled water   0.8 g
4. Sodium acetate trihydrate   0.023 g
   Succinic acid   0.055 g
   Malic acid   0.60 g
   Dissolve in 100 ml distilled water
5. Glucose, in 20 ml distilled water   1.8 g
6. Yeast extract, Yeastolate (Difco), in 50 ml distilled   1.36 g
   water
7. Lactalbumin hydrolyzate**   13.6 g
   Dissolve in 300 ml boiling water; then cool.

Mix these solutions in any order, except for Solution 3 which must be added at the very end (in order to avoid calcium precipitation).

Add the complement of Grace's vitamins (see Appendix 1.B)***: If the frozen solution is 500×, as suggested, add 2 ml.

Complete with distilled water up to 1 liter, adjusting to pH 6.6 with 1 N NaOH.

A pH indicator (phenol red: 1 ml of a 1% [w/v] water solution) and antibiotics (penicillin 250,000 U/liter) are routinely added.

Filter sterilize [Millipore (Bedford, MA) membranes HA (0.45 μm) + GS (0.22 μm)], then aliquot into sterile bottles. Can be kept for months in the refrigerator.

---

* This is the final "avatar" of our original *D20* (see Echalier, 1976). See Chapter 1, Section II,A and Table 1.V.

** Several years ago we had some unexplainable problems with different batches of lactalbumin hydrolyzate. We currently use SERVA (Heidelberg) enzymatic hydrolyzate.

*** As discussed in Chapter 1, Section II.A, it was never verified whether such a vitamin supplement is required in the presence of yeast extract.

As usual, D22 medium must be supplemented with 2–10% of fetal bovine serum* (except for a few sublines progressively adapted to grow without serum, e.g., *Kc0*).

1.B. Stock Solution of Grace's Vitamins

This solution is based on Grace's medium for *Antherea* cell line (1962). The stock solution is 500× concentrated, i.e., per 1 liter:

Thiamin hydrochloride	10 mg
Riboflavin	10 mg
Calcium pantothenate	10 mg
Pyridoxine hydrochloride	10 mg
Folic acid	10 mg
Niacin	10 mg
Inositol	10 mg
Biotin	5 mg
Choline chloride	100 mg
(*p*-Aminobenzoic acid	10 mg**)

Divide in small aliquots, which can be kept frozen for years.

1.C. Kakpakov's *C-46* Medium

This medium (Kakpakov, 1989) is commercially available from the Vavilov Institute of General Genetics (Prof. Kakpakov) in Moscow.

Salts		g/liter	mM
	NaCl	4.0	68.4
$NaH_2PO_4 \cdot 2H_2O$		0.5	3.2
	KCl	1.6	21.5
$CaCl_2 \cdot 6H_2O$		0.05	0.23
$MgCl_2 \cdot 6H_2O$		0.25	1.23
	$NaHCO_3$	0.35	4.2
	Tris	3.0	24.8

---

* FBS batches have to be carefully tested for toxicity against *Drosophila* cells (which was not always found to correspond to the commercial guarantees given toward vertebrate cells). We generally prefer batches that can be used without heat inactivation (1/2 hour in a 60°C water bath) (see Chapter 1, Section IV,I).

** *p*-Aminobenzoic acid was accidentally forgotten in our first *D20* formula and deliberately omitted thereafter.

Amino acids

α-Alanine	0.5
Arginine hydrochloride	0.4
Aspartic acid	0.25
Cysteine hydrochloride monohydrate	0.25
Glutamic acid	0.8
Glutamine	0.6
Glycine	0.4
Histidine hydrochloride monohydrate	0.4
Isoleucine	0.2
Leucine	0.25
Lysine hydrochloride	0.8
Methionine	0.5
4-Oxyproline	0.5
Phenylalanine	0.2
Serine	0.35
Threonine	0.5
Tryptophan	0.1
Tyrosine	0.4
Valine	0.3

Vitamins

	mg/liter
Yeast extract	2000.0
Niacinamide	0.1
Thiamin	0.02
Riboflavin	0.02
Calcium pantothenate	0.02
Pyridoxine	0.02
Inositol	0.02
Folic acid	0.02
Inositol	0.02
Choline hydrochloride	0.2

Sugar

Glucose	5.0 g/liter

Indicator

Phenol Red	10 mg/liter

Final pH is pH 7.2; osmotic pressure equals $-0.87°C$.

1.D. Robb's *Drosophila* Phosphate-Buffered Saline

This saline (DPBS) is from Robb (1969).

Stock solutions (2×)		g/liter
1.	NaCl	6.08
	KCl	5.96
	$CaCl_2 \cdot 2H_2O$	0.29
	$MgCl_2 \cdot 6H_2O$	0.49
	$MgSO_4 \cdot 7H_2O$	0.59
	Sucrose	36.0
	Glucose	1.8
2.	$Na_2HPO_4 \cdot 2H_2O$	0.7
	KH_2PO_4	0.1
	NaOH	4ml of $0.1N$ NaOH (pH 6.8)

Stock solutions are autoclaved separately, stored at room temperature, and mixed in equal volumes (1:1) just before use.

1.E. Chan and Gehring's Balanced Saline

This solution is prepared according to Chan and Gehring (*1971*).

	g/liter
NaCl	3.2
KCl	3.0
$CaCl_2 \cdot 2H_2O$	0.69
$MgSO_4 \cdot 7H_2O$	3.7
Tricine buffer (pH 7)	1.79
Glucose	3.6
Sucrose	17.1
+ Bovine serum albumin (BSA, fraction V)	1.0

Sterilize by filtration and store frozen

1.F. Minimum Hemolymph-like Solution

This solution (HL3) is taken from Stewart *et al.* (1994).

	mM
NaCl	70
KCl	5
CaCl$_2$ · 2H$_2$O	1.5
MgCl$_2$6H$_2$O	20
NaHCO$_3$	10
Trehalose	5
Sucrose	115
HEPES	5

Final pH of pH 7.2 and osmotic pressure of 343 $\pm$ 0.7

1.G. Ionically Matched Adult *Drosophila* Saline

This saline (IMADS) is prepared according to Singleton and Wood-ruff (1994).

Ionic Composition*

(Osmolarity:	255 mOsm)
Na$^+$	102 mM
K$^+$	25 mM
Mg^{2+}	15 mM
Ca^{2+}	5 mM
Cl$^-$	55 mM
SO$_4^{2-}$	5 mM
PO$_4^{3-}$	2 mM

Preparation

Sodium glutamate	100 mM
KCl	25 mM
MgCl$_2$	15 mM
CaSO$_4$	5 mM
Sodium phosphate buffer (pH 6.9)	2 mM

Glucose is added to bring the medium to 255 mOsm, measured with an osmometer.

* Derived from the adult hemolymph ionic composition reported by Van der Meer and Jaffe (1983) and from the adult hemolymph osmolarity measured by the authors.

1.H. *Drosophila* Extracts

1. FLY EXTRACT

The extract is prepared as in Currie *et al.* (1988):

"Two hundred well-nourished mature flies (at least 150 females) are homogenized in a tissue homogenizer (15ml capacity, 0.1 clearance), with 1.5 ml of *MM3* medium (see Chapter 1, Section II.C).

This homogenate is centrifuged at 1500*g* for 15 min in a refrigerated centrifuge.

The supernatant and the oily film above it are removed and heat treated at 60°C for 5 min.

This preparation is then spun at 1500*g* for 2 h in a refrigerated centrifuge.

The supernatant ("Fly Extract") is stored at −20°C prior to use.

Medium is resterilized by filtration (0.22 μm pore size) after the addition of fly extract."

2. DROSOPHILA PUPAL EXTRACT

This extract is given in Kakpakov *et al.* (1969, English translation).

"The extract is obtained by grinding 1g of pupae in 3 ml of salt solution *C-15* containing no Ca^{2+} or Mg^{2+}, in a glass Potter blender.

The homogenate is centrifuged in the cold for 20 min at 4000 *g*.

The supernatant fluid is heated at 60°C for 5 min and then frozen.

After thawing, the extract is centrifuged for 1 h at 40,000 *g*.

The supernatant liquid is sterilized by filtration through a G-5 glass filter and stored in the frozen state.

The extract contains 3–8 mg/ml proteins."

References

Abraham, M. (1991). The male germ cell protection barrier along phylogenesis. *Int. Rev. Cytol.* 130, 128–131.

Auclair, J. L., and Dubreuil, R. (1953). Etudes sur les acides aminés libres de l'hémolymphe des Insectes par la méthode de chromatographie sur papier filtre. *Can. J. Zool.* 31, 30–41.

Bacot, A. W., and Harden, A. (1922). Vitamin requirements of *Drosophila* I. Vitamins B and C. *Biochem. J.* 16, 148–152.

Baldwin, E. (1967). *The Nature of Biochemistry.* Cambridge Univ. Press, London.

Becker, H. J. (1959). Die Puffs der Speicheldrüsenchromosomen von *Drosophila melanogaster. Chromosoma* 10, 654–678.

Becker, J., and Landureau, J. C. (1981). Specific vitamin requirements of Insect cell lines (*P. americana*) according to their tissue origin and *in vitro* conditions. *In Vitro* 17, 471–479.

Begg, M. (1955). Osmotic pressure of *Drosophila* larval haemolymph. *DIS,* 29, 105.

Begg, M., and Cruickshank, W. J. (1963). A partial analysis of *Drosophila* larval haemolymph. *Proc. R. Soc. Edinburgh* **68**, 215–236.

Benz, G. (1955). Quantitative Veränderungen der Aminosäuren und Polypeptide während der Entwicklung von *Drosophila melanogaster*. *Archiv. der Julius Klaus-Stiftung fur vererbungsforschung* (Zurich) **30**, 498–505.

Boche, R. D., and Buck, J. B. (1938). Hydrogen-ion concentration of Insect blood. *Anat. Rec.* **72**, Suppl. 108–109.

Boche, R. D., and Buck, J. B. (1942). Studies on the Hydrogen-ion concentration of Insect Blood and their bearing on *in vitro* cytological technique. *Physiol. Zool.* **15**, 293–303.

Brooks, M. A., Tsang, K. R., and Frieman, F. A. (1980). Cholesterol as a growth factor for insect cell lines. In *"Invertebrate Systems in vitro"* (E. Kurstak, K. Maramorosch, and A. Dübendorfer, eds.). pp. 67–77. Elsevier/North Holland Biomed. Press, Amsterdam.

Budnik, V., Martin-Morris, L., and White, K. (1986). Perturbed pattern of catecholamine-containing neurons in mutant *Drosophila* deficient in the enzyme Dopa decarboxylase. *J. Neurosci.* **6**, 3682–3691.

Burnet, B., and Sang, J. H. (1968). Physiological genetics of melanotic tumors in *Drosophila melanogaster* V. Amino acid metabolism and tumor formation. *Genetics* **59**, 211–235.

Chadwick, L. E. (1947). The respiratory quotient of *Drosophila* in flight. *Biol. Bull. Woods Hole* **93**, 229–239.

Chan, L. N., and Gehring, W. (1971). Determination of blastoderm cells in *Drosophila melanogaster*. *Proc. Natl. Acad. Sci. U.S.A.* **68**, 2217–2221.

Chen, P. S. (1962). Free amino acids in Insect. In *"Amino-acid Pools"* (Holden, J. T., ed.). pp. 115–135. Elsevier, Amsterdam.

Chen, P. S., and Hadorn, E. (1954). Vergleichende Untersuchungen über die freien Aminosäuren in der larvalen Haemolymphe von *Drosophila, Ephestia* and *Corethra*. *Rev. Suisse Zool.* **61**, 437–451.

Chen, P. S., and Hanimann, F. (1965). Ionenaustausch-chromatographische Untersuchungen über die freien Aminosäuren und deren Derivate in *Drosophila melanogaster*. *Z. Naturf.* **20B**, 307–312.

Christie, N. T., Gosslee, D. G., Bate, L. C., and Jacobson, K. B. (1983). Quantitative aspects of metal ion content and toxicity in *Drosophila*. *Toxicology* **26**, 295–312.

Clements, A. N., and Grace, T. D. C. (1967). The utilization of sugars by insect cells in culture. *J. Insect Physiol.* **13**, 1327–1332.

Cohen, L. H., and Gotchel, B. V. (1971). Histones of polytene and non-polytene nuclei of *Drosophila melanogaster*. *J. Biol. Chem.* **246**, 1841–1848.

Cooke, J., and Sang, J. H. (1970). Utilization of sterols by larvae of *Drosophila melanogaster*. *J. Insect Physiol.* **16**, 801–812.

Croghan, P. C., and Lockwood, A. P. M. (1960). The composition of the haemolymph of the larva of *Drosophila melanogaster*. *J. Exp. Biol.* **37**, 339–343.

Crone Gloor, von der, U. (1959). Quantitative Untersuchungen der freien Aminosäuren und Polypeptide während der embryonalentwicklung von *Drosophila melanogaster*. *J. Insect Physiol.* **3**, 50–56.

Dulbecco, R., and Vogt, M. (1954). Plaque formation and isolation of pure lines with poliomyelitis viruses. *J. Exp. Med.* **99**, 167.

Eagle, H. (1959). Aminoacid metabolism in mammalian cell cultures. *Science* **130**, 432.

Ephrussi, B., and Beadle, G. (1936). A technique of transplantation for *Drosophila*. *Am. Nat.* **70**, 218–225.

Evans, C. L. (1952). *Principles of Human Physiology*, Churchill, Ltd, London.

Faulhaber, L. (1959). Biochemische Untersuchungen zum eiwess-stoffwechsel der letalmu-tante *lethal giant larvae (lgl)* von *Drosophila melanogaster. Z. Verbungslehre* **90**, 299–334.

Florkin, M., and Jeuniaux, Ch. (1964). Hemolymph: Composition. In *"The Physiology of Insects"* (M. Rockstein, ed.), Vol. 3, pp. 109–152. Academic Press, New York.

Garcia-Bellido, A., and Nöthiger, R. (1976). Maintenance of determination by cells of imaginal discs of *Drosophila* after dissociation and culture *in vivo. Roux's Arch. Dev. Biol.* **180**, 189–206.

Geer, B. W. (1965). A new synthetic medium for *Drosophila. DIS,* **40**, 96.

Geer, B. W., Dolph W. W., Maguire, J. A., and Dates, R. J. (1971). The metabolism of dietary carnitine in *Drosophila melanogaster. J. Exp. Zool.* **176**, 445–460.

Gey, G. O., and Thalimer, W. (1924). Observations on the effects of insulin introduced into the medium of tissue cultures. *Am. Med. Assoc.,* **82**, 1609.

Glassman, A. B., Rydzewski, R. S., and Bennett, C. E. (1980). Trace metal levels in commer-cially prepared tissue culture media. *Tissue Cell* **12**, 613–617.

Gloor, H., and Chen, P. S. (1950). Über ein Analorgan bei *Drosophila*-Larven. *Rev. Suisse Zool.* **57**, 570–576.

Grace, T. D. C. (1962). Establishment of four strains of cells from insect tissues grown *in vitro. Nature* **195**, 788–789.

Hadorn, E., and Garcia-Bellido, A. (1964). Zur proliferation von *Drosophila*-Zellkulturen in Adult-milieu. *Rev. Suisse Zool.* **71**, 576–582.

Hevert, F. (1974). Zur Regulation der IonenKonzentration von *Drosophila hydei* Na$^+$, K$^+$, Ca^{2+} and Cl$^-$. *J. Insect Physiol.* **20**, 2225–2245.

Hunt, V. (1970). A qualitatively minimal amino acid diet for *Drosophila melanogaster. DIS,* **45**, 179.

Jacobs-Lorena, M. (1989). (pers. comm.) in Ashburner (1989) *Drosophila, a laboratory manual,* Appendix. L, 379, Cold Spring Harbor Lab. Press, New York.

Jan, L. Y., and Jan, Y. N. (1976). Properties of the larval neuromuscular junction in *Drosophila melanogaster. J. Physiol.* **262**, 189–214.

Keith, A. D. (1967). Fatty acid metabolism in *Drosophila melanogaster*: interaction between dietary fatty acids and *de novo* synthesis. *Comp. Biochem. Physiol.* **21**, 587–600.

Landureau, J. C. (1970). Rôle biologique de la vitamine B12: analyse de l'exigence vitamin-ique stricte d'une lignée cellulaire d'Insectes *in vitro. C.R. Acad. Sci. Paris* **270**, 3288–3291.

Landureau, J. C., and Lenoir-Rousseaux, J. J. (1988). New culture media for insect cells. In *"Invert. Fish Tissue Culture"* (Y. Kuroda, E. Kurstak, and K. Maramorosch, eds.). pp. 23–27. Jap. Sci. Soc. Press and Springer-Verlag, Tokyo and Berlin.

Landureau, J. C., and Steinbuch, M. (1969). Cyanocobalamine as a support of the *in vitro* cell growth-promoting activity of serum proteins. *Experientia* **25**, 1078–1079.

Landureau, J. C., and Toulmond, A. (1980). Insect hemocytes: cells adapted to anaerobiosis. *Experientia* **36**, 966–967.

Larrivee, D. C. (1979). A biochemical analysis of the *Drosophila* rhabdomere and its extra cellular environment. Ph.D. thesis, Purdue University.

Levenbook, L., and Hollis, W. W. (1961). Organic acid in insects—I. Citric acid. *J. Insect Physiol.* **6**, 52–61.

McQuilkin, W. T., Evans, V. J., and Earle, W. R. (1957). The adaptation of additional lines of NCTC clone 929 (strain L) to chemically defined protein-free *J. Natl. Cancer Inst.* **19**, 885.

Meneses, P., and Ortiz, M. A. (1975). A protein extract from *Drosophila melanogaster* with insulin-like activity. *Comp. Biochem. Physiol.* 51, 483–485.

Mitchell, H. K., Chen, P. S., and Hadorn, E. (1960). Tyrosine Phosphate on paper chromatograms of *Drosophila melanogaster*. *Experientia* 16, 410.

Mitchell, H. K., and Lunan, K. D. (1964). Tyrosine-O-Phosphate in *Drosophila*. *Arch. Biochem. Biophys.* 106, 219–222.

Mitchell, H. K., and Simmons, J. R. (1962). Amino acids and derivatives in *Drosophila*. *In* "*Amino Acids Pools*" (Hoden, J.T., ed.). pp. 136–146. Elsevier, Amsterdam.

Nagasawa, H., Mitsuhashi, J., and Suzuki, A. (1988). Growth factors in the lactalbumin hydrolysate to the cells of the Flesh Fly, *Sarcophaga peregrina*. *In* "*Invert. Fish Tissue Culture*" (Y. Kuroda, E. Kurstak, and K. Maramorosch, eds.) pp. 29–32, Jap. Sci. Soc. Press and Springer-Verlag, Tokyo and Berlin.

Nishida, Y., Hata, M., Nishizuka, Y., Rutter W. J., and Ebina, Y. (1986). Cloning of a *Drosophila* cDNA encoding a polypeptide similar to the human insulin receptor precursor. *Biochem. Biophys. Res. Commun.* 141, 474–481.

Paul, J. (1973). Cell and Tissue Culture. 4th ed. Churchill Livingstone, Edinburgh, London.

Petruzelli, L., Herrera, L., Garcia, R., and Rosen, O. M. (1985a) The insulin receptor of *Drosophila melanogaster*. *In* "Cancer Cells, Vol. 3 Growth Factors and Transformation." pp. 115–121. Cold Spring Harbor Lab. Press, New York.

Petruzelli, L., Herrera, R., Garcia-Arenas, R., and Rosen, O. M. (1985b) Acquisition of insulin-dependent protein tyrosine kinase activity during *Drosophila* embryogenesis. *J. Biol. Chem.* 260, 16072–16075.

Rapport, E., and Yang, M. K. (1981). Effect of food deprivation on larval amino acid pools. *DIS,* 56, 109–110.

Sang, J. H. (1956). The quantitative nutritional requirements of *Drosophila melanogaster*. *J. Exp. Biol.* 33, 45–72.

Sang, J. H. (1978). The nutritional requirements of *Drosophila*. *In The Genetics and Biology of Drosophila* (M. Ashburner and T. R. F. Wright, eds.). Vol. 2a Chap. 3, pp. 159–192. Academic Press, New York.

Schneider, I. (1969). Establishment of three diploid cell lines of *Anopheles stephensi* (Diptera: Culicidae). *J. Cell Biol.* 42, 603–606.

Shaw, E. I. (1956). A glutamic acid-glycine medium for prolonged maintenance of high mitotic activity in grasshopper neuroblasts. *Exp. Cell Res.* 11, 580–586.

Shinoda, T. (1964) (in japanese). Biochemical studies on *Drosophila melanogaster*. 1. Free amino-acids and pteridins. *Seikagu J. Japanese Biochem. Soc.* 36, 816–819.

Stumm-Zollinger, E. (1954). Vergleichende Analyse der Aminosäuren und Peptide in der Hämolymphe des Wildtypus und der mutante *letal-translucida (ltr)* von *D. melanogaster*. *Z. Verbungslehre* 86, 126–133.

Szabo, K., Hollos, K., and Bartalis, E. (1967). Protein and free amino acid changes in different heterozygotes of *Drosophila melanogaster* Meig. in the course of ontogenesis. *Acta Biol. Hung.* 18, 403–419.

Szollosi, A., and Marcaillou, C. (1977). Electron microscope study of the blood-testis barrier in an Insect: *Locusta migratoria*. *J. Ult. Res.* 59, 158–172.

Treherne, J. E. (1974). *Insect Neurobiology* (Treherne, J. E., ed.) North Holland Pub. Co, Amsterdam.

Van der Meer, J. M., and Jaffe, L. F. (1983). Elemental composition of the perivitelline fluid in early *Drosophila* embryos. *Dev. Biol.* 95, 249–252.

Waymouth, C. (1965). Construction and use of synthetic media. *In* Wilmer's *"Cells and Tissues in Culture"* Vol. 1, pp. 99–142. Academic Press, New York.

Widmer, B. (1973). Untersuchungen zur Synthese und zum Metabolismus von Aminosäuren in Larven des Wildtypus und der Letalmutanten *l(3)tr* und *l(2))me* von *Drosophila melanogaster. Insect Biochem.* 3, 181–203.

Wigglesworth, V. B. (1949). The utilization of reserve substances in *Drosophila* during flight. *J. Exp. Biol.* 26, 150–163.

Wyatt, G. R. (1961). The biochemistry of Insect hemolymph. *Ann. Rev. Entomol.* 6, 75–102.

Wyatt, G. R., and Kalf, G. F. (1957). The chemistry of Insect hemolymph II / Trehalose and other carbohydrates. *J. Gen. Physiol.* 40, 833–847.

Wyatt, G. R., and Wyatt, S. S. (1976). The development of an Insect tissue culture medium. *In "Invertebrate Tissue Culture,* Applications in Medicine, Biology and Agriculture" (E. Kurstak and K. Maramorosch, eds.). pp. 249–255. Academic Press, New York.

Wyatt, S. S. (1956). Culture *in vitro* of tissue from the Silkworm *Bombyx mori. J. Gen. Physiol.* 39, 841–852.

Yamane, I., and Murakami, O. (1973). 6,8-dihydroxypurine: A novel growth factor for mammalian cells *in vitro,* isolated from a commercial peptone. *J. Cell. Physiol.* 81, 281–284.

Zwicky, K. (1954). Osmoregulatorische Reaktionen der Larve von *Drosophila melanogaster. Z. Verg. Physiol.* 36, 367–390.

2

Primary Cell Cultures of *Drosophila* Cells

69

The term "primary culture" corresponds to the settlement *in vitro,* the subsequent more or less prolonged survival and the possible proliferation, through a limited number of subcultures, of tissues or cells that have been newly explanted from an organism.

Primary cultures are in themselves an interesting experimental material for the study of the various processes of differentiation, if one is dealing with embryonic cells, or indeed any other problem of cell biology. *In vitro* conditions provide a relatively well-controlled environment, away from the complex interferences of the organism as a whole. Moreover, primary cultures are the compulsory first step toward the establishment *in vitro* of continuous cell lines.

Both aspects have been developed in cultures of *Drosophila* cells, but, as will be discussed below, the setting of proliferating primary cultures, on the one hand, and, on the other, the rather lengthy maintenance *in vitro* of living material until the possible "spontaneous" emergence of continuous lines occurs, require different methods. Primary cultures, as such, are surveyed in this chapter; Chapter 3 will deal with the establishment of permanent cell lines.

Shortly after the first undeniable successes in *Drosophila* cell culture (Echalier *et al.*, 1965; Kakpakov and Gvozdev, 1968; Gvozdev and Kakpakov, 1968), the exploitation of primary cultures has been continued mainly by two groups of investigators. Since 1968, Seecof and many collaborators have analyzed extensively the *in vitro* differentiation of larval-type muscles and neurons, in all their morphological, electrophysiological, and biochemical aspects; their results have been published in some 30 papers. Independently and in parallel, Sang and colleagues carefully studied the form and mode of development, not only of the same two predominant cell categories, but also of many other differentiated types which appear in cultures from early embryos. In close collaboration with this latter laboratory, Dübendorfer focused on the analysis of metamorphosis, after hormonal treatment *in vitro* or after transplantation into pupating larvae, of imaginal disc cells growing in such embryonic cultures amid larval cells.

Lastly, certain observations by Kuroda should also be quoted, although most of them are scattered about in multiple and difficult-to-find brief notes (more often than not, written in Japanese) (see Bibliography at end of this volume).

I. SETTING UP PRIMARY CULTURES

A general preparation procedure, derived from not only our laboratory but also from valuable suggestions from other laboratories, is given in Appendix 2.A.

We will discuss the successive technical steps necessary for the initiation and maintainance of primary cultures of *Drosophila* embryonic cells. The reader is also referred to two excellent reviews: Schneider and Blumenthal (1978) and Sang (1981). Furthermore, detailed descriptions of culture methods for *Drosophila* embryonic cells may be found in the TCA Manual (Donady and Fyrberg, 1975, 1977; Fyrberg *et al.*, 1977); they are reproduced here (Appendix 2.D) with the kind permission of Professor Donady and the Society for *In Vitro* Biology.

A. Embryonic Cell Material

With rare exceptions, e.g., a few cultures of larval hemocytes and, more recently, successful cultures of dissociated disc cells (see Chapter

3), most *Drosophila* cell cultures have been made from embryos. It was indeed assumed, as a result of the long experience of vertebrate tissue culture, that embryonic material is more plastic and more readily adaptable to artificial conditions and that the high mitotic rate which characterizes early embryogenesis favors *in vitro* proliferation. Another obvious practical advantage for culture is the easy sterilization of the egg surface.

The relative merit of any embryonic stage intended as primary explant is still questionable. The choice seems to depend mainly on the experimental goal: if one wishes to study differentiation of larval tissues, it is logical to start with early embryos (up to 6 to 8 hours old). In contrast, fragments from late embryos (20 to 24 h old) will display abundant growth of imaginal* material.

With cell cultures made from single dissociated embryos whose developmental stages** can be accurately defined, Seecof and colleagues (Seecof and Alleaume, 1968; Seecof *et al.*, 1971; Seecof and Donady, 1972) established that cells from blastulae do not survive, whereas, after initiation of gastrulation, the number of cells capable of differentiating *in vitro* increases progressively; a near maximum is reached with 30-min-older embryos. Therefore, in all later works, they used embryos less than 40 minutes old after the onset of gastrulation (signaled by the appearance of the ventral furrow). This corresponds approximately to embryos 4 h after fertilization (at 25 to 26°C). At this stage, a *Drosophila* embryo contains about 4000 cells which are not yet morphologically differentiated. It is noteworthy, however, that the same group (Petersen *et al.*, 1977) started with older embryos (up to 8 h and even 15 h of development) when growing continuous cell lines.

Sang's laboratory used embryos that had been incubated at 25°C for 6 to 8 h after oviposition (Shields and Sang, 1970; Dübendorfer *et al.*, 1975). In individual embryo cultures and for a more precise comparison between wild-type and mutant strains, Cross and Sang (1978a,b) selected embryos in the postgastrulation stage (45 min at 22°C, after the appearance of proctodeal invagination).

We (Echalier *et al.*, 1965) obtained proliferating cells from mid-embryogenesis eggs (6 to 12 h after oviposition, at 25°C). It must be

---

* It must be borne in mind that hatched larvae of holometabolous Diptera harbor segregated groups of cells that will form adult structures at metamorphosis. Cultures from embryos, therefore, contain two kinds of cells with quite different destinies ("larval" and so-called "imaginal," i.e., presumptive adult cells).

** The developmental stages of dechorionated eggs can be readily determined through the transparent vitelline membrane, under a binocular microscope. See detailed Stages of Embryogenesis in Campos-Ortega and Hartenstein (*1985*).

pointed out that our aim was primarily to grow continuous lines, so we were therefore looking for embryonic material from which mechanical dissociation would release cell clumps of a proper size (Echalier and Ohanessian, 1969, 1970; see Chapter 3).

In contrast, Schneider (1972) preferred to initiate cultures from much later embryos (20 to 24 h old, at 22°C). Schneider and Blumenthal (1978) specified that embryos should, at a minimum, exhibit well-defined tracheal tubes. Neonate larvae (1 to 2 h after hatching) were equally suitable: Chakrabartty and Schneider (unpublished, quoted in Schneider and Blumenthal, 1978) could grow a cell line, called *WR72-Di-1,* from newly hatched larvae of *Drosophila immigrans* (see Chapter 3, Section III. F).

All these late stages required a slight tryptic dissociation and resulting primary cultures were principally characterized by the growth of cell spheres, issuing from cut ends of embryonic fragments and obviously deriving from imaginal discs (see Section V.A).

B. Cell Dissociation

Efficient setting up of cell cultures from any tissue requires previous loosening of its intercellular attachments in order to release discrete and viable cells, or, at least, small cell clumps. In the case of vertebrate tissue, methods of cell dispersion are essentially based on an enzymatic treatment (predominantly with trypsin, but sometimes, for tissues particularly rich in collagen, with collagenase).

Conversely, most *Drosophila* cell cultivators agree with our opinion that early embryonic cells tolerate tryptic treatment poorly (Echalier and Ohanessian, 1970; Shields and Sang, 1970), and therefore mechanical dissociation techniques have been widely adopted. These latter take two main forms.

For mass cultures, several hundred dechorionated whole embryos are physically disaggregated in a small conical glass (or glass–Teflon) homogenizer (proper clearance around 0.2 mm), with a few gentle strokes of the pestle (Echalier *et al.,* 1965) (see details in Appendix 1.A).

In the case of cultures from individual embryos, each egg at the selected stage is, after dechorionation, placed on a dry surface and cleanly punctured with a micropipette (internal tip diameter of 50 to 70 μm) filled with culture medium and equipped with mouth-adapted tubing. The entire contents of the egg, or a specific embryonic region extruded from a localized puncture, is sucked up, then expelled into the culture drop and pipetted up and down, once or twice, to ensure good cell dispersal (Seecof *et al.*, 1971, 1973; Cross and Sang, 1978).

Schneider (1971, 1972) used much older embryos (20 to 24 h old). After cutting them into 2 to 3 fragments, it was necessary to expose them to the slightly dispersive action of 0.2% trypsin* (for 20 to 45 min at room temperature). As a matter of fact, this does not result in true cell dissociation, but, presumably, in sufficient loosening of the intercellular matrix. In their 1978 review, Schneider and Blumenthal noticed that collagenase (*Clostridium*, Sigma), used at the same 0.2% concentration in Rinaldini's solution but for 1 h at room temperature, was found to be more effective than trypsin and hyaluronidase; Pronase damaged too many cells.

Similarly, the isolation of viable cells from imaginal discs undeniably requires some sort of enzymatic treatment. Wyss (1982a) praised the use of *proteinase VII* (Sigma): some 50 to 300 dissected wing discs were transferred to 1 ml of synthetic *ZW* medium to which was added 50 μl of a 5 mg/ml water solution of proteinase VIII. After incubation for 1 hour at 25°C, the discs were washed in fresh medium, then dissociated by vigorous pipetting.

Using the same material, Currie *et al.* (1988) returned to a more classic tryptic treatment in a Ca^{2+}- and Mg^{2+}-free buffered saline, preceded by treatment with Dispase and EDTA, and followed by Vortex shaking and pipetting (see Appendix 2.B). Let us point out, however, that the final trypsin concentration remains rather low (0.1%) and that, at the recommended room temperature, the enzyme is not very active.

C. Methods of Culture

All the methods used in primary cultures of *Drosophila* cells are ruled by three basic considerations: (1) the small size of explants, (2) the cell concentration that must necessarily be ensured in the cultures (around 0.5 to 1 × 10⁶ cells per ml), (3) their possible observation under the highest magnification of a phase-contrast microscope.

A variety of technical solutions have been devised, according to the volume of medium used and the required cell confinement.

---

* In Rinaldini's Ca^{2+}- and Mg^{2+}-free saline solution (*1958*): NaCl, 800 mg; KCl, 20 mg; $NaH_2PO_4 \cdot H_2O$, 5 mg; $NaHCO_3$, 100 mg; $Na_3C_5H_6O_7 \cdot 3H_2O$, 68 mg; glucose, 100 mg; distilled water 100 ml.

1. COLUMN-DROP METHOD

Sang's school developed a modification of the classical hanging-drop technique that provides better conditions for phase-contrast microscopy. Shields and Sang (1970) described this so-called column-drop method in the following terms: "The depression slide has a cavity of 15 mm diameter and 0.8 mm depth. The drop of cell suspension, laid on the coverslip, is made sufficiently large and compact to make contact with the bottom of the depression slide when this latter is lowered onto it, so that a column of liquid, generally quite stable, is formed. Slides are kept inverted for 1 hour after mounting to allow the cells to settle out and adhere to the coverslip, but no longer, to avoid settling out of subcellular debris. Medium changes, generally performed every 3 days, could be accomplished, without loss of too many cultures, by carefully swiveling the coverslip (only sealed with Vaseline) to free one of its corners, lifting by this to break the column and free the coverslip completely, removing the half of the drop left in the depression slide and placing an equal volume of fresh medium on the half left on the coverslip and reforming the column by replacing the slide in the original way."

This column-drop method was significantly improved by using a new glass chamber (Shields *et al.*, 1975; Dübendorfer and Eichenberg-Glinz,

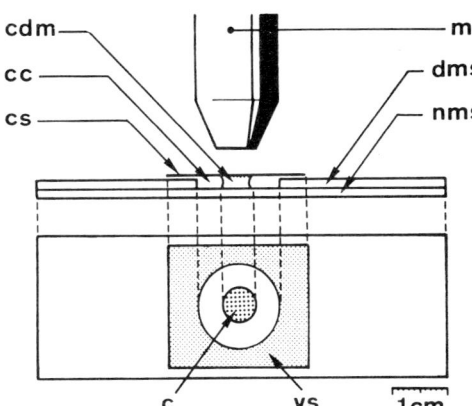

FIGURE 2.1 Chamber for "column-drop" method of culture. Seen from the side and from above. The cells are attached to the coverslip, i.e., to the roof of the chamber: c, cells; cc, culture chamber; dms, drilled microscope slide; nms, normal microscope slide; cdm, column drop of medium; cs, coverslip; ml, microscope lens; vs, petroleum jelly seal. From Dübendorfer and Eichenberger-Glinz (1980); courtesy of Andreas Dübendorfer.

1980): Two ordinary microscope slides were stuck together with silicone rubber, the top one having a hole drilled through it to make a chamber of 15 mm diameter and 1 mm depth (see Fig. 2.1 and handling procedure in Appendix 2.C). As before, cultures are mounted as column drops. The greater depth of this new slide chamber permits a greater volume of medium and, thus, longer intervals between medium changes (every 4 to 7 days instead of every 3 days).

2. MICRODROPS UNDER PARAFFIN OIL

The method devised by Lwoff and colleagues (*1955*) for collecting the viral yield from single vertebrate cells is very suitable for growing a reduced number of cells.

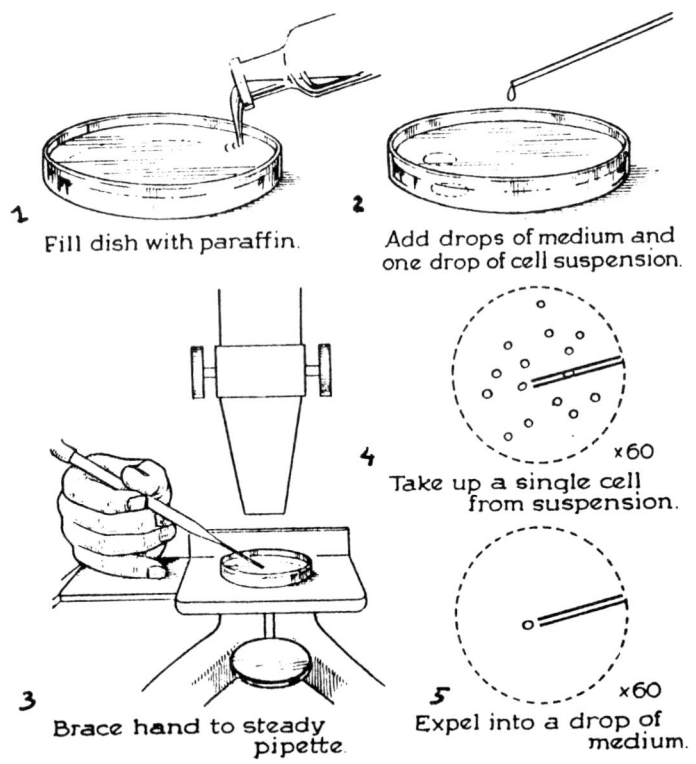

1 Fill dish with paraffin.

2 Add drops of medium and one drop of cell suspension.

3 Brace hand to steady pipette.

4 Take up a single cell from suspension.   ×60

5 Expel into a drop of medium.   ×60

FIGURE 2.2   Microdrops under paraffin oil. From J. Paul, *1973*. With permission.

Medium-saturated liquid paraffin* is poured into a small sterile glass or plastic petri dish (to a depth of 1 to 2 mm). Under a binocular microscope and with a fine capillary pipette attached to a mouth tube and filled with medium, a few individual droplets (5 to 10 μl) are expelled directly in contact with the dish bottom, so that they adhere separately to the glass (or treated plastic) (Fig. 2.2). It is easy, by further pipetting, to introduce the cell suspension into these drops and, periodically, to change the medium partially. The cells, which spread and proliferate on the dish floor, must be observed through an inverted microscope. Ink squares may be drawn under the dish, in order to facilitate the identification of each drop.

This method enabled Ohanessan and Echalier (1967) to subculture *Drosophila* embryonic cells successfully and they could demonstrate the multiplication, in such minimonolayers, of the *Drosophila* hereditary virus *Sigma* (see Chapter 11). In similar conditions, Seecof et al. (1971, 1973) could grow, under paraffin oil, cells from a single embryo.

3. VAGO-HIRUMI FLASKS

For careful monitoring of primary cultures or subcultures, we (Echalier et al., 1965; Echalier and Ohanessian, 1969, 1970) have extensively used a convenient device designed by Vago (personal communication, 1963) and improved by Hirumi (*1963*): it is a small Carrel-type glass flask in the bottom of which a central hole (15 mm diameter) has been cut out (see Fig. 2.3). This hole is closed by a coverslip, affixed with a 1 : 1 mixture of paraffin and petroleum jelly. In Hirumi's variant, three small legs prevent its accidental displacement. Sitting drop cultures are established on the internal side of the coverslip. The medium can be easily changed through the neck of the flask, and the cover glass provides a high-quality optical surface for observation of the culture with an inverted microscope.

4. MICROWELL PLATES

When isolation of a few cells in a reduced volume of medium (about 150 μl) is needed for primary cultures (Currie et al., 1988) or cell cloning (see Chapter 3), most investigators prefer to use the small wells of a "95-microwell plate" (especially treated for cell culture). For short cultures of a few days, evaporation can be minimized by keeping the plates in a humidified incubator, but we found it necessary for longer cultures (for

---

* *Paraffinum liquidum* (viscosity 125/135) does not need to be sterile, but must be saturated with medium. In a sterile bottle, paraffin is shaken with a small volume of medium, then left in contact with the latter for several days in the refrigerator, before use.

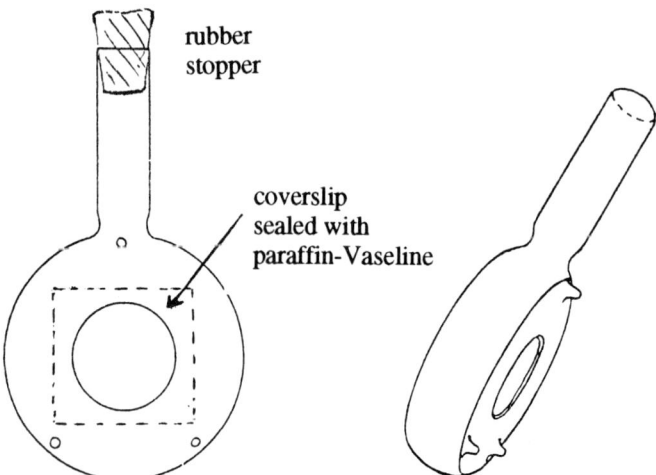

rubber
stopper

coverslip
sealed with
paraffin-Vaseline

FIGURE 2.3 Vago–Hirumi tissue culture flask. Ventral view at left shows position of hole and legs and sealed coverslip. From Hirumi (1963).

instance for cloning, see Chapter 3) to seal the plate cover tightly with an impermeable sticky tape.

5. Petri Dishes

For any larger cultures, for instance mass cultures initiated from several hundred *Drosophila* embryos, as well as monolayers of continuous cell lines, a great variety of commercially available culture vessels exists (Petri dishes or flasks). These are, nowadays, almost exclusively disposable, made from sterilized plastic, and their bottom surface is specifically treated for cell culture (thus providing strong cell adhesion and nontoxicity). Falcon (Bector Dickinson Labware, Franklin Lake, NJ) and Nunc (Nunc, Ltd., Roskilde, Denmark) products are the most extensively used commercial brands.

Small culture dishes (35 mm diameter × 10 mm) have been widely used in primary cultures, especially by Seecof and colleagues. These culture dishes contain 1 to 2 ml medium and have to be incubated in a moist atmosphere. Because some cell types attach more readily to glass than to plastic, or in order to facilitate subsequent fixation and staining of the culture, a coverslip is inserted into the dish and the cells pipetted onto it (Seecof and Donady, 1972; Seecof *et al.*, 1972). When the culture is restricted to the few thousand cells from a single embryo, it is convenient to gather the cells in the center of the dish, by pipetting them over a circular area (some 2 mm diameter), and to let them attach before adding further medium; moreover, in order to facilitate their localization under the microscope, a card with an inked circle delimiting the cell area can be placed underneath the transparent dish (Donady and Seecof, 1972).

With the same purpose in mind, that is, to help small groups of cells to create, by themselves, a local environment favorable for growth (see cloning methods, Chapter 3), Kuroda (1974) suggested that the embryonic fragments be confined between a coverslip and the bottom of the glass dish or flask.

6. SCREW-CAPPED FLASKS

For cultures that have to be kept for long periods (several weeks or months) in order to develop continuous cell lines (see Chapter 3), it is highly advisable to use screw-capped flasks. These will limit the chances of contamination, which is particularly to be feared in the saturated atmosphere of incubators.

II. EVOLUTION OF PRIMARY CULTURES FROM DISAGGREGATED WILD-TYPE EMBRYOS: CHARACTERISTICS OF MAIN CELL TYPES

Almost immediately after preparation of the culture, individual cells and cell aggregates of moderate size settle down and adhere quickly to the bottom of the culture vessel. Some cells can be seen to flatten, whereas others, in the hours that follow, will begin to extend timidly from the periphery of the clumps.

At this early time, all the cells from young embryos seem to be fairly similar, even though their size may vary in diameter from 5 to 25 μm. When special care was taken with the developmental stage of donor embryos, it could be confirmed that cells newly dissociated from postgastrulation embryos do indeed uniformly display unspecialized ultrastructures (Seecof *et al.*, 1971). Under a light microscope (see Fig. 2.4A), they appear either to have a circular, tear-drop shape, or to be rather angular, some of them with short processes. Their clear and relatively large nucleus, with a prominent single nucleolus, contrasts with a densely granular cytoplasm.

A

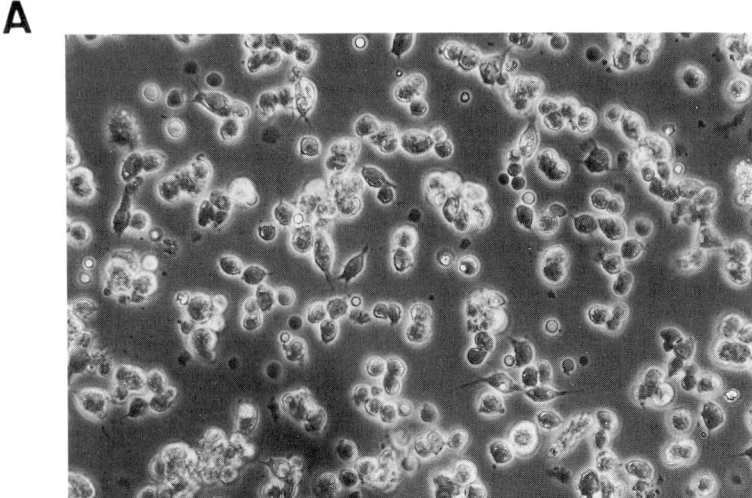

B

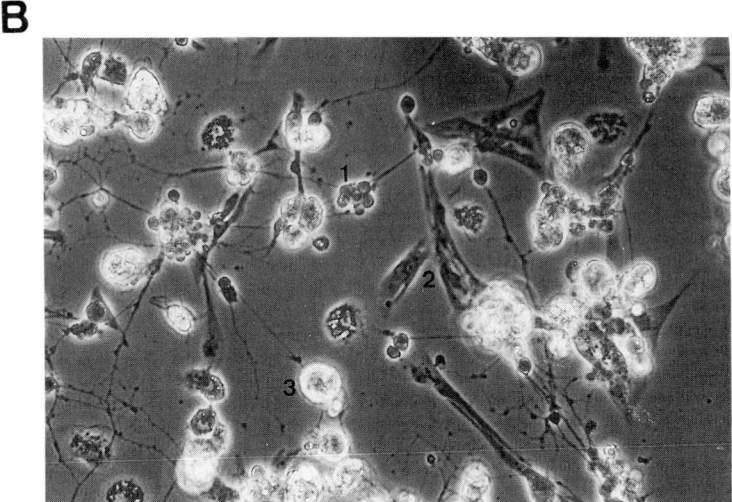

FIGURE 2.4   Primary cultures from *Drosophila* embryonic cells (early gastrulae). Magnification: ×1400. (Courtesy of A. Dübendorfer) (A) First hour after initiation of the culture. (B) Third day of development *in vitro*: differentiated nerves (1), muscles (2), and fat body (3) cells.

A variable proportion of these primary cells die rapidly, perhaps because they have been injured during the disaggregation process. They round up and detach. Thus, despite some apparent cell multiplication (division figures are, however, not very easy to distinguish), the cultures have usually thinned out significantly by 24 h. Remaining cells seem healthy and some of them will quickly enter various differentiation pathways.

Several distinct cell types become clearly identifiable, mainly by morphological criteria. Their respective characteristics and mode of development are described below, essentially according to the detailed descriptions given by Seecof and colleagues, for muscle cells and neurons, and by Sang and colleagues, for all other cell types. It seems meaningful to regroup them into three main categories, according to their early or late development and to their spontaneous or hormone-induced evolution (Fig. 2.4B).

It is fair to say that, despite improved methods, the developmental pattern of primary cultures of *Drosophila* cells still varies greatly. This, as pointed out by Shields *et al.* (1975), is possibly due to multiple factors that are difficult to control, namely, "the degree of dissociation and possible damage caused to the cells, the initial cell density, the size and timing of medium changes, and quality of the serum" used for supplementing the medium.

The following descriptions, of course, refer to the most successful cultures, and are directly inspired by the papers of the different authors.

A. Early Differentiating Larval Cell Types

1. MUSCLE AND NERVE CELLS

The differentiation of muscle and nerve cells is almost complete on the first day of culture. Some cells rapidly elongate and adopt a spindle shape. Usually grouped in a small bundle, they may also fuse into "sheeted" or "tube-like" syncytia. After having differentiated cross-striated fibrils, they very soon contract, at a slow rhythm or convulsively. Their muscle nature is obvious. At the same time, other cells, after rapid and characteristic divisions (see below), grow long axonal processes that ramify and spread over the bottom of culture flasks, often joining muscle fibers. They are functional neurons.

The eye-catching bustle produced by such numerous pulsatile foci and the complexity of the nerve network that connects cell clumps are among the more typical features of successful primary cultures, as observed during the first and second weeks.

Because of the special interest in both cell types, Sections III and IV are devoted to their *in vitro* differentiation.

2. *LARVAL FAT BODY CELLS*

These cells develop from small groups, usually created by the spreading of small cell clusters, of large polygonal cells that possess a central nucleus with its typical nucleolus, and whose cytoplasm is heavily charged with dark granules. At this stage, they were initially misinterpreted as being epithelial-type cells (Shields and Sang, 1970). In fact, as was soon recognized (Shields *et al.*, 1975; Dübendorfer and Eichenberger-Glinz, 1980), fat cell differentiation starts after one day in culture and it is marked by a rounding up of those cells and their progressive filling with bright refractile droplets. Red staining of such droplets with the dye Oil-Red-O indicates that they contain neutral fats (Seecof and Dewurst, 1974).

This process of lipid droplet accumulation and concomitant cell enlargement continues until the eighth to twelfth day of culture and many fat body cells may thus reach about four to six times their initial diameter (up to 50 μm, which corresponds approximately to a 1000-fold increase in volume) (Fig. 2.5). Moreover, they tend to become more loosely attached to the flask bottom, even though they remain together in translucent groups, especially abundant in some cultures and closely resembling the fat body lobules from 2nd or 3rd larval instars. Electron microscopy confirmed this similarity, showing large lipid vacuoles up to 6 μm in diameter, little glycogen, and no protein granules (Dübendorfer and Eichenberger, 1985).

It is noteworthy, and very likely in correlation with the crucial metabolic role played by the fat body in the insect organism, that the differentiation of numerous adipocytes seems to precede, in the cultures, the appearance of a series of new cell types. As suggested by Shields *et al.* (1975), fat body cells may have a beneficial conditioning effect on the medium or perhaps release some specific, although unidentified, growth factor(s). Moreover, after addition of ecdysone, they are very probably responsible for its conversion into the active hormone, 20-hydroxyecdysone (Dübendorfer, 1988, 1989).

3. *HEMOCYTES*

Their precursors are rather difficult to recognize and might correspond to colonies of proliferating rounded and granular cells which eventually

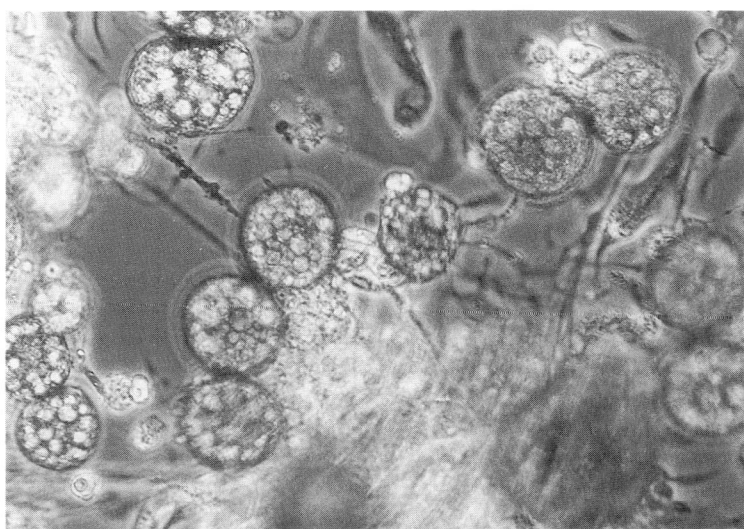

FIGURE 2.5 Differentiation of fat body cells. Primary culture of *Drosophila* embryonic cells (24th day *in vitro*). From Dübendorfer and Eichenberger (1985); courtesy of Dübendorfer.

reveal their nature by developing small collarettes of transparent cytoplasm or frail pseudopods.

Within a few days, several differentiated types can be identified, which are reminiscent of the various categories of hemocytes found in *Drosophila* larval hemolymph (see the classification proposed by Rizki and Rizki, 1984). Their identification *in vitro* was due to Sang's school (Shields and Sang, 1970; Shields *et al.*, 1975; Cross and Sang, 1978), whereas Dübendorfer and Eichenberger-Glinz (1980) published convincing pictures of transmission and scanning electron microscopy (Fig. 2.6A,B). The *plasmatocytes* (or macrophage-like cells) are the basic form, according to Riski's classification, and the most frequently observed in cultures. They correspond to the small groups, described above, of round cells (diameter ca. 10 μm), with a granular cytoplasm and short pseudopodial processes. Transmission electron microscopy (TEM) examination shows a variety of inclusions, such as endocytotic vesicles, lysosomes, or residual bodies. *Podocytes* are variant forms, characterized by long and filiform extensions of their cytoplasmic surface. *Lamellocytes* seem also to derive from plasmocytes. Losing their granulations, they flatten out considerably and

A

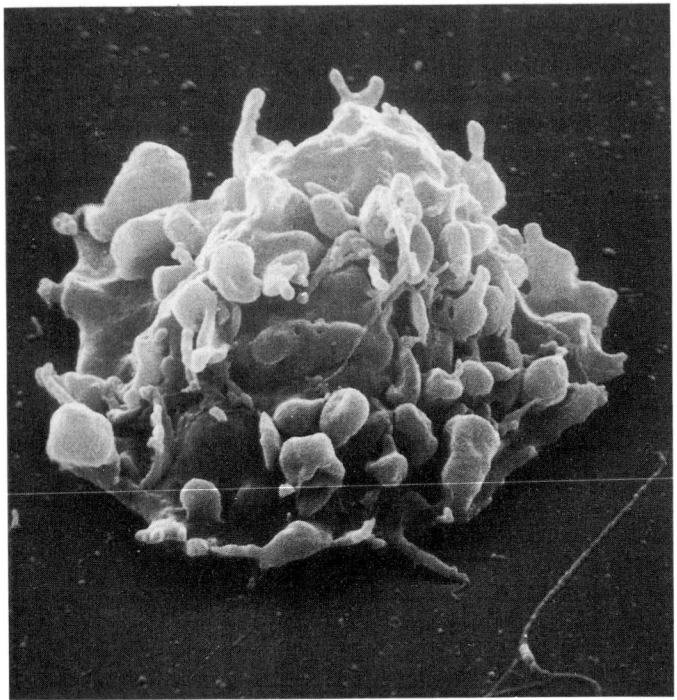

FIGURE 2.6   Hemocytes in primary cultures of *Drosophila* embryonic cells. Scanning electron micrograph (SEM) of (A) a plasmatocyte (×800) and (B) a lamellocyte (×100). From Dübendorfer and Eichenberger-Glinz (1980); courtesy of Dübendorfer. (*continues*)

become typical giant disc-shaped and clear cells (reaching 60 to 100 μm in diameter and less than 0.2 μm in thickness at their outermost borders). Because of their small dark perinuclear area, contrasting with the large transparent and pleated aureola of the peripheral cytoplasm, they have been compared to young medusae spread out on the bottom of the culture flask.

Furthermore, Sang (personal communication, as quoted by Dübendorfer and Eichenberger-Glinz, 1980) observed a distinct type of hemocytes, namely *crystal cells* with paracrystalline inclusions.

These hemocytes, especially the flattened forms, show a tendency to melanize. Sometimes they do so *en masse*, which should always be interpreted as a sign of deterioration of the culture (Cross and Sang, 1978).

B

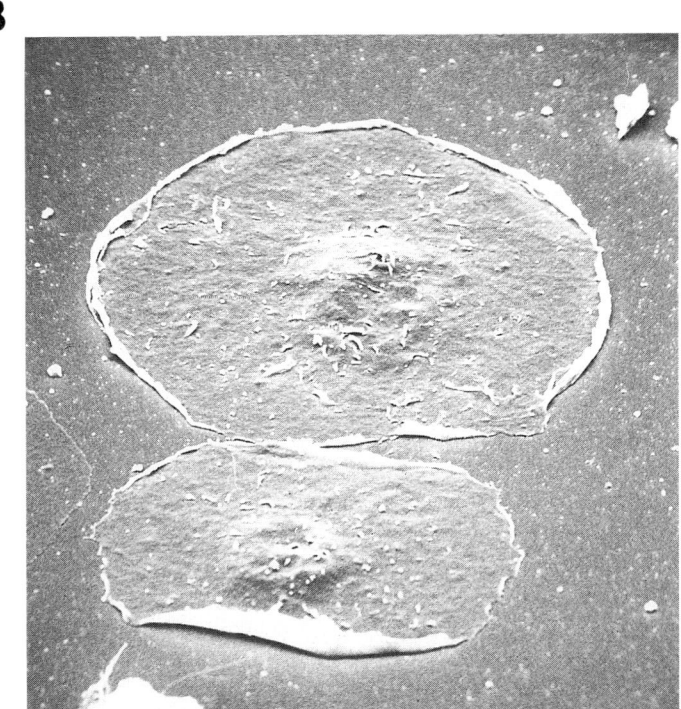

FIGURE 2.6   *(continued)*

4. *Chitin-Secreting Cells: Larval Epidermis*

In 48-hour-old cultures, bright refractile bodies can be observed in the center of isolated cells or of tight clusters. The containing cavity progressively swells up, but it is only on or around the tenth day that chitin is secreted as a continuous layer lining its inner surface. One or two more strata may be successively laid, forming free concentric capsules, in a characteristic onion-like structure. The process is very similar to ecdysis, more especially because one of the first responses of primary cultures to ecdysteroid treatment (Dübendorfer and Eichenberger-Glinz, 1980; Sang, 1981) was shown to be the secretion of additional layers by the larval epidermal cells (see also Section V.D).

Chitin layers are either clear and plain or somewhat textured and melanized. Sometimes, they present a few stubby hairs or more complex tooth-like structures. Sang (1981) suggested that transparent and thin capsules are possibly secreted by fore- and hindgut cells, whereas pigmented capsules with hairs might correspond to larval integument.

5. TRACHEOBLASTS

Occasionally, in a few cultures and after the third week, compact sheets of small clear cells extend several tongues of tissue, whose intermediate segments roll up into monolayered tubes, and a chitinous lining is secreted. Growth and branching of these tracheal tubes may form small networks.

B. Late Proliferation and Migration Waves of "Undifferentiated" Cells

Usually not before the end of the second week, new waves of cell migration burst out from some of the embryonic cell clumps that still adhere to the culture flask, and they may assume two main patterns (Fig. 2.7A,B): (a) The first pattern corresponds to large droves of roundish cells. They may, however, display an epithelium-like appearance, if they are more crowded; and (b) In the second possible pattern, more or less compact rows of spindle-shaped cells (fibroblast-like) form relatively vast meshes.

In both cases, they are small cells (8 to 10 μm diameter) with a large and clear nucleus and a conspicuous single nucleolus.

These cells spread during the following weeks and, in most successful cultures, can cover a significant area of the flask bottom. Such a proliferation is primarily due to cell multiplication; a few mitoses can be observed (Echalier and Ohanessian, 1970), either on living material with phase contrast microscopy (although they are difficult to observe because of the relatively slow division rate, the small size of the cells and the probable rapidity of the mitotic process*) or after treatment with colchicine and subsequent staining with orcein (Dolfini, personal communication).

These first waves of cell proliferation are generally doomed to deterioration and extinction within a few weeks, but several other similar cell outbursts may be periodically repeated during the following months. Finally, at least in a small percentage of the cultures, they may colonize the whole flask and can then be successfully transferred. It is obvious that most of the established cell lines derived from one or the other of these

---

* It should be remembered that in early embryonic stages of *Drosophila* DNA replication and nuclear division cycles succeed one another every 10 min.

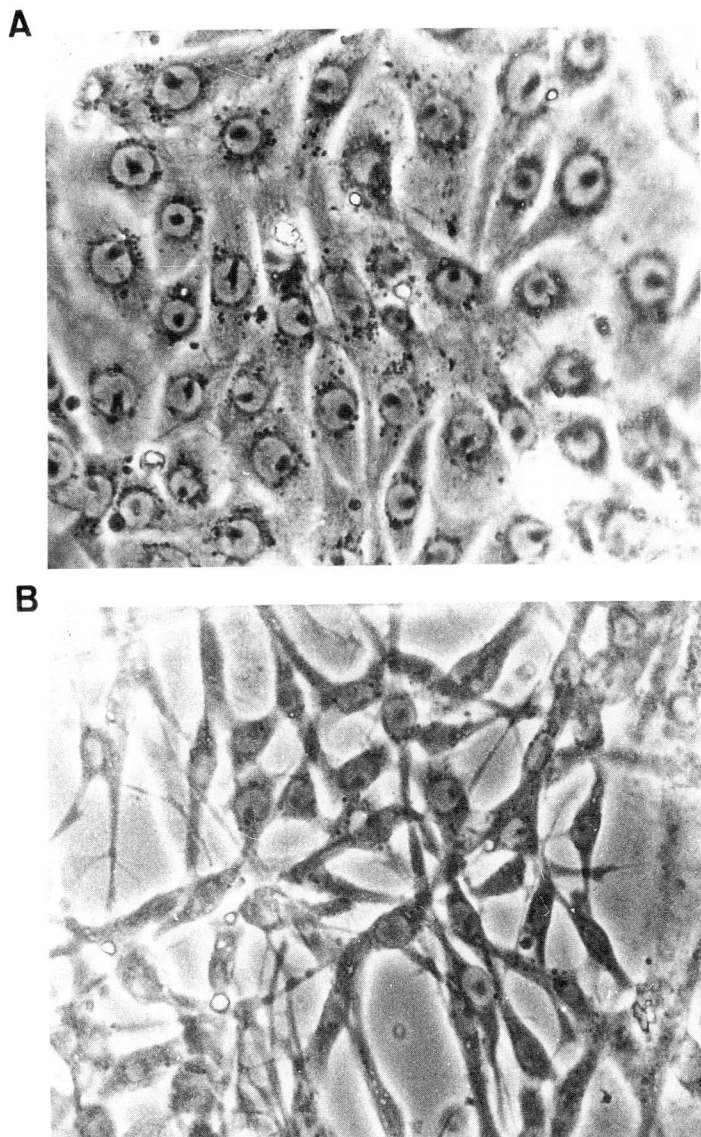

FIGURE 2.7   Waves of cell migration and multiplication in primary cultures of *Drosophila* embryonic cells. They display two main patterns: (A) "epithelium-like" or (B) "fibroblast-like."

cell types (or from hemocyte precursors, as suggested by Dübendorfer, personal communication, 1993).

It must be pointed out that it was not possible to determine the embryonic tissue from which such proliferating cells originated. Their rather banal morphology is of no help for tissue identification (see discussion in Chapter 3).

C. Development and Differentiation of Imaginal Cells as Induced by Hormonal Treatment

In holometabolous higher Diptera, such as *Drosophila,* the prospective adult insect, or "imago," is represented, from early embryonic stages onwards, by discrete small groups of cells that are sheltered by the larval organism until metamorphosis. So, in primary cultures of embryonic cells, two kinds of cells, with quite distinct fates, cohabit (larval and imaginal cells).

In cultures derived from embryos that are almost ready to hatch, Schneider (1972) observed, during a period of several months, continuous proliferation, from the cut ends of embryonic fragments, of hollow cellular spheres. After transplantation into 3rd instar larvae, those vesicles metamorphosed with their host and gave rise to adult cuticular structures, which demonstrated their epidermal "imaginal disc" origin. Later on, Dübendorfer, in collaboration with Sang's laboratory, carried out a thorough analysis of the *in vitro* growth and hormonal control of differentiation of such adult epidermis cells; the study was extended to several other imaginal tissues (muscles, fat body cells). Because of its particular interest, Section V of this chapter is devoted to this research field.

III. PRIMARY CULTURES OF MUSCLE CELLS

The attention of early *Drosophila* cell culturists was immediately riveted by the occurrence, and sometimes the long survival in cultures, of contractile or pulsatile material (Lesseps, 1965; Echalier *et al.,* 1965; Shields and Sang, 1970). Seecof and colleagues had the special merit of demonstrating, particularly by setting up cultures from single embryos at strictly defined developmental stages, that the differentiation of these muscle cells does indeed take place *in vitro* (Seecof and Unanue, 1968; Seecof and Alléaume, 1968; Seecof *et al.,* 1971). Then, as already mentioned at the beginning of this chapter, Seecof's school extensively studied all aspects of such differentiation *in vitro* of *Drosophila* larval muscle cells (as well as that of neruons, see Section IV). The achievement of mass cultures made thorough biochemical investigation possible, while the use of *Drosophila* mutations affecting myogenesis offered further opportunities for analysis. Finally, a fruitful experimental model was developed, comparable to that constituted by cultures of vertebrate myogenic cells.

A. Typical Sequence of Myogenesis *in vitro*

Dissociated blastula cells cannot acquire competence *in vitro*, probably because this requires specific cell interactions (see Chapter 6). On the other hand, in cultures made of embryos that had passed the stage of gastrulation initiation (marked by the appearance of a ventral furrow), myogenic cells become capable of differentiating autonomously (Seecof and Alléaume, 1968; Seecof *et al.*, 1971; Seecof and Donady, 1972).

Only their subsequent fate distinguishes *myoblasts* (the term being used in its classical sense) from all other cells obtained from disaggregated gastrulae and which appear to be fairly simple in gross morphology and ultrastructure (Seecof *et al.*, 1971; Gerson *et al.*, 1976). Myoblasts possibly represent about 10 to 15% of the entire cell population at gastrulation. Their defined sequence of differentiation was carefully described by Seecof, Gerson *et al.* (1973).

Myoblasts (some 5 μm in diameter when not flattened) divide once at about 5 h, thereby yielding two small *myocytes*. This unique division cycle may have been preceded by a few others before the setting of the culture. Some hours later (around 12 h), the daughter cells will begin to elongate, eventually adopting bipolar configurations. They frequently aggregate with neighboring myocytes into parallel arrays, and some of them, by aligning themselves and fusing, form syncytial *myotubes*. This fusion is limited to a 15 h-period and, around the 25th hour, myogenesis is virtually complete (see Table 2.I).

About a quarter of myocytes become contacted by one or several axons emanating from the many neurons that simultaneously differentiate in the culture (Seecof *et al.*, 1972; see following section).

The first sporadic contractions can be observed at 13 to 14 h and the number of foci of movement increases to reach a maximum around 24 h. It should be noted that some of the muscle cells can pulsate without any visible nervous connection. So, muscle differentiation proceeds *in*

TABLE 2.I   Events in *Drosophila* Myogenesis *In Vitro*[a]

Developmental event	Time (h) from fertilization
Fertilization	0.0
Initiation of gastrulation	3.5
Setting of the cultures	4.0
DNA synthesis	4.8–6.8
Myoblast division	8.5–6.8
Myocyte elongation	14.0–16.0
Fusion into myotubes	14.0–30.0

[a] After Donady *et al.* (1980).

vitro with a notable synchrony and at approximately the same rate as in intact embryos.

Muscle cells that spontaneously differentiate, in primary cultures, can be classified into three groups:

(a) The first one consists of spindle-shaped and mononucleated *myocytes* whose long processes attach to the substrate or to clumps of tissue. They are present either singly or in small parallel clusters. Their mean length is 35 to 50 μm.

In addition, there are a few "large myocytes" (100 μm) which might represent a distinct type and seem to arise from nondividing myoblasts (Seecof, Gerson *et al.,* 1973). Under electrophysiological stimulations, they respond to shorter pulse trains than "small myocytes" (Seecof *et al.,* 1972; see Section IV.A.2). Moreover, this distinction might also correlate with the two kinds of myocytes, with electron-dense or light cytoplasm, respectively, that can be seen by electron microscopy (Gerson *et al.,* 1975).

(b) Some muscle cells tend to spread into triangular or polygonal forms and their cytoplasm is very flat and almost granule free. They may become bi- or polynucleated. Such large *sheet-muscle cells* maintain their movement for up to about 2 weeks (Shields and Sang, 1970).

(c) *Myotubes* are strip-like elements, containing from 2 to 15 peripheral nuclei, usually aligned in an orderly fashion; a diffuse cross-striation can be seen by polarized light and phase contrast microscopy. They may achieve lengths of 100 μm or more, and during contraction are able to shorten their maximal size by 50 to 70% (Dübendorfer *et al.,* 1978) (Fig. 2.8).

In representative cultures made from individual embryos, about 200 myocytes and a dozen myotubes will develop from the initial 4000 gas-

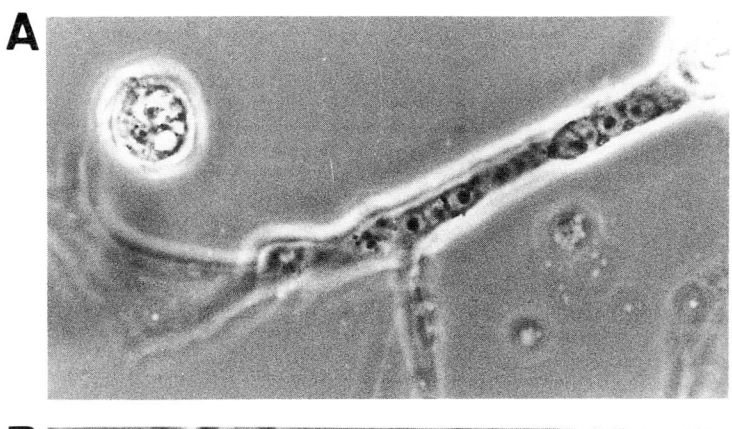

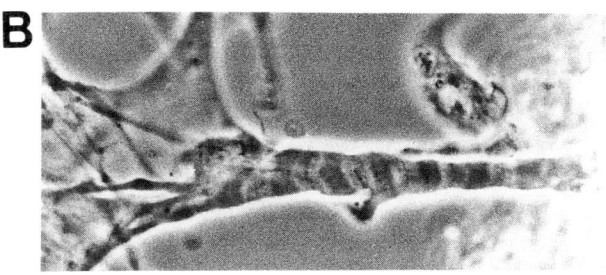

FIGURE 2.8 Differentiation *in vitro* of larval muscles from *Drosophila* embryonic cells: myotubes. Large nuclei are peripherally arranged (A), generally in the focal plane above the myofibrils themselves (B). From Dübendorfer *et al.*, 1978; courtesy of Andreas Dübendorfer.

trula cells. Only about 10% of those muscle cells pulsate spontaneously (which means approximately 20 foci of movement per culture), but nearly all of them contract if stimulated electrically (Seecof *et al.*, 1972).

Donady *et al.* (1980) investigated the effect of myoblast density on fusion to form myotubes: the unexpected relatively high fusion rate observed even at the lowest cell concentration might be due to "directed" movements of myocytes.

In other respects, Seecof and Dewhurst (1974) established that when cultures are supplemented with bovine insulin (see Chapter 1 and Chapter 6) the differentiation of myotubes is positively influenced: an optimal concentration of 3.7 mU/ml led to 8-fold increase in the number of myotubes per culture (i.e., a 14-fold increase in the ratio of myotubes to myocytes).

Electron microscopy confirmed that all cells, from early gastrulae onwards, have a relatively simple ultrastructure and that typical thin (actin) and thick (150 Å diameter; myosin) myofilaments are not detected in muscle cells before 10 to 11 h of culture. In the following hours, they become more numerous and begin to organize themselves into parallel bundles, so that some striation is visible at around 18.5 h (Gerson *et al.*, 1976): Z, A, H and I bands can be definitely identified, as well as T-system tubules and associated sarcoplasmic reticula. Transverse sections show 7 to 12 actin filaments surrounding each myosin filament. It is noteworthy that contractility may precede the full alignment of myofilaments.

In addition, desmosomes are commonly seen between myocytes, and, at the points of contact with terminal axons, the junctional structures bring to mind typical insect neuromuscular synapses (Seecof *et al.*, 1972) (see Section V.A.2).

This overall organization is clearly characteristic of larval (skeletal and visceral) muscles (Seecof *et al.*, 1971; Seecof, Gerson *et al.*, 1973; Dübendorfer *et al.*, 1978; Kuroda and Shimada, 1988).

B. Isolation and Mass Cultures of Myoblasts

In order to maximize the proportion of myogenic cells, in microdrop cultures, it is possible, with the tip of a thin tungsten needle, to puncture the ventral midline of individual embryos some 35 min after the appearance of the ventral furrow (more precisely on about a quarter of the length from the posterior end). It causes a clump of some 300 cells, most of them deriving from the mesodermal rudiment (Seecof, Gerson *et al.*, 1973).

For biochemical studies, mass cultures are necessary and the above procedure is not, therefore, appropriate. Methods for separating myoblasts, from mixed gastrula cell populations, took advantage of their selective attachment to the substrate. It was incidentally observed, for instance, that there is greater differentiation of muscle cells (and less of neurons) in cultures grown in tissue culture plastic dishes than in those grown on glass coverslips (Donady and Fyrberg, 1975). Two techniques were independently devised: (a) The method of Bernstein *et al.* (1978) used the calcium chelator EGTA. During a 2-h period, cells were plated, in plastic dishes, in medium containing 3 mM EGTA. Then, the culture dishes were transferred to a gyratory shaker, in order to detach loosely adhering cells. When rinsed with normal medium, the cells that remained attached were primarily myoblasts. In optimal conditions, cultures with as many as 80% myogenic nuclei could be obtained. Practical details are carefully described in the TCA Manual (Fyrberg *et al.,* 1977), as reproduced in Appendix 2.D. A new description of the technique has been given by Fyrberg *et al.* (1994). (b) By pretreating plastic culture petri dishes with a 0.1 mg/ml solution of protamine, a polyamine that increases the positive surface charge of the substrate, Storti *et al.* (1978) were able to enhance the preferential adhesion of myogenic cells. So, commencing also with early gastrula cells, they could work with relatively pure cultures of muscle cells.

Hayashi and Perez-Magallanes (1994) described a simple and effective method for separating the different precursor cell populations in early embryos, based on their differential sedimentation through a Ficoll gradient. Smaller cells, in the last fractions, produced predominantly myotube-containing cultures (see also Section IV.3).

C. Biochemical Approach: Synthesis of Myogenic Proteins

Valuable information was, at first, provided by the use of different metabolites or inhibitors.

Labeling of muscle cell nuclei (with 20 min pulses of tritiated thymidine) at various times of culture indicated that the only peak of incorporation, that is the last S period of myoblasts, takes place between 1.3 and 3.3 h. This was confirmed by the fact that, when cells were challenged with 10^{-4} M bromodeoxyuridine (BrdU) (which is well known to take the place of thymidine when incorporated into DNA) during this first 3-h interval but no later, myogenesis was seriously inhibited, as shown by the reduced number of muscle cell nuclei or movement foci (Seecof and Dewurst, 1976).

Actinomycin D treatment (used in short 10 min pulses, at a concentration of 10 μg/ml) completely inhibited the differentiation of all cells in the cultures, when applied at zero time. Up to 12.5 h, which is the usual time for the first observed foci of movement, the inhibitor prevented the development of any contraction. On the other hand, after exposure at 24 h and even if actinomycin D was not removed thereafter, the morphology and pulsation of myocytes remained normal.

Autoradiographs confirmed that the inhibitor reduced labeled uridine incorporation by about 90%, whereas leucine or thymidine incorporation decreased by only 6% or less.

The most likely interpretation of these results is that transcription is necessary in myocyte differentiation up to the point when pulsations can begin (Donady and Seecof, 1975).

When mass cultures of *Drosophila* myogenic cells became available, it was then possible to investigate more thoroughly the specific protein synthesis during myogenesis and to compare the results with those obtained from well-studied vertebrate myogenic culture systems. Myogenesis in *Drosophila* cell cultures, unlike that in vertebrate cultures, exhibits a high degree of synchrony and is virtually complete by 24 h, which are notable advantages. This analysis was initiated almost simultaneously by two different laboratories.

Two-dimensional polyacrylamide gel electrophoresis showed that profound changes in the pattern of protein synthesis occur during muscle differentiation *in vitro*. At least 22 muscle specific proteins were detected and their high rate of synthesis was found to be correlated with the fusion of myogenic cells into multinucleated myotubes (Fyrberg and Donady, 1978, 1979; Bernstein and Donady, 1980; Donady *et al.*, 1980). For their part, Storti *et al.* (1978) confirmed the occurrence of substantial changes in proteins, qualitative as well as quantitative, essentially after myogenic fusion has occurred. They distinguished between three discrete classes of proteins: class A proteins, the most abundant ones are synthesized continuously throughout myogenesis. The synthesis of class B polypeptides is initiated during the developmental sequence and apparently continues in the differentiated culture; they can account for approximately 20% of the protein species synthesized by 18-h cultures. The synthesis of class C proteins is tightly coupled with specific stages of development.

Some of these myogenic specific proteins have been identified. A muscle specific form of actin (which is the most acidic and is designated form I) becomes a major protein after myocyte fusion and approaches 80% of total actin synthesis by 14 to 16 h (whereas it represents only 40% at 4 to 6 h). In nonmyogenic cell cultures, this form never constitutes more than 5% of total actin (Storti *et al.*, 1978; Fyrberg and Donady, 1979).

Myosin heavy chain has been well characterized (Fyrberg and Donady, 1978; Bernstein, 1979) and even purified from mass myogenic cultures. The amino acid composition of this 200 kDa protein is very similar to that of chick myosin heavy chain (Bernstein *et al.*, 1986).

Among other proteins from the contractile apparatus, three myosin light chains and two forms of tropomyosin have also been identified.

Autoradiographs were scanned with a computerized densitometer and it was clearly shown that the maximum synthesis rate of all these myofibrillar proteins is reached at or near the beginning of the myotube fusion process (Donady *et al.,* 1986) (see Fig. 2.9).

Parallel studies on pulse-labeled polyadenylated RNAs established that their sedimentation profiles, from prefusion and postfusion stage cultures,

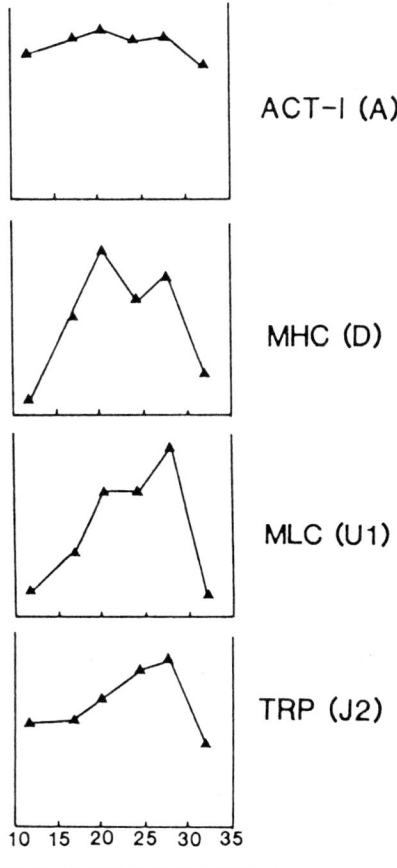

ACT–I (A)

MHC (D)

MLC (U1)

TRP (J2)

10 15 20 25 30 35

HOURS AFTER FERTILIZATION

FIGURE 2.9 Synthetic patterns of the major contractile proteins. *Drosophila* myogenic cell cultures were pulsed with [^{35}S]methionine for 2 h, at successive intervals. Proteins were separated by two-dimensional PAGE, and all gels were exposed for 7 days. Autoradiographs were scanned with a computerized densitometer. Only relative kinetics of synthesis can be compared, over developmental time. ACT–I, actin I; MHC, myosin heavy chain; MLC, myosin light chain; TRP, tropomyosin. From Donady *et al.,* 1986. With permission of Elsevier Science Publishers.

differ significantly (Bernstein, 1980). The products of their *in vitro* translation corresponded well to the proteins synthesized *in vivo* at both stages (Bernstein and Donady, 1980; Donady *et al.*, 1980).

The patterns of accumulation of specific mRNAs parallel the induction of corresponding muscle proteins, as was verified for actin I or myosin heavy chain messengers (Donady *et al.*, 1986; Bernstein *et al.*, 1986).

The data, including results of global inhibition by actinomycin D, established clearly that a family of muscle-specific genes is activated, in a strictly coordinated manner, at the fusion stage of *Drosophila* myogenesis. There are strong similarities with muscle differentiation of cultured vertebrate myoblasts and this confirms the value of such an experimental model (see Table 2.1).

IV. NERVE CELLS IN PRIMARY CULTURES

A. Neuroblasts in Embryonic Cell Cultures

Lesseps (1965) and Echalier *et al.* (1965) reported the abundance of axonal-like processes in primary cultures. Using younger embryos, Seecof and colleagues (Seecof and Unanue, 1968; Seecof and Alleaume, 1968; Seecof and Teplitz, 1971; Seecof, 1980), then Sang and colleagues (Shields and Sang, 1970; Cross and Sang, 1978a; see also Sang, 1981) were able to establish that neuroblast differentiation takes place *in vitro*.

Because of their rapid differentiation and easy visual identification, neurons are among the most characteristic cell types developing in early cultures of embryos.

Their small (some 5 μm in diameter), roundish, dark cell body is essentially filled with a barely distinguishable nucleus, and most of the cytoplasm extends in one or more axonal threads. These thin processes, which adhere closely to the substrate, can spread over 100 μm in length and are more or less richly and irregularly branched; typical varicosities are observed all along them. Numerous connections occur between the branches of one or more neurons, so that complex nerve networks can be formed. The tips of many nerve fibers clearly seem to be attached to neighboring myocytes.

Neurons are usually grouped in small, or larger ganglion-like clusters, which may be due to either a common origin or subsequent regrouping. Likewise, their axons frequently gather into straight bundles, connecting cell aggregates (Fig. 2.10).

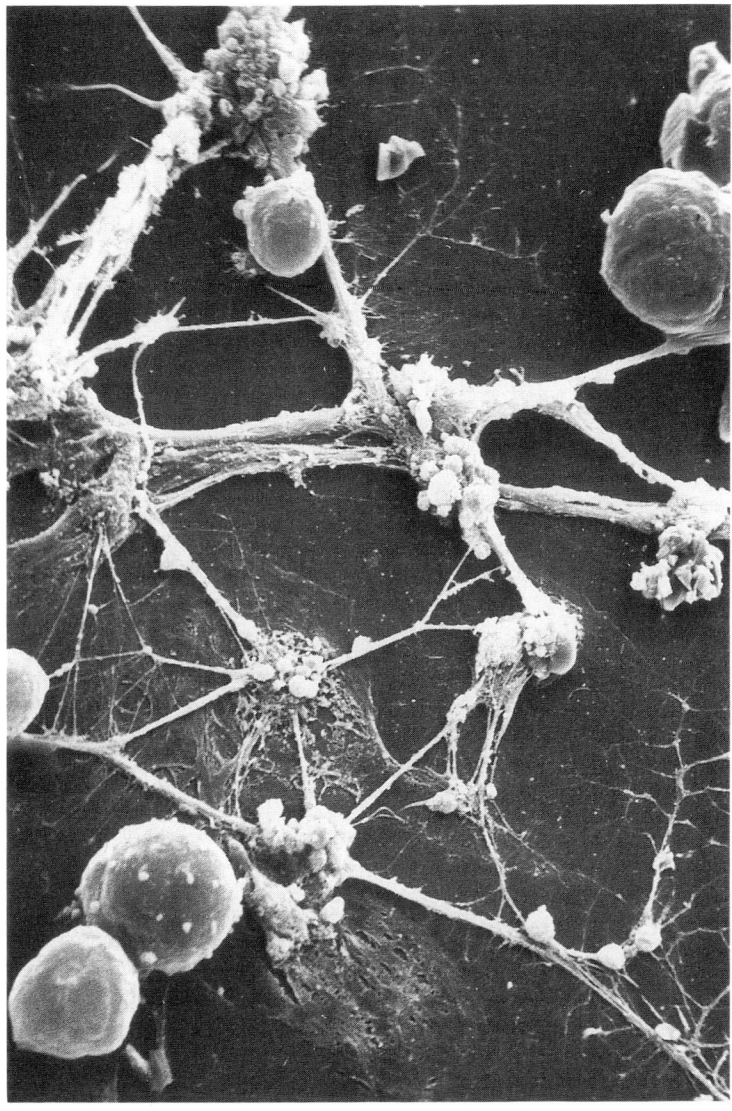

FIGURE 2.10   Neural net in 1-week-old culture from *Drosophila* embryonic cells. Note neural cell clusters and axon bundles, with muscle cells and early adipocytes (Scanning electron microscopy ×1000). Courtesy of Dübendorfer.

Electron microscopy confirmed that such cell processes have indeed the appearance of insect axons: with a diameter of 0.2 to 1 μm and bound by a membrane about 30 nm thick, they contain mitochondria, neurotubules, and neurosecretory vesicles. In 7-day cultures, the presence of mesaxons established the differentiation *in vitro* of glial cells (Seecof *et al.*, 1971, 1972). Preliminary observations of neuron clusters showed the typical organization of a miniature insect ganglion, with cell bodies at the periphery and a neuropile-like fiber association inside (Seecof *et al.*, 1973).

1. TYPICAL SEQUENCE OF NEURON DIFFERENTIATION IN VITRO

Successive steps in differentiation *in vitro* of nerve cells resemble those of normal events occurring in developing embryos [as described, over forty years ago, by Poulson (1950)].

Placing the dissociated cells from early gastrula embryos between the two coverslips of a Sykes–Moore culture chamber, Seecof and Teplitz (1971; see also Seecof, Donady and Teplitz 1973; Seecof, Donady and Fiorio, 1973) could monitor individual cells by phase contrast and photograph them at intervals.

Stem cells (*neuroblasts*) are large cells (about 12.5 μm in diameter), which are fairly similar to other blastoderm cells. Within the first few hours of culture, they undergo a sequence of about eight asymmetrical divisions, each time "budding off" a small *neurocyte* while the large cell continues to divide at intervals of about 30 min. A final division of all these daughter cells generates a three-dimensional clump of, theoretically, some 16 to 18 tiny *ganglion cells* (3 to 5 μm in diameter, that is, some of the smallest cells in the culture); however, division may sometimes halt before this total is reached.

Two or more bundles of axons grow from each cluster. The exact time of axon initiation is difficult to determine, because neuroblasts and ganglion cells already have short processes. The very first axons may appear at 4 to 5 h, but the most indubitable ones are visible by 10 h. In a 24-h culture, some of them attain a length of about 100 μm.

The thread network continues to develop during the fourth and fifth days and, after medium change, may reach extreme degrees of complexity.

Although neurons and axon-like fibers survive well during the first week, deterioration and progressive losses are observed later on, so that very few neurons persist beyond 7 weeks (Shields and Sang, 1970).

Under optimal conditions (Seecof and Donady, 1972) of pH (6.8) and temperature (25 to 27°C), a culture derived from a single wild-type gastrula yields about 200 to 300 neurons with axons longer than 50 μm. Their cell bodies are arranged in some 10 discrete clusters or hidden within heterogenous cell aggregates (Seecof, Donady, and Teplitz, 1973). By contrast with myocyte differentiation, addition of insulin had no obvious effect on the number or length of axons (Seecof and Dewurst, 1974). See Section VI for the influence of several *Drosophila* mutations.

2. NEUROMUSCULAR JUNCTIONS IN CULTURES

Many axons grow in contact with other axons or with nearby myocytes. It is not known whether any diffusing substance(s) or transmembrane factors influence the rate or direction of axon growth (see Chapter 6, Fasciclins).

Such myoneural attachments are visible under a light microscope (Seecof, Donady, and Fiorio, 1973), especially when the myocyte contracts, tautening the connected nerve thread.

Seecof *et al.* (1972) demonstrated by electron microscopy and electrophysiological techniques that such *in vitro* differentiated junctions are morphologically complete and functional.

When sections for electron microscopy are prepared from 24-h cultures, at each zone of contact between axon terminals and myocytes, a synaptic cleft (gap distance about 130 Å) is visible and is bridged by regions of electron dense material. Within the axon, clear synaptic vesicles (600 Å in diameter) are clustered against the presynaptic membrane. The general organization and dimensions of gaps and vesicles are very typical of insect neuromuscular junctions. As in intact insects, one muscle cell can be innervated by several axons or branches from a single axon.

When, in cultures, electrical stimuli were applied to axons, they caused the contraction of innervated myocytes. The investigators selected axons easily accessible to the electrode and with clear terminations on easily identifiable muscle cells. Axons of various diameters proved functional in 85% of the tested junctions. Two types of responses were observed: (1) small myocytes responded only to long trains of pulses (20 to 40 pulses, delivered at 30 to 40 Hz) and gave slow contractions (about 500 to 1000 msec), and (2) large myocytes (see above) responded to shorter pulse trains (4 to 10 pulses delivered at 30 to 40 Hz) and made faster twitch-like contractions. The "slow" or "fast" quality of the response seems, indeed, to depend on the identity of the myocyte.

3. PURIFIED NEURAL CELL CULTURE

By combining selective procedures of centrifugal elutriation (to fractionate embryonic cells on the basis of size) and differential adhesion of neuroblasts to the glass substrate, Furst and Mahowald (1985a) were able to prepare, on a relatively large scale, virtually pure cultures of developing neurons (see Appendix 2.E).

Elutriation is a counterflow centrifugation system which separates cells on the basis of their sedimentation rates and permits the handling of a large number of cells. Because it is a gentle procedure, without the use of potentially toxic agents, it has been widely used to isolate a variety of mammalian cell types.

Monodispersed *Drosophila* embryonic cell suspensions were prepared, as usual, by mechanical dispersion of early gastrulae (4 to 6h) in buffered saline solution. Whereas the majority of gastrula cells are small (most are smaller than 8 μm in diameter), neuroblasts are among the largest (in the range of 10 to 13 μm in diameter).

Some 10^9 cells were then injected into the elutriator system and, with an optimal rotor speed (around 2800 rpm), were separated into seven fractions. [More practical details about this elutriation of gastrula cells have been given by Mahowald (1994).]

The largest cell type (class III, diameter > 10 μm), after 1-day culture, showed typical clusters of small round ganglion cells and produced the greatest number of neurites. It was estimated that more than 1 in 5 cells was a neural precursor, whereas in total gastrula cell cultures the ratio was only 1 in 14 cells.

In a further step of enrichment, the authors took advantage of the good adherence to glass of neuroblasts and differentiated neurons. Myoblasts, too, can adhere firmly to substrate (see Section III.B), but they had been previously eliminated on the basis of size. So, after incubation in the culture flask for at least 2 h, the medium was aspirated and changed. The remaining cells gave rise primarily to neural clusters: an average of 72% of the cell clusters retained on the glass substrate contained nerve cells.

On the whole, after both combined procedures, some 2×10^7 neuroblasts were obtained from an initial elutriation of 1×10^9 gastrula cells; and, because of the early and multiple divisions of neuroblasts (see above), contrasting with the limited proliferation of contaminant cell types, more than 2×10^8 neurons could be derived in this manner, so that the proportion of neural cells could finally reach 95% in overnight cultures.

A few neural clusters, although usually much smaller (often a pair of neurons), were present in cultures from size class I (and II). They might, of course, result from contamination of the fractions by larger cells, but the authors made the interesting suggestion that they may perhaps correspond to a distinct neuroblast subclass, the specific "midline precursors" (MP), well identified by insect neurobiologists; they are smaller cells and undergo a single symmetrical division.

Such purified neuroblasts were isolated in microdrop cultures and monitored by time-lapse cinematography (Furst and Mahowald, 1985b). They could divide and differentiate in the absence of any influence from other cells, and it was thereby demonstrated that most of the typical neural cell clusters arise from single precursor cells (Fig. 2.11) (even though, in whole gastrula-derived cultures, NCC are frequently associated with a small number of flattened cells, possibly glia). After [³H]thymidine labeling, a neuroblast cell division cycle averaged approximately 1.5 h (versus 30 min, *in situ*, as estimated by Poulson, *1950*).

Krasnow and colleagues (1991) reported a general method for purifying virtually any cell type from cell suspensions of whole *Drosophila* early embryos, and they applied it to the selection of neural cells.

Cells harboring and expressing a *lacZ* transgene can be fluorescently stained with the β-galactosidase substrate, fluorescent

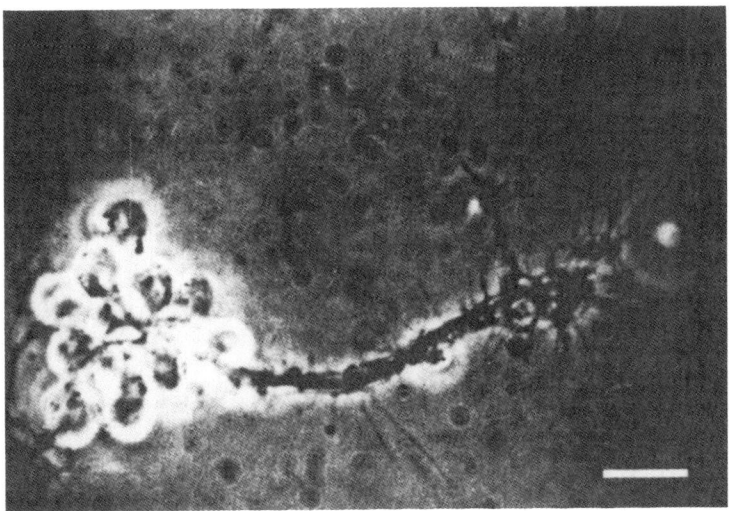

FIGURE 2.11   A clonal neural cell cluster developed *in vitro* from a single precursor cell. Overnight culture of *Drosophila* gastrula cells at 26°C. Living cells were photographed in a wet-mount preparation, using phase-contrast optics. Cell bodies surround neurites, which extend as a fascicle. Bar: 10 μm. From Furst and Mahowald (1985b). Reproduced with the kind authorization of Prof. Mahowald and Springer-Verlag.

di-β-D-galactopyranoside (FDG), and still remain viable. The analytical power of a fluorescence-activated cell sorting system makes it possible to separate them from neighboring cells. It is important to point out that this staining is specific to the *Escherichia coli* β-galactosidase.

Now, a number of transgenic *Drosophila* strains, with *lacZ* under the control of tissue-specific promoters, are available. For instance, using their method, the authors succeeded in purifying neural cells from *Drosophila* embryos of a strain in which *lacZ* expression was restricted to neuroblasts and derived cells. In 24-h primary cultures, up to 41% of the selected cells displayed typical neuron morphology with long neurites; moreover, immunocytochemistry revealed the presence, on their surface, of a characteristic neuronal antigen. The observation of clusters suggested that such purified neuroblasts could divide and differentiate in the normal way.

This purification technique is obviously applicable to most cell types. If the cell of interest represents a reasonable proportion of the embryo, some 10^5 to 10^7 cells can thus be purified in a few hours.

A new description of this fluorescence-activated cell sorting (FACS) method has been given by Cumberledge and Krasnow (1994).

Taking advantage of the differences in size between precursors of neurons and myocytes among early gastrula cells, Hayashi and Perez-Magallanes (1994) have developed a simpler and fast method of separation, by using a shallow linear Ficoll gradient. Some 3×10^7 cells can be handled per operation and their basic differentiation patterns, in culture, remain normal.

Cells from early fractions (over 10 μm in diameter) give pure neuronal cultures, as indicated by morphology and immunological tests, and they represent about 35% of the gastrula population.

Much smaller cells, in the last fractions, differentiate predominantly into myotubes (65% of the population).

With the kind permission of the authors, this method is reproduced in Appendix 2.F.

4. Biochemical Approaches

a. Nucleic acid synthesis inhibitors

The use of classical metabolic inhibitors permitted the determination of certain critical steps in the differentiation process.

Early treatment with BrdU blocked the typical sequence of unequal divisions of neuroblasts (Seecof and Dewurst, 1976).

In their study of the development of neural cell clusters from isolated neuroblasts (see above), Furst and Mahowald (1985b) showed that hydroxyurea, at 2 to 4 mM, is sufficient to inhibit cell division, but still allows differentiation, thus making it a very convenient way to truncate neuroblast cell lineages at specific times (Huff *et al.,* 1989).

In contrast, actinomycin D prevented differentiation of neuroblasts, but only when applied prior to axon initiation (Donady *et al..,* 1975); later on, neurons are no longer sensitive to the drug. It was confirmed that uridine incorporation is thereby inhibited, so that transcription seems necessary in neuron differentiation up to the stage of axon budding. This transcription concerns essentially messenger RNAs, because neuron differentiation was shown to remain normal in cultures from embryos without an rDNA region (see Section VI.2).

b. Acetylcholine metabolizing enzymes

Choline acetylase (choline O-acetyltransferase) and acetylcholinesterase, are, at least to some extent, characteristic of the central nervous system in *Drosophila* and may be used as indicators of functional activity.

Dewhurst and Seecof (1975) compared the pattern of development of both enzymes in whole embryos and in cultures, in correlation with morphological events.

Acetylcholinesterase appeared first, as it does *in vivo*, and was detectable in cell cultures from single embryos at 8 h, which represents a 2 h lag as compared to whole embryos and indicates that the enzyme is synthesized in neurons very soon after the growth of the very first axons is completed. Then, the rate of increase of acetylcholinesterase activity was approximately the same as that in embryos and, between 12.8 and 20.8 h, it increased more than sevenfold (up to 4 nmol per single embryo culture per hour). The authors verified that they were dealing with "true" acetylcholinesterase (AChE) activity, the latter being almost completely inhibited by *BW 284 C51* dibromide (at the classical concentration for this specific inhibitor, i.e., between 5×10^{-6} and $3 \times 10^{-5} M$).

About 5 h later (i.e., around 12.8 h), choline O-acetyltransferase (ChAT) could be detected, 3 h later than in whole embryos. After a slow increase during the following hours, its activity in both cultures and embryos increased more than fifteenfold (up to about 200 pM per embryo per hour).

Such a similarity of temporal sequences in cultured cells and intact embryos indicates that an identical program for enzyme appearance is present both *in vivo* and *in vitro* and that it can be expressed autonomously by the neurons.

Most of these results concerning the developmental program of AChE and ChAT in primary cultures of embryonic cells were confirmed by Salvaterra *et al.* (1987), with some additional data on the influence of various inhibitors or variations in culture conditions.

By using the "direct coloring" method for cholinesterase, devised by Karnovsky and Roots (*1964*), Furst and Mahowald (1985a) could also show that most of the clusters from their purified neural cell cultures (see above) were cytochemically AChE-positive.

c. Dopamine and serotonin

Huff *et al.* (1989) focused on the production of two other neurotransmitters, using the clonal neural cell clusters that they were able to obtain with sparse cultures of their purified neuroblasts. In approximately 4% of these clusters, they observed 1 to 3 cells that were immunoreactive for the presence of serotonin, whereas 5 to 6% distinct clusters contained a single dopaminergic neuron. In all transmitter-positive cells, transcripts of dopa-decarboxylase (an enzyme involved in the synthesis of both dopamine and serotonin) were localized by *in situ* hybridization. Moreover, it could be shown that each transmitter appeared at a specific time during the development of a lineage and, more especially, that serotonin requires a set number of DNA replication. These results are in perfect agreement with *in vivo* observations of *Drosophila* embryos.

5. ELECTROPHYSIOLOGICAL STUDY OF IONIC CHANNELS

Standard embryonic cultures allowed Byerly (1985, 1986) and collaborators (Leung and Byerly, 1987; Byerly and Leung, 1988) to study the K^+, Na^+, and Ca^{2+} currents of *Drosophila* neurons with voltage–clamp techniques.

The most prominent voltage-dependent currents in these *Drosophila* embryonic neurons are K^+ currents and, at least, two different types can be distinguished (as confirmed in cultured myocytes, too, by Brainard *et al.*, 1987). The *Shaker* mutation (which eliminates the A-type current of *Drosophila* muscles) does not affect either the rapidly inactivating or the slowly inactivating K^+ current of those embryonic neurons.

Strong Ca^{2+} currents, probably of multiples types, are present in both embryonic neurons and myocytes. Their peak amplitude varies from 0 to 100 pA. Calcium, as well as sodium channels seem to be primarily located in the process membranes, whereas the membrane of the neuron soma contains mainly potassium channels.

6. CULTURE OF ECTODERMAL NEUROGENIC PRECURSORS FROM *DROSOPHILA* GASTRULAE

The regulation of the neural/epidermal dichotomy in *Drosophila* embryos is very complex (see *Notch* and other proneural genes, in Chapter 6). In order to distinguish clearly the intrinsic capabilities of cells of the presumptive truncal ectoderm from the external influences of the many factors involved, Lüer and Technau (1992) succeeded, with technical advice from Dübendorfer, in culturing these precursor cells in isolation. The cells were delicately explanted from well-defined sites of early gastrula donors, with a thin beveled micropipette, and grown individually, according to the column-drop method (see Section I.C.1). In more than 90% of the cases, isolated cells divided and gave rise to basically three types of clones: neural (Fig. 2.11), epidermal, or mixed clones. Therefore, at the onset of gastrulation, ectodermal cells are already endowed with the intrinsic capacity to develop as epidermoblasts or neuroblasts, and their preferences are clearly correlated with their original positions along the dorsoventral axis.

Let us point out that, *in situ,* only 25% of the cells of the ventral neurogenic region develop as neuroblasts, which seems in apparent contradiction with these observations *in vitro*. It is thereby revealed that, as was previously hypothesized, ventral neurectoderm cells have a primary neural fate, but other influences are superimposed, forcing about 75% of them onto the epidermogenic pathway.

B. Cell Cultures from Larval Central Nervous System

Drosophila neurobiology is a very active research field, once again because a number of single-gene mutants altering the morphology or physiology of neurons are available. Cell culture systems from differentiated nervous tissue would facilitate the experimental analysis of normal neuronal properties and of their altered functions in neurological mutants.

However, except for a brief allusion to a limited proliferation *in vivo* of neuroblasts from invasive brain tumors of *lethal(2) giant larva* mutants (Sang, 1981), no report of successful cell culture of postembryonic neurons of *Drosophila* existed, until the paper of Wu and colleagues (1983).

They dissociated brain and ventral ganglia from mature 3rd instar larvae, that is, a developmental stage at which the CNS initiates the important remodeling of metamorphosis, and could observe division of neuroblasts and active growth of axons. Moreover, they demonstrated the suitability of such isolated cells, from wild-type as well as from mutants, for electrophysiological, immunological, or pharmacological approaches.

Cell dissociation was obtained with a 1-h collagenase treatment under the conditions recommended by Schneider and Blumenthal (1978) for loosening late embryos, followed by mild pipetting. The brain and ventral ganglion of a single larva yielded about 10^5 isolated cells, from which some 20% adhered to the cell culture dish and formed a monolayer (see details of technique in Appendix 2.G).

In 24-h cultures, a large proportion of the cells developed long processes which ramified and formed apparent contacts with other cells.

Such cultures could be routinely kept about 1 week and some of them survived up to 40 days. Figure 2.12 gives the curves of cell survival and neurite development in typical cultures.

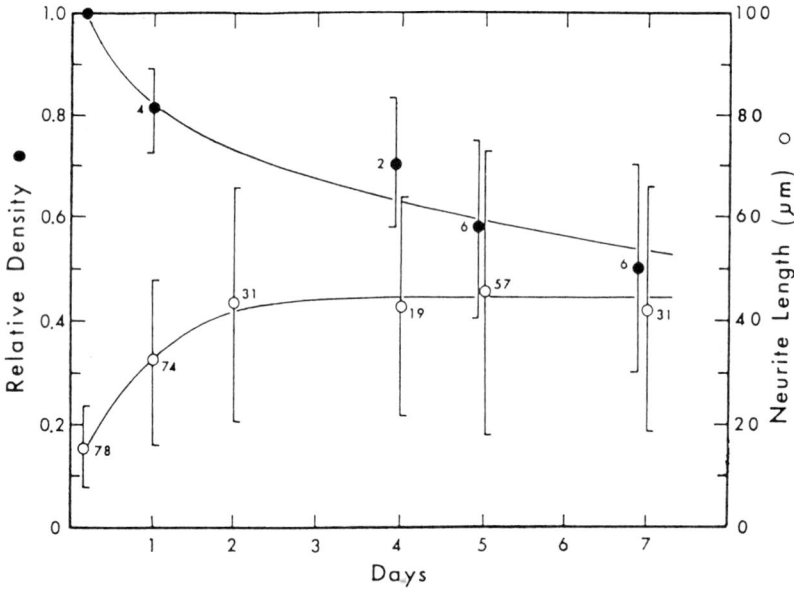

FIGURE 2.12 Viability *in vitro* of neurons developed from dissociated neuroblasts of *Drosophila* larval CNS. Filled circles (●) represent the cell density normalized to the initial value as determined in the number of cultures indicated. Unfilled circles (○) represent the average of the neurite length from the number of neurons indicated. Error bars represent standard deviation (SD). From Wu *et al.* (1983), by permission of Oxford University Press.

Three main categories of cells could be distinguished: Type I (some 10% of the population) were large (>8 μm diameter) neuroblast-like cells; they underwent several asymmetrical divisions, very like those observed in embryonic cell cultures (see above), and which generated small accompanying cells. The latter resembled Type II cells, that is, much smaller cells (2 to 3 μm) with axons longer than 20 μm, forming the majority of the cell population (40%). Type III cells (20 to 30%), whose derivation could not be determined, were oval or spindle-shaped cells of intermediary size (4 to 10μm), with one or several neurites. The remaining cells, in the cultures, were spherical, without any processes and seemed to correspond to degenerating neurons of all categories. Fibroblasts or glial cells were rarely encountered (<1%), probably in keeping with the low number of glial cells in the insect central nervous system (CNS).

In addition to their characteristic morphology, these cultured cells were shown to be neurons by several criteria: (1) Immunofluorescence, with a series of monoclonal antibodies raised against *Drosophila* CNS, revealed the presence of adult neuronal antigens; and (2) Acetylcholine is a major neurotransmitter in *Drosophila* CNS. After 1-h incubation in [³H]choline, 1 day cultures synthesized and accumulated [³H]acetylcholine, with an efficiency comparable to that of intact larval brains.

Electrophysiological methods were applied to these readily accessible cells: the largest types (I and III) could be penetrated with intracellular microelectrodes and they were also suitable for extracellular patch clamp recordings. The membrane resting potential was usually found to be below 25 mV. Single-channel currents induced by acetylcholine, as well as regenerative action potentials, were detected.

In the same paper, Wu *et al.* (1983) produced other convincing evidence that such a neuron culture system might be amenable to true pharmacological experimentation:

Veratridine is a neurotoxin highly specific to voltage-sensitive Na⁺ channels, leading to neuron lethality. The fact that cell viability decreased rapidly in mature 4-day cultures exposed to veratridine (500 μM) and that such a lethal effect could be specifically inhibited by tetrodotoxin, strongly suggests that Na⁺ channels are present in the cultured neurons.

nap^{ts} is a temperature-sensitive paralytic mutant of *Drosophila* in which nerve conduction fails at nonpermissive temperature. Since it was assumed that the mutation might affect Na⁺ channels, it was interesting to test, on isolated *nap^{ts}* neurons, their sensitivity to veratridine. It was indeed found to be greatly reduced, even at the permissive temperature (21°C), which is in accord with this hypothesis. In 1988, Wu reviewed the possibilities of this culture system.

After preliminary results (Ui *et al.*, 1988, 1989), continuous cell lines, displaying various characteristics of neural cells, were obtained from cultures of larval brain ganglia by Ui and colleagues (1994) (see Chapter 3).

V. *IN VITRO* DIFFERENTIATION OF IMAGINAL CELLS: HORMONAL CONTROL

A. Imaginal Disc Cells of Adult Epidermis

Imaginal disc cell spheres do not usually develop in primary embryonic cultures before 2 to 3 weeks. Small blebs form at the surface of adherent cell clumps and, by cell multiplication, swell up into hollow vesicles made of monolayers of flattened cells. They enlarge considerably, often containing thousands of cells, or bud off daughter vesicles which form characteristic bunches of little balloons floating in the medium (Fig. 2.13). Their proliferation may go on for months. Such spheres are particularly numerous (sometimes more than 100 in each flask) in cultures made from late embryos (Schneider, 1972).

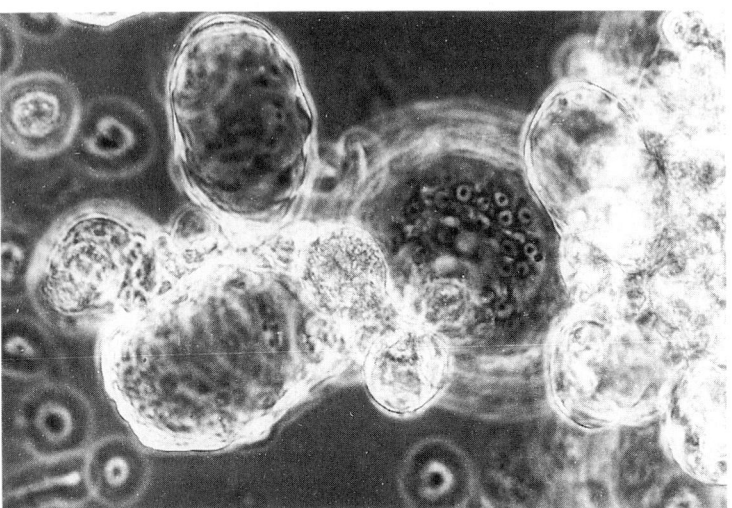

FIGURE 2.13   Imaginal disc cell vesicles, in primary *Drosophila* embryonic cell culture (third week *in vitro*). Courtesy of Dr. Andreas Dübendorfer, Zürich University.

1. IN VITRO *TREATMENT WITH* α- *OR* β-*ECDYSONE*

The insect growth and differentiation hormones, at concentrations of 10 μg/ml (i.e., 1.1×10^{-5} M) or 0.1 to 0.5 μg/ml, respectively, could induce metamorphosis of the vesicles, confirming their epidermal and adult nature (Dübendorfer *et al.*, 1974, 1975; Dübendorfer, 1976; Kambysellis and Schneider, 1975; Kuroda, 1986; Kuroda and Shimada, 1988). One to two days after exposure to hormone, the vesicles secreted chitin cuticle, either smooth or with simple hairs (trichomes), or even more complex bristles (including socket and bristle shaft which derive, as is well known, from a single precursor cell by differential cell division) (Fig. 2.14). One vesicle may contain hundreds of trichomes alone, or bristles (up to 20), or a mixture of both structures. More specific markers, however, such as claws, sensillae, or eye facets, never differentiated and the

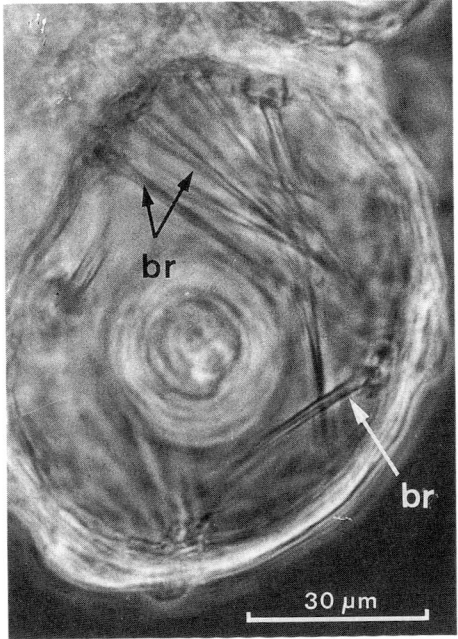

FIGURE 2.14 Differentiation *in vitro* of imaginal structures (bristles) from the cells of epidermal vesicles after ecdysone treatments. Four weeks in culture; the hormone was applied on the 13th and 22nd days. Phase contrast. From Dübendorfer (1976); courtesy of the author.

complex patterns that characterize certain regions of the adult body could not be identified.

As expected, a concomitant treatment with a strong juvenile hormone (JH) analog, *ZR515*, reduced the number of differentiating vesicles to 10% (Milner and Dübendorfer, 1982).

In a preliminary work on spheres from primary cultures of neonate larvae, Kambysellis and Schneider (1975) observed also, after ecdysteroid treatment, the formation of several cuticle-like structures, although they were intriguingly secreted at the periphery of a central cell aggregate.

2. TRANSPLANTATION INTO LARVAE

More complete adult structures could be formed, on the condition that the vesicles grown *in vitro* were transplanted into larvae and remained at least 2 days *in vivo* before the commencement of host metamorphosis (Dübendorfer *et al.*, 1974, 1975; Dübendorfer, 1976).

When the transplantation was made into ready to pupate larvae, the results did not differ from those obtained *in vitro*.

On the contrary, when the implant spent 2 days in the larva prior to metamorphosis, by transplantation into early 3rd instar larvae, easily recognizable patterns of the adult body (such as leg, antenna, or genital apparatus) could be recovered in some 8% of the samples.

The frequency of such specific patterns could even be greatly increased (up to 40%) if vesicles developed *in vitro* were, before their implantation into larvae, previously grown in an adult female abdomen for 9 days (according to the *in vivo* culture technique devised by the school of Hadorn, in Zürich; see Hadorn and Garcia-Bellido, 1964).

Those results suggest that some unknown factor(s), specific growth factor(s) or simple metabolite(s), ? present in the hemolymph but absent in synthetic culture media, may be required, concomitantly with ecdysteroids, for full expression of the differentiating capacity of *in vitro* grown imaginal disc cells.

B. Imaginal Muscles

Within a week after treatment with ecdysone, most of the larval "sheet" muscles and myotubes differentiated in primary cultures become granular and degenerate, while, usually 2 to 4 days after the second hormonal application that accompanies the medium change, two new muscle types appear. According to the careful observations of Dübendorfer and colleagues (Dübendorfer *et al.*, 1978; Dübendorfer and Eichenberger-Glinz,

1980), they seem to be formed by the assembly and fusion of some of the fibroblast-like cells, and their characteristics are very similar to those of the two main classes found in adult flies, namely "tubular" and "fibrillar" muscles.

1. TUBULAR FIBERS

These are the "ordinary" skeletal muscles: in cultures, they are often grouped in small bundles. Long and narrow, they contain, along their central axis, a single row of nuclei which are so dark that, when observed in polarized light, the fibers appear to be hollow tubes. Myofibrils lie at the periphery, with sharp cross-striations (Fig. 2.15). Their cross sections, in electron microscopy, are somewhat cuboid, each myosin filament being surrounded by a typical ring of 10 to 12 actin filaments.

2. FIBRILLAR MUSCLE CELLS

They correspond to adult indirect flight muscles. In cultures, they are represented by isolated fibers, with blunt ends and many small nuclei scattered throughout the sarcoplasma. Some of them are very short and

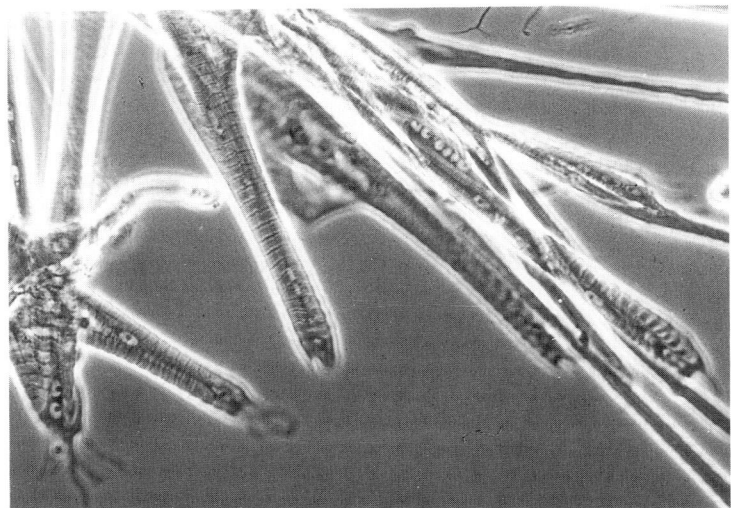

FIGURE 2.15 Adult muscles developed in *Drosophila* embryonic culture after treatment with ecdysone. Bundles of tubular muscles. Nuclei are seen in the center and cross-striations are very sharp. From Dübendorfer *et al.* (1978); courtesy of Dr. Andreas Dübendorfer.

almost rectangular, and others longer and more clearly fibrillar, with fine cross-striations. With electron microscopy, the cross sections of their myofibrils are seen to be round and a well-ordered array of 6 actin filaments encircles each myosin filament, as in adult flight muscles.

C. Adult Fat Body Cells

In vitro differentiation of imaginal fat body cells was studied by Dübendorfer and Eichenberger (1985).

The addition of ecdysteroids to 15-day-old primary cultures induced first a progressive lysis of differentiated larval fat body cells (see Section II.A.2). This was especially obvious after the second hormonal application and is perfectly consistent with normal degradation of larval fat body during metamorphosis. A little later, adult-type fat body cells differentiated from fibroblast-like precursors. They could be identified with electron microscopy by the presence of large reserves of glycogen and numerous, although small, lipid droplets, two features that are very typical of imaginal adipocytes.

The amount of adult fat body per ecdysone-treated culture was reduced by a concomitant exposure to the powerful JH analog Zoecon ZR 515 (Milner and Dübendorfer, 1982).

D. Differentiation of Other Tissues

The most immediate response of primary cultures to ecdysteroids is the secretion of additional chitin layers by larval epidermis cells (Dübendorfer and Eichenberger-Glinz, 1980; Sang, 1981) (Fig. 2.16). The persistence of these chitinous capsules makes it difficult to tell whether these larval cells will finally die, as an expected consequence of exposure to ecdysone.

Milner and Dübendorfer (1982) confirmed that the secretion of chitin layers is not inhibited but enhanced by the presence of JH analog, which is the normal situation during larval molting.

The development of tracheolar tubes is enhanced by ecdysterone treatment. Moreover, an identifiable chitinous lining is secreted, with the typical ribs of insect tracheae (Dübendorfer and Eichenberger-Glinz, 1980; Kuroda and Shimada, 1988).

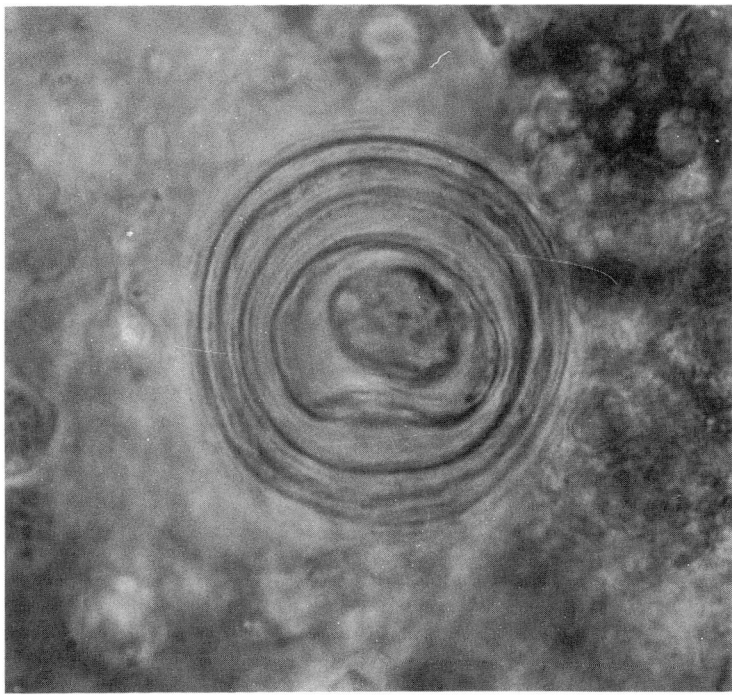

FIGURE 2.16 Secretion of several layers of chitin by larval epidermal cells. Primary *Drosophila* embryonic cell culture after ecdysone treatment. From Dübendorfer and Eichenberger-Glinz (1980); courtesy of Dr. A. Dübendorfer.

VI. CELL CULTURES FROM EMBRYONIC (LETHAL) MUTANTS

One of the major assets of *Drosophila melanogaster* as an experimental model is, once again, the availability of an exceptionally high number of mutants, affecting most of its developmental events. *In vitro* culture gives an added dimension to the analysis, since it makes it possible (1) to study the potentialities of individual cells in a situation where they are isolated from interactions of the whole organism, (2) to keep lethal mutant cells alive beyond the specific "kill period" (D. T. Suzuki) and to detect thus subtle defects which manifest themselves more clearly in the long term;

this is particularly true for cell lines permanently established from mutants (see Chapter 3), (3) to try to repair the defective gene action by preparing mixed cultures with wild-type cells or by adding specific metabolites to the culture medium.

As soon as it became possible to make cultures from individual gastrulae, because, in many cases of severe lathality, homozygous embryos have to be obtained from complex crosses and may represent only one class of the progeny, which excludes the use of mass cultures, Donady and Seecof (1971) investigated the effect of *Drosophila lethal myospheroid* mutation on differentiation *in vitro* of muscle and nerve cells. This approach has been developed by Sang and collaborators and also by Kuroda. It is obvious, however, that the potential of this technique has not yet been fully exploited.

1. LETHAL MYOSPHEROID (mys)

This mutation, *lethal myospheroid* (*mys*), which is sex-linked and recessive, causes death after mid-embryogenesis.

According to Donady and Seecof (1971, 1972), cell cultures derived from *mys* hemizygous embryos are more or less similar, after 24 h, to wild-type ones, except that they fail to produce any muscle cells and the number of axons is reduced by approximately 35% compared to normal cultures. Thus, because it causes defects in dissociated cells, the *myospheroid* mutation should be considered to be "autonomous," although axon failure observed *in vitro* might depend on the myocyte failure.

On the other hand, Cross and Sang (1978) found that, in cultures from *mys* embryos, muscle cells did differentiate but with an unusual form and weak pulsations, whereas neurons, fat body, and chitin-forming cells developed normally. A prevailing abnormality consisted in the loose attachment of all cell types to the glass substrate, so that the authors proposed a new interpretation of the *mys* syndrome. It might result from a general defect in the cell membrane rather than a specific defect in the basement membrane.

The conflicting conclusions from those two laboratories arise probably from the strains of flies used, or on different culture conditions. (For further development of this approach, see Chapter 6: integrins).

2. DEFICIENCY FOR RIBOSOMAL DNA

Using a deletion of the X-chromosome that includes the whole rDNA region, Donady and colleagues (1973) established that neuron and myocyte differentiation was normal in culture embryonic cells that contained

0, 1, or 2 regions of rDNA, respectively. As in the development of the "anucleolate" mutant of the toad *Xenopus,* the cells lacking rDNA had received ribosomal material from the ovum and this seems to be sufficient for protein synthesis, during the short culture period.

3. NOTCH

Extreme *Notch* alleles cause a dramatic hypertrophy of the nervous system in lethal embryos, to the detriment of hypoderm, and a complete failure of mesoderm derivatives.

Cross and Sang (1978) cultured embryonic cells from five distinct *Notch* mutants. In most cases, they observed some hypertrophy of neural tissue, revealed by a two- to three-fold increase in the number of large clusters of neuron cell bodies. Even so, the typical feature consisted of muscular abnormalities. Most individual myocytes failed to fuse into myotubes, so that their total number, as well as the number of myocyte bundles, appeared significantly higher than in normal cultures.

Data from mixed cultures of *Notch-8* and wild-type cells are consistent with the autonomous expression of muscle and nerve abnormalities. (See also further data and discussion in Chapter 6.)

4. *WHITE* DEFICIENCY LETHALS

Whereas the initial differentiation of cultured *white* mutant cells seemed normal, several deteriorations affected the three major cell types after the 3rd day (Cross and Sang, 1978). Muscle and fat body cells degenerated and neurons appeared swollen, with no new axon production. This situation reflects the late death of apparently complete embryos that characterizes this type of deficiency *in vivo.*

In mixed cultures from wild-type and mutant embryonic cells, these latter cells still displayed typical muscle and fat body abnormalities.

5. SHIBIRE[ts]

The development of these "conditional" mutants is normal at 22°C, whereas exposure at 29°C causes embryonic lethality or larval and adult paralysis.

During *in vitro* culture at a "restrictive" temperature, all cell types could differentiate, but their development was rapidly halted. This block could be removed by a simple downshift to the "permissive" temperature, even several days later. Conversely, when, at the end of the first day of culture at 22°C, the temperature was raised to 29°C, the development, until then normal, was arrested (Cross and Sang, 1978). (See also *Shi^ts* permanent cell lines, in Chapter 3, and, in Chapter 6, Section III.C.)

6. DEEP ORANGE (dor)

This is an X-linked recessive lethal gene which affects pteridine synthesis and provokes inexorably the death of homozygous embryos at various stages of embryogenesis.

It is surprising that, in cultures from *dor* mutant embryos that had developed beyond gastrulation, Kuroda (1974b) observed only limited abnormalities, which consisted essentially in a failure in myocyte fusion and the absence of epithelial vesicles.

Quite interestingly, addition of an extract from wild-type eggs to the culture medium resulted in the repair of the defective characters. One should bear in mind that *dor* embryos can be cured by the injection into eggs of wild-type cytoplasm.

To determine the exact time at which a *dor^+* gene introduced by a wild-type sperm expresses the necessary substance(s) in the zygote, the repair activity of extracts from embryos deriving from a cross between *dor* homozygous mothers and wild-type males was assayed, in culture, at various stages. The effective substances seem to be produced by the paternal *dor^+* gene between 3 and 5 h after fertilization (Kuroda, 1977a, 1982).

As pointed out by Sang (1981), this work, because of the quite interesting results it produced, merits repetition with more suitable culture media and techniques.

7. FUSED

It is also a X-linked recessive lethal mutation.

Cells from fused gastrulae showed quasi-normal growth and differentiation, although Kuroda (1976, 1982) noticed some defects in extension and branching of nerve fibers (?). These could not be cured with wild-type egg extracts.

8. RUDIMENTARY 39K

Located on the X chromosome, the rudimentary gene encodes a polyenzymatic macromolecule catalyzing the first three steps of the pyrimidine biosynthetic pathway (see Chapter 5).

In primary cultures from r^{39K}, Kuroda (1976b, 1978, 1982)* observed only an absence of pigmentation in what he called "epithelial cells" (?).

---

* Because of the medium used by the author, it would be advisable to consider these results with some caution.

APPENDIX 2

2.A. Setting up Primary Cultures from *Drosophila* Embryonic Cells

These procedures are based on our own experience and on several other sources.

COLLECTION OF EGGS

A large population (several hundred young adults) of well-fed flies is allowed to deposit eggs on small trays, or simple microscope slides, covered with agar which has been coated with an autoclaved paste of fresh yeasts.*

In order to eliminate eggs that are too old, the retention of fertilized eggs being frequent, especially in stressed flies, it is advisable to discard the first 2 h egg collection. The true collection, on fresh trays, will last some additional 2 h; then, trays are removed and incubated in a moist chamber, at 25°C, for the required time (see Chapter 2, Section I.A).**

The age of embryos may be roughly estimated from the midpoint of the egg laying period.

DECHORIONATION AND SURFACE STERILIZATION

Eggs are brushed off from the surface of the agar into a beaker containing distilled water. Then, they must be thoroughly rinsed with water, through two successive metallic or nylon sieves. The first one (500 μm mesh) retains large fragments of agar and dead flies, whereas the second one (110 μm mesh) will keep back eggs and let yeasts pass through.

After a final examination to make certain small debris and more especially newly hatched larvae (whose gut content might contaminate the whole culture) have been eliminated, 300 to 400 eggs are directly transferred, with a small brush, into the homogenizer conical tube containing a few milliliters of a freshly prepared and filtered 2.5% (v/v) sodium

---

* 2 g of Agar Noble (Difco) are dissolved in 100 ml distilled water (by stirring and warming up to boiling). After addition of 2.5 ml ethanol and 1 ml acetic acid, the mixture is carefully poured onto standard microscopy glass slides. After cooling and solidification of the agar, the slides are overlaid with an autoclaved yeast–water paste. Ready-to-use slides may then be kept for a few days in the refrigerator (within a moist box).

** Oviposition in *Drosophila* occurs preferentially at the beginning of the dark phase of the nycthemeral cycle. Fortunately, use of artificial light dispenses with unpleasant night work, which would result from the young embryonic stages selected for culture (see Section I.A). Using preferentially 6 to 12-hour-old embryos, we found it convenient to collect eggs in the late evening and then the embryos were ready for the next morning's work.

hypochlorite solution (or saturated calcium hypochlorite). Addition of a wetting agent [for instance, Nonidet P-40 (NP-40) 1 : 20,000], by reducing egg flotation, facilitates their handling.*

A 5 to 10 min exposure removes the chorion of eggs (this should be controlled under a binocular microscope) and achieves complete sterilization of their surface (so that any further treatment, for instance with 0.05% (w/v) mercuric chloride in 70% ethanol for 10 to 15 min, is unnecessary).

Wash thoroughly with serum-free culture medium, because hypochlorite is very toxic to cells.

When eggs have sunk to the bottom of the conical homogenizer tube, the quasi-totality of the liquid is carefully pipetted off and replaced with fresh medium; allow the eggs to settle again, then repeat washing 3 to 4 times.

Note: Whenever a more precise stage determination of the embryos is needed, it should be carried out, under a binocular microscope, after chorion removal and through the transparency of vitelline membrane. The procedure will differ slightly, since dechorionation and washings take place in a small test tube and, from this point on, all operations must be accomplished under sterile conditions, unless one prefers, after embryo selection, to resterilize the egg surface with a 10 min immersion in 70% ethanol, or a 5 min treatment with Zephiran chloride.

MECHANICAL EGG DISRUPTION AND CELL DISSOCIATION

Finally, the eggs are collected in a small volume of complete medium at the bottom of the conical homogenizer. By pressing them gently with the tip of the sterilized pestle without turning or slipping, all of the eggs are burst at once. This is the crucial step of the operation and difficult to control. It is a matter of knack. Two extremes must be avoided: if crushing is not sufficient, the large tissue fragments will not adhere to the culture vessels and will have to be removed from the culture; on the other hand, an excessive fragmentation may damage most of the cells.

---

* It was found very convenient, during this dechorionation step, when eggs, carried upward by small gas bubbles, show a tendency to float, to fill the tube completely and close it with a wetted rubber stopper, expelling carefully any imprisoned air (in order to eliminate the air–liquid interphase meniscus).

By pipetting a few milliliters of medium onto the pestle, any cellular material adhering to it is washed down to the bottom of the tube. If the homogenizer tube is not moved, the cell groups (containing approximately 50 cells*) settle to the bottom, while debris, such as vitelline granules, or isolated injured cells, continue to float in the liquid. Thus, the suitable cell clumps are easily collected and can be transferred into culture vessels at suitable concentrations (see Section I.C).

It might be advisable, according to some authors, to clear cells more completely from yolk debris by one or two washings with fresh medium and low-speed centrifugation.

2.B. Cell Dissociation Method for Imaginal Discs

This method is from Currie, Milner and Evans (1988) and reproduced with permission of *Development,* Company of Biologists, Ltd.

Thirty to 40 imaginal discs from late 3rd instar larvae are dissected in Shield and Sang *MM3* medium (see Chapter 1). They are cut into 3 to 5 fragments to facilitate dissociation.

These fragments are transferred to a 10-ml plastic centrifuge tube; as much of the medium as possible is removed and 0.8 ml of Ca^{2+} and Mg^{2+}-free Dulbecco's phosphate-buffered saline (CMF–PBS**) with 2 mg/ml^{-1} *dispase* (Boehringer Mannheim) and 4 mM EDTA are added.

The tube is then rotated at about 1 revolution per second on a bottle roller, at room temperature, for 45 min.

80 μl of 1% (w/v) *Bacto-trypsin* (Difco) in CMF–PBS is then added, giving a final concentration of 0.1% and is left for an additional 10 min.

Fragments are further dissociated by brief shaking on a Vortex mixer. Centrifugation for 4 min at 150 g in a refrigerated centrifuge.

The pellet is resuspended in complete medium, then cells are pipetted up and down 15 times to further dissociate remaining clumps, before seeding in wells of a 96-well plate (Nunc).

2.C. Column-drop Method

This method is based on that of Dübendorfer and Eichenberg-Glinz (1980 and personal communication from the authors).

---

* This optimum cell group size refers to primary cultures which are set up with the aim of developing continuous cell lines (see Chapter 3).

** Dissolve in one liter of distilled water: NaCl 8 g; KCl 0.2 g; Na_2HPO_4 1.15 g; KH_2PO_4 0.2. Autoclaved for 20 min. The pH should be between 7.4 and 7.5. Distribute in small flasks and store in refrigerator.

The slide chamber (see Chapter 2, Section I.C.1 and Fig. 2.1) is made by cementing two microscope glass slides on top of each other with silicone rubber; the top one having a hole of 15 mm diameter drilled through it.

The chamber must be siliconized (2% [v/v] dimethyldichlorosilane in CCl_4) in order to give a hydrophobic floor, and must be carefully washed before use.

After heat sterilization, some petroleum jelly is smeared, with a glass rod, around the culture well.

A drop of cell suspension is then placed in the center of the well. A coverslip (which has been previously washed free from any trace of grease and sterilized with a 1 : 1 mixture of ethanol and ether) is laid over it.

The petroleum jelly will seal the chamber tightly, when the coverslip is gently pressed down, and the drop of cell suspension will touch the coverslip and form a column. The drop has been made sufficiently large to give a column of about 5 mm in diameter, so that the air-to-medium volume ratio in the chamber is about 7.5 : 1).

The culture chamber is now kept upside down for 40 min, so that the cells can settle and adhere to the coverslip. When the culture is reverted, nonadhering debris settle to the floor of the chamber, whereas the adhering cells remain hanging on the coverslip and continue their development.

Medium changes, usually performed weekly, can be done by rotating the coverslip by 45 degrees and lifting it off the slide. Roughly, half of the medium is removed, using a sterile micropipette, and replaced by fresh medium. Then the chamber is closed again.

2.D. Isolation of Myoblasts from Primary Cultures of *Drosophila* Embryos

This protocol is reproduced from Fyrberg *et al.* (1977) from *TCA Manual*, copyright © 1977 by the Tissue Culture Association, with the kind permission of Professor Donady and the Society for *In Vitro* Biology.

MATERIALS

EGTA (Ethyleneglycol bis(β-aminoethyl ether)-N,N-tetraacetic acid), (Sigma, St. Louis, MO)

Shaker-incubator (No. C-24NB, New Brunswick Science Co., New Brunswick, NJ)

Stock solution (1 mg per ml; pH 3) of bovine insulin in Seecof (1971)'s saline
Clinical centrifuge and conical centrifuge tubes with cap
Tissue culture dishes (Falcon Plastics)

ISOLATION OF MYOBLASTS

(See also Appendix 2.A)
Collect eggs; clean; dechorionate and sterilize.
Under sterile conditions break the eggs in a tissue grinder.
Centrifuge the suspension to isolate and clean the cells.
Finally, plate the cell suspension in tissue culture dishes.
When isolation of myoblasts is desired make the following alterations in the general procedure for the plating of the cells:
Maintain 3 to 5 mM EGTA,* along with serum and insulin, in the initial plating medium. All subsequent mentioning of medium includes serum (18% per volume).
Decrease the cell density to 2.5×10^5 cells per mm^2 to reduce the incidence of contaminating cells attaching to myoblasts.
Incubate the culture at 26°C and observe periodically with inverted phase optics. One to two hours later, the majority of cells will have detached from the dish bottom. The cells that remain attached are primarily myoblasts.
Shake the culture for 10 min at 100 to 150 rpm** and 26°C. This will release additional cells.
Remove the medium from the side of the tilted dish with a Pasteur pipette. Released cells may be replated.
Rinse the culture dish three times with medium (omitting EGTA). Add and remove medium from the side of the dish.
Add final medium containing insulin (10 mU per ml) to facilitate differentiation. The remaining cells are greatly enriched for myoblasts (75 to 95% of differentiated cells are myogenic).
Due to their low proportion in mass cultures, myoblasts will now be sparsely distributed over the culture dish surface. If this is sufficient material, incubate the culture in a moist atmosphere at 26°C.

---

* Ranges of the optimal concentration of EGTA and the duration of exposure may differ depending the serum lot and age of serum.
** The ranges for shaker time are also related to the EGTA effect. The higher speeds produce purer preparations of myoblasts but at lower densities.

If higher densities of myoblasts are required, the cells may be concentrated by the following method.

CONCENTRATION OF MYOBLASTS

Remove the medium with a pipette (5 ml) and vigorously discharge the medium over the surface of the culture dish. Repeat this step several times until all cells are detached.

Pipette the cell suspension into a conical centrifuge tube and cap. Centrifuge at 360 g for 5 min. Discard supernate.

Resuspend cell pellet in a small, known amount of medium and determine cell number by standard hematocytometer method.

Dilute cell suspension to desired concentration. Plate cells. A final density of 1×10^6 cells per cm^2 will permit excellent differentiation.

Allow the cells to attach to the dish bottom and then replace medium to remove loose cells and debris.

2.E. Purified Neural Cell Cultures from *Drosophila* Embryos

This procedure is based on Furst and Mahowald (1985) and reproduced with the kind permission of Prof. Mahowald and Springer Verlag, Ltd.

MONODISPERSE EMBRYONIC CELL SUSPENSION

Eggs are collected for 2 h from fly population cages. After 4 h aging, nearly all embryos are between the earliest stages of gastrulation and germ-band elongation.

Dechorionation for 2.5 min with agitation in 2.5% (w/v) $NaClO_3$ in 50% ethanol. Surface sterilization for 10 min in 0.1% (w/v) alkyldimethylbenzyl ammonium chloride in 70% (v/v) ethanol. Wash three times in Chan and Gehring's BSS (see Chapter 1).

Mechanical disruption in a Dounce homogenizer. The cell suspension in BSS is centrifuged for 5 min at 1500 rpm, and washing is repeated 3 times.

SIZE FRACTIONATION BY CENTRIFUGAL ELUTRIATION

Approximately 1×10^9 cells are injected into a Beckman JE-6 rotor at 2800 rpm, at 10°C, with a flow of BSS at 12.5 ml/min.

The flow is increased by increments of 2.5 ml/min, and seven fractions of 100 to 200 ml are collected.

The cells are pelleted by centrifugation at 1500 rpm for 15 min, at 4°C, washed once in culture medium, and cultured. Cell recovery exceeds 90%.

FURTHER ENRICHMENT BY DIFFERENTIAL ADHESION

After overnight incubation of cell class III (fractions 4 to 6 are pooled) cultures, the medium is aspirated and replaced.

The remaining cells correspond to a virtually pure culture of developing neurons from the embryonic CNS of *Drosophila*.

2.F. Separation of Neuron and Myocyte Precursors from Early Gastrula Cells through Ficoll Gradient

This procedure is based on the work of Hashashi and Perez-Magallanes (1994) (*In Vitro Cell. Dev. Biol. Animal*, 30A, pp. 202–208; copyright © 1994 by the Tissue Culture Association, Inc., reproduced with the kind permission of Dr Salvaterra, and of the copyright owner. In Memory of Dr Izumi Hayashi.

PREPARATION OF CELLS

Drosophila melanogaster, Canton S, wild-type flies were maintained in population cages at 25°C under a 12-h light/dark cycle. Eggs were collected for 2 h from 1- to 2-week-old flies by placing standard corn meal-agar food plates in population cages.

The plates were incubated for 3 h at 25°C until the embryos approached early gastrula stage.

The embryonic cells were prepared for culture by a modification of the method of Seecof *et al.* (1971). Briefly, the embryos were harvested, washed extensively in water, and treated for 1 min with a 1 : 1 mixture of 3% (v/v) sodium hypochlorite and 95% (v/v) ethanol.

The sterile, dechorionated embryos were homogenized with a Dounce homogenizer in Schneider's modified *Drosophila* medium (Irvine Scientific, Santa Ana, CA) containing 5% (v/v) fetal bovine serum and 200 ng/ml bovine insulin (Sigma Chemicals).

The homogenate was passed through a Nytex mesh (25 μm) to remove debris and large cell aggregates, and the single cells were collected by centrifuging for 5 min at 1500 rpm, using a Beckman desktop centrifuge. The centrifugation was repeated twice under the same conditions, and the cell pellet was finally resuspended in Schneider's medium containing 5% FBS, 200 ng/ml insulin and 0.5% Ficoll (Sigma).

The cells were counted using a hemocytometer and the cell density was adjusted to 1×10^6 cells/ml and kept on ice until fractionation. (The total time required for the preparation of cells was about 30 min.)

MEDIA FOR FRACTIONATION

Schneider's modified *Drosophila* medium supplemented with 5% bovine calf serum and 200 ng/ml bovine insulin was used as the base medium for all fractionation operations. *Drosophila* cells do not grow or differentiate in the absence of fetal bovine serum, but the cells survive well in any serum for a short period. Because 1 liter of medium is necessary for a single operation, bovine calf serum was used for economic reasons.

For the Ficoll gradient formation, 450 ml of each of the base media, supplemented with 1.5% and 3% Ficoll, were prepared. The overlay was the base medium without Ficoll. The base medium was supplemented with 20% Ficoll to form a high-density cushion during cell separation. All media were sterilized by filtration through 0.2-μm filter units and kept at 4°C until use.

FRACTIONATION OF DROSOPHILA GASTRULA STAGE CELLS

Fractionation of the cells was performed using a Celsep apparatus (Beckman Instruments, Logan, UT). The apparatus, which can be operated either at a 30 degree angle or in a horizontal position, is basically a flat (1 in. height) cylindrical chamber with openings at the top and the bottom for the introduction of cell suspension and the Ficoll gradient, respectively. The cells are differentially sedimented through the flat cylinder in a short time and can then be collected into fractions after reorientation of the cylinder to 30 degrees.

Construction of the Ficoll gradient for Celsep was done according to the manufacturer's protocol. Briefly, the Celsep chamber (1 liter capacity) positioned at a 30 degree angle was filled from the bottom with the base medium containing a continuous gradient of 1.5 to 3% Ficoll. The gradient was formed by a gradient maker and delivered by a peristaltic pump at 30 ml/min. Following the gradient, approximately 100 ml of cushion medium was delivered to the chamber. The flow of the peristaltic pump was then reversed, and 30 ml of cell suspension containing a total of 3×10^7 cells was delivered onto the gradient at 10 ml/min, followed by the addition of about 50 ml of the overlay medium. Thus, at the completion of packing, the Celsep chamber contained, from the top, 50 ml of overlay medium, 30 ml of cell suspension, 900 ml of medium containing a continuous Ficoll gradient (1.5 to 3%), and about 20 ml of the cushion medium.

The apparatus was packed in ice and left for 2.5 h in a horizontal position to allow the cells to sediment. After sedimentation, the Celsep chamber was reoriented to a 30 degree angle and 20 fractions, each containing 50 ml of medium, were collected into sterile centrifuge tubes.

The tubes were centrifuged for 10 min at 1500 rpm to collect the cells. The cells were washed once in Schneider's modified *Drosophila* medium containing 20% fetal bovine serum, resuspended in the same medium and plated. The cultures were kept at 25°C. The Ficoll concentration in each fraction was determined by measuring the refractive index of the medium using a refractometer (Bausch and Lomb).

ANALYSIS OF FRACTIONATION OF DROSOPHILA EMBRYONIC CELLS

The size of cells in each fraction was determined by taking photographs of the Celsep-separated cells immediately after plating. A typical field contained 50 to 100 cells. A micrometer registering a 10 μm scale was also photographed using the same magnification. Cell size was determined by measuring the diameters of cells on photographic prints.

Cell number was determined by direct counting using a hemocytometer, both at the time of plating and 10 h after plating. This latter time coincides with the beginning of terminal differentiation, when the majority of cells have ceased to divide but myocytes have not fused completely and neurons have not yet formed tight clusters. For the 10 h cell count, the cells were plated at 2×10^5 10 mm well (of a 24-well plate) for all fractions to eliminate the effect of differing plating densities on cell growth. The final cell number for each fraction at 10 h was deduced from the percentage of cells used for each fraction for this experiment. At 10 h after plating, the medium was removed, the plates were rinsed once with phosphate-buffered saline, and the cells were dissociated with 0.25% trypsin and 5 mM EDTA in phosphate-buffered saline, and counted.

For the determination of differentiated cell types arising from each fraction, cultures were fixed for 30 min at 4°C in 4% paraformaldehyde in phosphate buffered saline 24 h after plating. Differentiation was observed using a phase contrast microscope, or, for neurons, by staining with antibodies to plant horseradish peroxidase (anti-HRP antibody), which are specific markers for insect neurons. For immunochemistry, the fixed cells were incubated with rabbit anti-HRP antibody (gift of Dr. Salvaterra,

City of Hope, CA) overnight at 4°C. Anti-HRP binding was visualized by detection of fluorescein-labeled goat antirabbit IgG secondary antibody, using a fluorescence microscope.

2.G. Primary Nerve Cell Cultures from Larval Central Nervous System

This culture is based on the work of Wu *et al.* (1983) and reproduced by permission of Oxford University Press.

Mature 3rd instar larvae are surface-sterilized by brief immersion in 70% ethanol and then rinsed three times with sterile distilled water.

The brain and ventral ganglion are dissected in Schneider's culture medium and torn into small fragments with fine dissecting needles.

Incubation for 1 h in a saline similar to Rinaldini's Ca^{2+}- and Mg^{2+}- free solution (see footnote on page 74) containing 0.1 to 0.5 mg/ml of collagenase (Type I, Sigma).

The fragments are rinsed three times with culture medium supplemented with 10% fetal bovine serum and then dispersed into single cells in the same medium by repeatedly flushing through a siliconized glass pipette.

To obtain monolayer cultures, 80 to 200 μl of cell suspension are added into a plastic petri dish containing 1 ml of culture medium or directly onto round glass coverslips (10 to 20 mm diameter) in dry petri dishes to form minicultures* confined by surface tension of medium. The coverslips and the Petri dishes have been precoated with poly(L-lysine) (M_r 10,000, Sigma).

Cultures with densities ranging from 0.5 to 5 $\times$ 10^4 cells/cm^2 are incubated in air atmosphere at room temperature. To minimize the problem of evaporation, drops of additional culture medium in separation from the minicultures are added in the petri dish, and the dish is covered and kept in a moist chamber.

References

Ashburner, M. (1989). *Drosophila, a Laboratory Handbook*, Cold Spring Harbor Laboratory Press, New York.

Campos-Ortega, J. A., and Hartenstein, V. (1985). *The Embryonic Development of Drosophila melanogaster*, Springer-Verlag, Berlin.

* Coverslips can be easily transferred and are very convenient for electrophysiological work.

Hadorn, E., and Garcia-Bellido, A. (1964). Zur Proliferation von *Drosophila*-Zellkulturen im Adult-milieu, *Rev. Suisse Zool.* **71**, 576–582.

Hirumi, H. (1963). An improved device for cultivating cells *in vitro* and for observations under high power phase magnification. *Contr. Boyce Thompson Instit.*, **22**(2), 113–116.

Karnovsky, M. J., and Roots, L. (1964). A "direct-coloring" thiocholine method for cholinesterase, *J. Histochem. Cytochem.* **12**, 219–220 (cf in Furst & Mahowald, 1985a).

Lwoff, A., Dulbecco, R., Vogt, M., and Lwoff, M. (1955). Kinetics of the release of poliomyelitis virus from single cells, *Virology* **1**, 128.

Paul, J. (1973). *Cell and Tissue Culture,* 4th ed., Churchill Livingstone, Edinburgh, London.

Poulson, D. F. (1950). Histology, organogenesis, and differentiation in the embryo of *Drosophila melanogaster* Meigen. In "Biology of Drosophila" (M. Demerec, ed.). pp. 168–274, Hafner, New York.

Rinaldini, L. M. F. (1958). The isolation of living cells from animal tissues. *Int. Rev. Cytol.* **7**, 587–647.

Rizki, T. M., and Rizki, R. M. (1984). The cellular defense system of *Drosophila melanogaster*. In "Insect Ultrastructure" (R. C. King, and H. Akai, eds.). Vol 2, pp. 579–604. Plenum Press, New York.

3

Drosophila Continuous Cell Lines

The terms continuous cell line, permanent cell line, and established cell line are synonymous and, in the terminology proposed by Hayflick and Moorehead (1961), refer to any line that "can be cultured for such a long time that it has apparently developed the potential to be subcultured indefinitely *in vitro*" (J. Paul, 1973). In contrast, primary cultures and cell strains are capable of only a limited number of serial passages.

As a general rule, a line of mammalian cells must, arbitrarily, have been subcultured at least 70 times, at intervals of 3 days, before being considered as fully established. Schneider and Blumenthal (1978) suggested that, for invertebrate cell lines, a comparable timetable should be around 40 subcultures at 3 to 5 days intervals, or some 120 cell division cycles (Simcox and Shearn, 1985).

The use of such continuous cell lines offers obvious advantages, particularly for biochemical approaches: one may utilize theoretically unlimited amounts of quite homogeneous cellular material, especially when the line is periodically subcloned.

When well established *in vitro, Drosophila* cell lines are easier to grow than mammalian cells and their doubling time, even at an optimal temperature of 22 to 25°C, is quite comparable (about 24 h): (1) there is no need of sophisticated incubators, as the cells actively divide at room temperature and most culture media do not contain a bicarbonate buffer requiring a controlled gas phase; (2) the cells tolerate wide variations of pH or osmolality (see Chapter 1); (3) when grown as monolayers, their transfer does not usually require any enzymatic treatment as the cells are detached from the vessel by simple scraping; additionally, several lines have been adapted to suspension cultures; (4) most lines can be kept frozen in liquid nitrogen for a very long time. For instance, *Kc* cell samples which had been frozen in our laboratory on July 1969 were successfully thawed for the 25th anniversary of the initiation of their primary culture.

There are, nevertheless, two disadvantages, when compared to mammalian cell systems: (1) establishing new *Drosophila* cell lines remains a long and delicate task; and (2) the majority of available lines derived from dissociated *Drosophila* embryos and their precise tissue origin are unknown. Only two lines from larval prohemocytes and, more recently, several lines from various imaginal disc cells have been established *in vitro* (see Section IV).

I. ESTABLISHING EMBRYONIC CELL LINES

Most available *Drosophila* cell lines derived from the long-term evolution of primary cultures made of mechanically dissociated young embryos (6 to 12 h). As indicated in our preparation protocol (Chapter 2, Appendix 2.A), when primary cultures are set up with a view to developing cell lines, the fragmentation of embryos must give rise to many cell clumps of an adequate size (that is, comprising some 50 to 100 cells), so that they can firmly adhere to the vessel bottom and remain alive for the long transition period (see below).

A. Usual Sequence in the Development of a New Cell Line

As was stated in Chapter 2, after the *in vitro* differentiation of many tissue cell types (myocytes, neurons, adipocytes, etc.), and not before the end of the second week, there appears, in the most viable embryonic cell cultures, one or a few outbursts of new multiplicative cell types. Surrounding well-anchored embryonic fragments, they constitute large swarms of roundish cells—sometimes more crowded and rather "epithelial-like" (see Chapter 2, Fig. 2.7A). Alternatively, they may form extensive networks of multilayered bundles of spindle-shaped "fibroblasts" (see Chapter 2, Fig. 2.7B). Spreading results from migration and cell division and, within the first month, the cell colonies may cover a limited area or be widespread on the culture vessel bottom. Both types of proliferating cells are small (*ca.* 8 to 10 μm in diameter), with a relatively large and clear nucleus containing one typical nucleolus. Their rather common morphology precludes any usefulness in identifying their tissue origin (see below).

Usually this first wave of cell multiplication is doomed to degeneration and its usual lifespan, like that of all differentiated cell types in the cultures, does not exceed 6 weeks. However lifeless most of the culture may then look, many cells do remain alive within adhering cell clumps, as proved by subsequent outbursts of new cell growth during the following months. Their appearance is very similar to that of the first wave. Likewise, few will survive longer than a few weeks and, in most cases, will not withstand subculturing.

During such alternating phases of cell proliferation and inactivity, one must be extremely patient. See, for instance, the anecdotal, although meaningful, story on establishing the *Kc* cell line (Echalier, 1989). For nonscientific reasons, in May 1968 in Paris (during what was called the "student revolution"), a series of cultures were kept for several months, with their medium periodically renewed. In all likelihood they would have been discarded, owing to their miserable look, had we taken enough time to look at them; but, as professors, we were rather busy at that time. Finally, when the turmoil quieted down, some months later, a few lines started proliferating vigorously. The moral of the story is that one should always keep a number of ongoing cultures, change their medium punctually (and perhaps partially) and watch them carefully. Rather abruptly, without apparent reason, and in many cases not before 4 to 6 months, the situation may change: in a limited number of cultures, a decisive new wave of cell multiplication takes place, which attracts attention due to the extensiveness of the growth. Soon thereafter, it becomes necessary to reduce the cell density of the original culture and subculturing will then be successful. Subsequently, the cells can be transferred gradually into culture vessels of increasing size.

The uncertainties, the prolonged latency, and the erratic nature of early growth that, according to our own experience, mark the initiation of a new cell line might well be reduced, to some extent, by improvements in the culture media and increased skill in monitoring the primary cultures (as first suggested by Schneider, 1972). As a matter of fact, several *Drosophila* cell culturists reported successful subculturing after a much shorter lag phase, e.g., a few weeks instead of a few months. It was often observed, however, that the first few transfers still correspond to a period of transition, as revealed by a progressive acceleration of the division rate in successive passages. So, it seems fair to say, as acknowledged by Schneider in her review with Blumenthal (1978), that the establishment of a new *Drosophila* cell line and its complete adaptation to *in vitro* growth may require from 6 months to 1 year and still remains a delicate task. Sang (1981) spoke of "something of an art" and added that patience might well be the crucial ingredient in the matter. In spite of the difficulties, however, it should be noted that tens of *Drosophila* cell lines have been established by many groups from wild-type fly stocks as well as from various mutant ones (see Sections III and IV).

The above description does not fully apply to cultures initiated from late embryos, according to the protocol recommended by Schneider (1972) (see Chapter 2, Section I.A). In such primary cultures, the picture is characterized by the repeated, but once again erratic, growth of cellular vesicles from the cut ends of the embryonic fragments. Relatively early subculturing of the spheres resulted in the development of cells, growing on the bottom of the vessels, in and around the collapsed and fragmented spheres. It is likely that the three established lines, i.e., Schneider's *L1, L2,* and *L3,* did not derive from the imaginal epidermis cells of such initial vesicles, but from different types of cells physically associated with the spheres, as they were no longer able (as were the vesicles themselves) to form cuticular structures after their *in vivo* transplantation into metamorphosing larvae. However, as an alternative interpretation proposed by Schneider, the loss of a vesicular organization might perhaps account for their incapacity to form recognizable structures.

B. Tissue Origin of Embryonic Cell Lines

One of the major limitations of using *Drosophila* embryonic cell lines in developmental biology studies comes from our ignorance of their exact tissue derivation. It should be stressed that every culture is made of dissociated and mixed cells from several hundred embryos of various ages. Moreover, the proliferating cells, from which any line will ensue, display a common "fibroblastic" morphology that precludes a precise cell type identification.

As was amply documented in mammalian cells, the ordinary "undifferentiated" shape and structures adopted by most cells under culture conditions should not obscure the fact that many of them are, nonetheless, still able to retain *in vitro,* to some extent, the specialized metabolic activities of the tissue from which they were derived.

In order to detect such "residual" differentiated programs in *Drosophila* cell lines, several criteria have been sought.

1. ISOZYMIC AND ALLOZYMIC PATTERNS

With this aim in mind, Debec (1974, 1976) made a thorough electrophoretic analysis of the isozyme and allozyme patterns of *Drosophila Kc* cells and of a score of related lines or clones. He compared them with the large amount of data previously collected by many laboratories on the particular enzymatic properties of most tissues of the organism *Drosophila* during successive developmental stages. The study covered some 25 different enzyme systems, randomly distributed in very distinct biochemical pathways (see Chapter 6, Table 6.I). The author pointed out that all of the enzymatic "profiles" of the cultured cells differed greatly from the overall patterns obtained with whole extracts of embryos, larvae, pupae, or adults, but rather resembled the properties specific of tissues. He finally concluded that "there was a good correlation between the general patterns of the studied cell lines (in particular, *Kc* and *C* lines) and the enzymatic profiles of imaginal discs and brain tissue of 3rd instar larvae" (Fig. 3.1). As a provisional hypothesis, the line *Kc* (with its rounded cells) might be identified with imaginal disc cells, and the line *C* (fusiform cells interwoven in nets) with nervous tissue.

In a similar comparison carried out between two independent lines, *Kc* and Schneider's *L2*, Alahiotis and Berger (1977) observed a few clear-cut differences that might be used as genetic markers. With regard to the "functional state" of these cultured cells, the α-glycerophosphate dehydrogenase (α-GDH) system allowed an unambiguous assignment, as both lines showed a larval isozyme pattern and a total absence of the adult-specific band. As for hexokinase-2, only one of the larval-specific isozymes could be detected but, again, none of the adult bands was seen. Additionally, the alkaline phosphatase isozymes that are involved in the larval-to-adult transition were lacking.

2. ANTIGEN SPECTRUM

Another attractive approach utilized an extensive collection of antisera in an exhaustive immunophoretic characterization of antigenic proteins

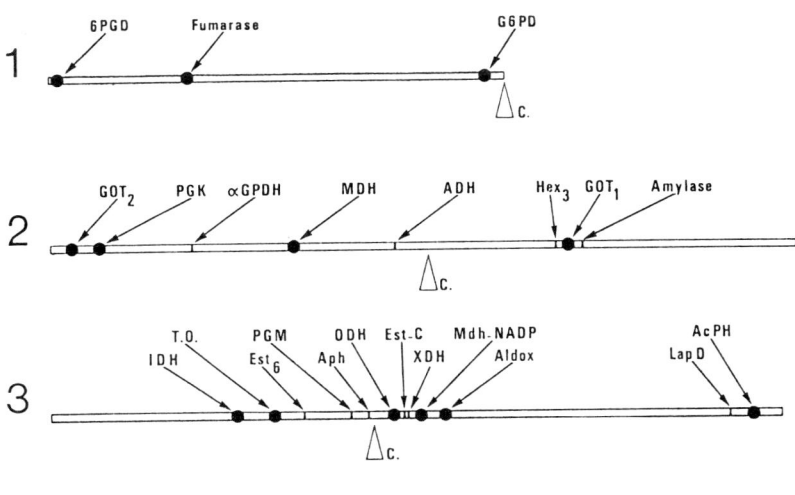

FIGURE 3.1 Isoenzyme "Program" of Drosophila *Kc* Cell Line. The loci corresponding to enzymes, the activity of which was demonstrated in *Kc* cells are marked (black circles) on the genetic map of *Drosophila melanogaster*. Those concerning undetected enzymes are represented by thin lines. On each pair of chromosomes, an arrow (c) shows the location of the centromere. From Debec (1976); courtesy of the author and the publisher Springer Verlag.

present in 10 distinct cell lines (namely Schneider's *L* lines, Dolfini's *Gm* lines, Echalier's *Kc* line, and subclones). They were compared with the specific antigenic patterns of many tissues and developmental stages of *Drosophila* (Moir and Roberts, 1976). Among the 85 detectable proteins (see Table 3.I), some 50 were found in the cell lines. Interest was focused on the 18 "luxury" proteins which, by contrast with the "essential" proteins present in all cell types, do vary in their qualitative distribution and so reflect the differentiated functions of the cells. Statistical comparisons with five tissues showed that all cell lines resembled in their serological properties both imaginal disks and salivary glands (whose intriguing resemblance, on antigenic grounds, had been previously shown by the same authors). Because these two types of structures derive from the lateral embryonic ectoderm, it was suggested that this was the most likely common origin of all studied cell lines.

TABLE 3.1 Distribution of Antigens in *Drosophila* Cell Lines, Developmental Stages, and Tissues [a,b]

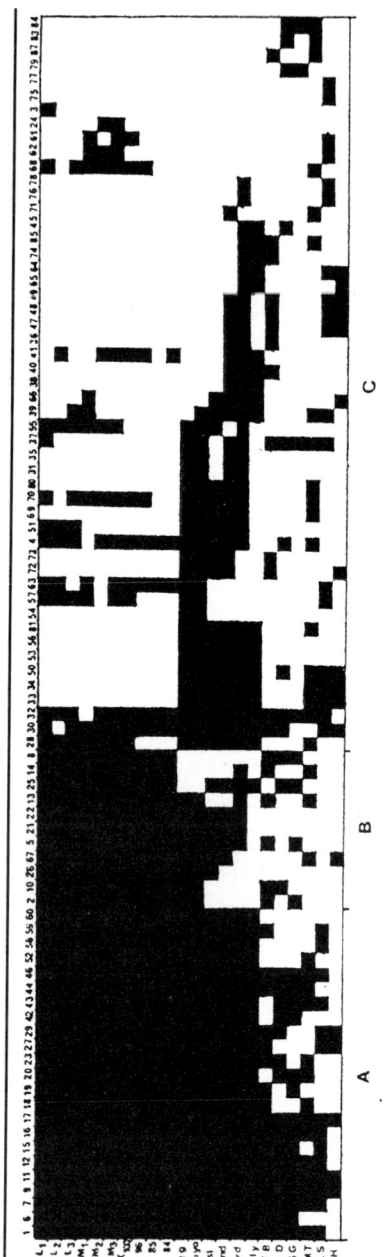

[a] (A) Proteins found at all developmental stages and in each cell line, the essential proteins, 23; (B) Proteins found in each cell line but not at all developmental stages, the proteins necessary for the cells to cope with the culture conditions, 11; (C) Proteins distributed unequally between the developmental stages and the cell lines, the "luxury proteins," 52. FB, fat body; ID, imaginal discs; SG, salivary glands; MT, Malpighian tubules; LS, larval serum; H, hemocytes.

[b] From Moir and Roberts (1976). Reprinted with the kind permission of Professor Roberts, copyright (1976), with kind permission from Elsevier Science Ltd., OX5 1GB, U.K.

3. SPECIFIC RESPONSES TO HORMONES

According to Sang (1981), it should be possible to distinguish at a minimum two broad categories among *Drosophila* established lines on the basis of their response to ecdysteroids. Some of them are rapidly killed by physiological concentrations of 20-hydroxyecdysone and might correspond to larval material (for instance, three of the many new lines established and tested by Simcox and Shearn, 1985). Many others, in particular *Kc* cells and *SL1* line, stop dividing and significantly change their morphology and metabolism (see Chapter 8, Figs. 8.1 and 8.2), i.e., do metamorphose when exposed to the hormone, as would be expected from putative imaginal cells.

In a careful study of the responses to ecdysteroids of a *Kc* subline (named *Kc-H* for its special sensitivity to the hormone), Cherbas *et al.* (1977) confirmed the initial observations of Courgeon (1972a,b) with regard to the morphological transformations induced by the hormone. Particularly impressive were the neuron-like shapes displayed by the treated cells and the growth of very long processes, sometimes exceeding 40 μm in length (see Chapter 8, Fig. 8.2). Moreover, the observed induction of acetylcholinesterase activity suggested that *Kc-H* cells might be precursors of neurons or glial elements, although the level of specific activity remains 50 to 100 times lower than in *Drosophila* brain ganglia. However, the occurrence of low AChE activity has now been reported in various other *Drosophila* tissues and can no longer be considered as a valuable neural marker. In addition, Berger *et al.* (1978) pointed out that choline acetylesterase, another enzyme diagnostic of neural elements, was not induced by 20-hydroxyecdysome, at least in *S3* cells (which, however, do display the same morphology and show the same AChE induction as *Kc* cells, after hormonal treatment). Therefore, Cherbas *et al.* (1980) became, later on, much more cautious about the conclusive value of AChE induction as a neural standard.

The enzyme DOPA decarboxylase is physiologically active in epidermal and neural tissues during the development of *Drosophila,* so that the possibility of its induction by ecdysteroids in *Kc* cells (Spencer *et al.,* 1982, see Chapter 8) might be advanced as an additional argument in favor of the ectodermal origin of this cell line.

Recently, new arguments, such as the tissue-specific regulation of *Eips* (i.e., hormone early inducible proteins, see Chatper 8), or the expression in *Kc* cells of type IV basement membrane collagen (Lunstrum *et al.,* 1988), led Andreas and Cherbas (1992) to retract their former hypothesis and to propose that *Kc* cells might not be derived from ectodermal material but rather from late embryonic hematopoietic cells.

We have detailed all these conflicting opinions about the actual tissue origin of one of the most extensively studied cell line, *Kc,* to clearly illustrate the difficulties and frailties of such *a posteriori* diagnoses.

C. Mechanisms of Cell "Immortalization"

The very process of establishing a continuous cell line from a primary culture, with its often prolonged period of "latency", has remained mystifying for a long time. Two schools of thought have been advanced: (1) either there is a gradual selection of a few cells that, for unknown reasons, are more "adaptable" to *in vitro* conditions; or (2) somatic mutations of various kinds may occur, perhaps favored by the altered environment, and the modified cells thereby become capable of indefinite multiplication and rapidly overtake the entire culture. The frequent aneuploidy and/or chromosomal accidents displayed by many mammalian cell lines seem to support the latter alternative. However, it should be emphasized that many *Drosophila* cell lines retain an euploid or, at least, a paradiploid karyotype (see Chapter 4).

Current studies emphasize the crucial role of some deregulation of a few cell protooncogenes (especially c-*myc* and related genes whose products can be localized in the nucleus) in the process of "immortalization" of cultured vertebrate cells. One of the best analyzed causes of such deregulation is the *cis*-activation of protooncogenes by integration of a retroviral provirus in their chromosomal vicinity, according to the "promoter-insertion" model proposed by Hayward *et al.* (*1981*) for explaining the growth-activating action of "slow" retroviruses.

As for *Drosophila,* although no *myc*-like gene has thus far been demonstrated in its genome, many other sequences homologous to the main categories of vertebrate oncogenes have been identified (see Chapter 6). Besides, we have shown (Echalier and Junakovic, 1988; Junakovic *et al.,* 1988) that amplification and transposition of several retrotransposons (see Chapter 10), which characterize most established *Drosophila* cell lines, seem to occur essentially within the first weeks (or months) of the culture. If one notes the striking structural similarities of those mobile elements with the integrated form of vertebrate retroviruses and, more particularly, the promoter strength of both their Long Terminal Repeats (LTRs), it is extremely tempting to correlate this initial burst of transposition, in early cultures, with the spontaneous establishment of continuous cell lines. Some moving transposons might haphazardly integrate into the genome in the proximity of some protooncogene (for instance a *myb*-related gene or any other) and, by enhancing the expression of this latter, might lead to continuous cell growth ("immortalization").

II. MAINTENANCE OF CELL LINES

A. General Management of Monolayer Cell Cultures

Well-established *Drosophila* cell lines are easily cultured in fluid medium, directly on glass or specially treated plastic surfaces of standard tissue culture vessels.

The following must be pointed out: (1) Cell adhesiveness varies greatly from one line to another and may be modified, quite often by unintentional selection, via the particular handling schedule of each laboratory. In any event, cell attachment remains rather tenuous compared to that of most mammalian cells and, in full-grown cultures, many cells often detach and float in the medium. (2) Cells rarely form a true, complete and adherent monolayer, although the populations of a few well-adapted lines may become so dense that cultures appear heavily crowded. (3) There is obviously no "contact inhibition," and the cells usually show a marked tendency to aggregate and pile up. This can be avoided by frequent subculturing and by reducing the percentage of serum in the medium, or mitigated to some extent by careful selection.

Each cell line is usually grown in the very medium with which it was established, but many sublines have been "adapted" to other media formulations (see Chapter 1, Section IV). The three most widely used culture media for *Drosophila* cells (i.e., Schneider's medium, Echalier's and Ohanessian's *D22*, and Shields' and Sang's *M3*) are now commercially available. However, media prepared in the laboratory ("home-made") remain far more economical, they are neither difficult nor take very long to prepare, and they may be stored in the refrigerator for prolonged periods (more than 1 year, for our *D22* medium, according to our experience). Freezing media must be avoided due to precipitation of some components on thawing.

With rare exceptions, supplementation with fetal bovine serum (2 to 20%) is required. Different serum batches must be tested for toxicity, the latter manifesting itself in slow growth, aberrant cell shape/size, detachment of cells from growing surface or even cell death.

Because most of the buffering of *Drosophila* culture media does not rely on a CO_2–bicarbonate system (see Chapter 1, Section IV.B), it is not necessary to control the gas phase of the cultures by introducing a controlled CO_2–air mixture into the flasks, as is done for most mammalian cell cultures.

The optimal temperature for growing *Drosophila* cell lines is around 22°C (between 19 and 25°C), e.g., normal room temperature. So there is no need of sophisticated incubators, although the temperature may need to be closely watched during the summer in order to avoid heat shock (see Chapter 7).

For ease and economy (careful cleaning and sterilization of glassware is both tedious and time consuming), most laboratories today use disposable plastic (polystyrene) vessels (flasks or petri dishes). Their bottom surface has been especially treated for cell culture and they are commercially available, in sterile condition, after sterilization with gamma rays.

For maintenance of all lines, our laboratory utilizes preferentially 25 cm² (50 ml) flasks with plug seal screw caps. When large quantities of cells are needed, we prefer to increase the number of replicated cultures rather than to handle larger vessels.

In the 3 ml of medium (*D22* + 5% (v/v) fetal bovine serum) with which we routinely grow our serial subcultures, each (25 cm²) flask can accommodate, at the end of one week's culture, as many as 1.5×10^8 cells of our standard *Kc* line (sometimes up to 2×10^8 cells). The cell yield in all other cell lines that we have had the opportunity to grow was found to be significantly lower, i.e., around 7.5×10^7 cells per flask in 3 ml of medium.

It is quite convenient to subculture the cells only once a week, even though, under the usual conditions of seeding, their exponential phase of growth seems to be limited to the first 3 to 5 days after a transfer (see Chapter 4).

Also, it is advisable as insurance against contamination or any other accident to keep each line in two independent series of cultures which are routinely subcultured on different days of the week.

After cautiously aspirating off about 2 ml from the used medium of the culture to be transferred, the cells are readily brought into suspension by simple scraping from the flask bottom and flushing, with the bent tip of a glass pipette. Cell clumps are then dispersed by gentle pipetting and an aliquot of the homogeneous suspension is directly transferred to a new culture flask (25 cm²), previously provided with 3 ml of fresh medium.

The optimal inoculation density remains rather high for most *Drosophila* cell lines. Even for long-established *Kc* cells, new subcultures must be seeded with approximately one-third to one-tenth of the whole yield of the parental culture. However, Schneider (personal communication, 1993) recommends a 1:20 split for *S2* cell cultures. In one of Debec's *1182* lines, each culture flask could be readily divided into 100 subcultures.

For experimental replicates, counting the cells with a hemocytometer together with a rapid estimation of their viability (see Section V) allows for more accurate control of the inoculum size.

If, for any reason, cells have to be washed and resuspended in another medium before transferring, very low rates of centrifugation (500 to 1000 rpm) should be used.

See a brief summary of our routine transfer protocol for monolayer cultures in Appendix 3.A.

B. Suspension Cultures

Suspended cell culture is commonly considered ideal for growing the large quantities of cells required for biochemical studies; but its main advantage is perhaps to facilitate the periodical sampling of experimental cell populations during kinetic studies.

The fact that *Drosophila* cells from most established lines attach very loosely to culture vessels and sometimes float freely in the medium, suggests that they may be able to grow efficiently in suspension.

Various systems have been developed for propagating mammalian cells in agitated suspensions. Most of them were more or less successfully adapted to *Drosophila* cells.

1. ROLLER BOTTLE CULTURES

Cylindrical bottles are constantly rotated horizontally on motor-driven rollers.

With the object of bulk isolation of metaphase chromosomes (see Chapter 4), Hanson and Hearst (1974) put 50 to 180 ml of Schneider's *L3* suspension into 500-ml glass screw-cap bottles (purchased from GIBCO Grand Island, NY). The cells came directly from monolayer cultures and the minimum initial density, in Schneider's medium + 15% FBS, was 5×10^5 cells/ml. The rotation speed was 30 rpm. Logarithmic growth was routinely observed, up to 8×10^6 cells per ml (exceptionally 2×10^7), and the doubling time varied from 16 to 24 h.

2. GYRATORY SHAKING

A gyratory shaker of the Brunswick type (for instance, New Brunswick Co. Inc., type G-10) can accommodate many Erlenmeyer flasks of various capacities, and produces a swirling motion of the cell suspensions.

Miyake *et al.* (1977) defined the optimal parameters for growing Mosna and Dolfini's *GM2* cells in Echalier's medium + 13% FBS: the cells grew well in 5- or 10-ml Erlenmeyer flasks (Tyson's type), with an inoculum size of 1 to 5×10^5 cells/ml and a rotation of 180 rpm. After a slight lag time, the population doubling size was approximately 16 h and the saturation density reached 5 to 10^6 cells per ml. Cell viability remained at the more than 0.95 level until the fifth day after inoculation, then decreased gradually. This system was also applicable to other lines (*GM1, SL2*).

3. MAGNETIC STIRRER UNITS: "SPINNER" FLASKS

The most commonly used method employs a plastic-enclosed magnet which is rotated, on the bottom of any kind of flask, by means of a magnetic stirrer outside the vessel. A range of spinner culture flasks have been specially developed by Bellco Glass Inc. (Vineland, NJ) in order to ensure uniform agitation of cell suspensions and allow easy sampling.

Lengyel *et al.* (1975) were among the first ones who succeeded in efficiently growing *Drosophila* cells in spinner bottles. After the adaptation to slightly modified Dulbecco's medium for Vertebrate cells (with the view of specific biochemical studies, see Chapter 5), Schneider's *L2* cells floated quite freely in the medium and seemed rather prone to grow in suspension. Yet, their adaptation was delicate and required previous passages in roller bottles (at very low speed, 2 to 3 rpm), then, after their transfer into spinner flasks, some alternate periods of spinning and stationary growth. Finally, the cells grew exponentially at densities of 1 to 4×10^6 cells/ml, with a doubling time of 30 h. Densities as high as 1×10^7/ml could be obtained. Unfortunately, the authors did not elaborate on the characteristics of the spinner bottles, nor their rotation speed.

Even *Kc0* cells, this subline capable of growing in serum-free *D22* medium, which is particularly economical, have been maintained in spinner flasks, for instance by O'Connor's group (Spencer *et al.*, 1983; Landon *et al.,* 1988). Cells were serially passed, two to three times a week, when they reached densities of approximately 6×10^6 cells per ml.

4. BIOREACTORS

See Appendix 3.B for a protocol for mass culture of *Drosophila* cells kindly communicated by Dr. Sondergaard (University of Copenhagen).

III. CATALOG OF ESTABLISHED *DROSOPHILA* EMBRYONIC CELL LINES

It is rather difficult to make an accurate count of all *Drosophila* cell lines that have been established throughout the world in the last two decades. In his 1980 survey, Sang considered their number to be more than 60. The list published by Ashburner (1989) in his *Drosophila* Laboratory Handbook approximates 100.

One should be aware, however, that most cell lines were developed for precise purposes, then were frozen by their laboratory of origin, and, more often than not, are no longer available.

In actual fact, a very small number of *Drosophila* cell lines have been extensively used: in the western scientific world, Echalier and Ohanessian's *Kc* line (and its serum-exempted *Kc0* derivative), Schneider's *S1*, *S2*, and *S3* lines and, to a lesser degree, Mosna and Dolfini's *GM1* and *GM3* and Dübendorfer's *D1* line. Both the *Kc* and *S2* lines are available from the American Tissue Culture Collection (Rockville, MD) and the European Collection of Animal Cell Cultures (Salisbury, UK). In the laboratories of the now dissolved Soviet Union, the preferred cellular material has been, of course, Kakpakov and Gvozdev's *67j25* sublines. Of special interest are a few metabolic mutant lines, as well as the haploid *1182* lines of Debec, Gateff's *l-mbn* lines developed from malignant larval hemocytes, and, more recently, established lines from imaginal disc cells (from Miyake's laboratory and Currie and Milner's). For useful comparisons, a few lines were grown from other *Drosophila* species.

Nevertheless, it seems worthwhile to give a tentatively exhaustive catalog of the various *Drosophila* cell lines which have been established, so far, particularly from mutant stocks. Most investigators, rather lost in the "maquis" of arbitrary appellations, tend to ignore the complex and yet meaningful derivations of the sublines they use in their experiments.

In order to facilitate their identification, all cell lines in the following compilation have been divided into several major categories and, within each division, the chronological order of their establishment *in vitro* has been observed.

A. Most Commonly Used *Drosophila melanogaster* Embryonic Cell Lines

1. UNIVERSITY OF PARIS, ECHALIER AND OHANESSIAN Kc LINE (AND ITS DERIVATIVE KcO) (FIG. 3.2) AND LINE C

(Both lines were simply named from the alphabet letters marking the primary cultures flasks from which they were derived)

References: Echalier and Ohanessian (1969, 1970).
Origin: Disaggregated young embryos (8 to 12 h old) of
                    Drosophila melanogaster. They corresponded to F$_2$
                    generation from a cross between *ebony* and *sepia*
                    stocks.
Medium: D22 (last "avatar" of our original D20 medium, see
                    Chapter 1) + 2 to 5% FBS (with the exception of
                    KcO sublines).

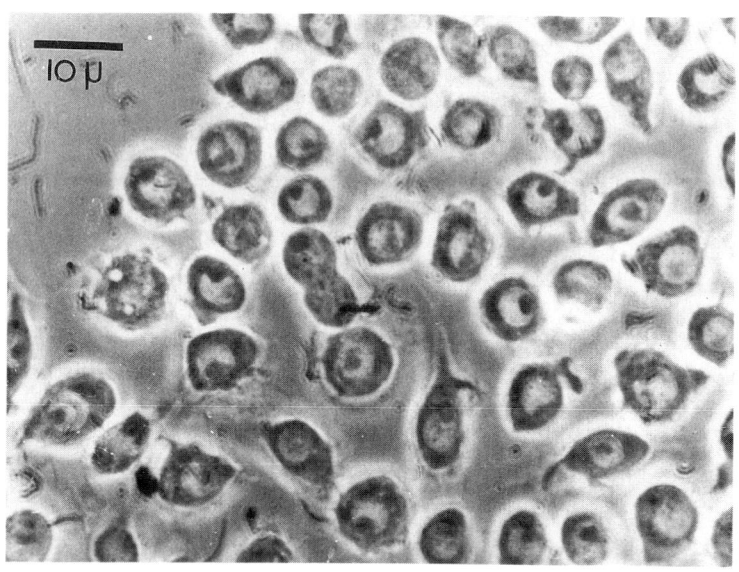

FIGURE 3.2   Kc cell line. Phase-contrast microscopy: in the large and clear nucleus, observe the single dark nucleolus.

Doubling time: In *Kc*, 18 h at 25°C (Dolfini *et al.*, 1970).

Karyotypes: *Kc* (a successful subline from the original *K* line) is fundamentally diploid (up to 90%), with a female chromosomal set (XX), but one single IVth punctiform chromosome (see Fig. 3.3).
C: male karyotype (XY).
(see also Chapter 4, Section I.E).

Interest: *Kc* is currently one of the most widely studied *Drosophila* cell line.

Available from American Tissue Culture collection (Rockville, MD) and the European Collection of Animal Cell Cultures (Salisbury, UK)* (catalog number 90070550).

Kc0: derivative from *Kc*, was adapted to serum-free culture and, therefore, has been extensively used. It is uncertain whether all cell sublines grown, throughout the world, under the same *Kc0* designation do derive from our originally adapted subline (because 0-serum sublines can be easily selected from standard *Kc* cells by simple gradual decrease of serum concentration in *D22* medium).

Kc-H (for "Hormone"): a clonal subline selected by the Cherbas (Cherbas *et al.*, 1980) for its peculiar sensitivity to ecdysteroids (see Chapter 8).

Kc 167: derived from a sample of our original *Kc* sent, at the time of its 167th passage, to Tissiere's laboratory, in Geneva. It has been claimed that this subline might be more efficient for transitory transfection, although our own experience did not confirm this assertion.

2. KURCHATOV INSTITUTE, MOSCOW, (SUB-) LINES 67j25 A AND 67j25D

(The latter one is sometimes abbreviated "D" in some Russian papers)

References: Kakpakov *et al.* (1969); Gvozdev and Kakpakov (1970).

---

* ECACC, Depositor: Dr. J. Sinclair, Department of Medicine, Addenbrookes Hospital, Cambridge, UK

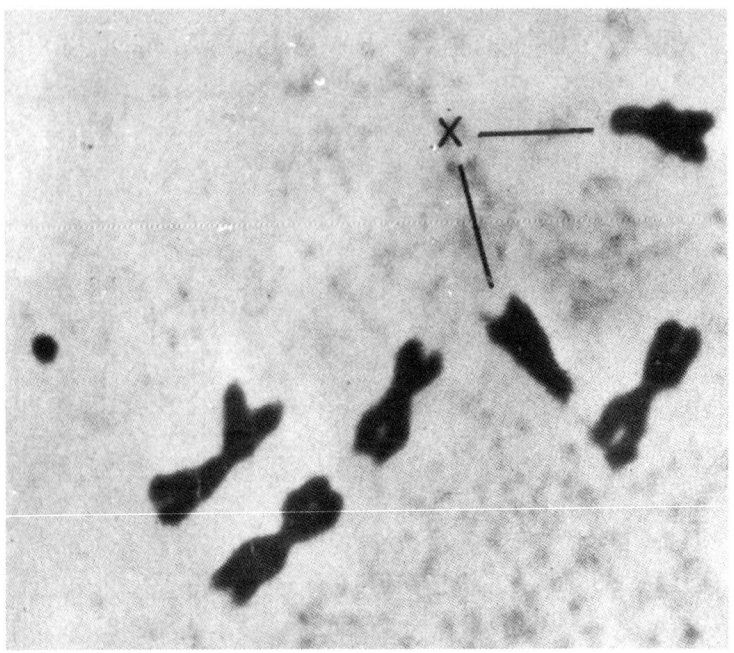

FIGURE 3.3   Typical karyotype of *Kc* cells. Orcein-stained preparation: "Female" diploid (XX) with a single dotlike 4th chromosome. Courtesy of Dr. S. Dolfini-Faccio, University of Milan, Italy.

Origin:	Embryos of *Drosophila melanogaster* (Oregon R-C).
Medium:	C-15 + 15% FBS and 10% heat-treated pupal extract. After establishment of the parent line, the pupal extract could be removed.
Cell cycle:	24 h at 28°C.
Karyotypes:	Female
	67j25 D: 85 to 90% diploid, but haploid IV.
	67j24 A: 60 to 70% aneuploid.
Sublines of interest:	Among the very first lines to be established, *67j25 D* has been the most widely used one in Soviet Union laboratories. (Available from

Professor Kakpakov, N. I. Vavilov Institute of General Genetics, Russian Academy of Sciences, Gubkin str.3, 117809 Moscow).

67j D-G and *67j24 Dbs:* adapted to serum-free media (quoted in Hinks' 1980 compilation).

67j25 T: 75% tetraploid, spontaneously formed from *67j25A* (Kakpakov *et al.,* 1971; Polukarova *et al.,* 1975).

Ta-6: triploid (3X + 6A), by cloning from T (Kakpakov *et al.,* 1971).

DaT (X;3): diploid subline with a spontaneously arisen translocation of a substantial portion of the X chromosome to an autosome of the 3rd pair. Advantage in growth and some resistance to ecdysone (Gvozdev *et al.,* 1974).

3. WALTER REED ARMY INSTITUTE RESEARCH, WASHINGTON, SCHNEIDER'S LINES S1, S2, AND S3

(*SL1, SL2,* and *SL3;* more generally known as *Dm1, Dm2, Dm3,* or, even more simply, *S1, S2,* and *S3*) (see Fig. 3.4).

Reference: Schneider (1971, 1972).
Origin: Oregon-R embryos, on the verge of hatching.
Medium: Schneider's medium (in its version with 5 g/liter bacteriological peptone, see Chapter 1) + 15% inactivated FBS.
Generation time: About 18 h at 25°C.
Karyotypes: (in 1972)
                      S1: both XX and XY.
                      S2: female, 60 to 80% tetraploid.
                      S3: XX and XY, 90% diploid.
Interest: *S2,* and, to a lesser degree, the two other lines are among the most widely used *Drosophila* cell lines all over the world.
                      Available from the American Tissue Culture collection, (Rockville, MD) and the European Collection of Animal Cell Cultures (catalog numbers "SL2": 90070554; "Dm1": 90070555: "DM3": 90070548).

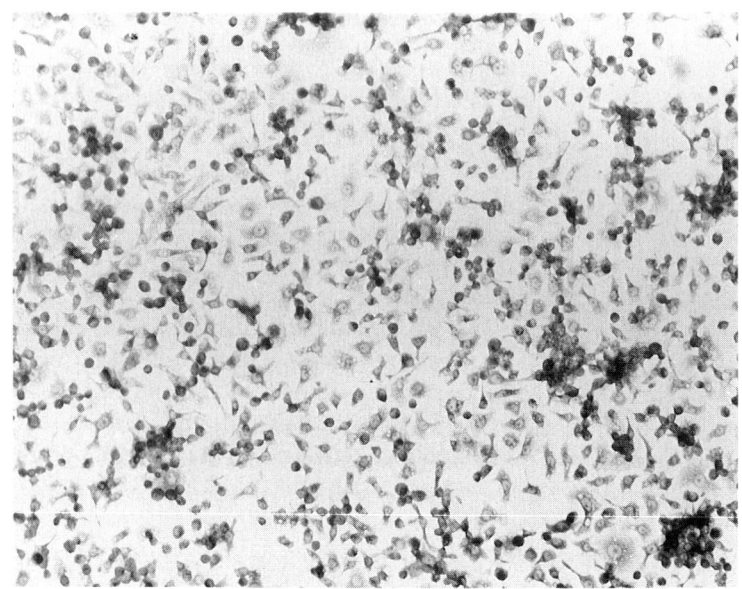

FIGURE 3.4   Schneider's *S2* cell line. Courtesy of I. Schneider.

4. *Department of Genetics, University of Milan,* GM1, GM2, *and* GM3 *Lines*

Reference:	Mosna and Dolfini (1972a,b).
Origin:	Wild-type stock "Varese." 12 to 15 h embryos.
Medium:	Echalier and Ohanessian's *D22* + 20% FBS.
Karyotypes:	*GM1:* high percentage (75%) of cells with chromosome X and heteropycnotic fragment of Y, haploid IV.
	GM2: 50% paradiploid formula: XO and presence of 2 telocentric chromosomes, probably deriving from the division of one of the metacentric II.
	GM3: diploid XY (73%), haploid IV.
	(see also Chapter 4 Section E).
Interesting features:	All three lines could be grown in serum-free medium (Mosna, 1972). For instance, *GM3 WS*

is a "without serum" subline (Dolfini, 1974).

GM1 thermosensitive clonal sublines *ts15* and *ts58* (Nakajima and Miyake, 1978) EMS-induced mutants. Their thermosensitive lesions are different, so that they can be mutually complemented in hybrid cells.

5. UNIVERSITY OF ZÜRICH, DÜBENDORFER'S LINES D1 AND D2

References: Dübendorfer (unpublished?); quoted in Schneider and Blumenthal (1978) and compiled by Hink (1980) and Sang (1981).

Origin: Oregon-S dissociated gastrulae (5 to 8 h) for *D1*, alternate passages *in vitro*, then *in vivo* and finally *in vitro* (Dübendorfer and Shields, 1972).

Medium: Shields and Sang's *M3* + 15% FBS.

Generation time: At 25°C.
              D1: 21 to 24 h; *D2*: 29 h.

Karyology: Both lines are 90% diploid and exclusively XY (the Y chromosome is "probably incomplete" in *D2*).

Contaminants: "Free of viruses" when examined by N. Plus in 1977.

B. Other Wild-Type Embryonic Cell Lines

1. INSTITUTE OF GENETICS, MILAN, LINES 0.57, 1.59 ETC.

(Several other lines *1.35, 1.56, 1.XII, 3.XIII, 3.38, 3.3* derived from the same series of primary cultures)

References: Mosna and Barigozzi (1976); Mosna (1979, 1983).
Origin: Wild stock "Varese" embryos.
Medium: Echalier and Ohanessian's *D22* medium + 15% FBS.
Karyology: *3XIII*: diploid 89%, XY (fragment), translocation between the 3rd chromosomes, no chromosome IV.
             0.57: tetraploid 52%, diploid 48%, XY (fragment), no IV.
             1.59: diploid 67%, tetraploid 21%, XY (fragment), no chromosome IV.

Interest: *MTX*ʳ subline *0.57:* methotrexate resistance (Mosna *et al.*, 1984). (see Chapter 5).

2. CITY OF HOPE MEDICAL CENTER (DUARTE, CA) SEECOF'S "OREGON-R" LINE

Reference: Petersen *et al.* (1977).
Medium: Schneider's medium (as modified by Seecof and Donady (1972) + 18% FBS + Insulin.
Doubling time: 48 h.

3. SUSSEX UNIVERSITY (UK) SHIELDS' G1 LINE

Reference: Shields, personal communication, as quoted in Schneider and Blumenthal (1978).
Origin: 6.5 to 8.5 h old Oregon-S embryos
Medium: *M3* + 10% FBS.
Cycle time: 24 h at 25°C.

4. BIOZENTRUM (BASEL), LINES EOR AND ECS

Reference: Bernhard *et al.* (1980).
Origin: 4 to 20 h embryos from wild-type stocks (Oregon-R and Canton-S, respectively).
Medium: *M3* (β-alanine omitted) + 10% inactivated FBS.
Karotypes: Diploid
               EOR: 90% XY.
               ECS: 80% XX.

5. JOHNS HOPKINS UNIVERSITY (BALTIMORE, MD), SIMCOX ET AL., WT-CS5 AND 3, WT-LM6

Reference: Simcox *et al.* (1985).
Origin: 3 to 14 h embryos, wild-type stocks (Canton S and LM).
Medium: *M3* + 10% FBS.

C. Cell Lines from Fly Stocks with Chromosomal Markers

1. DEPARTMENT OF GENETICS, BUDAPEST, PARADI'S CELL LINE IH₁

Reference: Paradi (1972, 1973).

Origin: 12 to 14 h embryos from *T(y:2)C/Cn³* mutant with a translocation of the Y chromosome.

Medium: *D22* + 20% FBS.

Karyotype: Polyploid (dominant octoploid); centric Y chromosome fragment; 0 to 7 IVth chromosome.

2. BIOZENTRUM (BASEL), GEHRING'S LINE G

Reference: Gehring, personal communication, as reported by Schneider and Blumenthal (1978).

Origin: 6 to 10 h embryos from a cross between *ywf³⁶ᵃ* and *In(3LR)C269,emwh.*

Note: Alternate passage, i.e., first *in vivo* culture, then *in vitro.*

Generation time: 22 to 24 h at 25°C.

Karyology: Near diploid.

3. CITY OF HOPE MEDICAL CENTER (DUARTE, CA), SEECOF'S LINE B26ywf

Reference: Petersen *et al.* (1977).

Origin: F1 embryos from females Y chromosome + attached X chromosome marked with *ywf* and males with reciprocal translocation between X and Y (one translocation marked with *Bˢ* and the other with *y⁺*).

Medium: Modified Schneider medium + 18% FBS + insulin.

4. KURCHATOV INSTITUTE (MOSCOW), LINE 75j23pe34

Reference: Kakpakov, personal communication (1977); indexed in Hink's compilation (1980).

Origin: Embryos from a *Dp(1;f)R/Df(1)Pgd-Kz* fly stock.

Medium: Shields and Sang M3.

Cell cycle: *ca.* 24 h at 28°C.

Karyology: A small fragment of the left end of X chromosome, *Dp(l;f)R*, present in 60% diploid cells.

5. BIOZENTRUM (BASEL), LINE EFM1

Reference: Bernhard *et al.* (1980).

Origin: Embryos from the balanced stock *Df(1)N8(Notch)/FM1* with an identifiable 1st chromosome inversion.

Karyology: Heteroploid, XX or X0; one or two FM1 balancer
             chromosomes (whose inversion can be cytologically
             recognized).

6. INSTITUTE OF GENETICS (MILAN), LINES P80 AND LINES P102

(Six lines were established from each mutant stock)

References: Halfer *et al.* (1980).
Origin: 12 to 15 h old embryos from mutants with a similar
              translocation of Y: *T(y:3)P80 Ubx/bx³⁴ᵉ* and
              T(y:3)P102 D³ bx³⁴ᵉ/Ubx(Me,Tm8).
              (see also Chapter 4, Section I.E).
Medium: Echalier and Ohanessian's *D22* + 20% FBS.
Karyology: Heteropolyploidy.
              P80: 14-20 chromosomes in 94% of the cells.
              P102: 14-23 chromosomes (92%). The original
              translocations of the Y chromosome are no longer
              present but a series of new marker chromosomes have
              appeared.

D. Cell Lines from Mutants of Interest

Resistance to *Drosophila* Sigma Virus

1. INSTITUT DE GÉNÉTIQUE (GIF/YVETTE, FRANCE), LINES B, G, I, AND J

References: Richard-Molard (1975).
Origin: 6 to 12 h embryos mutant fly stocks resistant (or
                   not) to the multiplication of *Drosophila* Sigma
                   virus: line *B* is homozygous for *ref(2)P⁰*, and
                   three other lines for the *Pᵖ* allele of the same
                   gene (see Chapter 11).
Medium: *D22* + 10% FBS.
Generation time: 48 h at 20°C.
Karyotypes: *B* and *J*: 90% diploid XO.
                   I: aneuploid.

Note: A series of additional lines were established from Oregon-M fly
stocks with deifferent mutatilons of the same *ref* gene involved in the

multiplication of Sigma virus lines *75A* to *75G, 75L, 75PE39, 77O$_{M3}$* [Ohanessian and Contamine, personal communications, as quoted by Schneider and Blumenthal (1978) and by Hink (1980)].

Also, another line was established from a stock (*ultra-rho*) containing a defective Sigma virus (Richard-Molard, 1973) (see Chapter 11).

Thermosensitivity

2. SUSSEX UNIVERSITY (UK), SANG'S GROUP, LINE S4

Reference: Simcox and Sang, as quoted in Sang (1981); Sang *et al.* (1984).
Origin: Embryos homozygous from the mutation *shibire^{ts}*.
Medium: *M3* + FBS.
Karyology: Tetraploid.
Interest: Cells grow normally at 22°C, but are killed by long-term exposure to a temperature of 29°C; can be used in a selective system (see Chapter 9, Section V.F).

3. SIMCOX ET AL., THIRTY-THREE LINES FROM VARIOUS TEMPERATURE-SENSITIVE MUTANTS

(*TS:403-1, 4,* and *11; TS:1006-1, 2,* and *8; TS:1704-2* and *6; TS:2366-1, 2, 3,* and *11; TS:2588-2, 5, 8,* and *10; TS:3803-4, 8,* and *9; TS:4875-3; TS:5697-4, 5,* and *7; TS:6255-8; TS:UC 13-9; TS:UC 32-1, 5, 6, 7,* and *8; TS:UC 88-6; TS:UC 259-8*)

Reference: Simcox *et al.* (1985).
Origin: Embryos homozygous or hemizygous for one of 14 temperature-sensitive recessive lethal mutations.
Medium: Sang *M3* + 10% FBS.
Interest: All lines express a temperature-sensitive phenotype: at the restrictive temperature (30°C) they died more or less rapidly.

4. UNIVERSITY OF CALIFORNIA, SANTA CRUZ, POODRY'S SHIBIRE LINE AND CLONE EH34A3

References: Huala and Poodry, unpublished; quoted in Woods and Poodry, 1983, and Sater *et al.*, 1984).
Origin: Late embryos, mutation *shibire^{ts1}*.
Medium: *M3* + 5% FBS + Bacto-peptone (6 mg/ml) + insulin (0.05 unit/ml).

Interest: Several clones, for instance *EH34A3*, responding to
                   ecdysteroids (Cherbas and Cherbas, 1991).

**5. BIOZENTRUM (BASEL), BERHNARD's EE 12ts LINE (SEE
BELOW)**

**6. MITSUBISHI-KASEI INSTITUTE (TOKYO), MIYAKE's CLONAL
SUBLINES ts15 AND ts58 (SEE LINE GM1)**

Metabolic Mutants

**7. BIOZENTRUM (BASEL), PYRIMIDINE AUXOTROPHIC LINES Er[1]
(a, b, c, d1, d2, e) AND Er[36]**

References: Regenass and Bernhard (1979); Berhnard *et al.*
                   (1980).
Origin: 4 to 20 h embryos from mutant stocks for the
                   rudimentary alleles r^1 or r^{36}.
Medium: *M3* (for establishing the lines, it was enriched with
                   uridine 0.5 m*M*) + 10% FBS.
Doubling time: Approximately 2 to 3 days.
Karyotypes: Primarily diploid (60 to 90%),
                   XX: *Er[1]a* (50% X0), *Er[1]d2*.
                   XY: *Er[1]b, c, d1, e*, and *Er[36]*.
Interest: Those mutant cell lines do not grow in pyrimidine-
                   free minimal medium (see Chapter 5).

**8. BIOZENTRUM, BASEL, "MARROON-LIKE" (Emal[1],
EE12[ts]mal[1], Emal[F3]) AND "ALDEHYDE OXIDASE" (Ealdox[n1])
OR "ALCOHOL DEHYDROGENASE" (Eadh[n1]) MUTANTS LINES**

Reference: Bernhard *et al.* (1980).
Origin: 4 to 20 h embryos from mutant stocks homozygous
                   for alleles of marroon-like, aldehyde oxidase, or
                   alcohol dehydrogenase.
Medium: *M3* (without β-alanine) + 10% FBS.
Doubling time: 1 to 3 days.
Karyotypes: *E ml[1]*: diploid XX (97%).
                   E ml[F3]: XY.
                   EE12[ts]: heteroploid.
                   E aldox[n1]: aneuploid (100%).
                   E adh[n1]: XX (72%).

Interest: As expected, *ml* lines do not express the enzymes
                xanthine oxidase, pyridoxal oxidase, and aldehyde
                oxidase.
                The *aldox* cell line expresses only residual amounts
                of aldehyde oxidase.
                EE 12^{ts} (derived from a stock homozygous for
                y l(1)E12^{ts} ml^1) exhibits a strict thermosensitive
                lethality: at the permissive temperature (22°C), the
                cell cycle is 72 h, whereas no proliferation is
                observed at the restrictive temperature (29°C).
                E adh^{n1} is derived from *Adh-null* flies.

Mutagen Sensitivity and DNA Repair Deficiency

(See Chapter 5 for a discussion of this topic.)

9. University of California, Davis* (Boyds group):
UCD-Dm-mei9-1 and 2, UCD-Dm-mei-41-1,
UCD-Dm-mei-218-1
UCD-Dm-mus103-1 and UCD-Dm-mus-104-1

References: Boyd *et al.* 1976; Hirschi and Boyd (1981); (see
                Chapter 5, Section IV.A).
Origin: Embryos homozygous for distinct X-linked
                mutations confering hypersensitivity to mutagens
                (*mus*), or meiotic mutants (*mei*) affecting genetic
                recombination.
Medium: Schneider (+2 mM glutamine) + 16% FBS + insulin.
Doubling time: From 24 h for both *mei-9* lines to 47 h for *mei-218*.
Karyotypes: General evolution to hypotetraploidy, with
                preferential loss of Y and IVth (dot) chromosomes.
Interest: Both *mei9a-1* and 2 lines display a similar DNA
                excision repair deficiency, which is reflected by
                their hypersensitivity to UV irradiation.
                In *mei41* and *mus104* lines, rates of DNA chain
                growth are reduced. "Unexpectedly, however,
                these established lines failed to respond to UV or
                inhibitor treatment under conditions in which
                primary cultures from the same mutants exhibited
                a strong reduction in postreplication repair."

---

* Nomenclature includes the laboratory of origin, a genus-species identification (Dm),
the genetic locus and a cell line number.

> Incidentally, *mus104* possesses a slow variant of the
> enzyme phosphoglucomutase.
> The expression of *mei218* being restricted to the
> germ line, the deriving somatic cell line can be
> employed as a control.

Available from the European Collection of Animal Cell Culture (Catalog
Number: Mei9a: 90070553).

**10. *Mitsubishi-Kasei Institute (Tokyo), Miyake's Group,
UV-Hypersensitive Line ML82-8***

Reference: Todo *et al.* (1985).
Origin: 3 to 7 h old embryos from the homozygous strain
            mus201^{D1}/mus201^{D1}.
Medium: *M3*(BF) (see page 19) + 10% FBS.
Interest: Hypersensitivity to UV and various chemical mutagens.
            Lacks excision repair for pyrimidine dimers induced by
            UV, although normal resistance to X-rays.

Other Mutants

**11. *City of Hope Medical Center (Duarte, CA), Seecof's
Line lgl^4***

Reference: Petersen *et al.* (1977)
Origin: Embryos whose parents were heterozygous for *lgl^4*
            (neuroblastoma).
Medium: Modified Schneider's medium + 18% FBS.

Note: theoretically, one-fourth of the disaggregated embryos of the primary culture were homozygous, but, according to the authors themselves,
"we have no reason to believe that cells homozygous for *lgl^4* dominate
the culture."

**12. *Kurchatov Institute (Moscow) Kakpakov's Group,
Many "Vestigial" Cell Lines (75e 7vg1 to 75e 7vg7)***

Reference: Kakpakov *et al.* (1977).
Origin: 1 to 18 h embryos from various *vestigial* mutant flies
            (from *vg1* to *vg7*).
Medium: Shields and Sang's *M3* + 15% FBS + 2% larval
            Drosophila extracts. After 10 to 20 passages, *75e
            7vg4* became able to grow without larval extract.

Karyotypes: Diploid female: *75e7vg5*.
Diploid male: *75e7vg7*.
Haploid X, haploid IV: *75e7vg2*.

13. UNIVERSITY OF CALIFORNIA, DAVIS (BOYD'S GROUP):
UDC-Dm-w-1

References: Boyd *et al.* (1976); Hirschi and Boyd (1981).
Origin: Embryos from a *white* stock.
Medium: Schneider's "revised medium," (see page 17) insulin, 18% FBS.
Doubling time: 26 h; S phase estimated to last about 10 h.

E. Haploid Cell Lines

(For additional details regarding haploid cell lines, see Chapter 4, Section III)

1. UNIVERSITY OF PARIS, DEBEC'S LINES *1182-1* TO *1182-7*

References: Debec (1978, 1984) (see also Chapter 4).
Origin: Embryos from females homozygous for the sterile mutation *1182^{ts}* (today called *maternal haploid* [*mh1182*]) grown at the nonpermissive temperature 25°C.
Medium: A mixture, in equal proportion of Echalier and Ohanessian's *D22* and Shields and Sang's *M3* + 20% FBS.
Karyotypes: Six of those continuous lines (with the exception of *1182-3a* primarily diploid) show haploid metaphases in different percentages.
The line *1182-4* was stably haploid (80–90%) after 3 years of culture.
Interest: Cells of line *1182-4* are devoid of centrioles (Debec, 1982; Szöllösi *et al.*, 1986).
CdR200: a subline from *1182-6*, selected for its high resistance to cadmium (due to overproduction of metallothionein) (Debec *et al.*, 1985).

F. Cell Lines from Other *Drosophila* Species

Drosophila virilis

1. WALTER REED ARMY INSTITUTE RESEARCH (WASHINGTON, DC) SCHNEIDER'S *WR-75-Dv-1*

Reference: Schneider and Blumenthal (1978).
Origin: Minced neonate larvae of *Drosophila virilis*.

Medium: Schneider's medium + 15% FBS.
Doubling time: Approximately 40 h at 25°C.
Karyology: Primarily diploid (80%), both XX and XY.

2. VAVILOV INSTITUTE OF GENETICS (MOSCOW) BRAUDE-ZOLOTARJOVA'S LINE *79 f7 Dv 3g*

Reference: Braude-Zolotarjova *et al.* (1986).
Origin: 20 h embryos of *Drosophila virilis*.
Medium: Shields and Sang *M3* + 15% FBS + 2%
                         Drosophila virilis pupal extracts.
                         Later on, routinely grown in *M3* (or NS-10*
                         medium) + 5% inactivated FCS.
Karyotype: Male karyotype of *Drosophila virilis*; 80 to 85%
                         diploid.
Sublines of interest: *Dv0*: the original line could be grown without
                         adaptation and over 100 passages in NS-10
                         medium without serum.

Drosophila immigrans

3. WALTER REED ARMY INSTITUTE OF RESEARCH (WASHINGTON, DC) SCHNEIDER'S *WR-72-Di-1*

Reference: Chakrabarty and Schneider, quoted in Schneider
                     and Blumenthal (1978).
Origin: Neonate larvae of *Drosophila immigrans*.
Medium: Schneider's medium + 15% FBS.
                     Note: the cells attach firmly, so that they have to
                     be released with trypsin for subculturing.
Generation time: 60 h at 25°C.
Karyotype: Diploid, both XX and XY.

Drosophila hydei

4. NIJMEGEN UNIVERSITY (THE NEDERLANDS), LUBSEN'S *KUN-DH-14, 15, 33,* AND *47*

Reference: Sondermeijer *et al.* (1980).
Origin: 0 to 18 h embryos of *Drosophila hydei*.

---

* One of the many formulas derived from *S-15* (see Chapter 1, Section II.D).

Medium: *M3* + FBS.
Doubling time: 1 to 2 days at 25°C.
Karyotypes: Normal diploid (XY), with the exception of line
                 KUN-DH-14 (diploid X0 with Y(?)
                 heterochromatic fragment).
Interest: Can be grown in spinner (up to $5 \times 10^6 - 2 \times 10^7$
                 cells/ml).
                 Burke *et al.* (1985) noticed that *Drosophila hydei*
                 lines are particularly efficient for expressing a
                 transferred gene (see Chapter 9, Section III).

All three lines are available from the European Collection of Animal Cell Cultures (catalog numbers: "DH14": 90070551; "DH15": 90070549; "DH33": 90070559).

Drosophila simulans

5. MITSUBISHI KASEI INSTITUTE OF LIFE SCIENCE (TOKYO), ML83-8b LINE

Reference: Quoted by Yoshioka *et al.* (1992).
Origin: Embryos of *Drosophila simulans.*

IV. CELL LINES FROM LARVAL OR IMAGINAL TISSUES

As has been repeatedly pointed out, very few *Drosophila* cell lines have been established from defined tissues; only two lines were derived from neoplastic hemocytes of larval mutants and, more recently, two laboratories succeeded in developing a number of cell lines from imaginal disks of 3rd instar larvae and from larval neural ganglia.

The unquestionable interest of these various lines deserves special discussion.

A. Larval Tumorous Blood Cell Lines

I(2)mbn and *I(3)mbn* (*lethal malignant blood neoplasm*) are two recessive lethal mutations of *Drosophila melanogaster* that were induced with EMS (Gateff, *1977, 1978*).

Although they correspond to genes located on distinct pairs of chromosomes, both display a quite similar phenotype: an abnormal hematopoiesis results in a true leukemia. At the premortal stage (late 3rd instar), the body cavities of mutant larvae are filled with enormous amounts of free immature plasmatocyte-like cells, invading all tissues. Compared to the normal blood count of wild-type larvae, the number of blood cells in hemolymph is increased some 40 times in *l(2)mbn* and 150 times in *l(3)mbn*.

These tumorous blood cells could be easily aspirated by simple puncturing of the surface-sterilized integument of mature mutant larvae, and cultured *in vitro*; only a month and half later, permanent cell lines were established (Gateff *et al.*, 1980). The malignant nature of these cells probably accounts for the absence of any lag period between primary cultures and the establishment of lines.

FREIBURG UNIVERSITY (GERMANY), GATEFF'S MALIGNANT BLOOD CELL LINES *l(2)mbn* AND *l(3)mbn*

Synonymy: *Blood II* and *Blood III*

Reference: Gateff *et al.* (1980).
Origin: Tumorous hemocytes from late 3rd instar larvae of two recessive mutants of *Drosophila melanogaster*: *lethal malignant blood neoplasm*, *l(2)mbn* and *l(3)mbn*.
Medium: *M3* or Schneider's medium + FBS.

B. Imaginal Disc Cell Lines

It is useful to reiterate that imaginal discs are small autonomous islets of cells which, in holometabolous insects, are sheltered and nourished by successive larval instars but whose fate is "imaginal," that is, adult. At metamorphosis, they develop into defined parts of the adult organism (eyes, wings, legs, etc.).

In the active field of *Drosophila* developmental biology, imaginal discs constitute especially valuable experimental models. Although significant results have been obtained from short-term *in vitro* culture of whole discs,* it is obvious that the availability of permanent cell lines remaining

---

* As noted in the introductory chapter, maintenance *in vitro* of whole discs relies on quite distinct culture methods and aims which do not fall within the scope of this book. Relevant references, however, are listed in the Bibliography.

capable of hormone-induced differentiation should greatly facilitate bio-chemical investigations.

As surveyed in Chapter 2, several proliferating structures deriving from imaginal discs could be identified in primary cultures from disaggregated embryos and their response to ecdysteroids was fully analysed by Düben-dorfer and colleagues. As a matter of fact, in Schneider's (1972) primary cultures, grown from mature embryos on the verge of hatching, a continu-ous production of imaginal disc vesicles was even the dominant feature. Subsequent established cell lines (i.e., *S1* to *S3*), however, probably devel-oped from quite different cell types.

For a long time, nevertheless, repeated attempts to culture dissociated cells from larval imaginal discs, chiefly from mature 3rd instar larvae (for obvious reason of size), remained unfruitful. Davies and Shearn (1978) empirically designed a highly complex medium which would support the regenerative growth of cut imaginal discs, although not the multiplication of isolated disc cells. Wyss (1982) was the first to obtain, after dissociation with protease (type VIII) and mechanical shearing, limited cell prolifera-tion leading to the formation of colonies in a defined medium supple-mented with fly extracts, insulin and a low concentration of ecdysterone (see Chapter 1).

Several years later, and without any notable technical differences from previous studies, two laboratories reported almost simultaneously the successful establishment of a number of cell lines from various imagi-nal discs:

Ui *et al.* (1987, 1988), after protease dissociation (as recommended by Wyss) and using Cross and Sang *M3*(BF) medium + 10% FBS, partly "conditioned" by embryonic primary cultures and supplemented with insulin, succeeded in developing twenty lines from wing, leg, and antennal discs, or a mixture of them.

Independently, Currie *et al.* (1988) dissociated leg and wing discs with classical trypsin treatment and grew the cells in *MM3* medium, supplemented with FBS, insulin, ecdysterone, and a fly extract.

With the exception of either "conditioning" the medium or its enrich-ment with fly extracts, there was no special "trick" readily explaining these successes; rather they may have depended on more careful handling of the cultures.

Unfortunately, these imaginal disc cell lines seem quite heterogeneous, surprisingly even in clonal colonies (according to Peel and Milner, 1990). So, they might not have been derived exclusively from imaginal epithelial cells. It is indeed well known that several cell types are present under the basal lamina of discs (especially adepithelial cells that give rise to adult muscles) and that, moreover, many hemocytes may adhere to the disc surface during dissection.

Nevertheless, if the formation of typical vesicles and their hormone-induced differentiation into adult cuticular structures was restricted to primary cultures, Currie *et al.* reported the spontaneous secretion, in established cultures, of two types of cuticle-like material: thin, untanned sheets, released from the surface of colonies and resembling apolysed pupal cuticle, or thicker, tanned material on the outer surface of individual cells, similar to imaginal cuticle.

1. Mitsubishi-Kasei Institute (Tokyo), Miyake's Group, Imaginal Disc Cell Lines ML-DmD1 to 20

Reference: Ui *et al.* (1987, 1988).
Origin: Mature 3rd instar larvae from the mutant stock *y v f mal.* Dissociated cells from each category of disc (wing, halter, legs, antenna, eye-antenna) or from mixtures of them.

2. University of St. Andrews (UK), Milner's Group, CME W1 and 2, CME L1 and 2

(Named from the initials of the authors ?)

References: Currie *et al.* (1988), Reel and Milner (1990), Cullen and Milner (1991).
Origin: Oregon-R mature larvae. Dissociated wing discs (lines *W*) and leg discs (*L*).
Medium: *MM3* (see Chapter 1) + 2% FBS + fly extract (see Chapter 1, Appendix 1.F) + insulin + 20-hydroxyecdysone.

C. Larval Central Nervous System Cell Lines

1. Mitsubishi-Kasei Institute (Tokyo), Miyake's Group, Larval CNS Cell Lines ML-DmBG1 to 8

Reference: Ui *et al.* (1994).
Origin: Dissociated brain and ventral ganglia from mature 3rd instar larvae of *Drosophila melanogaster y v f mal* strain.

Medium: *LM3*(BF) (i.e., modification of Cross and Sang's
                    M3(BF) in which Sigma Trizma was substituted
                    for Bis–Tris) + 10% FBS + insulin.
Karyology: Most of them are diploid.
                    Clone BG2-c5 is tetraploid.
Doubling time: 27 to 53 h.
Interest: Six tested clones could be immunostained with anti-
                    Horseradish Peroxidase (HRP). Variation of
                    acetylcholine content, and morphology may reflect
                    the heterogeneity of cells composing the CNS.

V. CELL COUNTING AND VIABILITY DETERMINATION

A. Determination of Cell Density

The level of accuracy with which the concentration of cells in any one culture has to be estimated mainly depends on the ultimate objective of the cell count. For routine serial subcultures of a well-established cell line, a standardized dilution is adequate (see Appendix 3.A), whereas for any experimental series the inoculum must be held to within certain limits.

The simplest method consists in counting *in situ* the cells of an un-crowded monolayer, that is, to say when cells, alive or fixed and stained, are still attached to the flask bottom. The grid of a micrometer disc, placed in the 10× eyepiece of a microscope, delimits a well-defined area. Due to the frequent heterogeneity of the cultures, counting must be repeated in several different regions of the flask. It is possible to monitor for several days, and with a reasonable precision, the multiplication or the death of cells subjected to some treatment (see, for instance, Courgeon, 1972a). This method may also be convenient for a quick estimation of the percentage of cells that, among a population, express some histochemi-cally identifiable products (for instance, after transfection with some expression plasmid; see Chapter 9).

In most cases, however, and for greater accuracy, all cells of a culture should be carefully scraped or pipetted from the vessel and suspended in a strictly defined volume of medium or buffered saline. A vigorous pipetting ensures a satisfactory homogeneous cell suspension, with the dispersion of cell clumps. For specific purposes, trypsin and EDTA have been recommended for dissociating cell aggregates (Besson, 1987). Yet, with the notable exception of Schneider who used trypsin repeatedly without special damage (personal communication, 1993), most investigators consider *Drosophila* cells to be particularly sensitive to trypsin and avoid such drastic treatment when the cells have to be subsequently subcultured. Besides, suspension cultures allow easy sampling of the cell suspensions and thus are especially suitable for kinetic quantitative studies.

However that may be, an aliquot of the cell suspension is serially diluted in order to reach a cell concentration suitable for counting. During all manipulations, great care must be taken for not losing cells (possibly stuck on pipette or tube walls; coating with silicone is advisable) and keeping the suspensions quite uniform.

The counting itself may be done on a small scale, or with automatic devices. It depends on the objectives and the financial capacities of the laboratory.

1. HEMOCYTOMETER

A hemocytometer slide consists of an optically flat counting chamber with a strictly defined depth (1/10 mm) on which a rigid coverslip is overlayed. On its bottom is engraved a small grid delimiting a series of tiny squares of accurate size (their distribution and dimension may vary according to the model of counting chamber used).

As soon as the cells have spread over the chamber floor, their number within a defined area (and thereby a defined volume) is counted under a low magnification microscope. The field iris is slightly closed to enhance the contrast and to see the grid and the cells clearly.

The original cell suspension must be appropriately diluted in order to count between 100 and 500 cells per mm^2. The cell concentration of the culture will be easily deduced from the direct cell count with regard to the volume scored and the serial dilutions made.

2. ELECTRONIC CELL COUNTERS

Electronic devices for particle counting allow an automatic and rapid quantitation of cell cultures and, possibly, concomitant cell sizing.

Their principle is rather simple: a defined volume of a well-homogenized cell suspension in buffered saline is sucked through the minute orifice of a glass tube. One electrode is placed within the tube and another outside of it in the suspension. Every time one cell passes through the hole, it changes the resistance of the current flowing through this orifice by an amount proportional to the volume of the cell. This generates a pulse which is amplified and counted. More sophisticated machines may directly print a size distribution histogram.

Counting accuracy is excellent because of the high number of cells counted. On the other hand, the apparatus is unable to distinguish between live and dead cells (or fragments) or between single cells and cell clumps; so that the dispersion of the cells has to be carefully verified under a microscope before counting.

Note: The use of fluorescence-activated cell sorters will be discussed in Chapter 4, Section II.

B. Cell Viability Tests

There are many experimental situations in which it is important to know the exact percentage of the live cell population: for instance, after the freezing-thawing of a liquid nitrogen-preserved sample or after dissociation of embryonic cells for primary cultures or else during the adaptative phase of a given subline to suspension culture, etc.

Short-term viability tests rely on the detection of lethal deteriorations of the cell membrane as determined either by the uptake of a dye to which living cells are normally impermeable or the reverse, e.g., by the release of a dye normally taken up and retained by viable cells.

The most extensively used stain, for "dye exclusion tests," at least for *Drosophila* cells, has been trypan blue:

One drop of a dilute solution of trypan blue (Sigma: 0.4% in water) is mixed with two drops of the cell suspension (previously diluted to a proper cell concentration for counting) just before introducing it into the counting chamber (hemocytometer). After a few minutes, only dead cells, not living ones, are stained. Cell viability is usually expressed as the percentage of unstained (i.e., living) cells.

Note: Quantitative estimations of the metabolic activity and growth rate of cultured cells require quite distinct methods, essentially the measure of uptake and incorporation rate of labeled amino acids or tritiated thymidine. Another indicative parameter of proliferating cell populations is the mitotic index (see Chapter 4).

VI. CELL CLONING

Most, and perhaps all, established cell lines are heterogeneous from the beginning, their polyclonal origin proved by the fact that both female and male karyotypes can be observed in some of them (see Chapter 4). Moreover, in such huge populations of rapidly dividing cells, genetic drift (by mutations or chromosomal accidents) is continuous. In order to work with homogeneous material, a prerequisite at least for genetic investigations, it is highly advisable to periodically develop new sublines from single cells (= cell "cloning").

Unfortunately, solitary cells from higher organisms cannot grow *in vitro* without great difficulty,* and this seems particularly true for *Drosophila* cells, probably because culture media remain suboptimal.

Long ago, Eagle pointed out with respect to mammalian cells that cells grown in culture, as opposed to their counterparts *in vivo*, are more or less drowning in an ocean of fluid. Even though each cell in a dilute culture may individually be able to synthesize all its metabolites, some of the latter may passively leak through the permeable membrane. Hence, intracellular concentrations will always remain at a subpar level; only when the cell population attains a certain density, will the cells "assist" each other by "conditioning" the culture medium. Besides, it is well known today that neighboring cells do exchange essential stimulating signals (growth factors, etc.; see Chapter 6).

Methods of cloning, all derived from the experience acquired with mammalian cells, aim at providing the isolated cell with a microenvironment conducive to optimizing its surroundings, e.g., medium components and pH, with or without the metabolic support of fellow cells.

The first condition is fulfilled by isolating the cell in a tiny volume of medium (some 25 to 50 μl), for instance, in one of the small cavities of a 96-well microtiter plate (tissue culture grade). An alternative procedure consists in dispersing the single cells of a very low density culture within a semisolid overlay [culture medium slightly gelled with a low concentration (0.3%) of bacteriological agar]. Compared to the simple "dilution technique", ineffective with *Drosophila* cells, this semisolid agar procedure offers the additional advantage of reducing any possible contamination between the clones, which is especially important for *Drosophila* cells that loosely attach to their substrate, and of facilitating the final collection of the small colonies (Echalier, 1971).

Useful metabolic components may be supplied with the use of conditioned medium, that is medium in which a healthy culture has been grown for a short while. Kakpakov *et al.* (1971) isolated clones with different karyotypes from their original 67J25 line by inoculating flasks at relatively

---

* For cultured cells, just as it is said in Scriptures for men, "it is not good to live alone."

low cell concentrations and with the stipulation that their pupal extract-enriched medium comprised 50% of what they termed "adapted medium." Nakajima and Miyake (1977) determined the optimal conditions for preparation and use of such "conditioned" medium: when the cells, from the same line as those to be cloned, have become confluent, the medium is replaced by an equal volume of fresh medium and harvested 24 h later. The pool of collected culture fluids must be freeze-thawed (in dry ice and acetone, then in water bath), treated at 56°C for 30 min and, finally, filtered (in order to ensure that no cells found their way into the conditioned medium as would be possible if clarification of the medium was attempted only by centrifugation). This medium stock could be kept at −20°C without decreasing activity for at least 10 days. As distinguished from "used medium," that is medium simply collected from several day-old cultures and which would be much less efficient, such a so-called *MC* medium significantly increased colony formation efficiencies in very low density cultures of *GM1* and *GM2* cells. For instance, even when only two *GM1* cells were inoculated per 60 mm petri dish with 100% *MC* medium, the plating efficiency was 62.5%. This efficiency, however, was much lower with *GM2* cells (only 5% with an inoculum size of 100 cells per dish).

Another method (feeder-layer technique), devised by Puck and Marcus (*1955*) for mammalian cells, consists of distributing the cells to be cloned among like cells whose capacity for division has been definitely blocked by heavy doses of ionizing radiation (X-rays or γ-rays). As an alternative, these "feeder" cells may be previously treated with mitomycin C. Their irradiation, or chemical treatment, must be accurately controlled, so that they can still survive for one week or so and support colony forming by their few scattered healthy companion cells. The quantitative aspects of this technique, when applied to *Drosophila* cells, was carefully analyzed by Richard-Molard and Ohanessian (1977; see also Koval, 1983): the resistance of *Drosophila* cells to irradiation is strikingly high (between 10 and 25 kilorads (Kr), instead of 1 to 4 Kr for mammalian cells), and varies widely from one line to another and even among individual cells of a same line. Mosna (1983) reported the exceptional case of some cells starting to multiply 2 months after irradiation of the culture with as high as 60 Kr. Thus, for each cell type, the appropriate radiation dose has to be empirically and accurately defined in order to give, in control cultures, less than 1 clone per 10^9 irradiated cells (Cherbas L. and P. in Ashburner, 1989).

The cloning protocols currently adopted may differ somewhat from one laboratory to another but each of them is usually a combination of several of the above mentioned growth enhancing procedures: (1) feeder layer and agar-solidified overlay (Bernhard and Gehring, 1975; see Cherbas' protocol in Ashburner, 1989; Brown *et al.,* 1981); (2) as above, but using agarose,* instead of agar (Mosna, 1983); (3) feeder layer using 96-well microtiter plates (Echalier, 1976; Richard-Molard and Ohanessian, 1977); (4) conditioned medium in microtitration plates (Lindquist's protocol as described in Ashburner, 1989).

Our own standard cloning procedure is described in Appendix 3.C (see also Fig. 3.5). Similar protocols, communicated by the Cherbas and by Lindquist, respectively, can be found in Ashburner's Laboratory Manual (1989; Protocols 61 and 62).

In all cases, plating efficiencies remain usually low, from 1 to 20%, depending on the cell line. A few authors, however, claimed that they obtained a better yield.

Note: More recently, Peel and Milner (1990) described a new device that allows one to periodically renew the feeder layer. This latter seems not to be irradiated (?), but is grown on the bottom of a 5 cm plastic petri dish, whereas the cells to be cloned are plated on the underside of the bottom of a smaller (3 cm) dish which is, itself, hanging and stuck below the lid of the larger disc [for details, see figure in Peel and Milner (1990)]. Thus, the growing clones, with the lid that carries them, may be easily transferred to a fresh culture of feeder cells. Unfortunately, this system assumes a strong adherence of the cells to their substrate, which is probably the case for the imaginal disc cell lines established by the authors, but rarely true for most available embryonic cell lines. It might perhaps be safer to use a Transwell insert device (Costar Corp., Cambridge, MA) in which one type of cell is grown on the bottom of the chamber and the other on the porous membrane of the insert.

VII. STORAGE BY FREEZING

Preservation of cell lines by very low temperatures is widely utilized. It spares one the effort of maintaining in culture lines/sublines/clones that are not in continuous use, reduces the number of subcultures employed as well as possibility of losing cultures through contamination.

---

* The melting temperature of agarose is significantly lower than that of Noble agar (38°C instead of 45 to 50°C), so that it replaces it avantageously for solidifying the medium. The risk of heat-shocking or even killing the cells, when its hot stock solution is mixed with the cell suspension, is thus attenuated.

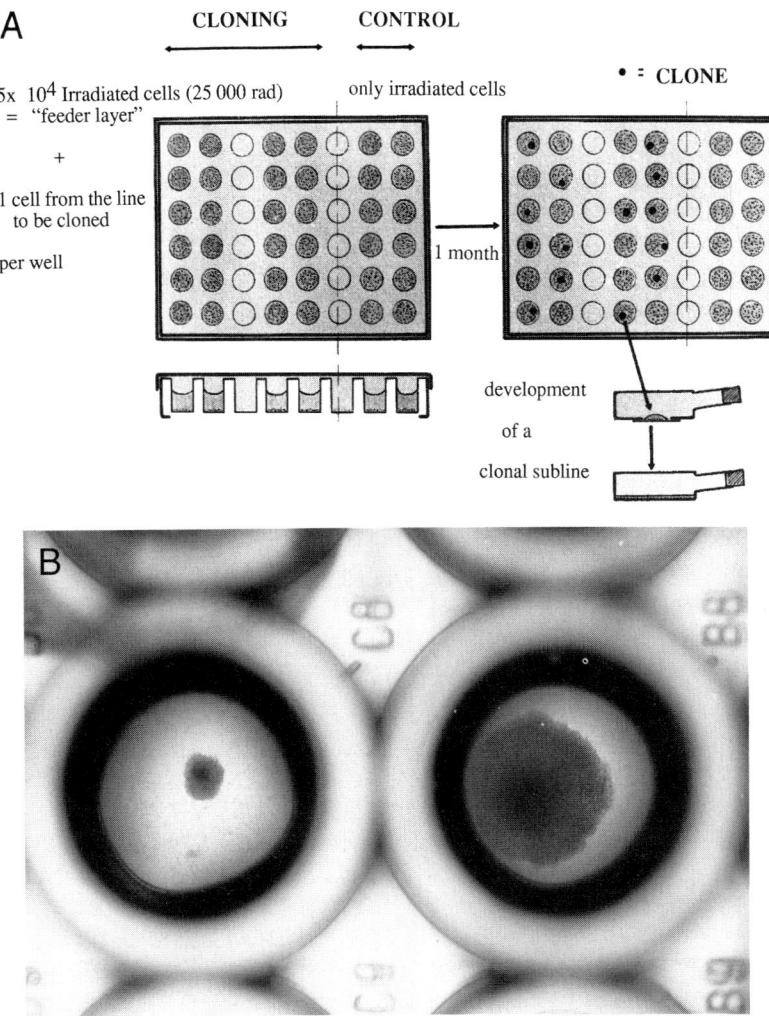

FIGURE 3.5 Cell cloning (feeder-layer technique in microwell plate). (A) General procedure. From Richard-Molard and Ohanessian (1976); reproduced with the kind authorization of the authors and Springer Verlag); (B) Clonal colonies in individual wells (photo Desrosiers, University P. et M. Curie, Paris).

Moreover, it allows useful comparisons between samples of the same line that were frozen at different stages of its evolution. For experimental series which are staggered over long periods (months or years), it is always advisable to periodically renew the cultures by thawing a series of aliquots (all frozen at the same date) in order to get rid of any possible "drift" in cultured cell populations.

For limiting cell injuries due to the formation of ice crystals during freezing, a protective agent must be added to the cell suspension and the suspension frozen very slowly. Rapid thawing seems essential for maximum survival. Our standard laboratory protocol for freezing and thawing *Drosophila* cells is described in Appendix 3.D.

The recommended protective agent is dimethyl sulfoxide (DMSO), rather than glycerol, at a concentration of close to 10% (v/v). According to J. Paul (*1973*), it might diffuse more readily in and out the cells, thus causing less osmotic damage when added to medium or diluted out.

Complex and expensive devices have been developed for freezing at a strictly controlled rate, but, for most routine laboratory work, much simpler protocols are sufficient. The tightly closed ampules containing the cell suspensions with DMSO (or glycerol), are gradually cooled by their successive transfer to a 4°C refrigerator or 0°C melting ice, a freezer (−20°C), then a dry-ice chest. As an alternative, ampules are wrapped in cotton wool, enclosed in an polyfoam box and directly placed into a deep freezer.

Finally, frozen samples are stored in a deep-freeze cabinet (−70°C, −80°C), or, even safer, in a liquid nitrogen container (−196°C). Many suitable containers are commercially available: duly labeled ampules are clipped, end to end, in channeled metal rods (or "racks") which are distributed among several deep "canisters."

As has already been pointed out, thawing must be rapid (a few minutes, for instance in a warm hand), and the cells are quickly transferred into a culture flask with a large volume of culture medium and serum (in order to lessen, by dilution, any toxic effect of DMSO). Moreover, as soon as the cells are firmly adhering to the flask, the medium is replaced.

As is well known, the viability of frozen cells depends on the cell line and on the freezing conditions. As an example, it is noteworthy that a sample from our original *Kc* line, which had been frozen in 1969 and kept in liquid nitrogen, was thawed and successfully grown 25 years later!

VIII. CELL FUSION AND SELECTION OF SOMATIC HYBRIDS

Genetics of somatic cells started with the initial observation of Barski *et al.* (1960) that, when two recognizable strains of murine cells were grown in the same culture vessel, a few cells could accidentally fuse their cytoplasm, then their nuclei, and give rise to viable "hybrid" cells.

Spontaneous cell fusions occur most probably in *Drosophila* cells as well. Fortuitously, Dolfini (as reported in Barigozzi, 1971) happened to observe, in *Kc* cell cultures, karyological pictures which were reasonable evidence of such an event: they were tetraploid metaphases with three X and one Y chromosome, which can only be explained by the fusion, amid a mixed population, of a "female" cell with a "male" one.

Such accidental events remain too rare for experimental work, so several methods have been devised to considerably increase chances of cell fusion.

Note: Treatment with UV-inactivated virions of Sendaî parainfluenza virus or some other vertebrate viruses that had been commonly used to induce polykaryocytosis in mammalian cells was found, not surprisingly, to be quite ineffective in *Drosophila* cells* (Echalier, 1971). However, given the report by Suitor and Paul (1969) that infection of mosquito cells by dengue 2 virus may generate large syncytia, Hannoun and Echalier (1971) tested the possible fusing capacity, on *Drosophila* cells, of this arbovirus and of 18 others of the same sero Group B, but without success (see Chapter 11).

A. Cell Fusion Methods

It was shown that various types of substances, active on cell membranes, can also efficiently fuse mammalian cells, and these biochemical treatments are now preferentially adopted, as they suppress the risk of infection from incompletely inactivated virions.

Three types of such agents can be successfully applied to *Drosophila* cells.

1. LECTIN-MEDIATED FUSION

Becker (1972) demonstrated that the phytohemagglutinin *concanavalin A* (ConA) (from the Jack Bean), not only agglutinates *Drosophila*

---

* It might result from a lack of terminal sialic sites which serve as receptors for the paramyxovirus (Nakajima and Miyake, 1978).

cells, as extensively shown for mammalian cells, but induces a rapid cell fusion and the formation of apparently viable synkaryons.

The technique is quite simple: cells at a high concentration (3×10^6 cells/ml) are kept in suspension via a slowly rotating magnet in the presence of 100 μg/ml of ConA (Calbiochem, Los Angeles, CA) for 20 min. $CaCl_2$ and $MgSO_4$, both at $10^{-4} M$, are also present as these bivalent cations are well known for enhancing ConA activity. The pH should not go below pH 6.8. The cells are then seeded into the usual culture flasks and, as soon as they adhere, are carefully washed twice with fresh medium.

The efficiency of the method, at least for our *Kc* line, was very high: half of the population assembled in pairs or larger groups and between the partners a complex web of interpenetrating microvilli could be seen by phase contrast microscopy (Fig. 3.6). Fusion occurred within the following 10 h resulting in the formation of di- or polykaryons and, 24 h later, of synkaryons.

Autoradiographic pictures of dikaryons, formed between one subline whose nuclei had been labeled with tritiated thymidine and another unlabeled subline, readily demonstrated that they did derive from cell fusion. Moreover, Becker showed that cell aggregation could be completely suppressed by adding the specific hapten inhibitor (α-methyl-D-mannopyranoside, 0.1 *M*) during the ConA treatment.

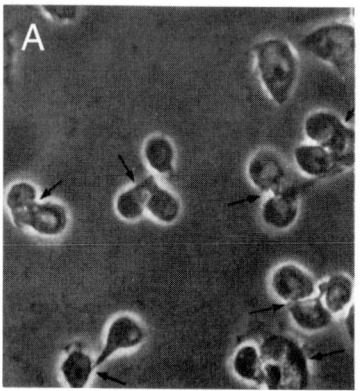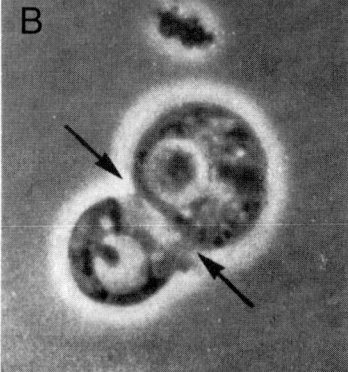

FIGURE 3.6   Concanavalin A-mediated *Kc* cell aggregation and fusion. (A) Many cell couples are forming (phase contrast; magnification: $\times$ 500); (B) First step of cell fusion (25th min of treatment) magnification: 1800$\times$. From Becker (1972). With permission.

Becker's data were confirmed by Gvozdev *et al.* and Gehring's group (personal communications), then by Riszki *et al.* (1975) and Halfer and Petrella (1976). (See also Chapter 6, page 282.)

In addition, Rizki and colleagues (1975) established that another lectin, wheat germ agglutinin (WGA) can also, at much lower concentrations, cause morphological alterations and fusion of Schneider's *S1* cells: when cells were grown on coverglasses in Leighton tubes in the presence of WGA (5 to 10 μg/ml, in a buffered saline) the typical surface filaments, observable with scanning microscopy, disappeared in the first minute and, after 3 min, cells were flattened and fused. The presence of the inhibitor N-acetylglucosamine (NAGA) did not block this phenomenon.

2. LYSOLECITHIN-INDUCED CELL FUSION *

The fusion capacity of lysolecithin, widely demonstrated in mammalian cells, was tested on *Drosophila* cells by Halfer and Petrella (1976). A high percentage of binucleate cells (*ca.* 45%) was observed in an *IB5* (a clone from *Kc* cells) line immediately after 30 to 45 min exposure of a cell suspension to a 200 μg/ml concentration of LL (Calbiochem); higher concentrations caused lysis of most cells. Another cell line (*GM1*) was found to be much more sensitive and yielded fewer fusions. Compared to ConA, the activity of this chemical seems more rapid but also more toxic.

3. POLYETHYLENE GLYCOL-MEDIATED CELL FUSION

Polyethylene glycol (PEG) is a potent fusing agent for mammalian cells and plant protoplasts. Bernhard (1976) was the first to demonstrate its efficiency on *Drosophila* cells, and the use of PEG has currently become the favorite method for fusing *Drosophila* cells.

According to Bernhard, after 10 min treatment with PEG 4000 (50 mM in a buffered saline) the frequency of heterodikaryons formed between two lines, one of which was labeled with tritiated thymidine, was estimated by autoradiography to be around 5% of the population, 24 h later.

Using a TAM selection (see below and Chapter 5), Wyss (1979) was able to recover proliferating synkaryons produced by PEG fusion between a *Kc* subline and *S3* or *D1* cells. Their hybrid nature was proved by their unambiguously hybrid isozyme pattern for isocitrate dehydrogenase; then

---

* Although used for quite different purposes, see a procedure for liposome-mediated delivery of antibody to Drosophila Kc cells in Walter *et al.* (1986).

it was confirmed on a larger scale by comparing their whole peptide synthesis pattern with those of both parental lines (Berger *et al.,* 1980b).

Our routine protocol for PEG-mediated fusion is described in Appendix 3.E.

B. Selection of Hybrid Cells

Even induced fusions involve only a low percentage of the treated cell population, so that somatic hybrid cells have to be sorted out by some selection.

Among the many systems of selection that were developed for mammalian cells, a few could be applied to *Drosophila* cells. They correspond to two main categories.

1. When two distinct enzymatic deficiencies affect respectively the two parental lines, they can be cured in the hybrids by reciprocal complementation. This is the principle of Littlefield's selective system, a method universally used for mammalian cells: one partner is deficient for a key enzyme (hypoxanthine–guanine phosphoribosyltransferase, HGPRT) of the purine "salvage" pathway, whereas the other lacks one critical enzyme (thymidine kinase, TK) of the pyrimidine "salvage" pathway. Thus, both parental lines die when their neosynthesis routes to nucleotides are blocked with the drug aminopterin (an analog of folic acid), even if they are supplied with exogenous purines and pyrimidines. Only hybrid cells are able to multiply in this selective hypoxanthine–aminopterin–thymidine (so-called *HAT*) medium.

Because of the special features of the purine metabolic pathways in *Drosophila*, as first established by Becker (1974 to 1980) (see Chapter 5), an orthodox HAT system could not be directly adapted to *Drosophila* cells. In particular, the purine salvage mechanism relies, here, on the activity of an adenosine phosphoribosyltransferase.

Yet, Wyss isolated an adenine-salvage-deficient variant from a *Kc* subline. It was called *MDR3* due to its double resistance to two adenine analogs: 6-methylpurine and diaminopurine. Unlike "wild-type" cell lines, this mutant was not able to grow in a medium supplemented with thymidine, adenine, and the folic acid analog, methotrexate (TAM medium). Conversely, probably because they are nutritionally more demanding, Schneider's *S3* or Dübendorfer's *D1* lines could not survive in this synthetic medium (*ZH1*, see Chapter 1, Section II.F) in which the *Kc* subline was grown. Therefore, only hybrid cells between *MDR3* and any one of the other lines, resulting from PEG fusion, were able to proliferate in a TAM synthetic medium (Wyss, 1979, 1980a).

Additionally, using a semiselective procedure, as devised by Davidson and Ephrussi (*1965*) to recover mammalian hybrid cells between established line cells and freshly explanted cells, Wyss (1980b) succeeded in obtaining hybrids of *MDR3* cells with primary culture embryonic cells; only synkaryons had a chance to continuously proliferate in TAM medium.

A quite different system, although also implying a mutual complementation between both parental lines, was utilized by Nakajima and Miyake (1978) for sorting out hybrids between two distinct thermosensitive sublines: *ts15* and *ts58* (two Ethylmethane sulfonate (EMS)-induced mutants from *GM1* cells) formed colonies at 23°C but not at 30°C, when inoculated, in conditioned medium, at concentrations less than 10^5 cells per 60 mm petri dish. After PEG treatment of mixed monolayers, a few hybrid colonies developed at the nonpermissive temperature, because the temperature-sensitive lesions of each of the partners were different and could thus be reciprocally complemented.

2. A second type of selection is based on the utilization of dominant markers, for instance enzymatic mutations or overexpression, conferring an unusual level of resistance to a given drug, which may be transmitted to the hybrid cells.

Such sublines with dominant markers have been selected from several *Drosophila* cell lines, and they might be conveniently utilized in efficient selection procedures.

Oubain resistance (the drug is known to inhibit the plasma membrane Na^+,K^+-ATPase) behaves in mammalian cells as a codominant trait and was used for selecting hybrids between ouabain-resistant and ouabain-sensitive cell lines.

Spontaneous ouabain-resistant *Kc* cells could be isolated (Echalier, 1976) but, apart of the fact that their Na^+,K^+-ATPase activity was not measured, they were useless for hybrid selection because they no longer fused after ConA treatment (at that time, the only fusing agent at our disposal).

After EMS mutagenesis, Sugino and Nakayama (1980) recovered mutants from Schneider's *S2* line, with high levels of resistance to the antibiotic aphidicolin (a specific inhibitor of DNA polymerase-α in eukaryotic organisms). In one of them (*aph-10*), the polymerase was much more resistant to inhibition by the drug (its apparent K_i was more than 100 nM aphidicolin, instead of 12 nM for the wild-type enzyme), while for another mutant (*aph-13*) the resistance was due to an overproduction of DNA polymerase (an eightfold higher level).

By gradually increasing the concentration of the drug, cells resistant to methotrexate were obtained from two different *Drosophila* lines (Mosna *et al.*, 1984). *Cl82 MTX^r* (from *Kc* cells) and *0.57 MTX^r* (from Milanese *0.57* line) could resist $2 \times 10^3 M$ and $2 \times 10^5 M$ methotrexate, respectively. Unlike methotrexate-resistant mammalian cells, no "double minute" chromosomes were observed nor could the presence of "homogeneously stained regions" be established, since *Drosophila* chromosomes are not susceptible to G-banding (see Chapter 4).

It is fair to say that none of these potentially interesting codominant markers has actually been used, up to now, in efficient selective systems for *Drosophila* cell hybrids.

Note: It is noteworthy that *Drosophila* cell culture has never been involved in the yet active research field of somatic cell genetics, most probably for reasons of chronology. When the first *Drosophila* cell lines were established, quite different molecular approaches had just been developed for analyzing eukaryotic genomes. Moreover, it is obvious that *Drosophila* geneticists had at their disposal, for several decades, much better tools for localizing genes on chromosomes (the most common utilization of mammalian cell hybridization in the early 1960s).

Nevertheless, it is quite clear that this technique of cell fusion and the ways it might expand functional analyses of the hybrid cells have been somewhat underutilized in *Drosophila* cells. Particularly worthwhile might be the fusion of cells from established lines with a few categories of differentiated tissue cells (as for neurons, see, for instance, Wu *et al.*, 1981; Suzuki and Wu, 1984), even though it has often been observed in mammalian cells that a large portion of the specialized program of the differentiated partner is "extinguished" in such hybrids.

IX. LIVING CELL TRANSPORTATION

Because temperate room temperatures (19 to 22°C) are optimal for *Drosophila* cells, most lines may be easily mailed from one laboratory to another.

The cultures to be sent should be rather densely seeded, 2 to 3 days before forwarding, and their appearance carefully checked. The flask should be completely filled with culture medium to avoid partial desiccation of the cell monolayer, in case the parcel is upended along the way, and to eliminate any air bubbles that might mechanically injure adhering as well as floating cells.

It is safer to wrap the vessel in an insulated package (small polystyrene boxes for ice cream are quite convenient). If possible, the cells should be mailed during mild weather conditions. Do not forget that *Drosophila* cells are sensitive to heat shock (that is, above 30 to 32°C); they survive for a long time at relatively low temperature (*ca.* 10°C).

On arrival, the cells should be permitted to settle down and reattach to the flask bottom for a few hours before most of the supernatant is aspirated. Recovered medium should be carefully stored for use in subsequent subcultures, especially if the cells have difficulty in adapting to the batch of serum used in the new laboratory.

Frozen aliquots of cells may also be successfully shipped throughout the world, although at a much higher price than that of flasks of growing cells. The frozen vials are surrounded by dry ice pellets and shipped in insulated polyfoam containers with suitable shipping labels indicating the perishable nature of the contents. On arrival the vials can be placed in liquid nitrogen or the cells immediately thawed by the usual method and placed in culture (Schneider, personal communication, 1993).

APPENDIX 3

3.A. Subculturing Monolayer Cultures

This is a brief summary of our routine schedule.

Once a week (on a definite day, for convenience), the cultures to be transferred are carefully inspected for healthy appearance and lack of (always possible!) contamination.

After pipetting off, with caution, a large part of the used medium, the cells are simply detached with a bent glass pipette (or a rubber spatula), then dispersed by gentle pipetting.

One fraction of the suspension (one-third to one-twentieth, according to the lines) is transferred to a new culture flask (25 cm²; 50 ml) already containing 3 ml of fresh medium supplemented with serum and previously brought to room temperature.

After a very brief flaming of the plastic flask neck, the cap is screwed tight.

Some gentle rocking of the flask ensures a homogeneous distribution of the cells that will rapidly adhere to the bottom.

Cultures should be kept stationary, in the dark, at a temperature of 22 to 25°C.

3.B. Growing *Drosophila* Cells in a Bioreactor

This procedure is from Dr. L. Sondergaard (University of Copenhagen, Denmark).

The following procedure is used for *S2* cells grown in a Brown Biostat M bioreactor equipped with a 2 liter vessel, but should be applicable to all bioreactors designed for growing mammalian cells.

1. BIOREACTOR AND ITS PREPARATION

The Biostat M is equipped with a pH electrode, O_2 electrode and O_2 control unit, a sampling/harvesting tube, thermocouple, a ring-shaped sparger, and a silicone tube device for aeration, consisting of approximately 5 m of thin-walled silicone tube wrapped around steel rods fixed to the periphery of the lid. Agitation is provided by a slow-running (25 to 300 rpm) propeller.

The sampling/harvesting stainless tube is used to fill and empty the vessel as well as to withdraw samples for counting cells. The tube is inserted through the lid of the fermentor and reaches the bottom of the vessel. The top end is fitted with a short silicone tube which is closed by inserting a small disposable, sterile filter (of the kind used for syringes). The sparger also carries a sterile filter inserted into a short silicone tube. The filter is connected to compressed O_2. The silicone tubing device for aeration is connected to the in-house compressed air (alternatively, a high capacity aquarium pump can be used) via a reduction valve and a carbon filter to trap oil vapor in the compressed air. Downstream of the carbon filter a sterile filter is mounted to trap particles escaping from the carbon filter.

Before use, the assembled bioreactor vessel is half-filled with cell culture grade distilled water and autoclaved. After cooling to 30 to 35°C, the bioreactor is ready for use.

2. PROTOCOL

The following steps are done in a sterile bench.

Using an autoclaved silicone tube connected to the sampling/harvesting tube, the water is pumped out of the vessel by means of a peristaltic pump (we use a Millipore pump equipped with a Masterflex pump head). Care should be taken to flame sterilize ends of tube fittings, etc., before connections are made. For this purpose, a tubeless bunsen burner (for instance, Flame-boy, Technorama, Switzerland) is very convenient. The outlet of the silicone tube is fitted onto a sterile 5-ml glass pipette (long type; not the short type normally used for cell culture).

To fill the bioreactor, the inlet of the silicone tube is fitted onto a new sterile 5-ml pipette and the outlet is attached (after flaming) to the sample tube. The pipette is flame sterilized and inserted into the bottle containing the sterile medium. Fresh medium is then pumped into the vessel. Principally we use 1 to 1.5 liters for a 2 liter vessel to allow some headspace in case of foaming. When half the medium has been pumped into the vessel, the pipette is removed from the bottle containing the medium.

After flame sterilization, the same pipette is used to pump the cell suspension directly into the fermentor vessel. (Care should be taken not to overheat the medium in tubes and pipettes during flaming.) After the cells have been added to the fermentor, the rest of the medium is pumped into the vessel.

The vessel is then connected to the main system of the fermentor. We set the agitation to 75 rpm. However, to resuspend cells sedimented at the bottom, the initial agitation could be set at 175 or 275, but should be decreased to 75 rpm after 1 to 2 min to minimize shear damage.

The silicone tube aeration device is connected to compressed air, the sparger to the O_2 supply, and the DO_2 control unit is set to 50% air saturation. The sparging time is set to 2 sec with a minimum interval of 10 sec. A relatively long interval between sparging periods is necessary to let foam and bubbles settle. This setting is sufficient to keep DO_2 in even rapidly growing cell cultures at 50% air saturation without foaming. If DO_2 drops below 45%, the sparging time must be increased.

After a few minutes, the culture is mixed sufficiently for the cells to be evenly distributed in the liquid, so a sample is drawn for counting. (Failure in sterile handling during this process is the most frequent cause of contamination; special attention should be paid to flame sterilizing and not touching the ends of the tube in order not to transfer microorganisms to the interior of the fermentor.) Since there are a few milliliters of "dead volume" in the sample tube, we usually blow about 10 ml of air, using a disposable syringe, through the sterile filter at the top end of the sample tube before removing the filter and withdrawing 5 ml of liquid from the fermentor, again using a 10-ml disposable syringe. After appropriate flame sterilization, a new disposable filter is fitted to the silicone tube.

Since pH changes of both Schneider's and *D22* media are only very small during cell growth, a pH control system is not necessary.

Although a final cell concentration of 1 to 2×10^5 cells/ml can be used to start the fermentor, such a low concentration usually results in a lag period of 2 to 3 days before cell growth begins. Therefore, we normally start with a concentration of 1 to 2×10^6 cells/ml. To obtain this concentration, we harvest cells (by swirling or washing, not scraping!) from six large (73 cm^2) cell culture flasks. Just prior to use, the suspensions from five flasks are collected in the last flask.

The cells to be used in the fermentor should have a history of several generations of vigorous growth in cell culture flasks without reaching a point of superconfluency. They should be used for the fermentor at a stage of approximately 75% confluency to ensure initiation of growth after only a minimal lag period.

After 3 days, the cell density should be approximately 1 to 1.5×10^7 cells/ml. At this point, the sucrose in the medium will be exhausted, and, in combination with a buildup of toxic waste products, this will cause an increasing inhibition of cell growth.

At this point, two approaches of continuation are available: the "fed batch mode or the "discontinuous batch mode."

In the fed batch mode, the cells are harvested, resuspended in fresh medium and pumped back into the fermentor and grown for 24 to 36 h to 2 to 3×10^7 cells/ml. Then the medium will again be depleted and has to be changed to let the cells grow to 5 to 7×10^7 per ml.

At these cell densities, the medium has to be changed every 24 h, and very often it is not possible to obtain cell densities beyond 5×10^7 per ml. The risk of contamination is high during the process of pumping cells out of the fermentor, recovering the cells by centrifugation and resuspending the cells in fresh medium (so that we prefer to use the "discontinuous batch" growth). If cells have to be recovered under sterile conditions, we use a centrifuge with removable buckets which has a tight lid and which can be sterilized inside with ethanol. Sterilizable centrifugation bottles of 500 ml or 1 liter are most useful.

During "discontinuous batch" growth, batches of cells are harvested at cell densities of 1 to 2.5×10^7 cells/ml. Enough cell suspension is left in the fermentor to allow a concentration of 1 to 3×10^6 cells/ml when fresh medium is supplied to the fermentor. In practice, we remove 1000 ml cell suspension if the fermentor contains 1100 ml and replenish the fermentor with 1000 ml of fresh medium.

After 3 days, the cell density has increased to 1 to 1.5×10^7 cells/ml and the medium needs to be changed; so again 1000 ml is removed and the fermentor is replenished with 1000 ml of fresh medium. In this way, the fermentor can be run for up to 10 batches, during which the doubling time of the cells is 24 h. However, during growth, there appears to be a buildup of toxic compounds which will inhibit growth, probably arising from dead cells as well as from metabolic waste compounds. Apparently the concentration of such compounds can be sufficient to reduce growth rate after just three or four batches. If this is the case, the fermentation should be stopped and the bioreactor thoroughly cleaned, autoclaved, and restarted using a fresh batch of cells.

Using the discontinuous batch mode, a 1.5 liter fermentor may yield 8×10^{10} cells in just 2 weeks. This should be sufficient for the isolation, for instance, of proteins expressed at even low levels.

3.C. Cell Cloning Procedure

This procedure is based on Echalier (1976) and Richard-Molard and Ohanessian (1976). See also Chapter 3, Fig. 5.

1. IRRADIATION AND PREPARATION OF "FEEDER" CELLS

Healthy cultures from the same line as the cells to be cloned are irradiated with an X-ray or gamma-ray source.

In our Institute, we have at our disposal a very convenient cobalt source, with a small compartment in which the plastic culture flasks (containing some 10^8 cells in their normal medium) can be directly introduced for an allotted time ensuring an adequate and accurate dose of gamma rays.

When using an X-ray machine, all the different parameters of an exposure must be rigorously standardized: size of the petri dish in which the cells have been transferred and volume of medium (in other words, the depth of the liquid layer), target-to-sample distance, type and conditions of utilization of the X-ray generator (dose rate, thickness of Al and Cu filters). Be sure that the temperature never rises above 25°C.

The irradiation dose must be empirically and accurately determined for each type of cells: the appropriate dose should be just enough to completely suppress any capacity of multiplication, and this must be carefully verified with cultures of "feeder" cells alone. As an example, we found 26 Kr to be a safe dose for most of our *Kc* sublines.

Irradiated cells are then scraped off the vessel and dispersed, in the medium, with gentle pipetting. An aliquot of the suspension, after appropriate dilution, permits rapid cell counting by means of a hemocytometer.

A final and important dilution is made with fresh medium supplemented with 5 to 15% FBS so that 0.1 ml of the suspension will contain approximately 50,000 cells. This volume (in practice, two drops of the final dilution) is poured into every small cavity of a 96-well microtiter plate (tissue culture grade), and those cells will constitute the feeder layer. Note: To limit the possibility of contamination, empty wells are alternated with those containing cells throughout the plate (see Fig. 3.5A)

2. CELLS TO BE CLONED

The cell suspension, coming from an almost confluent healthy culture, is vigorously pipetted, to disperse any cell clumps. Be careful, however, not to mechanically injure the cells. The efficiency of this monocellular dispersion, as well as cell viability (as estimated from the trypan blue exclusion test, see Section V) will be assessed via cell counting with a hemocytometer.

Serial dilutions with fresh medium are then made, until the cell density averages 1 cell per 0.05 ml. This volume corresponds approximately to one drop of the suspension and this is the quantity to be then added to each well of the plate.

As a result, one healthy cell is theoretically seeded into each well among some 50,000 "feeder" cells. However, because the probability of survival of individual cells may vary greatly from one line to another and the possibility of injury due to handling the cells, it is advisable to seed a series of wells with two or three cells, respectively.

3. COLONY FORMATION AND GROWTH OF CLONAL SUBLINES

In order to minimize evaporation, the lids of the plates must be securely sealed with a truly impermeable adhesive tape, as the microcultures have to be kept undisturbed for several weeks at 22°C.

Plates are periodically inspected for colony formation. Under a binocular microscope, their tiny whitish spots are easily identifiable on the bottom of the wells, among the sparse fragments of the dead feeder layer (Fig. 3.5B). As soon as the first colonies are detectable, and this is rarely before the fourth to sixth weeks, every well containing a single colony is carefully labeled with a pen.

For most cell types, the cloning efficiency does not generally exceed a few percent.

Each clone is allowed to grow further, for an additional 1 or 2 weeks, until it occupies a large portion of the well bottom. Adding a drop of fresh medium, at this stage, may help but is usually not necessary (and handling of the plate lid is often hazardous for sterility).

Finally, each clone, collected by simple pipetting, is gradually transferred into culture vessels of increasing sizes: 4-well plates, then small petri dishes, and finally screw-cap culture flasks.

As a general rule, in order to make absolutely sure of the clonal nature of the isolated cell sublines, it is highly advisable to repeat this cloning process twice. In addition, it is worthwhile to verify the homogeneity of karyotypes.

3.D. Freezing Cell Lines for Storage

1. FREEZING

Healthy cell cultures are prepared 3 to 4 days beforehand with a high inoculum size in order to be relatively dense (some 5×10^7 cells per flask).

Most of the medium is aspirated off and the cells are scraped or pipetted off, as usual, with a glass pipette and suspended in the remaining 0.5 ml of medium.

Transfer to small plastic ampules especially designed for freezing and with a tight screw cap [for instance cryotubes (NUNC), Roskilde, Denmark]. Each of them is carefully labeled for future identification with a special cryopen.

Add the protective agent, glycerol or dimethyl sulfoxide: 0.5 ml of a (Millipore-filtered) 2X solution that is, 20% (v/v) in medium supplemented with serum, will provide a final concentration of 10%.

Note: DMSO should be recently purchased to avoid toxic sulfone derivatives (detectable by their unpleasant smell).

After tightly screwing their caps, the ampules are successively transferred, for gradual cooling, into a 4°C refrigerator (or in melting ice) for 30 min, a −20°C freezer compartment for 30 min, and a dry ice chest for 1 h.

Then, the frozen ampules are quickly affixed, end to end, onto suitable metal tracks (channeled rods with a number of clips) and these are stocked in clearly numbered canisters of a special liquid nitrogen container (−196°C).

2. THAWING

When cells are to be recovered, thawing must be rapid. The simplest technique is to warm them in the hollow of your hand. Pipetted cells are quickly transferred into a culture flask, with a relatively large volume of medium (5 to 10 ml). As soon as the viable cells have attached to the flask (about 1 h), the medium is replaced in order to eliminate the toxic DMSO.

Subcultures usually become possible after a few days.

Notes: (*a*) It is always advisable to thaw one or two aliquots, a week or so after every series of cell freezing, to verify the viability of the frozen cells before discontinuing the culture of the corresponding cell line. Some cell lines (for instance, our *KcO* subline), at least under the above conditions of cooling, cannot be preserved by freezing. (*b*) A simplified protocol has been currently introduced into our laboratory routine use and seems quite efficient: Ampules, after careful wrapping in cotton wool, are enclosed in a small polyfoam box which is readily placed into a deep freezer cabinet ($-70°C$). This simple procedure apparently ensures a sufficiently low rate of freezing.

3.E. Polyethylene Glycol-Mediated Cell Fusion

1. *Materials*

PEG 1000 (Merck) 40% (v/v) in Dulbecco's modified Eagle's medium* Because of its strong damaging effect on any cell membrane, including microbial ones, it is not necessary to autoclave it for sterility. Melt PEG in a 70°C water bath. Using a sterile pipette with a wide tip (PEG is very viscous), mix 4 ml of the melted PEG with 6 ml of Dulbecco's medium.

D22 medium and its hypotonic version: adding 3 ml of sterile distilled water to 10 ml *D22* decreases the osmolarity to 275 mOsm.

Coverslips 22 × 22 mm (sterile).

Plastic petri dishes 60 mm diam (sterile).

2. *Protocol*

On the day preceding the experiment, a mixture of the two lines to be fused is seeded as a dense monolayer onto a small sterile coverslip placed in the bottom of a 60-mm petri dish. After the cells have settled, add 4 ml of standard culture medium.

About 2 h before the fusing treatment, the medium is replaced with 4 ml of diluted *D22* [or simply add a suitable volume (see above) of distilled water]. This hypotonic shock seems to help in fragilizing the cell membranes.

Remove hypotonic medium and pour a small drop of the *PEG* solution (40% in Dulbecco's medium) on the coverslip and the cells. Treat for 30 sec only.

---

* PEG solution in this medium designed for mammalian cells was fortuitously found to be more efficient than in *D22* medium. Whether the effectivness depends on its ionic balance and/or the higher pH is unknown.

Rapidly wash the cells to dilute and remove PEG (which is very toxic): for instance, wash successively four times with 4 ml diluted *D22* (twice for 2 min, then twice for 5 min).

Finally, replace hypotonic *D22* with normal medium (but without serum).

On the next day, the medium will be normally supplemented with 5 to 10% fetal bovine serum.

Many dikaryons or polykaryons can be observed immediately, although their number will continue to increase within the following hours.

It is advisable to allow the cells to recover, for 1 or more days, before using the selective system.

Note: As an alternative, according to the original technique recommended for mammalian cells, Wyss (1977) prefers to treat the cells with PEG when they are in suspension:

"PEG 1000 (Merck), after its autoclaving, is mixed with the serum-free and yeast extract-free ZH1 medium (see Chapter 1): 35% (w/v);

A mingling of the two lines to fuse (about 5×10^6 to 10^7 cells of each) is pelleted;

After removal of the supernatant, the pellet is stirred up and mixed with 0.3 ml of the PEG solution;

After 30 sec, 20 ml of serum-free medium is added and the cell suspension is centrifuged for 10 min at 600 *g* at room temperature;

Cells are then resuspended in fresh medium (or selective medium) and seeded into culture flasks."

This latter technique probably allows a more rapid elimination of the toxic PEG, but our treatment using a monolayer might prove especially suitable for fusing cells of an established line with a few differentiated cells previously grown on the coverglass.

References

Ashburner, M. (1989). *Drosophila: a Laboratory Manual,* Cold Spring Harbor Laboratory Press, Cold Spring Harbor.

Barski, G., Sorieul, G., and Cornefer, F. (1960). Production dans des cultures *in vitro* de deux souches cellulaires en association, de cellules de caractère "hybride." *CR Acad. Sci. Paris* **251**, 1825.

Davidson, R., and Ephrussi, B. (1965). A selective system for the isolation of hybrids between L cells and normal cells. *Nature (London)* **205**, 1170.

Gateff, E. (1977). Lethal malignant blood neoplasms: *l(2)mbn* and *l(3)mbn*. *Drosophila Information Service*, **52**, 4–5.

Gateff, E. (1978). Malignant and benign neoplasms of *Drosophila melanogaster*. *In "The Genetics and Biology of Drosophila"* (M. Ashburner and T. R. F. Wright, ed.). Vol 2b pp. 181–265, Academic Press, New York.

Hayflick, L., and Moorehead, P. S. (1961). The serial cultivation of human diploid cells. *Exp Cell Res.* **25**, 585–621.

Hayward, W. S., Neel, B. G., and Astrin, S. M. (1981). Activation of a cellular *onc* gene by promoter insertion in ALV-induced lymphoid leukosis. *Nature (London)* **290**, 475–480.

Paul, J. (1973). *Cell and Tissue Culture,* 4th ed., Churchill Livingstone, London.

Puck, T. T., and Marcus, P. I. (1955). A rapid method for viable cell titration and clone production with Hela-cells in tissue culture: the use of X-irradiated cells to supply conditioning factors. *Proc. Natl. Acad. Sci. U.S.A.* **41**, 432–437.

Suitor, E. C., and Paul, F. J. (1969). Syncytia formation of Mosquito cell cultures mediated by Type 2 Dengue Virus. *Virology* **38**, 482.

4

Karyotype and Cell Cycle

Appendix 4
A. Preparation of Well-Spread Chromosomes from Cultured Cells for Karyotype Analysis
B. Conventional Chromosome Staining with Aceto-Orcein
C. Chromosome Staining with Giemsa: C-banding
D. Fluorescent Banding of Chromosomes: Q-banding: Staining Procedure with Quinacrine Dihydrochloride
E. Bulk Isolation of Metaphase Chromosomes
F. Buffer A for Preserving Architectural Integrity of Isolated Mitotic Chromosomes
G. [³H]Thymidine Pulse Labeling for Analysis of Cell Cycle
References

The relative simplicity of its genotype is one of the major assets of *Drosophila melanogaster:* its karyotype is characterized by (1) a low number of chromosomes, (2) clear-cut differences between pairs, and (3) a wide choice of genetic and chromosomal markers. All these undeniable advantages are even more valid as far as genetic studies at the cellular level are concerned.

I. KARYOTYPE ANALYSIS

Cultured *Drosophila* cells can be used to investigate many basic problems, such as the detailed behavior of each chromosome pair during mitosis, DNA replication, heterochromatin function, maintenance or variation of ploidy, sister chromatid exchanges, etc., with a resolving power unparalleled in other systems.

A. Chromosome Complement of *Drosophila melanogaster*

The chromosomes of *Drosophila melanogaster* are particularly well-known from the classical studies of Kaufmann (*1934*), Cooper (*1950*) and many others. The normal diploid set is made up of only 4 pairs of chromosomes, which are easily recognizable on the basis of their relative sizes, the shapes and locations of primary and secondary constrictions, and whose correspondence with the four genetic linkage groups has long been established (see Fig. 4.1).

The 1st pair of so-called "sex chromosomes" is heteromorphic (XX in females and XY in males).

The X-chromosome is a rod-like element (approximately 4 μm long, at metaphase) with a subterminal spindle fiber attachment (centromere); in addition to this "primary constriction," a secondary constriction" marks a nucleolar organizer (NOR).

Chromosome Y is J-shaped (submetacentric: its short arm is about 1 μm in length and the long one 2 μm). It also carries a nucleolar organizer, detectable as a "secondary constriction" near the middle of the shorter arm.

Two large pairs of "autosomes" (II and III) are more or less equal-armed (V-shaped, i.e., metacentric) and their total length is approximately 5 μm, although the 2nd pair is slightly, albeit noticeably, smaller. In addition, experienced observers can identify this 2nd pair by the presence of a prominent secondary constriction, lying on its left arm and close to the primary constriction.

The IVth pair corresponds to tiny dotlike chromosomes (0.5 μm) with a subterminal primary constriction.

A large proportion of this chromosomal complex (about 30%) is heterochromatic, as revealed by positive heteropycnosis in the prometaphase

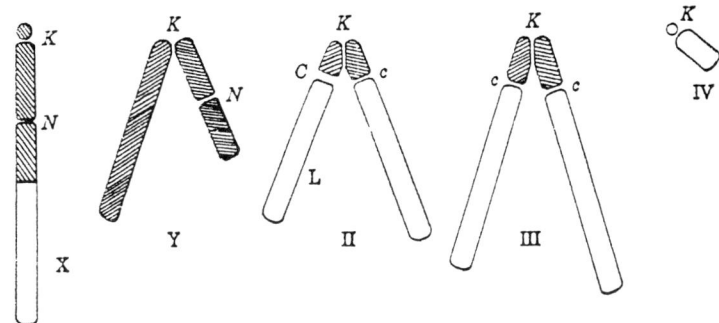

FIGURE 4.1 Diagrammatic representation of the chromosome complement of *Drosophila melanogaster*. Heterochromatin (cross-lined); K, kinetochore, primary constriction; N, nucleolar constriction; c, secondary constriction (the prominent secondary constriction, C, said to mark 2L, is tentatively represented in the shorter arm. From K. Cooper (*1950*) in Demerec, *Biology of Drosophila*, (1964 reprint) chap. 4, fig. 1, Reprinted by permission of Prentice-Hall Inc., Upper Saddle River, New Jersey).

chromosomes: constitutive heterochromatin occupies the whole length of chromosome Y, the proximal half of the X, and the centromeric regions of large autosomes II and III; as for the tiny chromosome IV, which was not generally considered to be heterochromatic, it was however shown to fluoresce with fluorochromes and to be late replicating (see below). Heterochromatic regions are not uniform, but are made up of a series of blocks of differential staining capacities. A detailed description of this banding pattern may be found in Cooper (*1959*) and in Pimpinelli *et al.* (*1976*) (see also summarizing diagram in Ashburner (*1989*), after Pimpinelli's communication).

B. Karyological Methods as Applied to *Drosophila* Cultured Cells

Chromosome preparations can be obtained from cells cultured under conditions that favor the occurrence of many mitoses, that is, during the exponential phase of growth (for instance, the second day of a subculture from established line cells).

In order to accumulate metaphases, colchicine,* which blocks spindle formation, may be added, at a final concentration of 0.04 μg/ml, for about 6 to 8 h before chromosome preparation. Because the drug causes further chromosome condensation, it should not be used when chromosomes are to be examined for banding.

1. PREPARATION OF WELL-SPREAD CHROMOSOMES

The standard method of Rothfels and Siminovitch (*1958*), devised for mammalian cells, aims at spreading metaphase chromosomes, so that they are well flattened and clearly distinguishable without any overlapping. It combines (a) a hypotonic treatment and swelling of cells in mitosis, which causes chromosomes to separate from one another, and (b) air-drying after fixation, which arranges the chromosomes in one plane.

See the operating procedure for *Drosophila* cells in Appendix 4.A.

2. STAINING METHODS

Figure 4.2 shows differential staining patterns for chromosomes.

a. Conventional staining

The most widely used dye is aceto-orcein (see staining technique in Appendix 4.B). It is perfectly compatible with [^{3}H]thymidine labeling and

---

* Colchicine or vinblastine sulfate (Sigma): 2 to 4 μg/ml for about one generation time.

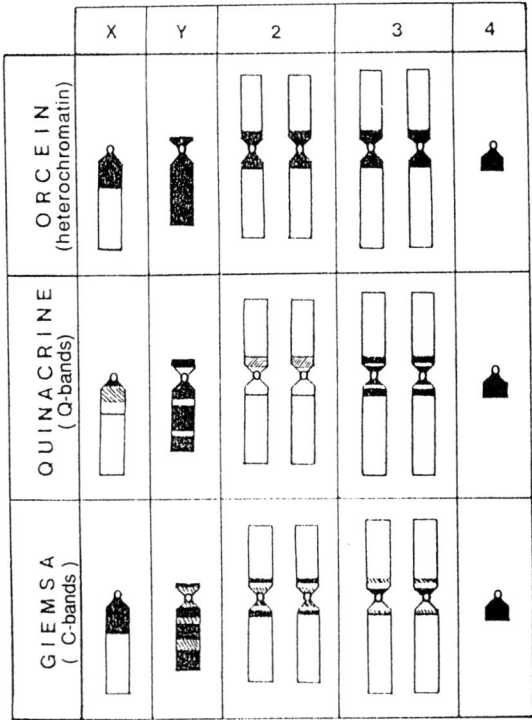

FIGURE 4.2 Differential staining patterns of the chromosomes of *Drosophila melanogaster*. Karyotypes of *GM3* cell line: orcein, quinacrine, and Giemsa staining. From Dolfini, 1974, 1976; courtesy of the author.

autoradiographic process, and can also be used on chromosomes from which a fluorochrome has been washed off after fluorescence observation.

b. C-banding: differential staining of heterochromatin

After DNA denaturation with an alkaline solution and annealing in warm saline solution, the darker staining with Giemsa (*C-bands*) corresponds to the heterochromatin; only the C-band localized on the X chromosome is reduced, in comparison with the corresponding heteropycnotic material on prometaphase.

Staining procedure is given in Appendix 4.C; the length of NaOH treatment is critical.

Note: N-banding. Another type of Giemsa staining consists in pretreating chromosomes with 5% trichloroacetic acid (95°C for 30 min), then with 0.1 N HCl (60°C for 30 min) (see Funaki *et al., 1975*). The deeply stained regions or N-bands are always located on heterochomatin, but their pattern differs from that of C-bands and in *Drosophila*, in contrast with the observations made on several other materials, they do not correspond to the nucleolar organizing regions.

c. Fluorescence observation after fluorochrome staining: Q-banding

Unstained chromosomes show no fluorescence, but some fluorescent dyes (Hoechst 33258, or quinacrine dihydrochloride) have a specific affinity for defined chromosome regions. Therefore, when excited by the light of a shorter wavelength (UV) under a fluorescence microscope, metaphase chromosomes stained with fluorochromes appear to be banded. Such bands are called Q-bands or H-bands, for quinacrine or Hoechst staining, respectively. Whereas euchromatin fluoresces very faintly and homogeneously, this method gives an excellent longitudinal differentiation of heterochromatin, revealing areas of bright or weak fluorescence characteristic of each chromosome pair (see Fig. 4.2). Both types of staining coincide roughly. In the X chromosome, however, only a proximal portion of the large Hoechst-fluorescent block is quinacrine bright.

It is reasonable to assume that this cytological heterogeneity corresponds to a heterogeneity at the structural and/or molecular level.

The staining procedure with quinacrine hydrochloride is described in Appendix 4.D. It should be remembered that fluorescence fades away rapidly on exposure to UV, and that the discrimination of photographs is far superior to that of direct eye observation.

C. Bulk Isolation of Metaphase Chromosomes from Cell Cultures

The only successful method for mass preparation of mitotic chromosomes from *Drosophila* cells grown *in vitro* was reported by Hanson and Hearst (1974) who used suspension cultures of line S2. As the authors observed, the cohesiveness of arrested metaphase chromosome spreads in *Drosophila melanogaster* cells seems to greatly exceed that of other cells.

Their operating procedure is reproduced in Appendix 4.E (see also Hanson, 1978).

Note: Belmont *et al.* (1987, 1989) devised a proper buffer (see Appendix 4.F) for preserving the architecture of such isolated mitotic chromosomes. They could observe chromatin structural domains approximately 130 nm in width.

D. Cultured Cells and Fine Structural Analysis of Somatic Chromosomes

Cell cultures are today widely recognized as being the best tool for obtaining large quantities of well-spread mitotic chromosomes for high resolution observations.

As already mentioned, the karyotype of *Drosophila melanogaster* is particularly favorable for such studies. Barigozzi's group, in Milan, investigated the structural distinction and the physiological meaning of constitutive heterochromatin, using primary cultures as well as established cell lines [see reviews by Dolfini (1971) and Barigozzi (1971, 1972); see also Faccio Dolfini and Bonifazio Razzini (1983)]. Moreover, several distinctive aspects of the mitotic process in Diptera, such as chromosome pairing in somatic cells and the peculiar behavior of the Y chromosome during prophase synapsis (Halfer and Barigozzi, 1973, 1977), or the intriguing phenomenon of sister chromatid exchange (Faccio Dolfini, 1978) could also be conveniently studied.

1. DIFFERENTIAL DNA REPLICATION PATTERNS OF EUCHROMATIN AND HETEROCHROMATIN

By the means of labeled precursors of DNA and subsequent autoradiographic demonstration of the synthesis patterns of different chromosome segments, it was established, in mammalian cells, that heterochromatin duplicates relatively late, near the end of the S period. (See technique of [^{3}H]thymidine labeling in Appendix 4.G).

Barigozzi and colleagues (1966, 1967) clearly demonstrated this asynchrony of replication of heterochromatin in very short-term primary cultures of *Drosophila* embryonic cells. Cultures were exposed to tritiated thymidine (0.5 μC/ml) for the last 6 h and $3\frac{1}{2}$ h, respectively, before chromosome preparation. Whereas all chromosomes were uniformly labeled in the 6 h series, the preferential location of silver grains, as observed in the $3\frac{1}{2}$ h series, revealed the late replication of all heterochromatic segments: the Y chromosome was obviously the last to terminate synthesis along its entire length; but all regions of the other pairs characterized by stable heteropycnosis (namely the proximal half of the X, the centromeric sections of II and III) and also the whole dotlike IVth pair, were shown to be late replicating (Fig. 4.3).

The existence, in *Drosophila melanogaster,* of many viable fly stocks with detectable chromosomal translocations offered special opportunities for a fuller analysis: for instance, in the sterilizer *(szw)* stock, the centric part of the Y forms a ring, whereas the acentric fragment is translocated to the X heterochromatic region. In other cases, various segments of the heterochromatic Y may be translocated to different regions of the large autosomes *(Y:2* or *Y:3).* It was interesting to determine whether their replication time had been modified thereby.

In primary cultures derived from embryos of these different stocks, it could be shown (Halfer *et al.,* 1969; Barigozzi *et al.,* 1969) that the rule of late replication of heterochromatin remains generally valid; yet, when, for instance, the Y is split into two portions, a clear asynchrony arises. Such a change in the duplication time of translocated segments, which, by the way, was not accompanied by a change in their heteropycnosis, seemed more or less independent of the hetero- or euchromatic nature of adjoining sections; the phenomenon could be interpreted as a position effect without any spreading effect. Moreover, the observations suggested that some factor(s) that specifically controls the late replication timing of the whole chromosome might exist in the central part of the Y chromosome.

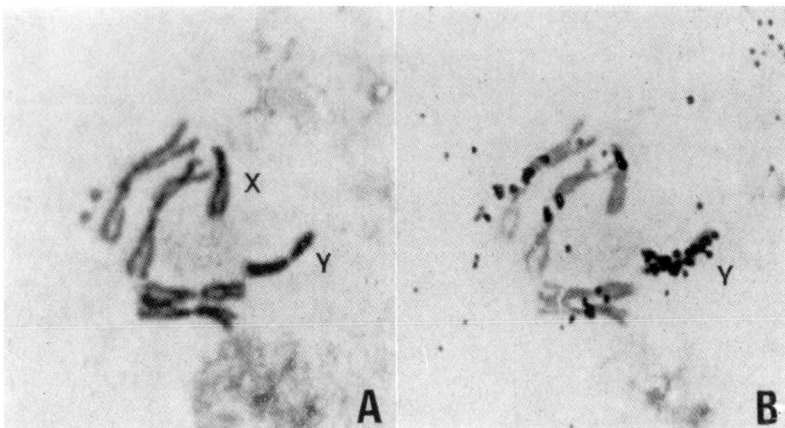

FIGURE 4.3   Late replication of heterochromatin (A) Metaphase of *Drosophila* male embryonic cell *in vitro* (orcein); (B) The distribution pattern of silver grains in an autoradiograph shows the late DNA replication of all heterochromatic portions. From Halfer and Barigozzi (1976); courtesy of the authors.

Using a similar technique of [³H]TdR (tritiated thymidine) short labeling, Ananiev *et al.* (1977) confirmed, in *67j25* cells, that heterochromatin and euchromatin replications occur in different periods of the 10 h long S phase. From their label distribution curves as a function of time, it could be concluded that, during the first 4 h of S phase, the synthesis concerns only euchromatic regions; then, just before the synthesis oᶠ euchromatin is complete, that of heterochromatin starts and will last for 6 h until the end of S phase.

2. BANDING PATTERNS AND IDENTIFICATION OF REARRANGED CHROMOSOMES

The development of various banding methods, and especially the use of fluorochromes, gave rise to more detailed investigation, at the structural level, of the differentiation between hetero- and euchromatin.

Becker (1970) was the first to demonstrate, in our established line *Ca* (comprising 80 to 95% of XY cells), that, after staining with DNA-binding quinacrine, the Y chromosome could be identified in interphase nuclei of *Drosophila* "male" cells, as had just been reported for human male cells (Pearson *et al.,* 1970). This chromosome appears as a small (*ca.* 0.4 μm) but bright fluorescent body, in contact with the single nucleolus (see Fig. 4.4). Since nothing similar was observed in cells with a female

FIGURE 4.4 Quinacrine fluorescent staining of the interphasic nucleus of a "male" *Drosophila* cell. Cell line *Ca:* the bright corpuscle is close to the nucleolus. From Becker (1970). Courtesy of the author, with permission of Gauthier-Villars, Ltd., Paris.

karyotype, this specific fluorescence might constitute a convenient marker in cell populations.

A short while before, Vosa (1970) had worked out the discriminating fluorescence patterns, on squashed larval brain ganglia, after quinacrine staining, of all the chromosomes of *Drosophila melanogaster* (Figs. 4.2 and 4.5). These data were then extensively exploited by Barigozzi's group, especially for identifying chromosome fragments or rearrangements in primary cell cultures deriving from various stocks, as well as in several established cell lines.

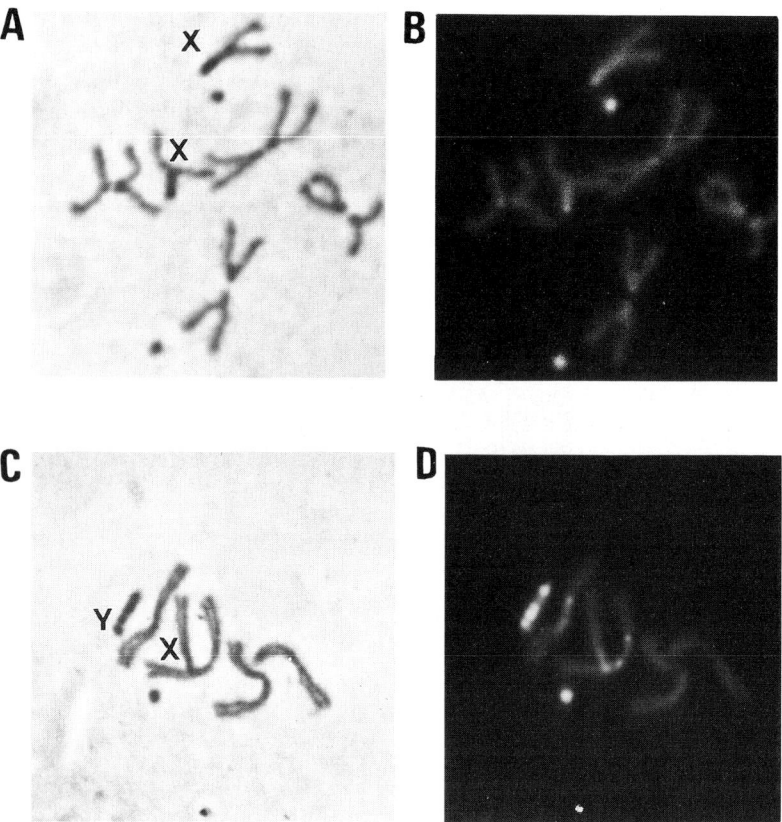

FIGURE 4.5   Female (A,B) and male (C,D) karotypes of *Drosophila melanogaster* cultured cells. Orcein (A,C) and fluorescent quinacrine (B,D) staining. Courtesy of S. Dolfini-Faccio, University of Milan, Italy.

The quinacrine fluorescence of the Y chromosome is quite characteristic, showing four bright segments distributed along its whole length: one involving the terminal portion of the short arm and three in the long arm (Fig. 4.5D). Therefore, it was relatively easy, in cultured cells from embryos of the sterilizer (*szw*) stock, to identify the ring chromosome, with three bright fluorescent sections, to the long arm of the Y, while the portion translocated to the X corresponds to the short arm. It should be pointed out, however, that the fluorescent intensity of specific chromosome segments may change somewhat when they are transferred onto other chromosomes (Zuffardi *et al.,* 1971). This finding calls for some caution in the interpretation of chromosomal rearrangements.

Nevertheless, this fluorescent method, possibly in association with other banding techniques, was very useful for identifying several abnormal chromosomes, as they occurred in some established cell lines (see Section E that follows). For instance, in the *GM1* line, a high percentage of cells contained, besides the autosomes and one normal X, a small centric and heteropycnotic fragment; its two bright fluorescent blocks might correspond to the sections of the Y-chromosome proximal to the centromere (Zuffardi *et al.,* 1971).

E. Stability and Variations of Karyotypes of Cells Cultured *in Vitro*

The establishment *in vitro* of permanent cell lines in mammalian cells is generally correlated with the development of heteroploid states. Interestingly, this does not seem to be the rule in *Drosophila* cell lines and many of them remain diploid or nearly diploid (see Chapter 3, Section III, the predominant karyotype of most established lines). However, throughout the years of growth *in vitro,* significant quantitative or qualitative rearrangements of the chromosome set can occur and still remain compatible with viability at this monocellular level. Various environmental conditions can exert powerful selective pressures on these huge cell populations (a standard culture flask may contain some 10^8 cells or more).

Accurate monitoring of the genome is not easily accomplished with mammalian cells because of their numerous chromosomes, which are difficult to classify, whereas the simple karyotype of *Drosophila* offers an almost unique opportunity for studying the dynamics of the genome in cultured cell populations. Serial karyological analyses have been carried out on several *Drosophila* cell lines, essentially by Barigozzi's group in Milan. Independently, additional information, relative to their own cell lines, was contributed by Gvozdev, Kakpakov, and collaborators. The particular case of haploid cell lines, examined by Debec, and which possibly depends on somewhat different regulation mechanisms, will be discussed in Section III.A.2.

In short-term primary cultures set up under the precarious conditions of Horikawa's medium, and in contrast with the earlier assertions of Horikawa and Fox (1964), Dolfini and Gottardi (1966) observed 12% heteroploid metaphases after only 12 h of growth *in vitro,* and as much as 80% after 72 h. This rapid poly- and aneuploidization proves only the unsuitability of this culture medium since, as soon as more fitting media were devised for *Drosophila* cells, Gvozdev and Kakpakov (1968) reported that embryonic cells retained diploid chromosome complements for up to 70 days *in vitro.* Subsequently, their continuous cell subline *67j25D* preserved a normal female karyotype (85 to 90% diploid cells + a few percent of tetraploids) until, at least, the 48th passage, which corresponds to several tens of cell generations (Kakpakov *et al.,* 1969). Another subline (*67j25A*), however, was isolated with an aneuploid formula (primarily 6X + 7A, no account being taken of the IVth pair, which is difficult to detect).

Let us now examine a few examples of long-term karyological observations.

1. The first permanently established *Drosophila* cell line, namely the *Kc* line, was monitored by Dolfini (1971) for more than one year. At the start of this study (that is, 15 months after the beginning of the initial primary culture), an XX diploid karyotype was already predominant (71.5%), although about 10% of XY cells could be observed. This latter category had completely disappeared 6 months later, and today, after more than 25 years of culture, this widely used cell line is still characterized by its female paradiploid state (Fig. 3.3, page 146).

However, the percentage of true XX cells was rapidly reduced (by about one-half) in favor of X0 and principally X + one centric fragment formulas. This heterochromatic fragment, at first interpreted as being a segment from the Y, was subsequently shown to have a quinacrine fluorescence pattern very similar to that of the proximal portion of the X. Moreover, a few cells had two X's of unequal lengths (XX_L); later on, Privitera (1980) was able to conclude, having used a rather delicate silver staining procedure considered as specific to the nucleolar organizing region (NOR), that the length difference was due to an elongation of the NOR. As for the autosomes, the line was, from the beginning, marked by the loss of one dotlike IVth chromosome and by a typical shortening of one arm of one of the major pairs. Another important point was the slow rise in the number of tetraploid cells (from 1 to 10%).

Rather interestingly, a further study (Dolfini, 1973) of two clonal sublines issued from this *Kc* line showed that both of them had lost their theoretical homogeneity and reflected, finally, the same karyological variations that characterized the parental line.

2. Over 3 years, Kakpakov *et al.* (1971) followed the evolution of the karyotypes of different sublines from their basic *67j25* line. A further evaluation was made after 7 years of culture (Polukarova *et al.*, 1975).

The diploid subline *67j25D* remained essentially diploid, with two X chromosomes and four large autosomes (2X + 4A), after 7 years of growth *in vitro* (350 passages; *ca.* 1400 cell generations). In time, the proportion of cells with only one detectable chromosome IV reached 80%. Tetraploid cells did not exceed 5 to 15%.

A tetraploid subline *T,* which arose spontaneously during the passage of the aneuploid subline *A,* and the triploid subline *TA-6,* obtained by mere chance from cloning, were also capable of retaining their own karyotypes for a long time (more than 100 passages). Similarly, about 10% hyperploids (that is hexaploids in the triploid line and octoploids in the tetraploid one) were noted. Moreover, a few percent of aneuploids occurred, especially hypoploid as for the X chromosomes; stable clones with 5X + 8A or 4X + 6A could even be isolated.*

3. Barigozzi's collaborators studied, with the greatest care, the chromosome complements of the various cell lines that they had established (Mosna and Dolfini, 1972; Faccio Dolfini, 1974, 1976, 1977; Faccio Dolfini and Halfer, 1976, 1978; Mosna, 1979).

Line *GM2* had a peculiar marker: resulting from a misdivision of the centromere of one of the metacentric autosomes (IInd pair), there were two "new" telocentric chromosomes; their derivation could be clearly confirmed by their quinacrine fluorescence patterns. Within the first year of culture, a doubling of the ploidy occurred and, during the three following years, many karyotypic changes were observed (most probably under the "protection" of this tetraploid state). Among these changes was a progressive loss of the telocentric chromosomes corresponding to the left arm of chromosome II. A similar breakage, at the level of the centromeric heterochromatin of two of the four chromosomes of the third pair could be observed in many of these tetraploid metaphases, followed by the disappearance of the entire euchromatic arm.

---

* An evaluation covering 26 years of uninterrupted growth *in vitro* of Russian cell lines was published by Kakpakov and Kakpakova (1993) in a Russian book (see Bibliography). It could not, unfortunately, be reviewed here.

On the other hand, *GM3* line cells were quite stable and, if one excepts the rapid loss of one, and sometimes two, chromosomes IV (although they seemed rather to "jump" from one chromosome to another), most cells still showed, 4 years later, their original diploid XY karyotype.

The report, by Halfer (1980), relative to seven embryonic lines deriving from stocks marked by reciprocal translocations between the Y and chromosome III, is of special interest. Their karyotypic behavior differed markedly from that of wild-type lines: they were exclusively polyploid. Moreover, if the original translocation was no longer found, a series of new aberrations took place, involving all the major chromosomes.

It is also worth mentioning a systematic and long-term (more than 5 years) study by Halfer (1978) concerning our line *Ca* (prevalently diploid male cells), because she noted the occurrence, after several years of reasonable stability, of an intriguing 1-year period of crisis. The line then underwent a polyploidization process, with broken chromosomes, correlated with changes in the cell morphology. This crisis was followed by a no less strange return to a diploid or quasi-diploid condition, accompanied, however, by various karyological transformations (for instance, an inversion in the Y chromosome).

It is particularly interesting that, in a number of cell lines derived from various *Drosophila* mutations (such as *mei, mus;* see Chapter 5) that affect DNA metabolism and, more especially, repair mechanisms, the karyotype evolution was found to be very similar to that observed in other cell lines (Hirschi and Boyd, 1981). They show a general tendency to hypotetraploidy, with a preferential loss of Y and small IVth chromosomes.

A series of general conclusions may be drawn from all these detailed studies (see also a short review by Barigozzi, 1982). First, a remark concerning the methodology should be made: such karyological analyses require the most careful sampling, because karyotypic variations, with regard to their relative percentages, may differ strikingly from one slide to another, each slide representing one Leighton tube, that is a small individual subculture (Barigozzi, 1972).

Theoretically, if large cell populations grown *in vitro* are genetically heterogeneous (and this poses the problem as to whether the established lines are of monoclonal or polyclonal origin), they should be submitted to powerful selective pressures, tending to change their primitive genetic constitution. In reality, a few variations were observed, at the karyological level, when a stable line, such as *67j25D,* was successively grown in media as different as *S-15* medium or Schneider's medium (Polukarova *et al.,* 1975). Similarly, the karyotypic changes in sublines adapted to serum-free media, as compared to the parental lines, remains controversial, even after the systematic checking of *GM2* and *GM3* lines carried out by Faccio Dolfini (1976, 1977). It seems, however, that our serum-free adapted subline *Kc0* shows much larger variations of ploidy than the standard parental *Kc* line, even though our opinion is based on only a few occasional verifications. In all events, it is clearly established that mediocre conditions of culture, such as the use of unsuitable media (Dolfini and Gottardi, 1966) or transfers with inocula that are too dilute (Kakpakov *et al.,* 1971), do induce a rapid aneuploidization. Nevertheless, in most cases, the diversification of the genomes in cultured cells appeared to be better explained by some random sampling throughout the multiple passages.

The contrast between the basic stability of most cell lines derived from wild-type embryos and the high frequency of chromosomal aberrations observed in those obtained from *Drosophila* stocks already marked by karyotypic rearrangements (Paradi, 1973; Halfer *et al.,* 1980) suggests a possible correlation between the chromosomal behavior of cultured cells and the genetic background of the parental fly stock.

Spontaneous numerical and structural variations concern essentially the X, Y, and small IVth chromosomes. Moreover, breakages, giving rise to deletions or translocations, occur predominantly in heterochromatin, and it may be pointed out that these latter three chromosome pairs are, to a large extent, heterochromatic. On the other hand, the autosomal pairs II and III are rarely involved, probably because no resulting gene imbalance is compatible with growth *in vitro*.

The chromosome Y is often lost or fragmented, with possible translocations.

Variations of the X number are also frequent, so that a X0 formula is unexceptional in "female" as well as in "male" lines. With regard to the important problem of the ratio of sex chromosomes to large autosomes (1 : 2 in the normal formula 2X : 4A), Kakpakov *et al.* (1971) reported a selective predominance, among polyploid sublines grown in precarious conditions, of aneuploids with abnormally high ratios (approximating 1; see, for instance, observed 6X7A or 5X5A formulas). On the other hand, Polukarova *et al.* (1975) could isolate, from hyperploid sublines, stable clones with a relative hypoploidy with respect to the X's (such as 5X8A), although no clone was ever obtained from cells with only 3X8A.

The disappearance of one of the dotlike IVth chromosomes is undoubtedly the most common karyological modification observed in established cell lines. As a matter of fact, even at the level of the organism, this haploid IV formula is not lethal. Moreover, detection by fluorescence revealed that in many cases, and at least when both IVth chromosomes seemed to have been lost, these minute chromosomes were, in fact, translocated onto other pairs; and because they apparently kept their centromere, they were able to "jump" from one chromosome to another*.

A doubling of the chromosome complement frequently occurs in *Drosophila* cell lines, as it does in cultured mammalian cells. Let us point out that true polyploidy is quite unusual in normal tissues of *Drosophila,* whereas polyteny is the rule in several larval structures. This latter situation, however, has, so far, never been reported in cell cultures *in vitro* from early embryos.

Tetraploids, and more rarely octoploids, are more probably due to endomitosis than to spontaneous cell fusions (even if such events have been documented in cultured cells; see Chapter 3, Section VIII), and the cause and mechanisms of the phenomenon remain unknown. Their percentage does not usually exceed 10 to 15% of the population and data do not suggest that polyploidy, *per se,* offers any selective advantage. On the other hand, it is an important source of variability. Chromosomal rearrangements obviously tend to occur more freely in polyploid cells because the increased gene dosage possibly acts as a "buffer", so to speak, against any imbalance resulting from abnormal karyotypes (Halfer *et al.,* 1980).

To conclude this long survey concerning the dynamics of the genome of cultured *Drosophila* cell lines, it is clear that no specific chromosomal change, at least at the cytological level, can be correlated with the capacity for continuous proliferation *in vitro.* Most of the observed chromosomal shifts seem rather to correspond to random events.

II. CELL GROWTH AND DIVISION CYCLE

Each cell subculture progresses through a characteristic growth sequence: after a brief lag period of adaptation to the fresh medium, the cells undergo a quick series of multiplication cycles, during the exponential or

---

* *Kc* cells seem to be truly haploid IV, as may be deduced from Illmensee's experiments (1976, 1978) of nuclear transplantation into early embryos and from the typical structure of bristles in the resulting chimeric epidermal areas (see the legend of Fig. I.1, in the introductory chapter of this book).

log period; a few days later, there is a deceleration, then an arrest of this proliferation (stationary period or plateau).

Typical growth and division cycle of a eukaryotic cell is composed of four successive phases: succeeding to a previous division, the so-called interphase starts with a Gap1 phase, or G_1, during which the cell recovers its biosynthetic activities. Yet, DNA synthesis and resulting chromosome duplication are restricted to the following S phase (S = synthesis). After the completion of this genome doubling, there is another interval, or Gap2 phase (G_2), before the cell enters the division processes, globally designated M phase (M, mitosis, that is, nuclear division followed by a cytoplasmic division, or cytokinesis). As a matter of fact, the division events, although spectacular, occupy only a small portion (about one-tenth) of the total cycle time. Moreover, in most organisms, although, as will be shown, this does not appear to be always the case in *Drosophila* somatic cells, differentiating cells retire from the proliferation cycle during G_1 and remain blocked in what is then called G_0.

Current investigations on the complex mechanisms of control of this cell cycle have emphasized the existence of 2 main decision points: one is late in G_1 and controls the initiation of DNA synthesis (designated "Start" or "Restriction Point," because it might be the principal control point in most cell types) and the other, at the end of G_2, corresponds to a mitotic induction.

Such a general description of the growth pattern of a subculture cell population as well as the proliferation cycle of individual cells, remains correct, of course, for *Drosophila* cultured cells, although a few adjustments are necessary.

For instance, populations of "normal" mammalian cells, which are usually well spread and strongly attached to the culture flask, will "plateau" as soon as they reach confluence. The early concept of "contact inhibition" could not be applied to *Drosophila* cells, which adhere loosely to the substrate, never form complete monolayers and frequently pile up at the stationary phase. Current interpretations of the phenomenon, in terms of depletion of available nutrients (especially growth factors), and its new and wider appellation of "density limitation of growth" are much more appropriate to *Drosophila* cells, which, similarly, reduce rapidly their proliferation when reaching a certain density.

Besides, as will be shown hereafter, there may be, in *Drosophila* cells, some significant peculiarities in the relative proportions of the different phases of their growth cycle (essentially G_2 versus G_1).

A. Generation Time and Lengths of Successive Phases of Cell Cycle

These parameters must be measured during the exponential phase of cultures, that is, in standard conditions of growth of *Drosophila* established lines, between the second and fourth to fifth days after seeding.

An easy way to monitor the proliferation rate of a culture is to count the total population periodically (see techniques in Chapter 3, Section V) or to measure its total mass. "Growth curves," depicting the increase in cell number throughout a given period, furnish precious information about the behavior of the culture, under various experimental conditions (see, for instance, Kakpakov *et al.,* 1969; Courgeon, 1972a). Nevertheless, the "doubling time", i.e., the time taken for the culture to increase twofold, should not be confused with the "generation time" (or "cell cycle time"), because it is only an average figure, describing the net result of the wide range of capacity for growth existing between different lineages of any population (even in theoretically homogeneous cloned sublines). Similarly, the "mitotic index", that is, the percentage of mitoses that can be directly observed under a microscope, at a given time, in the entire stained population, only provides a rough estimate of the duration of the M phase (the total generation time is simply divided by this percentage of mitotic cells).

Accurate figures require more elaborate methods.

1. PULSE LABELING METHOD AND TIMING OF THE CELL CYCLE

This method consists in fixing samples of an asynchronous cell population (during its exponential phase of growth) at various intervals after a short exposure to tritiated thymidine, then plotting the percentage of labeled metaphases after autoradiography (see protocol in Appendix 4.G). The duration of the total cell division cycle and of its different main phases may be deduced from the curve depicting these percentages as a function of the time [for discussion, see Quasler (*1963*)].

After preliminary results (Dolfini and Tiepolo, 1967) on short-term primary cultures, Dolfini *et al.* (1970) analyzed, according to this method, the cell cycle of *Kc* line cells. A typical biphasic curve was obtained (Fig. 4.6): the shape of such a graph can be explained by imagining that the group of cells in phase S at the time of the pulse move slowly toward M. After completion of G_2, the first labeled mitoses appear, and their number increases rapidly to reach a maximum, then decreases. The second wave in the curve corresponds to the progeny of the labeled cells, after

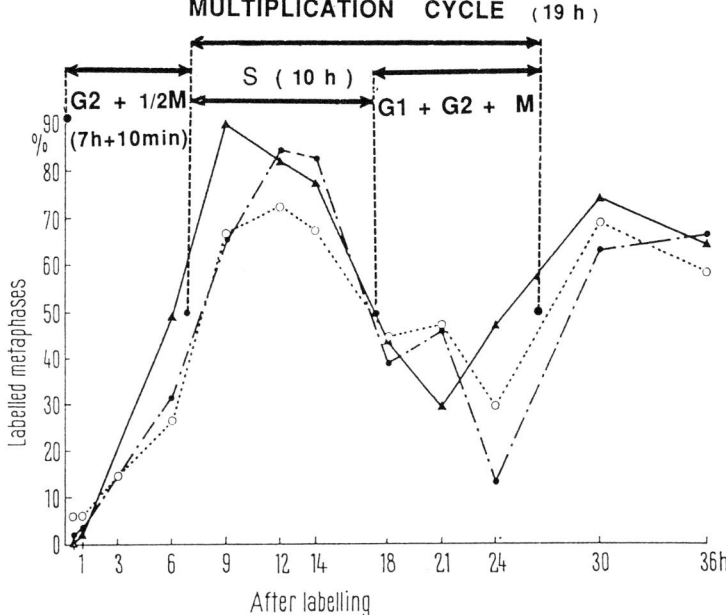

FIGURE 4.6 Cell cycle of *Kc* cell line. Percent of labeled metaphases plotted against the time after [³H]thymidine pulse, in three different experiments. How the lengths of the successive phases can be deduced from the curves is discussed in text (*D22* + 20% FBS; 26°C.). Modified from Dolfini *et al.* (1970) and Quastler (1963).

their passage through the next G_1, S, and G_2. The total cell cycle time [Generation Time (GT)] is measured by "the time between the 50% points of the two ascending limbs of the biphasic curve." Similarly, the S period may be calculated from "the distance between the two midpoints in the first wave of labeled metaphases", whereas G_2 is determined as "the time interval between the pulse labeling and the midpoint of the ascending limb of this first wave". The length of G_1 is indirectly obtained by subtracting the three other periods from the total duration of the cycle, but, in the case of *Drosophila* cells, some approximation of the M phase is necessary: the usual measurement of M, by direct observation *in vivo* of the mitotic chromosomal "ballet," is somewhat difficult in tiny *Drosophila* cells; and even the distinctive "rounding up" of cells in division is not easy to score in cell populations adhering so loosely to the flask. Only the mitotic index (see above) is informative.

According to Dolfini *et al.* (1970), the average durations of the total multiplication cycle and of G_1, S, G_2 phases were thus estimated to be 18.8, 1.8, 10.0, and 7.2 h, respectively, in *Kc* cells grown at 26°C in 20% serum-supplemented *D22* medium.

Using the same method, Ananiev *et al.* (1977) found that G_1, S, and G_2, in *67j25* cells (grown at 20°C in *C-39* medium + 10% FCS) last 6, 10, and 6 h, respectively. They emphasized the fact that cell populations can be extremely heterogeneous with regard to their division rates: for instance, already 16 h after the pulse, a few metaphases were observed with a label characteristic of the second wave of mitoses (i.e., with only one labeled chromatid), which means that some cells could complete their cycle within 8 h.

Similarly, Rizzino and Blumenthal (1978) published the detailed parameters of the cell cycle of Schneider's *SL2* line: the population doubling time, in monolayer or in suspension cultures, ranged from 15 to 40 h (at 25°C, in Schneider's medium + 15% FBS). A conventional interpretation of the kinetics of labeled mitoses after a [³H]TdR pulse, in suspension cultures, gave the following values: 16.1 ± 0.8 h for S phase and 6.4 ± 0.3 h for G_2 + 1/2 M.

Moreover, the total cell cycle length could be directly measured by time-lapse photography of individual cells, and varied from 6.3 to 27.8 h (mean: 15.1 ± 4.4 h; mode: 14 h). The same technique allowed also a rough estimation of the *M* phase (time between the loss of visibility of nuclear content and cytokinesis): 35 ± 49 min (mode: 20 min), a value in agreement with that calculated from the mitotic index (i.e., 25 min).

These figures call for a few remarks:

(a) As is reported in the brief description of several of the established lines indexed in the preceding chapter (Chapter 3, Sections II and IV), the "doubling time" of most *Drosophila* cell lines ranges from 20 to 48 h.

(b) It is noteworthy that this GT of *Drosophila* cells, even though the latter are cultured at temperatures ranging from 20 to 25°C, does not differ dramatically from the values observed in many mammalian cells grown at 37°C.

(c) The unusual length of G_2, as compared with other phases and especially with G_1, was in good agreement with results from experiments of continuous labeling, in which the first labeled metaphases did not appear before 7 to 8 h. This peculiarity of the *Drosophila* cell cycle will be discussed in the next section.

Note: In a series of cell lines derived from mutants with DNA repair deficiencies (*mei, mus,* see Chapter 5), Hirshi and Boyd (1981) estimated their cell doubling time to be within the usual range of 24 to 47 h. From autoradiographic analyses, after a 15 min exposure to [³H]thymidine, their labeling indices varied from 24 to 38%.

2. ANALYSIS OF CELL CYCLE BY FLOW CYTOFLUOROMETRY

The distribution of the cells of an asynchronously growing population, among the different phases of their proliferation cycle, can be easily determined, at any given time, by using an electronic fluorescence-activated cell analyzer.

After fixation, the cells are stained with a fluorescent dye (such as ethidium bromide) which binds to the DNA and confers on them a fluorescence whose intensity is directly proportional to their DNA content. A proper dilution of the cell suspension, so that individual cells travel in a single file, is then projected through the fine nozzle of the apparatus. When each cell briefly passes a tiny window where a laser beam excites its fluorescence emission, this emission is measured by a photomultiplier tube and recorded. Thousands of cells can thus be rapidly monitored.

The DNA fluorescence histograms typically show two peaks (Fig. 4.7): the first one corresponds to the lowest amount of DNA (cells with a *2C* DNA content, i.e., in G_1) and the second one to a double DNA content (cells with a *4C* DNA content, i.e., G_2 + M cells); and because the mitotic index is rather low, it can be assumed that this latter peak consists essentially of G_2 cells. Cells with intermediate DNA amounts (between the two peaks) are in S phase.

Using such an electronic device, Stevens and the group of O'Connor (Stevens *et al.,* 1980; Stevens, 1981; Stevens and O'Connor, 1982; O'Connor, 1985) observed that the G_2 phase is much longer in cultured *Drosophila* cells than in other studied cell lines as inferred from the near equivalence of the G_1 and G_2 peaks in fluorescence histograms obtained from logarithmically growing *Kc* cells. Moreover, they were the first to notice that the division arrest, induced by a 20-hydroxyecdysone treatment of hormone-sensitive *Kc* sublines (Courgeon, 1972a), occurs in phase G_2 (see further discussion in Chapter 8 and Fig. 8.5).

These results were confirmed by Besson *et al.* (1987) through a systematic study of the growth kinetics of two different *Kc* sublines; a clonal subline (*8-9K*), growing in normal 5% serum-supplemented medium, and the serum-independent *Kc0* subline (see Fig. 4.7). During the growth

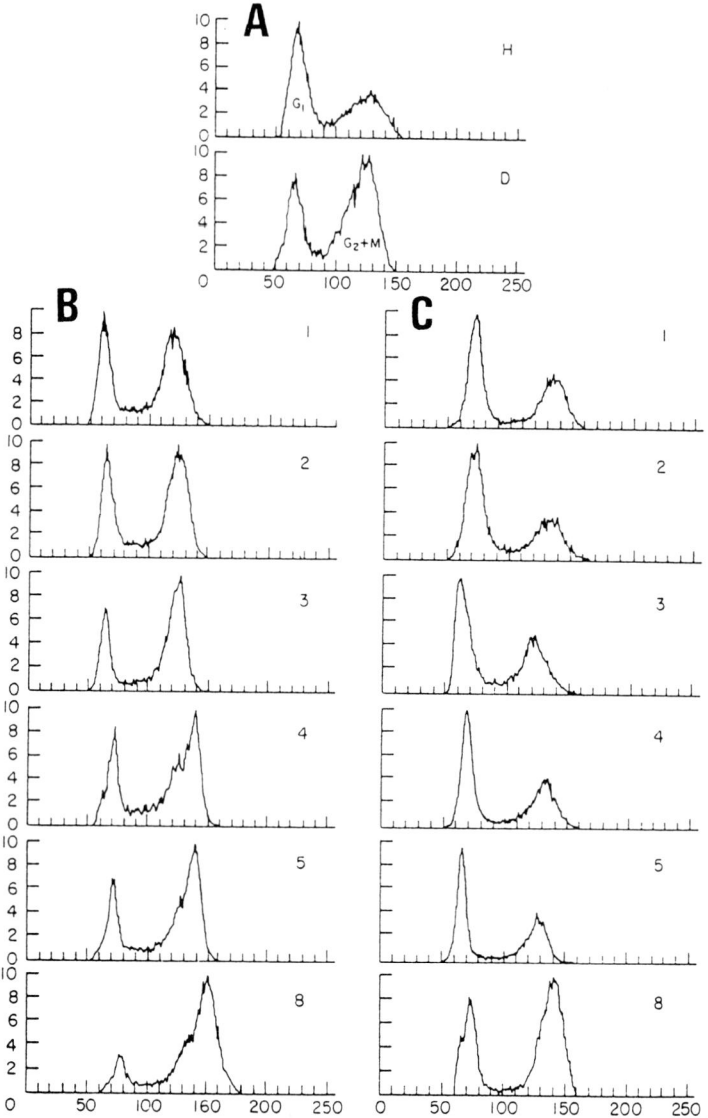

FIGURE 4.7 DNA fluorescence histograms of *Kc* and *KcO* cells. (A) *Kc* growing cells after hydroxyurea (200 μg/ml^{-1}, 18 h) (H) and demecolcin (10^{-6} *M*, 18 h) (D) treatments in order to calibrate G$_1$ and G$_2$ + M peaks; (B) A *Kc* clone during an 8-day culture (time in days, from 1 to 8); (C) *KcO* subline. (For more detail see text.) From Besson *et al.* (1987). Courtesy of the author.

phase of the *8-9K* clone, the percentages of G_1 and G_2 cells were found to be nearly equal, which means that G_1 and G_2 phases were of similar durations. When the cultures reached the stationary period (from Day 5), the G_2 percentage became predominant (73%), suggesting an increase in the length of G_2 and finally a G_2 arrest. In *Kc0* cultures, the general pattern was similar, although the percentage of G2 cells remained lower during the "log" period; moreover, the accumulation of G_2, in the stationary period (3 days later than in *Kc*), did not exceed 50%. Furthermore, continuous exposure to ecdysteroids blocked, within 3 days, most of the cells in G_2 (95% in *Kc* cells).

All these data emphasize the particular importance of phase G_2 in *Drosophila* cells and suggest that the main "control point" of the cell cycle might be located not at the boundary between G1 and S, as in most organisms, but rather at the end of G2, thereby being decisive in the triggering of mitosis.

This assertion is corroborated by current investigation of the control of division cycles in early *Drosophila* embryos (carried out in parallel by the groups of O'Farrell in San Francisco and Glover in Scotland). Ending an extraordinary fast succession of synchronous and strictly nuclear replications and divisions (every 8 min), the 14th interphase initiates a new era in postcellularization embryos (about 2 h after fertilization): henceforth, cell division rates become much slower and then differ greatly from one tissue to another, according to a precise spatiotemporal pattern; this regulation seems primarily due to the appearance and differential lengths of a G_2 phase. The experimental blocking of protein synthesis in this 14th cycle results in G_2 arrest, suggesting that new factors are then required to initiate mitosis.

It should also be remembered that Delachambre and collaborators (Dijon, France; personal communication in the early 1980s) have shown that the active cuticular synthesis by the epidermis of another Insect, the coleopteran *Tenebrio*, is associated with G_2-blocked cells. Perhaps the polytenic state of several larval differentiated tissues, in the *Drosophila* organism, corresponds to a somewhat analogous situation.

B. Synchronization of *Drosophila* Cell Cultures

For a number of biochemical approaches, it would be advantageous to have at one's disposal large pools of cells that are homogeneous with respect to their phase in the division cycle. Unfortunately, among the many methods developed for synchronizing the growth of mammalian cell cultures, very few could be successfully used with *Drosophila* cells.

For instance, the common and simple method ("mitotic selection"), based on the fact that dividing cells round up and tend to detach from the monolayer, and which consists in collecting selectively these mitotic cells by merely shaking them off the culture dish, cannot be applied to *Drosophila* cell populations because they all adhere so poorly to the substrate.

Besides, let us emphasize that in any synchronized cell culture and whatever the technique used, if a high degree of synchrony can sometimes be achieved in the first cycle, it is already much lower in the second cycle, because of the heterogeneity of most cell populations. It becomes usually close to random by the third cycle.

1. DNA Synthesis Blockade

Many procedures of synchronization use metabolic inhibitors blocking rapidly and reversibly the DNA synthesis, so that the cells accumulate at the beginning of phase S; as soon as the drug is removed, the proliferation cycle starts again synchronously in a large proportion of the population. Nevertheless, because their action might be incompletely reversible, such exogenous inhibitors are never desirable, especially for studies of DNA synthesis and related problems.

Ananiev *et al.* (1977) reported that fluorodeoxyuridine (FdU), at a concentration of 1 μg/ml, can block DNA synthesis in 67j25 cell cultures within 5 min. After a 12 h treatment, some 50% of the cells were arrested between the G_1 and S phases, and were subsequently able to resume their growth cycle.

2. Nutritional Deprivation

It had been shown, in mammalian cells, that the growth arrest of a culture at the stationary phase is the result of some specific depletion of medium components [such as isoleucine and glutamine, for instance, in the case of Chinese hamster ovary (CHO) cells]; upon subsequent addition of fresh medium, cells resume their life cycle in synchrony.

This simple procedure, whereby the cells of suspension cultures, when arrested for 3 to 4 days in stationary phase, are pelleted and then resuspended in fresh culture medium, produced only partial synchrony in *Drosophila Kc* and *SL2* cells (Rizzino and Blumenthal, 1978): after an 8 to 14 h time lapse, some 60 to 80% of the cells entered S phase. However, the duration of the peaks of DNA synthesis (24 to 29 h after release from stationary phase), was 1.5 to 3 times longer than the length of S phase estimated for either *Kc* or *SL2* cells, implying that the cells were only in a partially synchronous S phase (which might be due to the karyotype heterogeneity of the studied sublines).

C. DNA Replication Rate and Replicon Size Determination

Labeled DNA molecules, isolated from cultured cells incubated for various times with [³H]thymidine, are analyzed by radioautography and reveal grain tracks which extend in length as pulse durations increase. The average rate of this track growth can be directly measured and, for the estimation of the "replication rate," the fact that replication in eukaryotes is bidirectional should be taken into account. The average "replicon" size can also be deduced from the distances between centers of adjacent tracks aligned along the same molecule.

In Schneider's *S2* cells grown in a spinner, Blumenthal *et al.* (1974) established that the average rate of DNA synthesis per replication fork was equivalent to or greater than 2.6 kb/min^{-1}/fork^{-1} (or 53 μm/h), while the replicon sizes seemed to be divided into two classes: 28 kb and 57 kb (that is 9 μm and 19 μm).

The results of a similar study, carried out by Ananiev *et al.* (1977a and b) on semisynchronized *67j25* cells, differed significantly: the replication rate at a fork was estimated to be only 12.5 μm, and the mean replicon size was around 70 μm.

Such discrepancies could perhaps be explained by differences in methodology; moreover, an interference of the FdU used by the latter authors for synchronizing the cells cannot be dismissed.

D. Regulation of DNA Replication

The onset of eukaryotic DNA replication requires the highly coordinated expression of many replication-related genes and little is still known, in animal cells, about the *cis*-elements and transacting factors that are involved in such a strictly regulated polygenic induction.

After the isolation and sequencing of two important *Drosophila* genes involved in DNA replication, namely, DNA polymerase α (Hirose *et al.*, 1991) and PCNA (proliferating cell nuclear antigen or cyclin) (Yamaguchi *et al.*, 1990, 1991), Hirose *et al.* (1993) noticed the presence, in their upstream regulatory regions, of a common palindromic sequence

(5'-TATCGATA-3'). This sequence is even repeated three times in the DNA polymerase α gene. CAT constructs, including different lengths of these upstream sequences or related oligonucleotides, were transfected into *Kc* cells and it was clearly demonstrated that such 8 bp motifs are required for high CAT expression. They were, therefore, named DRE (for DNA replication regulatory elements).

In nuclear extracts from *Kc* cells, a specifc binding factor (DREF) could be identified (with gel mobility shift assay, DNA foot-printing analysis, and UV cross-linking method) and it was finally purified with DRE-oligonucleotide-immobilized affinity latex particles. Native DREF seems to be a homodimeric form of a 86 kDa polypeptide (which is in good agreement with the palindromic organization of the recognition site).

Further transient expression assay, again in *Kc* cells, using various mutations of the PCNA upstream region (base substitution or internal deletions), confirmed that the 8 bp motif (from nucleotide position −93 to −100) is essential for activation as it is for the binding to DREF *in vitro* (Yamaguchi *et al.*, 1995).

Moreover, by cotransfection with a *zerknüllt* (*zen*)-expressing plasmid, the same laboratory (Yamaguchi *et al.*, 1991; Hirose *et al.*, 1994) established that the *zen* homeodomain protein represses expression of DNA replication-related genes, probably by reducing DREF.

An independent approach refers to the mammalian cycle-regulatory transcription factor *DRTF1/E2F* (well known to be a heterogeneous complex that coordinates the expression of genes during the progression into S phase):

After cotransfection into *Drosophila* cells, Bandara *et al.* (1993) showed that two mammalian components, *E2F-1* and *DP-1*, are present *in vivo* in a DNA binding complex and, as a heterodimer, act synergistically in E2F site-dependent transcriptional activation.

Dynlacht *et al.* (1994) cloned two *Drosophila* cDNAs encoding proteins with significant homologies to human *E2F* family members and to human *DP,* respectively. They termed them *dE2F* and *dDP.*

By cotransfection in *S2* cells, they could demonstrate that overexpression of *dE2F* greatly increased (up to 100-fold) the transcription of a reporter plasmid containing *E2F*-binding motifs. Moreover, as is the rule in mammalian cells, in which *E2F* and *DP* associate in a stable complex, bacteria-expressed *dE2F* and *dDP* proteins were shown to bind strongly to each other, in *in vitro* conditions.

The discovery of this important *E2F* control pathway in *Drosophila* should allow a finer analysis of regulatory mechanisms of DNA replication, and perhaps lead to the identification of retinoblastoma-related protein(s) (which, in mammalian cells, inhibits E2F by direct binding to a region of the protein which is, as a matter of fact, strongly conserved in *dE2F*).

III. HAPLOID CELL LINES AND ACENTRIOLAR STATUS OF ONE OF THEM

The diploid constitution of eukaryotic cells is clearly an obstacle to recovery of recessive mutants from cultured animal cell lines, whereas this can be easily accomplished in bacterial population. Therefore, the *Drosophila* haploid cell lines established by A. Debec (1978) are of special interest for genetics.

By using natural or experimental parthenogenesis, a few authors succeeded in growing cells deriving from haploid (or partially haploid) embryos, in particular from cockroach, frog, and even mouse. Unfortunately, haploid cells were rapidly overgrown by diploid variant cells. Only Freed and Mezger-Freed (*1970*) could maintain, for 200 generations, two relatively stable haploid cell lines originating from androgenetic embryos of *Rana pipiens,* an animal that is unsuitable for genetic studies.

As for the *Drosophila* genus, spontaneous parthenogenesis was described in several species, although not in *Drosophila melanogaster,* but karyotypic regulation occurs during embryogenesis, resulting in diploid (sometimes triploid) flies. On the other hand, haploid/diploid mosaics could be detected in a few mutants of *Drosophila melanogaster* and *D. simulans,* for instance, as early as *1925* by Bridges in the *Minute* mutation. These latter observations proved, at least, that *Drosophila* haploid cells are not cell lethal and can even differentiate.

A. The Mutation *mh1182* and Haploid Cell Lines Established from Its Lethal Embryos

Looking for X chromosome genes that would be active mainly or exclusively during oogenesis, Gans *et al.* (*1975*) induced with ethylmethane sulfonate (EMS) and isolated about 100 "female-sterile" mutants of *Drosophila melanogaster:* this rather ambiguous denomination refers to mutations which have no direct effect on the growth and viability of homozygous individuals themselves, although homozygous females lay eggs that will develop abnormally, even when fertilized by wild-type

sperm. The lack of some specific maternally synthesized product inter-
rupts, more or less precociously, the morphogenetic processes.

Seven of these mutations gave rise to haploid embyros, or at least to
embryos comprising large haploid sectors (Zalokar *et al.*, *1975*). Among
them, *maternal haploid* (*mh*) *1182* was the most favorable for attempting
cell cultures, because homozygous females were normally fecund, and
defective embryos, although seeming entirely haploid, reached the blastoderm stage or beyond.

It must be pointed out that the mutation *mh1182*, which was, initially,
temperature sensitive and whose "penetrance" was not complete even
when the mothers had been carefully grown at nonpermissive temperature
(29°C), was modified in some unknown way in the fly stock kept in Paris,
so that it has now become totally lethal at any temperature (Santamaria
and Gans, *1980*).

1. ESTABLISHING HAPLOID CELL LINES

Cultures were set up according to a protocol similar to our standard
procedure (see Chapter 2): the homozygous *mh 1182/mh 1182* mothers
had to be carefully sorted out. The culture medium used was an 1 : 1
mixture of *D22* and *M3*, supplemented with 20% FBS.

As usual, during the first few days, the differentiation of various cell
types (muscle and nerve cell, etc.) could be observed, but no systematic
comparison was made with wild-type cultures. The development and
survival of most of these primary cultures did not exceed 1 or 2 months.

In a few flasks, however, a small islet of cell proliferation appeared
after several weeks, and finally seven continuous cell lines were established
and named *1181-1* to *1182-7b* (Debec, 1978). The morphological heterogeneity of each line probably implies a nonclonal origin, but many cells
had a diameter (7 μm) clearly smaller than that of normal diploid cells;
and six of these lines revealed various proportions of haploid metaphases
(Fig. 4.8).

2. LONG-TERM EVOLUTION OF KARYOTYPES OF HAPLOID CELL LINES

If one excepts the line *1182-3a* whose karyotype was primarily male
diploid (see tentative explanation in the following paragraph), all other
lines showed, initially, a high percentage of haploid cells, with the typical
constitution: chromosomes X/2/3/4. The other metaphases were mostly
diploid or hypodiploid (frequently monosomic for X and/or chromosome 4).

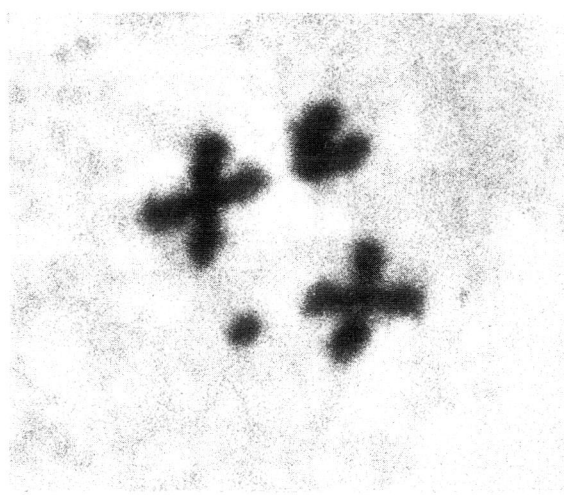

FIGURE 4.8   Haploid metaphase of *1182* cell lines. Courtesy of A. Debec.

Their karyology was monitored for 2 ½ years (Debec, 1984), and their evolution varied greatly from one line to another.

Lines *1182-2, 1182-3b,* and *1182-7b* showed a rapid decline in the percentage of haploid cells, which had completely disappeared after 9 months.

In lines *1182-1* and *1182-6,* the decrease in the number of haploid cells, although inexorable, was only a few percent per month.

Only line *1182-4* was quite stable. In 3 years of culture, involving some 150 transfers, it retained a very high percentage (80 to 90%) of haploid cells. A transitory drop to 50% was, however, observed, but this was followed by a quick reversion to the initial level. Out of ten haploid clones isolated from this line, six showed a comparable chromosome stability. A subline could be adapted to a serum-free medium and still keep the same karyotypic features.

3. MATERNAL ORIGIN OF HAPLOID GENOME OF THESE CELL LINES

In order to recognize the parental origin of the cultured cell haploid genome, Debec made primary cultures from embryos resulting from crosses between females *mh1182/mh1182* and males carrying an easily identifiable ring X chromosome. After a few days of culture, 98% of observed metaphases were haploid with a normal rod-shaped X chromosome; 2% were diploid but possessed exclusively rod-shaped X chromosomes. Such data clearly demonstrate the maternal origin of the haploid genome and strongly suggest that diploid cells could only result from diploidization of preexisting haploid cells. This conclusion agrees with the observation of Santamaria (unpublished) that, in fertilized eggs from *mh 1182* homozygous mothers, the sperm is found in a condensed form within the cytoplasm.

Notes: the perfectly homozygous condition of such "isodiploids" might partially explain the fact that they do not have, in culture, any clear selective advantage over haploid cells (as seems to be the rule in other cell systems involving "heterozygous diploids").

The development of a male diploid line (*1182-3a*) might correspond to the fact that the first cultures were set up by Debec at a time when the mutation *1182* was not yet fully "penetrant" (so that a few embryos, from which the line probably derived, were normal).

B. Potential Interest of Haploid Cell Lines in Somatic Genetics

As already emphasized, haploid cell lines should theoretically be particularly suitable for the selection of recessive mutants from the huge population of a culture.

Two examples illustrate their potential interest.

1. COMPARATIVE FREQUENCY OF 20-HYDROXYECDYSONE-RESISTANT CELLS IN HAPLOID LINES

The differentiation responses of cultured *Drosophila* cells to 20-hydroxyecdysone imply an arrest of division (see Chapter 8). Among hormone-sensitive populations, a few "resistant" cells escape this block.

It is well known that the resistance of a cell to a selective agent depends on various and complex mechanisms and may often require more than a simple mutation.

Debec (1986), however, attempted a comparison between haploid and diploid lines exposed to the hormone, by measuring their resistance frequency. He observed, in all cases, that lines comprising haploid cells showed a much higher number of resistant colonies than diploid lines. For instance, with a high hormonal concentration, he found a 1×10^{-3} frequency in the *1182-4* line, instead of 1×10^{-7} in the *Kc* line. These

significant differences are roughly in agreement with the theoretical estimates for a recessive mutation. Moreover, all studied resistant clones
were almost exclusively haploid (up to 99% of the population), which
confirms the preferential occurrence of resistants in this cell category.

2. Selection of Lines Resistant to Cadmium by Overexpression of Metallothionein

Heavy metals are highly toxic for cells, and their detoxification, up to
a certain level, is ensured by small proteins, named metallothioneins
because they are characterized by an exceptionally high cysteine residue content.

In cultured *Drosophila* cells, the lethal concentration of cadmium was
shown to be close to 20 to 30 μM.

Using increasing concentrations of cadmium chloride, Debec *et al.*
(1985) could isolate highly resistant sublines from a stable haploid clone
(derived from *1182-6* line), whereas no resistant colony was ever obtained
in a diploid clone from *Kc* cells. By successive steps, a cell population
(*Cd200* subline) could finally be grown at a concentration of 200 μM
$CdCl_2$. Suspecting that the low concentration of cysteine (1 μM) in the
culture medium limited the synthesis of metallothionein and thereby
cadmium resistance, the authors added five times more cysteine and
could, thus, select a "super-resistant" subline (*Cd800*, which grew
with 800 μM $CdCl_2$, in the presence of 5 μM cysteine).

This cadmium resistance was due to the overexpression of a metallothionein: *Cd200* cells contained 22 times more metallothionein than fully
cadmium-induced normal cells, and the super-resistants even four-fold
more.

Thus, the group of Wegnez (Mokdad *et al.*, 1987) could isolate and
characterize a new metallothionein (MTo), markedly different from that
(called MTn) previously identified in *Drosophila* by Maroni and Watson
(*1985*). After cloning of its cDNA, it was possible to show, on Southern
blots from *Cd200* cell DNA, that cadmium resistance seems to correlate
with a single duplication of the gene; the additional sequence, however,
is obviously located in a distinct genomic environment and might, thereby,
be under the control of some particularly strong "enhancer". Cadmium-
resistant lines are now diploid, but such subsequent "isodiploidization"
is not unusual, as mentioned above.

Note: Probably without connection with its haploid condition, it must
be remembered, at this point, that Saunders *et al.* (1989) claimed that
the *1182-4* line had an exceptional capacity for transient transfection

(which they found to be 100 times more efficient than in any other *Drosophila* cell line), even with naked DNA without any "facilitator" (see Chapter 9). We could not confirm their results, which means that some "drift" perhaps occurred in this cell line after it left our laboratory (Echalier and Fourcade-Peronnet, 1994).

C. *Drosophila* Cell Line without Centriole

In *Drosophila,* as in all Metazoa, the centrosome is a constant cyto-plasmic component of the cell. Observed with electron microscopy, this organelle comprises a pair of centrioles (*sensu stricto*), small cylinders typically formed by nine triplets of microtubules and surrounded by an electron dense cloud of pericentriolar material (PCM). Permanently present in the vicinity of the interphase nucleus, the centrosome splits in two, in early prophase, and its identical derivatives, at the center of radiating asters, then occupy the two poles of the mitotic spindle. Al-though it is well established today that the centrosome, or at least its PCM, acts as the "microtubule organizing center" of the cell, its precise functions, especially those of centrioles, in cell division are poorly under-stood; and the fascinating question as to its frequently assumed "genetic continuity" still remains unsolved. Therefore, the availability of an ac-tively dividing *Drosophila* cell line completely devoid of centriole might facilitate significant progress in this currently developing field.

A few karyological abnormalities observed in *mh 1182* line cells, for instance, the relative frequency of multinucleate cells or, in diploidized cells, the occurrence of aneuploidy concerning the second and third chro-mosomes (which is extremely rare in normal lines, see Section I,E), led Debec *et al.* (1982) to suspect some peculiarity in their mitotic apparatus. As a matter of fact, even with the light microscope, mitotic spindles appeared much less stretched out and more barrel-like than those of normal *Drosophila* cells.

A systematic and irksome exploration with electron microscopy estab-lished that, in *Kc* cells, centrioles were clearly recognizable in 5.5% of the 2000 cell profiles examined at random and showed a classical structure (although the cylinders appear to be particularly short: 0.16 μm in length and 0.2 μm in diameter). In contrast, no centriole could be found in any of the 4000 random cell profiles prepared from the *1182-4* line, or in the 3000 profiles from its diploidized subline. Moreover, serial reconstruc-tions of three whole metaphases of haploid cells proved also the absence of centrioles at the spindle poles or elsewhere in the cytoplasm; the spindle microtubules were seen to converge toward a large area in which several small foci of dense filamentous material were scattered.

This total lack of centriole in *1182-4* line cells was further confirmed by a high voltage EM study, which allows the examination of thicker sections (Szollosi *et al.,* 1986).

Finally, the use of a monoclonal antibody (*Bx63*, prepared by Dr. Frasch) that recognizes the *Drosophila* centrosome permitted a relatively rapid screening, by immunofluorescence, of large cell populations (Debec and Abbadie, 1989). Not only was the absence of a centriole in all *1182-4* cells reaffirmed, but a certain proportion of acentriolar cells were also detected in other *1182* lines.

This latter finding suggests that the phenomenon is indeed related to the special origin of these *1182* lines, but, at the same time, it cannot be merely explained by a modification (or absence) of some *1182* gene product, as centrioles are present in many cells of these lines and could also be observed in several *1182* lethal embryos.

An attractive hypothesis, based directly on Boveri's classical assumption that, in embryogenesis, centrioles derive exclusively from the spermatozoon head, is that, at least in a few *1182* embryos, not only are the paternal chromosomes kept aside (perhaps by some defect in the male pronucleus swelling, as noted by Santamaria, personal communication, 1982), but also the sperm centriole is not, accidentally, able to join the female pronucleus (Debec, 1986).

Whatever may have been the precise mechanism generating its anucleolar state, this *1182-4* line should provide an unique opportunity for re-evaluating the true role of the centrosome, and of its distinct components, in the control of animal cell division. As a matter of fact, there is no equivalent example of animal cells undergoing multiple cycles of division (in this case, for years) in the total absence of centrioles, even though this is the rule in higher plants.

Debec has already undertaken a variety of immunocytological and experimental studies; for instance, using immunofluorescence staining, with an O'Farrell's monoclonal antibody *F2F4* that is specific for *Drosophila* cyclin B, combined with confocal laser scanning microscopy, he could establish that a large part of the cyclin B pool, in normal *Kc* cells, is associated with the centrosome, at the prophase and metaphase, while the signal disappeared during interphase. This suggested association is yet, unexpectedly, totally absent in the acentriolar *1182-4* line (Debec and Montmory, 1992). Moreover, the same group (Debec *et al.,* 1990; Marcaillou *et al.,* 1993) reported, during heat shock or other stresses, the transient disappearance of centrioles, as detectable with specific mono-clonal antibodies, and their functional "eclipse" as centers of microtubule nucleation. More recently, the location of γ-tubulin with regard to the poles of the mitotic spindle could be established (Lajoie-Mazenc *et al.*, 1994; Debec *et al.*, 1995).

Obviously, many more studies will be necessary, in correlation with the current and important advances in the area of molecular control of cell division in order to better understand the biology of this still puzzling organelle. The availability of such an acentriolar cell line, as a control, and more especially because it is derived from *Drosophila*, should be a valuable asset.

APPENDIX 4

4.A. Preparation of Well-Spread Chromosomes from Cultured Cells for Karyotype Analysis

The procedure was provided by S. Faccio-Dolfini (Department of Ge-netics and Microbiology, University of Milan, Italy).

It is convenient to grow the cells on the coverslips* of Leighton tubes (or on microscope glass slides accommodated in 10 cm diameter petri dishes), because, 24 to 48 h later, spreading and staining of chromosomes can be carried out directly on those coverslips or slides.

HYPOTONIC TREATMENT

The coverslip, with adhering cells, is transferred to a petri dish and covered with a prewarmed (25°C) hypotonic solution [sodium citrate $(C_6H_5Na_3O_7 \cdot H_2O)$, 1% (w/v) in distilled water] for 30 to 45 min. The optimum time may vary significantly from one cell line to another.

FIXATION

Half of the hypotonic solution is substituted by an equal volume of fixative (freshly prepared: 1 part of glacial acetic acid + 3 parts methanol). After 10 min, the coverslip is transferred twice to a petri dish containing fresh fixative (each passage, again, for 10 min).

---

* For careful cleaning of coverslips and slides, they are immersed in a solution of washing detergent, rinsed in tap water, placed in an alcohol–ether mixture, then passed through absolute ethanol, and dried with a clean cloth.

AIR-DRYING

The fixative is entirely drawn off, and the coverslip, with its cell mono-layer, is left uncovered, at room temperature, until the cells are dried.

4.B. Conventional Chromosome Staining with Aceto-Orcein

The procedure was provided by S. Faccio-Dolfini (University of Milan, Italy)

1. PREPARATION OF ORCEIN SOLUTION

Dissolve 2 g of natural orcein (G. T. Gurr, London; or Fluka, Ronkonkoma, NY) in 60 ml of glacial acetic acid, by gently shaking the vessel. Use a round bottom flask with a narrow neck, equipped with a vertical condenser (water-refrigerated Pyrex helix). Add glass beads to the liquid for tempering the turmoil.

Boil for 30 min, under a hood.

Let cool down to 50°C and then add 40 ml of distilled water.

Filter and store in a closed bottle.

Each time before using, refilter the orcein solution.

2. STAINING PROCEDURE

Air-dried cells on coverslips (see Appendix 4.A) are covered with a few drops of filtered aceto-orcein solution, for 20 min, at room temperature.

Rinse off the stain by dipping twice the coverlips into 95% (v/v) ethanol.

Expose to 100% ethanol for 5 min, then air-dry.

Mount in Euparal (Chroma–Gesellschaft, Schmid and Co., Stuttgart, Germany)

4.C. Chromosome Staining with Giemsa: C-banding

This preparation is based on Dolfini (1974), according to Hsu (*1971*) and the modified method of Drets and Shaw (*1971*).

The air-dried preparations (see Appendix 4.A) are incubated, for 1 min, in 0.07 N NaOH in 0.112 M NaCl (pH 12) (i.e., 2.8 g NaOH + 6.2 g NaCl in 1 liter deionized water).

Rinse three times in 2× SSC [saline–citrate solution (6×SSC) is prepared from 52.6 g NaCl + 26.4 g trisodium citrate in 1 liter deionized water; adjust to pH 7 with 0.1 N HCl].

Incubate in 6 × SSC at 65°C for 24 h.

Pass through two changes of 70% (v/v) ethanol and two changes of 95% (v/v) ethanol (3 min each).

After air-drying, the slides are stained for 15 to 30 min in a buffered Giemsa solution (4% Merck Giemsa in 0.01 *M* Sorensen buffer, pH 7).

Rinse briefly in deionized water, air-dry, then mount in Euparal.

4.D. Fluorescent Banding of Chromosomes: Q-banding: Staining Procedure with Quinacrine Dihydrochloride

This procedure has been provided by S. Faccio-Dolfini (University of Milan), following essentially the method of Pearson *et al.* (1970).

The slides (with air-dried cells) are immersed for 7 min in a 0.5% aqueous* solution of quinacrine dihydrochloride (Atebrin, Gurr, London; or Mepacrine, K and K Laboratory, Plainview, NY) in doubly distilled water, without exposing to light.

They are washed in running tap water for 30 sec and wet-mounted with a few drops of doubly distilled water. The coverslips are sealed with nail varnish.

The preparation should be immediately examined under a fluorescence microscope equipped with specific filters which allow visualization of the wavelength emitted by the fluorescent dye (for a Leitz microscope, the series of filters is BP 436/7, RKP 475, LP 490; for a Zeiss microscope, the series is BP 436, FT 460, LP 470).

Because the intensity of the fluorescence rapidly decreases on exposure to light (any particular field kept in focus with full illumination will fade within a few minutes), only a rapid evaluation is possible directly under the microscope. A more detailed analysis of chromosomes can usually be carried out only on photographs, because small differences in fluorescence intensity, that may not be perceived by the eye, will be enhanced by the photographic process. It is advisable to photograph a sufficient number of metaphase plates.

Preparations stained with fluorochromes may subsequently be stained with another dye: the slides, with their coverslips, are directly immersed

---

* Gatti *et al.* (1976) recommend the use of an alcoholic solution of quinacrine: "The slides are soaked for 5 min in ethanol, then stained for 10 min in a 5% solution of quinacrine HCl in absolute ethanol, rapidly washed in alcohol and air dried. They are mounted in distilled water."

for 30 min in 95% ethanol, which removes the coverslips and dissolves quinacrine hydrochloride without damaging the cells.

After air-drying, any other staining, for instance with aceto-orcein, can be applied, and the chomosomes (and if possible the same metaphases) should be rephotographed.*

4.E. Bulk Isolation of Metaphase Chromosomes

This preparation is reproduced from Hanson and Hearst (1974), with permission of Cold Spring Harbor Lab. Press.

1. METAPHASE ARREST

Metaphase arrest of suspension cultures of *Drosophila* cells is obtained by addition of vinblastine sulfate (Velban, Eli Lilly Co.) to a final concentration of 2 to 4 μg/ml.

The drug is inoculated into a rapidly dividing culture in two equal doses, one half-generation time apart, with the harvest occurring one generation time (*ca.* 20 h) after the initial dose.

2. HYPOTONIC SWELLING

Induce the hypotonic swelling of the arrested cells in a buffered saline containing 0.05 to 0.10 *M* sucrose and precisely adjusted to pH 10.0. The buffer must be freshly prepared (use it within 1 day)

> 0.05 *M* sucrose
> 0.0013 *M* CaCl$_2$
> 0.001 *M* cyclohexylaminopropanesulfonic acid (CAPS, Calbiochem, La Jolla, CA)

The pH must be adjusted to 10.0 with solid Ca(OH)$_2$, before adding 0.33 *M* 2-methyl-2,4-pentanediol (hexylene glycol).

The cells are washed twice at room temperature in 3 to 4 volumes of buffer. After washing, the cell pellet is resuspended in a one-half volume of buffer, previously cooled to 0°C.

3. DETERGENT TREATMENT AND MECHANICAL HOMOGENIZATION

Add an equal volume of cold buffer containing 1.5% (v/v) Nonidet P-40 (Sigma)

---

* Black and white films: Kodak (Rochester, NY) Plus-X Pan, Ilford FP 4, Kodak Recordak AHU Microfilm 5460; color slides: Kodak Ektachrome 160).

Immediate mechanical homogenization is accomplished at 0°C with 20 to 30 strokes, in a cold, siliconized glass Dounce homogenizer fitted with a size "B" (tight) pestle.

4. CHROMOSOME ISOLATION

The contaminating interphase nuclei are removed from homogenates by several passages through a 3 μm Nuclepore filter (General Electric Co.)

A single filter can accommodate the homogenate from approximately 10^8 cells.

Centrifugal pelleting of relatively purified chromosomes (3000 rpm for 30 min).

Large autosomes and sex chromosomes are visually distinguishable in the isolate (with a phase-contrast high magnification microscope).

4.F. Buffer A for Preserving Architectural Integrity of Isolated Mitotic Chromosomes

This preparation is taken from Belmont *et al.* (1987).

KCl	80 mM
NaCl	20 mM
EDTA	2 mM
EGTA	0.5 mM
PIPES buffer	15 mM
2-Mercaptoethanol	15 mM
Spermidine	0.5 mM
Spermine	0.2 mM
Turkey egg white protease inhibitor (Sigma)	10 μg/ml

Adjust to pH 7.0

4.G. [^{3}H]Thymidine Pulse Labeling for Analysis of Cell Cycle

The procedure has been provided by S. Faccio Dolfini (Department of Genetics, University of Milan).

On the second day of culture in Leighton tubes, tritiated thymidine (Amersham Int., Buckinghamshire, U.K., specific activity 3Ci/mmol) is added to the medium, at a final concentration of 0.5 μCi/ml and for a 30-min pulse.

After washing, the cells are incubated, at 26°C, in fresh medium supplemented with "cold" thymidine at a concentration 100 times higher than that of [³H]thymidine.

At successive intervals (1/2, 1, 3, 6 h, etc.) the cells, still attached to the slides, are fixed and prepared for chromosome spreading and staining (aceto-orcein). Autoradiography is then performed using the Kodak NTB-2 nuclear track emulsion (approximately 15 to 20 day exposure). For a detailed description of the method, see Pardue (1986) in *Drosophila: a pratical approach* (D. B. Roberts, ed.) pp. 11–137, IRL Press, Oxford and Washington D.C.

About 50 metaphases were scored on each slide, a metaphase being considered as labeled when the ratio between the number of exposed silver grains over the dividing cell and that of background grains (over an equivalent area) is higher than one.

References

Ashburner, M. (1989). *Drosophila, a laboratory handbook*, CSH Lab. Press, Cold Spring Harbor.

Bridges, C. B. (1925). Haploidy in *Drosophila melanogaster*. *Proc. Natl. Acad. Sci. U.S.A.* **11**, 706–710.

Cooper, K. W. (1950). Normal spermatogenesis in *Drosophila*. In *Biology of Drosophila* (Demerec, M., ed.) (1965 edition), pp. 10–18, Hafner Publishing, New York.

Cooper, K. W. (1959). Cytogenetic analysis of major heterochromatic elements (especially Xh and Y) in *Drosophila melanogaster*, and the theory of "heterochromatin." *Chromosoma* **10**, 535–588.

Drets, M. E., and Shaw, M. W. (1971). Specific banding of human chromosomes. *Proc. Natl. Acad. Sci. U.S.A.* **68**, 2073–2077.

Freed, J. J., and Metzger-Freed, L. (1970). Stable haploid cultured cell lines from frog embryos *Proc. Natl. Acad. Sci. U.S.A.* **65**, 337–344.

Funaki, K., Matsui, S., and Sasaki, M. (1975). Localization of nucleolar organizers in animal and plant chromosomes by means of an improved N-banding technique. *Chromosoma* **49**, 357–370.

Gans, M., Audit, C., and Masson, M. (1975). Isolation and characterization of sex-linked female sterile mutants in *Drosophila melanogaster*. *Genetics* **81**, 683–704.

Gatti, M., Pimpinelli, S., and Santini, G. (1976). Characterization of *Drosophila* heterochromatin. I. Staining and decondensation with Hoechst 33258 and quinacrine. *Chromosoma* **57**, 351–375.

Hsu, T. C. (1971). Heterochromatin pattern in metaphase chromosomes of *Drosophila melanogaster*. *J. Hered.* **62**, 285–287.

Kaufmann, B. P. (1934). Somatic mitoses of *Drosophila melanogaster*. *J. Morphol.* **56**, 125–155.

Maroni, G., and Watson, D. (1985). Uptake and binding of Cadmium, Copper and Zinc by *Drosophila melanogaster* larvae. *Insect Biochem.* **15**, 55–63.

Pearson, P. L., Bobrow, M., and Vosa, C. G. (1970). A technique for detecting the presence of Y chromosomes in human interphase nuclei. *Nature* **226**, 78–80.

Pimpinelli, S., Santini, G., and Gatti, M. (1976). Characterization of *Drosophila* heterochromatin. II C- and N-banding. *Chromosoma* **57**, 377–386.

Quastler, H. (1963). The analysis of cell population kinetics. *In Cell Proliferation* (L. F. Lamerton and R. J. M. Fry, eds.), pp. 18–34, Philadelphia, F. A. Davis.

Rothfels, K. H., and Siminovitch, L. (1958). An air-drying technique for flattening chromosomes in mammalian cells grown *in vitro*. *Stain Technol.* **33**, 73–77.

Santamaria, P., and Gans, M. (1980). Chimaeras of *Drosophila melanogaster* obtained by injection of haploid nuclei. *Nature (London)* **287**, 143–144.

Vosa, C. G. (1970). The discriminating fluorescence patterns of the chromosomes of *Drosophila melanogaster*. *Chromosoma* **31**, 446–451.

Zalokar, M., Audit, C., and Erk, I. (1975). Developmental defects of female-sterile mutants of *Drosophila melanogaster*. *Dev. Biol.* **47**, 419–432.

5

Biology and Biochemistry of Cultured Cell Lines: 1. Nucleic Acids

227

I. NUCLEIC ACID METABOLISM: BIOSYNTHESIS OF NUCLEOSIDE TRIPHOSPHATES

In the synthesis of their nucleic acids, animal cells are generally able to build up both the purine ring system and the pyrimidine nucleus of their constitutive nucleotides. As an alternative to these *"de novo"* synthesis routes, most cells can also make use of preformed bases or nucleosides, whenever they are available. These so-called "salvage" pathways would be better named "spare" pathways.

The satisfactory growth of cultured *Drosophila* cells in minimal media deprived of exogenous purines and pyrimidines, either without any addition of yeast extracts (*MX3:* Sinclair *et al.,* 1983; see Section I.B.3) or with only a pool of high molecular weight fractions from Yeastolate (*ZH*1%: Wyss, 1977; see Section I.B.1), proves that most established cell lines are capable of endogenous synthesis. There is, however, a notable exception: the malignant blood cell line *l(3)mbn* (see Chapter 3, Section IV) was found to be devoid of any measurable *de novo* synthesis (Becker, 1980). In all other cases studied, the efficient incorporation of [^{14}C]formate (which is the customary test for verifying the correct functioning of the purine ring synthesis), and the efficient blocking capacities of many classical inhibitors (such as folic acid analogs, or azaserine; and, on the other hand, PALA; see below) strongly suggest that the routes of endogenous synthesis of purines and pyrimidines, in *Drosophila* cells, are the typical biochemical ones followed by most eukaryotic organisms.

Rather infrequently in eukaryotes, several enzymes catalyzing successive steps of the nucleotide synthesis may be encoded by a common locus, which ensures their coordinated regulation. This was shown to be the case in *Drosophila* cells and in mammals for the three enzymes that control the first three reactions of pyrimidine biosynthesis (Jarry, 1976; see Section I.B.4). Similarly, by transferring the entire GART (phosphoribosylglycinamide formyltransferase) locus into *S2* cells, Henikoff *et al.* (1986) observed the overproduction of a large multifunctional protein (150 kDa) displaying not only a GART activity, but also GARS (phosphoribosylglycinamide synthetase) and AIRS (phosphoribosylaminoimidazole synthetase) enzymatic activities.

On the other hand, when looking for selection systems that could be applied to cultured *Drosophila* cells, Becker noticed a series of unusual characteristics in their salvage pathways, which led him to investigate purine metabolism in *Drosophila* extensively (Becker, 1974a to 1980).

A. Purine Interconversion Routes

Becker's data, concerning primarily the *Kc* line and many of its clonal sublines, but also confirmed in milanese cell lines *GM*, are summed up in Table 5.I and Fig. 5.1.

The most striking result was the nondetectability, under usual conditions of culture, of any hypoxanthine–guanine phosphoribosyltransferase (HGPRT) activity, i.e., the enzyme which generally allows the recovery of exogenous guanine or hypoxanthine bases, by combining with phosphoribosylpyrophosphate (PRPP). On the other hand, the salvage route for adenine, through APRT activity, was found to be very active (Becker, 1974a,b). Besides, the only nucleoside kinase that seemed to be operational was adenosine kinase (Becker, 1975).

Autoradiographic experiments confirmed that neither [^{14}C]hypoxanthine nor guanine could enter the cells, whereas they became heavily labeled after addition of [^{14}C]adenine. Now, the isolation (through their resistance to 10 μg/ml 8-azaadenine) of APRT-deficient sublines, in which radiolabeled adenine did not enter, suggests that the transformation of the base into its mononucleotide is essential for cytoplasmic penetration. Moreover, normal cell lines were found to be spontaneously resistant to high doses (>200 mg/ml) of various analogs, such as 8-azaguanine, 6-mercaptopurine, and 6-thioguanine.

This lack of HGPRT activity (and ensuing insensitivity to 8-AG and 6-MP) as well as the presence of significant APRT activity (and, thereby, sensitivity to adenine analogs) was verified by Moiseenko and Kakpakov (1974) in their *67j25D* cell line.

TABLE 5.I   Activities of Main Enzymes
of Purine Metabolism in
Drosophila melanogaster[a]

Enzyme	Activity[b]
Phosphoribosyltransferases	
APRT	12 ± 2.4
HGPRT	undetectable
Nucleoside kinases	
Adenosine kinase	9 ± 1.8
Guanosine kinase	undetectable
Deaminases	
AMP deaminase	5.5 ± 1.1
Adenine deaminase	3.8 ± 0.7
Adenosine deaminase	7 ± 1.2
Guanine deaminase	2.6 ± 0.5
5'-Nucleotidases	traces
PNPases	
Adenosine phosphorylase	undetectable
Inosine phosphorylase	1.7 ± 0.2
Guanosine phosphorylase	1.8 ± 0.2
GMP reductase	undetectable
Xanthine oxidase	
larval extracts	traces
fly extracts	traces
Kc cell extracts	undetectable
IMP dehydrogenase and	traces
XMP aminase	

[a] From Becker (1974b). Data reproduced with
the kind permission of the author.
[b] Expressed in nmol of the product/min/μg proteins.

In fact, Becker (1978) could establish that HGPRT activity is "conditional" in *Drosophila,* that is to say, that the enzyme, although always
present in the cells, has no detectable activity unless the *de novo* purine
biosynthesis is blocked. Purine nucleotides appear indeed to be inhibitors
of HGPRT, so that the enzyme should be regulated by its end products,
a quite common situation in metabolic control: (a) Purine bases, and
especially hypoxanthine (0.1 mM) and its analog allopurine, were found
to activate HGPRT, apparently by lowering the cell PRPP content, which
leads, by competition, to a decrease in the endogenous nucleotide synthesis. Similarly, when well-known PRPP depletors (orotate or glutamine)

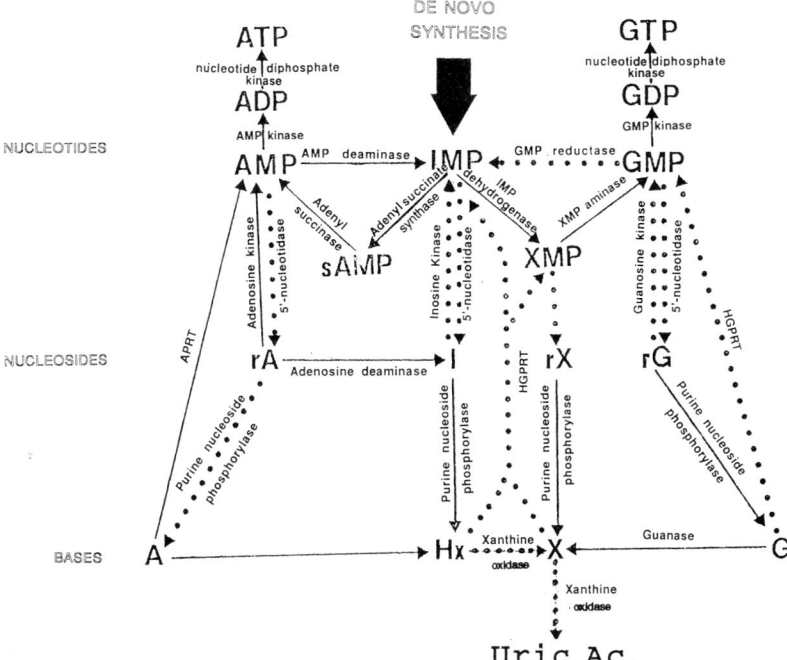

FIGURE 5.1   Purine interconversion routes and salvage pathways in cultured *Drosophila melanogaster* cells (*Kc* line). From J. L. Becker (1974, 1975, 1978); courtesy of the author.

or usual inhibitors of the *de novo* synthesis routes (for instance, methotrexate, 5×10^{-5} *M*) were added to the culture media, HGPRT activity was recovered, along with an important 5′-nucleotidase activity. (b) When nucleotides and derivatives were eliminated from cell extracts, either by an overnight dialysis or by treatment with activated charcoal, the measured activities of both HGPRT and 5′-nucleotidase were high. In such nucleotide-free cell extracts, the HGPRT activity could be inhibited by addition of purine mononucleotides (IMP and GMP), while 5′-nucleotidase could be inhibited by nucleotide triphosphates (dTTP or ATP).

Moreover, heating of cell cultures or crude cell extracts resulted in the recovery of both enzymatic activities. This fact might be explained by conformational changes and possible desensitization (with respect to its inhibitors) which affects the 5′-nucleotidase only; but since the enzyme hydrolyzes the pool of mononucleotides, the block on HGPRT activity would also be removed.

This assumed relationship between the two enzymes is consistent with the fact that, in the *l(3)mbn* hemocyte line, which was shown to be devoid of purine *de novo* synthesis, although without HGPRT and 5'-nucleotidase detectable activities, the only analog which can induce both activities is Azaserine (probably via an inhibition of CTP synthase and resulting depletion of pyrimidine nucleotide triphosphates which are themselves inhibitors of 5'-nucleotidase).

B. Selection Systems Involving Nucleotide Metabolic Pathways

1. *TAM SELECTION SYSTEM*

A universally used method for selecting mammalian somatic hybrid cells, directly inspired by bacterial genetics, was devised by Littlefield (*1964*). It consists of "crossing," by various cell fusion techniques, a cell line deficient in an enzyme of the purine salvage pathway (usually HGPRT) with another line which lacks the main enzyme of the pyrimidine salvage pathway [thymidine kinase (TK)]. In the presence of a folic acid analog (aminopterin or methotrexate) which blocks the neosynthesis routes to nucleotides, and provided that the medium is supplemented with a purine and a pyrimidine (for instance, hypoxanthine and thymidine), hybrid cells only are able to grow, by complementation, in this so-called HAT medium, not the parental ones. Such "auxotrophic" lines are, theoretically, easy to isolate, on the basis of their resistance to various guanine and thymidine analogs, respectively.

Unfortunately, several difficulties were encountered when adapting this method to *Drosophila* cells, the major one being the "conditional" activity of HGPRT (see above) and the resulting spontaneous resistance of most cell lines to purine analogs.

Only Wyss (1979) has described a successful "semiselective" system based on the same general principles, that is, the use of an adenine salvage-deficient variant, and could thus recover hybrid cells.

A subline from *Kc* cells was adapted to grow in a semidefined medium (called *ZH1%,* see Chapter 1) in which the yeast extract was depleted of purines; thus, after blocking the endogenous nucleotide synthesis with methotrexate, Wyss (1977) tested the capacities of various purines and pyrimidines to support cell proliferation. Thereafter, using adenine analogs (by successive resistance to 6-methylpurine and diaminopurine), he could isolate, albeit in conditions poorly explained, a few clones (especially *MDR3*) that were deficient in adenine salvage and remained so, even after several months without selective agent. Their APRT activity was indeed hardly detectable (less than 1% of the value measured in the parental line).

Taking advantage of the fact that two other lines (Schneider's *S3* and Dübendorfer's line *D1*) were more nutritionally demanding and unable to proliferate in the semisynthetic medium, Wyss could, after fusion with polyethylene glycol (see Chapter 3, Section VIII.B), recover cells that were hybrids of *MDR3* and either *S3* or *D1* cells in a selective medium [named TAM and consisting of ZH1% supplemented with methotrexate (10^{-6} M), adenine, and thymidine (10^{-4} each)].

The hybrid nature of growing cells was demonstrated with several isoenzymic markers and by comparing the peptide synthesis patterns of the cells with those of parental lines (Berger *et al.*, 1980). The system also allowed an analysis of the response to ecdysteroids of hybrid cells between hormone-sensitive and resistant lines (Wyss, 1980a) (see Chapter 8).

2. *Methotrexate-Resistant DHFR and* phGCo *Vector for Cell Transformation*

Current methods of gene transfer make it possible to confer on a few cells in a culture the capacity to resist some growth-blocking drug, or to utilize some unusual metabolite. The acquisition of such a "dominant" trait by the transformed cells is a convenient basis for an efficient selection (see Chapter 9).

It is well known that folic acid, in its activated form, tetrahydrofolate, participates as a coenzyme in several major steps of purine and pyrimidine biosynthesis. Therefore, its analog methotrexate, by blocking the conversion of dihydrofolate to tetrahydrofolate, normally accomplished by the enzyme dihydrofolate reductase (DHFR), can rapidly halt cell proliferation. The effective concentration of methotrexate for *Drosophila Kc* cells is between 10^{-6} and 5×10^{-5} (Wyss, 1977; Becker, 1978). It is noteworthy, however, that Mosna *et al.* (1984) could, using serial selection, isolate two sublines resistant to 2×10^{-3} and 2×10^{-5} M methotrexate, respectively.

With the hope of cloning the *Drosophila Dhfr* gene from a cell line in which it would be amplified (which is the traditional approach), Hao *et al.* (1994) selected methotrexate-resistant cells over a 4-year period from *S3* line cells. Unexpectedly, their highly resistant subline (200 μM MTX) did not overexpress DHFR nor its messenger. They succeeded, nevertheless, in cloning the gene of the *Drosophila* enzyme. Its organization differs significantly from that of its mammalian counterpart (its kinetic properties had previously been studied by the same group, and it was found to show characteristics of both prokaryotic and eukaryotic DHFRs (Raucourt and Walker, *1990*).

By putting the structural gene for a methotrexate-resistant DHFR enzyme of prokaryotic origin under the control of a strong *Drosophila* promoter (e.g., a Long Terminal Repeat [LTR] of the retrotransposon *copia*), Bourouis and Jarry (1983) could confer a high level of methotrexate resistance (up to 4 μg/ml) on transformed *Kc* cells. It was, incidentally, the first report of gene transfer into cultured *Drosophila* cells. Their selection plasmid (named *pHGCo*; see Chapter 9, Fig. 9.5A), in fact directly derived from the original construct of O'Hare *et al.* (*1981*), has been widely used for introducing nonselective genes by cotransfection.

3. Ecogpt Selection System

Escherichia coli possesses a xanthine guanine phosphoribosyltransferase (called *Ecogpt*, or *gpt* for short) that enables the bacteria to make use of the purine base xanthine, whereas animal cells are unable to do so in their salvage pathways.

Sinclair *et al.* (1983; Burke *et al.* 1984a) constructed several vectors in which the *gpt* gene was positioned downstream of the two tandem LTRs of extrachromosomal *copia* circles (see Chapter 9, Fig. 9.5B).

Transformed *Drosophila melanogaster D1* cells or *Drosophila hydei* cell lines could grow in a selective medium (*MX3*) containing methotrexate (for blocking endogenous synthesis) and in which xanthine was the sole purine source.

This system was useful for cotransfecting, for instance, nonselective heat-shock genes (Burke *et al.*, 1984b; Sinclair *et al.*, 1985a) (see Chapters 7 and 9).

4. Cell Resistance to PALA, Amplification of rudimentary Locus

It has long been known that *rudimentary*, one of the earliest mutations to be described in *Drosophila melanogaster* (Morgan, *1910*), is associated with various disturbances in the metabolism of pyrimidines. As a matter of fact, this locus *r* in *Drosophila* (*15A1* on chromosome X) encodes a macromolecular entity (called CAD) carrying the three enzymatic activities that catalyze the first three reactions of the pyrimidine biosynthesis route (Jarry and Falk, *1974*), namely the carbamoyl-phosphate synthetase

(CPSase), the aspartate transcarbamylase (ATCase), and the dihydroorotase (DHOase) activities.

In S30 supernatant from *Kc* cells, the three enzymatic activities were found to be two to three times higher than in corresponding extracts from *Drosophila* second instar larvae. Moreover, the purification of ATCase, via precipitation with ammonium sulfate and chromatography on hydroxylapatite columns, resulted in the concomitant purification of DHOase and CPSase. The molecular mass of this complex, determined in sucrose gradients, was close to 800,000 Da. The complex was unstable and spontaneously dissociable at low temperature into at least two units possessing ATCase and DHOase/CPSase activities, respectively (Jarry, 1976).

Starting from embryos of various *rudimentary* alleles, Bernhard and collaborators (Regenass *et al.*, 1979; Bernhard *et al.*, 1980) were able to establish *in vitro* several pyrimidine auxotrophic cell lines (see Chapter 3, Section III). These lines grew, at usual rates, in regular *M3* medium, but not in a minimal medium (i.e., without yeast extract), unless it was supplemented with dihydroorotate (5 m*M*).

The auxotrophic phenotypes of these lines correlate well with enzymic data: if CPSase was found to be normal in most r^1 cell lines, the levels of DHOase activity was drastically reduced (3 to 7% of the wild type) and ATCase activity was only 20 to 60% if compared with the wild type. In the r^{36} cell line, which could also be saved by addition of carbamoylaspartate (5 m*M*) to the minimal medium, the ATCase activity was reduced to only 10%, while the two other enzymes functioned at their normal levels.

In order to study possible gene amplification in *Drosophila,* Giorgi *et al.* (1983; Laval *et al.*, 1986; Azou and Laval, 1993) isolated, by serial selection, different cell sublines resistant to PALA [N-(phosphonoacetyl)-L-aspartate]. This specific inhibitor of the second enzyme of pyrimidine synthesis, i.e., ATCase, is well known to induce, in mammalian cells, an amplification of the gene coding for the multifunctional protein CAD.

The selection could not proceed beyond a concentration of 10 m*M* PALA. The three independent resistant sublines (from an Oregon-R line, 77 OM3) could be maintained for at least 18 months in the absence of the drug and still display a coordinated increase in the levels of their ATCase, CPSase, and DHOase activities (5- to 12-fold higher, compared to the parental line).

The physical and kinetic parameters of the aspartate transcarbamylase were apparently unmodified in resistant cells, and it could be clearly demonstrated, by immunotitration and immunoblotting, that this PALA-resistance was indeed due to an overproduction of the multienzymatic protein CAD. Moreover, Southern blot analysis showed that this could be accounted for by an amplification (about five to six times) of the *rudimentary* gene and its surrounding regions (at least 90 kb).

Nevertheless, karyological studies did not reveal any typical chromosomal alteration or the presence of additional minute chromosomes, as is usually the rule for gene amplification in mammalian cells.

Note: Provided that the methotrexate resistance which Mosna *et al.* (1984) obtained with a serial selection was also correlated with gene amplification, which still remains to be proved, it is worth noting that they did not observe any characteristic karyological alterations in their resistant *Drosophila* cell sublines either.

II. RIBOSOMAL RNAs

Eukaryotic ribosomes are composed of RNAs (approximately half of their total weight), in close association with a large variety of polypeptides. Their small subunit contains one molecule of 18S RNA, whereas the larger one comprises not only the so-called 28S RNA (more precisely, 26S RNA in *Drosophila*) and its hydrogen-bonded derivative(s) 5.8S RNA (with an additional 2S RNA in *Drosophila*; see Jordan, 1974), but also one molecule of 5S RNA.

Ribosomal genes occur in many copies per haploid genome, in order to meet the considerable cell requirements, especially during certain developmental periods. They are organized in tandem arrays along the chromosomes, each repeat being separated by spacer DNA. The sequences encoding 26S and 18S RNA molecules are closely linked and belong to a common transcription unit. Thus, the specific RNA polymerase I synthesizes a precursor pre-rRNA from which a complex processing will yield mature rRNAs. Their chromosomal site constitutes what is called a nucleolar organizer region (NOR). The 5S RNA genes are clustered in their own separate multigene family.

In *Drosophila melanogaster,* there are some two-hundred forty 18S/26S genes per diploid nucleus (i.e., approximately 0.25% of nuclear DNA), with significant variations between fly strains (not to mention the *bobbed* mutants which correspond to partial deletions of rDNA). These multigenes are distributed into two clusters: one on the X chromosome (in the proximal centromeric heterochromatin) and the other on the short arm of the Y chromosome (see Chapter 3).

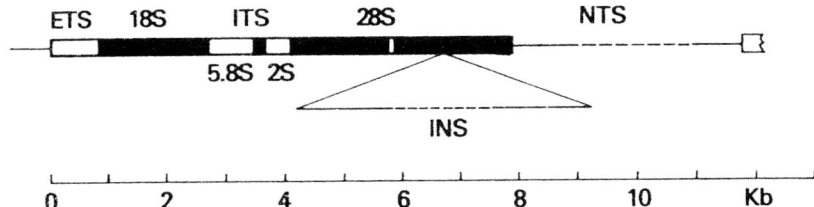

FIGURE 5.2   Map of repeating units of ribosomal genes (18S-26S) of *Drosophila melano-gaster*. Coding regions are shown as solid bars. Transcribed spacers, i.e. regions transcribed into nonconserved segments of pre-rRNA, are shown as open bars; they are distinguished as external (ETS) and internal (ITS); the latter one is thought to include the coding regions for 5.8S and 2S (this last form is only present in *Drosophila*). Nontranscribed spacers (NTS), which separate the repeating units, are shown as lines; NTS and insertions (INS) occur in variable length as indicated by broken lines. From Long and Dawid (*1980*). Reproduced, with permission, from the Annual Review of Biochemistry, Volume 49, © 1980, by Annual Reviews, Inc.

Transcription units (8 to 13 kb) alternate with spacer regions traditionally referred to as nontranscribed spacers (NTS) (see below), to distinguish them from transcribed but nonconserved segments that will be processed out from mature rRNAs and are known as transcribed spacers (subdivided into external ETS and internal ITS). In fact, within this latter group, Jordan and Glover (1977) were able to locate the sequences that are complementary to 5.8S and 2S rRNAs. Moreover, in about half of these repeating units of both X and Y chromosomes, the 26S RNA coding sequence is interrupted by an insertion of various types,* ranging from 0.5 to 6 kb) (Fig. 5.2). It could be shown that such interrupted genes are expressed either very poorly or not at all.

As in all other organisms, *Drosophila* 5S RNA genes form a quite distinct cluster, located on the right arm of chromosome 2 (in the region 56E-F). There are about 160 genes per haploid genome, interspersed with nontranscribed spacers, the modest length heterogeneity of which would be accounted for by variable numbers of short repeating blocks.

The possibility, presented by the cultures of *Drosophila* cells, of purifying large amounts of the various species of rRNA and of radiolabeling them to a high activity was quite helpful for defining the functional organization of ribosomal genes, as summarized above (see, for instance, Glover *et al.*, 1975; Glover, 1977; Glover and Hogness, 1977; White and

* They correspond to Line-like transposable elements (Jakubczak *et al.*, *1990*). See Chapter 10.

Hogness, 1977; Dawid and Botchan, 1977). Incidentally, Jordan (1974) was thus able to discover and then to sequence (Jordan *et al.*, 1976), a novel small ribosome RNA component (2S RNA) which seems to be peculiar to *Drosophila*. Similarly, having unraveled the nucleotidic sequence of 5S RNA from *Kc* cells (Benhamou and Jordan, 1976), the same group extended the analysis to the various tissues and developmental stages of *Drosophila* (Benhamou *et al.*, 1977). It was established that this latter ribosomal RNA species remains strikingly homogeneous throughout its development, which contrasts with the situation reported in *Xenopus* in which two separate sets of genes encode an oocyte type and a somatic type of 5S RNA.

The use of cultured cells was particularly valuable for monitoring the maturation steps of rRNA transcripts as well as for experimental analysis of the *cis*-acting sequences involved in transcriptional regulation.

A. Transcriptional Control of Ribosomal RNA Genes

It is worth remembering that, in eukaryotic cells, a substantial fraction of their transcriptional activity, perhaps half of all transcription, corresponds to the synthesis of rRNA and that this activity requires the intervention of a specific RNA polymerase, named polymerase I. Thus, the determination of the *cis*-acting DNA sequences involved in its regulation is of particular interest, and *Drosophila* rRNA genes have been among the most extensively studied models.

At first, a core promoter could be defined in a transcription system reconstituted *in vitro*. Kohorn and Rae (1982a,b) developed, from extracts of *Kc* cells, a cell-free *Drosophila melanogaster* RNA polymerase I transcription system. It is reliable, in the sense that transcription *in vitro* starts at (or within one nucleotide of) the normal initiation site *in vivo* and also that there is no observable polymerase II and III activity. Moreover, no detectable transcription was obtained with *Drosophila virilis* templates, which is in good agreement with the general observation that regulatory regions of ribosomal genes differ significantly from one species to another, so that rDNA genes are only poorly expressed in heterologous systems. The extract, however, could not discriminate between rDNA repeat units with or without an insertion in the 26S coding sequence, even though interrupted genes do not seem to be normally expressed *in vivo*.

A minimal rDNA segment (several tens of base pairs long on either side of the pre-RNA start site), capable of directing proper polymerase I-dependent transcription, could thus be defined.

Recombinant rDNA–CAT genes were constructed, in which the first 34bp of a *Drosophila* rDNA transcriptional unit, preceded by various lengths of nontranscribed spacer sequences, was joined to the "reporter" gene CAT. Their transient expression in transfected S2 cells confirmed that the −180/+34 segment is sufficient to promote a correct transcription, although the presence of a complete NTS greatly enhanced (about 10 times) the transcriptional efficiency (Grimaldi and Di Nocera, 1986). CAT transcripts could even be 20-fold more abundant than those driven, in similar conditions, by a *copia* promoter, even though this promoter is one of the most potent promoters in *Drosophila* (see Chapters 9 and 10). Moreover, it is noteworthy that, whereas the hybrid transcripts were easily recognizable with adequate probes, no CAT enzyme activity was measurable, as is usual with this reporter gene, after 48 h. It must be remembered that rRNA transcripts are synthesized by polymerase I and, even though it is not known whether the transfected hybrid DNA templates were transcribed in the nucleolus, like the endogenous ribosomal genes are, their transcripts might be poor substrates for the necessary posttranscriptional messenger modifications (namely capping and polyadenylation).

It was then noted that the *Drosophila melanogaster* rDNA spacer (so-called nontranscribed spacer) is composed of three adjacent repetitive regions (I to III) (see Fig. 5.3). Each region consists of a tandem array of homologous repeats, which are 95, 330, and 240 bp long, respectively.

In a more systematic analysis, Grimaldi and Di Nocera (1988) established that, while regions I and II play minor roles, the polymerase I transcriptional efficiency was linearly correlated with the number of

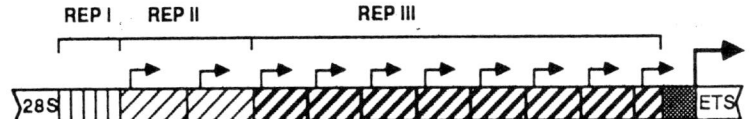

FIGURE 5.3 Structural organization of the *Drosophila melanogaster* rDNA "nontranscribed spacer" (NTS) region. Rep I, II, and III refer to three repetitive regions, composed of repeats of 95, 330, and 240 bp, respectively. Region III contains seven integer repeats and a truncated one preceding the pre-rRNA promoter region (filled box). ETS means "external transcribed spacer." Thick and thin arrows indicate the pre-rRNA start site and promoter duplications within 330- and 240-bp repeats, respectively. From Grimaldi *et al.* (1990). With the kind permission of Professor Di Nocera and the American Society of Microbiology.

240 bp modules of region III remaining in the constructs. Moreover, these repeats act in an orientation dependent manner. This stimulating effect is abolished if these repeats are oppositely oriented in the construct. Now, it should be pointed out that each repeat of regions II and III includes a sequence homologous to the pre-RNA core promoter. As a matter of fact, Miller *et al.* (1983) in embryos and then Murtif and Rae in *Drosophila melanogaster* and *D. virilis* cultured cells had previously demonstrated (through S1 nuclease protection experiments and Northern analysis) that such multiple promoters of the so-called nontranscribed spacer support the production of transcripts. Spacer transcripts could be identified in nuclear RNA, but not in the cytoplasm, and they may be 100-fold rarer than normal pre-RNA transcripts.

This possibility of spacer transcription is in some way connected with the report by Tantz and Dover (1986) that transcription termination of rRNA in *Drosophila* cells does not always occur just beyond the end of the 26S coding sequence; transcription might proceed through the spacer, up to the second 240 bp repeat.

Using artificial spacer arrays, built from various segments of the 240 module, Grimaldi *et al.* (1990) could finally demonstrate that the sequence of each repeat that is able to enhance transcriptional activity strictly coincides with the copy of the gene core promoter that it contains.

In conclusion, it is assumed that the multiple promoter copies carried by the upstream NTS repeats might help to attract and concentrate Pol I as well as transcription factors, some of them occurring perhaps in limited concentration, in order to convey them efficiently down to the true gene promoter.

Notes: (a) The fact that 18S/26S RNA genes carrying a Line-like insertion in their 26S coding sequence (i.e., about half of the gene copies in *Drosophila melanogaster*) do not seem to be expressed *in vivo,* or only at a very low level, raises a basic problem. It provided Dawid and Reppert (1986) with the opportunity for testing the hypothesis according to which activatable genes are in a special chromatin configuration. When cultured S2 cells were exposed to chloroquine (25 to 50 mM), a DNA intercalating drug that introduces some torsional stress, the abundance of type I insertion (but not type II) transcripts was indeed dramatically increased (up to 60-fold), but without any substantial modification of total RNA synthesis.

(b) The availability of series of established cell lines with different karyotypes (various levels of ploidy and, in particular, different sex chromosome numbers; see Chapter 4) led Kubaneishvili *et al.* (1983) to study the correlation between the number of ribosomal RNA genes and the rate of rRNA synthesis, a problem widely documented in *Drosophila* (especially in *bobbed* mutants and X0 flies):

Diploid and polyploid cell lines retained the usual percentage of rDNA (0.3% of total DNA), even though, according to the rather surprising data of these authors, the haploid genome of cultured cells contained four to five times more DNA than the amount usually reported for *Drosophila melanogaster* somatic and diploid cells(!).

On the other hand, in X0 cells, some compensation occurred, as had already been established in X0 flies, and the normal diploid number of ribosomal genes was restored.

Within the limits of the technique used for characterizing rRNA, the rate of synthesis of the latter was found to be proportional to the ploidy level and, more precisely, to the number of X chromosomes.

In X0 cells, however, and in spite of the recovery of a normal ribosomal gene number, the rate of rRNA synthesis was 50% lower than in diploid cells, which is in good agreement with Tartof's assumption that additional ribosomal genes are inactive in X0 flies.

(c) Let us recall that Donady *et al.* (1973) showed, in primary cultures (see Chapter 2, Section VI.2), that neurons and myocytes can differentiate normally in *Drosophila* embryonic cells hemizygous for the X chromosome and containing, after appropriate crosses, a single nucleolar organizer region or none at all. It is clear that these embryonic cells had inherited functional ribosomes from maternal oogenesis and that the situation of such *Drosophila* embryos completely lacking rDNA was very similar to that of the "anucleolar" mutants of the toad *Xenopus*.

B. Processing of 18S/26S Gene Transcripts

Kc cells grown in suspension culture could be labeled with ^{32}P to quite high activity: up to 1 mCi of ortho[^{32}P] phosphate per ml was added to low phosphate *D22* medium (i.e., deprived of its usual sodium phosphate component; see Chapter 1, Appendix 1.A). It was then verified that cell doubling remained unaffected during the first 24 to 36 h. Jordan and collaborators (Jordan, 1975; Jordan *et al.*, 1976; Jordan and Glover, 1977) used pulse experiments to characterize the successive steps of the maturation of 18S, 28S, 5.8S, and 2S rRNAs from their common 38S precursor (see Fig. 5.4).

The processing of the pre-rRNA, detectable in small amounts at the end of the 90-min labeling period, gives rise to 18S and 26S molecules. The 26S molecule appears first as a continuous polynucleotide chain, but, while it is still in the nucleus, a central cleavage induces the "hidden break" that has been widely reported in its mature structure. Then, the

DNA

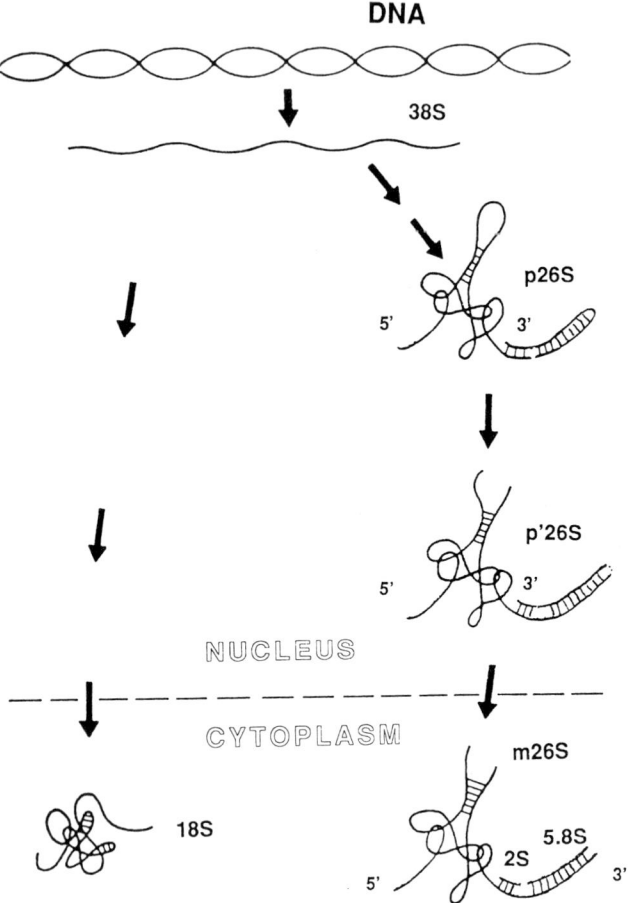

FIGURE 5.4   Maturation of *Drosophila* 18S and 26S rRNAs. Redrawn from Jordan *et al.* (1976) and Jordan and Glover (1977).

molecule is transported into the cytoplasm where late, and apparently simultaneous, cleavages generate 5.8S and 2S RNAs. Both small derivatives remain hydrogen-bonded to the 3' extremity of 26S RNA. Because the shorter one seems to occur only in *Drosophila*, the two together (i.e., 120 plus 30 nucleotides) possibly correspond to the classical 5.8S RNA (150 to 160 nucleotides) found in other eukaryotes.

C. Maturation of 5S rRNA

The occurrence of a precursor, with 15 additional nucleotides at the 3' end, was revealed by its abnormal accumulation in heat-shocked *Kc* cells (Rubin and Hogness, 1975). Since the formation of 5.8S is also reduced, there would seem to be some general block in rRNA processing during heat-shock (see Chapter 7).

Dingermann *et al.* (1981) devised a transcription system *in vitro*, with cell-free extracts from *Kc* cells. It enabled Reiser and collaborators (Preiser, 1990; Preiser and Levinger, 1991a,b; Levinger *et al.*, 1992) to investigate the effects of stem I and loop A transversions, selected additions and deletions, on the processing *in vitro* of 5S rRNA. *Note:* Preiser *et al.* (1993) have characterized, in *Kc* cells, a 50 kDa protein, binding specifically to the 3' terminus of *Drosophila* pre-5S RNA, which may protect it for customary processing and transport.

D. Morphofunctional Organization of Nucleolus of Cultured Cells

The nucleolus, where ribosomes are manufactured, develops around the chromosomal sites of the arrays of rRNA genes [nucleolar organizer regions (NORs)]. Since it is the expression of their transcriptional activity, this organelle may be considered to be, to some extent, analogous to the puffs of polytene chromosomes. Yet, in addition to the DNA templates and polymerase I transcription complex, it comprises many ribosomal protein components converging from other parts of the cell.

In phase-contrast observation of cultured *Drosophila* cells, there is typically a single large nucleolus which appears as a dark pupilla in the clear nuclear compartment and which may cover up to one-third of the nuclear area (see, for instance, fig. 2.4 in Chapter 2).

Knibiehler *et al.* (1982a,b) carried out ultrastructural studies on *Kc0* cells, based on a combination of classical electron microscopy with autoradiography, after tritiated uridine incorporation.

Always attached to the nuclear envelope by its associated chromatin, the nucleolus displays various morphologies, depending on the cell cycle and conditions of culture. During normal exponential growth, two main components, albeit intermingled, can be distinguished: a multilobed fibrillogranular structure is flanked by two (or more) condensed fibrillar masses of heterochromatin. All around, there are clumps of nucleolus-associated chromatin, in which, primarily toward the nuclear border, the internal core extends a few large ramifications becoming more and more granular (Fig. 5.5).

Enzymatic digestions (with Pronase, DNase, RNase), as well as pulse labeling with tritiated uridine, demonstrated that the central part of the nucleolar core corresponds to the location of the NORs and to the transcriptional activity of their rRNA genes, although there are no so-called clear "fibrillar centers" as reported in a variety of organisms (see minireview in Sommerville, *1985*). Then, proteins bind to the transcripts in a ribonucleoprotein network that appears more and more granular at the

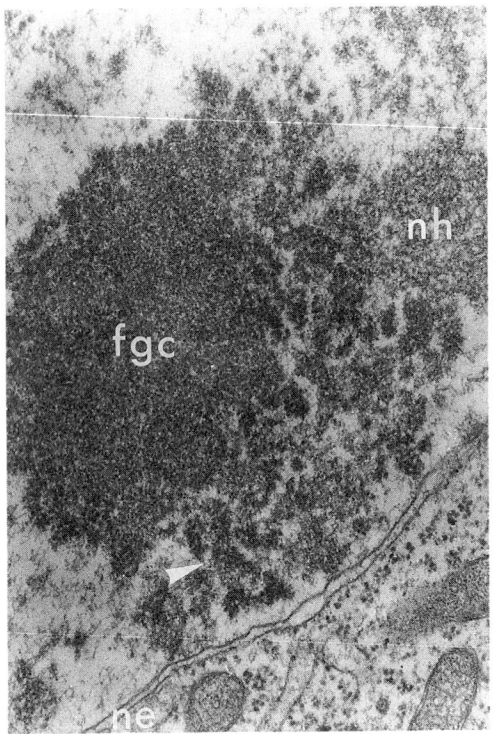

FIGURE 5.5   Nucleolar ultrastructural morphology in *Kc0* cells in exponential growth. Mass(es) of heterochromatin (nh), in relation with the nuclear chromatin flank the large central multilobed nucleolar core (fgc); this latter extends ramifications (arrowhead) toward the nuclear envelope (ne). From Knibiehler *et al.* (1982); courtesy of Professor R. Rosset, Univ. Marseille-Luminy, France.

periphery (which is the expression of the maturation of preribosomal particles). On the other hand, the dense homogeneous flanking masses, containing DNA fibers but not RNA molecules, should correspond to the inactive heterochromatin segments that surround the nucleolar organizer on the X chromosomes of *Drosophila* (it should be recalled that the original *Kc* line had a diploid female karyotype).

This functional organization of the nucleolus was confirmed by experiments of cell stimulation with ecdysteroid hormones. It was previously established (Rosset, 1978; see also Chapter 8) that β-ecdysone induces, in *Kc* cells, a transient increase of RNA synthesis. Autoradiographic observations, after pulse labeling with [³H]uridine, demonstrated that the nucleolar response is the most prominent (Knibiehler *et al.*, 1982b): the nucleolar size of hormone-treated *Kc0* cells enlarges significantly, and its uridine incorporation, between 0 and 8 h, might correspond to a 3-fold increase in the level of rRNA synthesis. Concomitantly, the maturation and transport processes were stimulated, as clearly shown by an obvious development of the nucleolus core and its fibrillogranular ramifications. *Note:* It should be recalled (see Chapter 4) that an improved silver staining method, devised by Goodpasture and Bloom (*1976*), allows detection of active nucleolar organizers on chromosomes. Thereby, Privitera (1980) and Kubaneishvili *et al.* (1983) showed that all the rRNA gene clusters take part in the rRNA synthesis of cultured *Drosophila* cell lines, irrespective of their number in the cell.

III. MITOCHONDRIAL GENOME

As in all other multicellular animals, *Drosophila* mitochondria contain their own decentralized genome in the form of one or several copies per organelle of a circular molecule (in this case a covalently closed duplex circle of approximately 19,000 base pairs). This mtDNA codes for only a small part (about 20%) of mitochondrial enzyme complexes, and also for specific ribosomal RNAs and transfer RNAs. So, there is an endogenous protein synthesis system* with a distinctive genetic code. All other structural, enzymatic, and regulatory polypeptides of mitochondria depend on nuclear genes and are synthesized by cytoplasmic ribosomes before being transported into the organelles.

* There is a specific RNA polymerase (*pol III*), although encoded by a nuclear gene. Likewise, a special type of DNA polymerase (called γ) is involved in mitochondrial DNA replication.

The mitochondrial (mtDNA) genomes of several *Drosophila* species (*D. yacuba:* Clary and Wolenstholme, *1985; D. melanogaster:* de Bruijn, *1983* and Garesse, *1988*) have been completely or partially sequenced and a high degree of homology was observed. The general organization (Fig. 5.6) is rather similar to that found in mammalian mtDNAs, although the arrangement of the genes and their distribution on the two DNA strands differ significantly. Moreover, there exists a large A + T-rich region (one-quarter of the mt genome in *D. melanogaster*), apparently without coding capacities but containing the origin of replication for both strands. Nevertheless, (a) the genes are strikingly similar to those identified in mammals (Chomyn *et al.* 1985), namely, two ribosomal genes (called 12S and 16S), 22 tRNAs and 13 structural genes corresponding to

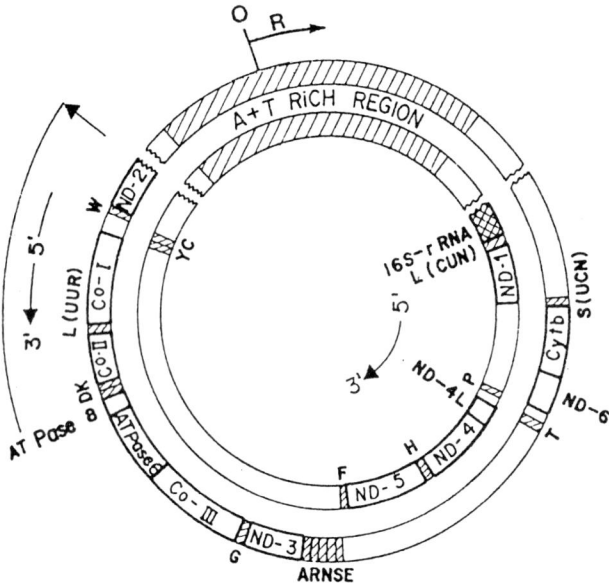

FIGURE 5.6 Map of the mitochondrial genome of *Drosophila melanogaster.* The protein genes are boxed, the tRNA genes hatched and the 16S rRNA gene cross-hatched at their corresponding position in the major coding strand (outside) or minor coding strand (inside). The (A + T)-rich region is also hatched and the different tRNAs identified with the one letter amino acid code; when necessary, the corresponding codon family is indicated in parentheses. The origin of replication, O, is arbitrarily located in the middle of the A + T-rich region. From Garesse (1988). Reproduced with the permission of the Genetics Society of America.

subunits of the enzyme complexes that are associated with the inner mitochondrial membrane; (b) they are intronless and tightly packed, i.e., with very short or no spacing between them; and (c) clusters of tRNA genes separate these protein-coding sequences.

A. Transcription Map of Mitochondrial DNA and Characterization of Mitochondrial Proteins

Cultured cells are very convenient material for testing the effects of various drugs, particularly because their concentration can be more easily controlled. Actinomycin D is well known for its blocking of normal nuclear transcription, and chloramphenicol is a specific inhibitor of mitochondrial synthesis.

Thus, *S2* cells treated with 5 μg actinomycin D per ml continued to incorporate tritiated uridine, albeit at a low level (Spradling *et al.*, 1977; Bonner *et al.*, 1978). Two different approaches—(a) isolation of the mitochondrial fraction after inhibiting nuclear RNA synthesis by camptothecin or after digestion of cytoplasmic RNA with RNase; (b) Northern hybridization to nuclear or mitochondrial DNA—were used to establish clearly that 12 to 13 electrophoretically separable poly(A)$^+$ RNAs were indeed products of the mitochondrial genome. Moreover, their synthesis was not affected by heat-shock (see Chapter 7, Section III.A: the so-called "group B" RNAs).

In parallel with extracts from flies or embryos, wide use was made of cell cultures for preparing mitochondrial DNA or highly labeled mtRNAs, thereby determining the transcription map of *Drosophila melanogaster* mtDNA, either by observation of RNA—DNA hybrids in the electron microscope, or by hybridization of purified transcripts to defined mtDNA fragments generated by restriction enzyme cleavage (Kuskas and Dawid, 1976; Bonner *et al.*, 1978).

By differential centrifugation onto a Percoll gradient, the mitochondrial fraction from *Kc0* cells could be isolated with a high degree of purification; measures of catalase and acid phosphatase activities may control any possible contamination by peroxisomes and lysosomes, respectively (see Appendix 5.A). With special care being taken for solubilization, because most proteins of the internal membrane are hydrophobic, bidimensional electrophoresis revealed about 300 to 400 polypeptides (Stepien, 1988), which is a reasonable number for the total components of mitochondria.

Moreover, a system of polypeptide synthesis *in vitro* in isolated *Drosophila* mitochondria was devised by Alziari *et al.* (1980; Stepien, 1988) (see Appendix 5.B). Ten major synthesized polypeptides, with molecular masses ranging from 6 to 40 kDa, and several minor bands were detected. This 1D electrophoretic pattern is very similar to that obtained with isolated mitochondria from embryos or with whole *Kc* cells in which cytoplasmic translation has been inhibited by cycloheximide. Figure 5.7 proposes a tentative identification of these mitochondrial translation products, according to Stepien (1988).

B. Juvenile Hormone and Mitochondrial Functions

Because *Drosophila* cultured cells are useful experimental models for studying the responses of cells to hormones (see Chapter 8), Stepien and collaborators (Stepien, 1988; Stepien *et al.*, 1988; Savre-Train, 1990) have investigated the effects of juvenile hormone (JH) on mitochondrial functions, in *Kc* cells. Even though JH does not elicit any morphological modifications in cultured cells (see Chapter 8, Section VI), metabolic or biosynthetic changes were expected, at the mitochondrial level, because earlier experimental data suggested that the *corpora allata,* the endocrine glands producing JH, stimulate general metabolism and oxygen consumption in insects. Three mitochondrial enzyme activities were monitored: cytochrome oxidase (its complex playing a key role in oxidative phosphorylations), malate dehydrogenase, and citrate synthase (both reflecting the activity of the Krebs cycle). Furthermore, mitochondrial protein synthesis was measured, after pulses of [^{35}S]methionine incorporation, in *Kc* cells exposed or not exposed to cycloheximide.

Juvenile hormone, and especially JH II, increased not only the labeling of mitochondrial proteins, this induction seemed transcriptional, but also the cytochrome oxidase activity (a 35 to 40% increase). If the two tested enzymes of the Krebs cycle remained unaffected, the stimulation might also concern other enzyme complexes of the oxidative phosphorylation pathway, as suggested by preliminary immunological titration of the subunit β of ATP synthase.

It seems paradoxical that higher effects were observed with JH II, since the normal hormone in Diptera is JH III. However, the structures of these two forms are very similar, and their different activities under these experimental conditions might merely be due to a better uptake of JH II by the cells or to its higher resistance to esterases.

Optimal concentrations were very low and ranged from 10^{-12} to 10^{-8} *M,* which is in good agreement with the 10^{-9} *M* physiological hormone concentration found in Diptera (Bownes and Rembold, 1987).

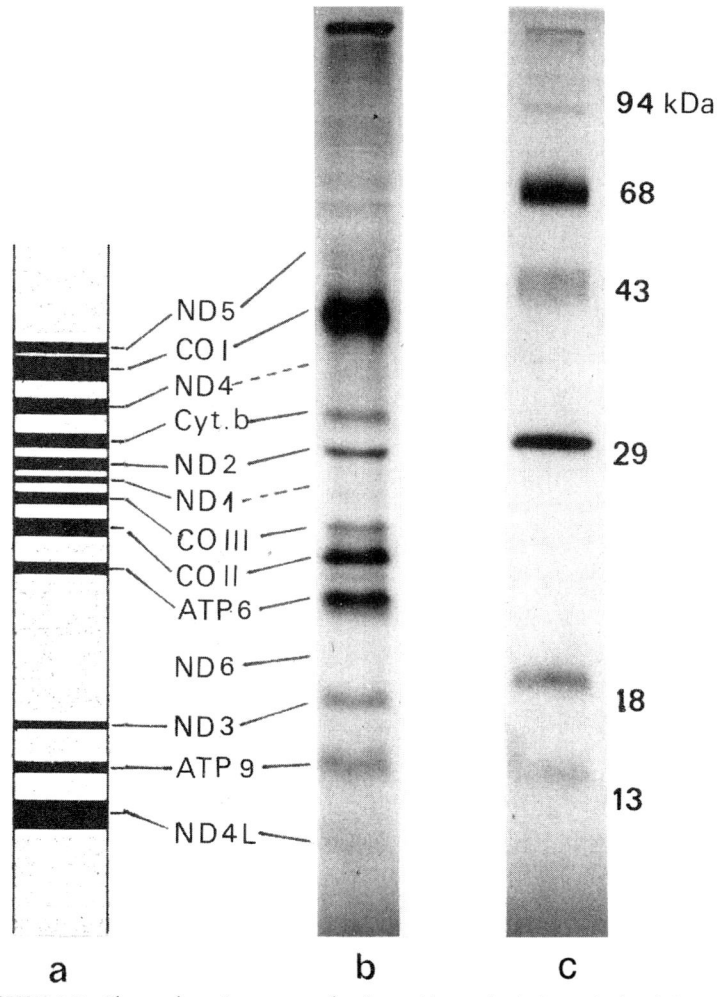

FIGURE 5.7 Electrophoretic pattern of polypeptide synthesis in an isolated fraction of mitochondria from *Kc* cells (b); Identification by comparison with human mitochondria (a) (from Chomyn *et al.* (*1985*); (c) standard markers. ND1–6 and ND4L, Subunits of NADH dehydrogenase; Cyt.b, cytochrome b; CO I–CO III, subunits of cytochrome oxidase; and ATP6 and 9: subunits of ATP synthase. From Stepien (1988). Courtesy of the author.

Kinetics studies (Fig. 5.8) showed that protein synthesis increased as early as the second hour of treatment and preceded, by 4 to 6 h, the increase in cytochrome oxidase activity. Maximal effects were reached after 12 h.

JH analogs, such as methoprene (ZR-515) and, to a lesser degree, hydroprene, were found to be active, but only at much higher (1000-fold) concentrations. On the other hand, JH antagonists (for instance, precocene) that are known to interfere with hormone biosynthesis in the organism, had no effect in this "target" cell system.

Incidentally, treatment with 20-hydroxyecdysone did not modify any of the monitored parameters, which is in striking contrast with other reports of a significant increase in the respiration rate of ecdysteroid-stimulated *Kc* cells (Peronnet *et al.*, 1986; see Chapter 8, Section II.E).

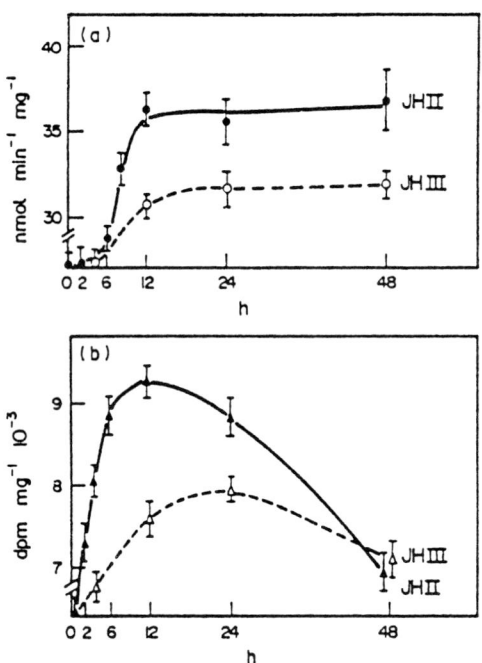

FIGURE 5.8   Effects of JH II and JH III on *Drosophila* mitochondrial activities. Cytochrome oxidase activity (a) and mitochondrial protein synthesis (b) were measured after 2 to 48 h treatment of *Kc0* cells with 10^{-9} M JH II or III. Reprinted from Stepien *et al.* (1988). Copyright (1988), with kind permission from Elsevier Science Ltd., The Boulevard, Langford Lane, Kidlington OX5 1GB, U.K.

The fact that JH had no detectable effect on isolated mitochondria, whereas mitochondrial biosynthesis could be increased in such an isolated mitochondria system by incubation in a postmitochondrial supernatant from JH-treated cells, strongly implies the intervention of some cytoplasmic factor and raises the crucial problem of the necessity for regulatory interactions between nuclear and mitochondrial genomes.

As a matter of fact, careful 2D-electrophoretic analyses and comparisons between cells stimulated or not stimulated by JH, and in which proteins had been labeled in the presence or absence of cycloheximide, showed that two polypeptides appeared reproducibly to be more strongly induced. After both Coomassie blue staining and autoradiography, the two corresponding spots were found to be increased by 100% (as estimated by densitometry). It could be demonstrated that they were encoded by nuclear genes. Unhappily for the simplicity of the interpretation, it is improbable (according to their molecular weight), that these two polypeptides correspond to cytochrome oxidase, even though they seemed to be located in the mitochondrion internal membrane.

C. Nuclear Genes Encoding Mitochondrial Enzymes

During an unsuccessful attempt to isolate and purify *Drosophila* cytochrome oxidase (Planques, 1988), it was observed that polyclonal antibodies, prepared against the whole complex of cytochrome oxidases from ox heart or rat liver, were able to recognize several proteins in the internal membrane of *Drosophila* mitochondria. Taking advantage of this implied cross-reactivity, the same laboratory (Stepien, 1988; Savre-Train, 1990) attempted to clone the nuclear genes encoding subunits of the *Drosophila* cytochrome oxidase, by screening a cDNA expression library from *Kc* cells, that had been built in the vector λgt11.* Several groups of "positive" clones, which supposedly corresponded to individual subunits of the complex, were recovered, but, surprisingly, further molecular analysis of one of them did not reveal any significant homology between its translation product and mammalian counterparts.

IV. DNA REPAIR AND SENSITIVITY TO MUTAGENS

Long nucleotidic chains are a target for a variety of physical and chemical agents, possibly present in the environment. Generally, to each

* cDNAs were synthesized from 10 μg of poly(A)⁺ RNA extracted from *Kc* cell cultures.

particular type of attack a specific kind of damage corresponds: for instance, deamination of base or base loss after heat exposure, single-strand breaks caused by X-rays, or formation of pyrimidine dimers following UV treatment. In order to recover the crucial integrity of their genome, all cells must be equipped with a complex array of DNA-repair mechanisms.

The sophistication of its genetics and especially the ease with which many repair-deficient mutants can be isolated in *Drosophila*, have made this organism, once again, an important model for the study of DNA repair functions in higher eukaryotes. To date, more than 30 genes seem to be involved in the different repair processes (see reviews by Boyd *et al.*, 1980, 1983, *1987*, 1988; Table 5-II) and, quite interestingly, some mutants display features which are comparable to those observed in several human hereditary repair-deficiency syndromes.

After mutagenization, three main screens were used to identify various repair deficiencies in *Drosophila,* and these criteria may sometimes overlap: (a) a small proportion of meiotic (*mei*) mutants, selected for a severe reduction (to about 8% of the control levels in *mei-9a*) in the genetic recombination, are actually defective in the DNA-exchange process itself; (b) mutagen-sensitive (*mus*) mutants show a hypersensitivity to a series of mutagens; (c) mutator strains, with an abnormally high frequency of spontaneous mutations, accompanied or not by chromosome instability, may also potentially lead to the characterization of repair-related genes.

For 20 years, Boyd and his laboratory (University of California, Davis) promoted the utilization of cultured *Drosophila* cells for extensive investigation of the different DNA repair pathways. This culture technique combines the large variety of available *Drosophila* mutants with the battery of biochemical assays developed for mammalian cells and, thereby, valuable comparisons are possible (see Boyd *et al.,* 1988). For practical reasons, most of the work has been conducted in primary cultures from homozygous mutant embryos, although a few cell lines could be permanently established from the most important mutant strains. More recently, Boyd's laboratory also praised the use of short cultures *in vitro* of dissected larval brain ganglia.

A. Identification of Defects in DNA Repair

1. NUCLEOTIDE EXCISION REPAIR (NER)

UV light is known to induce primarily cyclobutane dimers (or CPDs) and a few other photoproducts, resulting in structural distorsions of the

TABLE 5.II Major DNA Repair-Related Genes in *Drosophila*[a]

Repair function	Locus affected[b]	Map position	Hypersensitivity[c]	Meiotic effects	Mitotic chromosome stability
Photorepair	*phr*	2–57		NT	NT
Excision repair	*mei-9*	1–7	HN2, UV, X-rays, MMS, AAF, 4-NQO, BP, MNNG, DEB, EMS	Recombination reduced by 90%	Reduced
	mus201	2–23	UV, MMS, HN2, EMS	None detected	NT
Postreplication repair	*mei-41*	1–53	Hydroxyurea, UV, X-rays, MMS, AAF, 4-NQO, BP, MNNG, aflatoxin	Recombination reduced by 60%	Strongly reduced
	mus205	2–64	UV, MMS	None detected	NT
	mus302	3–45	MMS, HN2, X-rays, UV	None detected	Reduced
	mus310	3–47	MMS, X-rays, UV	None detected	Reduced

[a] From Boyd *et al.* (1987). Reproduced with the permission of the Company of Biologists, Ltd.
[b] Only those loci are included whose mutants exhibit the strongest repair defects.
[c] HN2, Nitrogen mustard; AAF, 2-acetylaminofluorene; BP, benzo[a]pyrene; MMS, methylmethane sulfonate; MNNG, N-methyl-N′-nitro-N-nitrosoguanidine; DEB, diepoxybutane; EMS, ethylmethane sulfonate; 4-NQO, 4-nitroquinoline oxide.

DNA duplex. First unraveled in prokaryotes, excision repair is a multistep enzymatic process: endonucleolytic breaks, adjacent to the modified bases, remove the CPD as an oligonucleotide; the gap is then filled by polymerization of nucleotides complementary to the undamaged strand.

As early as 1973, Trosko and Wilder measured, with two-dimensional chromatography, the pyrimidine dimers induced in wild-type *Drosophila* S2 cells by short-wave UV light (254 nm) and showed that, at low doses, 40% of the CPDs were excised after a 24 h dark incubation (in order to distinguish from the photoreactivating mechanism, see below). The process was not inhibited by caffeine posttreatment.

Two distinct genes, in *Drosophila,* have been implicated in such excision repair processes.

a. mei-9

The two "classical" alleles, *mei-9a* and *mei-9b,* of this locus (on chromosome X, map position 1-7) exhibit a strong meiotic effect; moreover, they are hypersensitive to all classes of chemical or physical mutagens, including X-rays.

The availability of established lines (named *mei-9-1* and *mei-9-2*; see details in Hirschi and Boyd, 1981; see also page 155) made it easier to demonstrate the existence of a deficiency in DNA repair (Nguyen and Boyd, 1977). In fact, a series of studies by Boyd and collaborators clearly established that the capacity of *mei-9* mutants to excise UV-induced pyrimidine dimers from DNA is severely reduced; most other repair functions seem to remain normal.

A method utilizing a specific endonuclease* that incises the DNA at the site of a CPD allows the monitoring, by subsequent sedimentation of the labeled DNA through alkaline sucrose gradients, of the number of endonuclease-sensitive sites, during increasing periods of cell incubation after UV-irradiation (5 Joules[J]/m2). Whereas, in a control *w* cell line, excision repair was complete after 1 day (96% recovery of a normal average molecular weight of DNA was achieved by 23 h, see Fig. 5.9), *mei-9* cell lines showed about a 16-fold reduction in the repair rate of detectable endonuclease-sensitive sites (Boyd *et al.,* 1976). Similar results were obtained in primary cultures.

The previous method could only detect 20% of the induced CPDs; therefore, a more efficient technique (alkaline elution procedure of Kohn *et al.,* 1976) for isolating and measuring the single-strand fragments

---

* A preparation from *Micrococcus luteus,* or purified T4 endonuclease V.

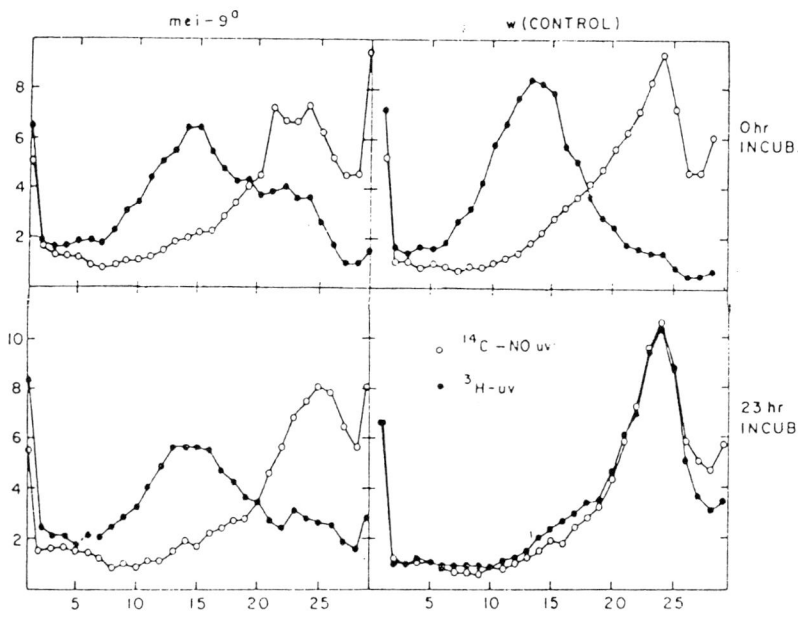

FRACTION NUMBER (SEDIMENTATION →)

FIGURE 5.9   Kinetic analysis of endonuclease-sensitive sites in the *Drosophila* established cell lines *mei-9a* and *w-1*. Cultures of each cell line were labeled overnight with either ³H or ¹⁴C. The ³H-labeled cultures received 5J/m2 of UV radiation and all cultures were further incubated for the specified time. Cells from parallel cultures were then mixed and assayed for the presence of endonuclease sensitives sites. Between 2000 and 7000 cpm of each isotope were layered on alkaline sucrose gradients. Each panel represents a single sucrose gradient. From Boyd *et al.* (1976). With the permission of the Genetics Society of America.

produced by the specific endonuclease, was adapted to *Drosophila* cells (Harris and Boyd, 1980). It confirmed that the *mei-9* mutation reduces the excision of pyrimidine dimers by 92%. Moreover, one can block the resynthesis phase of excision repair with various inhibitors (Cytosine-β-D-arabinofuranoside [araCyt], hydroxyurea), and thus force the accumulation of the transient strand interruptions that normally occur during repair. Now, in contrast with control cells, primary cultures of *mei-9^{D2}* failed to accumulate such breaks, either after UV exposure or during treatment with the carcinogen N-acetoxy-N-acetyl-2-aminofluorene (AAAF, which mimics CPD induction by UV).

Dusenbery *et al.* (1983) used an autoradiographic method to detect "unscheduled DNA synthesis," i.e., unusual [^{3}H]thymidine incorporation due to repair of DNA lesions induced by various insults (UV or X-rays; several mutagenic substances). In 2- to 3-day-old primary cultures and after 4 hours of [^{3}H]thymidine incorporation, the distribution of grains over non-S phase cells was estimated on autoradiographs (in standardized conditions, cells with more than 20 grains being scored as S-phase cells). Control cells, after exposure to UV or other mutagenic agents exhibited a clear dose-dependent increase in their median labeling index, whereas *mei9* or *mus(2)201* cells did not perform such a repair replication.

The cloning method on feeder layer and under agarose (see Chapter 3) was used to estimate the survival of *Drosophila* cells after irradiation with UV or ionizing radiations (Brown *et al.*, 1981). The lethal D_{37} doses (i.e., the dose reducing the population survival to 37%) were found to be, for the two *mei-9* cell lines, 0.7J/m2 in UV exposure and 230 rads in X-irradiation (versus 12J/m2 and 600 rads, respectively, in a repair-proficient wild-type cell line). Therefore, *mei-9* is unique, among excision-defective mutants of prokaryotes and eukaryotes, in its hypersensitivity to both UV and ionizing radiations.

Note: Among three newly isolated alleles of the same *mei-9* locus (induced by hybrid dysgenesis; see Yamamoto *et al.*, 1990), it is noteworthy that, although all were found to be deficient in DNA repair (on the basis of their sensitivity to methylmethane sulfonate), two of them remain proficient in female meiotic recombination. It was suggested that the insertion of the P-element might be located in some specific control region, thus affecting expression in somatic DNA repair, although not during the meiotic process.

b. *mus(2)201*

Another gene (located on the 2nd chromosome, at a map position of approximately 23) is involved in excision repair (Boyd *et al.*, 1982). It does not influence recombination proficiency, and was identified on the basis of an abnormal sensitivity of mutant larvae to various chemical mutagens (*MMS, HN2*) and UV light, but not to X-rays.

The biochemical assays, as described above, all demonstrated that primary cell cultures from homozygous *mus-201^{D1}* embryos are devoid of detectable excision repair and this defect seems to concern the initial incision step.

The special advantage of these mutants lies in a greater analogy, in all their characteristics of sensitivity, with excision-defective cells from the human hereditary disease *xeroderma pigmentosum*.

A permanent cell line (named *ML82-8*) was established from *mus-201^{D1}/mus201^{D1}* embryos by Todo *et al.* (1985) (see page 156). By measuring the inactivation of its colony-forming ability, the authors revealed the hypersensitivity of this mutant line to several chemical agents (for instance, a 7-fold higher sensitivity to methylnitrosourea, or a 3-fold higher sensitivity to mitomycin C, compared to the control *GM1* line) and to UV light (D_{37} values were 2J/m2 and 26J/m2 for *mus201* and wild-type *GM1* cells, respectively). On the other hand, resistance to X-rays was normal.

2. POSTREPLICATION REPAIR

This term refers to those "cellular mechanisms which permit synthesis of high molecular DNA to proceed on a damaged template." After a brief [³H]thymidine pulse, the increasing size of the labeled Okazaki-type DNA fragments can be monitored during the subsequent chase period, by centrifugation in alkaline sucrose gradients. Boyd and Setlow (1976) thereby established that, in primary cultures from a *Drosophila* repair-proficient strain, as well as in Schneider's *S2* line cells, the molecular weight profiles were quite similar to those observed in normal mammalian cells. After a 10J/m2 UV irradiation, the average molecular weight of newly synthesized DNA lagged only slightly behind that of unirradiated controls, which means that normal replication could proceed through the numerous UV-induced CPDs.

On the other hand, in embryonic cells from the mutant *mei-41^{D5}* whose synthesis was found to be normal in the absence of irradiation, the growth rate of pulse-labeled DNA was severely reduced after UV exposure (the average molecular weight was less than 60% of the control value) (Fig. 5.10).

Two other mutants, *mus101* and *mus104,* were also found to be defective in such postreplication repair, but their normal synthesis was also affected. Moreover, a caffeine treatment (0.3 mg/ml), that can inhibit, by 12%, the DNA increase after irradiation in control cells, had distinctive effects on these three mutants, which suggests the possible existence of at least two different postreplication repair pathways (one "caffeine sensitive" and another "caffeine insensitive").

Similar observations were extended to several other postreplication defective mutants, whose loci had been located on the 2nd and 3rd chromosomes (*mus205, mus302, mus304, mus308,* and *mus310*), either in primary cultures or in short organ culture of dissected larval brain ganglia (a procedure especially useful for strains with reduced fertility) (Boyd and Shaw, 1982).

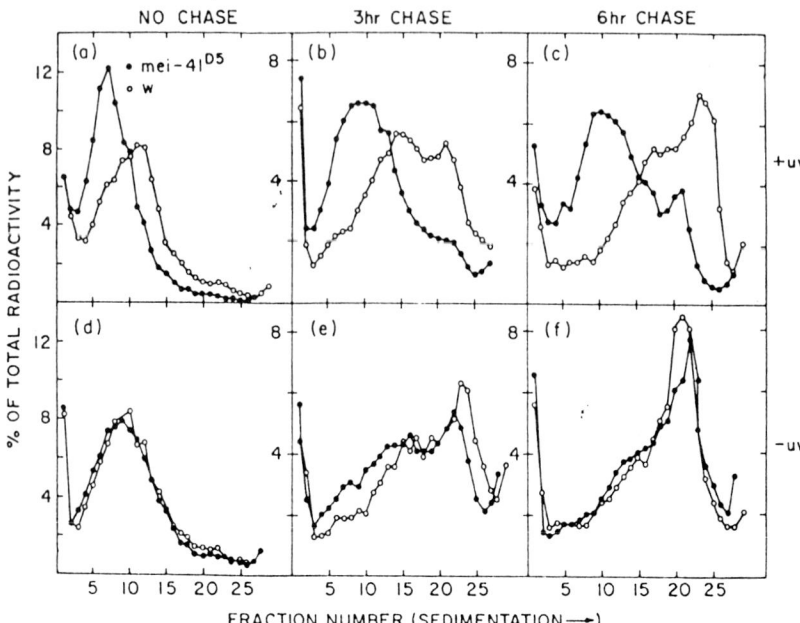

FIGURE 5.10   Kinetic analysis of postreplication repair and DNA synthesis in a *mei-41^{DS}*
cell line (black circles), compared to a control *w* cell line (open circles). All cultures were
labeled with [³H]thymidine for 30 min. Irradiated series have received 10J/m2 of UV
radiation 1/2 h prior to the pulse-labeling period. Whereas the rates with which unirradiated
cultures of the two strains incorporate [³H]thymidine are equivalent, the patterns observed
after irradiation are strikingly different: at each time investigated, the observed weight
average molecular weight of the mutant-derived DNA is less than 60% of the control value.
From Boyd and Setlow (1976). With the permission of the Genetics Society of America.

Brown and Boyd (1981) attempted a rather different approach, by
substituting a chase-pulse protocol for the previous pulse-chase procedure: i.e., the thymidine pulse was delayed throughout the incubation
period after irradiation, so that what is analyzed is the size of the newly
synthesized fragments at various times during the phase of recovery. In
control cultures, their molecular weight was significantly reduced just
after UV-treatment, but there was a gradual recovery to normal values
by 6 h. The thymidine incorporation rate, which was decreased concomitantly, was also restored within 10 h. In contrast, *mei-41* cells, and to a
lesser degree *mus101* and four other postrepair defective mutants, failed

to recover either thymidine incorporation or the ability to synthesize long nascent DNA fragments.

3. PHOTOREPAIR

Using the energy of light, a specific enzyme, called photolyase, is able to cleave the cyclobutane ring of abnormally induced pyrimidine dimers and, thereby, the regular configuration of the DNA duplex is restored.

Such a DNA photoreactivating enzyme, apparently very similar in its assay requirements to that of *Escherichia coli,* was purified from cultured *Drosophila* cells by Beck and Sutherland (1979; Beck, 1982). In fact, when primary cultures or established line cells of *Drosophila,* even those deriving from mutants (*mei-9*) which are defective in dark excision repair, were exposed to visible light (300 to 500 nm) just after UV irradiation, most of the induced pyrimidine dimers were rapidly split (Trosko and Wilder, 1973; Boyd *et al.,* 1976). Thus, the deleterious effect of UV is corrected: UV-irradiated cells treated with a fluorescent light for 40 min displayed an 8- to 12-fold photoreactivation of survival (Brown *et al.,* 1981).

Boyd and Harris (1985, 1987) isolated and characterized a photorepair-deficient mutant, termed *phr.** The gene could be mapped to position 56.8 on the 2nd chromosome. The monitoring of DNA endonuclease-sensitive sites, in primary cultures and in isolated brain ganglia or ovaries, showed that *phr* cells are entirely devoid of any photorepair (Fig. 5.11). Moreover, their dark excision repair system also appeared to be affected in some way. The situation is analogous to that of the complementation group B of *xeroderma pigmentosum* in human cells, in which a complete deficiency in photoreactivation is similarly associated with a partial defect in excision repair.

The *Drosophila ebony* strain was known to be more sensitive to UV-induced killing of larvae or embryos than wild-type Canton S, although this UV sensitivity is not directly correlated with the marker *ebony*. Using primary cell cultures and the usual bioassays, Ferro (1985) compared the repair abilities of both strains and found that photorepair of the endonuclease-sensitive-sites induced by UV-irradiation was indeed much slower in *ebony* cells. The construction of new strains of flies, with various combinations of chromosomes from Canton S and *ebony,* made it possible to conclude that repair deficiency was maximal in cells in which both major autosomal pairs derived from the *ebony* strain.

* This mutant also shows a partial defect in the incision step of excision repair (Harris and Boyd, 1993).

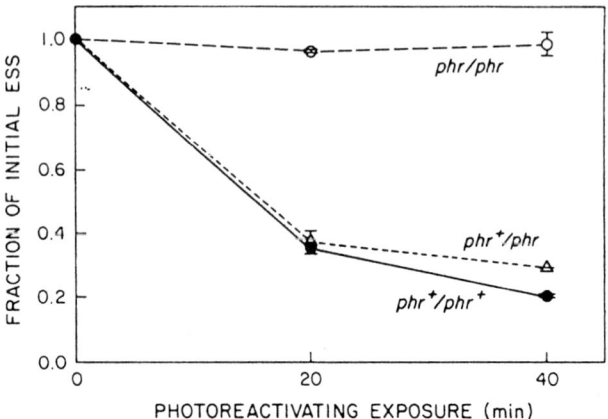

PHOTOREACTIVATING EXPOSURE (min)

FIGURE 5.11   Influence of mutant *phr* gene dosage on photorepair. Primary cell cultures of *Drosophila* embryos carrying 0, 1, or 2 copies of the wild-type *phr*⁺ allele were exposed to long-wavelength UV irradiation for 2 min. The cultured were subsequently exposed to fluorescent illumination for the indicated times. The remaining pyrimidine dimers were assayed (all data have been corrected for simultaneous excision repair which was assayed in parallel cultures incubated in the dark). ESS, endonuclease sensitive sites. From Boyd and Harris (1987). Reproduced with the permission of the Genetics Society of America.

Pyrimidine (6-4) pyrimidone photoproduct is the second most abundant UV photoproduct in DNA. Todo *et al.* (1993) demonstrated the occurrence in *Drosophila* embryo extracts of a photolyase specific for these photoproducts.

Kim *et al.* (1994) were able to characterize this specific photoreactivating activity in *Kc* cells.

4. REPAIR OF DNA SINGLE-STRAND OR DOUBLE-STRAND BREAKS

It is well known that X-rays induce numerous DNA single-strand or double-strand interruptions.

When *Drosophila* primary cultures were exposed to X-rays (10,000 rads, from a 50 kW source, through aluminum filter), the average molecular weight of their DNA was reduced from 270 × 10⁶ to 70 × 10⁶ Da. Within a few hours of incubation, wild-type cells, as well as most tested *mus* mutant cells, recovered an almost normal DNA size (Boyd and Setlow, 1976; Boyd *et al.*, 1976). If a few repair defective mutants, for instance *mei-9* (see above), are especially sensitive to X-rays (Brown *et al.*, 1981), this may be explained by the fact that X-rays can induce damage other than mere single-strand breaks.

During a study of the dose-dependent inhibition of *in vivo* transcription induced by X-rays in a variety of eukaryotic cells, including *Kc* cells, Luchnik *et al.* (1988) confirmed that DNA nicking was considerably repaired during post-irradiation incubation of the cultures.

The neutral elution procedure devised by Bradley and Kohn (*1979*) greatly facilitated the analysis of DNA double-strand break repair. Dezzani *et al.* (1982) adapted it to *Drosophila* cells and showed that relative elution is indeed a linear function of X-ray dose, between 5 and 10 kR. In control primary cultures, as well as in established cell lines, 88% of DSBs induced by X-rays (15 kR) were repaired within 3 h. None of the mutants tested (*mei* or *mus* primary cultures or cell lines) were found to be significantly deficient in DNA double-strand repair.

Note: Poly(ADP-ribose) synthase (i.e., a chromatin-bound enzyme which attaches polyanionic chains of ADP-ribose to nuclear proteins and whose function is unknown) was suspected to be involved in DNA repair. Nolan and Kidwell (1982) could demonstrate that DNA strand break repair was not impaired in *S2* cells, neither after the total inactivation of this enzyme by a brief heat-shock nor when its cellular amount had been reduced by specific inhibitors.

5. BASE EXCISION AND AP-SITE REPAIR

Loss of purine or pyrimidine bases from DNA is frequently induced by various mutagens. Moreover, cells can eliminate modified bases by enzymatic hydrolysis of the N-glycosylic bond. Thereby, AP (i.e., apurinic or apyrimidinic) sites are created and further processes are needed for completion of repair. Abasic sugars are removed by a specific AP-endonuclease.

Osgood and Boyd (1979, 1982) reported the presence of such a AP-nuclease in extracts from *Drosophila* primary cultures or established line cells, as well as from embryos or larval brain ganglia.

In only four alleles of *mei-9* and one of *mus201*, excision defective mutants, was this enzymatic activity found to be reduced. The correlation might be explained if AP-endonuclease was involved, in concert with a specific dimer glycosylase, in the initial endonucleolytic cut of the nucleotide excision repair. Besides, it is noteworthy that these mutants are hypersensitive to those mutagens, including X-rays, that can introduce AP-sites into DNA.

6. OTHER REPAIR PROCESSES

The X-linked *giant* (*gt*) locus was characterized in the 1920s. Not only is the 3rd instar larval period extended in mutants, but the presence of this mutation seems to correlate with a significant genetic instability.

Using, as a standard for comparison, the labeled DNA from a repair-proficient cell line, Narachi and Boyd (1985) established, with short cultures of brain ganglia, that the mutant *gt* synthesizes DNA of a molecular weight that is greatly reduced compared with the control and has an unusually high frequency of spontaneous single and double-strand breaks (the normalized molecular weight is 58% lower than that of the control). These phenomena, however, do not seem to be associated with a defect in the repair of X-ray induced breaks or with modified chromosome stability.

A nuclear extract system prepared from *Kc* cells could support efficient strand-specific mismatch correction *in vitro* within an artificial target made of a circular heteroduplex containing a single base-base mispair (Holmes *et al.*, 1990).

Koken *et al.* (1991) reported the cloning of *Dhr6*, a *Drosophila* homolog of the yeast DNA-repair gene *RAD6*; two transcripts were detected in poly(A)$^+$ RNA of *Drosophila* cell line *DM-2*.

B. Genomic Differential Repair

The possibilities of repair do not seem to be uniformly distributed along the genetic material, as has been documented in yeasts and mammalian cells. Transcriptionally active regions might be preferentially repaired and, moreover, repair rates might be higher in the transcribed DNA strand. Two explanations are possible: (1) either access to DNA of the repair machinery is easier through the "open" configuration of euchromatin than through the "closed" heterochromatin, or (2) a coupling of DNA repair and transcription systems occurs, perhaps because there is some common component? (see minireview by Buratowski, *1993*).

J. de Cock (1992) carefully investigated this putative preferential repair in two *Drosophila* permanent lines, *Kc* and *S2*.

A technique, developed by Bohr *et al.* (*1986*), permits the analysis of pyrimidine dimer excision repair in given sequences of DNA. Thereby, de Cock *et al.* (1991) compared the CPD removal, after exposure to 10 to 15 J/m2 UV, in three individual genes, two of them (*gart* and *notch*) being normally transcribed in the cultures, whereas the third one (*white*) is not expressed. In contrast with the results from other systems, the three

Drosophila genes were found to be repaired at the same rate and to the same extent (up to 80 to 100% CPD removal after 24 h incubation). Very similar observations were made with the excision of the other major type of UV photoproduct, i.e., 6-4 PPS (de Cock, van Haffen *et al.*, 1992).

When strand-specific probes were used, no difference could be observed in the repair kinetics of both DNA strands of these same three genes (de Cock, Klink *et al.*, 1992a).

The extensively studied responses of this *Kc* line to ecdysteroids (see Chapter 8) offer a variety of genes which can be specifically induced by the hormone. Such is the β-tubulin 60C gene, which, contrary to the constitutively expressed β-tubulin 56C, is only transcribed in 20-hydroxyecdysone-treated cells. Therefore, CPD removal could be monitored in the same gene either in the presence or in absence of transcription. Repair was not enhanced after hormonal induction, and, in both situations, the repair kinetics were very close to those of all other genes studied; again, there was no strand-specific repair (de Cock, Klink *et al.*, 1992b).

In order to account for such an intriguing absence of "preferential repair" in these *Drosophila* cell lines, the authors suggested that some repression mechanism, implying complex organization of the chromatin, might not yet have been activated in these embryo-derived cell lines.

V. TRANSCRIPTION APPARATUS AND REGULATORY FACTORS

Regulated transcription of eukaryotic protein-coding genes by RNA polymerase II requires the intervention of two distinct classes of transcription factors (1) A first set of ubiquitous general (or basal) transcription factors that bind to the core promoter elements (TATA box, start site or "initiator") recruit the polymerase and, together, catalyse a basal level of mRNA synthesis (initiation and elongation).

A second group of specific transcriptional regulation factors recognize and stick to typical short DNA "motifs" (or "response elements"), usually located in upstream regions of the promoter; through their "activation domain," they interact, in a way that is still poorly understood, with the basal transcription machinery, in order to modulate (i.e., activate or inhibit) the rate of transcription.

In this research area, *Drosophila* and its cultured cells in particular have been widely exploited (see Chapters 7 and 8 for examples of regulated transcription), more especially because there is a striking functional conservation of the basal transcription factors between *Drosophila* and mammalian cells.

A highly efficient *in vitro* transcription system was devised 12 years ago from isolated nuclei of *Kc* cells (Parker and Topol, 1984a,b) and reconstitution experiments made it possible to isolate and characterize many active components.

Moreover, as soon as cDNAs of various transcription factors were cloned, they could be overexpressed in transfected *Drosophila* cells (see techniques in Chapter 9). Thus, cotransfection experiments permitted fruitful analyses *in vivo* of the possible interactions between these different protagonists, especially between specific activators and the initiation complex. Even mammalian transcriptional regulation factors, such as *Sp1*, that do not exist in *Drosophila*, could be successfully studied in this heterogenous system (Courey and Tijan, 1988, and see below, Section V,C,3).

Most investigations of the transcriptional machinery of *Drosophila* have been carried out by Tijan and collaborators (University of California, Berkeley).

A. DNA-Dependent RNA Polymerase II

Whereas RNA polymerase I (or A, in European terminology) specializes in the transcription of major ribosomal RNAs and RNA polymerase III (or C) ensures that of tRNAs, the key enzyme in the synthesis of messenger RNAs, i.e., coding for proteins, is RNA polymerase II (or B). In eukaryotes, each one of these three distinct species is a large complex composed of 2 major subunits and about 10 small subunits.

The gene encoding the largest subunit (called "IIa"; 215 kDa) was identified in *Drosophila melanogaster,* on the basis of the resistance of several allelic mutants to α-amanitin (a toxin from the poisonous mushroom *Amanita phalloides* which is well known to inhibit specifically transcription by RNA polymerase II) (Greenleaf *et al., 1979*). It is referred to as *RpII215* (= RNA polymerase II, subunit 215 kDa) and was mapped at 35.66 on the X chromosome (polytene band region 10C). After preliminary information was obtained on the general organization of this gene (occurrence of 3 introns) and dimensions of its transcript (7 kb) (Biggs *et al., 1985*), Jokerst *et al.* (*1989*) cloned the locus, by P-element transposon tagging, and determined its complete sequence. The deduced amino acid sequence of this largest subunit, in *Drosophila*, exhibits N-terminal regions with striking homology to the largest component (β') of *E. coli*

RNA polymerase. In addition, like all other RNA polymerases from eukaryotes, it possesses an unusual carboxy-terminal domain (CTD) comprising multiple repeats of a heptapeptide consensus sequence (here, in *Drosophila*, there are 42 such tandem units).

The high efficiency of transcription of a preparation from *Kc* cell nuclear extracts allowed the characterization (Natori *et al.*, 1973) and chromatographic separation of *Drosophila* RNA polymerase II from several auxiliary, albeit essential, transcription factors (see following section). In such reconstructed transcription systems, Price *et al.* (1987) could compare the wild-type enzyme from *Kc* cells with the α-amanitin resistant RNA polymerase II prepared from mutant embryos. On an actin template, transcription by the normal enzyme was inhibited by the addition of 0.5 μg/ml α-amanitin, whereas, in identical conditions but with the mutant enzyme, transcripts were still detectable at inhibitor concentrations of up to 100 μg/ml.

It is interesting that, having cloned the *RpII 215* gene which encodes the 215 kDa subunit, Price (unpublished, but quoted in Jokerst *et al.*, 1989) used it as a template in the same transcription system *in vitro*, and observed only very low transcription signals.

A transformation vector, containing the mutant large subunit sequence from the α-amanitin-resistant allele "C4" inserted into the *XbaI* site of a "Carnegie 4" vector (Rubin and Spradling, 1983; see Chapter 9), was constructed by Jokerst *et al.* (1989, see above). This so-called *pPC4* plasmid can confer a high level of resistance on transformed flies, as well as on cultured cells (Thomas and Elkins, 1988). This is the basis for an efficient and widely used selection system which permits the recovery of cells that have been cotransformed with this selection vector and any additional nonselectable gene(s) (see Chapter 9).

B. Basal Transcription Complex

As in other studied eukaryotic models (primarily yeast and human cells), RNA polymerase II is assisted, in *Drosophila*, by at least seven general transcription factors (GTFs), designated TFIIA, B, D, E, F, H, J (an additional TFIIS might increase the efficiency of elongation). They assemble in a well-ordered pathway to form a functional transcription complex (or preinitiation complex), including the key enzyme (see review by Zawel and Reinberg, 1992).

Using their effective *Drosophila* transcription system *in vitro*, made of extracts from isolated nuclei of *Kc* cells, Parker and Topol (1984a,b) undertook its fractionation by stepwise chromatography. They were, thereby, able to distinguish what is now known as HSF (see Chapter 7), i.e., a typical sequence-specific transcriptional activator. On the other hand, two general factors (provisionally called A and B) were required for efficient initiation of transcription on *Drosophila* histone and actin (5C) genes, and were shown to be bound to their TATA box and start site regions. Price *et al.* (1987) developed this analysis, with the same *Kc* nuclear preparation, and, by reconstructions with partially purified fractions, could characterize seven activities (factors) affecting the transcription of four different genes; one of them (factor 3) seems to be the *Drosophila* counterpart of the TATA-binding factor previously identified in human cells.

The TATA-binding factor (TFIID, according to current nomenclature) plays a pivotal role in recognizing the core promoter and initiating the ordered assembly of the whole preinitiation complex. Whereas it is a single polypeptide in yeast, it corresponds in higher eukaryotes (Man and *Drosophila*) to a multisubunit complex consisting of the properly called TATA-binding protein (TBP) and at least 7 tightly bound TBP-associating factors (TAFs), ranging from 30 to 250 kDa (Dynlacht *et al.*, 1991; Yokomori *et al.*, 1993; see minireview by Pugh and Tijan, 1992 and Fig. 5.12). These TAFs may be involved in protein–protein linking with other

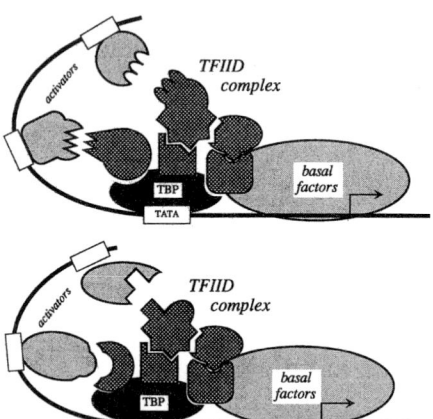

FIGURE 5.12 A schematic of transcription complex assembly on promoters with or without a TATA box. From Pugh and Tijan (1992). Reproduced with the kind authorization of Professor Tijan and the American Society of Molecular Biology.

components of the transcription machinery and some of them serve essentially, in the manner of "adaptor molecules," to mediate crucial interactions between sequence-specific activators and the basal transcription apparatus. They deserve, therefore, the epithet of "coactivator."

After the cloning of *Drosophila* genes or cDNAs encoding several of these factors (for instance, TBP: Hoey *et al.*, 1990; TAF 110: Hoey *et al.*, 1993), their transcripts could be identified in cultured *S2* cells and functional analyses could be carried out by transient cotransfection assay.

For instance, the over production of TBP in *Drosophila* cells, through the use of a proper expression plasmid, resulted in a substantial (more than 100-fold) and dose-dependent increase in the transcription of another transfected gene driven by a weak minimal TATA-containing promoter; this observation suggests that TFIID might be present in limited quantities in those *S2* cells (Colgan and Manley, 1992).

In similar conditions *in vivo*, Hoey *et al.* (1993) established that TAF110 functions, within the TFIID complex, as a specific target for the glutamine-rich activation domain of the human transactivator Sp1 (see below).

It is noteworthy, albeit intriguing, that a TIID complex containing TBP is essential for transcription even at a promoter that lacks a TATA box (Pugh and Tijan, 1991).

Furthermore, it is very likely that sequence-specific activators can also establish functional contacts with other general factors of the preinitiation complex and not only with components of TFIID. The transcriptional activating function of the product of *fushi tarazu*, which is a *Drosophila* developmental regulatory gene (see following paragraph), might be mediated by direct interactions with TFIIB, since the transacting effect of its glutamine-rich activation domain can be blocked by the coexpression of TFIIB truncated derivatives (Colgan *et al.*, 1993).

TFIIF (and its *Drosophila* counterpart, Factor 5) affects both the initiation and the elongation phases of the transcription by RNA polymerase II. Kephart *et al.* (1993, 1994) have cloned the cDNA encoding the large subunit of the *Drosophila* factor (*F5a*). Using an approach *in vitro* with a *Drosophila* RNA polymerase II purified from *Kc* cells (Price *et al.*, 1989; Kephart *et al.*, 1992), the authors could demonstrate that the addition of a chimeric factor, composed of bacteria-expressed *F5a* and *RAP30* (i.e., the small subunit of human TFIIF), was able to increase significantly the transcriptional elongation by the *Drosophila* enzyme (when assayed using a dc-tailed template). This was also true with human RNA polymerase II. Moreover, testing various truncated forms of *F5a* in the same system, they showed that the NH2-terminal region of the factor is necessary and sufficient for stimulation.

C. Sequence-Specific Transcription Factors

Since the "domain-swap" experiments of Ptashne, carried out on Yeast factors (see review in Ptashne, *1988*), it has been assumed that the organization of transactivating proteins is modular. They comprise, at least, one DNA-binding domain, which recognizes a specific DNA sequence motif. This is quite distinct from the activation domain(s) mediating productive interactions with some component(s) of the general initiation complex; other discrete segments may be involved either in the binding of some specific ligand (for instance, the hormone-binding domain in receptors) or in functional homo- or heteropolymerization.

On the basis of the different activation domains they possess, sequence-specific transcription factors may be subdivided into several classes. Three main types of amino acid sequence motifs, namely, acidic, glutamine-rich, or proline-rich, have thus far been identified.

Let us survey a few examples of transactivators that have been studied in cultured *Drosophila* cells.

1. HEAT-SHOCK FACTOR AND ECDYSTEROID RECEPTOR

The paradigm of such transactivating molecules, able to control the expression of an array of specific genes, is unquestionably the so-called heat shock factor. The predominance of the contribution of *Drosophila*, and more especially of cultured cells, to the analysis of this model of gene induction is discussed at length in Chapter 7.

The mechanisms of action of the ecdysteroid receptor (EcR-protein) are more complex, not only because the functional receptor corresponds to a heteromeric structure, but, moreover, the intervention of a specific ligand, the hormone, is required for a productive binding of the receptor to its target DNA motif, or EcRE (see Chapter 8).

2. MANY DEVELOPMENTAL GENES ACTING
AS TRANSACTIVATORS

Drosophila embryogenesis is currently the subject of an omnidirectional analysis, without any equivalent. Genetic approaches have established that embryonic pattern and specific structures are built in an orderly fashion by the successive interferences of cascades of genes which seem to control one another. In fact, the products of many of these developmental genes display a variety of potential DNA-binding domains and are very probably transcriptional transactivators.

Hence, a novel and fruitful strategy has been developed, and it is today adopted by most laboratories, for unraveling the strict functional hierarchy of all these genes [see a good description of this approach by Krasnow *et al.* (1989), as quoted in Chapter 9]. This strategy consists in introducing into *Drosophila* cells, by transient cotransfection, a "producer" plasmid encoding a supposed regulatory gene product, along with a second plasmid, or "responder, made of a reporter gene driven by a candidate target promoter (from another developmental gene). The putative regulatory activity is monitored by the expression levels of the reporter gene.

Many of these *Drosophila* developmental genes encode a highly conserved 60 amino acid sequence, referred to as the "homeodomain" which displays a specific DNA-binding activity. Using the cotransfection method, Jaynes and O'Farrell (1988) showed that the *fushi tarazu* (*ftz*) protein, as produced in cultured cells with an expression vector, was able to stimulate several hundredfold the expression of a reporter plasmid equipped with six tandem copies of a consensus homeodomain-binding site; the activation depends strictly on the presence of these specific DNA motifs, which behave like "enhancers" since they prove active in all promoters tested and in either orientation. In contrast, a vector producing another homeodomain gene product, *engrailed* (*en*), induced a slight, albeit reproducible, decrease in the expression of the same responder plasmids, and this effect was also dependent on the same binding sites. Moreover, in a triple cotransfection system, the stimulation by *ftz* was reduced 30- to 100-fold by the presence of *engrailed*. It was therefore suggested that these two homeodomain-containing proteins might compete for binding sites and influence transcription according to the activity of the resulting complex. Later on, the same authors (1991) refined their analysis of the repressor activity of *engrailed* and established, still in *Drosophila* transformed cells, that *en* could also block the stimulation promoted by several mammalian activators, although these latter bind to quite different DNA sites. The region involved in this so-called "active" repression by *en* product, as distinguished from the previous "passive" repression by competition, was mapped outside the homeodomain.

In similar cotransfection assays, Krasnow *et al.* (1989) demonstrated that, as predicted from genetic data, the family of *Ultrabithorax* (*Ubx*) proteins could repress an *Antennapedia* (*Antp*) promoter fusion construct, while they stimulated a reporter gene driven by their own *Ubx* promoter.

Interestingly, this double activity mimics, in cultured cells, the physiological shift between the functional states of these two promoters, as observed in the anterior thoracic segment (where *Ubx* is OFF and *Antp* is ON) and the posterior thoracic segment (*Ubx* ON and *Antp* OFF), respectively. Moreover, different constructs allowed a functional dissection of the target promoter sequences, as well as that of the involved domains of regulatory *Ubx* products. For instance, deletions affecting their homeodomain suppressed both types of regulation.

The transcriptional activating function of purified *Ubx* proteins was confirmed in a transcription system *in vitro* made from *Kc* cell nuclear extracts (Johnson and Krasnow, 1990).

On the other hand, in a similar *Kc* cell transcription system, especially deprived, however, of the *zeste* protein, it had been previously demonstrated that the product of this other *Drosophila* regulatory gene is involved in the *cis*-regulation of *Ubx* (Biggin *et al.*, 1988).

Similarly, the homeodomain proteins encoded, either by the homeotic gene *Antennapedia* or by the segmentation gene *fushi tarazu*, proved to be transcriptional activators in *Drosophila* cells for several specific promoters (Winslow *et al.*, 1989). For example, the *Antp* product stimulated (about 30-fold) the expression of an *Ubx* promoter plasmid. The most important *Ubx* sequence element was located downstream of the transcriptional start site, within the untranslated "leader" region.

These first few examples are fair illustrations of the rich possibilities offered by this cotransfection approach in cultured cells. *In vivo* functional analysis is today a routine method, in most laboratories, when a novel regulatory gene has been cloned and its sequencing reveals some putative DNA-binding structure. Cultured cell systems require much less toil than the obtention of cotransformed fly strains. In any case, they are very useful for preliminary bioassays of the constructed vectors.

The following list of works carried out with this strategy may, therefore, not be exhaustive. They have successively concerned: *bicoid* and *hunchback* (Driever and Nüsslein-Volhard, 1989; Driever *et al.*, 1989); *caudal* and *fushi tarazu* (Dearolf *et al.*, 1989); *dorsal* (Rushlow *et al.*, 1989); *fushi tarazu*, *paired*, *zen*, *engrailed* (Han *et al.*, 1989); general review on homeodomain factors (Hayashi and Scott, 1990); *fushi tarazu* and *engrailed* (Ohkuma *et al.*, 1990); *transformer*, *transformer-2*, *doublesex* and *sex-lethal* (Inoue *et al.*, 1990, 1992; Hoshijima *et al.*, 1993); Krüppel (Sauer and Jäckle, 1991, 1993); *Krüppel* and *hunchback* (Zuo *et al.*, 1991); *zerknüllt* and *Cyclin* (Yamaguchi *et al.*, 1991); *dorsal* and *twist* (Thisse *et al.*, 1991); *Ultrabithorax* (Gavis and Hogness, 1991);

c-rel/NF-κB/dorsal family (Lienhard-Schmitz and Baeuerle, 1991; Kerr *et al.*, 1993); *Antennapedia* (Oh *et al.*, 1992); *Polycomb* (Messmer *et al.*, 1992) *fushi tarazu* (Colgan and Manley, 1992, 1993); *even-skipped* (Han and Manley, 1993); *PRD, fushi tarazu* and *engrailed* (Anathan *et al.*, 1993; John *et al.*, 1995); *deltex* and *Notch* (Diederich *et al.*, 1994); *glass* and *opsin* (O'Neill *et al.*, 1995); *trithorax, Polycomb* and *Ultrabithorax* (Huang *et al.*, 1995); *dorsal* and *cactus* (Kubota and Gay, 1995); *Deltex, Notch,* and *Suppressor of Hairless* (Matsuno *et al.*, 1995).

3. STUDY OF MAMMALIAN TRANSCRIPTION FACTORS IN HETEROLOGOUS *Drosophila* CELLS

It may seem paradoxical that *Drosophila* cells have also been frequently used, in cotransfection systems, for functional analyses of mammalian transcription factors that do not even exist in this organism. In fact, because the basal transcription apparatus is highly conserved, Tijan and his group considered the null background of this heterogenous host cell system to be an obvious advantage. They adopted this approach for a thorough functional study of the structure and mechanism of action of several human transcription factors:

a. Sp1

Sp1 binds to GC-rich elements, or GC boxes, in a variety of viral and cellular promoters. In order to express it in *Drosophila* S2 cells, Courey and Tijan (1988) constructed expression vectors by linking, in frame, a *Drosophila Adh* or *actin 5C* promoter (see Chapter 9) to the whole ORF of the *Sp1* cDNA clone (or various deletion mutants of it). Reporter plasmids contained the bacterial *CAT* gene driven by the *SV40* early promoter, with its typical six tandem *Sp1* sites (or with a variable number of those GC-boxes). Whereas the *CAT* activity is very low in cells transfected with the wild-type *SV40* promoter plasmid alone, cotransfection with the *Sp1* expression vector $(pP_{ac}Sp1)$ stimulated *CAT* expression 100 to 500 times $(P_{ac},$ promoter actin). High levels of stimulation required multiple *Sp1*-binding sites close to the transcriptional start site of *CAT*. With the use of plasmids expressing a variety of N- or C-terminal deletions of *Sp1*, it was shown that, besides the DNA-binding domain with its 3 zinc fingers, four discrete regions of the factor, and especially two glutamine-rich domains (A and B) in the N-terminal half of the protein, are important for transcriptional activation.

A similar *Sp1* transient transfection system, in *Drosophila* cells, enabled McLachan and collaborators (Raney *et al.*, 1992; Zhang *et al.*,

1993) to demonstrate the regulatory importance of *Sp1* binding motifs in the promoters of the major surface antigen and the nucleocapsid of hepatitis B virus.

Because several known or suspected transactivators display similar glutamine-rich segments, Courey *et al.* (1989) were led to demonstrate, with the same cotransfection assay, that the glutamine-rich domain from the *Drosophila Antennapedia* product was also a potent activator when fused, as a chimeric protein, to the DNA-binding domain of *Sp1*. Moreover, they showed that, when the *CAT* reporter plasmid comprised GC-boxes, both close to the transcriptional start site and far away, the bound *Sp1* molecules had a synergistical effect; even finger-less variants of *Sp1* could interact with a normal DNA-binding form to "superactivate" transcription.

Further analysis of these different types of synergy (Pascal and Tijan, 1991), again carried out in cell transfection systems and combined with data from gel-shift and cross-linking experiments, strongly suggested that *Sp1* molecules, as bound on each GC-box, form multimers (primarily tetramers) by direct protein–protein interactions. Synergistic activation would correspond to the building of higher-order complexes between the *Sp1* tetramers bound to either adjacent or distant sites of the promoter. According to the authors' assumption, such large *Sp1* complexes might generate a more effective activation surface to interface with components of the general transcriptional machinery.

Hoey *et al.* (1993) established that it is more precisely TAF 110, within the TFIID tight complex, that is involved in specific connections with the activation domain of *Sp1*. When a TAF 110 hybrid construct was cotransfected with a DNA expressing the Gln-rich domains A and B of *Sp1*, a 60-fold increase in transcription of a reporter plasmid was observed in a *Drosophila* S2 bioassay.

An important aspect of transcriptional activation mechanisms is represented by the combinatorial interplay that must occur between the many different transactivators that bind to the same promoter. One recent paper (Seto *et al.*, 1993) deals with the interactions, studied, once again, in cotransfected *Drosophila* cells, between *Sp1* and *YY1*, a mammalian transcriptional "repressor" belonging to the *GL1-Krüppel* family.

Note: Because of this absence of endogenous *Sp1*, Chavrier *et al.* (1990) also chose *Drosophila* cells for the analysis, by cotransfection of the transcriptional activating capabilities of *Krox-20*, a mouse zinc finger protein which is expressed, in a segment-specific manner, in the embryonic nervous system and whose target nucleotidic sequence is closely related to that of *Sp1*.

b. AP-2

Drosophila Schneider's S2 cells also appear to be devoid of another human enhancer-binding factor, AP-2 (Perkins *et al.*, 1988). When its coding sequence, driven by the *Drosophila* ADH promoter (see Chapter 9), is cotransfected into these cells with a CAT reporter plasmid (equipped with the enhancer of either SV40 or human metallothionein), there is a significant induction (4-fold) of CAT activity, which clearly demonstrates that AP-2 indeed functions *in vivo* as a transactivator (Williams *et al.*, 1988).

c. CTF/NF-1

Turning to their account the fact that this transactivator is also lacking in *Drosophila*, Tijan and collaborators (Santoro *et al.*, 1988) studied, in cotransfected S2 cells, the mode of action of human CTF/NF-1, i.e., one of the families of activators that bind to the CCAAT recognition motifs (and which they had previously identified from HeLa cells). When a CTF expression plasmid is cotransfected with a reporter plasmid (containing the CAT gene linked to the promoter of human α-globin), it induces a significant (3- to 5-fold) increase in CAT activity; this was not observed with a reporter plasmid built with a truncated version (without CCAAT box) of the same promoter.

A systematic functional dissection (Mermod *et al.*, 1989) revealed that transcriptional activation requires, here, a C-terminal domain that is unusually rich (25%) in proline residues. This corresponds to a novel class of activators. As expected, constructed deletion mutants, without the N terminus of the protein (i.e., the DNA-binding domain), were also totally inactive in this assay *in vivo*.

d. P-53

P-53 is the product of the human tumor-suppressor gene. It inhibits the proliferation of several cell types, most probably because it can selectively activate or repress certain genes involved in the control of the cell cycle. Because DNA-binding assays *in vitro* showed some cooperation between *P-53* and TBP or TFIID, Chen *et al.* (1993) verified, in *Drosophila* cells (which do not normally contain any *P-53*-related gene), the activation by *P-53* of a reporter plasmid c-*fos*-CAT (i.e., containing a *P-53*-binding site located upstream of the TATA box). Then, in a triple cotransfection system, associating the reporter plasmid to both *P-53* and TBP expression vectors, they observed a plain combinatorial stimulation of the CAT expression.

APPENDIX 5

5.A. Preparation and Purification of Mitochondrial Fraction from *Drosophila* Cultured Cells

These procedures are given according to Alziari *et al.* (1985) and Stepien *et al.* (1988) (courtesy of the authors).

1. ISOLATION OF A CRUDE MITOCHONDRIAL FRACTION

Cultured cells, detached from the flasks by scraping, are collected by centrifugation (10 min, 130/1300 g).

Their membrane being more resistant than that of tissue cells, they are exposed to a 10 min osmotic shock in a hypotonic medium (HEPES–KOH 20 mM, pH 7.4; EGTA 1 mM). Then a normal osmolarity is restored by addition of 1 : 1 (v/v) complementary "double-isotonic" medium (i.e., hypotonic medium supplemented with mannitol 0.5 M).

First homogenization as carried out in an Potter-Elvehjem homogenizer fitted with a Teflon pestle (5 to 12 min; 700 to 900 rpm; 4°C).

A second homogenization is carried out in a Dounce homogenizer with a glass pestle (by hand; 30 sec to 2 min; 4°C).

By two successive centrifugations at 500 g (10 min; 4°C), nuclei and imperfectly crushed cells* are sedimented.

The supernatant is, then, centrifugated at 3700 g (10 min; 4°C), which allows the elimination of most lysosomes, peroxysomes, and membrane fragments.

After being resuspended in an isotonic medium (HEPES–KOH 20 mM; EGTA 1 mM; mannitol 0.25 M), the mitochondrial pellet is washed by a last centrifugation at 7000 g (10 min; 4°C).

2. FURTHER PURIFICATION OF MITOCHONDRIA

The separation of mitochondria (density = 1.075 to 1.085 g/ml) from contaminants such as lysosomes and peroxysomes (density = 1.05 to 1.06) is carried out in a preformed Percoll gradient.

Polycarbonate tubes are filled with the following mixture

    55%   Percoll-saccharose   0.5 M (final)
    45%   Isotonic extraction medium (see above)

---

* In such experimental conditions, about 80 of 100 cells have been crushed, as shown by staining with trypan blue.

and the gradient is preformed by 1 h centrifugation at 92,000 to 100,000 *g* at 4°C.

The crude mitochondrial fraction is laid onto the gradient and the separation is performed by 1 min centrifugation (100,000 *g*; 4°C). The band corresponding to mitochondria is sucked out and diluted into 5 volume of the isotonic extraction medium.

A final centrifugation (10,000 *g*; 5 min; 4°C) washed out the residual Percoll.

Control of purity of the isolated mitochondria may be performed by measuring catalase (an enzyme specific of peroxysomes) and acid phosphatase (characteristic of lysosomes) activities.

5.B. Incubation Medium I for Isolated *Drosophila* Mitochondria

This procedure has been provided by Professor Alziari (University of Clermont-Ferrand, France; personal communication, 1995).

1. STOCK SOLUTIONS (20×)

(Most of them may be kept frozen at −20°C)

Potassium acetate 2 *M*
Magnesium acetate 0.4 *M*
Cyclohexamide 4 mg/ml H_2O; must be freshly prepared
Amino acids: a mixture of the usual 19 amino acids, without
   methionine: 5 m*M*
GTP 50 m*M*
Dithiothreitol 100 m*M*
Spermidine 100 m*M*
Creatine kinase 1 mg/ml
Phosphocreatine 500 m*M*
PMSF (Phenylmethylsulfonyl fluoride) in DMSO (dimethyl
   sulfoxide) 500 m*M*
ATP 500 m*M*

2. PREMIX (2×)

To be prepared just before use: 30 to 50 µl of each ingredient is added in the order of the above list and will be diluted 10-fold by adding a Tricine buffer (100 m*M*, pH 7.8).

3. INCUBATION OF ISOLATED MITOCHONDRIA

Isolated mitochondria (see Appendix 5.A) are diluted to a protein concentration of 5 mg/ml

Pour 100 to 200 μg of the suspension of mitochondria into a tube and add the same volume of the 2× premix

Add [^{35}S]methionine (2 to 10 μCi/μl) and RNase inhibitor (RNasin, Boehringer, Mannheim, Germany) (200 U/ml)

Incubate at 28°C under shaking for 30 min

To end the incubation, add 9 volumes of cold ethanol (10 min at 0 to 5°C)

References

Biggs, J., Searles, L. L., and Greenleaf, A. L. (1985). Structure of the eukaryotic transcription apparatus: feature of the gene for the largest subunit of *Drosophila* RNA polymerase II. *Cell* **42**, 611–621.

Bohr, V. A., Okumoto, D. S., and Hanawalt, P. C. (1986). Survival of UV-irradiated mammalian cells correlates with efficient DNA repair in an essential gene. *Proc. Natl. Acad. Sci.* **83**, 3830–3833.

Bownes, M., and Rembold, H. (1987). The titre of Juvenile Hormone during the pupal and adult stages of the life cycle of *Drosophila melanogaster*. *Eur. J. Biochem.* **164**, 709–712.

Boyd, J. B., Mason, J. M., Yamamoto, A. H., Brodberg, R. K., Banga, S. S., and Sakaguchi, K. (1987). A genetic and molecular analysis of DNA repair in *Drosophila*. *J. Cell Sci.* **6**, Suppl., 39–60.

Bradley, M. O., and Kohn, K. W. (1979). X-ray induced DNA double strand break production and repair in mammalian cells as measured by neutral filter elution. *Nucleic Acids Res.* **7**, 793–804.

de Bruijn, M. H. L. (1983). *Drosophila* mitochondrial DNA, a novel organization and genetic code. *Nature* **304**, 234–241.

Buratowski, S. (1993). DNA repair and transcription: the helicase connection. *Science* **260**, 37–38.

Chomyn, A., Mariottini, P., Cleeter, M. N. J., Ragan, C. J., Matsuno-Yagi, A., Hatefi, Y., Doolittle, R. F., and Attardi, G. (1985). Six unidentified reading frames of human mitochondrial DNA encode components of the respiratory chain NADH dehydrogenase. *Nature* **314**, 592–597.

Clary, D. O., and Wolenstholme, D. R. (1985). The mitochondrial DNA molecule of *Drosophila yacuba*: Nucleotide sequence, gene organization, and genetic code. *J. Mol. Evol.* **22**, 252–271.

Dynlacht, B. D., Hoey, T., and Tijan, R. (1991). Isolation of coactivators associated with the TATA-binding protein that mediates transcriptional activation. *Cell* **66**, 563–576.

Garesse, R. (1988). *Drosophila melanogaster* mitochondrial DNA: gene organization and evolutionary considerations. *Genetics* **118**, 649–663.

Goodpasture, S. E., and Bloom, C. (1975). Visualization of nucleolar organizer regions in mammalian chromosomes, using silver staining. *Chromosoma* **53**, 37–50.

Greenleaf, A. L., Borsell, L. M., Jamachello, P. F., and Coulter, D. E. (1979). α-amanitin resistant *Drosophila melanogaster* with an altered RNA polymerase II *Cell* **18**, 613–622.

Jakubczak, J. L., Xiong, Y., and Eickbush, T. H. (1990). Type I(R1) and Type II(R2) ribosomal DNA insertions of *Drosophila melanogaster* are retrotransposable elements closely related to those of *Bombyx mori. J. Mol. Biol.* **212**, 37–52.

Jarry, B., and Falk, R. (1974). Functional diversity within the *rudimentary* locus of *Drosophila melanogaster. Mol. Gen. Genet.* **135**, 113–122.

Jokerst, R. S., Weeks, J. R., Zehring, W. A., and Greenleaf, A. L. (1989). Analysis of the gene encoding the largest subunit of RNA polymerase II in *Drosophila. Mol. Gen. Genet.* **215**, 266–275.

Kohn, K. W., Erickson, L. C., Ewig, R. A. G., and Friedman, C. A. (1976). Fractionation of DNA from Mammalian cells by alkaline elution. *Biochemistry* **15**, 4629–4637.

Littlefield, J. W. (1964). Selection of hybrids from matings of fibroblasts *in vitro* and their presumed recombinants. *Science* **145**, 709–710.

Long, E. O., and Dawid, I. B. (1980). Repeated genes of Eukaryotes. *Annu. Rev. Biochem.* **49**, 727–764.

Miller, R., Hayward, D. C., and Glover, D. M. (1983). Transcription of the "nontranscribed" spacer of *Drosophila melanogaster* rBNA. *Nucleic Acids Res.* **11**, 11–19.

Morgan, T. H. (1910). The method of inheritance of two sex-limited characters in the same animal. *Proc. Soc. Exp. Biol. Med.* **8**, 17–19.

O'Hare, K., Benoist, C., and Breathnach, R. (1981). Transformation of mouse fibroblasts to methotrexate resistance by a recombinant plasmid expressing a prokaryotic dihydrofolate reductase. *Proc. Natl. Acad. Sci. U.S.A.* **78**, 1527–1531.

Ptashne, M. (1988). How eukaryotic transcriptional activators work. *Nature* **335**, 683–689.

Pugh, B. F., and Tijan, R. (1992). (Minireview) Diverse transcriptional functions of the multisubunit eukaryotic TF IID complex. *J. Biol. Chem.* **267**, 679–682.

Raucourt, S. L., and Walker, V. K. (1990). Kinetic characterization of dihydrofolate reductase from *Drosophila melanogaster. Biochem. Cell Biol.* **68**, 1075–1082.

Rubin, G. M., and Spradling, A. C. (1983). Vectors for P-element mediated gene transfer in *Drosophila. Nucleic Acids Res.* **11**, 6341–6351.

Sommerville, J. (1985). Organizing the nucleolus. *Nature* **318**, 410–411.

Todo, K., Takemori, H., Ryo, H., Ihara, M., Matsunaga, T., Nikaido, O., Sato, K., and Nomura, T. (1993). A new photoreactivating enzyme that specifically repairs UV-light-induced (6-4) photoproducts. *Nature* **361**, 371.

Yamamoto, A. H., Brodberg, R. K., Banga, S. S., Boyd, J. B., and Mason, J. M. (1990). Recovery and characterization of hybrid dysgenesis-induced *mei-9* and *mei-41* alleles of *Drosophila melanogaster. Mut. Res.* **229**, 17–28.

Zawel, L., and Reinberg, D. (1992). Advances in RNA polymerase II transcription. *Curr. Opin. Cell. Biol.* **4**, 488–495.

6

Biology and Biochemistry of Cultured Cell Lines: 2. Proteins

I. CELL SURFACE GLYCOPROTEINS: ADHESION/RECOGNITION MOLECULES

The guiding of cell movement in morphogenesis, the necessary cohesion of multicellular organisms, and many other crucial cell interactions are mediated by a variety of transmembrane glycoprotein complexes that ensure recognition and/or cell adhesion. Mechanisms and molecules appear to have been highly conserved throughout phylogeny, so that *Drosophila,* because of the unique potentialities of its genetics and, thereby, its leading role in developmental biology, has become, once more, one of the major experimental systems in this research area. This is clearly emphasized in a recent review on "*Drosophila* cell adhesion molecules" (Bunch and Brower, 1993); see also Hortsch and Goodman (1991).

Until now, it seemed convenient, albeit somewhat arbitrary, to distinguish two main classes: cell adhesion molecules (CAMs), for direct cell–cell contacts, and substrate adhesion molecules (SAMs), mediating cell–matrix connections. Yet, it has become more and more obvious that a wide variety of other molecules may be involved in different processes of cell adhesion or recognition, so that there is, in fact, no clear distinction between adhesive and signaling functions.

Moreover, one should be fully aware that there is no real discontinuity between the system of integral membrane proteins, capable of transducing an exogenous signal into the cell machinery, and that of the so-called growth factors and their corresponding receptors (see Section II).

A. Preliminary Studies on Cell Surface Proteins of *Drosophila* Cultured Cells and Their Modification by Ecdysteroids

Cell clustering is considered to be one of the most typical responses of hormone-sensitive *Drosophila* cell lines to ecdysteroids. In pioneering work, A. M. Courgeon (1972a,b) reported that, in the days following stimulation by 20-hydroxyecdysone, *Kc* cells gather and form randomly distributed clumps of increasing sizes; finally, these latter detach from the flask bottom and float freely in the medium, as large aggregates (see Chapter 7, Fig. 7.1C). Later, it was shown with electron microscopy that clustered cells were held together by numerous contact points, and specialized junctions, including adhering zones and gap junctions, could be seen (Yudin *et al.*, 1982; Woods and Poodry, 1983; Rickoll *et al.*, 1986) (see Fig. 6.1). Such cellular adhesive interactions obviously reflect significant modifications in the cell surface components.

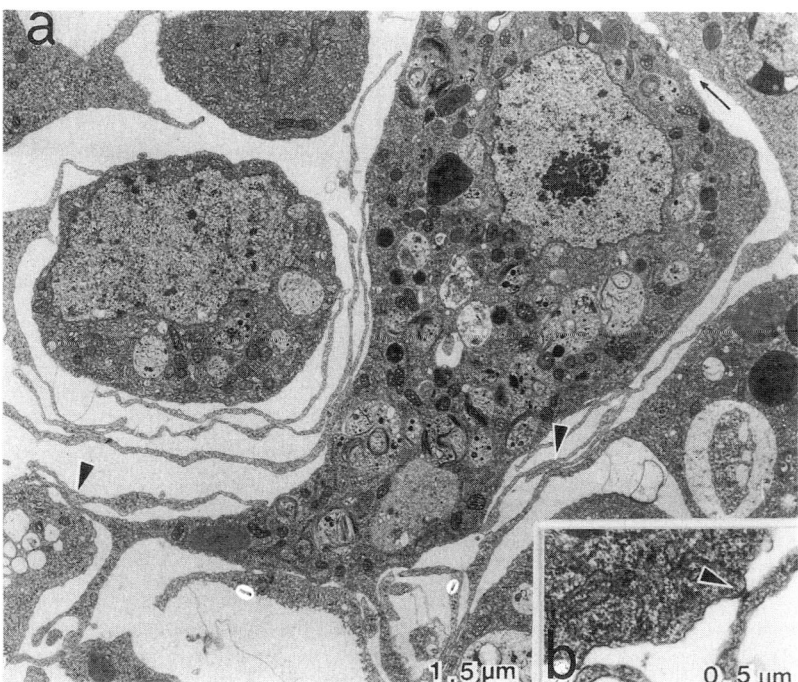

FIGURE 6.1   Hormone-induced aggregation of *Drosophila* cells. (a) *S3* cells cultured for 48 h with $10^{-6}M$ 20-hydroxyecdysone. Numerous filopodial contacts can be seen extending between the cells (arrowheads). Cells can also be seen to make contacts along their edges (arrow). (b) At some contact points, *macula adherens* (arrowhead) can be observed. From Rickoll *et al.* (1986); reprinted from *Insect Biochemistry*, **16**, 211, copyright (1986), with kind permission of Elsevier Science Ltd., The Boulevard, Langford Lane, Kidlington 0X5 1GB, U.K.

A first approach consisted of testing the cell binding capacities of various plant lectins, well known for their specificities with regard to different types of sugars and glycoproteins. Remember that, as early as 1972, Becker established that concanavalin A (ConA, the Jack Bean agglutinin) can not only agglutinate *Kc* cells but can also induce a rapid cell fusion. This observation was confirmed by several laboratories (see Chapter 3). Metakowsky *et al.* (1975) and then Dennis and Haustein (1982) showed that ecdysteroid treatment could modify the extent to which cells agglutinate in the presence of ConA, although opposite effects were sometimes observed in other cell lines. In a further examination, carried out on *Kc-H* cells exposed to 2×10^{-6} *M* 20-hydroxyecdysone, Johnson *et al.* (1983) noted that treated cells responded to wheat germ agglutinin at a lower concentration than control cells, whereas their agglutination by ConA required, conversely, a 100-fold greater concentration (1.25 mg/ml instead of the usual 12.5 μg/ml). Whatever the case, these data revealed important changes in the cell surface.

Preliminary analyses (Metakowsky *et al.*, 1975, 1978) established that the synthesis pattern of glycoproteins in *67j25D* cells, as identified from their incorporation of N-[^{14}C]acetylglucosamine or [^{14}C]galactose, was also markedly modified by hormonal treatment: the main electrophoretic fraction A showed a reduction in its average molecular mass (from 120 to 85 kDa), whereas fraction B (200 kDa) was slightly increased. The degradation of both major components by a mild pronase pretreatment of the cells confirmed the surface localization of these glycoproteins.

The use of lactoperoxidase (LPO)-catalyzed radioiodination (according to the method devised by Hynes (*1973*) for mammalian cells and adapted to *Drosophila Kc* cells by Dennis and Haustein, 1981) permits the labeling and, thereby, a clear distinction of exposed cell surface proteins. After two-dimensional electrophoresis and autoradiography, about 150 and 175 surface polypeptides could be resolved in *S2* and *Kc* cells, respectively (Woods and Poodry, 1983; Johnson *et al.*, 1983). Most of them corresponded to minor silver spots when the same gel was stained for total cellular proteins. Moreover, these hormone-induced surface proteins were found to bind to *Lens culinaris* agglutinin (LCA) in lectin affinity chromatography and are, consequently, glycoproteins.

After addition of 20-hydroxyecdysone, some 30 spots underwent reproductive, quantitative or qualitative, changes (see Fig. 6.2). This relatively high proportion of modified cell surface proteins contrasts with the fairly rare changes that could be detected in the synthesis of total proteins of ecdysteroid-treated cells (see Chapter 7). However, when cells were seeded at low density, in order to impede their aggregation, several

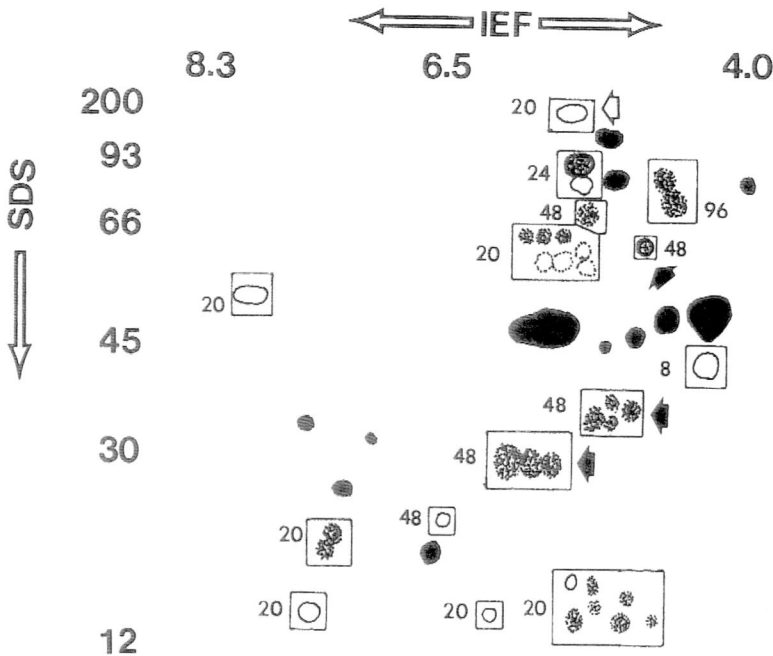

FIGURE 6.2   Summary diagram of radiolabeled cell surface proteins and their modulation by 20-hydroxyecdysone. (O) disappearing proteins; (✸) proteins induced by hormone; (●) reference proteins; (◌) glycoproteins; (◀) proteins which failed to appear at low cell densities; (◊) proteins which failed to disappear at low cell densities. Proteins are grouped and labeled based on the earliest time at which some type of change was observed (in hours after hormone addition). The molecular weight standards used were myosin, phosphorylase B, bovine serum albumin, ovalbumin, carbonic anhydrase, and cytochrome c. Woods and Poodry (1983) *Dev. Biol.* 96, 23, with the kind permission from Dr. Poodry and Springer-Verlag.

electrophoretic spots lacked their typical increase, which means that cell contact *per se* might be necessary for the induction of some of the proteins.

During a comparison of the effect of 20-hydroxyecdysone on the surface proteins of two cell sublines, which differed in their capacities for hormone-induced aggregation, Rickoll *et al.* (1986) observed, indeed, significant differences between their modulated polypeptides:

After their labeling with N-[³H]acetylglucosamine, Galewsky *et al.* (1988) found a large 110-kDa electrophoretic band in hormone-treated S3 cells only, i.e., a Schneider's line which is capable of forming large aggregates (see Chapter 8). This glycoprotein appeared to be identical to a major hormone-dependent protein secreted by the same cells, as previously detected by Rickoll and Galewsky (1988), and whose secretion can be inhibited with the ionophore monensin. Now, when monensin was added to cultures of 48 h hormone-treated S3 cells (10 mM monensin, 1 h), the cell aggregates could be easily dispersed by gentle agitation of the flasks. Moreover, in a nitrocellulose-blot cell-binding assay, hormone-treated cells displayed a preferential binding to the band corresponding to this 110-kDa glycoprotein. A polyclonal antibody raised against this so-called P110 was able to inhibit the reaggregation of hormone-treated S3 cells (Galewsky and Rickoll, 1989).

Because this glycoprotein is primarily a soluble extracellular component, it was deduced from these data that it might function as "an extracellular ligand increasing cell–cell association during hormone-induced aggregation."

Notably, very similar modifications in the composition of cell surface glycoproteins were observed in the presence of 20-hydroxyecdysone, during metamorphosis of imaginal discs *in vitro*. See, for instance, the data of Rickoll and Fristrom (1983) who isolated a membrane vesicle fraction (enriched for the plasma membrane enzyme Na^+, K^+-ATPase) or those of Woods *et al.* (1987) who used the radioiodination method: about 15% of the surface molecules underwent significant changes and it is likely that such modulations, via mediated cell–cell interactions, play a crucial role in the cellular rearrangements of disc morphogenesis.

At least some of these hormone-induced surface proteins from imaginal discs seem to be identical to those detected in hormone-induced cell lines (Rickoll and Galewsky, 1987; see also Roberts, 1975). This constitutes an additional argument in favor of considering established cell lines as valuable models for studying the mechanisms of hormone action.

Note: A thorough analysis of the cell surface antigens of 12 established cell lines, carried out by Spragg *et al.* (1982) with a large array of monoclonal antibodies, should be quoted in this context.

B. Major Cell Adhesion Molecules

1. Fasciclins and Other Cell Adhesion Molecules with Immunoglobulin-like Domains

According to the "labeled pathways hypothesis," today widely verified, the progression of axon fascicles in embryonic nervous systems is selectively guided by differential surface recognition glycoproteins. Mechanisms and molecules appear to be remarkably conserved throughout phylogeny.

The so-called fasciclins were first characterized, or rediscovered, in the *Drosophila* embryo. Their encoding genes have been cloned and corresponding mutations could be isolated. Most of these molecules (with the notable exception of fasciclin I) are members of the immunoglobulin (Ig) superfamily, that is to say they contain, in their extracellular domain, several repeats of a typical three-dimensional structure, the "Ig-fold" [a sandwich of two sheets made of antiparallel β-strands, stabilized by a disulfide bond, and providing a hydrophobic interior (Fig. 6.3)].

Fasciclin II displays great overall sequence homology with the well-known vertebrate neural adhesion molecule (N-CAM), while neuroglian closely resembles L1.

a. Fasciclins and neuroglian

In order to test the potential adhesive properties of vertebrate cell surface molecules, Tadeichi (*1988*) studied their expression, after transfection, in cultured cells. Snow *et al.* (*1989*) were the first to adapt this efficient strategy to *Drosophila* cells (Fig. 6.4).

S2 cells usually grow in suspension as single roundish cells, with low adhesivity. It must be noted that no fasciclin III transcripts could be detected in this cell line normally, or in other lines tested (*S1* or *Kc* cells). The authors constructed an expression plasmid in which the complete *fas III*-encoding cDNA was driven by a "conditional" promoter (*hsp70* promoter; see Chapter 9). A cotransfection system, associating this vector with an α-amanitin-resistant plasmid, allowed the selection of *fas III*-expressing cells. The low levels of constitutive transcription of fasciclin III and of its expression on cell surface could be greatly increased on a simple 15-min heat shock. When such heat-induced cells were gently agitated in suspension for 1 to 2 h, they formed large aggregates of tens to hundreds of cells, whereas control cells remained as single-cell suspensions. The use of a monoclonal antibody demonstrated that only cells involved in these clusters produced large quantities of fasciclin III; in confocal microscopy, the molecule appeared to be concentrated at intercellular junctions. This adhesive function of fasciclin III was further confirmed by adding to preformed aggregates, under standard conditions, either *fas III*-expressing or nonexpressing cells that had been stained with the lipophilic dye DiI; only expressing cells bound to the periphery of the clusters. The elimination of Ca^{2+} from the medium had no effect on the

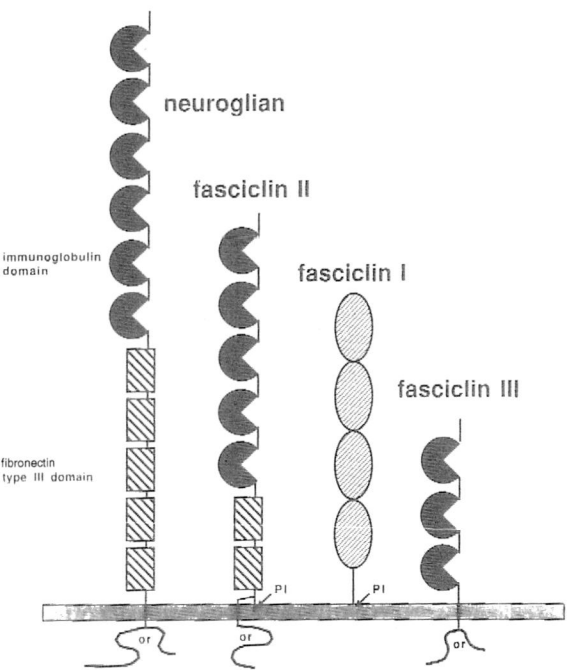

FIGURE 6.3   Schematic domain structure of the neuroglian, fasciclin II, fasciclin I, and fasciclin III axonal glycoproteins. Three are members of the immunoglobulin superfamily: Fasciclin II has five immunoglobulin C2-type domains followed by two fibronectin domains. It comes under different forms including one with a putative P1 membrane anchor. Neuroglian has six immunoglobulin C2-type domains followed by five fibronectin domains. Fasciclin III has three more divergent immunoglobulin domains and no fibronectin domains. Fasciclin I has a novel structure made up of four tandem domains. From Grenningloh et al. *(1990) CSH Symp. Quant. Biol.* **45**, 327, reproduced with the kind authorization of Professor Bieber and Cold Spring Harbor Laboratory Press.

aggregation. Moreover, when *fas III,* expressing cells were mixed with other transfected cells producing a quite different surface molecule (for instance the *Drosophila* EGF-receptor), they did not attach to one another. So, it is clear that cell aggregation mediated by fasciclin III is both homophilic and Ca^{2+}-independent.

These first experiments have been described in some detail because they plainly illustrate the wide possibilities offered by the method so that aggregation bioassays were, thereafter, extensively adopted by most investigators.

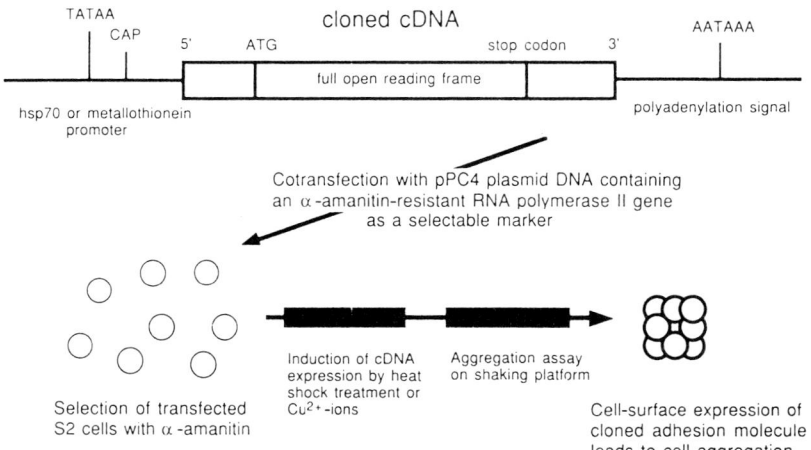

FIGURE 6.4   Strategy for examining the function of a putative adhesion molecule. Modified from Snow *et al.* (1989) *Cell* **59**, 313, reproduced with the permission of Professor Goodman and Cell Press, Inc. For more detail see text.

It was thereby established that fasciclin I, fasciclin II, and neuroglian can also function like homophilic cell adhesion molecules (Elkins *et al.*, 1990a,b; Grenningloh *et al.*, 1990; Hortsch *et al.*, 1995): for instance, *S1* cell lines (which normally express *fas I* endogenously), or *fas I* transfected *S2* cells aggregate in a specific manner, as proved by efficient inhibition of the process with antisera raised against fasciclin I (Fig. 6.5). The observation of a cell-type specific sorting into homogenous clusters, when *fasc I*, expressing cells were mixed, either with *fasc III*, or neuroglian, transfected cells, was of even greater interest.

These data are very suggestive of the putative role played by these molecules in "selective axonal fasciculation" during neurogenesis.

b. Drtk

A *Drosophila* gene was cloned owing to its striking homology with the tyrosine kinase family of mammalian neurotrophin receptors; its expression was indeed detected in selective tracts of the *Drosophila* embryonic nervous cord. The transfection of *S2* cells with a *hsp70* promoter-equipped vector (Pulido *et al.*, 1992) resulted in the synthesis of a surface glycoprotein with an apparent molecular weight of 160,000 (named *gp160^{Drtk}*). Heat-shock-induced expression caused cell aggregation in large clusters, whether or not Ca^{2+} ions were present. However, unlike

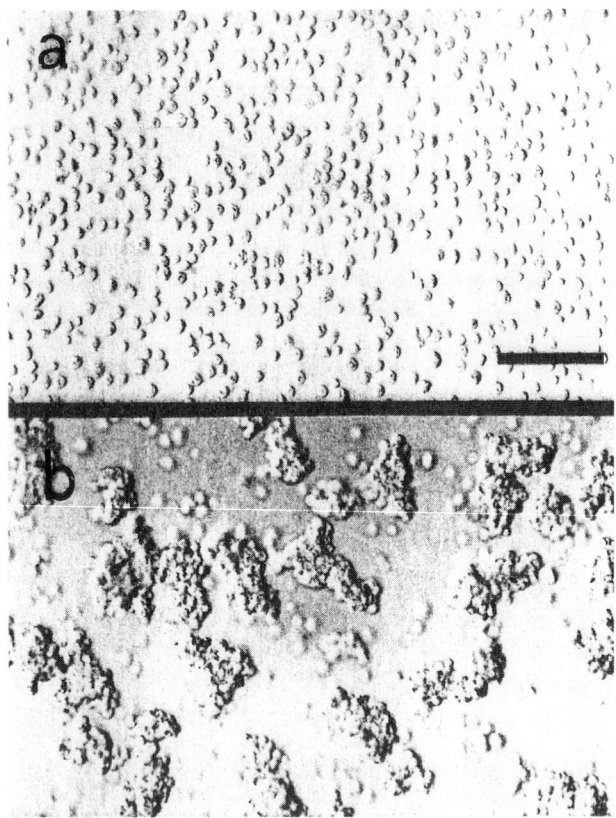

FIGURE 6.5   Aggregation of *S2* cells transfected with the fasciclin I cDNA (driven by hsp70 promoter). Cells were heat shocked for 20 min, and, after 90-min recovery, were agitated for 20 min. (a) Cell subline *5H6*, which does not express Fasciclin I; (b) Transfected cell subline *1B12* which expresses fasciclin I on the surface of almost all cells, after 20 min. From Elkins *et al.* (1990b); *Cell* **60**, 565, with the permission of Professor Goodman and Cell Press, Inc.

the above described molecules, *gp160^Dtrk* possesses, in addition to an extracellular domain with six Ig-like motifs and other noticeable sequences, a typical tyrosine kinase catalytic domain within its intracytoplasmic region. The authors were able to demonstrate that cell adhesion, as mediated by *gp160^Dtrk*, induces a phosphorylation of its own tyrosine residues, but not when heat-induced cells were prevented from aggregating; such an activation of the tyrosine kinase activity of this cell surface protein might thus play an important role in cell–cell communication.

c. Neuromusculin

A novel *Drosophila* cell adhesion molecule has been described (Kania *et al.*, 1993). The expression of the *nrm* gene is restricted to a few tissues during embryonic development (neural precursor cells and muscles), so that its product might function as a chemoattractant or substrate adhesion molecule for motoneuron growth cones. After cloning and sequencing, the deduced protein appeared to be a membrane-anchored molecule, with a short cytoplasmic tail, and belonging to the immunoglobulin superfamily (with nine putative Ig-like domains).

In *S2* cells transformed with a proper expression vector, two different forms of *nrm* were immunodetected: a 145-kDa membrane-linked protein and a shorter 135-kDa molecule secreted in the medium. This latter results from a proteolytic cleavage, as proved by the fact that another cell subline transfected with a *nrm* cDNA lacking the putative cleavage site no longer produced this lower molecular form.

Moreover, aggregation experiments with transformed cells showed that neuromusculin seems to act like an homophilic CAM (the soluble form being probably able to compete with the transmembrane form).

Note: As for other members of the immunoglobulin superfamily, it is noteworthy that cultured cells from *Drosophila,* a species in which conventional *major histocompatibility complexes* (MHC) have not been identified,* proved able, however, after proper cotransfection, to express and assemble at their surface, and in apparently normal conformation, large quantities of mammalian class I MHC molecules. Such a heterologous system made it possible to study empty class I heterodimers and to analyze the process of acquisition of extracellular peptides (Jackson *et al.*, 1992, 1994; Song *et al.*, 1994).

Similarly, stable cotransfectants from the *S3* cell line could produce large amounts of secreted MHC class II α and β chains, and characterization of the purified material showed that the molecules were correctly assembled into $\alpha\beta$ heterodimers (Wallny *et al.*, 1995).

---

* Shalev *et al.* (1983) reported serological evidence of the presence of β-microglobulin-like and H2-like epitopes on the surface of *Drosophila* cells (*Kc0* and *OM3* lines). Further attempts to isolate H2 or $\beta2\text{-}m$ homologous sequences from a *Drosophila* genomic library remained fruitless (Israel *et al.*, 1985).

2. CADHERINS

Cadherins are another important family of cell–cell adhesion molecules in vertebrates, and their name refers to the characteristic Ca^{2+} dependence of their homophilic adhesive activity. Since it is well known that the elimination of Ca^{2+} by various chelators greatly facilitates the dissociation of most tissues, cadherins should play a prominent role in the cellular cohesion of organisms. Moreover, different cadherins or combinations of various forms might confer adhesion specificities to all types of tissue cells (which would explain why cells sort themselves out, during normal morphogenetic movements or in experimental conditions when several cell types are mixed). See review in Tadeichi (1988).

Cadherins are transmembrane glycoproteins, with typical repeated domains (cadherin domains) in their extracellular region; the association of their intracellular domain with a group of cytoplasmic proteins (catenins, plakoglobin) and, via this complex, with cytoskeletal components, seems to be crucial for cell-binding functions.

There is increasing evidence that a similar Ca^{2+}-dependent cell adhesion system exists in *Drosophila,* although the molecules involved may be somewhat different.

Gratecos *et al.* (1990) reported that mechanically dissociated early embryonic cells of *Drosophila* formed pluricellular aggregates within 30 min when cultured in glass vials put on a roller (100 rpm); electron microscopy showed subsequent development of cell junctions of the *adherens*-type: (1) This binding was clearly Ca^{2+}-dependent and also temperature-dependent. (2) If established line cells (*S2* or haploid *1182*), labeled with a fluorescent stain, were mixed with the embryonic cell suspension, no chimeric cluster was formed, which suggests some cell-type selectivity. (3) When dissociated embryonic cells were treated with a low concentration of trypsin (0.01%; 5 min; 25°C) and on the condition that, at least, 1 mM Ca^{2+} was present, they retained their aggregability; conversely, after trypsin treatment in the absence of Ca^{2+}, they hardly aggregated at all. (4) Cell aggregation could be inhibited by an antiserum directed against cell surface components. These features are all very reminiscent of the properties of cadherins.

Some surface molecules of this type, as well as others which are known to interact with cadherins, have been identified.

One of the *Drosophila* segment polarity genes, *armadillo,* encodes a modular protein which is the homologue of the vertebrate adhesive junction proteins *β-catenin* or *plakoglobin* (Pfeifer and Wieschaus, 1990).

With the use of heterologous probes, the *Drosophila* α-*catenin* gene could also be cloned (Oda *et al.*, 1993); antibodies were thus prepared against a fusion protein and the immunostaining of *MLDmBG-1* line cells (from larval CNS, see page 162) revealed clear signals at cell-cell boundaries, but not at other points of the cell surface. Moreover, after cell labeling with [^{35}S]methionine, the electrophoresis of the immunoprecipitated materials showed several bands: a 110-kDa band recognized by an anti-α-*catenin* antibody, a 106-kDa band characterized by an anti-*armadillo* antibody, and a 150-kDa glycoprotein (perhaps some cadherin?).

In addition, two surface molecules with many cadherin-like domains have been characterized in *Drosophila*. Both are the products of two tumor-suppressor genes whose mutations result in neoplastic overgrowths of larval brain or imaginal disks: *lethal(2)giant* (Lützelschwabe *et al.*, 1987; Jacob *et al.*, 1987) and *fat* (Mahoney *et al.*, 1991). The former one has no transmembrane domain, although it seems to be associated with the cell surface, whereas the latter is a huge integral protein (more than 5,000 amino acids). They contain 8 and 34 tandem cadherin-like domains, respectively (compared to only four in vertebrate cadherin extracellular region).

However, there is, so far, no direct evidence of any adhesive activity in these *Drosophila* molecules,* even though it is currently suspected that, in vertebrates, some downregulation of cadherin expression might be implicated in tumor invasion.

In any case, mixed cells from different imaginal discs did have the ability of sorting out, during either *in vivo* or *in vitro* culture (Fausto-Sterling and Hsieh, 1987): Trypsin or collagen-dissociated cells were kept in suspension on a roller drum, and they formed large aggregates in which, after 40 h, tightly connected disc cells were sitting in small cystlike groups. In order to distinguish cells from different types of imaginal discs, or from differently determined regions of the same disc, the authors combined a cytochemical staining for acid phosphatase with the use, for one of the partners, of the acid phosphatase null mutant *acph*$^{n-11}$. Prospective wing disc cells, for instance, did sort out from leg cells, and also from prospective notum cells (although they derived from the same wing discs; see also Section I.C).

---

* According to further information, *l(2)gl* product should no longer be reasonably considered as a recognition molecule, since it appears not to be accessible from the outside and rather to be attached to the inner surface of plasma membrane, near cell contact sites (Strand *et al.*, 1991).

In this connection, another interesting observation was made in transient *in vitro* cultures by Gauger *et al.* (1985): when single cells or small clumps from defined imaginal discs were incubated, on a rocking shaker, with germ-band-shortening stage embryos (dissected from their vitelline membrane), they displayed a preferential binding for the epidermis of the embryonic segments from which they derived.

Note: A novel member of mammalian cadherins was recently characterized by cloning from rat liver and its adhesive properties were tested after heterologous expression in *Drosophila S2* cells (Berndorff *et al.*, 1994).

C. Integrins and Their Putative Extracellular Matrix Ligands

Members of the integrin family are the major transmembrane receptors whereby cells attach to the adjacent extracellular matrix (ECM), although they may sometimes also mediate direct cell–cell interactions. Through their intracytoplasmic domain, they form a link with the cortical cytoskeleton.

They are heterodimerized glycoproteins, comprised of two large (100 kDa) integral polypeptides, α and β, stabilized by Ca^{2+} ions. In vertebrates, the combination of a variety of both types of subunits generates a multitude of different, and functionally distinct, integrins. Their corresponding ECM ligands usually contain a typical binding site, the sequence Arg-Gly-Asp-Ser/Thr (called the RDGT motif).

The *Drosophila* integrins were, at first, identified by random immunocytological screening with monoclonal antibodies, during a search for cell surface antigens that would display a specific developmental and regional distribution. They are therefore known as position-specific (PS) integrins: for instance, in "mature" wing imaginal discs of a 3rd instar larva, PS1 integrin specifies the dorsal compartment, whereas PS2 integrin shows a complementary ventral preference. They obviously ensure the correct apposition of dorsal and ventral epithelia during wing morphogenesis, via direct or indirect contacts (Wilcox *et al., 1981, 1989;* Brower *et al., 1984;* Brower and Jaffe, *1989*). Moreover, during embryogenesis, PS1 integrin is mainly expressed in ectodermal and endodermal derivatives, whereas PS2 is found on mesodermal derivatives (particularly at sites of muscle attachment).

1. INTEGRINS

(a) Wilcox *et al.* (1984) established that the monoclonal antibodies that had revealed regionally restricted PS antigens in imaginal discs could detect the presence of similar antigens on the surface of many *Drosophila* line cells. Out of the 11 lines tested, *Emal¹* cells (Bernhard *et al.*, 1979, see Chapter 3, page 154) bound the highest levels of each antibody and, thus, provided a very convenient source of antigens.

SDS electrophoretic patterns of the antigens extracted by PS antibodies from cultured cell and from imaginal disc lysates were very similar (Fig. 6.6). Each antigen corresponds to a complex composed of a common glycoprotein of 110 kDa associated with one or more unique proteins

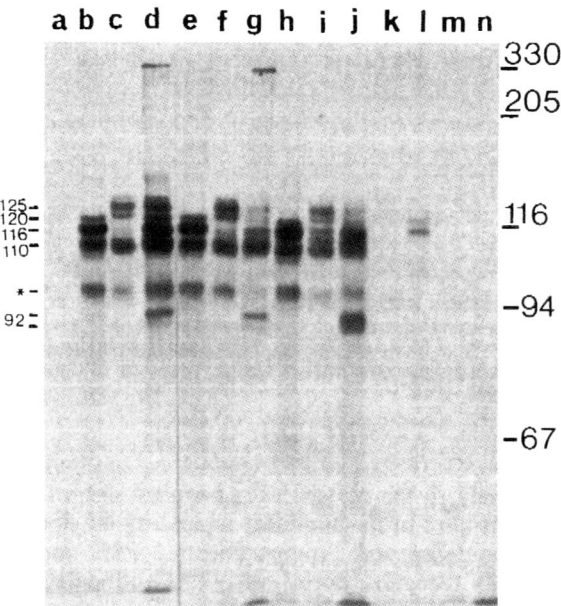

FIGURE 6.6 The components of the three PS antigens in *Emal-1* cells and imaginal discs. SDS-PAGE on 8% gels of antigens extracted from *Emal1* cell and imaginal disc lysates by PS antibodies covalently attached to Affi-Gel. Major glycoprotein components are indicated with their molecular weight values: Lanes a and k: controls (*Emal¹* cell lysates extracted by deactivated Affi-Gel). Lanes b-d: PS1, PS2 and PS3 antigens, respectively, extracted from a lysate of *Emal¹* cells. The PS3 antigen (lane *d*) contains (in addition to the unique *gp92*) components running identically to all the PS1 and PS2 components. The remaining lanes show products extracted sequentially, first by PS1, then PS2 and finally PS3 antibodies from lysates of *Emal¹* cells (lanes e-g), imaginal discs (lanes h-j) and *D1* cells (lanes l-n).

Each lysate was prepared from an equal volume of cells. After blotting to nitrocellulose, glycoproteins were detected by Con A/HRP staining. From Willcox *et al.* (1984); *EMBO J.* 3, 2307, with the permission of the European Mol. Biol. Organization.

(*gp116-120* in PS1, *gp120* in PS2, respectively). As proved by a treatment with cross-linking reagents, both components are closely bound on the cell surface, albeit in a noncovalent manner. The total number of these complexes was estimated to be about 2×10^4 per *Emal[1]* cell.

This heterodimeric organization is very reminiscent of that of vertebrate integrins. Further analysis, still carried out with cultured cells (Leptin *et al.*, 1987), showed that *Drosophila* PS antigens are indeed strikingly similar to integrins in their overall structure and biochemical properties. For instance, the common component *gp110* migrated more rapidly under nonreducing conditions than under reducing conditions, which suggests that it is, in its native state, densely compacted by intramolecular disulfide bonds (as vertebrate integrin β subunits are). On the other hand, the mobility of the variable and larger components was increased after reduction, which was due to the dissociation of disulfide-linked heavy and light chains (as in vertebrate integrin α subunits).

Because of the initial characterization of PS integrins, it was interesting to look for the integrin equipment and adhesive properties of cultured cell lines derived from imaginal discs. Peel and Milner (1992), who had established several of these lines (see Chapter 3), studied, by immunostaining, the expression of PS integrins in clonal sublines from wing and leg disc cells. These cells all show an obvious propensity to form large multicellular aggregates and the authors noticed a strong correlation between antibody staining and cell aggregation; single cells were poorly stained. Moreover, in monolayers, the antigen was clearly localized at points of cell–cell contact, whereas in large clumps, the staining was found to be heavier and more diffuse. Disappointingly, these imaginal disc-derived cells proved positive to both PS1 and PS2, even in clonal sublines, which suggests that they had lost, in the long run of culture, the identity of their "dorsal" or "ventral" origin.

See also a recent comparison, using two-dimensional electrophoresis with microsequencing, between proteins from imaginal discs and a wing disc-derived cell line, *CME W2* (see page 162) (Santaren *et al.*, 1993).

(b) There is a lethal mutation of *Drosophila*, called lethal *myospheroid* (*l(1)mys*, mapped at 7D1-5), in which, among other gross abnormalities, embryonic muscles detach from the body wall, at the time of the first contractions, and become spheroid. The cDNA cloning of this gene (McKrell *et al.*, 1988) revealed that it encodes a protein highly homologous to a β subunit of vertebrate integrins (45% of overall sequence homology, with a typical concentration and alignment of 56 cysteine residues).

In their pioneering works on differentiation *in vitro* of embryonic neuroblasts and myoblasts, Donady and Seecof (1971, 1972), then Cross and Sang (1978) observed failure, or at least important defects, in the differentiation of cultured myocytes from myospheroid embryos (see Chapter 2).

A more recent analysis (Volk *et al.*, 1990), using the same primary culture system, showed that, when myotubes were usually formed at around the 15th hour of culture, integrin could be primarily detected by immunofluorescence at cell–cell contacts. After 36 h, a sarcomeric pattern was visible and integrin antibodies prominently decorated the Z bands (in colocalization with an actin antibody). On the other hand, null *mys* cells from dissociated late gastrulae failed to stain with antibodies against β-PS integrin. Integrin-deficient myoblasts underwent fusion, however, but the cytoarchitecture of myotubes remained somewhat disorganized and lacked defined Z bands.

Note: The MP300 protein is typically expressed by muscle precursors at muscle–ectoderm and muscle–muscle attachment sites. Its sequence analysis revealed some homology with the repetitive domains of vertebrate α-actinin and spectrin.

Primary cell cultures of myotubes, seeded on laminin-coated coverslips, permitted the study of the subcellular localization of *MP300* (Volk, 1992). This latter was found to colocalize with integrin in small patches which may represent focal adhesion plaques. Cultures from integrin-deficient myospheroid embryos, in which the typical membrane location of *MP300* was no longer observed, served as a valuable control.

(c) The cloning of *Drosophila* integrin genes* made their functional analysis in transfected cells possible (Bunch and Brower, 1992): Schneider's S2 line cells were chosen, because they express very little endogenous integrin and grow as rounded and nonaggregating cells. In constructed expression vectors, the sequence coding for PS integrin β-subunit, or PS2 α-subunit, was driven by a *hsp70* promoter. As shown on immunoblots or by immunocytostaining, transformed cells expressed significant levels of α (or β) on their surface, and this was greatly enhanced after heat shock. As for their adhesive abilities, transformed cells, unlike control ones, could spread on uncoated plastic culture dishes, providing the medium contained a substantial amount (12%) of a proper batch of fetal

---

* In addition to the cloning of the *mys* gene, i.e., encoding the *Drosophila* integrin β subunit (McKrell, *1988*, see above), the cDNA coding for *PS2* α subunit was identified by Bogaert *et al.* (*1987*). Its deduced product presents extensive homologies with the heavy and light chain components of Vertebrate α-subunits. See also Wehrli *et al.* (*1993*).

calf serum (which suggests that the cells might accumulate some matrix component present in the serum, albeit, apparently, in variable concentrations).

2. EXTRACELLULAR MATRIX

Although the most common ligands of vertebrate integrins, which are present in the extracellular matrix, especially fibronectin (FN), laminin, type IV collagen, have been characterized in *Drosophila,* the specific ligands of identified PS integrins have still to be defined. [See reviews on *Drosophila* extracellular matrix by Fessler and Fessler (1989), and Fessler *et al.* (1994)].

A protein very similar to vertebrate fibronectin (with a subunit of molecular weight 230 kDa) was detected in *Drosophila* hemolymph with polyclonal antisera (Gratecos *et al.,* 1988). Moreover, from conditioned culture media of *Drosophila* established cell lines, Fessler and collaborators could purify several molecules that are usual components of basement membranes, namely, type IV collagen, laminin, high molecular weight proteoglycans (Lunstrum and Fessler, 1980; Fessler *et al.,* 1984). Labeling with [^{3}H]proline or [^{3}H]leucine made it possible to ascertain that these molecules were indeed neosynthesized and released by *Kc* cells (Fessler *et al.,* 1987), *S2,* and *E85* cells (Lunstrum *et al.,* 1988).

a. Fibronectin

Cultured embryonic cells can functionally make use of vertebrate fibronectin for spreading and differentiation (Gratecos *et al.,* 1988). In a culture medium supplemented with fibronectin-depleted fetal bovine serum, cells from gastrula-stage embryos could not spread normally or differentiate. Reintroduction of vertebrate FN, or addition of small amounts of *Drosophila* hemolymph, could restore both processes, at least to some extent. This effect seems to be mediated by the integrin system, as suggested by the consequences of simple addition of the Gly-Arg-Glu-Asp-Ser peptide (i.e., a specific competitor of the binding motif of most integrin ligands): following a short initial spreading, cells became rounded again and no differentiation occurred.

Hirano *et al.* (1991) tested the adhesiveness of an established cell line (*MLDmBG-1*)* to a variety of substrates: the cells spread very well on dishes coated with vertebrate vitronectin and, to a lesser extent, to those coated with fibronectin; however, addition of RDG peptides had no effect. On the other hand, the cells attach to laminin, but without spreading, and they did not even attach to collagen type I, type IV, or fibrinogen.

---

* From larval central nervous system; see page 162.

During the above mentioned transfection experiments, Bunch and Brower (1992) also studied the spreading capacities of integrin-overexpressing S2 cells on different substrates, after the cells' surrounding matrix had been removed by trypsin. They observed, too, an extensive spreading on vitronectin, and a moderate spreading on fibronectin. Moreover, the morphology of spreading cells appeared to be very dependent on the nature of the matrix molecules to which they attached themselves.

b. Laminin

Not only could laminin be isolated from Kc cell cultures, but, in differentiating primary cultures derived from 3-h-old embryos, some cell clusters secreted laminin ca. the 20th h, as detected by immunostaining (Fessler et al., 1987). This Drosophila laminin closely resembles its vertebrate homolog, with a three chain organization and a typical cross-shaped appearance.

When glass coverslips, on which Volk et al. (1990) grew dissociated late-gastrula cells in the absence of serum, were coated with various ECM components, only laminin (but, unexpectedly, not collagen IV) proved a good substrate for attachment and spreading. Although the peptide CDPGYGSR reduced the adhesion of muscle cells (but not of neurons) to laminin, the interaction does not require any known PS integrin, since mys mutant cells (see above, Section I.C.1) could also attach normally on laminin under serum-free conditions. However, in such an absence of serum, laminin-adhering myotubes, even from wild-type embryos, formed neither sarcomers nor Z bands. Further addition of serum or human plasma fibronectin induced the typical development of a sarcomeric architecture. It is obvious that some external ligand, other than laminin, is essential for myogenesis.

After the cloning of the B1 light chain of Drosophila laminin, Gow et al. (1993) showed that a pCAT plasmid containing 1000 bp of its 5′ flanking region was, as expected, strongly expressed in transfected S2 cells.

c. Type IV collagen

A collagen could be purified from cultures of S2 or E85 cells and, in larger amounts, from Kc cells in partial suspension. Antibodies raised against it were shown to recognize all locations where basement membranes are expected to be, in Drosophila embryos, larvae, or adults, so that this collagen can be considered to be the Drosophila equivalent of vertebrate type IV collagen (Blumberg et al., 1987; Lunstrum et al., 1988).

It is probably an homotrimer of three identical polypeptides (denoted *Drosophila pro-α1*). The amino acid composition of the extracted protein and, similarly, the complete cDNA sequence, cloned from a cDNA library made from *Kc* cells, confirmed the significant homologies with vertebrate basement membrane collagens (Blumberg *et al.*, 1987, 1988).

Observed with electron microscopy, monomeric *Drosophila* procollagen IV molecules looked like flexible threads (428 nm long) with a prominent knob at one end.

Pulse-chase labeling of cell cultures indicated that *Drosophila pro-α1* chains assemble relatively slowly into triple helices; then, these latter become disulfide linked to each other in a layer associated with the cell surface.

d. *Proteoglycan-like* papilin

Among the proteins secreted into the culture medium by *Kc* cells, Campbell *et al.* (1987) identified a large sulfated glycoprotein with many other characteristics of a proteoglycan.* O-linked carbohydrate chains were found to be relatively short, however, and correspond to neutral sugars and not amino sugars. Further immunolocalization of this molecule, named *papilin,* showed that it is a component of the basement membranes in the organism *Drosophila.*

Some *Kc* sublines synthesize primarily, or even exclusively (for instance, the *Br⁻* subline), the core protein. In SDS-gel electrophoresis, the apparent molecular mass of this core protein is about 400 kDa, whereas that of the completely glycosylated form approximates 900 kDa. As expected, the incorporation of radioactive sulfate was greatly decreased in the presence of a xyloside (i.e., a competitive inhibitor which can indeed block the addition of sulfated o-linked sugar chains to the core protein of many vertebrate proteoglycans).

e. *Glutactin*

Another much smaller sulfated protein was found to be present in medium conditioned by *Kc* cells and it was at first erroneously identified as *entactin* (Olson *et al.*, 1987, 1990). It was finally called *glutactin* because half of its carboxyl domain consists of glutamic acid or glutamine residues. The amino domain displays an intriguing homology with several

---

* It is notable that Parker *et al.* (1991) studied a crucial step of N-linked protein glycosylation in *DM3* cells.

In addition, an UDPglucose: glycoprotein glucosyltransferase (170 kDa) was recently characterized by Parker *et al.* (1995) in microsomes from highly secretory *Kc* cells.

serine esterases, such as acetylcholinesterase, although it lacks the catalytically crucial serine residue (see below neurotactin). Glutactin was immunolocalized in characteristic regions of the nervous system envelope and of muscle basement membranes and nothing is known of its potential enzymatic activity.

f. Gliotactin

Another transmembrane protein that shows, in its extracellular domain, a high degree of sequence similarity to the serine esterase family has recently been identified and named *gliotactin,* because it is transiently expressed on peripheral glial cells; it seems to be required for the formation of the peripheral blood–nerve barrier.

When its coding sequence, in an expression vector, was transfected into Schneider's cells, a 105 to 110-kDa protein could be characterized on Western blots from membrane proteins. No cell aggregation was observed, so that gliotactin does not appear to bind to itself, or to other proteins present on the cell surface of control Schneider's cells.

g. Syndecan

As in mammals, *Drosophila syndecan* is an heparan sulfate proteoglycan which is expressed at the cell surface and can be shed from cultured cells. The conservation of this class of molecule suggests that it can be used as a receptor or coreceptor for extracellular signals.

Syndecan could be immunocharacterized in conditioned media from *Kc* cells, but not from *S2* cells (Spring *et al.,* 1994).

h. Tiggrin

Fogerty *et al.* (1994) purified, from a *Kc167* subline and its 24 h conditioned culture medium, a novel *Drosophila* EMC protein, comprising the potential cell attachment motif RDG, and which is the first proven ligand of αPS2βPS integrins.

With a purified antibody, *tiggrin* cDNA could be cloned from an expression library of *Kc167* cells and it predicts a 255×10^3 molecular weight protein, with the integrin recognition motif RDG located near the C terminus and a large central domain composed of 16 repeats of a new type.

In *Drosophila* larval musculature, tiggrin was observed, by immunostaining, at the muscle–epiderm attachment sites.

The use of transformed *S2* cell lines, overexpressing αPS2 and βPS integrins (see above) made it possible to assay the capacity of a tiggrin-coated substrate to mediate cell spreading: whereas control *S2* cells did not spread, the so-called *HSP2(C)* cells showed extensive spreading, and it could be specifically inhibited by increasing concentration of soluble RDG peptide.

D. Intercellular Signaling Surface Molecules

1. *NOTCH, DELTA, AND SERRATE: MOLECULES WITH EGF-LIKE STRUCTURES*

The genetic control of neurogenesis has been extensively studied in *Drosophila*: the CNS derives from two symmetrical cell monolayer bands along the embryonic ventral ectoderm. Among the approximately 1600 cells, initially equipotent, of these "neurogenic regions," about one-quarter move inside and become neuroblasts, whereas the others remain at the surface and form ventral epidermal structures. This cell fate decision is controlled by a series of genes, collectively known as neurogenic (NG) loci, of which *Notch** is the best characterized—genetically and molecularly. In null mutations of any of these NG genes, homozygous embryos are inviable and display a hypertrophy of the nervous system at the expense of epiderm, which reflects a developmental misrouting of ectodermal cells into aneural pathway [as first described by Poulson, more than 50 years ago (*1940*)].

The resemblance of their mutant phenotypes implies that all these NG genes possibly act within a common regulation circuitry.

The products of two of them (*Notch* and *Delta***) have been shown to be integral proteins which, besides a typical membrane-spanning domain, contain, in their extracellular regions, multiple tandem EGF-like repeats, i.e., units 40 amino acid long with high sequence homology to mature vertebrate epidermal growth factor (especially a characteristic spacing of cystein residues). The enormous (300 kDa) N protein comprises as many as 36 such repeats, and the smaller *Delta* product only 9.

There is increasing evidence that neural/epidermal dichotomy relies on cell surface interactions between neighboring cells of the neurogenic area. The analysis of genetic mosaic clones and of direct transplantation of isolated mutant cells showed that N mutant cells display cell autonomy,

---

* The first mutation in this locus was reported by Mohr as early as *1919*. The name refers to the typical wing-nicking phenotype of heterozygous *Notch* mutant flies.

** *Notch* (Kidd *et al.*, *1983*; Wharton *et al.*, *1985*; Johansen *et al.*, *1989*); *Delta* (Kopczynski *et al.*, *1988*); *Serrate* (Fleming *et al.*, *1990*; Thomas *et al.*, *1991*).

but not Dl cells. This is consistent with the hypothesis whereby the *Notch* product functions as a receptor and *Delta* constitutes the signal molecule. See review by Artavanis-Tsakonas and Simpson (1991).

The complex phenotypic interactions observed between *Notch* or *Delta* alleles and another *Drosophila* gene, *Serrate,* which also encodes a transmembrane protein with 14 EGF-like repeats in its extracellular domain, evokes similar direct protein interactions and signaling at the cell surface.

Since the expression of so-called neurogenic loci, such as *Notch,* is not restricted to embryonic stages or to neural analagens, the possible intermolecular interactions of their products may be analyzed in transfected cultured cell systems.

Kc cells spontaneously express a normal *Notch* protein, but a small number of the detected molecules consists only of the EGF-like repeat domain, and this processing does not seem to be due to an artifact of long term culture (Fehon and Johansen, 1988; de Cock *et al.,* 1991, 1992b). On the other hand, *S2/M3* cells express an aberrant *Notch* messenger and no detectable *Notch* product. *Delta* protein was also absent in this latter line, which was therefore chosen for transfection assays.

Fehon *et al.* (1990) demonstrated that transformed *S2* cells expressing *Notch* on their surface, as confirmed by immunofluorescence with antibodies against the extracellular domain, aggregated with cells transfected to produce *Delta.* This heterophilic interaction, which required Ca^{2+} ions, is specific, since it could be inhibited by anti-*Notch* antibodies. In addition, while *Notch* cells did not bind to one another, *Delta* cells aggregated homotypically; this latter aggregation, however, was weaker than *Notch–Delta* aggregation, which suggests that *Notch* and *Delta* might compete for binding to *Delta,* on the cell surface.

According to unpublished observations (as quoted by Artavanis-Tsakonas and Simpson, 1991), in *Notch–Delta* cell aggregates, *Notch* cells displayed "vesicular structures containing both proteins, as if *Delta* had been internalized using *Notch* as the receptor."

Moreover, a variety of deletion constructs showed that *Notch* and *Delta* interact via their extracellular domains. Thus, a deletion of most of the long intracytoplasmic region had no effect in aggregation assays. Particularly noteworthy was the observation that, among the 36 EGF-like repeats of *Notch,* only two of them (repeats 11 and 12) were necessary and sufficient for *Notch–Delta* binding (Rebay *et al.,* 1991). As for *Delta,* the important structural element might be the N-terminal segment rather than the EGF repeats (Shepard and Muskavitch, as reported by Bunch and Bower, 1993).

It should be pointed out that such cell aggregations, as observed in experimental conditions, do not mean that the two products of these neurogenic loci function primarily as cell adhesion molecules; they might merely reveal a direct and efficient interaction of both molecules, which is, in fact, expected for their putative signaling function.

In a similar aggregation assay, *Serrate*-transfected cells, when mixed with *Notch*-expressing S2 cell, adhere to them, in a Ca^{2+}-dependent manner analogous to that of the *Notch–Delta* interaction. No aggregate, however, could be observed between *Serrate* cells themselves, or between *Serrate* and *Delta* cells (Rebay *et al.*, 1991).

When analyzing several alterations in the *Notch* protein that could generate gain-of-function "antineurogenic" phenotypes (i.e., underproduction of neuroblasts in transformed embryos), Lieber *et al.* (1993) discovered that one of these active truncated proteins, restricted to the sole intracytoplasmic domain of *Notch*, accumulated in nuclei. After transfection into cultured cells, this partial protein was also found to be located in nuclei (which allowed a functional dissection, in order to define two distinct nuclear localization signals). For all that, such an ectopic nuclear accumulation of its truncated segment, when it is untethered from the cell membrane, does not necessarily mean that *Notch* plays, *in vivo*, any role in the nucleus.

In recent studies carried out in cotransfected S2 cells, Artavanis-Tsakonas's group (Diederich *et al.*, 1994; Matsuno *et al.*, 1995) addressed the functional role of *Deltex* (*Dx*), another neurogenic gene, in the *Notch* signaling pathway. They propose that *Dx*, by antagonizing the interaction between *Notch* and *Suppressor of Hairless*, may prevent the sequestration of this latter factor in the cytoplasm.

Note: An aspartyl/asparaginyl β-hydrolase, similar to the enzyme that posttranscriptionally hydroxylates Asp and Asn residues within certain EGF-like modules of several vertebrate proteins and might thereby regulate protein–protein interactions, was characterized in S2 cells extracts (Monkovic *et al.*, 1992).

2. SIGNAL TRANSDUCERS WITH LEUCINE-RICH REPEATS (LRRs)

a. Toll

Among the dozen genes involved in the control of the dorsoventral pattern of the *Drosophila* embryo, the *Toll* protein, present in the oocyte membrane, appears to act like a receptor for an extracellular signal (which is probably the product of some other "maternally transcribed" gene, discharged in the perivitelline space of the egg); relying this signal to the cytoplasm, in the ventral regions of the embryo, *Toll* influences the nuclear gradient of the morphogen *dorsal* (i.e., a *Drosophila* member of the Rel/ NF-κB family of transcription factors).

Sequence homologies of the extracellular domain of *Toll* with the human adhesion factor platelet glycoprotein *Gp1* (more especially the presence of many "leucine-rich repeats"), whereas its intracytoplasmic domain is similar to that of the interleukin 1 receptor, led Keith and Gay (1990) to test the adhesiveness of the *Toll* product in transformed *S2* cells. Its coding sequence was put under the control of an *hsp70* promoter. Under conditions of gentle agitation, only heat-induced *Toll-S2* cells formed large aggregates. Mixing experiments with labeled control cells, which were found to participate in the clusters, suggested that the aggregation was, in fact, heterotypic. Moreover, immunofluorescence microscopy revealed that *Toll* protein accumulated in the plasma membrane at sites of cell–cell interaction. By comparison with the thrombin-induced platelet aggregation, which is not directly mediated by *Gp1*, the authors could not dismiss the possibility that *Toll* promotes cell adhesion indirectly through the activation of other surface adhesion molecules. Further analyses have been recently reported (Kubota *et al.*, 1993, 1995).

The specific enhancing influence of *Toll* on the intranuclear transloca- tion and activity of *dorsal* was analyzed by transient cotransfection of Schneider's cells [this *dl* activity being measured by its ability to increase the expression of the CAT gene put under the control of a *zen* promoter, which is a well-known target of *dorsal* (Rushlow *et al.*, 1989; Norris and Manley, 1992, 1995)]. Not only did the deletions of most of its extracellular domain not affect *Toll* activity, but the expression of its intracytoplasmic domain alone (restricted to *IL-1R* homology) was suffi- cient for enhancing *dl* activity.

Furthermore, among the three genes that are known to act downstream of *Toll*, in *Drosophila*, the *tube* gene was studied by the same authors with the same transfection system. *Tube* expression can also enhance *dl* transcriptional activity, although, apparently, in a proper manner: when *dl* protein is localized in the nucleus, so is *tube*, and this latter might act as a chaperon and coactivator of *dorsal*.

Moreover, because of its homology to the mammalian IL-1 receptor, the possible involvement of *Toll* in the immune response of *Drosophila* has been recently investigated by the group of Hultmark (Stockholm University), in a *Drosophila* hemocyte cell line (Rosetto *et al.*, 1995; see Section III.A).

b. Chaoptin

It was characterized as a specific surface antigen of photoreceptor cells and seems to be required for the development of rhabdomeric microvilli of these so-called R cells. This glycoprotein, which is merely attached to the cell membrane via a glycosylphosphatidylinositol linkage, is mostly composed of 41 tandemly arranged leucine-rich repeats (a motif well known to be important in a variety of protein–protein interactions).

Krantz and Zipursky (1990) engineered *chaoptin*-expressing S2 cells, by stable transformation with the corresponding cDNA driven by a *hsp* promoter. After heat induction, new 160 to 165-kDa electrophoretic bands were detectable within 60 min. Furthermore, the cell surface could be stained with a monoclonal antibody specific for chaoptin, and this extracellular location was confirmed by the fact that the staining pattern remained identical when the cells had been previously permeabilized with saponin (0.2%). Typical membrane linkage could be demonstrated by removal of the antigen by the phosphatidylinositol-specific phospholipase C.

Agitation in suspension culture showed that the expression of chaoptin caused aggregation of heat-induced *chp*-S2 cells. This binding is homotypic, as neither untreated *chp*-S2 cells nor heat-shocked parental S2 cells joined the aggregates. It is indeed mediated by chaoptin, as proved by its inhibition by either a monoclonal anti-*chp* antibody or pretreatment with phospholipase C.

c. Connectin

This molecule shows a typical distribution on the surface of a small subset of embryonic muscles and, concomitantly, of the neural growth cones, during the period of motoneuron outgrowth and innervation. The molecule might be attached to the cell membrane via a phosphotidyl (PI) anchor.

The *connectin* cDNA, linked to a metallothionein promoter, was transfected into S2 cells, the transformed cells being selected, as usual, by cotransfection with an α-amanitin-resistance plasmid (Nose *et al.*, 1992) or with a plasmid conferring a G418 resistance (Meadows *et al.*, 1994). In the presence of Cu^{2+}, i.e., a good inducer of the metallothionein promoter, *con*-S2 cells expressed a 62-kDa protein (i.e., close to the sequence-deduced value) recognizable by anticonnectin antibodies. Moreover, induced cells form very large aggregates (up to thousands of cells). To ascertain that connectin did not bind, in a heterotypic fashion, to some ligand endogeneously expressed by the S2 line, control S2 cells (labeled with the lipophilic dye DiI) and con-S2 cells were mixed and agitated together. Very few control cells were actually incorporated into the con-S2 aggregates, therefore the adhesion mediated by connectin should be considered to be homotypic.

3. OTHER PUTATIVE ADHESION OR SIGNALING SYSTEMS

a. Sevenless and bride of sevenless

These two transmembrane proteins constitute a communication system, by direct cell–cell contact, between two neighboring photoreceptor cells, during the development of the Drosophila retina: the commitment of the so-called R7 cell, in each ommatidium, depends on the interaction between the sevenless (sev) protein tyrosine kinase receptor, expressed in the R7 cell, and its ligand, encoded by the bride of sevenless (boss) gene and localized in the R8 cell.

The cultured cell aggregation strategy was used, once again, to demonstrate an efficient heterotypic aggregation between cell lines expressing sev and boss product, respectively (Krämer et al., 1991). The specificity of this interaction was revealed by its inhibition by either anti-sev or anti-boss antisera and the fact that this inhibition was completely reversed by blocking this latter antibody with a 40-fold molar excess of the boss peptide. A requirement for exogenous Ca^{2+} was suggested by the aggregation inhibiting effect of the chelator EDTA.

In this system, too, an internalization of the entire boss transmembrane molecule by sev-expressing cells was, surprisingly, observed (Cagan et al., 1992), and this process could be, subsequently, confirmed in vivo.

b. Wingless and engrailed

Intercellular signaling processes seem to be involved in the intrasegmental pattern of the Drosophila embryo. In 3-h-old embryos, engrailed (en) and wingless (wg) are expressed in adjacent rows of cells of the posterior half of each segment.

As proposed by Cumberledge and Krasnow (1993), wg might function as an extracellular signal for maintaining en expression in neighboring cells. To test this hypothesis, the authors inaugurated a novel strategy which consists of reconstructing, in culture, various associations between embryonic cells isolated by "whole animal cell sorting" (WACS, a promising technique previously devised by the same laboratory; see Krasnow et al., 1991).

When purified *en*-expressing cells (isolated by fluorescence-activated cell sorting from dissociated embryos carrying an *en* promoter–LacZ transgene) were cultured alone, the fraction of the cells that could be stained with an anti-*en* antibody rapidly decreased. On the other hand, coculture with *wg*-expressing embryonic cells (isolated, in parallel, in an analogous manner) prevented this loss. To ascertain that the stimulation is specific to the *wg* protein, *en*-embryonic cells were mixed with stably transformed S2 cells producing *wg* under the control of a *hsp70* promoter: only heat-induced cells could rescue *en*-expressing cells. Moreover, the activity of the conditioned medium from these *wg*-S2 cells was negligible, which suggests that *wg* is not, or is only poorly diffusible and that one is dealing with a true intercellular signaling process.

Investigation of the biological activity of the *wingless* product was carried out with transfected cells by van den Heuvel *et al.* (1993) and van Leeuwen *et al.* (1994). These latter authors developed a culture assay, using an imaginal disc cell line (*cl-8*, see Chapter 3) and measuring the effects on the *adherens* junction protein *armadillo* (which is a well-known genetic target for *wingless*). Wg protein, in the extracellular matrix (as seen in coculture) but also as a soluble extracellular molecule (by addition of medium conditioned by S2 cells overexpressing Wg protein), can increase, about 10-fold, the level of the *arm* protein, most probably by increasing its stability.

More recently, the functions of *dishevelled (dsh)*, another segment polarity gene that seems to be the earliest component in the *wingless* signaling pathway, could be studied in the same system. In *wg*-stimulated *cl-8* cells, the *dsh* product remained in the cytoplasm, but was hyper-phosphorylated and, concomitantly, *arm* protein was increased (Yanagawa *et al.*, 1995). Overexpression of *dsh* in transfected cells, even in the absence of Wg, resulted in the raising of *arm* levels, which made it possible to analyze the different domains of *dsh* product that are necessary for activity.

c. Neurotactin

The name of this transmembrane glycoprotein refers to the fact that, in late embryogenesis, it becomes restricted to cells in the central and peripheral nervous system, and several arguments suggest that it plays a role in cell adhesion. Its extracellular domain contains a region with an intriguing homology to cholinesterase, although the central block including the critical serine residue, here replaced by arginine, is not conserved (de la Escalera *et al., 1990*; Hortsch *et al., 1990*).

Amalgam, another *Drosophila* putative recognition protein, might be a possible ligand because its spatial distribution is strikingly similar to that of neurotactin.

Direct evidence of the adhesion/recognition functions of *neurotactin* was provided by cell transfection experiments (Barthalay *et al.,* 1990). The induction of neurotactin, whose cDNA had been put under the control of a heat-shock promoter, allowed the testing of partative adhesion properties of transformed *S2* cells (Fig. 6.7). Although the protein could be clearly detected by immunostaining on the surface of heat-shocked cells, these latters remained in a single cell suspension. Moreover, *Kc* or *1182* cells, which, unlike *S2* cells, spontaneously express high levels of neurotactin, did not aggregate either. This absence of homotypic adhesiveness led the authors to look for heterotypic potentialities. In fact, when dissociated gastrula cells (labeled with methylene blue) were poured onto a confluent layer of induced neurotactin transfectants, about one-third of these embryonic cells bound to *neur-S2* cells. It is thus suggested that only a subpopulation of the embryonic cells expresses a ligand. This heterotypic recognition is actually mediated by neurotactin, since it could be specifically inhibited by anti-neurotactin antibodies.

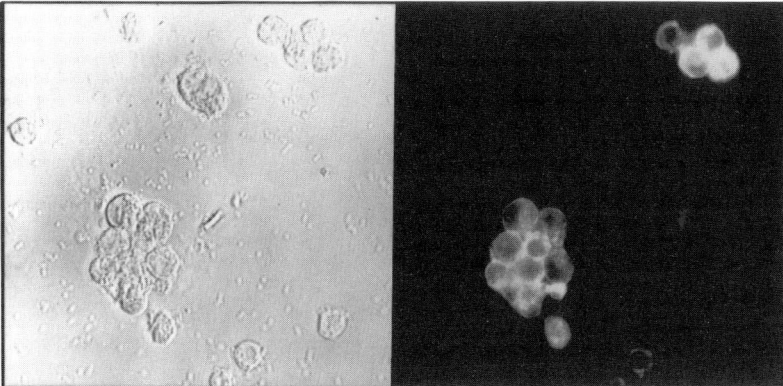

FIGURE 6.7   Aggregation of *S2* cells transfected with neurotactin cDNA. After cotransfection with the selection plasmid *pPC4* (see Chapter 9 and Appendix 9.F), transformed cells have been selected for resistance to α-amanitin. They are exposed to the 100,000 g supernatant from *Drosophila* embryonic esotracts, that contains a ligand activity. The presence of neurotactin is revealed by an ACm + a secondary antibody labeled with FITC (observation by phase contrast and immunofluorescence, respectively). It can be observed that unlabeled cells are not involved in the aggregates. Courtesy of Dr. Piovant, University of Marseille-Luminy, France.

d. Frizzled

The best studied of the tissue polarity genes that control hair polarity during *Drosophila* pupal wing morphogenesis is the *frizzled (fz)* locus. The expression of a *hs-fz* fusion gene in transfected cultured cells confirmed that the *fz* protein is a membrane protein (Park *et al.*, 1994).

e. Argos

The product of the *argos* gene is needed for cell fate decision, in the *Drosophila* eye morphogenesis.

Freeman (1994), using transfection into cultured cells, could demonstrate that the protein is indeed secreted in a soluble form. It was detected, by anti-*argos* antibody staining, in the Golgi apparatus of transfected S2 cells.

Note: Heterologous transfection of S2 cells and subsequent aggregation were used by Sap *et al.* (1994) to demonstrate the homophilic adhesive properties of a receptor tyrosine phosphatase (R-TPK-k) (i.e., a large family of transmembrane molecules with a protein tyrosine phosphatase homology domain).

II. GROWTH FACTORS, RECEPTORS, AND PROTOONCOGENES

Growth factors and protooncogenes appear more and more clearly to be components of common cellular regulatory pathways which control the proliferation rate and differentiation of the cells. Several of these circuits of regulation appear to have been highly conserved in *Drosophila*. See review by Hoffman (*1989*).

A. *Drosophila* Epidermal Growth Factor Receptor Homolog

When a *Drosophila* genomic library was screened with a v-*erb* probe containing the kinase domain, several *src* proto-oncogene homologs were identified (see below), but, in addition, an unique sequence that codes for a protein possessing three functional domains very similar to those of the human EGF-receptor was isolated by low-stringency hybridization (Livneh *et al.*, 1985; Wadsworth *et al.*, 1985). The gene of this *Drosophila* EGF-receptor homolog (or *DER*) is located on the right arm of the 2nd chromosome (position 57C); it was later shown (Schejter and Shilo, *1989*) that it is allelic to a locus essential for embryonic development and corresponding to various mutant phenotypes [designated *faint little ball (flb)* or *torpedo*]. See review by Shilo and Raz (*1991*).

Let us point out that the endogenous ligand of DER is not obvious, since no sequence homologous to EGF has been found in *Drosophila*. It is perhaps not necessary for this ligand to be soluble and several important surface proteins with many EGF-like repeats (see above paragraph) occur in *Drosophila* (reminiscent of the integral form of the mammalian EGF precursor protein).

Using Western blotting with polyclonal antisera prepared against denaturated human EGF-receptor, Thompson *et al.* (1984, 1985) detected, among membrane proteins of *S2* and *Kc* cells, a predominant electrophoretic band of approximately 190 kDa [a value in the molecular weight range of the sequence-deduced size (170 kDa) of the DER product]. Moreover, after cell labeling with [^{32}P]P$_i$, they could immunoprecipitate a phosphorylated protein of identical molecular weight, as can be done with the autophosphorylated human EGF receptor. Yet, this protein was not able to bind [^{125}I]-labeled EGF, so that DER probably more closely resembles the vertebrate protooncogene *neu* (i.e., a homolog of the true EGF receptor but whose ligand is still unknown).

The biogenesis of DER was studied by Zak and Shilo (1990) in *S2* cells, after their transfection with DER expression constructs. No endogenous DER protein or RNA could be detected in the parental subline of Schneider's cells they used.

Pulse-labeling of transfected cells showed the synthesis of a 170-kDa molecule which was specifically immunoprecipitated by anti-DER antibodies. Treatment with tunicamycin revealed its glycoprotein nature, while a mild trypsin treatment of the whole cells, before extraction, demonstrated that the mature form is partially exposed at the cell surface.

The monitoring of autophosphorylation of DER *in vivo* provided the authors with a sensitive assay for potential ligands: in fact, neither EGF or TGF-α, nor laminin or conditioned medium, had any triggering effect.

Interestingly, the same laboratory (Wides *et al.*, 1990) attempted to deregulate the tyrosine kinase activity of DER by engineering in the expression plasmid various structural changes that mimic the oncogenic alterations reported in its vertebrate counterparts. For instance, the highest increase in kinase activity (7-fold) of DER was observed after a single amino acid substitution (from valine to glutamic acid) in the transmembrane domain (i.e., the exact point mutation that transforms the vertebrate *neu* gene into a potent oncogene).

The *Drosophila Spitz* gene is known to encode a TGF-α homolog. S2 cells were transfected with a *Spitz* expression construct in which a stop codon had been introduced by PCR in the putative cleavage site of the coding sequence, so that the secreted form of Spitz was detectable, in significant amounts, in the culture medium. When DER-expressing *Drosophila* cells (see above) were incubated with this conditioned medium, Spitz triggered the DER signaling cascade: there was a dramatic increase in the level of tyrosine autophosphorylation of DER and, consequently, MAP kinase was activated (Schweitzer *et al.*, 1995).

Note: A posttranslational β-hydroxylation of specific aspartyl and asparagyl residues occurs within certain EGF-like modules of several mammalian proteins. Preparations derived from *Drosophila* S2 cells could catalyze this specific hydroxylation, in a manner quite similar to that of the purified mammalian aspartyl/asparagyl β-hydroxylase (Monkovic *et al.*, 1992).

B. *Drosophila* Insulin Receptor

Insulin-like peptides were immunocharacterized in *Drosophila*, as in other insects and many nonvertebrate groups (Seecof and Dewurst, 1974; Meneses and Ortiz, *1975;* LeRoith *et al.*, *1981*). Later, at all developmental stages of *Drosophila*, a large membrane-associated glycoprotein (ca. 400 kDa), which was able to bind ^{125}I-labeled insulin with high affinity and displayed an insulin-dependent protein tyrosine kinase activity was detected (Petruzzelli *et al.*, *1985a,b*).

When the sequence of the human insulin receptor became available, it became possible to clone a *Drosophila* homolog by low stringency hybridization to human probes (Petruzzelli *et al.*, *1986;* Nishida *et al.*, *1986*).

It should be recalled that the vertebrate insulin-receptor complex is a tetramer composed of two α subunits, which are external to the cell membrane and bind insulin, and two β subunits, traversing the membrane and possessing a tyrosine kinase domain for the transduction of the signal. Deriving from a common precursor, both types of polypeptides are glycosylated and linked to one another by disulfide bonds.

First, it must be remembered that the addition of mammalian insulin to *Drosophila* embryonic cell cultures was found to be beneficial for cell differentiation (Seecof and Dewurst, 1974), as had been recognized for long time in vertebrate *in vitro* cell culture. Therefore, medium supplementation with bovine or porcine insulin is today common practice in most *Drosophila* cell primary cultures (see Chapter 1, Section IV.I.2).

The use of established cell lines has greatly facilitated the determination of the overall structure and ligand specificity of the *Drosophila* insulin-receptor homolog (DIRH or, more simply, INR, according to the current terminology). In wheat germ agglutinin eluates from three cell lines (*Kc*, *S2*, and *S3*), antibodies prepared against peptide sequences characteristic of either α- or β- subunits of the human proreceptor could recognize two different forms of the tyrosine kinase component (β unit: 95 and 1760kDa) and one insulin-binding component (α unit: 110 kDa, albeit slightly larger in the *Kc* line (120 kDa) (Fernandez-Amonacid and Rosen, 1987). As for the ligand specificity of this DIRH, it is noteworthy that only insulin (for instance, porcine insulin, 100 nM), but not IGF-I, IGF-2, EGF, or even the insulin-like hormone purified from the silkworm, could induce the phosphorylation of the 95- and 170-kDa proteins, in extracts from *Kc* cells.

Further analyses conducted by the same group (Fernandez-Almonacid, 1992; Fernandez *et al.*, 1995), in *S2* cells transfected with various sequences of the *inr* ORF under the control of a metallothionein promoter, confirmed that the INR$_{\beta170}$ subunit may be cleaved into the β_{90} subunit and a C-terminal fragment of 60 kDa. Both possible forms of mature receptors [i.e., α_2-(β170)$_2$ and α_2-(β90)$_2$] can activate the Ras/MAP kinase [recall that Biggs and Zipursky (1992) have identified the *Drosophila* MAP kinase homolog (called DmERK-A)].

Note: In vertebrate cells, after internalization of the insulin-receptor complex, the breakdown of the hormone seems to be initiated by a nonlysosomal *insulin-degrading enzyme (IDE)*. A cytosolic enzyme, sharing both physical and kinetic properties with its vertebrate homolog, could be isolated from *Kc* and *S3* cells (Garcia *et al.*, 1988; Stoppelli *et al.*, 1988). Such a *Drosophila* IDE is identical to a 100-kDa protein (*dp100*) previously characterized in *Kc* and *S2* cytoplasmic fractions for its dual binding specificity for both insulin and EGF (or related molecules, particularly TGF-α). In fact, TGF-α, like insulin, but not EGF, can be degraded by *Drosophila* and mammalian IDE (Garcia *et al.*, 1989a,b).

C. Transforming Growth Factor β-like *Decapentaplegic* and *60A*

The product of the *Drosophila decapentaplegic (dpp)* gene is a member of the TGF-β superfamily of growth and differentiation factors.

The highly diverse phenotypes displayed by mutations of this large genetic unit (located on the left arm of chromosome 2) have revealed that it is implicated in many crucial developmental events, primarily in relation with pattern formation in early embryogenesis (establishment of a dorsoventral specificity) as well as in adult morphogenesis (for instance, in the proper formation of radial positional information in imaginal disc). Although five different overlapping transcripts are produced, with a complex temporal and spatial regulation, they all encode the same polypeptide (Padgett *et al., 1987;* St. Johnston *et al., 1990).*

In its 110 C-terminal amino acids, the *dpp* protein shows extensive sequence homology with the vertebrate members of the TGF-β family (especially with human bone morphogenesis proteins, BMPs), namely, a conserved motif of some seven cysteine residues and a putative cleavage site, right at the border of this region of high conservation. It must be remembered, indeed, that all vertebrate factors of this superfamily are first synthesized as large propolypeptides; the C-terminal region is cleaved, then a dimerization forms the mature secreted substance. See review by Gelbart (*1989*).

1. DECAPENTAPLEGIC

Kc line cells can be readily adapted to grow in a serum-free culture medium (see Chapter 1), which suggested that they might be autocrine, i.e., able to synthesize their own growth factor.

So, using the same isolation and activation procedure that had enabled him to detect a latent form of TGF-β in most normal mammalian tissues, Lawrence and collaborators, (Benzakour *et al.,* 1990; Benzakour, 1992) demonstrated the occurrence of a TGF-β-like activity in *Kc0* cell extracts. This bioactivity, like mammalian TGF-β, induced the anchorage-independent growth of rat *NRK-49F* cells (in the presence of EGF), whereas it inhibited [^{3}H]thymidine incorporation in the Mink epithelial cell line *CCL-64* (two classical bioassays for TGF-β). It was found to be acid- and heat-stable, but destroyed by dithiothreitol.

Moreover, at a concentration of 10 μg/ml, the *Kc0* extract, but not mammalian TGF-β1, produced also a 66% inhibition of [^{3}H]thymidine incorporation in another *Kc* subline, which would seem to indicate that it has more effect on homotypic cells than on heterotypic ones (only 40% inhibition in Mink *CCL-64* cells).

In order to obtain sufficient quantities of the *dpp* protein for the study of its structure and biogenesis, Panganiban *et al.* (1990) transfected *S2* cells with an expression vector (containing the whole *dpp*-encoding cDNA driven by a metallothionein conditional promoter).

In Northern analysis, no *dpp* mRNA was found in control *S2* and only small amounts in uninduced transfected cells, whereas, after copper induction of these latter cells, the level of *dpp* transcripts was increased 20- to 100-fold (depending on the subline).

With antibodies prepared against a *dpp* polypeptide made in bacteria, predominant 68- to 70-kDa proteins could be immunoprecipitated from cell lysates, but only from induced cells (the predicted *dpp* product is 65 kDa). In addition, two smaller proteins, 47 and 20 kDa, were recognized in both induced cell extracts and their conditioned culture medium. Partial amino acid sequencing established that they were cleavage products of the full-length protein. It was finally demonstrated that the smaller fragment, corresponding to the C-terminal domain, was secreted by the *S2* cells as a homodimer of 30 kDa apparent molecular weight (in nonreducing SDS-gels) and was rapidly adsorbed onto the surface of plastic petri dishes (from which it could be recovered by washing with a high concentration of NaCl and anionic detergent). Moreover, unlike the situation observed with the TGF-β1 released by a hamster cell line *CHO*, the mature secreted product was not associated, in a latent form, with the N-terminal 47 kDa component or other exogenous protein.

2. 60A

A new member of the TGF-β family, most closely related to human BMP5, 6, and 7, was recently identified in *Drosophila* by the PCR approach (Wharton *et al.*, 1991; Doctor *et al.*, 1992). It was named *60A* after its localization on polytene chromosomes.

The expression of *60A* cDNA in stably transformed *S2* cells was verified by immunostaining with antibodies prepared against a bacteria-made fusion protein. Immunological methods permitted the analysis of the processing and secretion of this *60A* protein (Doctor *et al.*, 1992).

A 48-kDa protein, which is probably the *60A* proprotein, could be detected. In conditioned medium, smaller polypeptides were recognized by pro-region-specific antibodies and C-terminal-specific antibodies; their respective molecular weights were 29 and 16 to 20 kDa. Under nonreducing conditions, this latter C-terminal fragment was recovered as an homodimer, with an apparent molecular weight of 32 kDa.

Note: It is particularly interesting that both *Drosophila* proteins of this TGF-β superfamily (*dpp* and *60A*), as produced in the *S2* expression system, were able to induce endochondral bone formation in mammals (Sampath *et al.,* 1993).

D. Basic Fibroblast Growth Factor Binding Protein

By incubating *Drosophila* culture cells (*S2* or imaginal disc-derived lines *D12* and *D20*) with human ^{125}I-labeled basic fibroblast growth factor (bFGF) or bovine ^{125}I-labeled acidic aFGF, Doctor *et al.* (1991) established, by competitive binding and cross-linking experiments, the presence of specific binding sites for basic FGF, but not for acidic FGF. The size (160 kDa) of the binding protein is close to that of vertebrate FGF receptors. Moreover, *Drosophila* cells were able to metabolize this growth factor, but not aFGF, in conditions strikingly similar to the degradation of FGF by vertebrate cell types.

E. *Jak/Stat* Signaling System

Several growth factor receptors lack intrinsic kinase activity, but, after their own stimulation by specific ligands, they can recruit a member of the so-called Janus tyrosine kinase (*JAK*) family (conversely devoid of a transmembrane domain). Activation of these latter enzymes, by autophosphorylation, led to the phosphorylation of downstream effectors, the *Stat* family of transactivators (i.e., "signal transducers and activaters of transcription"), which are then translocated in nucleus. This *Jak/Stat* pathway is well documented in mammals.

A Jak kinase homolog, *hopscotch,* has been characterized in *Drosophila* (Binari and Perrimon, 1994). A gain-of-function mutation (hop^{Tum-l}) causes hypertrophy of the hematopoietic organs and formation of melanotic tumors. Very recently, the same laboratory (Harrison *et al.,* 1995) showed that overexpression of *hop* or its allele, in transfected *S2* cells, results indeed in tyrosine phosphorylation of both normal and mutant *hop* proteins.

Independently, Sweitzer *et al.* (1995) have identified a Stat-like activity in *Drosophila S2* cells, after treatment with vanadate/peroxide (a procedure which, in various mammalian cell types, enables bypassing the normal receptor stimulation step). Electrophoretic mobility shift assay showed the formation, in treated cell lysates, of a GRR-binding complex, containing two phosphoproteins of 100 and 150 kDa.

F. *Drosophila* Protooncogenes

The use of vertebrate oncogene probes has allowed the characterization of several proto-oncogene homologs from the *Drosophila* genome, and they correspond to the three known major families, namely *src, ras,* and *myc/myb*. See reviews by Bishop *et al.* (*1985*) and Shilo (*1987*) (Fig. 6.8). Intriguingly, relatively few studies, so far, have used *Drosophila* cell lines, in this important research field.

1. *Src* AND *RELATED PROTOONCOGENES*

These protein-tyrosine kinases, with high homology in their catalytic domain with vertebrate counterparts, belong to two groups: those which, contrary to the above-mentioned receptors, do not cross the membrane, but are associated to it at their N terminus (for example, true *src*), and others which have no association with the cell membrane (for example, *abl*).

The patterns obtained with the various *src* probes tested were identical for DNAs extracted from flies or from cultured cells (Shilo *et al.,* 1981). Moreover, Simon *et al.* (1983) established that *Kc* cells possess tyrosine-

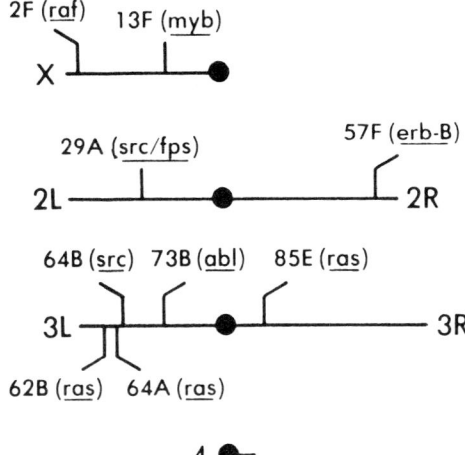

FIGURE 6.8   Protooncogenes of *Drosophila melanogaster.* They are here assigned to their approximate positions on polytene chromosomes. From Bishop *et al.* (*1985*); *CSH Symp. Quant. Biol.* 727, with the permission of Professor Bishop and Cold Spring Harbor Laboratory Press.

specific protein kinase activity in a fraction immunoprecipitated with antisera directed against RSV-induced tumors. Similarly, Gateff (1988) noticed that *pp60^{c-src}* kinase activity was significantly increased in blood cells of the tumoral mutant *l(2)mbn* and *l(3)mbn*, but only after their culture *in vitro*, and more especially during the exponential growth phase.

Kimchie *et al.* (1989) studied the transcription patterns of *Drosophila src, abl*, and *ras* homologs in Schneider's S2 cells (see Fig. 6.9). It must be remembered that these protooncogenes usually express one constitutive transcript in all developmental stages and, in addition, a subset of shorter transcripts differentially expressed in eggs and embryos. It is noteworthy that such maternal/embryonic specific transcripts were found in S2 cells, i.e., a line derived from dissociated embryos, which suggests that regulatory mechanisms have been conserved in the cultured cells.

a. src

Besides the great similarity of its tyrosine kinase domain, the product, called *p62^D* of the *Drosophila src* homologous gene *(Dsrc)* possesses, in common with the protein *p60^{c-src}*, several typical phosphorylation sites that are well known to be important for regulating the catalytic activity of the vertebrate protooncogene. In order to determine whether similar mechanisms regulate its *Drosophila* counterpart, Kussick and Cooper (1992) overexpressed the latter by transient transfection into S2 cells.

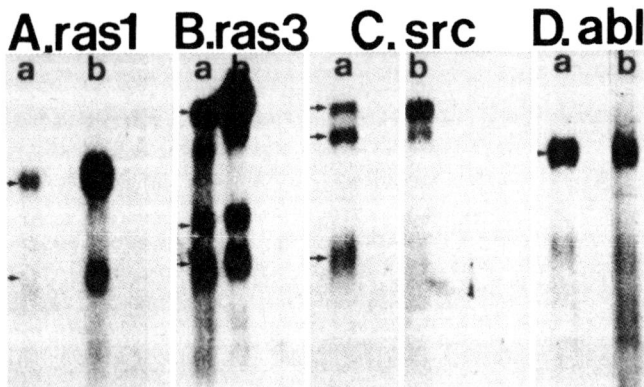

FIGURE 6.9   Some protooncogene transcripts in *Drosophila* tissue culture cells. Poly(A)⁺ RNA blots of (A, B, Da) 0 to 2 h embryos or (C, a) adult flies, and Schneider's S2 cells, were hybridized with DNA probes of (A) *ras1*, (B) *ras3*, (C) *src*, and (D) *abl*. From Kimchie *et al.* (1989) *Cell Diff. Dev.* **26**, 79 with the permission of Elsevier Science Publishers.

Immunoprecipitation revealed an 8-fold expression, compared to the endogenous $p62^D$. When transfected cells were metabolically labeled with $[^{32}P]P_i$, $p62^D$ was found, by tryptic phosphopeptide analysis, to be phosphorylated, not only at the C-terminal tyrosine but at several other sites; the same sites were also observed in autophosphorylation *in vitro*. The construction of an altered $p62^D$ in which Lys was substituted for Arg in the putative ATP binding site, i.e., a mutation that should render the enzyme inactive, demonstrated that the phosphorylation of such additional sites implies indeed an autophosphorylation. Thus, in contrast with the kinases of the vertebrate *src* family when they are overexpressed in mammalian fibroblasts, the overexpressed $p62^D$ appeared to be enzymatically active in S2 cells. Schneider's cells clearly differ from vertebrate fibroblasts in their incapacity to repress *src*-like kinases; and this could even be verified when S2 cells were transfected with a wild-type vertebrate $p60^{c-src}$.

b. abl

Genetic studies of the enhancing interactions of *disabled* (*dab*) and *abl* *Drosophila* mutations, which promote defects of embryonic neurogenesis, and also the characterization, in the *dab* product, of potential phosphorylation sites very similar to the major autophosphorylation sites in *abl*, led Gertler *et al.* (1993) to propose that the product of disabled might be a substrate for the *abl* protein-tyrosine kinase.

In S2 cells, the same authors showed, with specific antibodies, that both *abl* and *dab* are endogenously expressed and, moreover, that the *dab* protein is indeed immunoreactive with antiphosphotyrosine antiserum. Unfortunately, no physical association between *abl* and *dab* could be demonstrated in immunoprecipitates.

2. ras FAMILY

A monoclonal antibody reacting with the *p21* coded by the three known members of the vertebrate *ras* gene family could recognize *ras*-related products in a lysate from *Kc* cells (with a prominent band in the 21 kDa region of the gel) (Papageorge *et al.*, 1984). Yet, the fact that the protein was not detected by another monoclonal antibody known to be specific for the native form of the vertebrate *ras* product suggested that the conformation of the *Drosophila* protein is rather different.

3. myc/myb PROTOONCOGENES

As for the *myc/myb* family, whose encoded proteins reside and are active in the nuclear matrix, it seems that only a single *myb* homolog could be rigorously identified in *Drosophila* with vertebrate probes; this might be due to a lesser degree of sequence conservation in this family (Katzen *et al.*, 1985).

Madhavan *et al.* (1985) reported the localization, on polyribosomes of *Kc* cells, of two transcripts (1.7 and 2.2 kb) which hybridized with avian v-*myc*. Yet, they established that the several genomic or cDNA clones they could thereby isolate did not reveal any significant amino acid homology with the *myc* protein (in spite of a nucleotide sequence similarity sufficient for hybridization).

4. Fos

The product of the protooncogene *fos* complexes with another protooncoprotein, *Jun*, to constitute the transcription factor *AP-1*. The *Drosophila fos*-related gene (*D-fos*) was shown to be developmentally regulated (Perkins *et al.*, 1990).

Engel and Cornelius (1995) have recently shown in *S2* cells that *D-fos* can be stimulated by some serum factor, as in mammalian cells. After a 7-day serum depletion, the reintroduction of serum (10% FBS) in the culture medium elicited a significant *D-fos* mRNA increase within 30 min. Moreover, its stimulation by the phorbol ester PMA points to a role of protein kinase C in the activation pathway.

III. *DROSOPHILA* HUMORAL DEFENSES

Like many higher insects, *Drosophila* responds to a microbiological challenge, not only by phagocytosing or encapsulating the invader, but also by the rapid synthesis of a variety of bactericidal peptides (up to a dozen) that accumulate transiently in its hemolymph. These inducible antibacterial substances, which belong to several distinct types, have a wide spectrum of activity. They are mostly small cationic peptides which might create ion channels through the bacterial membranes. Such a cell-free immune response can be triggered by various agents, especially bacteria or components of the bacterial envelope, probably through the mediation of nonclonally distributed receptors able to recognize common prokaryotic structural motifs. Moreover, although a similar induction can be repeated at each aggression, there is no immune memory.

So, this defense system bears no resemblance with that of classical antibodies, but is strongly evocative of innate immunity (so-called acute phase response) in mammals.

See reviews by Boman and Hultmark (*1987*) and Hoffmann *et al.* (*1992*, 1993), that is, from the two laboratories that have played a leading role in this research field.

Since the observation that an embryonic cell line from the flesh fly *Sarcophaga* released constitutively several antibacterial peptides* (Matsuyama and Natori, *1988*), then the demonstration of the inducible immune response *in vitro* of *Drosophila* S2 cells (Samakovlis *et al.*, 1990), cultured cells have already significantly contributed to the analysis of the molecular mechanisms of induction of these bactericidal factors, and they should constitute an irreplaceable model system.

A. Cecropins

They were the first inducible antibacterial peptides to be characterized in insects. Their name was coined from the giant lepidopteran *Hyalophora cecropia* from the pupae of which they were isolated, in the early 1980s, by the group of Hultmark (Stockholm, Sweden). Later, the use of cDNA probes from the fly *Sarcophaga* major cecropin made it possible to identify, in *Drosophila,* four functional *cecropin* genes: *Cec A1, A2, B,* and *C*, which are all arranged, with two pseudogenes, in a compact cluster cytologically mapped at the chromosomal location 99E (Kylsten *et al., 1990;* Tryselius *et al., 1992*).

Cecropins are 4-kDa nonglycosylated peptides and are active against both gram-negative and gram-positive bacteria.

Looking for a convenient *in vitro* experimental model, Samakovlis *et al.* (1990) established, with Northern blotting, that the expression of cecropin genes can be induced in Schneider's S2 cells by mere addition of microbial produts, such as lipopolysaccharide or laminarin.

Cecropins being synthesized in the organism primarily by the fat body and a fraction of hemocytes, it was tempting to test the only cell lines whose the hemocytic origin was assured, i.e., Gateff's tumor blood cell lines (see Chapter 3, Section IV.A). Moreover, in the original description of these lines, Gateff *et al.* (1980) noted that, when chance contaminations occurred in the cultures, the cells were able to inhibit excessive bacterial growth. In fact, *mbn-2* cells responded more strongly to the various bacterial substances than previously studied S2 cells (Samakovlis *et al.,*

---

* As early as *1970*, in our laboratory, Landureau and Jolles discovered that two permanent cell lines established from the cockroach *Periplaneta* produced a factor that destroyed several kinds of microorganisms.

1992). The levels of the transcripts of the four *cecropin* genes were similarly increased within the first few hours (Fig. 6.10).

Lipopolysaccharide (LPS = a constituent of the membrane of gram negative bacteria), laminarin (an algal glucan similar to cell wall structures of fungi), but also flagellin (component of bacterial flagellae) proved to be very potent inducers (active at 0.1 ng/ml). Experiments with different inhibitors suggested that the cell response seemed to require some protein synthesis and might be mediated by a G protein.

In contrast, four other *Drosophila* cell lines (*Kc0, Emal, G,* and also an *S2* subline from another source) were found to be totally unresponsive. The fact that two sublines from the same original *S2* line can differ so strikingly leads to the idea that it is perhaps too simple to assume that the particularly efficient response of *mbn-2* cells has to be correlated with their hemocytic origin. It must be remembered that some transcription of cecropin genes could also be detected, *in vivo,* in various other tissues (Kimbrell, unpublished). I suggest, therefore, that it would be worthwhile, for further experimentation in this important research field, to compare systematically the responses of clonal sublines from a number of different cell lines.

The upstream regions of several immunity genes, including cecropins and diptericins (see below), contain κB-like binding motifs and, besides, it must be remembered that in mammals the regulatory factor NF-κB

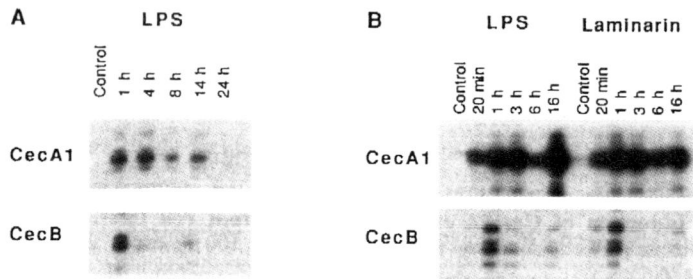

FIGURE 6.10   Induction kinetics of the cecropin *CecA1* and *CecB* genes in *Drosophila mbn-2* cells. RNAs were assayed by RNase protection at the indicated time after the addition of the elicitor at 100 μg/ml. Two independent experiments are shown: (A) expression was followed for 24 h after the addition of LSP; (B) for 18 h after the addition of LPS or laminarin. From Samakovlis *et al.* (1992) *Biochem. Biophys. Res. Commun.* **118**, 1169, with the permission of Professor Hultmark and the copyright owner.

participates in the triggering of the acute phase response. Now, there is in *Drosophila* an extensively studied member of this rel/NF-κB family of transcriptional regulators: the product of *dorsal* (*dl*) is involved in the dorsoventral patterning of the embryo and, like NF-κB, its highly controlled transport to the nucleus conditions its binding to several target promoters displaying the κB-like motif.

Engström *et al.* (1993) carried out a functional analysis of the regulatory region of a cecropin gene, by transient transfection of *mbn-2* cells with a vector containing a β-gal reporter gene under the control of upstream sequences of *Cec A1*. A −760-bp region, but not a deleted promoter with only −68 bp (that is, lacking the κB motif), ensured a high level of expression, even in the absence of induction by LPS. To test directly the involvement of the κB-binding motif, a trimer of the insect consensus κB-like sequence was ligated upstream to the inactive deletion construct; it proved indeed to be 10- to 20-fold more efficient than a similar construct with a trimer of slightly altered sequences.

The same authors characterized in *mbn-2* cells, using electrophoretic mobility shift assay, two factors that bound to a κB-like DNA probe. One of them was detected only after activation by LPS, both in the cytoplasm and nucleus. It was named DIF (*Drosophila* immunoresponsive factor) because it is putatively responsible for the induction of immunity genes, and it appeared to be very similar to a factor previously isolated from "vaccinated" *Hyalophora* pupae.

Ip *et al.* (1993) established that the true regulatory factor of the *Drosophila* immune response was, in fact, not the product of *dorsal* but that of another novel gene which they identified by low stringency screens with conserved regions of the *dorsal/rel* domain. They also named it *Dif*, although these initials mean here "dorsal-related immunity factor." The gene maps to the 36C chromosomal region, very close to *dorsal*. Its deduced protein (667 amino acids) contains a well-conserved *rel*-domain. Using a gel shift assay with nuclear extracts from LPS-induced *mbn-2* cells, in conditions similar to those reported by Engström *et al.*, they showed that the induced complex binding to the *Cec A1*-κB-like motif could be specifically disrupted by pretreating the nuclear extract with anti-Dif antibodies (prepared against an *Escherichia coli*-made fusion protein).

More recently, Petersen *et al.* (1995) brought the direct demonstration, by cotransfection assays in *S2* or *mbn* cells, that the *Dif* gene product transactivates the *Drosophila Cecropin A1* and that the κB-like motif, in the promoter of this latter gene, is imperatively required.

The well-known role of *Toll*, a *Drosophila* gene that enhances nuclear translocation and activity of the *dorsal* protein (see Section I.D.2), led

Hultmark's group to investigate the possible involvement of this gene in the immune response of *Drosophila*, especially because the *Toll* protein displays some sequence homology to the mammalian interleukin 1 receptor: Rosetto *et al.* (1995) cotransfected cells from the hemocyte line *mbn2* with plasmids containing wild-type *Toll* or a constitutively activated *Toll*[10B] mutant under the control of an *hsp* promoter, concomitantly with the *CeA1–lacZ* reporter construct. Overexpression of *Toll*[10B] gave maximal stimulation, and this stimulating effect relies on the same κB-like elements, in the cecropin promoter, as those that serve as binding sites for Dif.

B. Defensin

Insect *defensins*, thus named because they present a sequence homology with the defensins isolated from mammalian neutrophiles and macrophages, were first characterized in another dipteran, *Phormia*. They are also 4-kDa nonglycosylated peptides, but show three intramolecular disulphide bridges and are especially active on gram-positive bacteria.

An antibacterial molecule of this family has been recently purified from "immunized" *Drosophila* flies (Dimarcq *et al.*, 1994), and its microsequencing allowed the cloning of a *preprodefensin* gene (located at position 46CD on polytene chromosome).

Defensin transcripts could be detected in LPS-treated *mbn-2* cells, but, as in larvae or adults, the intensity of the signals was notably weaker than those observed for diptericin (see below) (Dimarcq *et al.*, 1994).

C. Diptericin

Another distinct type of potent inducible bactericidal peptides has been characterized by Hoffmann's group and, so far, only in Diptera.

The *Drosophila diptericin* gene was cloned (Wicker *et al.*, 1990; Reichhart *et al.*, 1992) and it maps at the position 56A, on the right arm of the second chromosome.

Concomitantly with similar experiments on transgenic flies, *mbn-2* cells were transfected with plasmids in which the chloramphenicol aminotransferase (CAT) gene was driven by different upstream sequences of the *diptericin* gene (Kappler *et al.*, 1993).

Normal *mbn-2* cells did not express diptericin, but its transcription could be rapidly (1 to 8 h) elicited by simple addition of LPS or bacteria (alive or heat-killed) (Fig. 6.11). This diptericin system in cultured cells was selected by Hoffmann's group as a model for a molecular analysis of the inducible antibacterial response.

In cells transfected with a construct containing the nucleotides −180 to −35 (upstream of the initiation site) and comprising, in particular, two 17 bp repeats harboring a decameric consensus κB-motif, the basal expression level was low, but could be enhanced 7-fold by LPS induction. Replacement of these repeats by random sequences completely abolished the inducibility. Inversely, there was a good correlation between the transcriptional activation and the number of repeats that had been put in tandem upstream of the reporter gene cassette. For instance, in the presence of eight repeat copies, a 40-fold increase of CAT activity in transfected cells was observed after LSP induction.

An electrophoretic mobility shift assay, with a radiolabeled probe made of 16 of the 17 bp of the diptericin κB-like motif, demonstrated that the extracts from *mbn-2* cells contained a major DNA-binding protein complex, 2 h after induction by bacteria or LPS. Experiments of competi-

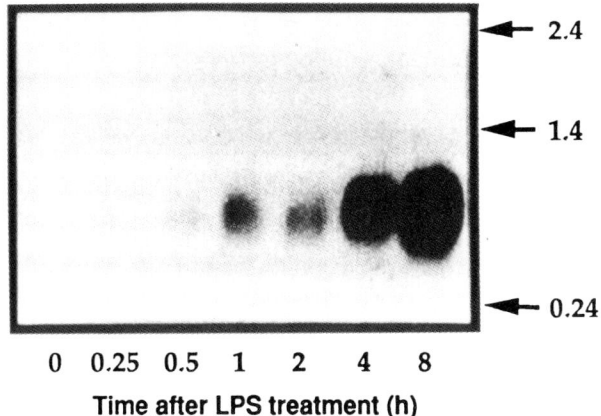

0   0.25   0.5   1   2   4   8
Time after LPS treatment (h)

FIGURE 6.11   Transcriptional profiles of the Diptericin gene following LPS treatment in *Drosophila mbn-2* cells. Poly(A)-enriched RNA (5 μg) was isolated from *mbn-2* cells at different times after LPS (Sigma, *E. coli* 55:B5) treatment and subjected to electrophoresis on a 1.5% agarose gel, blotted onto a nitrocellulose filter and hybridized with a nick-translated cDNA probe. Marker positions are indicated on the right of the figure. From Kappler *et al.* (1993) *EMBO J.* **12**, 1561, with the permission of the Eur. Mol. Biol. Organization.

tion with wild-type or mutant oligonucleotides confirmed the specificity of this binding.

Moreover, in the poly(A) RNA from bacteria-challenged *mbn-2* cells (as from "vaccinated" larvae or flies), a dorsal probe revealed the rapid intensification of a positive band at 2.8 kDa, which is the reported size for *dorsal* transcripts (Reichhart *et al.*, 1993). Likewise, a monoclonal anti-*dorsal* antibody recognized several bands in extracts from the same cells. According to the recent discovery by Ip *et al.* (1993; see above), this hybridizing RNA band, as well as the observed *dorsal*-like antigen, might in fact correspond to the novel *dorsal*-related *Dif* gene, which has the same transcript size and a high conserved *rel* domain.

The same Strasbourg laboratory (Georgel *et al.*, 1993), yet again using LPS-induced *mbn-2* cells, established by foot-printing that, in addition to κB-related motifs, another series of oligonucleotidic elements might control the expression of the diptericin gene. Interestingly, they are homologous to consensus sequences that are known to bind other different factors involved in the regulation of the acute phase response in mammals. In particular, within three repeated regions of the diptericin promoter and in close vicinity to NF-κB and NF-Il6 response elements, Georgel *et al.* (1995) have recently detected a GAAANN motif, which is a consensus response element in several mammalian interferon-stimulated genes. In transfecting *mbn-2* cells with appropriate constructs, they demonstrated that the new nucleotide motif cannot, by its own, permit the full induction by LSP of the immune response, but probably cooperatively interacts with the other response elements (as it is also true in mammals) to enhance the expression of the diptericin promoter, by about five times. Moreover, in extracts from activated *mbn-2* cells or larval fat body, a 45-kDa protein was characterized after its specific binding to the GAAANN sequence; it displays structural similarities to mammalian interferon regulatory factor 1.

D. Immunophilins

Immunophilins are cell receptors through which different immunosuppressive drugs, such as cyclosporin and FK 506, exert their effect on vertebrate lymphocytes.

A cDNA encoding a FK 506-binding protein, has been recently isolated from an expression library of *Drosophila* flies (Theopold *et al.*, 1995). Although the gene is expressed in *mbn-2* cells, their antibacterial response is unaffected by the corresponding immunosuppressive drugs.

E. Cellular Immunity: Phagocytosis

In spite of the fact that the phagocytic abilities of insect "blood" cells and their protective role against bacteria have been known for more than a century, very few contemporary studies on cellular immunity have been performed on *Drosophila* cells *in vitro*. This is probably due to the apparent fragility of newly drawn hemocytes in contact with glass or plastic substrates and to the rapid darkening of the larval hemolymph when exposed to atmospheric oxygen (resulting from the phenol oxidase activities of so-called crystal cells that burst out). Yet, many *Drosophila* cell lines have been suspected of deriving from hemocytes and there are, in any case, two lines, Gateff's *l(2)mbn* and *l(3)mbn*, which were directly established from the abundant plasmocyte-like cells of *lethal malignant blood neoplasm* mutant larvae (see Chapter 3). (See also, in Chapter 2, the description and pictures of hemocyte-like cell types differentiating in primary embryonic cell cultures.)

Rizki and Rizki (1988) defined suitable conditions for maintaining larval blood cells for several hours in sealed Sykes chambers. This system should prove very valuable for *in vitro* assays. The authors could thus demonstrate the selective destruction of lamellocytes by a "factor" (provisionally called *lamellolysin*) stored in the reservoir of an accessory gland associated with the female reproductive system of a small parasitoid wasp (*Leptopilina*). Thereby, the parasite eggs, which have to develop in *Drosophila* larval hemocoel, might escape the encapsulation response of the host (Rizki and Rizki, 1990). Surprisingly, this reservoir fluid contains very peculiar virus-like particles which can enter *Drosophila* blood cells *in vitro*. A better simulation of the physiological circulation of hemolymph was obtained when the cultures, in siliconized depression-slides, were placed on the table of a rotary shaker (70 rpm); and the bipolar shape of the *lamellolysin*-affected lamellocytes more closely resembled that of the capsule-forming hemocytes in parasitized hosts (Rizki and Rizki, 1991).

In mammalian defense systems, macrophages are able to endocytose or phagocytose a wide variety of molecules or cells, because of the unusually broad ligand specificity of their "scavenger receptors." These latter can, thus, bind a series of modified proteins, such as acetylated low density protien (or AcLDL). Using fluorescently or radioactively labeled derivatives of this experimental ligand, Abrams *et al.* (1992) looked for scavenger-like receptors in *Drosophila* cultured cells.

Primary embryonic cultures were concomitantly exposed to DiI–AcLDL and stained with X-Gal (to detect β-galactosidase activity). Most of the β-galactosidase-positive macrophages present in the cultures displayed the Dil-AcLDL red fluorescence.

Moreover, S2 cells (but not Kc) also exhibited scavenger receptor-mediated endocytosis, which means that they probably represent a macrophage-derived cell line. Like in mammalian macrophages, this endocytic pathway (monitored from the ^{125}I-labeled AcLDL degradation products, as measured in the culture medium) could be specifically inhibited by chloroquine. A cDNA that encodes a class C scavenger receptor (named dSR-C1) has recently been isolated, by expression cloning, from the same S2 cell line (Pearson et al., 1995). In *Drosophila* embryonic development, its expression is restricted to macrophage-like hemocytes.

Furthermore, Kirkpatrick et al. (1995) identified in supernatants from serum-free cultured S2 cells (and a few other cell lines) an abundantly secreted 47-kDa glycoprotein that is related to mammalian proteins produced in rheumatoid tissues and by activated macrophages. Present throughout *Drosophila* development, this so-called DS47 protein is primarily produced by fat body and circulating macrophage-like granular hemocytes, and it is secreted into the hemolymph.

Notes: (a) The endocytic abilities of S2 cells may perhaps explain their well-known high efficiency of transfection with calcium phosphate-mediated DNA (see Chapter 9). (b) According to Kosaka and Ikeda (*1983*), the reversible paralysis of *shibire*[ts] mutants, at the nonpermissive temperature of 29°C, seems to be due to some block in the recycling of synaptic vescicles, which corresponds, in actual fact, to a blockage in the process of endocytosis in general.* As primary cultures and several cell lines have been established from such mutants (see Chapters 2 and 3), it might be worthwhile investigating their capacity for endocytosis.

IV. OTHER PROTEINS OR ENZYMATIC ACTIVITIES

As has been repeatedly emphasized in this book, *in vitro* cultures may provide investigators with large amounts of a very homogeneous cellular material, especially when clonal sublines are grown.

For instance, the total protein yield of a stationary culture of Kc cells, approximates 2 to 3 mg per 25-ml Falcon plastic flask, which corresponds

---

* It was established that the *shibire* gene encodes a *dynamin*-like protein which is associated with vesicular traffic (Van der Bliek and Meyerowitz, *1991*).

to about 10^8 cells in the usual 3 ml of *D22* medium (see Appendix 1.A); it must be remembered that individual cells from this paradiploid line are indeed rather small, about 7 μm diameter. As many culture flasks as necessary can be easily prepared. See also (Appendix 3.B in Chapter 3) a protocol for growing *Drosophila* cells in mass culture with a bioreactor (reproduced by courtesy of Dr. Sondergaard).

A. Proteins

1. Two-dimensional electrophoresis of total cell extracts revealed about 300 to 400 distinct spots (see Chapter 7, Fig. 7.2A). Unexpectedly, only a very low fraction (about a dozen) of those polypeptides appeared to be quantitatively or qualitatively modified by treatment with 20-hydroxyecdysone (Berger *et al.*, 1978-1980; Couderc and Dastugue, 1980; Rollet *et al.*, 1986). This strikingly selective action of the hormone is comprehensively described in Chapter 8. Rabilloud *et al.* (1985) attempted a computer-mediated systematization of the variations of such electrophoregram patterns after various stimulations.

Another valuable method for characterizing cultured cell proteins and their possible variations consists in the production of large libraries of monoclonal antibodies. As for *Drosophila* cells, this approach was initiated by Roberts' group in Oxford (Roberts, 1975; Moir and Roberts, 1976).

2. See also a fine study of nuclear proteins by Huo *et al.* (1982)[*] and by Fuchs *et al.* (1983), and several papers by Fisher's group on nuclear *lamina* from cultured cell lines (Smith *et al.*, 1987; Smith and Fisher, 1989; Maus *et al.*, 1995). For the currently emerging picture of the role of chromatin packaging in the active and repressed state of genes, see Chapter 7, Section IX.B.4.

3. Heat-shock proteins and the conditions of their induction are discussed in Chapter 7.

4. Information concerning cytoskeletal proteins can be found in two sections. The synthesis of myogenic proteins was studied in mass cultures of purified myoblasts deriving from dissociated young embryos, and is discussed in Chapter 2. The regulation of actin and tubulin by ecdysteroids is dealt with in Chapter 8.

---

[*] A technique of preparation of *Drosophila* chromatin from cultured cells by Elgin and Miller is given in Volume 2A (1978) of "Genetics and Biology of Drosophila" (M. Ashburner and T. R. F. Wright, eds.), pp. 146–147, Academic Press.

TABLE 6.I   Main enzymatic activities[a] detected in *Drosophila* cell lines *Kc* and *S2*

Enzyme	Cell line	Control	Ecdysone	Remarks	Refs.
Carbohydrate metabolism					
α-Amylases	Kc, Ca	−			Debec (1976)
α-L-Fucosidase	Kc	−	−		Best-Belpomme (1981)
β-L-Fucosidase	Kc	−	−		Best-Belpomme (1981)
α-D-galactosidase	Kc	−	+++		Best-Belpomme (1981)
β-D-galactosidase	Kc	−	+++	see chapter 8	Best-Belpomme and Courgeon (1978)
α-D-glucosidase	Kc	+	+		Best-Belpomme (1981)
β-D-glucosidase	Kc	−	++		Best-Belpomme (1981)
α-D-mannosidase	Kc	+	++		Best-Belpomme (1981)
α-D-xylosidase	Kc	−	−		Best-Belpomme (1981)
β-D-xylosidase	Kc	+	+		Best-Belpomme (1981)
N-Acetylglucosaminidase	Kc	+++	+++		Best-Belpomme (1981)
Glucokinase (Hex 2)	Kc	+	+		Debec (1974)
	Ca, S2	+			Best-Belpomme 81
Mannokinase (Hex 1)	Kc	−	−		Alahiotis and Berger (1977)
	Ca, S2	−			Alahiotis and Berger (1977)
Fructokinase (Hex 3)	Kc, Ca	−			Alahiotis and Berger (1977)
	S2	(+)			Alahiotis and Berger (1977)
Phosphoglucoisomerase	Kc, S2	−			Alahiotis and Berger (1977)
Phosphoglucomutase	Kc, Ca	+	+		Debec (1976); Best-Belpomme (1981)
Glucose-6-phosphate dehydrogenase (G6PD)	Kc, S2	−and+			Alahiotis and Berger (1977)
6-Phosphogluconate dehydrogenase (6PGD)	Kc, Ca	+			Debec (1974)
α-Glycerophosphate dehydrogenase (α-GPDH)	Kc, Ca	(+)			Debec (1974)
	Kc, S2	+		Kc "slow"/S2"fast"	Alahiotis and Berger (1977)

Enzyme	Tissue			Notes	Reference
Phosphoglycerate kinase (PGK)	Kc, Ca	+			Debec (1976)
Alcohol dehydrogenase (ADH)	Kc, Ca, S2	−	−		Debec (1974); Best-Belpomme (1981); Alahiotis and Berger (1977)
Octanol dehydrogenase (ODH)	Kc, Ca	++	++		Debec (1974); Best-Belpomme (1981)
fumarase	Kc, Ca	+or−		according sublines	Debec (1974); Debec (1976)
malate dehydrogenase	Kc	+	+		Best-Belpomme (1981); Debec (1976)
NAD-dependent malate dehydrogenase (=malic enzyme)	Kc, Ca	++ and +			Debec (1976)
Lactate dehydrogenase (LDH)	Kc, Ca	++			Debec (1974)
NADP-dependent isocitrate dehydrogenase (IDH-NADP)	Kc, Ca	+			Alahiotis and Berger (1977); Debec (1976)
Aconitase	Kc, S2	+		"slower" form in Kc	Best-Belpomme (1981)
	Kc, Ca	+	+		

AMINO ACID AND PROTEIN METABOLISM

Aminopeptidases	Tissue			Notes	Reference
Val	Kc, Ca	−	−		Best-Belpomme (1981)
Asp	Kc, Ca	+	+		Best-Belpomme (1981)
Arg	Kc, Ca	+	+		Best-Belpomme (1981)
Glu	Kc, Ca	−	−		Best-Belpomme (1981)
Leu	Kc	−			Debec (1976)
	S2	+		major pupal isozyme	Alahiotis and Berger (1977)
trypsin	Kc	−			Best-Belpomme (1981)
chemotrypsin	Kc	−			Best-Belpomme (1981)
arginine kinase (AK)	Kc, Ca	+/−		according sublines	Debec (1974)

(continues)

329

TABLE 6.1 (continued)

Enzyme	Cell line	Control	Ecdysone	Remarks	Refs.
LIPID METABOLISM					
Lipases					
Caprylate	Kc	+	+		Best-Belpomme (1981)
Myristate	Kc	+	+		Best-Belpomme (1981)
Palmitate	Kc	+	+		Best-Belpomme (1981)
β-1-Hydroxy-acid dehydrogenase (β-1-OHDH)	Kc, Ca	−			Debec (1976)
3-Hydroxy-3-methylglutaryl-CoA reductase	Kc, S2	+			Gertler et al. 1988)
NUCLEIC ACID METABOLISM					
xanthine dehydrogenase (XDH)	Kc, Ca	−			Debec (1974)
topoisomerase II					see Chapter 7
DNA polymerase α					see Chapter 5
phosphodiesterase	Kc	−	−		Best-Belpomme (1981)
OTHERS					
acid phosphatase (Ac Ph)	Kc, Ca, S2	++	++	minor band, variable	Debec (1974); Best-Belpomme (1981)
alkaline phosphatase (A Ph)	Kc, Ca	−			Alahiotis and Berger (1977)
	Kc, S2	(+)			Debec (1976)
acetylcholinesterase	Kc, S3	++/−		see Chapter 8	Alahiotis and Berger (1977)
esterases	Kc, Ca	++/−		variations	Cherbas et al. see Chapter 8
	Kc, S2	+		only Esterase C & Esterase 6	Debec (1974) Alahiotis and Berger (1977)
tetrazolium oxidase (TO)	Kc, Ca	++			Debec (1974)

[a] +, ++, and +++, presence and relative activity; −, nondetectable activity.

5. Last, we remind the reader that glycoproteins of the cell surface have been surveyed, at the beginning of this present chapter, in the relevant context of recognition/adhesion molecules.

B. Enzymes

Enzymatic activities, because of their specificities, are excellent indicators of the functional capacities of a cell. Moreover, many enzymes give rise to very sensitive methods of detection and measure, which are priceless for quantifying the responses of the cell to any stimulation or insult. For instance, the inducibility of acetylcholinesterase and β-galactosidase by ecdysteroids in sensitive *Kc* cells (see Chapter 8) made it possible to demonstrate (a) the precocity of the cell reaction (within the first day), (b) the extremely low concentrations of hormone sufficient for eliciting a response (about one log lower than those required for promoting morphological transformations), and (c) the fair dose-dependence of the cellular responses.

Therefore, a number of data concerning a variety of enzymes are dispersed throughout the different chapters of this book. For example, enzymes involved in the biosynthesis of nucleotides, or in DNA repair, are discussed in Chapter 5.

Furthermore, many enzymatic activities correspond to several isoforms that may be typical of a particular tissue or developmental stage. So, it was tempting to look for the electrophoretic variants detectable in cultured lines, for two different purposes: (1) to have at one's disposal reliable cellular markers (for instance, in somatic hybridization experiments; see Chapter 3), or in order to distinguish the product of some transferred enzyme gene from its endogenous form (see Chapter 9). (2) The comparison of the isoenzymatic pattern of a certain cell line with the well-known characteristic enzymatic equipment of each *Drosophila* tissue, would, hopefully, give us some idea of the putative tissular origin of *in vitro* established cell lines, which, let us recall, were usually derived from hundreds of dissociated embryos (see discussion in Chapter 3).

It seemed useful to collate in table form (Table 6.I), even though this tabulation is certainly not exhaustive, most of the enzymes that have been assayed in the two most commonly used cell lines, i.e., *Kc* and *S2* cells. Their list covers a very wide spectrum of cell metabolic pathways.

References

Binari, R., and Perrimon, N. (1994). Stripe-specific regulation of pair-rule genes by *hopscotch*, a putative Jak family tyrosine kinase in *Drosophila*. *Genes Dev.* **8**, 300–312.

Bishop, J. M., Drees, B., Katzen, A. L., Kornberg, T. B., and Simon, M. A. (1985). Proto-oncogenes of *Drosophila melanogaster. Cold Spring Harbor Lab. Symp. Quant. Biol.* **50**, 727–731.

Bliek, van der, A. M., and Meyerowitz, E. M. (1991). Dynamin-like protein encoded by the *Drosophila shibire* gene associated with vesicular traffic. *Nature* **351**, 411–414.

Bogaert, T., Brown, N., and Willcox, M. (1987). The *Drosophila* PS2 antigen is an invertebrate integrin that, like the fibronectin receptor, becomes localized to muscle attachments. *Cell* **51**, 929–940.

Boman, H. G., and Hultmark, D. (1987). Cell-free immunity in Insects. *Ann. Rev. Microbiol.* **41**, 103–126.

Brower, D. L., and Jaffe, S. M. (1989). Requirements for integrins during *Drosophila* wing development. *Nature* (London) **342**, 285–287.

Brower, D. L., Willcox, M., Piovant, M., Smith, R. J., and Reger, L. A. (1984). Related cell-surface antigens expressed with positional specificity in *Drosophila* imaginal discs. *Proc. Natl. Acad. Sci.* **81**, 7485–7489.

Escalera (de la), S., Bockamp, E. O., Moya, F., Piovant, M., and Jimenez, F. (1990). Characterization and gene cloning of *neurotactin*, a *Drosophila* transmembrane protein related to cholinesterase. *EMBO J* **9**, 3593–3601.

Fleming, R. J., Scottgale, T. N., Diederich, R. J., and Artavanis-Tsakonas, S. (1990). The gene *Serrate* encodes a putative EGF-like transmembrane protein essential for proper ectodermal development in *Drosophila melanogaster. Genes Dev.* **4**, 2188–2201.

Gelbart, W. M. (1989). The *decapentaplegic* gene: a TGF-β homologue controlling pattern formation in *Drosophila. Development* **107**, (Suppl.) 65–74.

Hoffmann, F. M. (1989). Roles of *Drosophila* proto-oncogene and growth factor homologs during development of the fly. *Curr. Top. Microb. Immunol.* **147**, 1–29.

Hoffmann, J. A., Dimarcq, J. L., and Bulet, P. (1992). Les peptides antibactériens inductibles des insectes. *Médecine/Science* **8**, 432–439.

Hortsch, M., Patel, N. H., Bieber, A. J., Traquina, Z. R., and Goodman, C. S. (1990). *Drosophila neurotactin*, a surface glycoprotein with homology to serine esterases, is dynamically expressed during embryogenesis. *Development* **110**, 1327–1340.

Hynes, R. O. (1973). Alteration of cell-surface proteins by viral transformation and by proteolysis. *Proc. Natl. Acad. Sci. U.S.A.* **70**, 3170–3174.

Israel, A., Becker, J. L., LeBail, O., and Kourilsky, P. (1985). A sequence from *Drosophila* DNA cross-hybridizing with a mouse class I H-2 gene absent of relevant nucleic or amino acid sequence homology. *Biochimie* **67**, 1225–1229.

Jacob, L., Opper, M., Metzroth, B., Phannavong, B., and Mechler, B. M. (1987). Structure of the *l(2)gl* gene of *Drosophila* and delimitation of its tumor suppressor domain. *Cell* **50**, 215–225.

Johansen, K., Fehon, R. G., and Artavanis-Tsakonas, S. (1989). The *Notch* gene product is a glycoprotein expressed on the cell surface of both epidermal and neuronal precursor cells during *Drosophila* development. *J. Cell Biol.* **109**, 2427–2440.

Katzen, A. L., Kornberg, T. B., and Bishop, J. M. (1985). Isolation of the proto-oncogene c-myb from *Drosophila melanogaster. Cell* **41**, 449–456.

Kidd, S., Lockett, T. J., and Young, M. W. (1983). The *Notch* locus of *Drosophila melanogaster. Cell* **34**, 421–433.

Kosaka, T., and Ikeda, K. (1983). Reversible blockage of membrane retrieval and endocytosis in the garland cell of the temperature sensitive mutant of *Drosophila melanogaster, shibire*[ts1]. *J. Cell Biol.* **97**, 499–507.

Krasnow, M. A., Cumberledge, S., Manning, G., Herzenberg, L. A., and Nolan, G. P. (1991). Whole animal cell sorting of *Drosophila* embryos. *Science* **251**, 81–85.

Kopczynski, C. C., Alton, A. K., Fechtel, K., Kooh, P. J., and Muskavitch, M. A. T. (1988). *Delta*, a *Drosophila* neurogenic gene, is transcriptionally complex and encodes a protein related to blood coagulation factor and epidermal growth factor of vertebrates. *Genes Dev.* **2**, 1723–1735.

Kylsten, P., Samakovlis, C., and Hultmark, D. (1990). The *cecropin* locus in *Drosophila*: a compact gene cluster involved in the response to infection. *EMBO J.* **9**, 217–224.

Landureau, J. C., and Jolles, P. (1970). Lytic enzyme produced *in vitro* by insect cells: lysozyme or chitinase? *Nature* **225**, 968–969.

LeRoith, D., Lesniak, M. A., and Roth, J. (1981). Insulin in Insects and Annelids. *Diabetes* **30**, 70.

Livneh, E., Glazer, L., Segal, D., Schlessinger, J., and Shilo, B. Z. (1985). The *Drosophila* EGF receptor gene homolog: conservation of both hormone binding and kinase domains. *Cell* **40**, 599–607.

Lützelschwab, R., Klämbt, C., Rossa, R., and Schmidt, O. (1987). A protein product of the *Drosophila* recessive tumor gene, *l*(2)giant gl, potentially has adhesive properties. *EMBO J.* **6**, 1791–1797.

McKrell, A. J., Blumberg, B., Haynes, S. R., and Fessler, J. H. (1988). The lethal *myospheroid* gene of *Drosophila* encodes a membrane protein homologous to the vertebrate integrin β subunits. *Proc. Natl. Acad. Sci. U.S.A.* **85**, 2633–2637.

Mahoney, P. A., Webe, U., Onofrechuk, P., Biessman, H., Bryant, P. J., and Goodman, C. S. (1991). The *fat* tumor suppressor gene in *Drosophila* encodes a novel member of the cadherin gene superfamily. *Cell* **67**, 853–868.

Matsuyama, K., and Natori, S. (1988). Purification of three antibacterial proteins from the culture medium of *NIH-Sape-4*, an embryonic cell line of *Sarcophaga peregrina*. *J. Biol. Chem.* **263**, 17112–17121.

Meneses, P., and Ortiz, M. D. (1975). A protein extract from *Drosophila melanogaster* with insulin-like activity. *Comp. Biochem. Physiol.* **51A**, 483–485.

Mohr, O. (1919). Character changes caused by mutation of an entire region of a chromosome in *Drosophila*. *Genetics* **4**, 274–282.

Nishida, Y., Hata, Y., Nishizuka, W., Rutter, W., and Ebina, Y. (1986). Cloning of a *Drosophila* cDNA encoding a polypeptide similar to the human insulin receptor precursor. *Biochem. Biophys. Res. Commun.* **141**, 474–481.

Padgett, R. W., St Johnston, R. D., and Gelbart, W. M. (1987). A transcript from a *Drosophila* pattern gene predicts a protein homologous to the transforming growth factor-β family. *Nature* **325**, 81–84.

Perkins, K. K., Admon, A., Patel, N., and Tijan, R. (1990). The *Drosophila* Fos-related AP-1 is a developmentally regulated transcription factor. *Genes Dev.* **4**, 822–834.

Petruzelli, L., Herrera, L., Garcia, R., and Rosen, O. M. (1985a). The insulin receptor of *Drosophila melanogaster*. In "Cancer Cells 3/ Growth factors and transformation" (J. Feramisco, B. Ozanne and C. Stiles, eds.) *Cold Spring Harbor Lab. Press* 115–121.

Petruzelli, L., Herrera, R., Garcia-Arenas, R., and Rosen, O. M. (1985b). Acquisition of insulin-dependent protein tyrosine kinase activity during *Drosophila* embryogenesis. *J. Biol. Chem.* **260**, 16072–16075.

Petruzelli, M., Herrera, R., Arena-Garcia, R., Garcia, R., and Rosen, O. M. (1986). Isolation of a *Drosophila* genomic sequence homologous to the kinase domain of the human insulin receptor and detection of the phosphorylated receptor with an anti-peptide antibody. *Proc. Natl. Acad. Sci. U.S.A.* **83**, 4710–4714.

Pfeifer, M., and Wieschaus, E. (1990). The segment polarity gene *armadillo* encodes a functionally modular protein that is the *Drosophila* homolog of human plakoglobin. *Cell* **63**, 1167–1178.

Poulson, D. F. (1940). The effects of certain X-chromosome deficiencies on the embryonic development of *Drosophila melanogaster. J. Exp. Zool.* **83**, 271–325.

Reichhart, J. M., Meister, M., Dimarcq, J. L., Zachary, D., Hoffmann, D., Ruiz, C., Richards, G., and Hoffmann, J. (1992). Insect immunity: developmental and inducible activity of the *Drosophila diptericin* promoter. *EMBO J.* **11**, 1469–1477.

Schejter, E. D., and Shilo, B. Z. (1989). The *Drosophila* EGF receptor homolog (DER) gene is allelic to *faint little ball,* a locus essential for embryonic development. *Cell* **56**, 1093–1104.

Shilo, B. Z. (1987). Proto-oncogenes in *Drosophila melanogaster. Trends Genet.* **3**, 69–72.

Shilo, B. Z., and Raz, E. (1991). Developmental control by the *Drosophila* EGF receptor homolog DER. *Trends Genet.* **7**, 388–392.

St Johnston, R. D., Hopffmann, F. M., Blackman, R. K., Segal, D., Grimaila, R., Padgett, R. W., Irick, H. A., and Gelbart, W. M. (1990). Molecular organization of the *decapentaplegic* gene in *Drosophila melanogaster. Genes Dev.* **4**, 1114–1127.

Strand, D., Török, I., Kalmes, A., Schmidt, M., Merz, R., and Mechler, B. M. (1991). Transcriptional and translational regulation of the expression of the *l(2)gl* tumor suppressor gene of *Drosophila melanogaster. Adv. Enzyme Reg.* **31**, 339–350.

Tadeichi, M. (1988). The cadherins: cell-cell adhesion molecules controlling animal morphogenesis. *Development* **102**, 639–655.

Thomas, V., Speicher, S. A., and Knust, C. (1991). The *Drosophila* gene *Serrate* encodes an EGF-like transmembrane protein with a complex expression pattern in embryos and wing discs. *Development* **111**, 749–761.

Tryselius, Y., Samakovlis, C., Kimbrell, D. A., and Hultmark, D. (1992). *CecC,* a cecropin gene expressed during metamorphosis in *Drosophila* pupae. *Eur. J. Biochem.* **204**, 395–399.

Wadsworth, S. C., Vincent, W. S. III, and Bilodeau-Wentworth, T. H. (1985). A *Drosophila* genomic sequence with homology to human EGF receptor. *Nature* **314**, 178–180.

Wehrli, M., DiAntonio, A., Fearnly, I. M., Smith, R. J., and Wilcox, M. (1993). Cloning and characterization of α_{PS1}, a novel *Drosophila melanogaster* integrin. *Mech. Dev.* **43**, 21–36.

Wharton, K. A., Johansen, K. M., Xu, T., and Artavanis-Tsakonas, S. (1985). Nucleotide sequence from the neurogenic locus *Notch* implies a gene product that shares homology with proteins containing EGF-like repeats. *Cell* **43**, 567–581.

Wharton, K. A., Thomsen, G. H., and Gelbart, W. M. (1991). *Drosophila* 60A gene, another transforming growth factor β family member, is closely related to human bone morphogenetic proteins. *Proc. Natl. Acad. Sci. U.S.A.* **88**, 9214–9218.

Wicker, C., Reichhart, J. M., Hoffmann, D., Hultmark, D., Samakovlis, C., and Hoffmann, J. (1990). Insect Immunity. Characterization of a *Drosophila* cDNA encoding a novel member of the diptericin family of immune peptides. *J. Biol. Chem.* **265**, 22493–22498.

Wilcox, M., Brower, D. L., and Smith, R. J. (1981). A Position-Specific cell surface antigen in the *Drosophila* wing imaginal disc. *Cell* **25**, 159–164.

Wilcox, M., DiAntonio, A., and Leptin, M. (1989). The functions of PS integrins in the *Drosophila* wing morphogenesis. *Development* **107**, 891–897.

7

Experimental Models of Gene Regulation: 1. Heat-Shock Response of *Drosophila* Cells

The main objective of contemporary biological research is to unravel the extraordinarily complex mechanisms that control gene expression in higher eukaryotes. *Drosophila,* because of its small genome and sophisticated genetics, as has been discussed at length in the Introduction to this book, asserts itself as an ideal intermediary between bacteria, yeasts, and the higher vertebrates. Cell cultures have greatly facilitated the molecular approach. Studies have, of course, been concomitantly carried out on the whole organism, using its exceptional mutagenic and transgenic capabilities, but this survey will be limited to the data obtained from cell cultures.

In fact, two competitive, or more correctly, complementary, and widely used "experimental models" have been developed using *Drosophila* cell lines: (1) One of them is the "heat-shock" response. (2) The other is the specific induction of proteins by hormonal treatment and will be discussed in Chapter 8. These have become two of the paradigms for gene regulation in higher organisms.

I. THE HEAT-SHOCK SYSTEM

The heat-shock response was first discovered in *Drosophila* where it is especially intense. Studies carried out in this organism have enabled us to understand most of the fundamental mechanisms of the response.

A simple elevation of temperature, from 25 to 37°C, results in a rapid and profound alteration of the pattern of gene expression in nearly all types of *Drosophila* cells. Within a few minutes, the synthesis of a small number of new and specific proteins, the so-called heat-shock proteins (or *HSPs*), is vigorously induced, while most of the normal preexisting program of protein synthesis is inhibited.

This cell response, that is, the induction of a specific set of proteins after a brief exposure to supraoptimal temperatures, was shown to be universal, from bacteria to higher animals and plants. The main *HSPs* have been remarkably conserved throughout evolution, which testifies to the general importance of the phenomenon. Moreover, a variety of other stresses (e.g., perturbations of respiratory pathways and treatment with heavy metals) can elicit the same kind of response, so that the heat-shock response should be regarded as an adaptation of the cell to environmental stress and deserves, therefore, the more general denomination of stress response. Even though the specific role of individual *HSPs* remains poorly understood, accumulating evidence suggests that they have very general protective functions under stress conditions and, in addition, play a few other vital roles at normal temperature.

As early as *1962,* Ritossa noticed that short temperature shocks induced marked and well-defined variations in the "puffing" patterns of polytene chromosomes in the salivary glands of *Drosophila busckii* larvae. Similar puffs could also be triggered by various drugs known to interfere with respiratory metabolism. This latter aspect of the phenomenon was extensively explored in *Drosophila hydei* by Berendes and collaborators (review in Berendes, *1972*). Finally, Ashburner (1970), in a detailed analysis of experimentally induced puffing in *Drosophila melanogaster,* confirmed that a 40-min exposure at 37°C, either of whole 3rd instar larvae or of salivary glands isolated *in vitro*, resulted in the development of nine new characteristic puffs (at loci 33B, 63BC, 64F, 67B, 70A, 87A, 87B, 93D, and 95D), while there was a general regression of all other puffs active during development. These new puffs were also induced when larvae were brought back from anoxia to aerobic conditions.

The molecular understanding of this heat-shock response was initiated when Tissieres *et al.* (*1974*) and then Lewis *et al.* (*1975*) established (1) that radiolabeled uridine was indeed incorporated by each of these heat-shock puffs, (2) that this typical heat-shock puffing coincided with the neosynthesis of at least six to seven specific polypeptides, and (3) that such heat-induced proteins were not tissue specific, but were synthesized by every cell type tested (i.e., salivary glands, but also midgut, malpighian tubules, imaginal discs, and larval or adult brain).

Biochemical work was greatly facilitated as soon as it had been demonstrated that even cultured cells respond to heat shock in an identical manner (Lindquist McKenzie *et al.*, 1975; Linquist McKenzie, 1977; Moran *et al.*, 1977), and, thereafter, most investigations were quickly shifted to such *in vitro* material. HSPs accumulate in the cells and, after 8 h at 37°C, they can represent, for instance in *Kc* cells, several percentages of total cellular proteins (Mirault *et al.*, 1978) (Figs. 7.1 and 7.2).

Undeniably, this heat-shock response in *Drosophila* cell lines has provided one of the most valuable experimental models for studying the mechanisms regulating gene expression in higher eukaryotes.

II. MAJOR HEAT-SHOCK PROTEINS AND THEIR ENCODING GENES

As pointed out above, the fact that the set of proteins induced in cultured cells by temperature shock is identical to the few *HSPs* first characterized in salivary glands greatly promoted further studies (Lindquist McKenzie *et al.*, 1975; Spradling *et al.*, 1975, 1977; Mirault *et al.*, 1978).

From this abundant material, more amenable to biochemical analyses than dissected organs, it was possible to separate newly transcribed RNAs into distinct categories and, after translation *in vitro*, individual *hsp* messages could be correlated with the various types of *HSPs*. Moreover, their labeled complementary DNAs could be, by *in situ* hybridization on polytene chromosomes, attributed to each of the heat-shock puffing loci. Finally, the use of the same probes, for screening libraries of *Drosophila* DNA recombinant plasmids, enabled several laboratories to isolate the corresponding genomic clones. Thus, the general organization and nucleotide sequence of the main *hsp* genes could be extensively analyzed (*hsp70*: Schedl *et al.*, 1977; Livak *et al.*, 1978; Lis *et al.*, 1978; Mirault *et al.*, 1979; Craig *et al.*, 1979; Ingolia *et al.*, 1980; small *hsps*: Corces *et al.*,

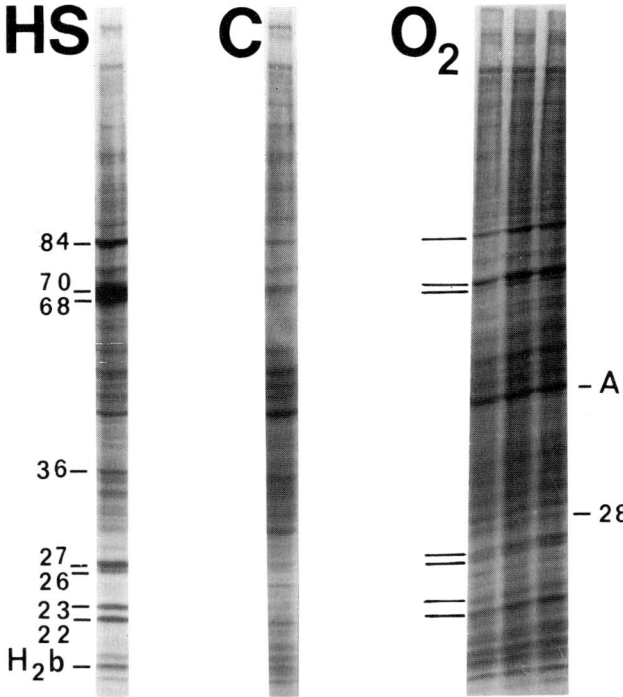

FIGURE 7.1   Induction of heat-shock proteins in *Drosophila Kc* cells. Monodimensional electrophoresis on polyacrylamide gel. Comparison between control cells (C) grown at 23°C, heat-shocked cells (HS) (37°C, 2 h), and cells in recovery (1, 2, and 3 h, 23°C) from a 24-h-period of anoxia (O_2). Protein 28 is specifically induced by oxygen. Courtesy of Dr. A. M. Courgeon, CNRS, Paris.

1980; Voellmy *et al.*, 1981; Ingolia and Craig, 1982). See Southgate *et al.* (1985) for more detail and a complete bibliography.

A careful two-dimensional gel electrophoretic analysis, carried out by Buzin and Petersen (1982) on both salivary glands and cultured cell lines, revealed that the complexity of the heat-induced proteins was much greater than had been previously suspected. The most prominent ones fell into six or seven broad molecular weight groups, but with slight, albeit distinguishable, molecular weight differences in each group and multiple isoelectric point variants in some of them. They corresponded, nevertheless, to the conventional HSPs, as identified by one-dimensional SDS–polyacrylamide gel electrophoresis (i.e., 83, 70, 68, 34, 26 to 27,

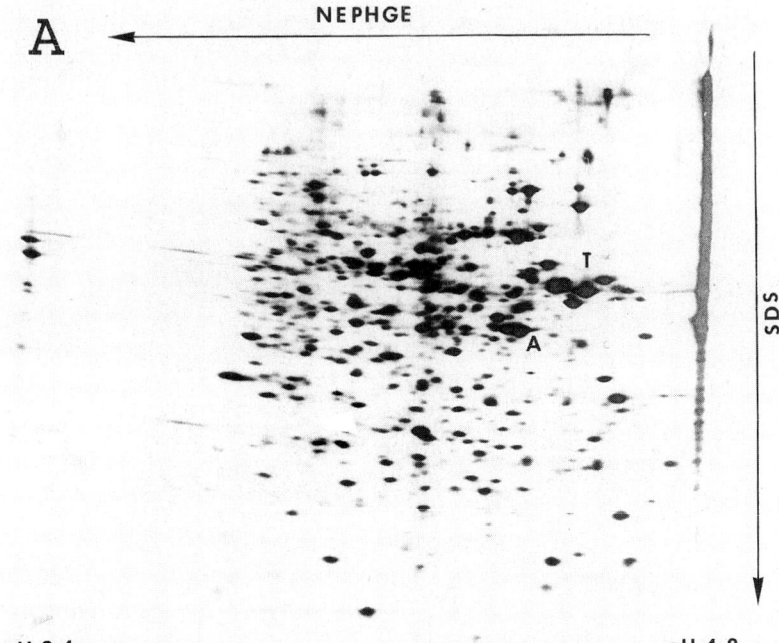

FIGURE 7.2 Analysis by 2D electrophoresis of heat-shock-induced protein synthesis in *Drosophila Kc* cells. *Kc* cells (clone 23) were heat shocked for 1 h at 37°C. Control (A) and heat-shocked (B) cultures were incubated with [³⁵S]methionine (80 μCi/ml for 1 hr). Fifty milligrams of total proteins were analyzed by 2D-NEPHGE (according to O'Farrell *et al., 1977*). IEF, isoelectrofusing, first dimension, and SDS, second dimension on SDS–polyacrylamide gel. *HSP84* (□) *HSP68-70* (◣) *HSP26-27* (✔) *HSP22-23* (✔) actin (A) tubulin (T). Observe in B the general arrest of protein synthesis contrasting with a rapid increase of a few typical heat-shock proteins. Courtesy of Dr. C. Maisonhaute, Université P. et M. Curie, Paris. (*continues*)

and 21–23-kDa types). In addition, a number of minor species were observed, ranging from 44 to 66 kDa (Fig. 7.1).

For a useful comparison with the heat-shock response in other organisms, it is convenient to classify *Drosophila* major *HSPs* into three families. Some other heat-inducible proteins will be considered later.

A. The HSP70 Family and Related Proteins

Seventy kDa proteins are the most abundant heat inducible proteins (they can incorporate, in heat-shocked *Drosophila* cells, one-fifth of the

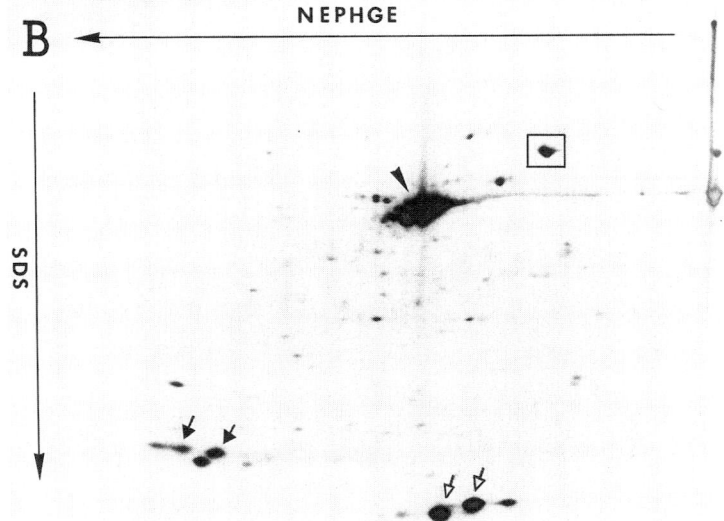

FIGURE 7.2 *(continued)*

total label of neosynthesized polypeptides) and, during evolution, they have also been the most highly conserved HSPs (human *HSP70* is 70% identical to the *Drosophila* homologous protein).

The heavily labeled 70-kDa band, as detected on monodimensional SDS–acrylamide gels, could be resolved into many distinct spots with two-dimensional electrophoresis (Mirault *et al.*, 1978). Buzin and Petersen (1982) could even distinguish some 40 components. These multiple isoforms may be explained not only by the occurrence of small variations among the different gene copies, but also by the existence of *HSP68* and *heat-shock cognate* genes and, very probably, by posttranscriptional modifications.

1. GENUINE *HSP70*

These proteins are coded for by a multigene family, containing usually five genes in *Drosophila melanogaster,* although some polymorphism in this copy number exists among fly strains. All *hsp 70* genes are distributed into two distinct clusters, at the cytological subdivisions *87A7* and *87C1,** respectively, on the right arm of the third chromosome (Fig. 7.3).

---

* The initially described major heat-puff 87B was, later, more accurately assigned to locus 87C1.

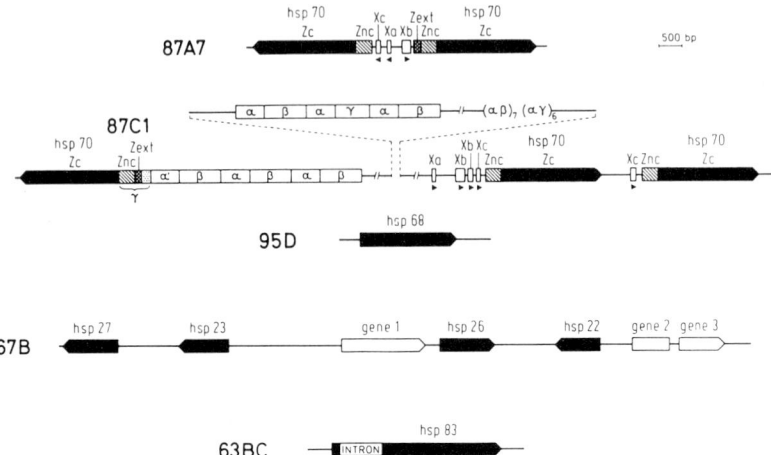

FIGURE 7.3   The arrangement of the *Drosophila melanogaster* heat-shock genes. Those regions complementary to heat-shock mRNAs are indicated by black arrows, the orientation of which gives the direction of transcription of each heat-shock gene. The top two maps portray the organization of the *hsp70* gene family at the cytological loci 87A7 and 87C1, respectively. The left-hand part of these maps is proximal to the centromere of chromosome 3R. The 2.2-kb Z_c element is complementary to *hsp70* mRNA. The 0.35 Z_{nc} regions (hatched boxes) are approximately 98% conserved at each locus and 90% homologous between the two loci. Additional areas of homology (X_a, X_b, X_c, and X_{ext}) are found to be very AT rich and well conserved. The α, β, and γ elements are arranged as either $\alpha\beta$ or $\alpha\gamma$ tandem arrays (their precise order is unknown) in the approximately 40-kb spacer which separates the single proximal *hsp70* gene from two distal *hsp70* gene copies at locus 87C1. The third map represents the *hsp68* gene at locus 95D. The fourth map gives the organization of *small-hsp* genes at the 67B locus. The map of the *hsp83* gene (at locus 63BC) shows the unique intron. From Southgate *et al.* (1985); reproduced with the permission of Professor Tissière (University of Geneva) and Academic Press.

The locus 87A7 comprises two copies, with diverging orientations. The organization of the 87C1 locus is much more complex: a single "proximal" copy (with respect to the centromere) is in opposite polarity and separated from two "distal" tandem genes by about 40 kb of DNA. In several *Drosophila* strains, and this is also the case in the *Kc* cell genome (Mirault *et al.*, 1979), there are three, instead of two, distal tandem copies. The large DNA interval contains a score of repeated units, the so-called αβ repeats, whose transcription is also heat inducible,* although it does not give rise to any protein.

Each individual *hsp70* gene is organized as a basic well-conserved unit Z, subdivided into a 2.2-kb Z-coding (Z_c) segment, which is complementary to the 70-kDa protein message, and, at its 5' end, a 0.35-kb Z-noncoding (Z_{nc}) segment. There is no intron, and the large ORF specifies a polypeptide with a calculated molecular weight of 70,270 (Ingolia *et al.*, 1980). Coding sequences are extremely well conserved among the different gene copies, and the degree of homology between the genes of a same locus is even higher. Additional regions of various sequence homologies, the so-called X elements, are present upstream of all *hsp70* genes (see Section IX.A).

The occurrence of fly mutants, with specific deficiencies of either locus, made it possible to demonstrate that both gene clusters are involved in the production of *HSPs70*, and it could be verified, at least for the locus 87A, that each individual copy is activated after heat-shock.

2. HSP68

HSP68 is encoded by a single gene at the chromosomal division 95D, which is also a heat-puff site. Its transcribed segment is 2.1 kb long, and heteroduplexes demonstrated a strong cross-homology with the *hsp70* genes (only some 15% divergence).

3. HEAT-SHOCK COGNATE GENES (HSCS)

In situ hybridization of polytene chromosomes, using an *hsp70* probe, revealed the presence in the *Drosophila* genome of several other genes (so far, seven have been identified), all closely related to *hsp70* genes (about 75 to 80% homology). They are not inducible by heat treatment, however, but they are normally expressed at various developmental stages of *Drosophila* and some of them at basal levels never reached by induced true *hsp70* (Craig *et al.*, 1983; Palter *et al.*, 1986); Rubin *et al.*, 1993). So, they were named heat-shock cognate genes, or *Hscs*. For instance, Hsc-1 (with one 1.7-kb intron) is located at chromosome subdivision 70C, Hsc-2 (with a 650-bp intron) at 87D and Hsc-4 at 88E.

Using monoclonal antibodies against the major *hsc*, Palter *et al.* (1986) observed by indirect immunofluorescence in primary culture of *Drosophila*

---

* The fact that αβ repeats are also present in the chromocenter of polytene chromosomes, although those copies are not inducible, suggests a recent transposition of such αβ repeats to the locus 87C1, in the vicinity of *hsp70* genes, with a possible usurpation of regulatory elements responsible for heat inducibility (Hackett and Lis, 1981).

embryonic cells a network of cytoplasmic fibers, densely concentrated around the nucleus.

B. Small Heat-Shock Proteins

The small molecular weight heat-shock proteins (or s-*HSPs*) comprise not only the four initially recognized HSPs (i.e., *HSP27, HSP26, HSP23,* and *HSP22*), but also the products of three closely linked genes called 1, 2, and 3 (Ayme and Tissieres, *1985*). These seven distinct genes are clustered in a 12-kb DNA stretch at chromosomal subdivision 67B (on the left arm of chromosome 3). They are all oriented differently, which indicates that, despite of being activated by common stimuli, they do not form a common transcription unit (Fig. 7.3).

Comparison of the coding sequences of the four conventional *s-hsp* genes revealed a high degree of homology over some 50% of their lengths and a remarkable resemblance between this region of homology and the similar region of mammalian α-crystallin (Ingolia and Craig, *1982*). It was assumed that this common domain was related to the ability of s-HSPs to form aggregates (see Section VIII.B.2). Furthermore, the 5' transcribed but untranslated sequences of *s-hsps* are unusually long (over 100 bp).

If, finally, all the seven genes of the locus 67B were found to be heat inducible, the four "classical" *hsp27, -26, -23,* and *-22* genes (but apparently not the other interspersed genes) were also shown to be regulated by ecdysteroids (Ireland and Berger, 1982; Ireland *et al.,* 1982), which explains, for the most part, their typical developmental expression patterns (see Chapter 8).

C. HSP83

HSP83 is the *Drosophila* representative of what is called, in other organisms, the *HSP90* group. These proteins, present at normal temperature and at fairly high levels, not only in cultured cells but in all tissues throughout development, might play significant roles by binding to various molecules of interest (see Section VII.D). They are further induced by heat, although moderately (a 5 to 6-fold increase in *Drosophila* cells, whereas, for instance, *HSP70* might be increased several hundredfold).

The *hsp83* gene is unique and located at the chromosomal division 63BC (Fig. 7.3). It is the only *Drosophila hsp* gene, if one excludes *hsp cognate* genes, to possess one intron; its transcription unit (3.7-kb long) is indeed interrupted by a 0.9-kb intervening sequence.

III. HEAT-SHOCK PROTEIN INDUCTION AT TRANSCRIPTIONAL LEVEL

The extraordinarily rapid (1 min for 87C locus) development of specific puffs, in salivary gland chromosomes, after the temperature shift, suggests that transcription is the primary level of regulation of heat-shock genes.

In cultured cells, within minutes of a temperature elevation, the specific program of protein synthesis, which is the characteristic cell response to heat shock, is initiated by the transcription of a completely new set of mRNAs, while the synthesis of most of the normal messages is suddenly curtailed.

This dramatic change, discernible from the modified size distribution of the polysome population, was first observed by Lindquist McKenzie *et al.* (1975) in heated *Kc* cells. In extracts from cells grown at 25°C, the sedimentation profile of polysomes, in a sucrose gradient, was rather complex, which correlates with the extreme diversity of the proteins normally synthesized by cells, but showed a major peak at around 12 to 13 ribosomes per message. In contrast, at 37°C, almost all polysomes disappeared after 10 min and, 1 hr later, a new population of polysomes was observed, mainly in the region of 20 to 30 ribosomes per message. This could be prevented by even a low concentration (0.04 μg/ml) of actinomycin D (added before the temperature was raised), although preexisting polysomes are dismantled as before (see the next section).

Independently, the laboratory of M. L. Pardue (Spradling *et al.*, 1975; Lengyel and Pardue, 1975) demonstrated, on the basis of their *in situ* hybridization pattern on polytene chromosomes, that cytoplasmic mRNAs from 37°-treated *S2* cells consisted of only a small number of RNA species, compared to the large and diversified population at 25°C (which might correspond to some 3000 to 4000 active genes). Similar analysis of hnRNAs suggested that the difference was principally due to a change in transcription rather than to an alteration in hnRNA processing.

Moreover, both laboratories established that most of the newly synthesized messages hybridized to the main heat-shock puffing sites of salivary gland polytene chromosomes.

All further studies have confirmed the differential transcription of *hsp* genes in heat-shocked cells. See, for instance, the clear results from *in vitro* transcription experiments carried out with extracts from normal or heat-shocked *S2* cells (Love and Minton, 1985).

More recently, the pattern of transcription during heat shock was carefully reinvestigated in *Kc* cells by Tissière's group (Vasquez *et al.,* 1993), using the nascent chain analysis technique, or "nuclear run-on". The method has the advantage of blinding the preexisting levels of mRNAs in the cells, since only nascent transcripts are labeled. This novel approach resulted in a more precise determination of the kinetics of transcriptional activity of heat-shock genes and transcriptional repression of other genes.

A. Main Types of Transcripts Present in Heat-Shocked Cells

1. The pioneering work of Spradling *et al.* (1977) and Mirault *et al.* (1978) showed that polysomal RNA from heat-shocked cultured cells, when labeled with tritiated uridine at 37°C, sedimented in sucrose gradients as two major peaks: a 20S RNA fraction, predominant in large polysomes and a 12-13S RNA fraction, corresponding to a minor population of smaller polysomes.

A further purification of individual messengers was obtained by polyacrylamide gel electrophoresis, and their identification was attempted by two different approaches: either translation *in vitro* in a cell-free system, or hybridization *in situ* to polytene chromosomes.

The 24 or so species of labeled RNAs thus isolated from heat-shocked cells could be arranged into three main classes:

Group A (according to Spradling's nomenclature) is detectable at significant levels in heated cells only and represents the great majority of poly(A) RNA synthesized at 37°C: The A1, A2, and A3 species (named 20-I, -II, and -III, in Mirault's paper), forming almost entirely the 20S fraction, correspond to heat-induced polypeptides 83, 70, and 68 kDa, respectively.

The A4 to A6 species, derived from the 13S fraction, as well as the two other groups, B and C, are the messengers of small hsps.

Group B is composed of 13 poly(A) RNA species (B1 to B13) and are the products of the mitochondrial genome. They are, of course, present in cells grown at normal temperature, but their synthesis obviously continues, although at a slower rate, after heat treatment. Their mitochondrial nature was confirmed by testing their complementary with mitochondrial

DNA and by the use of a specific inhibitor of nuclear and not mitochondrial RNA synthesis, the drug camptothecin (see Chapter 5, Section III). *Group C* is also found in normally grown cells. Most of these last RNA species correspond to histone messengers. Histone H2B synthesis even seems to be specifically increased in heat-shocked cells (more detail in Section VI.B).

Thus, an important point is made: although, broadly speaking, the normal program of gene expression is indeed interrupted during heat shock in *Drosophila melanogaster* cells, in fact, previous transcriptions are not all affected to the same extent.

2. Yost and Linquist (1986) established that severe heat shock produces a general block in the splicing of intervening sequences from messenger RNA precursors. As has been reported in Chapter 5, the normal processing of ribosomal 5S RNA is also disturbed by heat shock. Thus, the lack of processing of normal messages could be a significant part of the regulatory strategies of the heat-shock response (Yost *et al.*, 1990). *hsp* genes, with the exception of *hsp83*, possess no intron. Furthermore, HSPs themselves, when induced and accumulated in the cell during pretreatment at a moderately high temperature, will protect the splicing machinery during a subsequent severe heat shock.

Note: Dellavalle *et al.* (1994) reported that heat shock might disrupt other aspects of RNA processing: following a standard heat shock, 40% of *hsp70* transcripts in S2 cells were found to lack a poly(A) tail. This is due to a deadenylation and might play a significant role in regulating *hsp70* expression (deadenylated mRNA being translated with lower efficiency).

3. Preexisting messenger molecules are not degraded during heat shock, which could be plainly proved by the translation *in vitro* of total polyadenylated RNA from heated cells, with either heterologous (rabbit reticulocyte lysate: Mirault *et al.*, 1978; Storti *et al.*, 1980; wheat germ extracts: Sondermeijer and Lubsen, 1978) or homologous (nonheated *Drosophila* cultured cell lysate: Scott and Pardue, 1981; Kruger and Benecke, 1981) systems. The translation products included not only *HSPs*, but also a number of polypeptides that are normally synthesized at 25°C, and their amounts were equivalent to those obtained under similar conditions with RNA from control cells. Another indication of such a retention of "normal" messages in heat-shocked cells is the rapid "reactivation" of the full spectrum of normal protein synthesis, during the recovery from heat shock, even when new transcription is blocked with actinomycin (see Section V).

4. As already mentioned, the characterization from cultured cells of these individual messengers for heat-induced proteins allowed the isolation of their cDNA clones and then of corresponding genomic clones. In return, the availability of specific probes permitted a more refined analysis of the kinetics of synthesis of all types of *hsp* mRNAs. (Lis *et al.*, 1981; see the bibliography concerning the functional organization of *hsp* genes in Mirault *et al.*, 1985).

B. Kinetics of Transcription of *hsp* mRNAs

Lindquist (1980a) observed that, when actinomycin D is added up to 2 min after the temperature rise, the appearance of *HSP70* was blocked, whereas, if addition of the drug is delayed until the 4th min, a small quantity of this major *HSP* was synthesized. This means that the transcription of the first messages for *HSP70* is probably completed within the first 4 min of heat shock. Since synthesis of all *HSPs* could be detected at around 10–12 min, this allows some 5 to 8 min for total processing, transport to cytoplasm, and translation of the first messages.

According to Spradling *et al.* (1977), in S2 cell cultures shifted to 37°C, the incorporation of tritiated uridine into cytoplasmic poly(A)$^+$ RNA remained linear for at least 2 h [as opposed to 4 hr for poly(A)$^-$ RNA], but, thereafter, the total amount declined, with a half-life of about 2.5 h.

These rapid changes in *Drosophila* cell transcription were recently reestimated by O'Brien and Lis (1993), who monitored the RNA polymerase II association with a set of *Drosophila* genes at 30-sec intervals following an instantaneous heat shock. Transcription at the 5′ end of the *hsp70* gene could be detected within 30 to 60 sec after induction, and by 120 sec, the first wave of polymerase could already be detected near the 3′ end of the gene. In contrast, the nucleosome core protein *H1* gene was found to be rapidly repressed (about 90% within 300 sec of heat shock).

Lindquist (1980b) was able to estimate that the number of messenger molecules for large *HSPs* approximates 1.5×10^4 per cell and that some 1.7×10^7 molecules of *HSP70* and 2.9×10^6 molecules of *HSP83* can be produced per cell within 120 min of continuous 37°C treatment.

It is particularly noteworthy that the relative amounts of the various mRNA species in the cytoplasm depend critically on the temperature of the heat shock (Spradling *et al.*, 1977). This is in good agreement with the differences observed in the induction patterns of the different families of *HSPs* at various temperatures (Linquist, 1980a): *HSP70* is maximally induced over a narrow range of temperatures, centered around 37°C, while the four *small-HSPs* reach a maximal level at 35°C and *HSP83* at 33°C. Remember that the blocking in *HSP83* expression at higher temperatures is essentially due to an inhibition of the splicing of its premessenger, whereas its transcription rate remains unabated (Yost and Lindquist, 1986).

C. Inhibition of HSP Expression by Antisense RNAs

Recombinant plasmids, containing a segment of the *hsp26* coding sequence linked, in an antisense configuration, to an *hsp70* promoter, were transferred to S2 cells. Cotransfection with a selection plasmid (see Chapter 9) allowed the isolation of several stably transformed lines containing different numbers of integrated copies (McGarry and Lindquist, 1986). Whereas, in normal heat-shocked cells, *HSP26* is the most strongly expressed among the *small-HSPs*, on the other hand, in transformant lines it was poorly induced and the level of inhibition was proportional to the quantity of transcribed antisense RNA. Significant inhibition was thus achieved when the concentration of antisense RNA was 2 to 10 times the normal concentration of the *hsp26* message.

Similar results were obtained with *hsp70* antisense RNA (Lindquist *et al.*, 1988). After heat shock, the expression of *HSP70* was reduced in transformant lines, but it is noteworthy that, as if in compensation, other *HSPs* (especially HSP68) were found to be overexpressed.

Then again, Nicole and Tanguay (1987) showed that *in vitro* translation (in a rabbit reticulocyte lysate cell-free system) of mRNA coding for *HSP23* was specifically inhibited when *hsp23* transcripts in the antisense orientation were previously mixed with mRNA from heat shocked cells.

IV. TRANSLATIONAL CONTROL IN HEAT SHOCK

Since preexisting mRNAs are not destroyed in heat-shocked cells, it is very likely that the vigorous induction of *hsp* messages does not alone account for the drastic change in protein synthesis. The shut-off of normal protein synthesis cannot be explained by a simple model of competition between *hsp* mRNAs and control mRNAs, despite Jackson's assumption (1986).* As a matter of fact, even in the absence of *hsp* mRNAs, when

---

* The main merit of this work (see also Jackson, 1982) was to show that normal protein synthesis is perhaps not always so drastically and abruptly curtailed in *Drosophila* heat-shocked cells as was generally assumed.

all new transcription had been, for instance, blocked by actinomycin in heat-shocked cells, the translation of normal messages still present in the cells is stopped. It is obvious that an efficient translational control regulates the almost-exclusive synthesis of heat-shock specific proteins in *Drosophila melanogaster* cells maintained at 37°C.

1. A direct demonstration of such a translational discrimination between "normal" and heat-shock messages was provided by homologous *in vitro* translation systems [as devised by Scott *et al.* (1979) from *Drosophila* cultured cell lysates]. When supplied with polyadenylated RNA from heat-shocked cells, a cell-free translation lysate prepared from control nonheated cells produced both types of protein, even though HSPs predominated. In contrast, in lysates prepared from heat-shocked cells, the translation of 25°C mRNAs was almost completely suppressed, while the translatability of heat-shock messages was unaffected (Storti *et al.*, 1980; Scott and Pardue, 1981; Kruger and Benecke, 1981).

These *in vitro* systems provided additional information: (1) Preferential translation cannot merely be explained by direct temperature-induced changes in the secondary structure of "normal" mRNAs because no difference between the two types of messengers was observed when they were simultaneously translated at various temperatures in reticulocyte lysates (Lindquist, 1981); and (2) In a heat-shocked *Drosophila* cell lysate, the translation of "normal" messengers could be "rescued" by supplementation with a crude ribosome fraction from 25°C cells (but not from 36°C cells). Washed ribosomes from a 36°C cell extract recovered, however, some rescuing activity, so that it is not clear whether the discriminating property correlates with the presence of some soluble factor or is due to modifications in the ribosomal protein content or conformation (Scott and Pardue, 1981; Krüger and Benecke, 1981; Sanders *et al.*, 1986). It is relevant here to recall, for instance, that Glover (1982a,b) observed the complete and rapid dephosphorylation of one specific ribosomal protein (rP) in heat-shocked *Kc* cells.

2. Among the structural features of the *hsp* messages that might mediate their specific recognition, it was observed that they have unusually long 5' untranslated "leader" regions, which are particularly rich in adenosine residues. Moreover, some sequence elements seem to be conserved in the leaders of several *Drosophila hsp* genes (McGarry and Lindquist, 1985). Using transformant cell lines containing different modified constructs of the *hsp70* gene, these authors demonstrated that the heat-shock message was no longer translated at high temperature when its leader sequence was deleted, whereas, at normal temperature, it became

translatable. If the deletion is limited to discrete conserved sequences, however, there is no appreciable effect. The importance of specific stretches in the leaders of *hsp* messages for their preferential translation during heat shock was also confirmed in transgenic flies by Gehring and collaborators (Klemenz *et al., 1985;* Hultmark *et al., 1986*) (see Fig. 7.9).

Hess and Duncan (1994), in a UV cross-linking analysis, did not, however, succeed in detecting any specific protein binding to the 5′ untranslated leader of *hsp70* mRNA in heat-shocked S2 cell lysates.

3. The dramatic disappearance of the polysome population within a few minutes of the temperature shift, as first described by Lindquist-McKenzie *et al.* (1975), seriously suggested that the polypeptide initiation is the crucial step in this translational regulation. Further studies, however, by Ballinger and Pardue (1982, 1983, 1985), showed that even though there are undeniable and spectacular changes in the polysome population profile of heated cells, "normal" mRNAs remain ribosome-associated and the translational control involves, at least in part, a specific block in elongation of 25°C mRNAs in heat-shocked cells.

4. The currently emerging picture in this translational regulation of heat-shocked cells emphasizes the importance of the structural characteristics of the messengers and protein factors involved in mRNA binding to ribosomes. Sierra and Zapata (1994) reported some inactivation of the eukaryotic initiation factor *eIF-4E* in *Drosophila* heat-shocked cells and, even more recently, Duncan *et al.* (1995) indeed noted significant changes in the phosphorylation of this factor. Conversely, according to a previous study (Gordon *et al.,* 1987), the posttranslational conversion of a lysine residue into the unusual amino acid hypusine, normally observed in another "initiation factor" 4D, was unmodified in heat-shocked S2 cells.

In any case, this type of translational regulation seems to be a general feature of the heat-shock response in eukaryotic cells. It has been observed in a number of organisms, but was particularly well characterized in *Drosophila melanogaster* cells. Similar observations were made in *Drosophila hydei* by Sondermeijer and Lubsen (1978).

What could be the advantages for the cell or the whole organism of this regulation step?

1. It enables the stressed cell to focus on the rapid synthesis of HSPs, with their assumed beneficial intervention.

2. Functional sequestering, instead of destruction, of preexisting "normal" messengers will facilitate the recovery of the usual protein synthesis pattern, when the cell returns to normal temperatures. Moreover, let

us emphasize that in the whole organism the destruction of maternally inherited mRNAs by a simple elevation of temperature would cause severe damage to young embryos.

V. RECOVERY FROM HEAT SHOCK

When the heat shock is over, the expression of *hsp* genes comes to a halt quite quickly, and the full spectrum of normal protein synthesis is resumed within a few hours (Mirault *et al.*, 1978), even when new transcription is blocked with actinomycin (McKenzie, 1977; Storti *et al.*, 1980; Lindquist, 1981) (which demonstrates the retention of normal messages by heat-shocked cells). It must be pointed out that this return to normal patterns of macromolecular synthesis is a much more gradual process than the initial shift to heat-shock synthesis.

Arrigo (1980) was the first to demonstrate that the synthesis of *HSPs* during heat shock is necessary for the restoration of normal RNA synthesis at 25°C, and also that there is a correlation between this resumption of normal synthesis and the migration of *HSPs* back to the cytoplasm.

Lindquist and collaborators (DiDomenico *et al.*, 1982a,b; Lindquist *et al.*, 1982; Lindquist, 1985; Petersen and Lindquist, 1988) have carefully investigated these "recovery" processes, and their main results are the following:

1. The length of the time lag, during which heat-shock synthesis continues at normal temperature, greatly depends on the severity (both temperature and duration) of the preceding heat shock.

2. Whereas the repression of the different *HSPs* is asynchronous (*HSP70* is the first one to be repressed and *HSP83* the last) the reactivation of the whole cohort of normal proteins is uniform (Fig. 7.4).

3. It is possible that *HSP* repression is primarily transcriptional, although their individual messages are rapidly degraded at normal temperature. Incidentally, Simcox *et al.* (1985) noted that mRNA of *hsp40* (a mutant of the major *hsp70* with a deletion of its 3' end) was not actively destabilized after return to the control temperature, thus permitting a prolonged production of the mutant protein.

4. Restoration of normal synthesis seems to be coordinate with the kinetics of repression of *HSP70*, but not with those of the other *HSPs* (as can be seen in Fig. 7.4).

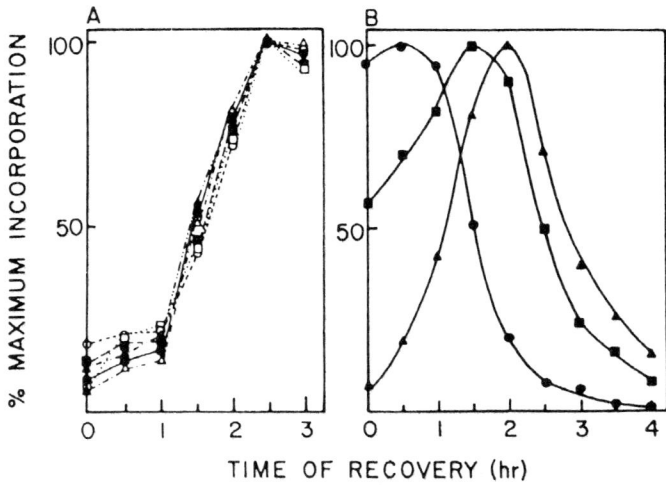

TIME OF RECOVERY (hr)

FIGURE 7.4   Restoration of normal protein synthesis and repression of HSP synthesis during recovery from heat shock. *Drosophila S2* cells were heat treated at 36.5°C for 30 min and then returned to 25°C for recovery. Protein synthesis was quantified by densitometry. (A) Profiles for actin and five randomly chosen 25°C proteins. (B) Profiles for three heat-shock proteins: HSP70 (●), HSP27 (■), and HSP82 (▲). From DiDomenico *et al.* (1982a); reproduced with the permission of Professor Lindquist.

5. A series of experiments (DiDomenico *et al.*, 1982b), whose purpose was to decrease (with actinomycin or cycloheximide) or to perturb (with the incorporation of amino acid analogs) the synthesis of functional *HSPs* during the preceding heat shock, demonstrated that a specific amount of *HSP70* must be accumulated by the cell before its transcription is repressed and its own mRNAs are destroyed at the time of recovery. In other words, there seems to be a self-regulation, which means that functional *HSP70* is required for repressing its own synthesis, both at the transcriptional and posttranscriptional levels, and this may also be true during the heat-shock period itself. Moreover, *HSP70* also seems to be involved in the concomitant release of the block in normal protein synthesis.

HSP70's predominant role in the regulation of this recovery period, and especially in the restoration of normal synthesis, has been confirmed by Solomon *et al.* (1991), who used transformant cell lines either underexpressing or overexpressing wild-type or altered versions of *HSP70*.

VI. OTHER TEMPERATURE-INDUCIBLE GENES

In addition to the better-known *hsp* genes, a number of genes can be activated by supraoptimal temperatures or other experimental insults (stresses).

A. *hsr-omega* Gene: Locus 93D

Among the largest puffs induced by a rise in temperature in the polytene chromosomes of *D. melanogaster* larvae, the locus 93D harbors a gene named *hsr-omega* (i.e., the locus encoding the omega set of heat shock RNAs), whose expression pattern is rather strange. Other *Drosophila* species also have one heat-inducible puff which shares the same distinguishing characteristics (for instance, *D. hydei* at locus 2-48B).

This *hsr-omega* gene does not yield any detectable protein product, despite the synthesis of quite abundant complementary RNAs, especially present in the nuclear fraction from heat-shocked S2 cells (Lengyel *et al.*, 1980; Hogan *et al.*, 1994). The smallest cytoplasmic transcript was found to be associated with polysomes of monosome and disome sizes. Moreover, the use of inhibitors of transcription or initiation, as well as transfection experiments with constructs in which a CAT reporter gene was linked to the *hsr-omega* promoter and its 5′ transcribed sequence, led Fini *et al.* (1989) to the surprising conclusion that one tiny ORF (see below) was possibly translated, even though no new protein of the appropriate size could be identified.

Puffing of 93D can be induced by several chemical agents, such as benzamide or colchicine, independent of all other loci (Lakhotia and Mukherjee, 1980). The same inducers led to elevated levels of *hsr-omega* transcripts in cultured cells and preferentially of the nuclear *omega-1* (see below) (Bendena *et al.*, 1989).

hsr-omega expression is constitutive in unstressed cells, but its transcription is significantly increased by heat shock. This is also true of *hsp83*.

hsr-omega is also ecdysone inducible, so that the changes in hormone titers during *Drosophila* development may play a role in the variations of *hsr-omega* transcript levels, as observed among the various stages and tissues (Bendena *et al.*, 1991). Hormone inducibility and typical developmental patterns of expression are two further features of the *small-hsp* family (see Chapter 8).

Let us now consider the molecular analysis of the 93D locus. Specific probes resulting from the microcloning of heat-induced 93D puffs cut out from salivary glands (Walldorf et al., 1984) or corresponding to cDNA prepared from RNAs of heat-shocked cultured cells (Garbe and Pardue, 1986; Garbe et al., 1986) allowed the screening of Drosophila melanogaster genomic libraries and the isolation of several clones of this 93D region. Similarly, genomic clones became available for the homologous locus of Drosophila hydei (i.e., hsr2-48B) (Peters et al., 1984).

This locus shows the same characteristic organization in both species (Rysek et al., 1987; Garbe et al., 1989; Bendena et al., 1989a,b). Their transcription units consist of a unique sequence (about 2.5 kb) followed by a long (8 to 10 kb) stretch of short direct tandem repeats (280 bp in D. melanogaster, and 115 bp in D. hydei, respectively) (Fig. 7.5).

Their transcription patterns (Fini and Pardue, 1988), too, are very similar. There are three major transcripts with the same start site. The major transcript (10 kb), called omega-1, is colinear with the entire transcription unit, and it is limited to the nucleus. omega-2 (1.9 kb) is produced by an alternative termination and is the nuclear precursor of the cytoplasmic transcript omega-3 (1.2 kb) (which implies the removal of a 700-bp intron). All three hsr-omega RNAs are produced constitutively and enhanced by a temperature shock, their relative levels changing in time during heat shock. Such striking similarity in gene structure and transcription pattern, despite a diverging primary sequence, suggests a conserved function among Drosophila species.

Surprisingly, the unique sequence present in hsr-omega loci of different Drosophila species shows only short ORFs; the largest one, which is quite well conserved, encodes a small polypeptide of about 3–4 kDa (34 amino

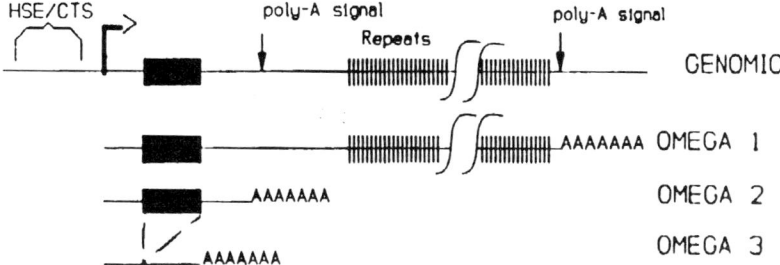

FIGURE 7.5   The genomic organization of the Drosophila melanogaster hsr-omega locus and the structure of its transcripts. HSE/CTS indicates a region containing heat shock elements and at least some of the constitutive transcription signals. Garbe et al. (1989); reproduced with the permission of the Genetics Society of America.

acids in *D. melanogaster*). As has already been pointed out, Fini *et al.* (1989; see also Bendena *et al.*, 1989a) provided a series of experimental results strongly suggesting that the cytoplasmic *omega-3* transcript is actually translated, although no corresponding small polypeptide could be detected.

The function(s) of this very peculiar heat-inducible locus remains, therefore, intriguing. It might be significant that, while it is constitutively expressed and developmentally regulated in most *Drosophila* tissues, its transcripts are absent from early embryonic cells and primary spermatocytes only (i.e., the two rare categories of cells incapable of mounting a heat-shock response) (Bendena *et al.*, 1991).

Besides, the differential patterns of the three transcripts after induction with various drugs and the fact that there is no evidence that *omega-1*, with its particularly long stretch of tandem repeats, is a precursor of *omega-3*, lead one to think that they might have quite distinct functions, one acting in the nuclear compartment and the other in the cytoplasm (Bendena *et al.*, 1989b). The notion that they might act together to coordinate some aspects of nucleocytoplasmic interactions is a good working hypothesis.

B. Histone H2B

Among the relatively few "normal" transcriptions that are maintained during heat shock, the messenger RNAs of *Drosophila* histones were expressly recognized, either by *in situ* hybridization on locus 39D of polytene chromosomes (Spradling *et al.*, 1975, 1977; McKenzie *et al.*, 1975) or, after separation on two-dimensional polyacrylamide gels, by cross-hybridization with sea urchin histone DNA (Burckhardt and Birnstiel, 1978).

Moreover, because the rate of synthesis (primarily of transcription) of histone H2B in *Kc* cells increases 3- to 4-fold in the first hour of heat shock, it was considered to be a heat-shock protein (see Fig. 7.1). In contrast, the synthesis of other histones (H1, H3, and H4) decreased (between 2- and more than 10-fold) (Sanders, 1981a,b, 1982; Tanguay *et al.*, 1983).

This accumulation of H2B in the nuclear fraction, at a time when the DNA synthesis is blocked by the temperature shift, is difficult to understand, and the noncoordinate synthesis of the different histones is even more surprising, because it is well known that the four components of the nucleosomal core (H2A, H2B, H3, and H4) are usually present in a molar ratio of 1:1:1:1 (which implies a well-regulated expression of their clustered genes). The hypothesis that H2B is a particularly unstable species at elevated temperatures and has to be rapidly replaced was ruled out by pulse–chase observations.

In order to correlate any possible modification of chromatin with the dramatic transcriptional changes elicited by heat shock, the methylation and acetylation patterns of the nucleosomal core histones were investigated (Camato and Tanguay, 1982; Arrigo, 1983; Tanguay and Desrosiers, 1988; for a general discussion see Tanguay and Desrosiers, 1990):

1. Histones H3 and H4, which are highly methylated at normal temperature, were no longer methylated in heat-shocked *Kc* cells, whereas H2B was methylated after 1 h at 37°C.
2. Concomitantly, the rise in temperature induced a drastic decrease in acetylation of the four core histones (to only 15% of normal values).

The biological significance of these postsynthetic modifications of histones remains speculative:

1. The histone acetylation and deacetylation cycles, which succeed one another rapidly in normal *Drosophila* cells (Munks *et al.*, 1991) are generally thought to be associated with variations in the packaging of DNA and possibly play a central role in transcriptional regulation. So, heat-induced modifications of histone acetylation might result in a gradual compaction of the chromatin, perhaps helpful to prevent chromatin damage or to reduce the expression level of normal (i.e., non-heat-shock) genes.

2. Similarly, Desrosiers and Tanguay (1985, 1986) showed that changes in the histone methylation pattern in heat shock seem to be closely correlated with the repression of normal genes rather than with the activation of *hsp* genes, as deduced from experimental situations in which the two effects of stress are unlinked (for instance, after treatment with ethanol which induces *HSPs* without impairing the normal protein synthesis) (Tanguay and Desrosiers, 1988; Desrosiers and Tanguay, 1988). The same authors (Desrosiers and Tanguay, 1989) established that inhibitors of topoisomerase II can induce changes in histone methylation that are similar to those observed in heat shock, which suggests that chromatin conformation is probably an important factor in the accessibility of histones to methyl transferases. Moreover, it was hypothesized that this hypermethylation of H2B ensures its protection from the

ubiquitin-mediated proteolytic system which is supposed to be activated in heat shock (see the following section).

C. Ubiquitin

As its name implies, *ubiquitin* is a small polypeptide, present everywhere in all eukaryotic cells. Its 76-amino-acid sequence was found to be remarkably similar in all animals analyzed, which suggests some important cellular function.

In each *Drosophila melanogaster* haploid genome, there are (probably) two ubiquitin genes mapping at two closely linked sites (about 60 kb apart) in the early ecdysone and heat-shock puff region 63F (90B in *Drosophila hydei*). Both are organized as multiple head-to-tail repeats (228 bp each), without any "spacer," and they encode a polyubiquitin precursor which is rapidly cleaved (Arribas *et al.*, 1986). This unusual type of organization, which is also the rule in other organisms, probably allows the production of large amounts of the protein within a short period.

Ubiquitin is induced by heat shock, strongly in vertebrate and yeast cells and more moderately in *Drosophila*. Among nine cDNAs homologous to poly(A)-containing RNAs whose levels were significantly higher in heat-shocked *Kc0* cells than in controls, Lis *et al.* (1981) characterized a clone hybridizing to locus 63F; the ratio of inducibility was estimated to be about 4.5. Arribas and collaborators (Izquierdo *et al.*, 1984), however, were surprised that they did not find a net increase in the 63F RNA levels of the whole organism, either after heat shock or after ecdysone stimulation (even when the corresponding polytene chromosomal region puffs undeniably, in both experimental situations, and despite the fact that the genes occupy the broadest center of this puff).

Ubiquitin participates in a nonlysosomal proteolytic pathway which, in normal cells, helps to get rid of short-lived proteins (some of which are important cell regulators) and to eradicate abnormal proteins. It becomes covalently ligated to the cytoplasmic or nuclear target proteins (via an unusual peptide linkage between its own terminal glycine carboxy group and the α- and ε-amino groups on the protein) and this "earmarking" serves as a signal for an attack by specific proteases. Owing to its similarity with a homologous yeast enzyme, the gene of one enzyme catalyzing the formation of ubiquitin–protein conjugates (called UbcD1) was cloned in *Drosophila* (Treier *et al.*, 1992).

It is obvious that environmental stresses can alter a series of cellular proteins and, in order to avoid overloading the cell with such damaged molecules, it is possible that this ubiquitin-dependent degradation system cooperates with the "chaperone" functions of conventional HSPs. As another possible connection with *HSPs*, Munro and Pelham (*1985*) even proposed (because ubiquinated proteins are not always doomed to destruction) that ubiquitin might be involved in the normal inactivation of the heat shock factor (see Section IX.C.4), whereas, under stress conditions, the affluence of abnormal proteins might result in a transient shortage of ubiquitin, which would release HSF.

D. "Prompt" Heat-Shock Proteins

As first discovered in mammalian cells, a strong elevation in temperature very quickly promotes the synthesis of a specific set of minor proteins distinct from the conventional *HSPs*. Fourteen new spots could be distinguished on two-dimensional gel electrophoregrams from heat-shocked *Drosophila* S2 cells (Ornelles and Penman, 1990); some of them might have been previously observed by Buzin and Petersen (1982). They were called *prompt HS* proteins and can be distinguished from the prominent and well-known *HSPs* by a series of characteristics:

1. They appear immediately after the temperature shift (as their name implies) and do not require any transcription, which suggests the translational regulation of preexisting inactive mRNAs.
2. Their inducing temperature is significantly higher (36°C), and they could not be induced by toxic agents, such as arsenite or cadmium.
3. They are found to be strictly associated with the nuclear matrix–intermediate filament complex, where they could be distinguished (despite their small amounts) from a series of other proteins which are not normally located in the nucleus, but shift to this NM-IF fraction at a high temperature (see Section IX.B.4). "These changes in composition of the NM-IF during heat-shock may reflect a fundamental role of the nuclear matrix in modulating gene expression in response to heat shock."

E. Glutathione S-Transferase 1-1

Glutathione S-transferases are a family of dimeric enzymes which are ubiquitous in eukaryotes and whose multiple roles may be essential in drug biodegradation and protection against peroxidative damage. In insects, the particular importance of *GSTs* stems from the fact that they might be involved in resistance to currently used pesticides, such as DDT, or organophosphorus compounds.

In *Drosophila melanogaster, GSTs* constitute an isozyme family composed of two or three distinct subunits encoded by different multigenic families. One cluster of GST genes is located at 87B on *Drosophila* polytene chromosomes near the heat-shock genes (Toung *et al., 1993*).

Cultured *Kc0* cells express at least two classes of GSTs, which may be identical to those detected in adult flies, and this permitted the molecular characterization and sequencing of one of the *Drosophila* GST genes, encoding subunit 1 (209 amino acids) (Toung *et al.,* 1990, 1991). This gene family was mapped by *in situ* hybridization to chromosome 3R at 87B (i.e., just between the two loci of the *hsp70* gene family, at 87A7 and 87C1), in a region covered by the heat-shock puff. The authors, therefore, looked for heat inducibility and, in *Kc0* cells, they indeed observed an increase, albeit a moderate one (less than two-fold) in GST activity (as assayed by 1-chloro-2,4-dinitrobenzene conjugation), after a 37°C heat shock.

VII. CELL RESPONSES TO DIFFERENT STRESSES

In his pioneering work published in *1962,* Ritossa observed that the puffing pattern of *Drosophila* polytene chromosomes that is inducible by an elevation of temperature can be reproduced by exposure *in vitro* of the salivary glands to 2,4-dinitrophenol (DNP; 10^{-3} M, 30 min) or salicylate (10^{-2} M). Then an identical set of puffs was observed during recovery of larvae from anoxia (Ritossa, *1964;* Ashburner, *1970*).

As a matter of fact, a large variety of experimental (or pathological) situations, in particular treatment with various substances which have apparently no chemical relationship (see Ashburner and Bonner (*1979*) for a partial compilation), are able to trigger the synthesis of typical *HSPs* in all kinds of cells. For instance, in primary cultures, Buzin and Bournias-Vardiabasis (1984; see also Bournias-Vardiabasis and Buzin, 1986, 1987) observed the induction of small heat-shock proteins by a variety of teratogens. Therefore, it would be more correct, despite the custom, to talk about a common cell response to stress, heat shock being only one of many stresses.

A. Inducing Agents or Conditions That Affect Respiratory Metabolism

Berendes and his Nijmegen laboratory in the Netherlands devoted a series of studies (see, for instance, Leenders and Berendes, 1972; Sin, 1975) to the puff-inducing activities of several drugs that are all known to interfere with mitochondrial respiratory pathways and oxidative phosphorylation (such as arsenite, DNP, salicylate, menadione), but did not succeed in characterizing any possible common denominator.

The heat-shock mimicking effect of arsenite was later used, for comparison with conventional heat shock, to study the synthesis and intracellular translocation of *HSPs* in *Kc* cells (Vincent and Tanguay, 1982) and the noncoordinated expression of histones (Tanguay *et al.*, 1983) or their modifications by acetylation and methylation during stress (Arrigo, 1983; Desrosiers and Tanguay, 1988) (see Section VIII.B).

Ropp *et al.* (1983) carefully investigated in cultured *Drosophila Kc* cells the return to normoxia, after a period of anaerobiosis (by replacing air phase with pure N_2). They observed a strong correlation between the synthesis of *HSPs* and the rapid rise in O_2 consumption (i.e., a doubling of the maximal rate of O_2 uptake, after a few hours of reexposure to a normal air phase) and pointed out that, in heat shock, there is a similar abrupt increase in respiration, simply due to temperature. Moreover, during this O_2 V_m overcharge, there was no rise in the enzymatic defences of the cell against the dangerous by-products of the partial reduction of oxygen (contrary to the situation in cell hormonal induction; see Chapter 6). This led the authors to formulate a general hypothesis: the highly reactive products of oxygen reduction (O_2 and H_2O_2) are the best candidates for the first links in the chain of chemical events leading to *HSP* induction (see Section IX.C.4). It was indeed shown, later, that exposure to extracellular hydrogen peroxide enhances the expression of *hsp* genes in *Drosophila* cultured cells within a few minutes (Love *et al.*, 1986; Courgeon *et al.*, 1988). Very short treatment (30 sec, 1 mM H_2O_2) of cell cultures or cell extracts was sufficient to induce the specific binding of HSF to the HSE nucleotide motifs (Becker *et al.*, 1990, 1991) (Fig. 7.6) (see Section IX.C.5).

A concomitant induction of cytoskeletal proteins (especially actin) was observed after H_2O_2 treatment, just as in ethanol stress or recovery from anoxia (Courgeon *et al.*, 1993); but one has to admit that this result contrasts with that of a classical heat shock, when actin transcription

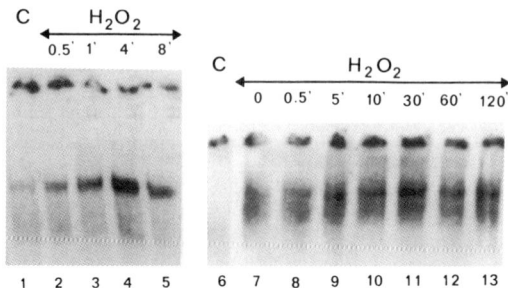

FIGURE 7.6   Kinetics of heat-shock element-binding activity in *Drosophila* cells following H_2O_2 treatment. *Drosophila Kc0* cell extracts were prepared and analyzed by gel-retardation assay (with a synthetic oligonucleotide bearing two overlapping heat-shock consensus sequences). Lanes 1 and 6, extracts from control untreated cells (C); Lanes 2 to 5, extracts prepared after 0.5, 1, 4, and 8 min of 1 mM H_2O_2 treatment; Lanes 7 to 13, cellular extracts of control cells were treated *in vitro* with 20 mM H_2O_2 and analyzed at time 0 and at 0.5, 5, 10, 30, 60, and 120 min after treatment. Arrowhead, HSE 2–protein complex. From Becker *et al.* (1990); reproduced with the permission of the *Eur. J. Biochem.*

is rapidly suppressed (5% of the 25°C value, according to Findly and Pederson, 1981).

It is fair, too, to quote the conflicting conclusions of Drummond and Steinhardt (1987). After a study of the thiol status (measure of total cellular glutatione and the ratio between reduced and oxidized glutathione) carried out on *Kc* cells, either exposed to heat shock or to reoxygenation, the authors considered that there was little evidence of cellular oxidative stress during a conventional heat shock. However, when examined more closely, their experimental data might not be as conclusive as they claimed (Best-Belpomme, personal communication, 1987).

B. HSP-Inducing Metal Ions

Such HSP inducers are of particular interest because some of them, for instance cadmium, are common industrial pollutants, with a wide range of toxic, teratogenic, or carcinogenic effects.

Courgeon *et al.* (1984) established that, as had been previously reported for vertebrate cells, cadmium, which is irreversibly lethal for *Drosophila* cultured cells in concentrations over 100 μM, induces heat-shock proteins; after 24 h of treatment with 10 μM CdCl$_2$, only *HSP70* and *HSP68* were induced, whereas 50 to 100 μM were needed to observe clearly the whole set of the main *HSPs*. Although the concomitant induction of metallothionein (see, in Chapter 3, the possible selection of Cd-resistant cell lines) guards the cells by binding Cd^{2+} ions, the protective role generally attributed to *HSPs* is more questionable.

Using the *in vitro* differentiation of *Drosophila* embryonic cells in primary cultures as an assay (see Chapter 2), Bournias-Vardiabasis *et al.* (1990) observed that most of the transition series metals lowered the numbers of terminally differentiated neurons and myotubes and, simultaneously, induced *HSPs*. Arsenite (5 μM), cadmium (200 μM $CdCl_2$), and mercury (10 μM $HgCl_2$) induced the entire set of *HSPs*, whereas, intriguingly, nickel (3 $\times$ 10^3 μM $NiCl_2$) and zinc (2 $\times$ 10^3 μM $ZnCl_2$) treatments resulted in the induction of only *HSP22* and *HSP23*. Let us point out that a similar induction, restricted to *HSP22* and *-23*, had been previously reported by the same authors after treatment with "classical" teratogens, such as thalidomide and diphenylhydantoin (Bujon *et al.*, 1982).

VIII. INTRACELLULAR LOCALIZATIONS AND PUTATIVE FUNCTIONS OF HEAT-SHOCK PROTEINS

The universality of the response to stress among living organisms and the extreme evolutionary conservation of *HSPs* themselves led to the notion that they might mediate some basic, although as yet poorly understood, function(s) in aggressed cells and perhaps, even, in normal cells.

It seems logical that *HSPs*, because of their rapid induction in various situations of stress, might have some, broadly speaking, protective functions; this is indeed suggested by a series of experimental data.

In this respect, it is important to define the extent to which *HSP* synthesis is restricted to situations of stress. Another useful clue in this search for *HSP* activities is their precise intracellular localization.

A. Presence and Activity of *HSPs* in Normal Cells

As we have just mentioned, a key point is to determine whether *HSPs* are strictly limited to stressed cells or preexist, and possibly act, in normally grown cells. The question is only relevant for the major *HSP70* family, because it is well known that *HSP83* is significantly expressed at normal temperatures, even though its level can multiply five- to six-fold during heat shock. As for the *small-HSPs,* they are also ecdysone inducible and their developmental expression patterns, in normal conditions, have been extensively studied.

It can be observed that, on one- or two-dimensional electrophoregrams, a few bands or spots corresponding to *HSP68/70* are already visible in extracts from control cells grown at 25°C, although their autoradiographic intensities reveal low levels of expression. This fact was observed by several authors; for instance, in *Kc* cells by Mirault *et al.* (1978), Savakis *et al.* (1980), and Sanders (1981). Similarly, Findly and Pederson (1981), when characterizing nuclear RNAs, confirmed the occurrence of a low but finite level of transcription of *hsp70* in untreated *S2* cells.

The discovery of the existence of several heat-shock "cognate" proteins, whose molecular properties are very similar to those of *HSP70* (see Section II.A.3) and which are abundantly synthesized by normal cells (without further induction by heat), required a serious reevaluation of all previous data.

Therefore, Lindquist's group (Velazquez *et al.,* 1983) devoted a careful study to this important problem. The use of specific monoclonal antibodies, binding only to the genuine *HSP70,* led to the conclusion that *HSP70* is indeed detectable, although at the limits of current immunological techniques, in *S2* cells maintained at 25 °C. Moreover, they could establish, from densitometric scanning, that the concentration of *hsp70* mRNA was about only one-thousandth of that attained at 35°C. After theoretical estimation, at 25°C there might be, at the most, 10^4 molecules of *HSP70* per cell.

The possibility remained, however, that induction might have been inadvertently caused by poorly controlled culture conditions or by experimental manipulations. Lindquist (1980) had previously shown that a variety of culture conditions can significantly affect the heat induction patterns of the various *HSPs:* for instance, the basal expression level of *HSP70* is increased when cells are grown with large volumes of media or in suspension cultures. Moreover, during the preparation of cell extracts, a transient induction occurs when the cells are pipetted too vigorously or pelleted too hard in centrifugation.

The same laboratory (Solomon *et al.,* 1991; Feder *et al.,* 1992) obtained transformant *S2* cell lines producing large amounts of *HSP70* at normal temperature, either constitutively or under the control of a "conditional" promoter. The only conclusion relevant to the present discussion was that abundant *HSP70* expression is detrimental to cell growth at optimal temperature.

In any case, it must be concluded that the transcriptional activation of *hsp70* genes by heat shock does not represent a true "off–on" situation.

As for the "cognate" gene products, it is legitimate to ask whether they perform, at normal temperature, the same roles as those played by genuine *HSP70* in stressed cells, or whether they have quite distinct functions.

B. Distribution of *HSPs* in the Cell

It was originally observed from microdissected salivary gland nuclei of *Drosophila* and other Diptera that induced *HSPs* have a propensity to concentrate into the nuclear compartment. The greater part of our information, however, was obtained from cultured *Drosophila* cells.

Two main approaches were combined: (1) classic cell fractionation techniques and (2) cytological studies, using either autoradiography of labeled *HSPs* or their more accurate characterization by specific antibodies and indirect immunofluorescence (Arrigo *et al.,* 1980; Velazquez *et al.,* 1980; Velazquez and Lindquist, 1984; Tanguay and Vincent, 1982; Vincent and Tanguay, 1982; Rabilloud *et al.,* 1985; see reviews by Tanguay, 1983, 1985).

1. One of the first conclusions was to rule out any preferential accumulation in mitochondria, as had been suggested from the heat shock mimicking effect on puffing of a variety of inhibitors of oxidative phosphorylation and electron transfer (see preceding Section VII). Another important result was the discovery of the differential behavior of the three main groups of *HSPs,* and even, very likely, of individual *HSPs,* which might imply functional differences. Their distribution patterns in the cells can be summed up as follows.

With the exception of *HSP83*, which remains primarily cytoplasmic with an increased concentration at the periphery of the cell during heat shock (Carbajal *et al.,* 1986, 1990), a major proportion of the newly synthesized *HSPs* is rapidly transported to the nucleus*:

After 1 h at 37°C, about 36% of *HSP70* and *HSP68* and over 80% of *small-HSPs* were found to be associated with the nuclear fraction of *Kc* cells (Arrigo *et al.,* 1980).

---

* As for *cognate HSP70* proteins, Carbajal *et al.* (1993) more recently established, using specific affinity-purified antibodies and colloidal gold immunoelectron microscopy, that *hsc4* is present in the cytosol, the nucleus and, interestingly, in the mitochondria (close to the inner membrane) of S3 cells and that heat shock induced some redistribution with enrichment of the nucleus. See also Palter *et al.* (1986).

During recovery at 23°C, most of the *HSPs* that were previously present in the nucleus moved back to the cytoplasm within 5–10 hr, and there seems to be a close correlation between this shuttle of *HSPs* and the resumption of normal RNA synthesis (Arrigo, 1980).

However, the precise intranuclear distribution of *HSPs,* and even their *bona fide* nuclear location, remain rather controversial.

According to Arrigo *et al.* (1980) and Velazquez *et al.* (1980), *HSPs* are associated with (soluble) chromatin and nucleoli. However, the attractive hypothesis that they might thereby block the transcription of "normal" cell messages is contradicted by the fact that the observed repression occurs too rapidly (within a few minutes) after the temperature shift.

In contrast, Sinibaldi and Morris (1981) as well as Levinger and Varshavsky (1981) established that after nuclease solubilization and then further extraction with 2*M* NaCl (two treatments that remove most of the DNA and histones) *HSPs* remained in an insoluble fraction with components of a nuclear scaffold or matrix. A structural role was postulated, perhaps "to preserve the spatial organization of transcriptionally active chromatin." Note, too, the relatively recent discovery of *prompt HSPs,* a novel set of heat-inducible proteins which associate specifically with the nuclear matrix–intermediate filament complex (Ornelles and Penman, 1990) (see Section VI.D).

Kloetzel and Bautz (1983) explored quite a different track: using an *in vivo* UV cross-linking procedure, they were able to demonstrate that *HSPs* are bound to heterogeneous nuclear RNA, thus forming RNP complexes in heat-shocked culture cells, but possibly also in cells grown at normal temperature. Independent studies (Wieben and Pederson, 1982; Mayrand and Pederson, 1983) revealed that heat shock alters the normal assembly of hnRNPs in *Kc* cells, as evidenced by a greatly decreased protein content of these nuclear particles in cesium density gradients. The authors, however, wondered whether *HSPs* were involved in the phenomenon, because the latter was observed within 10 min of temperature elevation, that is, when the translation of *HSPs* was just beginning. Similarly, Schuldt *et al.* (1989) noted significant changes in the general organization and interaction of the hnRNA population, which might be an important part of the survival strategy of the cell. In connection with the above, let us recall here that Yost and Lindquist (1986; see also Yost *et al.,* 1990) demonstrated that the general block on RNA processing produced by a severe heat shock (see Section III.A) could be alleviated when the cells had previously accumulated *HSPs* during a previous mild heat shock, which suggests that *HSPs* help to protect the splicing machinery.

In all those conventional cell fractionation methods, the same question recurs: Are HSPs *bona fide* nuclear constituents or do they only associate with structures that (artifactually?) cosediment with the nuclear pellet? Tanguay and Vincent (1982; Vincent and Tanguay, 1982) reported that the presence of *HSPs* in the nuclear fraction, after heat shock, could be correlated with the translocation and phosphorylation of a major cytoskeletal protein with a molecular weight of 45,000. Falkner and Biessmann (1980) had described, in the same *Kc* cells, the shift to heat-shocked nuclei of a cytoplasmic protein of almost identical molecular size (46,000); monoclonal antibodies prepared against it cross-reacted with vimentin and related components of intermediate-sized filaments of vertebrate cells. Thus, indirect immunofluorescence revealed the existence, in cultured *Drosophila* cells, of a cytoplasmic filamentous network that, after a brief heat shock (5 min at 37°C), rearranged itself and formed juxtanuclear cap structures (Falkner *et al.*, 1981) (the same phenomenom occurs in baby hamster kidney cells). Leicht *et al.* (1986) later confirmed that the cofractionation of the vimentin-like intermediate filament proteins with nuclei is indeed explained by the collapse of this IFP network against the nucleus upon heat shock and it thus seems possible that cofractionation of *small-HSPs* is due to the same mechanism.

Yet, a further series of data, obtained using specific monoclonal antibodies, seem rather to corroborate the actual nuclear location of some *HSPs* under heat-shock conditions. *HSP70* accumulation in nuclei obviously correlates with the induction of tolerance to heat or anoxia, which is very suggestive of a protective function (see Section VIII.C) (Velazquez and Lindquist, 1984). *HSP23* was found to be concentrated in the nucleolus, as well as in cytoplasmic granules (see below) (Duband *et al.*, 1986). As for *HSP27*, it was immunolocalized in the perinucleolar regions and in dense nuclear structures of heat-shocked cells (Beaulieu *et al.*, 1989).

A quite pertinent question, especially with respect to the potential functions of *HSPs* in the cell, was raised by Tanguay and Vincent (1982; also Vincent and Tanguay, 1982): Do *HSPs* have a compulsory nuclear localization *per se*, or does their translocation depend on a perturbation, or even physical alteration, caused by elevated temperature? It is, in any case, noteworthy that when the major *HSP70* was induced in *Kc* cells by an arsenite treatment (see Section VII.A) it remained concentrated in the cytoplasm. Moreover, after a subsequent temperature rise to 37°C, a high proportion of this previously labeled *HSP70* was recovered in the nuclear fraction. Likewise, *small HSPs* can be induced over a certain range of temperature (between 33 and 37°C), and they were progressively enriched in the nuclear pellet as the induction temperature was raised. Furthermore, Berger and collaborators (Ireland *et al.*, 1982), who discovered that these *small-HSPs*, but not *HSP70*, are also ecdysteroid inducible (see Section II.B and Chapter 8), noticed a striking difference, in the intracellular distributions of heat-shocked and ecdysone-stimulated cells. In the latter case, the four *small-HSPs* remained almost entirely within the cytoplasm (as opposed to 80% in the nucleus during heat shock). These data do not exclude, however, the potential role of *HSPs* in the nucleus of a heat-shocked cell, but only suggest that their translocation might imply some heat-dependent process, as distinct from their induction mechanism. Beaulieu *et al.* (1989) confirmed that the strength of *HSP27* association with nuclear constituents can differ greatly, depending on whether the cells were subjected to supraoptimal temperatures. During a severe heat shock, *HSP27* was tightly bound in a detergent-resistant form whereas after hormonal induction it was readily solubilized by the same detergent. Moreover, as had previously been reported for their mammalian counterparts, ATP might be involved in the reversible interaction between nuclear components and the major *HSP70*. The large quantities of the multiple members of this *HSP70 -72* family (as well as the major "cognate" protein), which concentrated in the nuclei of S3 cells under heat shock, could be easily released from the nuclear pellet by ATP (Beaulieu and Tanguay, 1988). Such ATP-binding properties might be related to their role in the stress response.

2. Although in the preceding paragraph we focused mainly on the nuclear accumulation of *HSPs*, it should not be forgotten that during the recovery period the majority of nuclear *HSPs* migrate back to the cytoplasm and seem to be useful for the resumption of normal cell synthesis.

The *small HSPs* (*HSP23, -26, and -27*) are present in cytoplasmic aggregates, sedimenting at 20–30S (Arrigo *et al.*, 1980). The absorption spectrum suggested their association with nucleic acids (Arrigo and Ahmad-Zadeh, 1981).

Schuldt and Kloetzel (1985) established that similar 19S RNP particles preexist any stress in the cytoplasmic supernatant from S3 cells grown at normal temperature as well as in the postribosomal fraction from wild-type 10 to 20-hr-old embryos constantly maintained at 23°C. In negatively stained EM preparations, the particles possess a hollow ring-shaped morphology (outer diameter, 12 nm; hollow core, 3 nm). These RNP complexes (approximate RNA–protein ratio of 1 : 8) contain small RNA molecules (60–200 nucleotides) and a characteristic set of some 16 polypeptides in the molecular mass range of 35 to 23 kDa. *Small-HSPs* (particularly *HSP23*) were immunologically identified among the latter (see also

Kloetzel and Haass, 1988). This typical structure and composition were very reminiscent of the so-called "prosome" (today called "proteasome"), a small cytoplasmic RNP particle discovered in vertebrate cells (Schmid *et al., 1984*) and of controversial significance.

Independently, Arrigo *et al.* (1985) further characterized this normal RNP constituent of *Drosophila* cells, but showed that, despite its copurification and related properties, it can be distinguished from the small-HSP ribonucleoprotein complexes observed in cells recovering from a heat shock. Finally, Rollet (1988; Martins de Sa *et al., 1989*), in collaboration with Scherrer's group who had first discovered the prosome, succeeded in separating, on sucrose gradients, the two types of RNP complex: the prosome sedimented at about 19S, whereas the *HSP* complex was isolated in the 16S and 12S regions of the gradient. She could, thereby, demonstrate that none of the prosomal proteins is a genuine *HSP*.

With regard to *HSP70*, it is noteworthy that, even during an intense heat shock, about half the total amount of the induced proteins is retained in the cytoplasm.

While they were investigating the consequences of expressing *HSP70* at normal temperature in transformant cell lines (with the use of *hsp70* constructs driven by constitutive or conditional heterologous promoters; see Chapter 9), Feder *et al.* (1992) discovered that a certain proportion of the cells that, after several days, had escaped the growth block elicited by a *HSP70* overproduction no longer displayed a diffuse immunofluorescence staining, but rather distinct and bright granules dotted all over the cell. As a matter of fact, closer scrutiny of wild-type cells in the late recovery period from heat shock also revealed staining restricted to discrete granules. Because the authors could not remobilize *HSP70* from these granules by heat shock (or, in the case of wild-type recovering cells, by a second repeated heat shock), and because high concentrations of such granules had no effect on thermotolerance, it is clear that they are not storage structures, but rather sites of sequestration and inactivation (primarily distinct from lysosomes). They might represent an additional mechanism for control of *HSP70* activity. This posttranscriptional regulation of *HSP70* and its correlation with the maintenance of acquired thermotolerance (see following section) have been recently reexamined in *S2* cells by Li and Duncan (1995).

C. HSPs and Thermotolerance

The striking rapidity with which aggressed eukaryotic cells launch out into the massive production of a few highly conserved proteins is very suggestive of a general system of protection. Experimental data, from a variety of organisms, have supported this assumption.

As was reported from various developmental stages of *Drosophila*, as well as from other organisms or cell lines, a preexposure of *Drosophila* cultured cells to a mild stress inducing *HSPs* (for instance a 30-min treatment at 35°C) greatly improves their ability to withstand the usually lethal effect of a subsequent more-severe insult (1 h at 42°C). This increased level of resistance is closely correlated with the presence of *HSPs*. See a review by Parsell and Linquist (1993).

Furthermore, thermotolerance may also be elicited by any experimental treatment that can induce *HSPs* (see Section VII); and the reverse is often true (i.e., a moderate heat shock will protect against the damaging effects of such experimental stressors: anoxia, heavy metals), which confirms the widespread protective nature of this emergency response. Yet, intriguingly, Love *et al.* (1986) observed that a mild heat-shock treatment of *S2* cells did not provide the protection expected against a subsequent H_2O_2 exposure, but instead enhanced the toxicity of a high dose of H_2O_2. Moreover, Corell *et al.* (1994) reported that chemical induction of heat-shock proteins, which leads to a general survival thermotolerance, does not, however, allow cells subjected to a serious heat treatment to splice correctly their mRNAs (whereas this remains possible with a previous mild-heat pretreatment).

Various *HSPs* have been attributed the role of major effector in thermotolerance and Berger & Woodward (1983) took advantage of the hormone inducibility of *small-HSPs* to demonstrate that a prior 3-day treatment with β-ecdysone could significantly protect *S3* cells against the deleterious effects of a 42°C shock (as measured from their [³H]uridine incorporation rates and their survival curves with the test for Trypan blue exclusion). On the other hand, the hormone-insensitive *F6* line did not display any similar protection. As a more direct evidence of the activity of this *s-HSP*, expression of *Drosophila HSP27* was shown to confer thermal resistance to Chinese hamster 023 cells (Rollet *et al.*, 1992) and to COS cells (Mehlen *et al.*, 1993).

It would be nevertheless paradoxical if *HSP70*, which is by far the most abundantly synthesized protein in heat shock, was not involved in this thermotolerance. Velazquez and Lindquist (1984) showed that a previous induction of *HSP70* and its return to the nucleus, when cells are heat-shocked again at 40°C, correlated with tolerance to this second and more intense stress; this was true, too, for cell resistance to anoxia.

More recently, the same laboratory initiated a more thorough study, with the use of transformant cell lines, either underexpressing or overexpressing *HSP70* (Solomon *et al.*, 1991). As expected, *S2* cells transformed with antisense constructs (which led to a slower accumulation of *HSP70*; see Section III) showed a sharp decrease in their survival rate after a severe heat shock (45 min at 41.5°C), whereas the integration of extra-copies of the wild-type *hsp70* construct had a positive effect on survival (see Fig. 7.7). The results indicate clearly that *HSP70* plays a major role in thermotolerance. Moreover, the use of mutated versions (deletions or frameshifts) of the transferred *hsp70* gene should permit an analysis in depth of the different functional domains of the protein.

It must be admitted that the molecular mechanisms of this thermoresistance remains poorly understood, although it is likely that they are related to the so far purely speculative functions that the different groups of *HSPs* fulfill in normal cells.

D. Current Views on Putative Functions of HSPs

It is widely held (see, for instance, reviews by Pelham, *1989*; Gething and Sambrook, *1992*) that the heat-shock proteins perform, both during environmental insults and in normal cells, and in all cellular compartments,

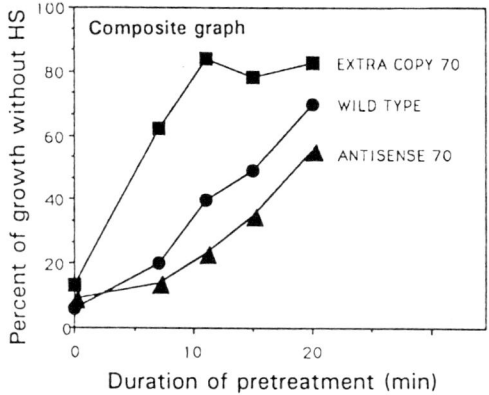

FIGURE 7.7 Thermotolerance in *Drosophila* cell lines that under- or overexpress *HSP70*. The ability to resume growth after a severe heat shock provides a qualitative assessment of thermotolerance: after a 35°C preheat treatment, *S2* cells were given a severe heat shock (45°C) for 45 min, then the growth was measured on Day 5. Comparison between control and *S2* cells transformed either with antisense *hsp70* constructs or with extra copies of the wild-type *hsp70* construct. From Solomon *et al.*, 1991; reproduced with the kind permission of Professor Lindquist and Saunders, Ltd.

what has become known as a protein "chaperoning" role. Chaperones are well-conserved cellular proteins which are able to recognize and stabilize unfolded polypeptides, i.e., newly synthesized protein intermediates as well as the abundant denatured proteins created by stress (see, for instance, Westwood and Steinhardt, 1988). They protect these fragile molecules against damaging, mostly hydrophobic, interactions and catalyze their correct folding and assembly.

It is noteworthy that, by the transient stabilization of some target regulatory proteins in an inactive or unassembled state, they can participate in important regulatory processes. For instance, *HSPs* of the so-called *HSP90* family have been shown, in vertebrate cells but not in *Drosophila*, to link to a steroid receptor and, thereby, prevent, in the absence of the hormone, its intranuclear translocation. See also Section IX.C, where, according to a current model (Fig. 7.11), *HSP70* is shown to maintain the heat-shock factor in a monomeric inactive form, until a stress denatures many proteins that will then compete with HSF for association with *HSP70*.

IX. REGULATORY REGIONS OF *hsp* GENES AND TRANSCRIPTIONAL ACTIVATORS

It should be borne in mind that the cell response to heat shock was primarily exploited as an attractive experimental model for analyzing coordinate gene regulation in higher eukaryotes, even if it appeared later on to be, *per se,* an important and universal system of cell protection against manifold stresses. So far, transcriptional regulation of *hsp* genes is unquestionably, in parallel with hormonal induction by steroids, the model eukaryotic system that has been the most extensively studied at the molecular level; and *Drosophila* cells continue to play a predominant role in this research area.

It is well known that eukaryotic genes comprise, in the 5' position with respect to their transcribed sequences, quite long stretches of DNA (the so-called "promoters," in a broad acceptation) that control gene expression via the specific binding of transcriptional activator proteins to characteristic nucleotide "motifs." Such regulatory regions are conveniently divided into several functional components: (1) the transcription initiation element, including the TATA box (or equivalent sequences) and initiation site, which might be recognized by RNA polymerase II and associated factors and which would define the position at which transcription starts, several tens of basepairs downstream; (2) upstream elements, with typical binding motifs for specific regulatory proteins, vary in position and promote the efficient use of this polymerase-initiation complex; (3) enhancers are also regulatory cis-acting elements that, after their occupation by cellular factors, can greatly increase the level of transcription, frequently according to tissue-specific patterns; they are usually located at much greater distances (sometimes several thousand basepairs away) and can act in both orientations.

As soon as the main *Drosophila hsp* genes had been cloned and sequenced, their regulatory elements were analyzed by various approaches. (1) Computer searches revealed frequent homologies in the 5' upstream regions of most heat-inducible genes, not only in *Drosophila* but also in other organisms. (2) The construction of recombinant plasmids, in which variable lengths of the "upstream" sequences of any *hsp* gene were linked to the coding region of a "reporter" gene (see Chapter 9), then their transfer, either into *Drosophila* embryos or cultured cells, allowed an accurate delimitation and characterization of the promoter elements that are essential for heat inducibility. In such a "functional dissection" of the promoter, it is obvious that transformant flies (in the genome of which one or several copies of the fusion gene have been permanently integrated) provide irreplaceable information about the expression patterns of the *hsp* gene at the different developmental stages or in various tissues. However, transient transfection of cultured *Drosophila* cells has been successfully utilized, since it was demonstrated that the expression of transferred *hsp* genes was very similar to that of corresponding endogenous genes (Larocca, 1984; Morganelli and Berger, 1984; Morganelli *et al.*, 1985). For instance, the last authors observed that, in *S3* cells transfected with any *hsp* promoter–thymidine kinase (tk) hybrid gene, heat shock induced a 40- to 60-fold increase in the relative abundance of *tk* mRNA.

On the other hand, the level of chromatin compaction, in the vicinity of activated *hsp* genes and the accessibility of the regulatory signals for the polymerase complex or transactivating factors have also been carefully investigated.

A. Heat-Shock Responsive Element

Gene transfection methods were devised for mammalian cells a few years before they became available for insect cells (see Chapter 9). Therefore, several groups, at first, transferred *Drosophila hsp70* gene sequences into various mammalian cells and were able to show their heat-shock inducibility, although at the temperature that normally elicits a heat-shock response in the recipient cells (i.e., 42.5°C for mammalian cells). This important observation emphasizes the high degree of conservation of transcriptional regulation of the heat-shock response among higher eukaryotes.

  1. By deletion analysis of the *Drosophila hsp70* promoter, carried out in COS cells (a line derived from green monkey), Pelham (*1982*) identified an essential region upstream from the TATA box (between residues −44 to −66, with regard to the transcription start site). An obvious feature of this rather short segment of DNA is an inverted repeat centered at the junction −51/−50, as previously observed by Ingolia *et al.* (*1980*). The existence of similar inverted repeats with high homologies in most of the other *Drosophila hsp* gene promoters led Pelham to propose a palindromic 14-bp "heat-shock consensus element," C--GA--TTC--G (currently known as "Pelham's element") (Fig. 7.8). He could indeed demonstrate (Pelham and Bienz, *1982*) that synthetic oligonucleotides similar to this consensus sequence were sufficient in COS cells to confer temperature inducibility on the Herpes simplex virus thymidine kinase gene when they were placed upstream from the TATA box.

  Yet, P element-mediated transformation of flies (Dudler and Travers, *1984*) and, finally, the possible utilization of *Drosophila* cell lines for

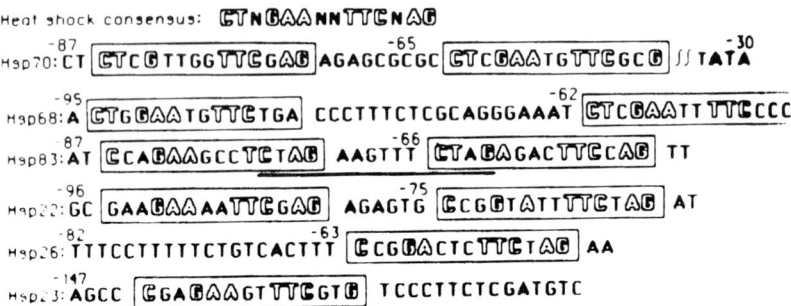

FIGURE 7.8   Location of copies of the heat-shock consensus sequence in several *Drosophila* heat-shock genes. The top line shows the 14-bp heat-shock consensus sequence derived by Pelham (*1982*). Nucleotide sequences in regions upstream of the TATA boxes of five heat-shock genes are shown. Matches to the consensus at 8 of 10 positions or better are enclosed in boxes (with matching nucleotides in outline type) except for the third *hsp83* match, which is underlined. Nucleotides are numbered relative to the transcription start site. From Simon *et al.* (*1985*); reproduced with the kind permission of Professor Lis and Cell Press.

transfection (Lawson *et al.*, 1984; Amin *et al.*, 1985a) rapidly showed the limits of these heterologous experimental systems, albeit without questioning the value of the Pelham's consensus sequence. For an efficient heat induction, a longer 5' flanking sequence is required in *Drosophila* cells than in monkey COS cells.

It was established from promoter deletion analyses (Lawson *et al.*, 1984; Amin *et al.*, 1985a, 1987, 1994; if one restricts the bibliography to studies carried out on cultured cells), combined with searches for sequence homologies and, subsequently, confirmed by the localization of binding sites for HSF (Topol *et al.*, 1985; see Section IX.C) that the *Drosophila hsp70* promoter actually comprises four distinct "heat-shock regulatory elements" (or HSEs). These sequences display various degrees of homology with respect to the canonic Pelham sequence and are all located within the −275 bp upstream 5' region (Fig. 7.9). A single HSE is not sufficient in homologous *Drosophila* cells, but two of them (for

HSP 70

HSP 27

FIGURE 7.9   Regulatory regions of *Drosophila hsp70* and *hsp27* genes, upstream from the TATA box. HSE, heat-shock consensus sequences (gray boxes); EcRE, ecdysteroid responsive element (empty box). Arrows indicate the start site (+1) of transcription. In the "leader" of *hsp70*, thinner black boxes correspond to sequences probably involved in the translational control at high temperature.

instance, the two proximal ones or a duplication of any one), in either orientation, can promote a high level of heat-shock induction, very similar to that observed with the complete gene. It was shown that they have a synergistic rather than additive effect on promoter activity.

As for the *hsp27* gene, it was established (Riddihough and Pelham, 1986, 1987) that the regulatory elements for heat shock and ecdysone induction lie in two separate clusters, respectively, some 300 and 500 bp upstream from the transcription start site. The heat-shock region contains three HSEs, spreading over a DNA stretch of about 100 bp (between -369 and -276); the two proximal ones, called HSE1 and HSE2, show a pattern of two matches to the consensus overlapping each other (Fig. 7.9). Full induction requires the presence of at least two of these HSEs. *Note:* see also binding sites for GAGA protein (Lu *et al.*, *1993*; Sandaltzopoulos *et al.*, *1995*) and discussion in following Section IX.B.4; moreover, for ecdysone regulation, see Chapter 8).

Similarly, in the 5' upstream region of the *hsp23* gene, there are four distinct segments containing a functional heat-shock responsive element, the closest one being at a distance of 101 bp from the TATA box (Lawson *et al.*, 1985b; Mestril *et al.*, 1986). Within the same DNA stretch Pauli *et al.* (*1981*) identified at least seven homologies to the consensus sequence. It is clear that multiple HSEs are required for normal gene expression.

The *hsp83* gene differs from other *Drosophila hsp* genes in that it already presents substantial levels of tissue-general expression at normal temperature. Xiao and Lis (1989) demonstrated in *Kc* cells transfected with deleted *hsp83–lacZ* fusion gene that the overlapping array of typical heat-shock consensus sequences (residing between -88 and -46) is sufficient for mediating heat-shock inducibility, although not constitutive expression.

Let us point out, however, that, in addition to HSEs, the promoters of heat-shock genes contain other control elements which are used during development or, at least for some of them, after hormonal activation.

2. Further investigations and especially the findings that (1) in recombinant flies, two perfect matches to Pelham's 14-bp HSE conferred only 27% of the wild-type induction on a *hsp70–lacZ* hybrid gene, and (2) variations in the nucleotide sequences flanking the HSEs further reduced levels of induction, led Xiao and Lis (*1988*) to propose a quite novel interpretation of the structure of the heat-shock element: all heat-shock regulatory regions include sequences beyond the conventional 14-bp HSE and would be "better described as a dimer of a 10-bp sequence: **nTTCnnGAAn**; actually, each of those 10-bp units is itself a

dyad of a 5-bp inverted repeat "which seems to be, finally, the basic site of interaction for the heat-shock transcriptional factor. This was indeed directly proved by studying the binding of HSTF (see Section IX,C) purified from heat-shocked *Kc* cells (Shuey and Parker, 1986a,b; Perisic *et al.*, 1989). Figure 7.10 shows the correlation between the regulatory regions of the main *Drosophila hsp* genes, as identified by deletion analysis and centered on the Pelham consensus, and the distribution of the newly defined 10-bp motif.

B. Structural Changes in Chromatin and Accessibility to Transcriptional Units

The periodical winding of DNA around small cores of histones and further packaging of these so-called nucleosomes into coiled 250–300 Angström fibers, which is the typical organization of chromatin in eukaryotes, would not give an RNA polymerase (and its transcription complex) a chance to reach and correctly exploit its template, nor would transacting regulatory factors be able to recognize and bind to the specific DNA sequence "motifs" of the gene promoters.

So, based on such theoretical considerations only, it may be assumed that transcribed genes, or ready-to-be activated ones, should show a significant disruption of the basic chromatin structure.

Studies carried out on *Drosophila* heat-shock genes, since they are particularly easy to switch on, have greatly helped to establish and popularize this notion of "open" chromatin conformation in the vicinity of activated eukaryotic genes.

In experiments of *in vitro* transcription carried out with *Escherichia coli* RNA polymerase, Biessmann *et al.* (1978a,b) showed that heat-shock-specific sequences were transcribed from chromatin isolated from Schneider *S2* cells with a 100-fold higher efficiency when cells had been previously heat-shocked. Identical results were obtained with isolated nuclei from uninduced and heat-induced *Kc* cells (Craine and Kornberg, 1981b)

1. Such an alteration in the organization of chromatin could be directly assessed by a relative increase in the susceptibility of activated sequences to various endonucleases. Wu and colleagues (1979; Wu, 1980) showed, in chromatin isolated from *S2* cells, that DNase I hypersensitive sites were present at the 5' end of *hsp70* genes and that such 5' sensitivity was already detectable before induction. After a 15-min heat-shock (35°C), the whole *hsp70* coding sequence also became sensitive to DNase I. The

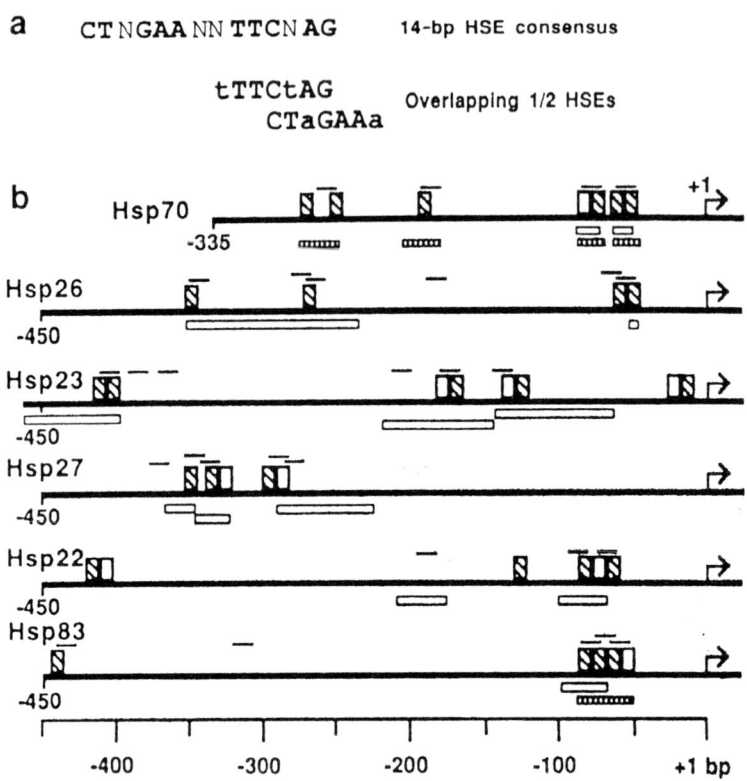

a   CT NGAA NN TTCN AG   14-bp HSE consensus

tTTCtAG   Overlapping 1/2 HSEs
CTaGAAa

FIGURE 7.10   Current interpretation of HSEs: (a) Correlation between the 10-bp heat-shock consensus sequence and the classical HSE. (b) Distribution of the best 10-bp elements (diagonally hatched boxes) in the heat-shock regulatory regions of *Drosophila hsp* genes. They are often found in tandem arrays, and less stringent matches to this consensus (open boxes) are shown when they immediately flank these best matches. Matches to the 14-bp HSE are indicated by the solid lines above each map. Regions containing upstream regulatory elements identified by deletion analyses are designated by the open bars under the maps. Regions of *hsp70* that bind heat-shock transcription factor or of *hsp83* that bind activator protein are designated by the dashed bars under the maps. From Xiao and Lis (*1988*); reprinted from *Science* with the kind permission of Professor Lis and American Society for the Advancement of Science, copyright (1988).

organizational integrity at the level examined was restored when the cells recovered from heat shock.

Similar observations were made on all other *hsp* genes; for instance, the four *small-hsps* (Keene *et al.*, *1981*; Costlow and Lis, 1984; Cartwright and Elgin, 1986). Also noteworthy is the observation that *hsp83* chromatin already displayed a slight sensitivity to DNase I at normal temperature, which is in good agreement with the low level of constitutive expression characterizing this *HSP* (Wu, 1980).

This sensitivity to DNase I obviously corresponds to important changes in the nucleosomal structure and/or higher levels of conformation of the chromatin, which could be demonstrated in several other ways. (1) Microccocal nuclease, known to cleave the linker DNA between nucleosome cores, digested the active *hsp70* chromatin from heat-induced *Kc* cells 30 times more rapidly than that of repressed genes, and the sizes of the fragments were significantly different (Levy and Noll, 1981). (2) Using a "protein-image" hybridization technique, Karpov *et al.* (1984) did not find any histone in the 5'-terminal DNase I hypersensitive region of *hsp70* chromatin, even in non-heat-shocked *Kc0* cells. After induction, they observed a preferential removal of first H1 and then of core histones from the coding region. (3) In possible relation to this latter observation, Levinger and Varshavsky (1982) had previously established that "ready-to-be-transcribed" chromosome regions from nonshocked *S2* cells, and in particular *hsp70* genes, are greatly enriched in ubiquitin–H2A histone. Remember that ubiquitin is itself heat inducible (see Section VI.C) and that such ubiquitin conjugates might become substrates for a proteolytic system.

2. As has already been pointed out, local discontinuities in the nucleosomal array should allow the penetration and binding to DNA of a variety of nuclear factors involved in the transcription process or its regulation. Therefore, it was of major interest to show that, within the promoter region of hypersensitivity to DNase I, a few discrete DNA stretches were occupied by nonhistone proteins and, furthermore, that additional specific sites became occupied (and thus protected from the enzyme) when the gene was in an active state.

For instance, convergent exonuclease III, "nibbling" from definite units on either flank of the DNase I hypersensitive site of the *hsp70* promoter, revealed a strong blockade between positions −40 and −12 in chromatin isolated from nonstressed *S2* cells (Wu, 1984a). In the induced state, this first resistant site is retained, but a novel weaker barrier (or series of barriers) was encountered from positions −108 to −40. If one refers to the nucleotide sequence of this promoter (see Fig. 7.9), it may be concluded that resistant site I surrounds the TATA box area, while site(s) II covers the heat-shock control region and probably corresponds to the presence of a "heat-shock activator protein," or HAP (now called HSF, see Section IX,C).

Similar results were obtained for the *hsp83* promoter (site I between −39 and −17; site II from −86 to −50). Wu (1984b) further demonstrated that when *Drosophila* embryo nuclei were incubated with extract from heat-shocked *S2* cells, but not from normal cultured cells, a distinct exonuclease resistance appeared at site II of the *hsp83* promoter and this HAP binding could be repressed by competition with short DNA fragments containing typical heat-shock elements (see following Section IX,C).

Later, using a high-resolution indirect end-labeling technique, Thomas and Elgin (*1988*) analyzed the genomic footprints on the *hsp26* promoter and similarly observed that a TATA box binding factor was present before and after heat-induction, whereas three of the seven possible heat-shock consensus sequences were occupied but only after heat shock. Besides, in two other *small-hsps* (*hsp27* and *hsp22*), whose hormonal inducibility should be recalled, an ecdysone stimulation of *S3* cells elicited obvious perturbations of the sensitivity to DNase I in upstream regions that coincided with consensus hormone-responsive DNA motifs (Kelly and Cartwright, 1989).

It is now well established that the transcriptional machinery is constitutively bound to the TATA box and adjacent downstream sequences of the *hsp* genes.

The RNA polymerase II itself had previously been localized by indirect immunofluorescence in the puffs of salivary gland polytene nuclei (Jamrich *et al.*, 1977). In *Drosophila* cultured cell, Lis and collaborators (Gilmour and Lis, 1985; Rougvie and Lis, 1988), using protein–DNA photo-cross-linking, could demonstrate not only that this key enzyme of transcription is indeed associated *in vivo* with the 5′ end of induced *hsp70* genes (more precisely between nucleotides −12 and +65), but also that it is already present in the uninduced state. Prior to any induction, the enzyme is even transcriptionally engaged and has already formed a nascent RNA chain of about 25 nucleotides. Thus, the resumption of elongation by this paused polymerase might be a crucial rate-limiting step in the control of *hsp70* transcription (O'Brien and Lis, 1991; Giardina *et al.*, 1992; Rasmussen and Lis, 1993; Giardina and Lis, 1993; O'Brien *et al.*, 1994).

This key enzyme is accompanied by all components of the "basal initiation complex" (see Chapter 5).

3. A series of other proteins, which seem to be associated with the activated state of *hsp* genes, has been immunologically characterized in the chromatin of induced cells (Saumweber *et al.*, 1980; Frasch and Saumweber, 1989). One of these, a 52-kDa protein (designated B52) is

clearly associated with boundaries of transcriptionally active chromatin (Champlin *et al.*, 1991), and it has been suggested that it might participate in the condensation or decondensation of chromatin, perhaps at the level of the 30-nm fiber.

Since the transient unfolding of the chromatin fiber and the passage of the polymerase complex along the coding DNA strand is, *a priori,* greatly facilitated by some "swivel" activity, it is noteworthy that topoisomerases could be detected *in vivo,* in association with the highly transcribed genes of heat-shocked *Drosophila* cells.

UV-cross-linking, and immunocharacterization combined with the use of a specific inhibitor, camptothecin, allowed Gilmour and collaborators (Gilmour *et al.*, 1986; Gilmour and Elgin, 1987) to demonstrate that topoisomerase I (1) is concentrated in *hsp* gene regions during heat shock (found to be 20-fold more abundant on *hsp70* transcription units when they are in induced state), (2) is concentrated in the transcribed region but not in nontranscribed flanking sequences, and (3) interacts with specific sites on each strand.

Similarly, by using the drug VM26 (teniposide), which interferes with the breakage–reunion reaction of topoisomerase II, Rowe *et al.* (1986) showed that multiple interaction sites occur at both the 3' and 5' boundaries of the *hsp70* genes and that the intensity of some of these cleavage sites is significantly increased on heat-shock induction. Rather interestingly, the antibiotitic novobiocin, which was shown to inhibit bacterial topoisomerase II *in vitro,* was able to block the heat-shock response of *Kc* cells, provided that it was added to the culture (150 μg/ml) prior to the induction. Moreover, it prevented the alterations in chromatin organization of the *hsp70* genes that normally accompany heat-shock induction (Han *et al.*, 1985).

4. Since the early 1990s spectacular progress has been made in our understanding of how stable changes in chromatin packaging can account for early embryo "determination," i.e., the mechanisms by which genes can be either permanently "locked up" or remain potentially active in subsequent differentiated tissue cells. Once again, *Drosophila* has proved to be the leading animal model for this approach (see two recent reviews by Singh (*1994*) and Moehrle and Paro (*1994*). In particular, two groups of regulatory genes, the *trithorax* and *Polycomb* groups, play a crucial role "in maintaining, respectively, the active and repressed state of homeotic genes"; this may be true as well for other genes.

The easy inducibility of *Drosophila hsp* genes remains a great advantage in this type of analysis.

The GAGA factor (so called because it binds to GA/CT sites present in the promoter of many *Drosophila* genes and can transactivate them in an *in vitro* transcription system derived from embryo or *Kc* nuclear extracts [Biggin and Tijan, 1988; Sandaltzopoulos *et al.*, 1995]) is encoded by the *trithorax-like* gene (Farkas *et al.*, 1994) and seems to be involved in the formation of an accessible chromatin structure. After reconstruction of the chromatin structure of the *hsp70* promoter, using an *in vitro* nucleosome assay system, Tsukiyama *et al.* (1994) showed plainly that addition of the GAGA protein led to an ATP-dependent nucleosome disruption.

In direct relation with the above-mentioned constitutive binding and then inducible progression of RNA polymerase II along the transcription units of *hsp* genes, we should cite a recent and stimulating paper by O'Brien *et al.* (1995) on the *in vivo* distribution of GAGA protein along these genes (as observed in normal and heat-shocked *Kc* cells, with an *in vivo* UV cross-linking method). Prior to heat shock, GAGA protein was shown to be bound constitutively to the promoter region of *hsp70* genes. Upon heat shock, the factor was rapidly recruited by the transcription unit, with a kinetics of distribution strikingly coincident with that of RNA polymerase II. For instance, GAGA protein was detected on the 3' half of the gene, about 120 sec after initiation of the heat shock. This fact strongly suggests that GAGA factor, by decondensing chromatin, can "open the way" to the polymerase. Concordant observations were made with *hsp26* and *hsp83* genes, but also with genes which are constitutively expressed in these cultured *Kc* cells; for instance, actin *5C* and the histone genes *His3* and *His4*.

5. At a higher level of structural organization, it was suggested some specific association existed between transcribed genes and the nuclear matrix. Small *et al.* (1985) showed in control and induced *Kc* cells that heat-shock genes were indeed matrix associated, both before and during heat shock, and this phenomenon did not apparently depend on high levels of transcription. (See also more recent investigations carried out by Käs *et al.* (1993) in *Kc* cells on the possible cooperative binding of histone H1 and scaffold-associated regions (RARs).

Note: To conclude this rapid survey of the importance of chromatin structural changes in the activation of eukaryotic genes, as results from analysis of this *Drosophila* heat-shock model, it should be pointed out that the "open" conformation of chromatin (ensuring great accessibility to the regulatory regions of the *hsp* genes, even in the inert state of nonstress conditions) in addition to the fact that RNA polymerase is already poised at the transcription site, is perhaps peculiar to these *hsp* genes and reflects their special aptitude for abrupt activation by any environmental aggression.

C. Heat-Shock Transcription Factor and Its Activation

Once again, it was in *Drosophila* cells that the activator protein specific for *hsp* genes was originally discovered.

1. As early as 1977, Compton and Bonner reported that heat-shock puffs could be induced in microdissected polytene nuclei (from *Drosophila* larval salivary glands) by their *in vitro* incubation with the cytosol of heat-shocked *Kc* cells, whereas they remained uninduced in a control cytosol (see also Compton and McCarthy, 1978; Bonner, 1981). Likewise, cytoplasmic extracts from heat-shocked cells were able to activate *hsp* genes in nuclei prepared from unstressed cells and a partial purification of the "activator" factor was, at first, obtained (Craine and Kornberg, 1981a); it was protease sensitive and heat labile.

Several other attempts (Jack *et al., 1981;* Wu, 1984a,b) finally resulted in the characterization (Parker and Topol, 1984b), from nuclear extracts of *Kc* cells, of a transcription factor specific for the heat-shock genes (abbreviated HSTF or, more simply, HSF). In footprint analyses, it was found to be quite distinct from the many general components of the basal initiation complex and to bind to a 55-bp sequence 5' upstream from the TATA box of the *hsp70* gene, i.e., the same region that is protected from DNase I cleavage and comprises the regulatory consensus motifs HSE. This HSF was also present in nuclear extracts from non-heat-shocked cells, although it was less active in transcription assays than the equivalent factor obtained from heat-shocked cells, which suggests that a preexisting factor has to be activated, in some way, by the thermal stress.

The use of sequence-specific DNA-affinity chromatography, based on HSE oligomers, allowed Parker and collaborators (Wiederrecht *et al.,* 1987) and Wu *et al.* (1987) to purify *Drosophila* HSF to near homogeneity from liters of *Kc* or *S2* cell cultures (overall purification close to 250,000-fold). The rigorous sequence-specific binding of the highly purified protein was confirmed by both DNase I footprinting and competition experiments. Its ability to activate the transcription of the *hsp70* gene was directly demonstrated, either in an *in vitro Drosophila* extract transcription system (with an observed 25- to 50-fold stimulation) or by coinjection of the target gene and the activator protein, into nonshocked *Xenopus* oocytes (with a 30-fold increase of the transcripts).

2. Some discrepancies were reported in the initial estimates of the apparent molecular mass of HSF, as measured by SDS–gel electrophoresis: 70 kDa according to Parker's group versus 110 to 115 kDa according to Wu's laboratory. However, having microsequenced partial peptides from HSF and used the corresponding oligonucleotides to screen a cDNA library, Clos *et al.* (1990) were able to isolate cDNA clones for *Drosophila* HSF. The latter revealed a single open reading frame of 2073 nucleotides encoding a 691-amino-acid peptide whose calculated molecular mass is 77,300 Da. Moreover, DNA gel blot analysis showed that the HSF gene is a single-copy gene in the *Drosophila* genome. *In situ* hybridization on polytene chromosomes indicated the position 55A.

Despite the high degree of phylogenetic conservation of the heat-shock system and its regulation among eukaryotes, and more especially, notwithstanding the remarkable similarity of the HSF-binding nucleotidic sites, the deduced amino acid sequence of the *Drosophila* heat-shock factor diverges surprisingly, at least for a large portion, from that of other known HSFs (for instance, yeast HSF). There are, nevertheless, two major and two minor local regions of significant homology: (1) The DNA-binding domain, near the N-terminal area, is the most highly conserved one and shows an intriguing resemblance to the helix-turn-helix binding motif of bacterial *sigma* factors.* (2) An oligomerization domain, lying C-terminal to the first one, comprises several heptad repeats of hydrophobic residues ("leucine-zipper motifs") which would allow the formation of a three-stranded α-helical coiled coil. (3) Two minor homology regions with conserved serine and threonine residues, which are possible sites for phosphorylation (see below).

3. HSF is already present in the nuclei of unstressed cells, although in an inactivated form, in order, probably, to minimize the cell reaction time to stress. The factor was indeed, as already mentioned, isolated from normal *Drosophila* cultured cells (Parker and Topol, 1984b; Wu, 1984b; Wu *et al.*, 1987, 1988) and could even be detected in their nuclei by immunofluorescence (Westwood *et al.*, 1991). Yet Zimarino and Wu (1987) observed a distinct increase (20 times more) in the specific HSE-binding activity of cell extracts within 30 sec after transfer of the cells to a 37.5°C bath; a plateau was reached after 5 min. Furthermore, the

---

* It is noteworthy that, in *E. coli*, the transcriptional induction of heat-shock genes is mediated by an alternative *sigma* factor, σ^{32}.

kinetics and extent of this variation were unaffected by the presence of a protein synthesis inhibitor, such as cycloheximide or puromycin, showing that HSF is preexistent.

In this posttranslational regulation of HSF activity, there is a marked difference between lower and higher eukaryotes. In the yeast *Saccharomyces cerevisiae*, sc-HSF is constitutively bound to its target DNA and only its transcriptional activity is modified in heat shock, probably through phosphorylation. On the other hand, in *Drosophila* and mammalian cells the transcription factor does not bind to the HSE until a stress is applied. The nuclease hypersensitivity of the regulatory region of *hsp70* genes, at normal temperature, whereas it is protected in heat-shocked cells, had been clearly demonstrated by Wu (1984).

This conversion of HSF into a more efficient binding form in animal cells does not exclude further step(s) of activation; for instance, by phosphorylation. It is relevant here to recall the presence of conserved serine and threonine residues (i.e., putative phosphorylation sites) in two short regions of homology between yeast and *Drosophila* HSFs (Clos *et al.*, 1990).

4. Careful studies of the molecular interactions between purified HSF and its specific binding site, carried out in the light of the new interpretation put on the HSE (see above; i.e., a variable number of the five-base module nGAAn (so-called "GAA boxes"), led to the conclusion that HSF binds to DNA as homotrimers, each subunit contacting a 5-bp repeat (Shuey and Parker, 1986a,b; Perisic *et al.*, 1989; Xiao *et al.*, 1991; Weswood and Wu, 1993). This trimerization is inferred by the observation that the electrophoretic mobilities, and therefore sizes, of HSF complexes with synthetic DNA containing increasing numbers of inverted 5-bp units varied stage by stage with the addition of further three GAA units. Moreover, after *in vitro* cross-linking of purified HSF, the predominant complex had indeed the mobility predicted for a HSF trimer. So, the activation of HSF in stress conditions implies a transition from a monomeric to an oligomeric state.

The fact that recombinant *Drosophila* HSF synthesized in *E. coli* displayed full binding activity, whereas, when overexpressed in *Xenopus* oocytes, it did not acquire any significant binding activity without a 10-min heat treatment of the oocytes, strongly suggested that in normal animal cells the preexisting HSF is under some negative control, this inhibition being relieved after heat shock (Clos *et al.*, 1990). Thus, it was assumed that the monomeric form of HSF could be "sequestered" in unstressed cells by its association with a "chaperoning" protein. Western blotting on nondenaturing gels revealed indeed the presence of a 220-kDa HSF complex in cytoplasmic extracts of *S2* cells (Westwood *et al.*, 1991). Ubiquitin (see Section VI.C) was at first proposed to be this hypothetic inhibitory protein. The large pool of abnormal proteins resulting from any stress would compete with HSF for ubiquitin, thus liberating a nonubiquinated active factor (Munro and Pelham, *1985;* Bienz and Pelham, 1987). It is now widely admitted, although not directly proved, that the negative regulators are in fact the *HSPs* themselves, especially *HSP70*. In this model, too, the abundant misfolded proteins produced by the stress compete for association with *HSP70,* and the released HSF subunits can then assemble in homotrimers and bind to DNA (Fig. 7.11).

An attractive hypothesis, which is not necessarily in contradiction with the preceding one, has been put forward in our laboratory by Best-Belpomme and Courgeon (Ropp *et al.*, 1983; Courgeon *et al.*, 1988, 1990; Becker *et al.*, 1990, 1991) and is supported by solid experimental arguments. It aims to identify a common "signal" that would trigger *HSP* synthesis via the activation of HSF in all kinds of stress situations.

It has long been known that reoxygenation after a period of anoxia can induce heat-shock-like puffing in the polytene chromosomes of *Drosophila* salivary glands (see Section VII). It was confirmed that such a reoxygenation, without any temperature elevation, was sufficient to promote typical *HSP* synthesis in *Kc* cells (Ropp *et al.*, 1983). The sudden increase in oxygen consumption, resulting from this recovery from anoxia, generates by-products of the partial reduction of oxygen, i.e., the so-called "reactive forms of oxygen" [or RFOs, such as the superoxide ions (O_2^-) or hydrogen peroxide (H_2O_2)]. Among the numerous and varied experimental "stressors" that can induce *HSPs,* most might also act via accumulation of RFOs or, more generally, via modifications of the redox potential of the cells. (1) For instance, several inducing agents (arsenite, dinitrophenol) interfere with the respiratory metabolism; (2) the drug menadione is known to elevate the intracellular concentration of superoxide ions; (3) metal ions, such as cadmium (Courgeon *et al.*, 1984), react with sulfhydryl groups; (4) finally, H_2O_2 itself induces some or all *HSPs* in bacteria, *Drosophila* (Courgeon *et al.*, 1988), and mammalian cells.

Ames and collaborators (Lee *et al.*, *1983*) have shown that adenylated nucleosides such as AppppA (P^1,P^3-diadenosine 5'-triphosphate) accumulate rapidly in *Salmonella* and *E. coli* after heat shock or under conditions of oxidative stress. They speculated that these molecules might be the "alarmones," a quite suggestive name (i.e., signal molecules that alert the cell to the onset of stress and lead to *HSP* synthesis). This latter

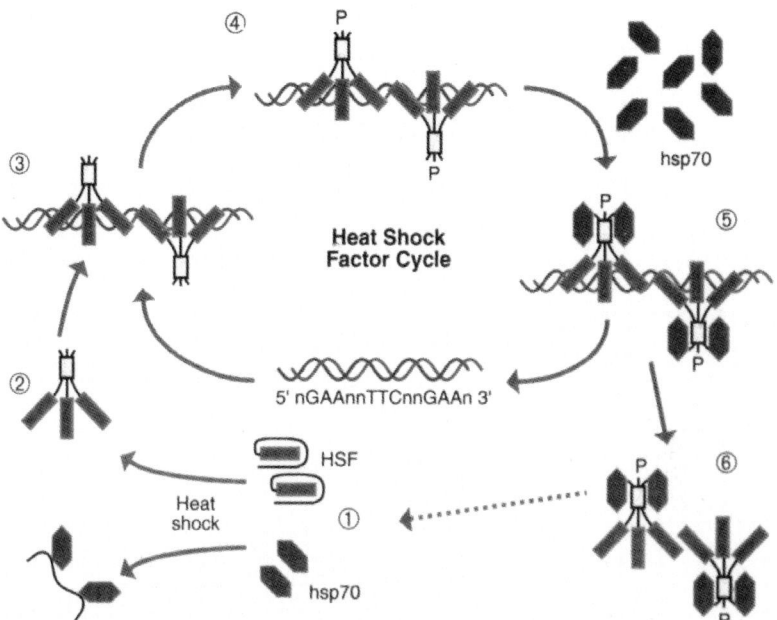

FIGURE 7.11 A current model of HSF regulation. In the unstressed cells, HSF is maintained in a monomeric, non-DNA-binding form through its interactions with *HSP70* (1). Upon heat shock or other forms of stress, HSF assembles into a trimer (2), binds to specific sequence elements in heat-shock gene promoters (3), and becomes phosphorylated (4). Transcriptional activation of the heat-shock genes leads to increased levels of *HSP70* and to formation of an HSF–*HSP70* complex (5). Finally, HSF dissociates from the DNA and is eventually converted to non-DNA-binding monomers (6). From Morimoto (*1993*); reprinted from *Science* with the permission of Professor Morimoto and American Society for the Advancement of Science, copyright (1993).

hypothesis was ruled out, at least in *Drosophila*, by the fact that in stressed *Kc* cells (Brevet *et al.*, 1985), there are no such rapid variations of AP4A and other dinucleosidephosphates. So, Best-Belpomme and Courgeon proposed that it is the reactive forms of oxygen alone which are the intracellular induction signals common to the different stresses.

By mimicking the usual situation in heat shock, in which the accumulation of RFOs, resulting from the elevation of temperature and the correlated sharp increase in respiration, must be quite rapidly resorbed by the specific enzymatic defences of the cells (see Chapter 8), Courgeon *et al.* (1990) were able to demonstrate that a brief (2 to 5 min) H_2O_2 (1 mM)

treatment of *Kc* cells was sufficient to trigger the synthesis of *HSPs* and histone H2B. The H_2O_2 effect was even greater when associated with aminotriazole (a specific inhibitor of catalase activity), which suggests that hydrogen peroxide itself, and not oxygen, is the inducer (Becker *et al.*, 1991).

The action of H_2O_2 seems to be direct, either on the heat-shock factor itself or on some activator (kinase?), as inferred from *in vitro* assays; whereas the specific binding to an oligonucleotide comprising two HSE sequences is weak in extracts from unstressed *Kc* cells, it could be significantly increased after a 5-min incubation of the same extracts with 1 mM H_2O_2 (Fig. 7.6) (Becker *et al.*, 1990; Becker, 1994). Similar results were obtained with murine cell extracts. It must be remembered that in their *in vitro* puffing system Compton and McCarthy (1978) could also stimulate puff induction by simply adding hydrogen peroxide to the cytosol of nonheated *Kc* cells in which the polytene nuclei were incubated.

Moreover, the fact that quite different treatments (heat or low pH) could also enhance the DNA-binding activity of HSF in extracts from unshocked *Drosophila* cells (Zimarino *et al.*, 1990) might be reconcilable with this "RFO hypothesis."

However, as a possible alternative, it cannot be ruled out that highly reactive forms of oxygen might possibly rapidly damage cellular proteins, which would be in agreement with the current prevailing HSF–*hsp70* complex model.

5. The precise molecular mechanism whereby the heat-shock factor, as soon as it is bound on HSE sequences, can activate transcription remains, as of the mid-1990s, unexplained. But this can also be said of most eukaryotic transcriptional activators. Remember that the RNA polymerase and its initiation complex are already bound to the promoters of *hsp* genes and are even transcriptionally engaged (see above, Section IX.B), so that it is rather a block on elongation which has to be released.

Careful studies by Shuey and Parker (1966a) established that HSF binding promotes conformational changes in the protein–DNA complex of the *hsp70* promoter and, in particular, results in the introduction of a specific DNA bend. Such structural modifications can be immunologically recognized, so much so that the activation of HSF that could be observed in extracts of unshocked cells, after treatment with polyclonal antibodies prepared against the activated factor (Zimarino *et al.*, 1990), may be reasonably explained by a specific antibody-mediated stabilization of the active form of HSF. Thus, in the presence of this antibody, the thermodynamic equilibrium between the inactive and active forms present in the cell extracts might be irreversibly shifted toward the active form.

References

Arribas, C., Sampero, J., and Izquierdo, M. (1986). The *ubiquitin* genes in *Drosophila melanogaster:* Transcription and polymorphism. *Biochim. Biophys. Acta* **869**, 119–127.

Ashburner, M., and Bonner, J. J. (1979). The induction of gene activity in *Drosophila* by heat shock. *Cell* **17**, 241–254.

Ayme, A., and Tissières, A. (1985). Locus *67B* of *Drosophila melanogaster* contains seven, not four, closely related heat shock genes. *EMBO J.* **4**, 2949–2954.

Berendes, H. D. (1972). The control of puffing in *Drosophila hydei. In* "Developmental Studies on Giant Chromosomes" (W. Beermann, ed.), Springer-Verlag, Berlin, pp. 181–207.

Dudler, R., and Travers, A. A. (1984). Upstream elements necessary for optimum function of the *hsp70* promoter in transformed flies. *Cell* **38**, 391–398.

Farkas, G., Gausz, J., Galloni, M., Reuter, G., Gyurkovics, H., and Karch, F. (1994). The *Trithorax-like* gene encodes the *Drosophila* GAGA factor. *Nature* **371**, 806–808.

Garbe, J. C., Bendena, W. G., and Pardue, M. L. (1989). Sequence evolution of the *Drosophila* heat shock *hsrω*. I. The nonrepeated portion of the gene. *Genetics* **122**, 403–415.

Gething, M. J., and Sambrook, J. (1992). Protein folding in the cell. *Nature* **355**, 33–45.

Hackett, R. W., and Lis, J. T. (1981). DNA sequence analysis reveals extensive homologies of regions preceding *hsp70* and *αβ* heat shock genes in *Drosophila melanogaster*. *Proc. Natl. Acad. Sci. U.S.A.* **78**, 6196–6200.

Hultmark, D., Klemenz, R., and Gehring, W. (1986). Translational and transcriptional control elements in the untranslated leader of the heat-shock gene *hsp22*. *Cell* **44**, 429–438.

Ingolia, D., and Craig, E. (1982). Four small *Drosophila* heat shock proteins are related to each other and to mammalian *α-crystallin*. *Proc. Natl. Acad. Sci. U.S.A.* **79**, 2360–2364.

Ingolia, D., Craig, E., and McCarthy, B. J. (1980). Sequence of 3 copies of the gene for the major heat shock induced protein and their flanking regions. *Cell* **21**, 669–679.

Jack, R. S., Gehring, W. J., and Brack, C. (1981). Protein component from *Drosophila* larval nuclei showing sequence specificity for a short region near a major heat-shock protein gene. *Cell* **24**, 321–331.

Jamrich, M., Greenleaf, A. L., and Bautz, E. K. F. (1977). Localization of RNA polymerase in polytene chromosomes of *Drosophila melanogaster*. *Proc. Natl. Acad. Sci. U.S.A.* **74**, 2079–2083.

Keene, M., Corces, V., Lowenkaupt, K., and Elgin, S. (1981). DNAse I hypersensitive sites in *Drosophila* chromatin at the 5' ends of regions of transcription. *Proc. Natl. Acad. Sci. U.S.A.* **78**, 143–146.

Klemenz, R., Hultmark, D., and Gehring, W. J. (1985). Selective translation of heat-shock mRNA in *Drosophila melanogaster* depends on sequence information in the leader. *EMBO J.* **4**, 2053–2060.

Lakhotia, S. C., and Mukherjee, A. S. (1980). Specific activation of puff *93D* of *Drosophila melanogaster* by benzamide and the effect of benzamide treatment on the heat-shock induced puffing activity. *Chromosoma* **81**, 125–136.

Lee, P. C., Bochner, B. R., and Ames, B. N. (1983). AppppA, heat-shock stress, and cell oxidation. *Proc. Natl. Acad. Sci. U.S.A.* **80**, 7496–7500.

Leenders, H. J., and Berendes, H. D. (1972). The effect of changes in the respiratory metabolism upon genome activity in *Drosophila*. I. The induction of gene activity. *Chromosoma* **37**, 433–444.

Lewis, M., Helmsing, P. J., and Ashburner, M. (1975). Parallel changes in puffing activity and patterns of protein synthesis in salivary glands of *Drosophila*. *Proc. Natl. Acad. Sci. U.S.A.* **72**, 3604–3608.

Lu, Q., Wallrath, L. L., Granok, H., and Elgin, S. C. R. (1993). (CT)n · (GA)n repeats and heat shock elements have distinct role in chromatin structure and transcriptional activation of the *Drosophila hsp26* gene. *Mol. Cell. Biol.* **13**, 2802–2814.

Mehlen, P., Briolay, J., Smith, L., Diaz-Latoud, L., Fabre, N., Pauli, D., and Arrigo, A. P. (1993). Analysis of the resistance to heat and hydrogen peroxide stresses in COS cells transiently expressing wild type or deletion mutants of the *Drosophila* 27-kDa heat-shock protein. *Eur. J. Biochem.* **215**, 277–284.

Moehrle, A., and Paro, R. (1994). Spreading the silence: Epigenetic transcriptional regulation during *Drosophila* development. *Dev. Genet.* **15**, 478–484.

Morimoto, R. I. (1993). Cells in stress: Transcriptional activation of heat-shock genes. *Science* **259**, 1409–1410.

Munro, S., and Pelham, H. (1985). What turns on heat shock genes? *Nature* **317**, 477–478.

O'Farrell, P. Z., Goodman, H. M., and O'Farrell, P. H. (1977). High resolution two-dimensional electrophoresis of basic as well as acidic proteins. *Cell* **12**, 1133–1142.

Pauli, D., Spierer, A., and Tissières, A. (1981). Several hundred base pairs upstream of *Drosophila hsp23* and *hsp26* genes are required for their heat induction in transformed flies. *EMBO J.* **5**, 755–761.

Pelham, H. R. B. (1982). A regulatory upstream promoter element in the *Drosophila hsp70* heat-shock gene. *Cell* **30**, 517–528.

Pelham, H. R. B. (1989). Heat shock and the sorting of luminal ER proteins. *EMBO J.* **8**, 3171–3176.

Pelham, H. R. B., and Bienz, M. (1982). A synthetic heat shock promoter element confers heat-inducibility on the herpes simplex virus thymidine kinase gene. *EMBO J.* **1**, 1473–1477.

Peters, F. P., Lubsen, N. H., Walldorf, U., Moormann, R. J. M., and Hovemann, B. (1984). The unusual structure of heat-shock locus 2-48B in *Drosophila hydei*. *Mol. Gen. Genet.* **197**, 392–398.

Ritossa, F. (1962). A new puffing pattern induced by temperature shock and DNP in *Drosophila*. *Experientia* **18**, 571–573.

Ritossa, F. (1964). Experimental activation of specific loci in polytene chromosomes of *Drosophila*. *Exp. Cell Res.* **35**, 601–607.

Rollet, E., Lavoie, J. N., Landry, J., and Tanguay, R. M. (1992). Expression of *Drosophila*'s 27 kDa heat shock protein in rodent cells confers thermal resistance. *Biochem. Biophys. Res. Commun.* **185**, 116–120.

Rubin, D. M., Mehta, A. D., Zhu, J., Shoham, S., Chen, X., Wells, Q. R., and Palter, K. B. (1993). Genomic structure and sequence analysis of *Drosophila melanogaster hsc70* genes. *Gene* **128**, 155–163.

Rysek, R. P., Walldorf, U., Hoffmann, T., and Hovemann, B. (1987). Heat-shock loci *93D* of *Drosophila melanogaster* and *48B* of *D. hydei* exhibit a common structural and transcriptional pattern. *Nucleic Acids Res.* **15**, 3317–3333.

Sandaltzopoulos, R., Mitchelmore, C., Bonte, E., Wall, G., and Becker, P. B. (1995). Dual regulation of the *Drosophila hsp26* promoter *in vitro*. *Nucleic Acids Res.* **23**, 2479–2487.

Schmid, H. P., Akhayat, O., Martins De Sa, C., Puvion, F., Koehler, K., and Scherrer, K. (1984). The prosome: An ubiquitous morphologically distinct RNP particle associated with repressed mRNPs and containing specific scRNA and a characteristic set of proteins. *EMBO J.* **3**, 29–34.

Simon, J. A., Sutton, C. A., Lobell, R. B., Glaser, R. L., and Lis, J. T. (1985). Determinants of heat-shock induced chromosome puffing. *Cell* **40**, 805–817.

Sin, Y. T. (1975). Induction of puffs in *Drosophila* salivary glands by mitochondrial factor(s). *Nature* **258**, 159–160.

Singh, P. B. (1994). Molecular mechanisms of cellular determination: Their relation to chromatin structure and parental imprinting. *J. Cell Sci.* **107**, 2653–2668.

Thomas, G. H., and Elgin, S. C. R. (1988). Protein/DNA architecture of the DNAse I hypersensitive regions of the *Drosophila* hsp26 promoter. *EMBO J.* **7**, 2191–2201.

Tissières, A., Mitchell, H. K., and Tracy, U. M. (1974). Protein synthesis in salivary glands of *Drosophila melanogaster*: Relation to chromosome puffs. *J. Mol. Biol.* **84**, 389–398.

Toung, Y. P. S., Hsieh, T. S., and Tu, C. P. (1993). The Glutathione S-Transferase D genes: A divergently organized, intronless gene family in *Drosophila melanogaster*. *J. Biol. Chem.* **268**, 9737–9746.

Treier, M., Seufert, W., and Jentsch, S. (1992). *UbcD1* encodes a highly conserved ubiquitin-conjugating enzyme involved in selective protein degradation. *EMBO J.* **11**, 367–372.

Tsukiyama, T., Becker, P. B., and Wu, C. (1994). ATP-dependent nucleosome disruption at a heat-shock promoter mediated by binding of GAGA transcription factor. *Nature* **367**, 525–532.

Walldorf, U., Richter, S., Rysek, R. P., Steller, H., Edström, J. E., Bautz, E. K. F., and Hovemann, B. (1984). Cloning of heat shock locus *93D* from *Drosophila melanogaster*. *EMBO J.* **3**, 2499–2504.

Xiao, H., and Lis, J. T. (1988). Germline transformation used to define key features of heat shock response elements. *Science* **239**, 1139–1142.

8

Experimental Models of Gene Regulation: 2. Cell Responses to Hormone

Ecdysteroid* hormones play a crucial role in the control of growth and differentiation in insects. Several cultured *Drosophila* cell lines appear to be quite sensitive to physiological concentrations of 20-hydroxyecdysone and display within a few hours or days of treatment a characteristic array of morphological and metabolic modifications. This experimental model, although it is more delicate to handle than the heat-shock system (see Chapter 7), is perhaps even more productive, since it combines the modulations of the best known genome of all higher organisms with the mechanisms of action of a steroid hormone. As will be discussed below, a set of so-called small-heat-shock proteins are also induced by the hormonal treatment, which creates an interesting "catwalk" between the two models.

A. M. Courgeon (1972a,b) was the very first to show that *Kc* line cells are sensitive "targets" for ecdysteroid hormones. In her two short basic papers and a few later ones, she established most of the main features of the "model": (1) strict specificity of the hormonal stimulus and its dose dependency (in particular, the fact that β-ecdysone is about 500-fold more active than α-ecdysone); (2) stereotyped sequence of dramatic morphological changes displayed by the treated cells, which flatten and emit long pseudopodia (Fig. 8.1); (3) increased cell mobility, resulting in a final aggregation into large and floating clumps; (4) concomitant arrest of division (Fig. 8.4); and (5) isolation of "resistant" cells. She studied the differential sensitivity of several cell lines or clones to ecdysteroids (1975a) and noted the antagonistic effect of the juvenile hormone (JH) (1975b). Then, in collaboration with Best-Belpomme, she reported a preliminary characterization of "saturable" ecdysone-binding factor(s) in fractions from "sensitive" *Kc* cells and their absence in "resistant" cell sublines (Best-Belpomme and Courgeon, 1975). They also observed a

---

* Ecdysteroid is a generic name for the arthropod molting hormones and related molecules and was first proposed by J. Kuhlman. The term ecdysone, in its broadest sense, is currently used for all biologically active ecdysteroids. Strictly speaking, however, ecdysone (α-ecdysone) is the prehormone, while the active form of the hormone is 20-hydroxyecdysone (20-HE, β-ecdysone, ecdysterone).

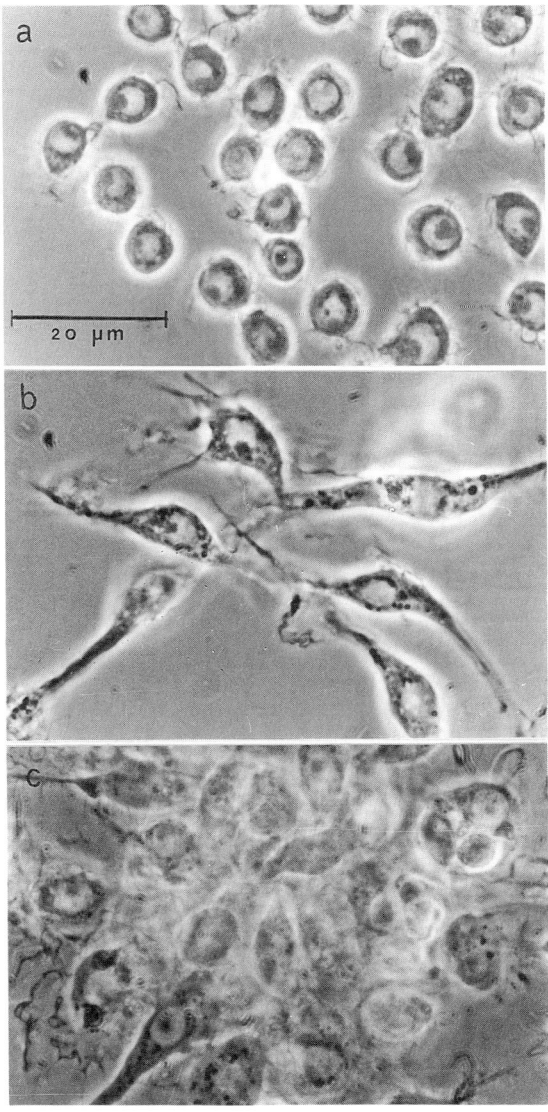

20 µm

FIGURE 8.1   Morphological changes of *Drosophila Kc* cells induced by ecdysteroids. Cells of line *Kc* were cultured in Leighton tubes, with or without 0.006 µg/ml of β-ecdysone. The coverslip bearing the living cells was turned upside down on a slide to be photographed under phase contrast. (a) Control *Kc* cells. (b) 1 to 2 days of hormonal treatment: the cells become spindle-shaped and seem to flatten onto the bottom of the culture vessel. (c) 3 to 4 days of treatment: cells aggregate into clumps of increasing size. Courtesy of Dr. Anne Marie Courgeon, Université P. et M. Curie, Paris.

specific protein induction in hormone-treated cells (Best-Belpomme and Courgeon, 1976a,b).

Soon after, several other laboratories adopted this model of hormonal regulation and have actively participated to its enrichment and popularization.

First, Cherbas' group (at Harvard University, Cambridge, MA, then at Indiana University, Bloomington, IN) provided two major contributions. (1) They discovered in hormone-treated *Kc* cells the specific induction of an enzymatic activity (acetylcholinesterase), which allowed for an easy and precise quantification of the cell response (Cherbas *et al.*, 1977). (2) They focused on the study of two early inducible polypeptides whose synthesis starts within the first hour of treatment (EIPs, i.e., ecdysteroid-inducible proteins, see Sekeris *et al.*, 1980). The analysis was continued to the cloning of the corresponding genes, then to the identification of the cis-acting sequences involved in their hormonal regulation (Cherbas *et al.*, 1991; Cherbas, 1993). It contributed greatly to the definition of a consensus motif for ecdysone-responsive elements.

Of the few detected proteins whose synthesis is modified after ecdysteroid exposure, Couderc and collaborators (University of Clermont-Ferrand, France) chose to study the various aspects of hormonal regulation of the main cytoskeletal proteins, actin (Couderc *et al.*, 1982) and tubulin (Sobrier *et al.*, 1986).

Similarly, Berger (Darmouth College, Hanover, NH) played an important part in the development of the model. After several attempts in collaboration with Wyss (Zurich) to understand, by somatic cell hybridization experimentation, the true nature of cell resistance to the hormone, he, too, became interested in the structural proteins responsible for cell morphological transformation and mobility. Later, the intriguing assumption that the so-called "small" heat-shock proteins can also be induced by ecdysteroids (Ireland *et al.*, 1982) led him to devote most of his work to this problem of "dual regulation." His laboratory was one of the first to use the method of transfection of "chimeric" constructs in *Drosophila* cells for a "functional dissection" of the regulatory flanking regions of these genes (Morganelli and Berger, 1984; Morganelli *et al.*, 1985) (see also Chapter 7).

The early observations of the Russian group (Kurchatov Institute, Moscow) concerning the hormone-induced synthesis of macromolecules (Gvozdev *et al.*, 1974), and more particularly the changes in pattern of cell surface glycoproteins (Metakovsky and Gvozdev, 1978), should not be forgotten in this rapid bibliographic survey of the pioneering works.

I. HORMONAL STIMULUS

A. Hormone Specificity

It was essential to ascertain that the dramatic transfiguration of the cells elicited by β-ecdysone was not due to an unspecific pharmacological effect of steroid compounds.

The specific nature of the stimulus had already been established by Courgeon (1972b), who could not discern any activity of vertebrate steroids (male and female hormones, corticoids) or cholesterol, even at concentrations as high as 100 μg/ml. Moreover, as she pointed out, *Kc* cells respond to a dose as low as 10^{-8} to 10^{-7} M β-ecdysone, a concentration lower than that employed by the different authors in various *in vivo* or *in vitro* assay systems ("puffing" of polytene nuclei of salivary glands or evagination of imaginal discs); this value was later found to be even lower than the physiological hormone titer, as observed during the metamorphosis of *Drosophila*.* Another argument in favor of the nontoxicity of such low concentrations is the existence of "resistant" cells that were able to go on multiplying, even in the presence of 10^{-4} M β-ecdysone.

These observations were greatly extended by Cherbas and collaborators (1980). Apart from the fact that a number of unrelated compounds known to induce differentiation in a variety of vertebrate cell types (cyclic AMP, DMSO, several growth factors) failed to produce any visible effect on *Kc* cells, the activity of 60 different steroids (see the impressive list in the original paper) was carefully tested. Only those that had been previously reported to be active in other *in vivo* or *in vitro* insect systems induced the typical elongation of *Kc-H* cells, and their relative activities were shown to be quite similar to those already published (see Table 8.I). So, as the authors emphasized, "the (cultured) *Kc* cells have maintained the capacity of normal *Drosophila* cells to discriminate among closely related ecdysteroid compounds." Further characterization of specific ecdysteroid receptors in sensitive *Kc* and *S3* cells has fully confirmed this assertion.

B. Dose Dependence of Cell Responses and Relative Activities of Main Ecdysteroids

To correlate the intensity of the cell response with increasing concentrations of ecdysteroids, one needs a semiquantitative bioassay. This could

---

* At the time of puparium formation, this titer was estimated to be about 180 pg β-ecdysone equivalents/mg live weight (Hodgetts *et al.*, 1977; Pak and Gilbert, 1987).

TABLE 8.I   Relative Activities of Different Steroids[a]

| | Relative activity | | |
Compound	Kc-H cell morphology	Kc-H cell AChE induction	Imaginal disc eversion
Ponasterone A	8	9	20
Polypodine B	4	3	2
20-Hydroxyecdysone[b]	[1]	[1]	[1]
Cyasterone	0.5	No data	1
Inokosterone	0.1	0.06	0.8
5-Hydroxy-22,25-deoxyecdysone	0.03	0.006	0.03
2-Deoxy-20-hydroxyecdysone	0.03	0.005	0.1
Ecdysone	0.006	0.004	0.002
22,25-Deoxyecdysone	0.003	0.004	Not detectable
2,22,25-Deoxyecdysone	<0.002	<0.002	Not detectable
14,15-Epoxy-22,25-deoxyecdysone	<0.002	<0.002	Not detectable
14,22-Deoxyecdysone	<0.002	<0.002	Not detectable

[a] From Cherbas *et al.* (1980); reproduced with the kind permission of the Cherbas and of Springer-Verlag.

[b] Chosen for arbitrary unit.

be (1) the slowing down of the culture proliferation after exposure to the hormone (Courgeon, 1972a), (2) the rapidity of succession of the different stages of cell transformation at the population level (Courgeon, 1972b), or (3) the percentage of treated cells whose elongation under standardized conditions reaches an arbitrarily defined length (Cherbas *et al.*, 1980). As soon as several specific enzymatic inductions had been discovered, especially the appearance of acetylcholinesterase or β-galactosidase activities, the measure of which is particularly sensitive, the evaluation of the cell response became much more accurate (see below). In all cases, a dose-dependent effect was clearly established (see, for instance, Figs. 8.6 and 8.7).

Similarly, it became possible to compare the action of the main ecdysteroids. The great sensitivity of this cell system must be emphasized. When assayed in living insects, the various analogs are active at approximately the same concentrations, whereas, in cultured *Drosophila* cells, the range of their respective active concentrations may span more than three orders of magnitude.

As for 20-hydroxyecdysone (20-HE, alias β-ecdysone or ecdysterone), the physiological hormone,* the concentrations that elicit a half-maximal response, in its various forms, vary from 10^{-8} to 10^{-7} M, in Kc cells (Courgeon, 1972b; Cherbas *et al.*, 1980).

Ecdysone (α-ecdysone) was found to be 200 to 500 times less active (see, for instance, Fig. 8.3, in which the profile of the dose–response curve for α-ecdysone is almost identical to that of β-ecdysone, just as if it has only been transposed to the right along the abcissa of concentrations. Ecdysone activity might require its conversion to 20-HE, but, although its hydroxylation is well documented in insect organisms, it has not been reported in established cell lines from *Drosophila.***

Among the 60 analogs that they tested (see Table 8.I and footnote, page 437), Cherbas *et al.* (1980) found that polypodine B and, more especially, ponasterone A were more potent inducers than the natural hormone (about four- and eight-fold, respectively). This high biological activity must be correlated with the strong affinity to $Kc167$ cell extracts displayed by synthetic [^{125}I]iodoponasterone (Cherbas *et al.*, 1985) and justifies its use as a radioligand for the characterization of ecdysteroid receptors (see Section V). Surprisingly, at least for the induction of β-galactosidase in Kc cells, Best-Belpomme and Ropp (reported in Peronnet *et al.*, 1986) did not observe such a high activity with ponasterone; on the other hand, muristerone proved particularly efficient.

C. Differential Sensitivity of Main Cell Lines

Kc sublines offer a paradigm for ecdysteroid-sensitive cells, since most of the typical responses to the hormone have been described in these cells. Courgeon (1975) checked a number of clones isolated from a Kc responsive subline for their responsiveness to β-ecdysone. Each of them, with a great homogeneity among its cells, responded in its own character-

---

* It is noteworthy that Lafont and Best-Belpomme (as reported in Peronnet *et al.*, 1986), using high-performance liquid chromatography, observed that the major part (over 95%) of 20-OH ecdysone was not metabolized in hormone-sensitive Kc cells. This molecule is, therefore, likely to be the intracellular active hormone. Dinan (1985) confirmed that Gateff's *(2)bmn* cell line does not metabolize ecdysone, 20-hydroxyecdysone, or ponasterone.

** On the other hand, Dübendorfer (1988, 1989) observed such a conversion of ecdysone to 20-hydroxyecdysone in primary cultures of *Drosophila* embryonic cells; it gradually developed between the 7th and 15th day of culture, along with the differentiation of clusters of larval adipocytes.

istic manner, a fact that seems strongly in favor of the idea that the sensitivity to the hormone has a genetic basis. Moreover, she noticed that the sensitivity of any given clone remained stable for several months, then declined [(which might be correlated with some drift in the cell population (?)]. Thus, as a practical conclusion, it is advisable, when one is comparing experimental series, to work always with clonal cell lines and, if possible, within a 3- to 7-month period after their isolation. For ecdysteroid responses of various cell lines see Table 8.II (Cherbas and Cherbas, 1981).

Cherbas and colleagues (1977) used a cell strain derived from the *Kc* line and named *Kc-H* (or Harvard subline, because it had been grown for 2 years in C. Thomas' laboratory at the Harvard Medical School), which apparently gave more extreme morphological responses to β-ecdysone than the parent line.

According to the criterion of a rapid arrest of cell division, the Russian line *67J25D* and most of its derived tetraploid or triploid sublines respond to β- and α-ecdysone, although at relatively high concentrations (0.05 to 1 μg/ml and 5 μg/ml, respectively) (Gvozdev *et al.*, 1973, 1974).

Of the three Schneider's lines, *S3* cells show a sensitivity to the hormone which is rather similar to that of *Kc* cells (Wyss and Ellenberger, 1973; Berger, 1978). In the presence of 5×10^{-5} M β-ecdysone they displayed rapid and pronounced morphological transformations, whereas *S1* cells remained unchanged. As for *S2* cells, they simply increased their size then detached and formed small or (in about 10% of the experiments) very large aggregates. In all three lines, however, cell division stopped after 2 days, and the more significant changes in the spectrum of induced protein synthesis were observed in line *S2* (at least in the rare cases in which cells formed large aggregates).

Courgeon (1975) noticed that the genuine spindle-shaped appearance of our male diploid cell line *Ca* (established concurrently with the *Kc* line) was not modified by the hormone, although cell multiplication slowed down and stopped.

The Milanese cell lines *GM1* and *GM3* displayed a slow response, whereas *GM2* cells remained rounded and continued to multiply (Courgeon, 1975a).

Peel and Milner (1992a) tested the sensitivity of their newly established leg and wing imaginal disc cell lines (Chapter 3) to 20-hydroxyecdysone. Morphological responses to the hormone were, rather deceptively, similar to those of other cell lines: arrest of cell division, cell elongation and

TABLE 8.II Ecdysteroid Responses of Various *Drosophila* Cell Lines[a]

			Response			
Cell line	Elongation	Inhibition of proliferation	AChE induction	β-Galactosidase induction	Cell surface changes	EIP induction
Kc	+[b,c]	+[d,e,f]	+[g,h,i]	+[j]	+[k]	+[l]
Ca			−[h]	+[j]		
S1	−[i]	+[i]	−[i]			
S2	−[i]	+[i]	−[i]			−[l]
S3	+[i]	+[i]	+[i]		+[i]	+[l]
67j25D		+[m,n]			+[o,p,q]	
TA-2		+[m]				
Da		+[m]				
DaT(X;3)		+[m]				
GM1	+[r]					
GM2	−[r]					
GM3	+[r]					
D1		+[s]				
1182-6			+[t]			

[a] From Cherbas and Cherbas, 1981; reproduced from *Adv. Cell Cult.* with the kind permission of the Cherbas and Academic Press.

[b] Courgeon (1972a).

[c] L. Cherbas *et al.* (1980b).

[d] Courgeon (1972b).

[e] Wyss (1976).

[f] Rosset (1978).

[g] P. Cherbas *et al.* (1977).

[h] Best-Belpomme and Courgeon (1977).

[i] Berger *et al.* (1978).

[j] Best-Belpomme *et al.* (1978a).

[k] P. Cherbas *et al.* (1980).

[l] Savakis *et al.* (1980).

[m] Gvozdev *et al.* (1974).

[n] Kakpakov *et al.* (1974).

[o] Metakovsky *et al.* (1975).

[p] Metakovsky *et al.* (1977).

[q] Metakovsky and Gvozdev (1978).

[r] Courgeon (1975a).

[s] Wyss (1980a).

[t] Best-Belpomme *et al.* (1980).

aggregation, and possible recovery of resistant sublines. However, in a sensitive clonal line, C1.8, chitin synthesis, as measured from the incorporation of radiolabeled glucosamine, was drastically increased after exposure to 10 ng/ml of 20-HE.

Note: The problem of "ecdysteroid-resistant" clones, derived from sensitive lines, deserves special discussion (see Section IV).

II. REPERTOIRE OF CELL RESPONSES

A. Morphological Changes and Cell Aggregation

The most spectacular response of cells to the hormone lies in the dramatic modification of their morphology.

According to Courgeon's pioneering description (1972b), the overall sequence of events may require 5 to 12 days, depending on the nature and concentration of the ecdysteroid and on the sensitivity of the cell line.

When grown in monolayer culture, Kc cells are independent round cells (5–10 μm in diameter), loosely attached to the flask bottom. Within 1 day of exposure to β-ecdysone (threshold concentration 10 nM), the cells flatten on the substrate, thus looking bigger. They elongate, becoming spindle-shaped, and emit large membranous or thread-like pseudopodia. With such a fibroblast-like appearance, the nucleus and its nucleolus become clearly visible, while the cytoplasm exhibits various types of inclusions (Fig. 8.1).

The development of cell processes correlates with cell mobility, and the cells gather together to form many and randomly distributed small clumps. These latter become larger, by apparently attracting neighboring cells, a process particularly obvious with microcinematography,* and some of them may finally attain 500 μm in diameter. After a few days, most cell aggregates detach and float freely in the culture medium (even though the cells remain alive).

The Cherbas preferred to use suspension cultures and focus on individual cell transformation (Cherbas et al., 1977, 1980; Cherbas et al., 1980). They worked with a Kc-H subline whose hormone-induced morphogenesis seemed even more dramatic than that of the parental Kc line. Exposed to a concentration of 10^{-8} to 10^{-6} M β-ecdysone, the cells, normally spheric, elaborated very long processes which, after

---

* Long ago, the Cherbas kindly sent us a copy of a short movie which showed treated cells rapidly moving and converging into small foci. See also Berger et al. (1980).

2–3 days, extended 50 μm or more in length. The mean value of the total cell length could be used as a bioassay (see Section I.B and Fig. 8.3) and it increased with ecdysone concentration in a sigmoid fashion. Virtually every cell bore one or several long primary processes decorated with numerous thinner ramifications, along which dark inclusions could be seen to migrate. Their appearance was very "neural-like" (Fig. 8.2) and this, coupled with the fact that acetylcholine esterase activity was concomitantly induced by the hormone, led to the assumption that *Kc* cells might be of neuronal origin (see the discussion in Chapter 3).

The filopodia extended by treated cells are dynamic structures and participate in cell movement. When observed with a polarizing microscope, they show a moderate birefringence, indicating the presence of underlying linear elements. Electron micrographs of pseudopodia from 48-h 20-HE-treated cells revealed, as expected, an accumulation of microtubules parallel to their long axes (Sobrier *et al.*, 1992). Both the formation of cell processes and their maintenance were found to be sensitive to

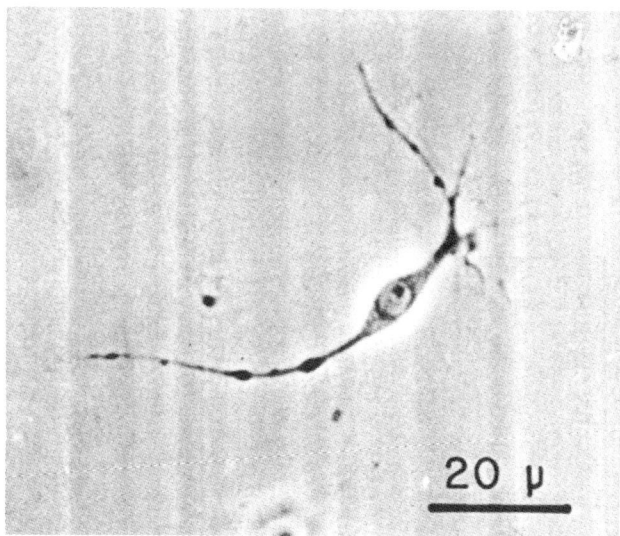

FIGURE 8.2 Neuron-like appearance of a *Kc-H* cell after treatment (4 days) with ecdysteroids. From Cherbas, L. and Cherbas, P. *et al.* (1980); reproduced with the kind permission of the Cherbas and Elsevier Science Publishers.

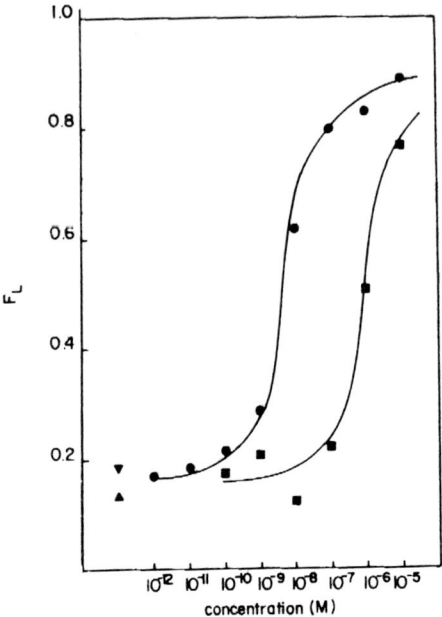

FIGURE 8.3   Assay of *Kc-H* cell elongation: dose–response curves for ecdysone (■) and 20-hydroxyecdysone (●). At least 1000 cells were scored for each point after 74 h treatment. Note that the ecdysone line is the same curve transposed to the right by a factor of 200. From Cherbas, L., Yonge *et al.* (1980); reproduced with the kind authorization of the Cherbas and permission of Springer-Verlag.

Colcemid (10 μg/ml), an inhibitor of microtubule assembly. Cytochalasin B (10 μg/ml), an inhibitor of microfilament formation, did not affect pseudopodia extension or stability, although it prevented motility (Berger *et al.*, 1980). This would mean that filopod formation involves microtubule assembly and that motility requires the additional establishment of microfilaments (see hormone-induction of cytoskeletal proteins, in Section II.D.1).

Coinciding with hormone-promoted cell aggregation, Courgeon and Cailla (1981, 1984) noticed a spectacular increase in cyclic GMP both in *Kc* cells and in their culture medium, whereas cyclic AMP remained constant. It is impossible to decide whether this high level of cyclic nucleotide was the cause or the consequence of the aggregation of the cells. We can only bear in mind the fact that chemotactic signaling between amebas of the slime mold *Dictyostelium* is carried out by the secretion of cAMP.

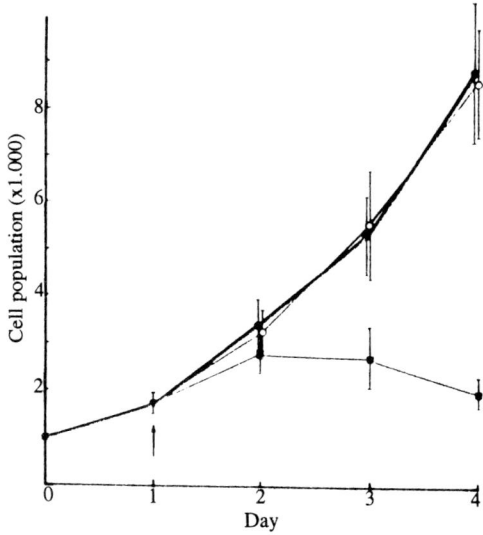

FIGURE 8.4   Arrest of *Drosophila* cell multiplication induced by β-ecdysone. *Kc* cells were cultured in *D22* medium supplemented with 20% FBS and counted daily (the density of cell population is expressed as the number of cells per mm² of culture vessel). (●) Control, untreated cells; (○) 0.0006 μg/ml β-ecdysone; (■) 0.006 μg/ml. The vertical lines represent 95% confidence limits. From Courgeon (1972a); reprinted with permission from *Nature*. Copyright (1972) Macmillan Magazines Limited, with the authorization of Dr. Courgeon.

B. Arrest of Cell Division

The most widespread consequence of an ecdysteroid treatment in all *Drosophila* responsive cell lines (Courgeon, 1972a,* 1975; Gvozdev *et al.*, 1973, 1974) is a rapid block in cell proliferation. Following the addition of 20-HE (10^{-8} to 10^{-7} M), the growth curve of a *Kc* cell culture reaches a plateau within 36–48 hr, which is concomitant with the induced elongation of the cells (Fig. 8.4).

* According to early observations which showed, on the contrary, a slight stimulation of the growth rate with 0.6 μg/ml α-ecdysone, Courgeon (1972a) thought, along with many other people at that time, that there might be some qualitative difference between the effects of α- and β-ecdysone: the first one promoting cell proliferation and the second one cell differentition. In reality, the activities of the two hormones are only quantitatively different, β-ecdysone being some 200-fold more efficient than α-ecdysone. So, even β-ecdysone at a low concentration (3 ng/ml) can cause a small and transient increase in cell proliferation (Cherbas *et al.*, 1980).

Since the initial population of cells has usually doubled in size at the time this blockade takes place, it was at first believed that, under continuous hormonal treatment, each cell divides once before elongating. Time-lapse films demonstrated that this is only an average value, with some cells multiplying two or three times, while others do not divide at all (Cherbas *et al.*, 1980). Therefore, no strict correlation appears to exist between cell division and differentiation, as can occur in other systems [see, for instance, the early steps of neuroblast differentiation in primary cultures of embryonic cells (Chapter 2)].

Using as a criterion the drop in cloning efficiency of hormone-treated cells after hormone withdrawal, Cherbas *et al.* (1980) established that the cells were committed at a very early stage to such a growth inhibition, long before the declining proliferation of the population could be detected. A 1 h hormone pulse was sufficient to decrease, by almost one-half, the cloning efficiency of a *Kc* subline. Such a commitment seems to require protein synthesis.

This interruption of cell multiplication is obviously accompanied by an arrest of DNA synthesis, as could be deduced from the measure of tritiated thymidine incorporation (Gvozdev *et al.*, 1973, 1974; Rosset, 1978; Cherbas *et al.*, 1980; Munsch and Cade-Treyer, 1982). In *Kc* cells grown in suspension, most of the DNA synthesis halted some 10 hr after addition of the hormone. Similarly, the level of histone mRNAs declined precipitously within 24 hr (Vitek and Berger, 1984).

It should be stressed that the block of cell division induced by ecdysteroids occurs in the phase G_2 of the cell cycle. This was first discovered by O'Connor's group (Stevens *et al.*, 1981; Stevens and O'Connor, 1982) during their analysis, using flow cytofluorometry, of the growth kinetics of suspension cultures of *Kc* cells (see Fig. 8.5). This fact was confirmed later by Besson *et al.* (1987). This arrest in phase G_2, which is moreover unusually long in the normal cell cycle of untreated *Kc* cells (as already discussed in Chapter 4), could be the rule for several insect differentiating tissue cells.

C. Enzymatic Inductions

The most convenient way to quantify the response of cells to a specific stimulus is to measure the possible induction of defined enzymatic activities, since their assay is fully standardized and particularly sensitive. In vertebrate endocrinology, the prototype for such an experimental system was the model developed by Tomkins and collaborators (1966), that is,

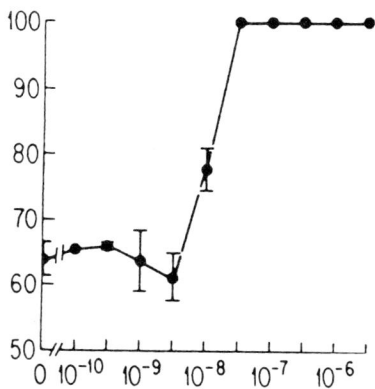

FIGURE 8.5 Ecdysone-induced block in G_2. Dose–response curve of *Drosophila Kc* clone 7C4 to 20-hydroxyecdysone (percent of cells in G2M). Cells were fixed after 48 h in the presence of hormone. After staining with a fluorochrome, they were analyzed for fluorescence in a Becton-Dickinson fluoresence-activated Cell Sorter. From Stevens *et al.* (1980); reproduced with the kind permission of Professor O'Connor and of Cell Press, Inc.

the induction by adrenal steroids of tyrosine aminotransferase in hepatoma cell cultures.

Some 40 enzyme systems, known to act along quite distinct metabolic pathways, were explored in *Kc* cells in the absence or presence of β-ecdysone. Most of them did not vary with the ecdysteroid treatment (see a detailed list in Best-Belpomme, 1980), which, by the way, is a strong argument against the criticism that the hormone might merely have a toxic effect, resulting in a rapid mitotic arrest and cell death. Only a few enzymatic activities were found to be modified by the hormonal treatment in *Drosophila* cultured cells; namely, acetylcholinesterase (Cherbas *et al.*, 1977; Best-Belpomme and Courgeon, 1977), β-galactosidase (Best-Belpomme *et al.*, 1978), dopa-decarboxylase (Kraminsky *et al.*, 1980), catalase (Best-Belpomme and Ropp, 1972), superoxide dismutase (Ropp *et al.*, 1986), and protein kinases (Rollet and Best-Belpomme, 1986; Rollet, 1988; Peronnet *et al.*, 1986). The fact that only a limited number of functions are modulated by the ecdysteroid appears to be a good indication of the specificity of its action (see also Chapter 6, Table 6.I).

Moreover, it is noteworthy that, owing to the great sensitivity of available assays, such enzymatic inductions can be detected very early and, usually, with hormone concentrations that are about one log lower than those required for eliciting morphological responses.

1. *ACETYLCHOLINESTERASE*

The resemblance to neurons displayed by hormone-treated *Kc-H* cells led Cherbas and colleagues (1977) to look for some possible tissue-specific "marker." Thus, they were the first to establish that physiological concentrations of β-ecdysone induced an eserine-sensitive acetylcholinesterase activity in *Kc-H* cells (Fig. 8.6). Practically indetectable in this subline before the treatment, AChE activity increased as early as the first day and reached a maximum (a 50-fold increase) after 3 days. Moreover, AChE histochemical staining seemed to be uniform within the cell population.

It was soon confirmed that such an enzymatic induction is a general feature of the *Drosophila* cell response to ecdysteroids, although each responsive cell line reacts in its own way (Best-Belpomme and Courgeon, 1977).

For instance, in *Kc0* cells, a subline adapted to growth without serum (which, by the way, excludes any possibility of confusion with the pseudo-

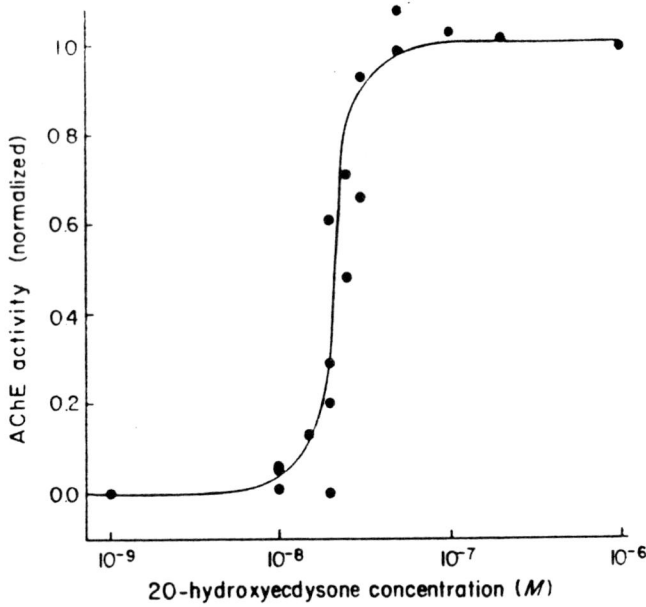

FIGURE 8.6   Dose–response curve for the induction of AChE activity in a clone of *Kc-H* cells. Enzyme activity in cell extracts after 3 days of treatment with 20-hydroxyecdysone. From Cherbas, P. *et al.* (1980); with permission of the Cherbas and of Springer-Verlag.

cholinesterase activity reported in fetal bovine serum), AChE activity became detectable after 1 day of treatment and rose linearly for 11 days. Untreated *1182-6* cells (a line derived from haploid embryos; see Chapter 4) showed a constitutive and relatively high AChE activity (2400 nmol/hr/mg proteins) but this activity could still increase six-fold within 5 days in the presence of 5×10^{-8} M β-ecdysone (Best-Belpomme *et al.*, 1980).

The biphasic response of *S3* line cells as observed by Berger and Wyss (1980) is quite unusual and difficult to explain. From a high constitutive level, AChE activity fell dramatically in hormone-treated cells, then, after this decline, the specific activity increased again at a linear rate for at least 3 days, reaching the initial value. The authors noticed some slight but reproducible differences in the electrophoretic mobilities of the uninduced and induced enzymes.

It can be deduced from lysate mixing experiments in which extracts from control cells did not inhibit the AChE activity of hormone-treated cells (Cherbas *et al.*, 1977) and from experiments with protein synthesis inhibitors (Berger *et al.*, 1984) that increased AChE activity in treated cells is due to a *de novo* synthesis of the enzyme.

The induced enzyme cannot be distinguished from the major AChE form extracted from *Drosophila* fly heads (Cherbas *et al.*, 1977; Best-Belpomme *et al.*, 1980) in its substrate and inhibitor (eserine sulfate, BW284c51) specificities and also in its physical properties (K_m , sedimentation constant (5.8S); electrophoretic mobility, although this point was questioned by Berger and Wyss, 1980, as discussed above). Moreover, the enzymatic activity appears to be linked to a membrane glycoprotein, since the extraction yield is better with high-ionic-strength detergent (Triton X-100) and the activity is totally precipitated by Concanavalin A.

In *Drosophila* cell lines, however, this specific AChE activity represents only one or, at most, a few percent ages of that observed in the *Drosophila* nervous system. It was estimated to be about 400 nmol of hydrolyzed acetylthiocholine/h/mg proteins in stimulated *Kc-H* cells, even though it reached a value 30 times higher in *1182-6* cells. Comparable low levels have been reported in a variety of fly tissues, so they can no longer be considered a specific neural marker. Besides, Berger *et al.* (1978) pointed out that choline acetyl transferase, another neural cell diagnosing enzyme, was not induced by β-ecdysone, at least in *S3* cells (see also discussion in Chapter 3, Section I.B).

AChE induction remains, however, a very useful biochemical indication of the cell response to the hormone.

2. β-GALACTOSIDASE

This enzyme is of special interest because of the extensive research that was carried out on the induction of β-galactosidase (alias *lacZ*) in the prokaryote *Escherichia coli,* and resulted in the famous model of gene regulation, the so-called "lactose operon" model, put forward by Jacob and Monod (*1961*). Moreover, the colorimetric assay of β-galactosidase is extremely sensitive.

Whereas untreated cells showed no detectable activity, in responsive *Drosophila* cell lines, β-galactosidase could be induced with the usual concentrations of β-ecdysone. Specific activity was revealed after 1 day, it rose for 2 to 3 days, then usually decreased after 4 days of treatment (Best-Belpomme *et al.,* 1978).

This induction is clearly dose dependent (Fig. 8.7); however, at higher concentrations (up to 0.5 to 1 μM β-ecdysone), induction levels are lower. We point out that this is also true for acetylcholinesterase induction. Such a "paradoxical" effect (well known in the action of vertebrate steroid

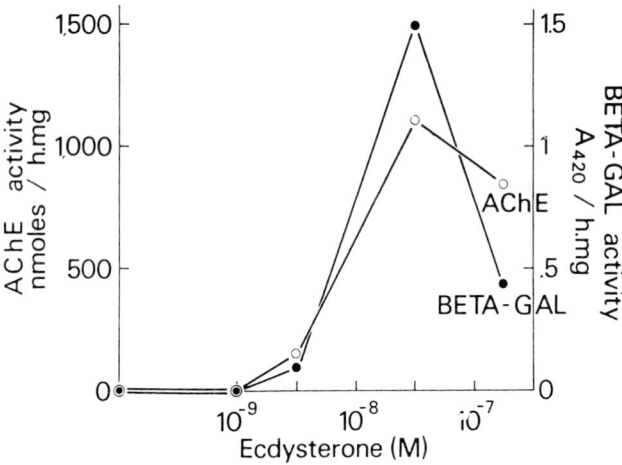

FIGURE 8.7   Enzymatic inductions (β-galactosidase and AChE) induced by ecdysteroids in *Drosophila Kc* cells. Each culture of clone 89K was treated for 4 days by different concentration of 20-hydroxyecdysone. Then the cellular extracts were prepared in T.S. buffer and measured for their enzymatic activities. A "paradoxical" effect can be observed at higher concentrations. Best-Belpomme *et al.* (1980); reproduced with permission of Elsevier Science Publishers.

hormones) must be correlated with the occurrence of lethal molts due to experimental "hyperecdysonism" in insects.

The characteristics of the β-galactosidase induced in treated cells and the optimal conditions of its measurement (pH 5 and 5 mM $MgCl_2$) have been defined with precision. The K_m for o-nitrophenyl galactoside is 0.3 mM and the K_i for lactose is 12 mM. Its sedimentation constant, as measured in a linear sucrose gradient, is 7S. The apparent molecular weight estimated by Sephadex G-2000 chromatography, is about 220,000, which seems to correspond to a tetrameric structure [as judged by the molecular weight (64,000) of its SDS subunits]. The figures are all given here because they are very useful in experiments of cell transfection when *E. coli lacZ* is used as "reporter" gene (see Chapter 9) in order to distinguish clearly the exogenous bacterial enzyme from the *Drosophila* endogenous form.

The physiological significance of such a β-galactosidase induction in hormone-stimulated *Drosophila* cells is unknown. It is clear that, as opposed to its normal role in *E. coli,* the enzyme here is unlikely to be involved in the cell's utilization of galactosides (which are absent in usual culture media). Bearing in mind that cerebrosides are β-galactosides, Best-Belpomme and collaborators suggested that it has a possible function in the assumed neural-like evolution of treated cells. This enzyme is generally considered in eukaryotic cells to be a lysosomal enzyme, which is consistent with its low optimum pH, and lysosomes might perhaps be involved in the structural cytoplasmic rearrangements that must accompany any program of differentiation. The fact that β-galactosidase activity decreases after the 4th day of exposure to the hormone seems, in all events, to contradict Cherbas' suggestion (1981) that it might be a manifestation of cell moribundity.

3. CATALASE AND SUPEROXIDE DISMUTASE

The products of O_2 reduction, such as the radical superoxide ion (O_2^-) and oxygenated water (H_2O_2), are very dangerous for aerobic cells because they are highly reactive, so that cells have to be protected by efficient enzymatic mechanisms of defense, especially catalase and superoxide dismutase activities. As will be reported below (Section II.E), O_2 uptake is significantly increased in ecdysteroid-treated cells and, in order to avoid the dangers of this accelerated O_2 reduction, the cells should concomitantly increase the levels of their enzymatic defenses.

Best-Belpomme and collaborators were thus able to show that catalase (Best-Belpomme and Ropp, 1982) and superoxide dismutase (Ropp and

Best-Belpomme, 1986) were indeed induced in *Drosophila*-sensitive *Kc* clones when the latter were exposed to β-ecdysone. On the other hand, no peroxidases, and more precisely no glutathione peroxidase, were detected.

The kinetics of catalase induction are quite similar to that reported for acetylcholine esterase or β-galactosidase although some increase may be observed even earlier (as soon as 8 hr after addition of the hormone). Likewise, the enzyme activity is not affected by mixing extracts from treated cells with those of controls and vice versa, which suggests that the induction does not reflect the existence of an activator of the enzyme or the release of an inhibitor, but, more likely the neosynthesis of catalase.

Besides, since it is well known that the peroxidative properties of catalase play an important role in the metabolism of ethanol in several *Drosophila* tissues, it is noteworthy that the addition of ethanol (at a low concentration, 1 : 300) to the culture medium can also induce a catalase activity: after 4 days, it is about four times higher than in control cells.

Last, let us point out that the hormonal induction of catalase in cultured cells may have a physiological counterpart *in vivo:* catalase activity was reported to be increased in late 3rd instar larvae near the time of the ecdysteroid peak (Baird *et al.,* 1971).

4. DOPA-DECARBOXYLASE

The early work of Karlson and Sekeris (*1962*) on the blowfly *Calliphora* established that dopa-decarboxylase (DDC) activity could be induced by 20-hydroxyecdysone in the mature larval epiderm.

In the *Drosophila* organism, the enzyme is physiologically active, mainly in two tissues: (1) in the epiderm, in which the conversion of dopa to dopamine allows the production of N-acetyldopamine, which is responsible for the hardening of the cuticle after each molt, and (2) in the neural tissue in which dopamine might act as a neurotransmitter. Moreover, the hormone peaks and DDC activity are clearly correlated, although there is usually a long lag (10 to 55 h) between their two maxima.

Therefore, Spencer *et al.* (1982) investigated the possible hormonal induction of DDC in *Kc* sublines. The specific enzyme activity appeared in β-ecdysone-treated cells, but not before a notable delay (48 hr) and then increased continuously. Despite similar substrate or inhibitor specificities, the DDC produced by cell lines seemed somewhat immunologically different from that found in the organism *Drosophila*. This might be due either to a rearrangement of the DNA during the many cell culture passages or to posttranscriptional modifications occurring in the cells. In any case, on Northern blots the authors observed mRNA transcripts that were apparently homologous to those of the *Ddc* structural gene. A further study (Swiderski and O'Connor, 1988) confirmed that the 26-fold increase in DDC activity (between 72 and 96 hr of continuous hormonal treatment) is coincident with a 6-fold increase in corresponding poly(A) RNA. Cell primary transcripts and mature messengers were found to be some 500 nucleotides longer than corresponding fly transcripts. Yet, both the *Kc* cell DNA and Canton-S fly genome contain a single copy of the *Ddc* gene, with a very similar restriction map. Interestingly, the authors pointed out that, according to recent data, *Ddc* RNAs found in the central nervous system of *Drosophila* would be the result of an alternative splicing and contain an additional exon. Even though it is not known whether the longer mRNA observed in cultured cells resembles this nervous system transcript, such a possibility would be a valuable argument to feed the controversy on the tissue origin of the *Kc* cell line (see Chapter 3). In any case, *in vitro* translation of hormone-treated cell RNAs resulted in polypeptides recognizable by a DDC antiserum and capable of binding a tritiated dopa analog.

Furthermore, let me point out that during a "differential screening" of a *D. melanogaster* genomic library with RNAs from hormone-stimulated and control *Kc* cells (in order to characterize β-ecdysone-responsive genes) Couderc *et al* (1984) isolated an "induced" sequence [i.e., corresponding to a poly(A)$^+$ RNA more abundant in treated cells] which, after *in situ* hybridization on polytene chromosomes, mapped at the *Ddc* locus (37A-B).

D. Specific Protein Induction

Considering the dramatic transfiguration of cells triggered by ecdysteroid hormones, one would expect some drastic changes in the synthesis of proteins in treated cells, more especially among cytoskeletal proteins.

Early experiments of labeled precursor incorporation (with [^{3}H]uridine and [^{14}C]lysine) showed that in various *67j25* sublines the total protein synthesis did not decrease after 48 hr of hormone action despite a substantial inhibition of the synthesis of total RNA (Gvozdev *et al.*, 1973, 1974). This suggests that the ecdysteroid effect is obviously not toxic, but might correspond to metabolic reorientations.

The maintenance of normal and linear rates of [^{35}S]methionine incorporation into TCA-insoluble material, in Schneider's cell line incubated for

4 days in the presence of 10^{-5} M β-ecdysone, was later confirmed by Berger *et al.* (1978).

The first specific induction of one class of soluble proteins after β-ecdysone treatment was demonstrated by Best-Belpomme and Courgeon (1976a,b) who used the rather rough method of one-dimensional electrophoresis in nondenaturing polyacrylamide gels. After simple staining of the protein bands with Coomassie Blue, they observed a strong reinforcement of one band (Rm 0.29), which became a major band in treated *Kc* cells only and exclusively in hormone-responsive clones. Rabbit antisera against this protein band were used in double immunodiffusion tests after labeling of the synthesized proteins (Best-Belpomme *et al.*, 1978; Cade-Treyer and Munsch, 1980), and it could be demonstrated that (1) the band is independent of the presence of calf serum in the culture medium (comparable induction in *Kc0* line), (2) it corresponds to an increase in the rate of synthesis (six to eight times higher after 48 h, but already detectable between 5 to 16 h after addition of the hormone), and (3) proteins immunologically similar to those induced in cultured cells are present *in vivo* throughout the development of *Drosophila*, especially in adult flies.

Such changes in the pattern(s) of protein synthesis were rapidly confirmed by Berger *et al.* (1978). The more sensitive autoradiographic method revealed the induction of six new protein bands in β-ecdysone-treated *S2* cells; altogether, they represented over 90% of total protein synthesis.

The use of O'Farrell's method of two-dimensional electrophoresis combined with autoradiographic or fluorographic visualization of the spots led, of course, to greatly improved resolution (Berger *et al.*, 1978 to 1980; Couderc and Dastugue, 1980).

Some 300 to 400 distinct polypeptides could thus be detected in normal cell extracts (see Chapter 7, Fig. 7.2A) and, despite a comparable general pattern, there were significant differences between cell lines. Individual spots could be quantitatively analyzed by densitometer scanning or radioactivity counting.

Only a few of these peptides were found to be modified by β-ecdysone treatment, which confirms the highly specific action of the hormone. For instance, in treated *Kc0* cells, changes were restricted to 11 spots: three of them were newly induced, two were repressed, and six showed an increased rate of synthesis.

Rabilloud *et al.* (1985) attempted a systematization by computer of the variations of protein electrophoregrams induced either by hormone or by heat shock.

A different approach consists in isolating mRNAs that are induced in cultured cells by ecdysteroids. Two methods were used.

"Hydridization substraction chromatography" (Vitek *et al.*, 1981) allowed an enrichment of the mRNA population in inducible mRNAs. In hormone-stimulated *S2* cells, it was estimated that they represented less than 0.2% of total cytoplasmic mRNAs [but further data, concerning specific mRNAs and resulting from Northern blot analysis, contradict this low value (see below)].

A "differential screening" of a *Drosophila* genomic library that compared the levels of hybridization with labeled poly(A)$^+$ RNAs prepared from treated and control cells resulted in the characterization of a few "inducible" or "repressed" sequences (Couderc *et al.*, 1984).

Of all these induced proteins, a few classes are of particular interest from various points of view, and their hormonal regulation has been carefully investigated by different laboratories. We will consider successively (1) the cytoskeletal proteins, actins, and tubulins (Berger *et al.*, 1980; Couderc and Dastugue, 1980); (2) the early inducible EIPs (Savakis *et al.*, 1980); and (3) the small-heat-shock proteins (Ireland and Berger, 1982) because of their dual regulation. Variations in cell surface glycoproteins will be considered in the more pertinent context of Chapter 6 (Section I.A).

1. INDUCTION OF CYTOSKELETAL PROTEINS

Cell elongation, emission of filopods and concomitant cell mobility, as triggered by β-ecdysone, imply the assembly of microtubules and microfilaments, which suggests that the synthesis of the main cytoskeletal proteins, namely actins and tubulins, is induced by the hormone.

Inhibitor experiments with Colcemid or cytochalasin B (specific inhibitors of microtubule and microfilament formation, respectively), but also with general inhibitors of RNA (actinomycin D) or protein (emetin) synthesis, supported this assumption (Berger *et al.*, 1980).

However, although indirect immunofluorescence (taking advantage of the binding of exogenous DNase I to actin) showed a homogeneous coloration throughout the cytoplasm of hormone-treated *Kc0* cells, it did not permit the detection of actin in defined filamentous structures (Couderc *et al.*, 1982).

a. Actins

In *Kc* cells and in the three Schneider lines, 2D -PAGE revealed the presence of two distinguishable forms of cytoplasmic actins, named actins

II and III (possessing the same molecular weight of 44,000, but different isoelectric points; 5.77 and 5.84, respectively) (Storti *et al.*, 1978; Berger and Cox, 1979). Moreover, the latter authors showed by pulse–chase experiments that actin III is a kinetic precursor of the stable actin II. This conversion involves an enzymatic acetylation near the N-terminus (Berger *et al.*, 1981).

As for hormonal induction, 2D electrophoresis autoradiograms of total cell extracts, in both the absence and presence of the hormone, suggested that actins (and tubulins) belong to the small category of polypeptides whose labeling is increased in S3 or Kc cells after β-ecdysone treatment (Berger *et al.*, 1980; Couderc and Dastugue, 1980).

After preliminary difficulties in demonstrating that this increase in the measurable actin content was indeed due to new synthesis, and not to the slowing down of its degradation rate or to a conversion from a hypothetic inactive pool (Berger *et al.*, 1981), it was, finally, clearly established that the transcription and translation of two actin genes are stimulated by ecdysteroids.

In Kc0 cells, actin represents about 4% of total proteins and only 46% of it is in the polymerized form ("filamentous" or F-actin); this could be estimated by an actin assay based on differential inhibition of DNase I activity by "globular" (G) and F-actins. After 3 to 4 days of exposure to β-ecdysone, the total actin pool rose to 9% of total proteins and the proportion of F-actin increased to 72%. Similarly it was deduced from densitometer scans that, in treated S3 cells, the stimulation of actin synthesis was already evident by 4 h and increased by a factor of five over controls between 12 and 18 h; the rate then decreased substantially (Vitek *et al.*, 1984).

The availability of specific molecular probes, corresponding to each of the six different actin genes that have been identified in D. *melanogaster*, permitted a more refined analysis at the transcription level. Only the two cytoplasmic actin genes (localized in loci 5C and 42A of polytene chromosomes), and none of the four others, are expressed in cultured cells (the second one at a much lower level). Yet, the six EcoRI fragments, which can be recognized on genomic blots from fly DNA (using as a probe the common actin coding sequence), are well conserved in Kc0 cells. This suggests that the six actin genes are still present in cultured cells and have not been reorganized after so many years of replication.

The hormone induces a significant increase in the steady state of the various transcripts of 5C and 42A actin genes (3- to 5-fold and 5 to 8-fold increase, respectively) (in Kc0: Couderc *et al.*, 1983, 1987; in S3:

Vitek *et al.*, 1984). On the basis, precisely, of its more abundant transcripts in treated *Kc* cells, actin 5C was among the few "inducible" sequences that could be isolated by "differential screening" of a genomic library (Couderc *et al.* 1984, see above).

Moreover, a nuclear run on assay (i.e., *in vitro* transcription in nuclear extracts from treated or control cells) demonstrated that this accumulation of actin content and transcripts is essentially due to a stimulation of the transcription rate of both genes. This was already evident after 4 h of exposure to β-ecdysone, whereas the general transcriptional activity of the *Kc* cells was not significantly modified (Couderc *et al.*, 1987).

Notes: (1) Let us point out, together with Vitek *et al.* (1984), an intriguing ambiguity. In the period 12 to 24 h after hormone addition, although mRNA levels remain high, the rate of actin synthesis clearly declines. (2) The analogous hormonal control of *Bombyx* cytoplasmic actin was more recently analyzed by transfection into *Drosophila* S2 cells (Abraham *et al.*, 1993).

b. Tubulins

According to the [^{3}H]colchicine binding assay (based on the specific binding of colchicine to tubulin dimers), the tubulin content appeared to be more than doubled in *Kc* cells exposed to β-ecdysone (at least on a per cell basis; the increase was only 70% on a per milligram protein basis) (Berger *et al.*, 1980). Yet, when studying the tubulin synthesis or breakdown rates as well as after indirect estimation of mRNA levels (through a reticulocyte translation system *in vitro*), the same authors had the utmost difficulties in finding any significant differences between untreated and treated cells.

Finally, it was demonstrated (Sobrier *et al.*, 1986) that tubulin induction by ecdysteroids specifically involves one isoform of β-tubulin (β_3). It should be recalled that tubulin has a dimeric structure, formed by the association of various α and β subunits, and that D. *melanogaster* possesses four different α genes and four distinct β genes (β_3 localized at the band 60C of polytene chromosomes).

This β_3-tubulin, practically absent in control *Kc* cells, had been reported as a new spot (molecular weight: 55,000; isoelectric point: 5.1) on 2D PAGE autoradiograms of total labeled proteins from treated cells (Couderc and Dastugue, 1980). An initial purification of tubulins, by precipitation with vinblastine, gave a much better resolution in electrophoretic blotting and enabled Sobrier (1986) to immunocharacterize two spots with a monoclonal anti-β-tubulin. Moreover, the induced protein comigrated with β_3-tubulin obtained from the translation *in vitro* of 10-to 13-h-old embryo mRNAs (a preferential stage for the synthesis of this isoform).

The use of nucleotide probes highly specific for the genes corresponding at loci 56D (encoding the β_1-tubulin isoform) and 60C (β_3 isoform) made it possible to show that the transcriptional level of the first gene is unmodified by the hormonal treatment, whereas a 2.6-kb transcript (characteristic of the 60C gene and virtually absent in control cells) was considerably increased in treated cells; it could even be detected after 4 h, preceding by a few hours the synthesis of the corresponding β_3-tubulin (Montpied *et al.*, 1988). At the height of induction, the level of this mRNA can be 200- to 400-fold higher than its initial value (Couderc, 1987).

Concomitantly, an accumulation of a nuclear 7-kb premessenger was observed. The fact that it was totally absent in controls combined with the results of transcription experiments *in vitro* strongly suggest a hormonal induction at the transcriptional level (Chapel *et al.*, 1993).

A functional analysis, through the classical approach of transient transfection with recombinant plasmids containing different potentially regulatory regions of the gene in phase with a CAT "reporter" gene (see Chapter 9), has revealed (1) the presence in the 5' flanking region of this β_3-tubulin gene (between -910 and -220 bp upstream from the transcription start site) of several cis-positive elements responsible for a constitutive expression, and (2) the crucial involvement in the hormonal regulation of a 360-bp segment of the first large intron, which contains both enhancer and silencer sequences. The hormonal regulation is the result of the combined activities of all the positive and negative elements of the intron in a dialogue with the promoter sequences (Bruhat *et al.*, 1990; Tourmente *et al.*, 1993; Chapel *et al.*, 1993).

Note: In direct connection with this hormonal induction of β_3-tubulin in *Drosophila* cultured cells, it should be noted that, *in vivo,* this isoform is obviously regulated by ecdysteroids. For instance, its expression is switched on at mid-embryogenesis as well as at the end of the 3rd larval instar, two stages which coincide with peaks of β-ecdysone.

Moreover, β_3-tubulin is sometimes considered as more or less characteristic of mesodermal tissue; but its presence has also been reported in pupal brain, so that its induction in *Kc* cells cannot be used as a conclusive argument for diagnosing the tissue derivation of this cell line.

2. ECDYSTEROID-INDUCIBLE PROTEINS

At the time when the ecdysteroid peak precedes and promotes the metamorphosis of *Drosophila* larvae, Ashburner (1972) observed, more than 20 years ago, the stereotyped succession on polytene chromosomes of a few "early puffing" genes, then of a number of "late puffing" ones. It seems, today, well established that the products of the genes which are the first to be expressed might be involved in the transcriptional triggering of the following ones (Ashburner *et al.*, 1974). So early induced polypeptides, in the responses of cells to steroid hormones, are, potentially, of particular interest.

Thus, the Cherbas and their collaborators focused their efforts on the characterization and regulation of two small EIPs whose synthesis can be detected within the first hour of exposure to the hormone* (Savakis *et al.*, 1980). Moreover, the fact that their mRNAs (see below) increase in the presence of inhibitors of protein synthesis (such as cycloheximide) suggests that EIP induction is indeed a "primary" response to ecdysteroids (Cherbas and Cherbas, 1981).

EIPs were designated 28, 29, and 40, respectively, according to their apparent molecular masses in kilodaltons) as deduced from PAGE autoradiograms (Fig. 8.8). Two-dimensional separation showed that each species consists of one major electrophoretic variant and one or two minor ones (Cherbas and Cherbas, 1981; Savakis *et al.*, 1984).

Although they were present in control *KcH* cells, their synthesis is significantly increased after addition of β-ecdysone. Induction is maximal by 4 hr (an approximately 10-fold increase), but *EIP28* synthesis is already elevated after half an hour (this is a minimum delay for detection because protein labeling is necessary before electrophoresis). As for *EIP29* and *EIP40*, increases are detectable after 1 hr. Similar observations could be made with other responsive lines, such as *S3* cells.

Cell-free translation experiments, showing that EIPs are synthesized at a much higher rate from poly(A)$^+$ RNAs extracted from hormone-treated cells than from controls, suggested that EIP induction is indeed caused by an increase in the levels of their translatable RNAs.

Two classes of cDNA clones could thus be isolated by differential screening. The first one selected mRNAs encoding the entire array of *EIP28* and *EIP29*. It hybridizes to a single site of polytene chromosomes. The other selects mRNAs that encode EIP40 and hybridized *in situ* to another chromosome.

Using them as specific probes on Northern blots, it was established that the hormonal induction at the RNA level is indeed quite similar to the effect on *EIP* synthesis; that is, a 10-fold increase, after 4 h, for *EIP28*

---

* AChE and β-galactosidase are not detectable before 1 day of hormonal treatment, although this is slightly earlier than the morphological responses.

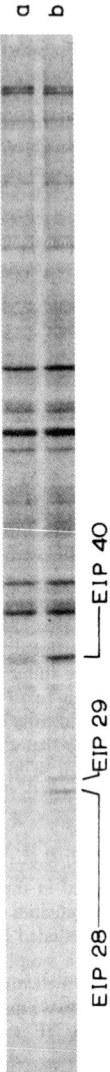

FIGURE 8.8 EIP 28-29 and EIP 40. Ecdysteroid-induced changes in the pattern of protein synthesis in *Kc-H* cells. Clone A4E6 cells, untreated (a) or exposed to 10^{-6} M 20-hydroxyecdysone for 4 h (b) were incubated for 20 min with [^{3}H] leucine. Extracts were separated by electrophoresis in a 8 to 15% linear gradient SDS-acrylamide gel, and the radioactivity localized by autofluorography. From Cherbas, L. *et al.* (1980); reproduced with the kind permission of the authors and Elsevier Science Publishers.

and a somewhat smaller increase for *EIP40*. *EIP28* RNA may represent about 1% of total poly(A)$^+$ RNA in treated *Kc-H* cells (as opposed to only 0.1% in controls) (Savakis *et al.*, 1984). Moreover, this transcriptional increase could be detected within 10 min after addition of the hormone (Bieber, 1986).

Finally, corresponding genomic clones were isolated, then sequenced, and their structures and transcriptional modalities were carefuly unraveled (Cherbas *et al.*, 1986; Schulz *et al.*, 1986).

It was plainly demonstrated that both families of EIP28 and EIP29 are encoded by a unique structural gene, but are generated by the alternative splicing of a common premessenger. This *Eip28/29* transcription unit, which resides in position 71CD of salivary gland polytene chromosomes, extends over some 2087 nucleotides and includes four introns. The "long" transcript (979 nt) is about three times more abundant than the "short" one (967 nt), and the ratio was not modified during hormonal induction. Their primary translation products correspond to the less acidic form of each family, from which the other isoforms derive by co- or posttranslational modifications

Eip40 is also a single-copy gene, located at position 55E3,4. It gives rise to a unique transcript (Bieber, 1984; Cherbas, Benes *et al.*, 1986). Sequence comparisons revealed no homology with the previous gene.

Furthermore, a variety of transcription units were identified in the immediate vicinity of both *Eip* genes, sometimes overlapping them (Schulz *et al.*, 1989). Some of these neighboring genes may be expressed at a low level in *Kc* cells [for instance, *gonadal (gdl)*], but they are unaffected by ecdysteroids. There is a notable exception: a transcript, overlapping the *Eip40* transcript but made from the opposite DNA strand, is present at a much lower level and is induced about 25-fold by the hormone.

The "functional dissection" of the flanking regions of the *Eip28/29* gene, by the usual method of transient cell transfection with DNA constructs associating differently deleted segments in phase with the coding sequence (itself tagged in its 3' untranslated region), revealed that the sequences involved in hormonal regulation are widely dispersed (Cherbas *et al.*, 1991). There is, at least, one significant element in the upstream region and, in addition three other elements are dispersed within the 3735 bp downstream from the polyadenylation site.

In all these regions, short sequences were identified and characterized as being putative "ecdysteroid response elements" (EcREs) by their ability (1) to bind ecdysteroid receptors and (2) to confer ecdysteroid inducibility on a reporter CAT sequence.

It must be emphasized that Cherbas's results on *Eip* regulation have greatly contributed to the general definition of the consensus ecdysteroid response element. It corresponds to an imperfect palindrome, composed of half-sites of the consensus hexamer **5′-TGA(AC)CY-3′**, with a single undefined nucleotide at the center of symmetry (see reviews by Cherbas, 1993, or Andres *et al.*, 1993).

The physiological function(s) of *EIPs* remain(s), unfortunately, unknown.* In the whole organism, *EIPs* are expressed throughout the developmental stages, although the maximal signals due to their transcripts on Northern blots never exceed 30–50% of those observed in untreated *Kc* cells. The tissue specificity of their expression conveys the complexity of their regulation more clearly and indicates close correlation with ecdysteroid titers; more especially, in the different "expression groups" that can be distinguished among the various tissues or organs, lymph gland and hemocytes show a sudden rise in *EIP28/29* expression during the interval between 96 hr and pupariation, to the level reached in *Kc* cells** (Andres, 1990; Andres and Cherbas, 1992, 1994).

3. HORMONAL INDUCTION OF SMALL-HEAT-SHOCK PROTEINS

Among the series of proteins whose synthesis is triggered by a sudden rise in temperature, i.e., the so-called heat-shock proteins (see Chapter 7), there are four polypeptides which can be distinguished, on PAGE autoradiograms, by their relatively low molecular weights (22,000, 23,000, 26,000, and 27,000, respectively). The four corresponding genes are clustered in a short DNA stretch (10 to 12 kb) in the 61B1 band of polytene chromosomes (see Chapter 7, Fig. 7.3).

Ireland and Berger (1982; see also Ireland *et al.*, 1982) discovered that these small-heat-shock proteins (or s-HSPs), but not the major ones, can also be induced by ecdysteroids. This intriguing dual regulation deserves special attention.

In hormone-responsive *S3* cells, for instance, this induction by β-ecdysone can be detected within 2 hr and shows the usual dose–response profile. It should be recalled that, during heat shock, the four s-*HSPs* are synthesized at approximately equal rates, whereas, after hormonal

---

* *EIP40*, however, has been recently identified as the enzyme γ-cystathionase (Andres and Thummel, 1992).

** An argument put forward by these authors to assume a possible hematopoietic origin of the *Kc* line (see discussion in Chapter 3).

stimulation, the [^{35}S]methionine incorporation into *HSP23* was about 20 times greater than that into *HSP22* and some 3 times greater than that into HSP26 (Vitek and Berger, 1984).

This heightened synthesis is accompanied and presumably caused by an increase in the abundance of *s-hsp* transcripts. As measured by the dot-blot hybridization technique, the level of *hsp23* mRNAs rose during the first 4 days, reaching a value about 50% of that observed in heat shock. In contrast, the level of *hsp22* messengers rose during the first 4 h, but reached a plateau at a much lower concentration (the ratio *hsp23/hsp22* mRNAs was approximately 6:1). The *hsp26* and *hsp27* mRNA increases were intermediary. Pulse–chase experiments indicated that this divergence in transcript levels, and consequently in HSP23 and HSP22 synthesis, was essentially due to a differential messenger stability: the *hsp23* mRNA appeared here to be some four times more stable, whereas in heat shock all *hsp* messengers remain equally stable when cells are kept at 35°C (Vitek and Berger, 1986).

Since this transcription could be initiated and continued for at least 2 hr in the total absence of protein synthesis (for instance, after addition of 10^{-4} *M* cyclohexamide), this *s-HSP* induction may be considered a "primary response" to the hormone. However, according to later observations by Amin *et al.* (1991), this might be true for *HSP27* but not for *HSP23*, which would imply rather different mechanisms of regulation.

In any case, these studies, at the transcriptional level, confirmed that none of the major heat-shock genes, nor the various small transcription units that are interspersed throughout the *s-hsp* cluster, are in the least affected by β-ecdysone treatment (Ireland *et al.*, 1982). The absence of response to the hormone of the so-called *gene1*, which lies between two highly susceptible genes, *hsp23* and *hsp26* (see Chapter 7, Fig. 7.3), is especially significant. This observation, coupled with the above-mentioned differences in the transcription rates of individual *s-hsp* genes, strongly suggests that, in this system, every single gene and its immediate flanking sequences form a regulation unit. Moreover, it is obvious that heat shock and hormone inductions rely on different mechanisms.

In order to identify the cis-acting sequences that are involved in either type of regulation, a functional analysis was undertaken by the transient transfection method (see Chapter 9). In a first step, Morganelli and Berger (1984), Lawson *et al.* (1984), and Larocca (1986) demonstrated the value of such an approach. The constitutive and induced (by heat-shock or by hormone) expression patterns of all transfecting *hsp* genes generally resembled those of their endogenous counterparts.

Then, in recombinant constructs, long or more or less deleted upstream regions from every *s-hsp* gene (as also from the major species *hsp70*) were ligated in phase to the coding sequence of a "reporter" gene; for instance, herpes virus thymidine kinase gene. In transfected S3 cells, heat shock elicited a 40- to 60-fold increase in the relative abundance of *tk* messengers, while β-ecdysone treatment produced a 6- to 10-fold increase with *s-hsp* promoters exclusively (Morganelli *et al.*, 1985). As expected, no hormonal induction was observed in a transfected ecdysteroid-insensitive line F6. Neither was there any induction in S3 cells when the promoter was too largely shortened.

a. hsp23

By progressively resecting the upstream sequences of the *hsp23* gene, Voellmy and collaborators (Lawson *et al.*, 1985b; Mestril *et al.*, 1986) showed that ecdysteroid regulation depends on elements quite distinct from those used for heat-shock activation.

While basic signals for heat shock are distributed throughout four regions within a relatively large (-500 bp) upstream sequence (each one corresponding to a functional HSE, see Chapter 7), the regulatory signals for ecdysteroids are differently located within the first -300 bp. Two putative "ecdysone responsive elements of about 20 bp each could be identified using constructs with specific internal deletions. Their sites are between -242 and -218, and between -200 and -181, respectively.

Later, exonuclease III protection assays and gel retardation experiments, carried out with a purified ecdysteroid receptor (Luo *et al.*, 1991), revealed that, even though these two proximal elements seem especially critical for hormone induction, there might be at least four receptor-binding sites in this *hsp23* promoter.

b. hsp27

Riddihough and Pelham (1986) entered on a similar "functional dissection" of upstream regions of the *hsp27* gene. (1) Flanking sequences of 782 bp are sufficient for full induction in both types of stimulation. (2) The regulatory regions for heat shock and hormonal responses lie in quite distinct clusters, some 300 and 500 bp, respectively, upstream from the transcription start site (see Chapter 7, Fig. 7.9). The region involved in heat shock resides between -369 and -273 and comprises three segments of homology with the consensus heat shock regulatory element

(HSE1 to HSE3).* On the other hand, the hormonal regulation depends on a 23-bp sequence with a dyad symmetry, located between -551 and -528 bp and which is reminiscent of the binding site for vertebrate steroid receptors; it was called HERE (this ecdysone responsive element is today known as EcRE).

Gel retardation and DNase footprinting experiments made it possible to identify a single protein-binding site corresponding precisely to this 23-bp hyphenated dyad. Moreover, a synthetic double-stranded oligonucleotide, reproducing this typical sequence, was indeed able in transfection experiments to confer a 20-fold ecdysteroid inducibility to a "reporter" gene (Riddihough and Pelham, 1987).

More recently, for a decisive demonstration that the newly isolated *EcR* gene does indeed encode the true ecdysteroid receptor (see below, Section V), Koelle *et al.* (1991) used a construct, including seven copies of this *hsp27* EcRE motif, inserted slightly upstream from the TATA box of the *Drosophila Adh* distal promoter; and the product of the cotransfected *EcR* gene, driven by a conditional promoter, could indeed greatly enhance *Adh* expression.

Independently, Voellmy's laboratory (Luo *et al.*, 1991) not only confirmed that the *hsp27* response element (like those from *hsp23*) can confer hormonal regulation on a basal promoter in transfected cells but also showed that the same sequences make transcription *in vitro* dependent on a purified ecdysteroid-receptor. This was the first direct evidence of a link between the hormonal regulation of *s-hsps* and an ecdysteroid receptor.

Further experimental analysis of this ecdysone responsive element of *hsp27* (and of the regulatory sequences of *Eip28/29*) led, almost simultaneously, the Cherbas (Cherbas *et al.*, 1991) and then Dobens *et al.* (1991) to assume that such EcREs function as both transcriptional repressors (in the absence of hormone) and activators (in its presence). As seems to be the case for the thyroid hormone receptor, the unliganded ecdysteroid receptor might already be bound by its target DNA site, and the critical function of the hormone should be to trigger some conformational changes, converting its repressor activity into an enhancer one.

E. Increase of Oxygen Uptake

By keeping *Kc* cells in suspension in the closed chamber of a Clark electrode, Ropp and Best-Belpomme (Ropp, 1983, 1986; see also Peronnet

---

* See also binding sites for GAGA protein (Lu *et al.*, 1993; Sandaltzopoulos *et al.*, 1995) and discussion in Chapter 7, Section IX.B.4.

et al., 1986) measured the kinetics parameters of their oxygen consumption.

The maximal rate of O_2 uptake (V_m) was found to be equal to 0.2 ± 0.05 fmol* min^{-1} $cell^{-1}$ at 25°C and the K_m was 4 ± 1 μM.**

Assuming that in aerobic metabolism 1 liter of consumed oxygen corresponds to the biological production of 4.7 × 10^3 cal (Jöbsis, 1964), it may be estimated that the cell energy production for such a V_m is equal to 2 × 10^{-11} calories per *Drosophila* cell during 1 min (corresponding to the synthesis of about 1.3 fmol ATP/min/cell).

After β-ecdysone stimulation the V_m could be increased up to threefold, whereas the K_m did not vary. This constancy of K_m suggests that the mitochondrial system of O_2 reduction remains the same (so that the use of the same correspondence between the volume of O_2 consumed and the production of calories appears justified). In other words, the energetic production of hormone-treated cells is also multiplied 3-fold (that is, about 6 × 10^{-11} cal min^{-1} $cell^{-1}$, corresponding to the synthesis of 3.9 fmol ATP/min/cell). After all, it seems logical that the new metabolic program triggered by the hormone should require additional energy, and the difference between the energetic production of treated and untreated cells should be a measure of the "energetic cost of cellular differentiation."

This assumption was confirmed by the fact that by varying the oxygen concentration in the gas phase of culture flasks it could be shown that enzymatic inductions by ecdysteroids (for AChE and β-galactosidase) were decreased by one-half in almost pure N_2, whereas they could be doubled in 50% (v/v) oxygen (Peronnet *et al.,* 1986).

Besides, as stated above (Section II.C.3), hormone-stimulated cells avoid the dangers of an acceleration of O_2 reduction by increasing, correlatively, the levels of their specific enzymatic defenses, i.e., by increasing superoxide dismutase and catalase activities.

III. "MATURATION" PHASE: A CRITICAL PERIOD FOR HORMONE ACTION

Vertebrate endocrinology has fully established that after an initial stimulation by a steroid hormone, cells become capable of responding

---

* 1 femtomole = 10^{-15} mol.

** It may be deduced from these parameters and from the rate of diffusion of oxygen through the culture medium that, in the usual culture flasks whose gas phase is atmospheric air, a monolayer of *Drosophila* cells will never be in hypoxia, as was sometimes assumed, so long as the depth of liquid over the cells does not exceed 1 cm.

more quickly and strongly to a second hormonal treatment. That is the case, for instance, for the secretion of ovalbumin by the oviduct of impubescent chicken, as induced by, first, a series of estrogen injections at high concentration followed after a few days by a second series of similar injections (Oka and Schimke, *1969;* Palmiter, *1972).*

Analogous observations were first reported by Cherbas *et al.* (1980) concerning acetylcholinesterase induction by ecdysteroids in *Drosophila Kc* cells. An initial stimulation by β-ecdysone (1 μM; 24 h) followed by hormone withdrawal (1 to 2 days), then by a second exposure (at the same active concentration) brought about a much more rapid enzymatic induction than a single continuous treatment (although the final level of activity was approximately identical).

The use of this convenient *in vitro Drosophila* model, not to mention the decisive advantage that lies in the fact that cultured cell lines are completely "naïve" with regard to the hormone, allowed Best-Belpomme and Courgeon (1980; Peronnet *et al.*, 1986) to go further in the analysis of this "primo-stimulation." They adopted the following protocol. In a first step, and for 2 days, responsive *Kc* clones were treated with a low, infrathreshold concentration of β-ecdysone (i.e., between 1 to 5 nM), which indeed did not induce any detectable response (despite the extreme sensitivity of enzymatic assays). Then, in a second step, and by simple addition of new hormone to the culture, the cells were exposed to the usual stimulating concentrations (50 to 250 nM). Following this preparation of the cells, which made them, in some unknown way, "competent" or "mature," the enzymatic inductions, both for acetylcholinesterase and β-galactosidase, were not only accelerated, but reached much higher levels than after direct stimulation of naïve cells. More precisely, for AChE, the activity increased linearly without any lag after addition of the active concentration of β-ecdysone, the initial induction rate being multiplied 6-fold and the final level doubled. As for β-galactosidase, there is a short 12-h lag, but the initial rate of induction was multiplied 2-fold and the final activity level 1.7-fold (Fig. 8.9).

So, the initial effect of a steroid hormone, even at subliminal concentrations, is to increase the capacity of the target cells to respond later, to an efficient concentration of the same hormone. The "competence" of a cell is, of course, the result of multiple processes, but the term "maturation" was proposed by Best-Belpomme and Courgeon to describe this specific and critical step which is promoted by the hormone itself during its first contact with the cells.

It seems particularly interesting to point out that this experimental sequence of a period of low hormonal concentration followed by a high

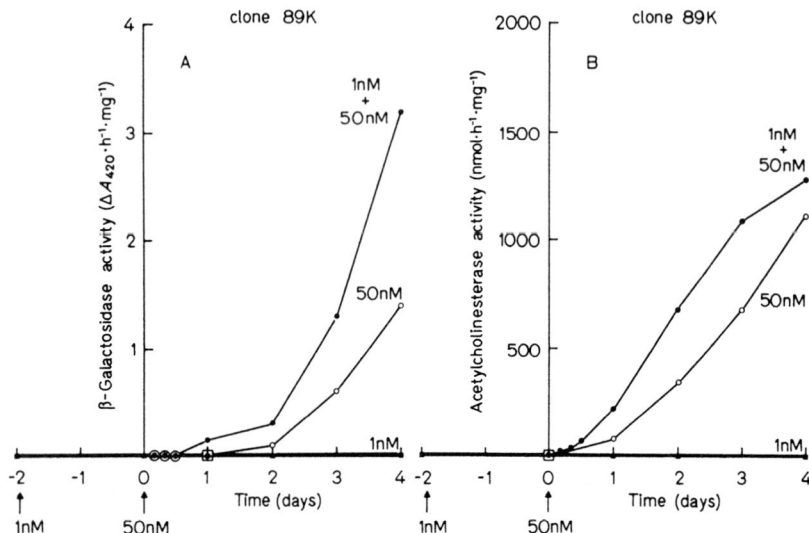

FIGURE 8.9   Effect of "maturing" infrathreshold concentration of 20-hydroxyecdysone on the enzymatic inductions obtained by a subsequent stimulating concentration of the same ecdysteroid. After a "maturation" period [1 nM ecdysterone for 2 days; between the arrows], *Kc* cells of clone 89K are treated with a stimulating concentration (50 nM) of the same hormone. β-galactosidase (A) and acetylcholinesterase (B) activities are more rapidly induced and to higher levels than in cells without pretreatment. From Best-Belpomme and Courgeon (1980); reproduced with permission of the *Eur. J. Biochem.*

concentration phase mimics the physiological conditions encountered *in vivo* by the tissues of the insect. Published curves of ecdysteroid titers in the hemolymph of *Drosophila* clearly show that the hormonal peaks that trigger moulting and differentiation occur in a background of low, albeit continuous, hormonal concentration.

The molecular mechanisms of this cell maturation remain completely obscure. One may speculate that the process allows the completion of the cellular equipment in specific receptors. Since, in the mid-1990s, the receptor gene *EcR* has been cloned and antibodies against EcR–protein are available (see Section V), it would be worthwhile to try to test this simple hypothesis. Another possibility is that the maturation phase modifies the conformation of the chromatin and, for instance, might have something to do with the setting up of DNase I hypersensitive sites in the vicinity of inducible genes.

IV. CELL RESISTANCE TO ECDYSTEROIDS

Because the arrest of cell multiplication is the most generalized response of *Drosophila* cell lines to ecdysteroids, the isolation of "resistant" sublines is quite easy. As shown by Courgeon (1972b) in *Kc* cell cultures continuously exposed to the hormone and in which the responsive population becomes detached within a week or so, a few clonal colonies (from 0 to 20 per flask, which corresponds to a frequency of about 0.3×10^{-6}) will develop after a few days to 1 month. These resistant cells appear identical to untreated cells, in shape and growth rate, even in a concentration of β-ecdysone as high as $10^{-4} M$ (that is, 10,000 times greater than the usual threshold dose); they will finally overgrow the whole culture. It was subsequently confirmed that resistant cells are also lacking in all other aspects of the hormonal response (Cherbas *et al.*, 1980).

This refractoriness to the hormone appears to be due to mutation(s). (1) The progeny of resistant cells remain insensitive to β-ecdysone, even after growing for many generations (several months) in the absence of the selective agent. Courgeon (1975) and Cherbas and Cherbas (1980), however, reported a few reversions. (2) The demonstration that the occurrence of these variants is independent of the presence of the hormone was obtained from a "fluctuation analysis" (Wyss, 1980b). (3) Mutagenesis (Cherbas *et al.*, 1980) allowed the recovery of resistant sublines, although the authors did not clarify the extent to which it was facilitated.

The simplest explanation of such an insensitivity to ecdysteroids is that resistant cells are deficient in specific receptors, as had already been well established in vertebrate systems. Indeed, during their preliminary characterization of saturable β-ecdysone-binding sites in extracts from responsive *Kc* cells, Best-Belpomme and Courgeon (1975) could not find any similar site in resistant clones. This nondetectability of receptors in resistant cells was confirmed by all subsequent research (for instance, O'Connor and the Cherbas, as quoted in Cherbas *et al.*, 1980). In the early 1990s, after the molecular cloning of the receptor gene *EcR*, it has been clearly demonstrated by immunoblot analysis (using antibodies against the EcR protein) that the concentration of receptor molecules was greatly reduced in ecdysteroid-resistant *S2* sublines. Furthermore, expression of EcR by transfection into such resistant cells could restore their ability to respond to the hormone (Koelle *et al.*, 1991).

This does not mean, however, that the presumed mutational event which is responsible for resistance necessarily affects the structural receptor gene itself. The situation is perhaps more complex, as was concluded from somatic hybridization experiments between sensitive and resistant cells (Wyss, 1980a,b; Berger *et al.,* 1980). In many cases, resistance to the hormone behaved as a codominant trait in the somatic hybrids, which suggests the likely intervention of epigenetic mechanisms. It seems, moreover, obvious that different types of ecdysteroid resistance might occur. It is relevant here to note that Beckers (personal communication to Wyss, 1980b) observed one resistant strain with apparently normal hormone-binding properties.

Note: The controversial assertion (Stevens *et al.,* 1980; Stevens, 1981; Stevens and O'Connor, 1982) that responsive cells, when continuously exposed to the hormone, might escape the G2 block at some time between 120 and 160 h and resume a normal cell cycle needs to be discussed in the present context.

According to O'Connor's laboratory, all hormone-treated cells would rapidly become, phenotypically resistant, so to speak, to the hormone in all aspects (unchanged morphology and division rate, receptor deficiency, etc.); even after hormone withdrawal, they maintain for months this refractoriness to subsequent hormonal stimulation.

These conclusions, based principally on fluorescent flow cytometry experiments with suspension cultures, are in complete contradiction to the observations of other laboratories, which all support the hypothesis of a selection, in the presence of the hormone, of a low percentage of spontaneous resistant cells present among the population. For instance, in stationary cultures, the clonal growth of a few resistant colonies is evident (Courgeon, 1972b, 1975). Likewise, in suspension cultures, careful cell counting showed a late and slow resumption of cell proliferation, perfectly compatible with the overcoming of a new population of resistants (Cherbas and Cherbas, 1981). This selection hypothesis was supported by a later flow fluorocytometry study (Besson *et al.,* 1987) which revealed, after a clear hormone-induced block in G_2, a slow and progressive reappearance of a few G_1 cells, although not before the second week.

V. ECDYSTEROID RECEPTORS

By analogy with the general mechanism of action of steroid hormones, as had previously been established in vertebrates, it was soon assumed that ecdysteroid stimulation is mediated through a hormone–receptor complex able to bind specific "target" DNA sequences in the promoters of hormone-regulated genes.

1. Best-Belpomme and Courgeon (1975, 1976) were the very first to show, albeit in rather preliminary experiments, the presence of "saturable" β-ecdysone-binding sites (i.e., putative receptors) in an insoluble fraction from hormone-responsive Kc cells, whereas they were absent in "resistant" cell clones.

The availability of tritiated ponasterone A (PNA), whose biological efficiency in cultured cells is greater than that of β-ecdysone (see Section I.B) and which proved also to be a more potent high-specific ligand, allowed Maroy et al. (1978) to confirm clearly the existence of an ecdysteroid binding moiety within the cytosol and mostly in nuclear extracts from Kc cells. The number of such binding sites was estimated to be about 1000 to 2000 per cell.

2. A series of studies, from O'Connor's group and participating laboratories (Beckers et al., 1980a,b; Stevens and O'Connor, 1982; Sage and O'Connor, 1985), established the main characteristics of this ecdysteroid-binding activity in cultured cells.

It is temperature-dependent. Kc cells accumulate and bind [^{3}H]PNA about 12 times more slowly at 0°C than at 20°C. At this latter temperature, the uptake kinetics indicate a multiple-stage process, with an initial rapid phase followed by a slower one which reaches a plateau after 60 min. Most of the ecdysteroid taken up by the cells was then located in the nucleus (about 70%).

The saturability and specificity of this binding was determined in the presence of increasing concentrations of unlabeled competitors (Table 8.III). A homogenous class of binding sites was observed by either Arrhenius or Scatchard analysis, and both cell compartments (nuclear and cytoplasmic) seemed to exhibit quite similar characteristics.

It may be concluded from these data that this ecdysteroid-binding activity of Drosophila cells presents all the classical attributes of steroid-receptor molecules: saturability, high affinity, specificity (with good correlation between K_d values and biological activities of the different tested ligands), presence restricted to responsive cells, and ability to bind to DNA (and especially to putative target sequences).

Sage et al. (1982) pointed out the preexistence of "resident" receptors in nuclear preparations from "naive" Kc cells (that is, cells that have never been exposed to the hormone). The generally assumed translocation of cytoplasmic receptors into the nuclear compartment, after binding to the hormone might not be necessary here, as would seem to be the case in other systems of steroid stimulation in vertebrates.

TABLE 8.III Kinetic Properties of Cytosol and Nuclear Ecdysteroid Receptors[a,b]

Binding parameter	K_c cytosol	K_c nuclear extract
Saturation level [³H]ponasterone A	$1.2 \times 10^{-8} M$	$1.8 \times 10^{-8} M$
Binding sites/cell	1200	800
Equilibrium dissociation constants [K_d]		
Ponasterone A	$3.4 \times 10^{-9} M$	$4.2 \times 10^{-9} M$
20-OH ecdysone	$2.4 \times 10^{-7} M$	$2.0 \times 10^{-7} M$
Ecdysone	$3.6 \times 10^{-5} M$	$2.7 \times 10^{-5} M$

[a] Cytosol receptor preparations were made from frozen cell populations. Scatchard [G. Scatchard, *Ann. N.Y. Acad. Sci.* **51**, 660 (1949)] analyses of the saturation kinetics of [³H]ponasterone A binding to ecdysteroid receptor preparations were used to derive the values for binding sites/cell and the K_ds of ponasterone A. The equilibrium dissociation constants for 20-OH-ecdysone and ecdysone were calculated from the ecdysteroid inhibition of [³H]ponasterone A receptor binding.

[b] From Sage and O'Connor (1985); reproduced with the kind permission of Professor O'Connor and Academic Press.

3. The purification of receptor molecules met with technical difficulties.

The use of another phytoecdysteroid, muristerone A (see Cherbas *et al.*, 1982), whose complex with the receptor was not as sensitive to dissociation in high salt buffers as other ecdysteroid-receptor complexes, permitted, at first, a 750-fold enrichment (Landon *et al.*, 1988); and the molecular weight could be estimated to be 120,000.

Furthermore, Cherbas *et al.* (1988) reported the synthesis of an analog, 26-[¹²⁵I]iodoponasterone, with both a very high specific activity (2175 Ci/mMol) and a superior affinity for the receptor (correlated with its more powerful biological activity in the cells). It has greatly facilitated the counting and study of these receptors (see, for instance, Rickoll *et al.*, 1986).

Their successful isolation, however, was finally obtained by quite a different approach. After the precise identification in the promoters of ecdysteroid-regulated genes of the specific DNA "motifs" to which the hormone–receptor complex is supposed to bind, Voellmy and collaborators (Luo *et al.*, 1991) could "fish out" the receptor molecules from 20 to 40 liters of *S3* suspension cultures by using DNA affinity columns containing "ecdysone responsive elements" (as defined from *s-hsp23* and *s-hsp27*; see Section II.D.3). When added to an *in vitro* transcription system, this protein, purified to apparent homogeneity, was able to increase, up to 100-fold, the transcription of EcRE-containing promoters.

Moreover, it was noted that the treatment *in vitro* of this receptor with β-ecdysterone greatly enhanced its DNA-binding capability.

Finally, on the basis of the sequence homology of their putative DNA-binding domain, a series of genes whose products obviously belong to the large superfamily of steroid and thyroid receptors were cloned from *Drosophila* genomic libraries. The use of cultured cells, however, was crucial for determining, on the direct criterion of its *in vivo* functionality, the "true" ecdysteroid receptor (i.e., *EcR*) (mapped to the cytological location 42A) (Koelle *et al.*, 1991). Immunoblot analysis clearly showed that the abundance of *EcR*–protein is indeed greatly reduced in ecdysone-resistant sublines (from *Kc* or *S2* cells). Above all, the response to the hormone of such resistant cells could be restored by transfection of this *EcR* sequence. Moreover, overexpressing *S2* lines could be constructed for the production of *EcR*–protein by stable transformation with a plasmid containing the *EcR*-coding sequence driven by a "conditional" metallothionein promoter (see Chapter 9).

At a time when it is more and more obvious that many trans-acting proteins, and especially various so-called "orphan" receptors, interfere in a combinatorial way with the action of the ecdysteroid–receptor complex (Thomas *et al.*, 1993; Schräder *et al.*, 1993; Ayer *et al.*, 1993; Amero *et al.*, 1993; Imhof *et al.*, 1993), such *in vivo* tests of functionality in cotransfected cultured cell systems should reveal themselves to be of the utmost interest.

Thus, it is well established that functional receptors mediating the response to ecdysteroids are heterodimeric complexes of *EcR* and *Ultraspiracle* (*USP*) (Yao *et al.*, 1992, 1993) and that there are several isoforms of *EcR* with possible distinct roles (Talbot *et al.*, 1993). Sutherland *et al.* (1995) characterized another *Drosophila* hormone receptor (DHR38) which is a homolog of rat nerve growth factor-induced protein B (NGF 1-B); they showed that, quite unexpectedly, it could compete *in vitro* against *EcR* for dimerization with *USP*. Moreover, in transfection experiments in Schneider's cells, this second partner for *USP* could indeed affect ecdysone-induced transcription, which suggests an important physiological role in modulating the response to hormone.

VI. FUNCTIONAL INTERFERENCE WITH JUVENILE HORMONE

In addition to ecdysteroids, another quite different type of hormone, the so-called *juvenile hormone*, is involved in the developmental control of the insect. So long as it is present in the insect organism, growth and moulting maintain a larval type, whereas its physiological disappearance during the last larval instar provokes the metamorphosis into pupa and adult. Therefore, its action has sometimes been described as being "antagonistic" to that of ecdysteroids, which is, of course, an oversimplification.

The natural juvenile hormone exists in three forms, namely JH I, JH II, and JH III (this last one being the predominant form in Diptera), but they are very similar in their sesquiterpenoid structures and physiological effects. Several synthetic JH analogs with a strong activity and better stability, one of which is the methoprene (*ZR-515*, Altosid SR-10), have been widely used in experimental conditions.

Early observations with farnesol or its derivatives (which are relatively poor JH mimics) demonstrated that, at concentrations which are difficult to estimate because of the insolubility of these molecules in water, but which might be close to 10^{-8} *M*, the habitual effects of β-ecdysone on the morphology of *Kc* cells (Courgeon, 1975) and also on their proliferation (Wyss, 1976) were counteracted to some extent, and in a dose-dependent manner, by the addition of JH analogs. Interestingly, the cell "maturation" (see Section III), induced by a primostimulation with a subliminal concentration of β-ecdysone, could also be abolished by addition of 1 nM of the juvenoid Altosid (Best-Belpomme and Courgeon, 1980). If Yudin *et al.* (1982; Chang *et al.*, 1982) observed only a minor effect on ecdysteroid-induced cell aggregation, this might be due to the high instability of the natural JH III they used.

A further analysis was carried out by Cherbas *et al.* (1989). The analog methoprene was chosen because of its better metabolic stability and relatively high water solubility and also because it is known to be particularly active in Diptera. This juvenoid, even at concentrations as low as 10^{-10} to 10^{-8} *M*, inhibited most of the *Kc-H* (or *S3*) cell responses to β-ecdysone (i.e., morphological changes, proliferation arrest, and its early commitment, AChE induction) with, however, the notable and unexplained exception that early induction of EIPs was not affected. The natural JH I caused qualitatively comparable effects. In both cases, the inhibition was never complete, even after several hours of preincubation with methoprene. For instance the final level of AChE induction was only about 20% that of cultures exposed to β-ecdysone alone. Dose-dependence curves support the hypothesis that JH has specific saturable receptors. Moreover, results from experiments in which the ratio of JH to β-ecdysone concentrations ranged from 10^{-4} to 10^{2} indicate that the observed antagonism between ecdysteroids and JH involves a mechanism other than a simple competition for binding sites on the same receptor.

More recently, Berger *et al.* (1992) established that the ecdysteroid-induced expression of *small-HSPs* could be abolished by the presence of methoprene (10^{-5} *M*) or JH III (although this inhibiting effect requires a pretreatment of at least 2 h). It is particularly noteworthy that the juvenoid did not significantly affect the synthesis, after heat shock, of the same proteins, which excludes the possibility of a purely toxic effect. These results were confirmed in *S3* cells transiently transfected with CAT-reporter plasmids in which the CAT coding sequence is driven by the *hsp27* promoter. Surprisingly, a similar inhibition was observed with a construct containing only a tandem of *hsp27* ecdysone responsive elements placed upstream from a herpes virus minimal promoter. This would suggest that JH, or its receptor complex, "might interfere with the binding of the ecdysteroid–receptor complex to its target EcRE sequence, unless, as a distinct hypothesis, it prevents some step in the transcriptional activation process that follows ecdysteroid–receptor binding."

With tritiated JH I and a dextran-coated charcoal assay, Chang *et al.* (1980) characterized in the cytosol of *Drosophila Kc* cells a macromolecule which binds comparably to the three forms of JH, with specificity, saturability, and high affinity ($K_d = 1.56 \times 10^{-8} M$). The number of such binding sites was estimated to be about 2500 per cell. This protein is, of course, a good candidate for the role of JH receptor. Let us emphasize the total absence of binding competition by ecdysteroids. Facilitated by the use of a labeled photoaffinity analog, [³H]EFDA (Prestwich *et al.*, *1984*), further analysis (Chang, 1985; Chang *et al.*, 1985) permitted a more accurate estimation of the molecular weight of this JH binding protein. It might be a dimer, consisting of identical subunits (each with an approximate molecular weight of 24,600). Moreover, JH-binding sites were also identified in a nuclear fraction.

Yet, more recently, Shemshedini *et al.* (1990) isolated from fat body of newly hatched *Drosophila* flies a single major JH III-binding protein with a much higher molecular weight subunit (85,000), which proved capable of stimulating protein synthesis in male accessory glands cultured *in vitro*.

VII. VALIDITY OF MODEL AND ITS CURRENT INTEREST

The value of this experimental model for the analysis of gene regulation in higher eukaryotes lies in its four main characteristics:

1. Its specificity, with regard to the nature and physiological concentration of the hormonal stimulus (see Section I) as well as to the selectivity of the cell responses (for instance, the synthesis of relatively few specific proteins, out of hundreds, is modified in treated cells; see Section II.D).

2. Its robustness; that is, the fact that cell responses are easily reproducible, remain rather similar in monolayer or suspension cultures, and are not modified, at least in their essentials, by cell concentration or any of the other usual variations in culture conditions.

3. Cultured cell lines are far more amenable to biochemical analysis than whole flies or dissected tissues, more especially since these *in vitro* cells are "naïve" with regard to the hormone and, thus, all parameters of stimulation can be easily controlled (see Section III).

4. As has already been emphasized, one is dealing with the action of a steroid hormone on the one genome that has been indisputably the most thoroughly explored, by far, in all higher animals.

The large repertoire of cell responses elicited by the hormone obviously resembles the metamorphic transformations of imaginal cells in holometabolous insects, so that it seems fair to consider it as a cell attempt, more or less abortive, to differentiate. Because the true tissue derivation of these established cell lines remains totally obscure, it is impossible to find any precise *in vivo* counterpart. In any case, all experimental data have confirmed that the steroid hormone, after binding to a specific receptor, is acting in a normal manner. Thus, the molecular mechanisms involved in such a genetic reprogramming of the cells appear to be physiological, which is the crucial point for the validity and interest of this experimental model.

As for future prospects, it is more and more evident that the characterization of receptors and corresponding response-element sequences is only the first step in our understanding of the molecular mechanism of action of steroid hormones. The cloning, especially in *Drosophila,* of a series of enigmatic "orphan" receptors (such as the *ultraspiracle* protein, or the products of "early puffing" genes), as well as current investigations on JH and binding molecules, have confirmed the constitution of heteromeric complexes with the "true" ecdysteroid receptor. Moreover hormonal regulation is only one aspect of developmental regulation in higher eukaryotes. Many trans-activating factors, especially tissue-specific ones, must interact, in a combinatorial manner, with the regulatory regions of each gene, in order to stimulate RNA polymerase and its transcription complex. It is obvious that the problems of tissue or stage specificity of gene expression can only be tackled, at least in their globality, in the organism as a whole; and *Drosophila,* because of its leading role in developmental biology, and the high efficiency of available gene transfer methods (through the vector P element; see Chapter 10) appears particularly well suited for such studies. Nevertheless, the utilization of cultured cells remains extremely valuable: either for overproducing specific factors in transformed cell lines or for testing, pair by pair in cotransfection systems, the complex interactions of the many molecular protagonists.

The *Drosophila* cell model also has great potential for the development of a new generation of insecticides, through the synthesis of molecules capable of competing with the binding, or antagonizing, in any other possible way, the action of ecdysteroids (Spindler-Barth and Spindler, 1987).* As a good example, Wing (1988, 1990) reported that a nonsteroidal substance, the insecticide RH 5849 (1,2-dibenzoyl-1-*tert*-butylhydrazine), could mimic perfectly the effects of β-ecdysone in *Kc* cells. Its activity, however, as estimated from acetylcholinesterase induction, was about 150-fold less potent. Moreover, RH 5849-resistant cells did not respond to β-ecdysone and vice versa. Competition experiments with ponasterone A showed that both molecules share a common binding domain in the hormone receptor.

References

Baird, M. B., Samis, H. V., and Massie, H. R. (1971). Changes in *Drosophila* catalase activity associated with pre-adult development. *Drosophila* Information Service, **47**, 81.

Hodgetts, R. B., Sage, B., and O'Connor, J. D. (1977). Ecdysone titers during postembryonic development of *Drosophila melanogaster. Dev. Biol.* **60**, 310–317.

Jacob, F., and Monod, J. (1961). Genetic regulatory mechanisms in the synthesis of proteins. *J. Mol. Biol.* **3**, 318–356.

Jöbsis (1964). *In* "Respiration Handbook of Physiology," Vol. 3(1), pp. 63–124. Am. Physiol. Soc., Washington, DC.

Karlson, P., and Sekeris, C. E. (1962). Kontrolle des Tyrosin-Stoffwechsels durch Ecdyson. *Biochem. Biophys. Acta,* **63**, 489–495.

Lu, Q., Wallrath, L. L., Granok, H., and Elgin, S. C. R. (1993). (CT)n · (CA)n repeats and heat shock elements have distinct roles in chromatin structure and transcriptional activation of the *Drosophila hsp26* gene. *Mol. Cell. Biol.* **13**, 2802–2814.

* In this connection, Dinan *et al.* (Exeter University, U.K.) reported, at the last (XIth) Ecdysone Workshop (Ceske Budejovice, Czech Republic, June, 1994), the results of impressive screens for ecdysone and antiecdysone activities from plant sources, using an assay based on the morphological response of one of Gateff's cell lines (see bioassay in Clement *et al.,* 1993).

Oka, T., and Schimke, R. T. (1969). Interaction of estrogen and progesterone in chick oviduct development. II. Effects of estrogen and progesterone on tubular gland cell function. *J. Cell Biol.* **43**, 123–137.

Pak, M. D., and Gilbert, L. I. (1987). A developmental analysis of ecdysteroids during the metamorphosis of *Drosophila melanogaster. J. Liquid Chromatogr.* **10**, 2591–2611.

Palmiter, R. D. (1972). Regulation of protein synthesis in chick oviduct. *J. Biol. Chem.* **247**, 6450–6461.

Prestwich, G. D., Singh, A. K., Carvazlho, J. F., Koeppe, J. K., Kovalick, G. E., and Chang, E. S. (1984). Photoaffinity label for insect juvenile hormone binding proteins: Synthesis and evaluation *in vitro. Tetrahedron* **40**, 529.

Sandaltzopoulos, R., Mitchelmore, C., Bonte, E., Wall, G., and Becker, P. B. (1995). Dual regulation of the *Drosophila hsp26* promoter *in vitro. Nucleic Acids Res.* **23**, 2479–2487.

Tomkins, G. M., Thompson, E. B., Hayashi, S., Gelehrter, T., Granner, D., and Peterkofsky, B. (1966). *Cold Spring Harbor Symp. Quant. Biol.* **31**, 349–360.

Yao, T. P., Forman, B. M., Jiang, Z., Cherbas, L., Chen, J. D., McKeown, M., Cherbas, P., and Evans, R. M. (1993). Functional ecdysone receptor is the product of *EcR* and *Ultraspiracle* genes. *Nature* **366**, 476–479.

Yao, T. P., Segraves, W. A., Oro, A. E., McKeown, M., and Evans, R. M. (1992). *Drosophila ultraspiracle* modulates ecdysone receptor function via heterodimer formation. *Cell* **71**, 63–72.

9

Gene Transfer
into Cultured
Drosophila Cells

439

The introduction of exogenous DNA sequences into eukaryotic cells (or transfection) and the subsequent analysis of their expression have proved a very efficient method for studying *in vivo* gene function and regulation in higher organisms.

As for *Drosophila*, the elegant method devised by Spradling and Rubin (*1982*) for transferring any gene into the germinal line, thereby making transgenic fly strains, exploits the natural mobility of a *Drosophila* transposon (P element, see Chapter 10). It is an unrivaled approach to the analysis of the tissue and stage specificities of expression of any transfected gene.

Concurrently, the transformation of *in vitro* cultured *Drosophila* cells retains, however, several obvious advantages. (1) *In vitro* culture avoids the many complexities of the organism and its developmental context and thus permits an easy quantification of the levels of gene expression in a single cell type. (2) The putative interactions of two or even more genes can be experimentally reconstructed and analyzed by their cointroduction into the same cell (see Sections I.A.4 and II.6). (3) Stably transformed cell lines in mass culture can be used for the overproduction of specific and rare gene products (for instance, regulatory proteins, in fundamental research, or any protein of commercial interest). (4) The transcriptional and translational apparatus of *Drosophila* cells appear to be similar to that of vertebrate cells, so they offer an attractive alternative to mammalian cell expression systems, even for the analysis of heterologous vertebrate genes.

In the mid-1990s gene transfer is most probably one of the main applications of *Drosophila* cell culture.

Two general systems of transfection have been developed: (1) those which rely on short-term expression of recently introduced genes (transient expression), and (2) those necessitating the selection of cell lines in which the foreign genes are duplicated with the rest of the genome during cell multiplication (stable transformation).

To be efficiently expressed, any transferred gene must retain its own upstream regulatory regions. Alternatively, to obtain higher expression, current techniques of molecular biology can link the gene coding sequence to a stronger promoter. Moreover, if one wants to "dissect" the regulatory regions of any given gene, different segments of its upstream sequences may be ligated to the coding sequence of a so-called reporter gene, that is, another gene whose product is easily identifiable and/or measurable.

The first successful transformation in *Drosophila* cells was reported by Wyss (1981) in a short abstract. A *Kc* subline, deficient in the purine nucleotide salvage pathway and therefore unable to grow on a selective medium TAM (thymidine, adenine, methotrexate) (see Chapter 5.I.B), could be transformed and thereby rescued by exposure to wild-type *Kc* DNA previously complexed with poly(L-ornithine).

Since then, a variety of gene transfer methods have been proposed, most of them initially devised for mammalian cells but rapidly adapted to *Drosophila* cell lines. Their different potentialities and respective efficiencies in each experimental system should be compared (see also a review on *Drosophila* cell transformation techniques by Cherbas *et al.*, 1994).

I. CELL TRANSFORMATION

In transformed cell lines, input DNA sequences are maintained within the host cells by a regular replication at each division cycle, either after their integration into chromosomal DNA or, occasionally, as extrachromosomal plasmids. Because transformation events are relatively rare and haphazard in a transfected cell population, transformants have to be selected on the basis of some new ability, usually resistance to a specific drug (see Section V), conferred by the exogenous gene. It is easy to understand that, when a transformed cell line is grown in the absence of selection, integrated copies of the transgene can be stably inherited (stable transformation), whereas extrachromosomally replicating plasmids may be lost by dilution (unstable transformation).

Both situations were observed, respectively, in the first two reports of successful transformation of *Drosophila* cells, which made use of dominant selectable markers.

Bourouis and Jarry (1983) showed that the acquired methotrexate resistance (see Section V.A) displayed by several transformed cell lines could be correlated with the presence, in their high-molecular-weight DNA (supposedly chromosomal DNA), of a very large number of copies of their DHFR vector plasmid in the form of large head-to-tail oligomeric structures (from 1000 to 3000 copies per cell, which represented 2–6% of total DNA).

In contrast, when using a *gpt* selection system (see Section V,B), Sinclair *et al.* (1983) established that their *gpt* vector was propagated extrachromosomally in the form of rearranged plasmids and that there was a maximum of two copies per cell. As expected, growth in a nonselective medium rapidly led to the loss of this *gpt* gene. Moreover, the authors suggested that, owing to this small number of plasmids, a spontaneous segregation of cells lacking *gpt* could also occur under selective conditions, which might account for the apparent slow doubling rate of their transformed cell lines.

It is important to distinguish these two quite different types of situations.

A. Stable Transformation

In most transfection experiments with *Drosophila* cells, stable transformation (that is, integration into the host genome) was found to be the rule.

As in the pioneering work of Bourouis and Jarry, Southern blot analyses of high-molecular-weight DNA from transformed cell lines indeed revealed, in most cases, a high number of vector plasmid copies, randomly integrated into the genome. They did not usually display any rearrangement and were present in the form of long tandem head-to-tail arrays in one or just a few sites in the host cell chromosomes.

Integration was conclusively demonstrated, at least in one example, by the subcloning in lambda (λ) phages of the transduced DNA from a transformed cell line, then by the hybridizing of the cloned fragments with polytene chromosomes. It was shown that some of the hybridization sites corresponded to flanking sequences acquired by integration (Burke *et al.,* 1984).

The multicopy arrangements, which were presumably the result of homologous recombination, could be revealed by the fact that enzymes cleaving only once in the vector plasmid gave a single prominent band with the exact size of the linearized plasmid; on the other hand, digestion with other enzymes that do not cleave within the vector gave rise to a very high-molecular-weight band (several tens of kilobases).

1. INTEGRATED GENE COPY NUMBER

Reconstitution experiments (i.e., comparisons with plasmid DNA standards; see Potter *et al., 1980*) permit a rough estimation of the average copy number of integrated genes. It may vary enormously: from 1 to 2 copies per haploid genome (Maisonhaute and Echalier, 1986); 10 to 20 (Rio and Rubin, 1985); 3 to 50 (Burke *et al.,* 1984); 20–40 (Allday *et al.,* 1985); 50 (McGarry and Lindquist, 1985); 30 to 80 (Henikoff *et al.,* 1986); 40 to 120 (Van der Straten *et al.,* 1989); 50 to 200 (Sinclair *et al.,* 1985); 500 (Johansen *et al.,* 1989); 700 to 1000 (Martin and Jarry, 1988); and up to several thousands (Bourouis and Jarry, 1983; Moss, 1985; Moss *et al.,* 1985).

Although the mechanisms of gene integration are still poorly understood, such variations in integrated gene copy number seem to depend on multiple parameters: (1) the vector construction and especially the strength of the driving promoter (Sinclair *et al.,* 1986); it is indeed well documented in vertebrate cells that a good correlation exists between stable transformation frequency and the level of transient expression of the vector; (2) the nature or the physiological state of recipient cells (see Section III); (3) the procedures of transfection and of selection (see following paragraphs): a striking example concerns a *gpt* vector which replicated extrachromosomally when transfected with wild-type DNA as a carrier, whereas it was integrated when cotransfected with an *hsp* vector (Burke *et al.*, 1984).

2. CORRELATION BETWEEN INTEGRATED GENE COPY NUMBER AND EXPRESSION LEVEL

In transformed cell lines, it was generally observed that a significant correlation exists between the expression level of a transfected gene, as estimated from the quantity of its transcripts or its protein product, and its copy number in the genome. This RNA/DNA ratio, however, seems to drop with an increasing number of copies (see, for instance, Van der Straten *et al.*, 1989), which might perhaps reflect some limitation in the cellular contents of specific transcriptional activators.

The general conclusion that a high copy number usually corresponds to a high expression level is well illustrated by the results of Moss (1985). The copy number of integrated plasmids was estimated to be up to several thousands (i.e., represented 8% of the cell DNA) and thereby their transcripts accounted for as much as 20% of total cell mRNA.

These considerations should be borne in mind for the future construction of stably transformed *Drosophila* cell lines intended for overproduction of proteins of interest.

Note: Another important question is the relative level of expression of transfected genes with respect to that of corresponding endogenous genes. To make this comparison, one must be able to distinguish the transfected gene product with the aid of a modification in the coding sequence. For example, McGarry and Lindquist (1985) used a variant (*hsp44*) of an *hsp70* gene (see Chapter 7). Thus, they could establish that, although the integrated copy number of these modified genes was, in some of their transformed lines, about 10 times higher than that of endogenous *hsp70* genes, nearly equal quantities of HSP70-kDa and HSPp44-kDa proteins were produced after heat shock. This means that after induction the expression level of each transfected gene reached only 10% of the expected value.

3. INFLUENCE OF CHROMOSOME ENVIRONMENT

It should never be forgotten that in such transformed cell systems the chromosomal environment of integrated genes may have a crucial influence on their expression efficiency. This fact has been strongly documented in *Drosophila* transgenic fly strains in which *in situ* hybridization on polytene chromosomes makes a precise localization of the transgene possible. It must be emphasized, nevertheless, that in most cell transformation experiments, and despite the probable ectopic location of integrated genes, these genes were efficiently expressed and even correctly regulated (see, for instance, the systematic study of transduced *hsp* genes by Larocca, 1986).

There is in the literature at least one documented example of the possible influence of the insertion site in a cultured cell system. To their great surprise, Burke *et al.* (1984) observed, in their first cotransformed cell lines, that integrated *hsp* genes (up to 50 copies) remained inactive. They suspected that their inactivity might merely be due to the fact that plasmid arrays were located in heterochromatin; it proved a likely hypothesis since they demonstrated, at least in one case, that the flanking sequences that they had subcloned hybridized with the chromocenter of polytene chromosomes.

As suggested by Burke *et al.* (1984), systematic comparisons between the activity of a transiently transfected plasmid and its activity after integration (see a later study by Martin and Bourouis, 1988) permit an analysis of the influence of the integration site and its chromatin structure on gene expression.

4. COTRANSFORMATION

Cotransfection, that is, the simultaneous introduction of two different genes into the same cell, is a universal method for transferring a nonselectable gene under the cover of a selectable one. It is indeed well known that the few cells that, in a culture population, are able to take up one exogenous DNA sequence can also accept concomitantly another gene.

Under selective conditions, both cotransfected plasmids are integrated into the chromosomes of the resistant cells, usually as separate concatemeric blocks; sometimes, however, probably after assembly by homologous recombination, they can be found in interspersed series (Van der Straten *et al.*, 1989).

Their respective copy number can be modulated by altering the relative amounts of the two plasmids in the transfection cocktail (Burke *et al.*, 1984; Moss, 1985; Sinclair *et al.*, 1985; Henikoff *et al.*, 1986; Van der Straten *et al.*, 1989). Van der Straten *et al.* (1989) made a systematic study of the phenomenon (see Table 9.I). With a 1 : 1 ratio, which is the usual proportion in cotransfection procedures; both plasmids were found

TABLE 9.I Correlation between DNA Ratio of Two Cotransfected Plasmids and Numbers of Integrated Copies in Transformed Cell[a,b]

Input plasmid DNA (μg/transfected culture)			Gene copy numbers (in hygromycin-resistant cells)	
pCodGalk	pCodhygro	Ratio	Galk	Hygro
20	0.02	1000/1	1050	1-2
20	0.2	100/1	600	10
18	2	10/1	450	70
10	10	1/1	150	135
2	18	1/10	43	420
0	20		0	400

[a] From Van Der Straten *et al.* (1989); copyright © (1989). Reprinted by permission of Wiley–Liss, Inc., a subsidiary of John Wiley & Sons, Inc.

[b] The selection plasmid is pCod-hygro which confers hygromycin resistance.

in nearly equal numbers in each individual transformed cell line, even though their total number could vary fourfold from one line to another.

5. POLYCLONAL ORIGIN OF MOST *Drosophila* TRANSFORMED CELL LINES

A decisive point, not sufficiently emphasized in most experiments with stably transformed *Drosophila* cells, is that one is usually dealing with polyclonal cell lines. This is due to the general difficulty encountered in cloning *Drosophila* cultured cells (see Chapter 3, Section VI), so that when the first transformant colonies appear under selective conditions, their poor viability and weak adhesion to the culture flask make impossible any accurate counting (to estimate transformation efficiency) or rigorous isolation and independent propagation of true clonal sublines. Resistant cells in each culture flask are, therefore, usually grown in mass culture and constitute what is called a transformed cell line. It is true that the low efficiency of most transformation methods guarantees that the number of founder cells for each line remains low; nevertheless, the number of integrated gene copies and their precise chromosomal location differ in each of these "initials."

It is not surprising, therefore, that some uncontrolled cell "drift" has been observed in such heterogenous populations. For instance, Van der

Straten *et al.* (1989) reported that the average number of *hygro* genes (see Section V.D) in their transformed lines decreased 3-fold over a 4-month period, which may be readily explained by the outgrowth of subpopulations containing a lower number of copies.

In any case, these facts throw a shadow of doubt on most of the above-mentioned estimations of the "average" copy number of integrated genes in transformed cell lines as well as any attempt to correlate too strictly this copy number with a level of expression.

Despite practical difficulties, it is highly advisable, at least for such purposes, to subclone each polyclonal transformed line shortly after the selection period (Martin and Bourouis, 1988).

B. Unstable Transformation

The retention of input DNA in *Drosophila* cells as autonomously replicating plasmids was first observed by Sinclair *et al.* (1983), but this situation of unstable transformation remained an exception in subsequent transformation experiments. Yet, these authors noted the same phenomenon (Sang *et al.*, 1984) when using the same type of vector plasmid, although in entirely different conditions of selection.

Three observations are relevant at this point.

1. Kurata and Marunouchi (1988) studied specific nucleotide sequences, called ARS (for autonomously replicating sequences), which are responsible for the initiation of DNA replication in *Saccharomyces* and may confer on a plasmid the ability to replicate as an episome in this yeast. Similar sequences, comprising two typical successive motifs, were found in the DNA from other eukaryotes, including *Drosophila melanogaster*. When such ARS were inserted into *copia–gpt* vectors (see Section V.B), these plasmids survived in *Drosophila* cells for several weeks. Moreover, a certain proportion of these plasmids (from 50 to 80%, depending on the construct) replicated autonomously in the transformed cells. This could be deduced from the extent of methylation of recovered plasmid DNA, because it is well known that introduced plasmids are methylated in *Escherichia coli*, whereas no methylation takes place in *Drosophila* (Uriel-Shoval and Gruenbaum, 1982). Nevertheless, after 2 months, only transformed cells in which the plasmids had been integrated into the chromosomes could survive under selective conditions.

2. There is also an interesting observation made by Allday *et al.* (1985) who, after cotransfecting *Drosophila* cells with an Epstein-Barr virus

(EBV) cosmid clone, detected the nuclear EBNA-1 antigen in the nucleus. This EBV antigen is apparently necessary in vertebrate cells for the maintenance of extrachromosomal EBV episomes (Yates *et al.*, *1985*). It is not known whether it functions in a similar way in transformed *Drosophila* cells, but, if it does, it would be of practical value to use EBV-derived recombinant plasmids for introducing genes into such EBNA-1 expressing lines and, thereby, maintaining them in an extrachromosomal situation (in order to avoid the hazards of an improper chromatin environment).

3. More recently, a series of plasmids containing various sequences from the *Drosophila* gene *Ultrabithorax*, as well as random human or *E. coli* genomic fragments, could still be detected 18 days after their transfer into S2 line cells (Smith and Calos, 1995). This implies a capacity for autonomous replication and a lack of specific sequence requirement for initiation (which seems to be a characteristic of animal cells, in contrast with the high sequence specificity observed in yeast).

II. TRANSIENT EXPRESSION

An alternative to stable transformation systems is offered by experiments in which expression of plasmid-borne genes is measured within the first few days following transfection with DNA (= transient expression).

The main advantage of transient assays is that they are considerably faster: a few days rather than several weeks (for stable transformation) are required. Moreover, the average level of gene expression can be measured from the entire population of transfected cells, instead of the considerable variability typically observed among individual clones of transformed cells. In addition, because the plasmids remain unintegrated during the short assay period, any potential influence of neighboring host chromosomal sequences is, *ipso facto,* avoided.

In order to estimate the expression level of transduced genes and to distinguish it without ambiguity from endogenous cell activities, wide use is made in these transient expression systems of fusions between the gene regulatory sequences and various "reporter" genes (Section VII), i.e., encoding heterologous enzymes whose activities are easy to characterize and to quantitate.

Since the pioneer work of Di Nocera and Dawid (1983), who applied the CAT assay system (see Section VII.A) to *Drosophila* cells, this transient expression method has been rapidly adopted by many groups (Martin, 1984; Burke *et al.*, 1984; Morganelli and Berger, 1984; Benyajati and Dray, 1984) and it is now routinely employed. Although CAT remains the most frequently used reporter gene, it is not the only possible one (see below).

1. TRANSFECTION EFFICIENCY

As in cell transformation systems, the proportion of cells from a transfected population which express the foreign gene and the average number of plasmids taken up by individual cells depend greatly on both the recipient *Drosophila* cell line (Section III) and the chosen transfection procedure.

A few data are available concerning this expression efficiency in a cell population:

1. The *copia–gpt* vector (Fig. 9.5b), which is strongly expressed in the *Drosophila hydei DH33* cell line, allows the cells to incorporate [^{3}H]guanine and thus to become identifiable by autoradiography. About 20% of labeled cells were observed in transfected cultures (Burke *et al.*, 1984).

2. Analogously, the marker *lacZ* (see Section VII.B) gives rise to a dark-blue staining of the individual cells that express β-galactosidase after *in situ* treatment with a chromogenic substrate (such as X-Gal, see Appendix 9.K) [Sinclair and Bryant (1987); Thummel *et al.*, 1988)].

3. Another method of estimation consists of using specific antibodies against the gene product and of characterizing expressing cells by an indirect immunocytochemical staining. For instance, a clonal anti-CAT mouse antibody was secondarily detected with an anti-mouse antibody conjugated with peroxidase (Davies *et al.*, 1986), and no more than 1% of the cells were shown to produce the CAT enzyme. In contrast, 48 h after transfection with an *Antp*-coding plasmid, 10 to 30% of the cells synthesized immunochemically detectable *Antp* protein (Winslow *et al.*, 1989).

In most cases, nevertheless, it seems that expression efficiency in transient systems does not exceed a few percentages of the exposed cell population. Such low percentages might reflect either (1) the presence in heterogenous cell lines of a small subpopulation especially suited to the taking up and expressing of exogenous DNA or (2) the rare distribution in such nonsynchronized cultures of some brief cell cycle stage that would be best fitted for transfection (see discussion in the following Section III).

2. USEFULNESS OF INTERNAL CONTROL

Whatever their causes, the aforementioned low transfection efficiencies might, at least partly, explain the wide variations that are typically observed from one experiment to another in transient transfection systems, even when using strictly standardized protocols. As a rule, identical assays have to be repeated several times and their results pooled.

Moreover, it is highly advisable to check the competence with which the cell batch takes up and expresses exogenous DNA. For this purpose, an aliquot from the cell culture may be transfected in parallel with a reference reporter plasmid (Martin and Jarry, 1988); or, even better, this reference plasmid may be used as an internal control by cointroducing it into the same cells. For instance, when Sinclair *et al.* (1986) tested deleted variants of *copia–gpt* vectors (see Section VI.A), they cotransfected the cells with their standard pCV2CAT plasmid, which permitted correction for variations from one experimental series to another. Similarly, Driever *et al.* (1989), like Sneddon and Flavell (1989), in concert with their complex experimental systems that were based on a CAT assay cointroduced an efficient *lacZ* fusion construct whose β-Gal expression could be measured in the same cell extract.

3. NONREPLICATION OF INPUT PLASMIDS DURING TRANSIENT ASSAYS

The demonstration that transfected plasmids do not measurably replicate during the first few days following transfection relies on the fact that, in contrast with *E. coli,* in which the plasmids have been prepared, *Drosophila* does not methylate its DNA (Uriel-Shoval and Gruenbaum, 1982). There are two restriction enzymes that are isoschizomers, i.e., cleave the same recognition sites in DNA: *Mbo*I does not, however, cut methylated adenines, whereas *Sau*3A does. Digestion of input plasmid DNA, about 60 h after transfection, showed that the plasmids were cut by *Sau*3A but remained uncut by *Mbo*I, which means that they had retained their initial bacterial methylation (Burke *et al.,* 1984).

Moreover, Morganelli and Berger (1984) reported that on Southern blots the significant amounts of intact plasmid DNA present within the cell 2 h after transfection were already reduced by 20 h.

4. OPTIMAL TIMING FOR TRANSIENT EXPRESSION ASSAYS

The expression of transfected plasmids may sometimes be particularly rapid: for instance, as early as 2 h after transfection with an *hsp22*-containing plasmid and following heat-shock induction (Morganelli and Berger, 1984). Likewise, after 4 h, substantial β-Gal activity was already measurable in cells transfected with an *hsp70–β-Gal* vector (Lawson *et al.,* 1984).

More often, however, the expression of a transfected gene cannot be detected before approximately 24 h; its level generally increases linearly during the following 20 to 40 h, up to a plateau which may be maintained for 3 to 5 days (see Fig. 9.1).

Therefore, after transfection with the standard DNA-calcium phosphate coprecipitate method (Section IV.A), expression assays are usually carried out at the end of a 48 h incubation period and sometimes even later (for instance, 90 h: Burke *et al.,* 1984; 4 days: Davies *et al.,* 1986).

This latency seems to be due to the time necessary for the cells to recover from the treatment (which can vary with different transfection procedures).

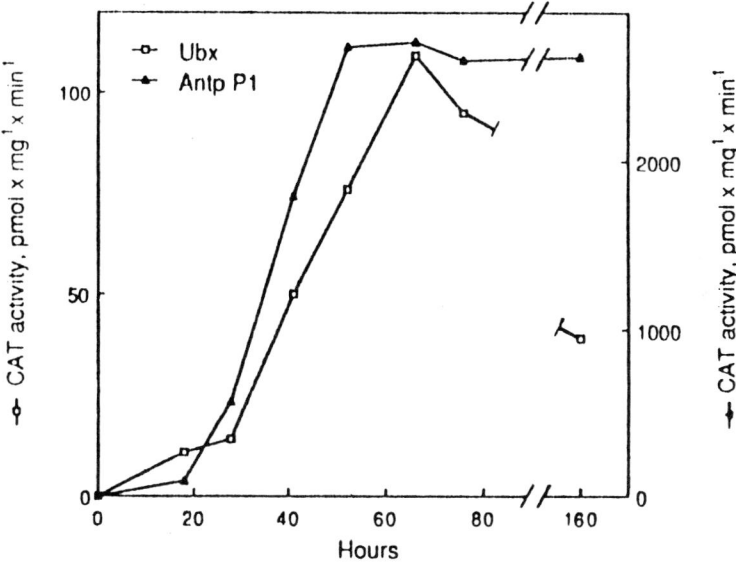

FIGURE 9.1 Time course of expression in *Drosophila* cells of two "reporter" plasmids in which the gene *CAT* is driven by the promoter sequences of two developmental genes, *Antp* and *Ubx*, respectively. Parallel cultures of *D. melanogaster S2* cells were transfected with 1 μg of either pP$_{Antp1}$CAT or pP$_{Ubx}$CAT DNA by the calcium phosphate technique. At the times indicated after transfection, individual cultures were harvested and cell extracts were assayed for CAT activity. From Krasnow *et al.* (1989), with the kind permission of Professor Hogness and Cell Press.

5. CORRECT OR ECTOPIC EXPRESSION OF TRANSFECTED GENES

In most transient expression systems, the constitutive as well as induced expression of transfected genes were found to be very similar to that of their endogenous counterparts. This has been especially well documented for *heat-shock* genes (Di Nocera and Dawid, 1983; Morganelli and Berger, 1984; Lawson *et al.*, 1984; Larocca, 1986) and supports the validity of such experimental cell systems for the analysis of the regulatory regions of a gene *in vivo.**

Nevertheless, one should not ignore the fact that a few genes, even though it is known that they display in the *Drosophila* organism strict stage or tissue specificities of expression, can be more or less efficiently expressed after transfection into undifferentiated cell lines. This was true for *alcohol dehydrogenase* (Benyajati and Dray, 1984); *larval serum protein 1,* normally synthesized only in larval fat body (Davies *et al.*, 1986); *Sgs3,* coding for a glue protein specific of larval salivary gland; or *yolk polypeptides 1* and *2* (Martin and Jarry, 1988).

The most surprising case is perhaps that of *ADH* which, despite its relatively wide tissue distribution in larvae and flies, is not normally expressed in most available *Drosophila* cell lines (Debec, 1974). How exogenous *Adh* sequences can be expressed in transfected Schneider's *line 2* cells, whereas their endogenous *Adh* genes are not, is a question that remains unanswered.

It was suggested (Davies *et al.*, 1986) that such ectopic expression of transduced genes could be a consequence of the uptake of naked DNA. The input plasmid-borne gene may not be condensed correctly into chromatin and therefore could be in a conformation particularly accessible to the transcription machinery.

6. COTRANSFECTION IN TRANSIENT SYSTEMS

Two or more genes can be simultaneously transfected and thereby transiently coexpressed in cultured cells.

The main advantage of such a cotransfection system, in transient conditions, is to allow an *in vivo* analysis of putative interactions between the transferred genes, which implies, of course, that both types of plasmids (or more) have entered the same cells.

In fact, with immunocytochemical detection of both products of two cotransfected recombinant plasmids, using fluorescein and rhodamine-conjugated secondary antibodies, respectively, Winslow *et al.* (1989)

---

* According to the meaningful pun made by Moreno and Nurse (*1990*), *In Vivo Veritas!*

could demonstrate that cells that produced one protein also synthesized the other. Incidentally, this observation suggests once more that in cultured cell populations a small subset of receptive cells can take up and express not only one but several different plasmids.

This cotransfection assay has been currently developed by several laboratories engaged in studies of the developmental biology of *Drosophila* in the hope of unraveling step-by-step the combinatorial interactions that occur among the many genes participating in embryonic pattern formation (as deduced from genetic data). For instance, and to limit these references to the first published papers, let us quote experiments with *fushi tarazu* and *engrailed* (Jaynes and O'Farrell, 1988); *bicoid* and *hunchback* (Driever and Nüsslein-Volhard, 1989); *Antennapedia* and *fushi tarazu* (Winslow *et al.*, 1989); and *Ultrabithorax* and *Antennapedia* (Krasnow *et al.*, 1989). In this last paper, the strategy is clearly stated by the Hogness group:

> A plasmid that directs expression of a developmental control protein (the *effector plasmid* or *producer*) is transiently transfected into cultured *Drosophila melanogaster* cells, along with a second plasmid containing a candidate target promoter coupled to a reporter gene (the *reporter plasmid* or *responder*). Regulatory function is monitored by changes in reporter gene expression

For a finer analysis, both partners, that is, the functional domains of the transacting protein and the corresponding binding sites of the target gene, can be systematically modified (Fig. 9.2).

In such relatively simple *in vivo* systems, at least, in comparison with the complexity of whole embryos, it might be possible to reconstruct large parts of the regulatory hierarchy of developmental genes by introducing each target gene together with its multiple putative regulators.

A variant method consists in stably transforming *Drosophila* cells with a developmental regulatory gene and then transiently introducing a series of potential target genes into these expressing cell lines. Many such cell lines, each constitutively expressing one of the main developmental regulatory genes, are already available for this purpose.

A similar cotransfection approach, which made intentional use of *Drosophila* cells as a heterologous *in vivo* system, was exploited by Tijan's group for the analysis of the functional organization of several mammalian transcription factors whose homologs do not exist in *Drosophila* cells (although they proved, in such experiments, to be compatible

A

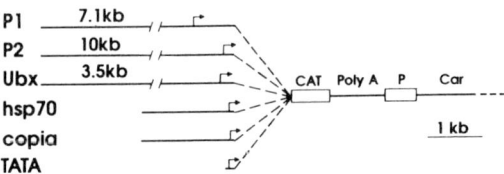

B

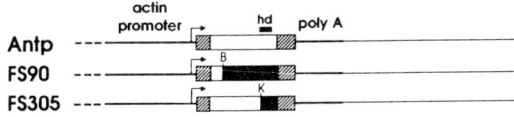

C

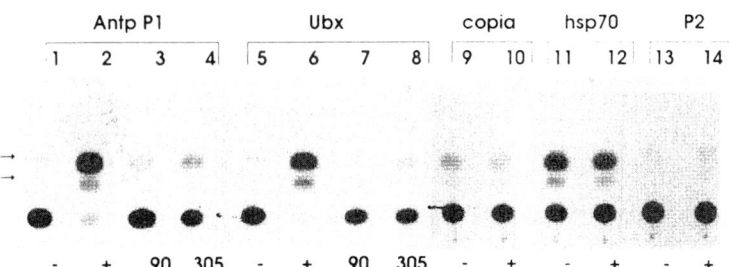

FIGURE 9.2   Cotransfection system for analyzing the influence of a developmental gene product on the expression of another gene, e.g., *Antp*-dependent stimulation of the CAT activity of different "reporter" plasmids. (A) Reporter plasmids contain the 5' regulatory regions and 5' nontranslated sequences from the indicated genes inserted into the vector pC4CAT upstream of the bacterial chloramphenicol acetyltransferase: P1 and P2, DNA fragments from the two *Antp* promoters; TATA, a short fragment from the *hsp70* gene (-40 bp to $+70$ bp); Open boxes indicate the coding sequence of the CAT gene and P element sequences; Poly(A), SV40 polyadenylation signal; Car, Carnegie 4P element transformation vector. The horizontal arrows indicate the start of transcription. (B) Protein expression plasmids. The coding (open boxes) and noncoding (hatched boxes) regions from

with their transcriptional machinery): for instance, *CTF/NF-1* (Santoro *et al.*, 1988); *Sp1* (Courey and Tijan, 1988) (see Chapter 5, Section V.C.3).

III. RECIPIENT CELL LINES

As has also been documented in mammalian cells, the transfection characteristics of the different *Drosophila* cell lines vary considerably. This widespread opinion is mostly based on unpublished observations and certainly deserves a finer experimental analysis. Only two papers have been devoted to such an important question. For a more detailed discussion of the transfection efficiency of different recipient cell lines see a review by Echalier and Fourcade (1994).

Burke *et al.* (1985) made a systematic comparison of several *Drosophila* cell lines by measuring their capacities to incorporate [³H]guanine as conferred by transfection with the bacterial gene *ecogpt* (see Section V.B). When using their *copia*-based *gpt* vector (pCV2gpt), they observed that none of the tested *Drosophila melanogaster* cell lines (*D1, S2, Kc*) incorporated [³H]guanine to the same high level as *D. hydei* cell lines (only 3 to 6% of the level measured with *Drosophila hydei* DH33 (see Table 9.II).

Such marked differences of expression could be at least partly accounted for by differences in plasmid uptake, as shown by Southern analysis: in *D1* cells, the uptake was estimated to be only one-tenth and in *S2* cells one-third to one-half that of *Drosophila hydei* cells.

However, when the same *gpt* gene was driven by an *hsp70* promoter, even though differences in plasmid uptake remained the same and, conse-

---

the *Antp* gene were inserted into the expression vector pP$_{ac}$. FS90 and FS305 are *Antp* constructs that contain frameshift mutations engineered at the *Bst*(B) or *Kpn*(K) sites. The light horizontal lines indicate *pUC18* sequences. The homeodomain is indicated by the solid bar above the construct. (C) Results of the CAT assays. The promoter-CAT construct is indicated at the top. The notation at the bottom refers to the protein-producing plasmids: −, the actin expression vector (pP$_{ac}$) with no coding sequence; +, the actin vector with the insertion of the full-length *Antp* protein; 90, the FS90 plasmid (see B); 305, the FS305 plasmid (see B); Arrows point to the acetylated forms of chloramphenicol. The full-length *Antp* protein causes a dramatic increase in CAT activity produced from the *AntpP1* (lane 2) and *Ubx* (lane 6) promoters, while no significant changes are observed from the *copia* (lane 10), *hsp70* (lane 12), or *AntpP2* (lane 14) promoters. The basal level of expression from the *hsp70* promoter was obtained without heat shock. From Winslow *et al.* (1989) with the kind permission of Professor Hogness and Cell Press.

TABLE 9.II   Differences in Plasmid Uptake and/or Expression of
Various *Drosophila* Cell Lines[a]

Cell type	[³H]Guanine incorporation (% cpm of DH33)
D1	5.75
SL2	3.3
Kc	6.2
DH14	80.0
DH15	66.0
DH33	100

[a] From Burke *et al.* (1985); with the permission of *EMBO J.*
Levels of [³H]guanine incorporation are compared between different *Drosophila* cells lines transfected with CV2gpt. *D1, SL2,* and
Kc cells are *D. melanogaster* cell lines, whereas *DH14, DH15,* and
DH33 are *Drosophila hydei* cell lines.
Approximately 90 h after transfection, cells were assayed for
[³H]guanine incorporation. The highest incorporation observed (i.e.,
in *DH33* cells) was 12,624 cpm, which is equivalent to 10.0 pmol
of guanine incorporated over the 20-hr labeling period.

quently, seem indeed to characterize each cell line, the levels of expression
after heat shock were much closer to one another. As before, *Drosophila
hydei DH33* cells were the most efficient, but *Drosophila melanogaster
DI* cells equalled the performance of the *Drosophila hydei DH14* line.
So, cell lines may also differ in their capabilities for transcription and
translation. These characteristics remain, nevertheless, contingent on the
construct to be expressed. For instance, a chimeric plasmid, with a human
cytomegalovirus promoter driving a CAT sequence, which was strongly
expressed in *Drosophila melanogaster* cells, was not expressed in the
D. hydei line *DH33*, in direct contrast with the preceding observations
(Sinclair, 1987).

Saunders *et al.* (1989) reported the exceptionally high expression efficiency of a haploid *Drosophila melanogaster* cell line, *1182* (Debec,
1978). These cells could transiently express transfected genes between
20 and over 100 times more efficiently than other *Drosophila* cell lines
(*D1, S3,* but also *DH33*).* Although *1182* cells could be efficiently transfected with naked DNA, even in the absence of calcium phosphate or

* Unfortunately, we could not confirm such a high efficiency with our own original
lines of *1182* cells.

any other facilitator, their high expression efficiency did not seem to be attributable to a preferential uptake, replication, or lesser degradation of the input DNA.

Because of its theoretical interest and practical implications, let us consider the various aspects of this problem. It must be remembered that one is measuring the global expression of a transfected gene by a cell population, and several parameters could account for the observed differences.

The percentage of expressing cells in a transfected culture is generally low (see Section II.1). It might correspond to the presence of a small specific subpopulation whose relative importance varies from one line to another and even between subcultures from the same cell line. For instance, some researchers report better performances of the subline *Kc167*, which is merely a derivative of our standard *Kc* line (Bourouis and Jarry, 1983).*

Another explanation could be the likely existence of a physiological stage that would especially favor transfection. If this stage is brief during the cellular cycle, the proportion of cells concerned at any time should be low in nonsynchronized cultures; again, this percentage could differ from one cell line to another. Moreover, culture conditions at the time of transfection, and especially the timing of the last cell seeding, would be of the utmost importance. Most transfection protocols recommend the use of nonconfluent cell cultures that have been seeded on the previous day.

In fact, every step of the gene expression process, from the penetration of exogenous DNA to its utilization by the transcriptional then translational cell machinery, must be subject to the intrinsic properties of the different cell lines.

1. As mentioned above, Burke *et al.* (1985) showed that each cell line displays its own characteristic level of plasmid uptake through the cell membrane and this level also greatly depends on the choice of transfection method.

2. As a second step, input DNA has to be safely transferred through the cytoplasm into the nuclear compartment.

Morganelli and Berger (1984) could deduce from Southern blots that the amounts of their *hsp22* vector, although abundant within the cells 2 h after transfection, appeared significantly reduced by 20 h. Their semiquantitative estimations, however, are vitiated by too many artifacts;

* We could never observe any differences between this subline and our standard *Kc* line.

in particular, it is impossible to distinguish between the simple release of DNA initially adsorbed to the cell surface and the actual degradation of already internalized plasmids.

It was claimed that some substances, such as chloroquine, could inhibit the degradation of DNA engulfed by mammalian cells, either by increasing pH value in lysosomes or by strongly binding to DNA and protecting it (Luthman and Magnusson, *1983*). A posttreatment with chloroquine diphosphate is therefore recommended as a final step in some transfection methods, although its beneficial effect is doubtful in *Drosophila* cells (Echalier, unpublished data); and it was in fact found to be extremely toxic (Sinclair and Bryant, 1987).

3. With regard to the entry of exogenous DNA into the nucleus, Sene and Nicolau (*1981*) assumed that it might be facilitated by the breakdown of the nuclear membrane at metaphase and its reformation during telophase. In this connection, Saunders *et al.* (1989) suggested that the high efficiency of expression that they observed in transfected haploid *1182* cells might be correlated with a disturbance in their cellular cycle, since some of these sublines were shown to be devoid of any centriole (see Chapter 4, Section III.C).

Moreover, Sinclair and Bryant (1987) reported that *Drosophila* cells preexposed to ecdysteroids for 48 h prior to transfection showed a 4- to 5-fold increase in the expression of transfected genes, and this was in keeping with an increase in the percentage of expressing cells in the population. Now, such hormonal treatment is well known to induce cell arrest in stage G_2 (see Chapter 8, Section II.B). Sinclair and Bryant's data might perhaps be explained (as suggested by Saunders *et al.,* (1989) by a removal of this hormonal block and entry into cell mitosis just before transfection.

4. It was demonstrated (see Section II.3) that in any *Drosophila* cell line there is no or very little plasmid replication during the first few hours after transfection. Differential replication of intranuclear plasmids would not, therefore, account for the differences in the levels of expression observed between cell lines.

5. In eukaryotes, transcription remains a crucial step for the regulation of expression. It has been well established that each cell type in higher organisms runs its own program of transcription, presumably controlled by specific combinations of transcriptional modulating factors. In fact, a good correlation usually exists between the final expression of a transfected gene and the levels of its specific mRNA(s).

Variations in any component of the cellular basal transcription machinery, as well as differences in amounts or in the relative proportions of the available trans-activating factors, can greatly influence the general transcriptional capacity of the cells or even the specific transcription rate of a given gene. It is clear that these characteristics may totally differ from one cultured cell line to another, particularly because the precise tissular origins of most *Drosophila* cell lines are unknown and might be quite different (see Chapter 3).

Moreover, they might vary from one *Drosophila* species to another, and one example seems to particularly demonstrate this: "Functional dissections" of the regulatory regions of *copia* LTRs, as conducted in transient expression systems with a CAT reporter gene (see following Section VI.A, and Chapter 10), led to rather different results in transfected *Drosophila hydei* and *Drosophila melanogaster* cell lines, respectively. A regulatory region downstream from the RNA start site was needed for optimal expression in the *Drosophila melanogaster Kc* line (Sneddon and Flavell, 1989), whereas it was not essential in *Drosophila hydei DH33* cells (Sinclair *et al.*, 1986). Incidentally, it must be remembered that the genome of this latter *Drosophila* species does not normally contain this *copia* transposon. Once again, the differences between these two cell types probably correlate with differences in the nature, or in the abundance, of the protein modulators that are liable to interact with *copia* LTR regulatory regions. This is true for any transfected gene, and it would not be surprising if each gene required a particular cell line for its optimal expression.

6. It is obvious that several posttranscriptional steps (such as mRNA processing or degradation, translation rate, etc.), even though they are much less studied, must have an influence on the levels of expression efficiency.

For instance, McDonald *et al.* (1988), when comparing the promoter strength of retrovirus or retrotransposon LTRs in *Drosophila* and mammalian cells observed that the relative levels of *cat* mRNAs were nearly identical in *Drosophila S2* cells transfected with human immunodeficiency virus (HIV)–CAT (and HIV–TAT), Rous sarcoma virus (RSV)–CAT, and *copia*–CAT plasmids. Yet the level of CAT enzymatic activity in *copia*–CAT-transfected *Drosophila* cells was more than 5 times higher than in HIV–CAT transfected ones. This clearly indicates that the transcripts promoted by RSV and HIV LTRs are posttranscriptionally repressed in *Drosophila* cells. The reverse was observed in mammalian cells.

As a general and practical rule, for each experimental system transfection assays must be optimized by the proper choice of recipient cells and appropriate transfection methods.

IV. TRANSFECTION METHODS

In order to facilitate the penetration of exogenous DNA into cells and its safe transfer into the nucleus and possibly into the genome, several methods, with multiple variant protocols, have been recommended; but, so far, the most universally used technique remains the DNA–calcium phosphate coprecipitate procedure, first devised by Graham and Van der Eb (*1973*) for mammalian cells.

However, the DEAE–Dextran method was successfully employed by several groups for transfecting *Drosophila* cells. The Polybrene technique was also found to work efficiently in *Kc* cells (Echalier, unpublished data) and some other cell lines (Walker *et al.*, 1989).

Last, lysosome-mediated transfection, which is simple and ubiquitously efficient, is currently gaining ground.

Electroporation, which ensures high transfection efficiencies in many mammalian cell lines, has recently been adapted to *Drosophila* cells (personal communication [1995] from the Cherbas'; see Section IV.F and Appendix 9.E).

It must again be emphasized that the success of any transfection technique can differ greatly from one *Drosophila* cell line to another, and the optimal conditions for operating procedures have to be defined for each one.

Note: As mentioned above, Saunders *et al.* (1989) reported that Debec's haploid *1182* line (see Chapter 3, Section III) might even show a high transfection efficiency with naked DNA without the help of any facilitation technique. (We could not confirm this assertion in the original line, and it might be peculiar to the subline these authors were using.)

A. DNA–Calcium Phosphate Coprecipitation Method

Studying the influence of divalent cations on the uptake of adenovirus DNA by human *Kb* cells, Graham and Van der Eb (*1973*) showed that the observed enhancement of infectivity correlated with the formation, in their experimental conditions, of a coprecipitate of DNA and calcium phosphate, which bound strongly to the cell monolayer.

With this new technique, the same group (Bacchetti and Graham, *1977*) achieved the first reported transformation of mammalian cells from a thymidine kinase-negative to a thymidine kinase-positive phenotype by using purified DNA from herpes simplex virus.

The optimization of the Ca procedure allowed Wigler *et al.* (*1977, 1978, 1979*) to transfer efficiently into cultured mammalian cells defined genes, not only of viral origin but also from complex vertebrate genomes. Their 1979 paper, published in the Proceedings of the National Academy of Sciences (USA), gives a detailed description of the DNA–calcium phosphate coprecipitation technique. This description is generally considered as the reference protocol for cell transformation.

The fine DNA–calcium phosphate coprecipitate is probably taken up by endocytosis through the cell surface, but its helpful role in DNA protection and genomic integration has not been fully elucidated (Loyter *et al., 1982*).

Several other conditions or agents might stimulate the DNA uptake by affecting the cellular membrane or facilitate its insertion by damaging chromosomal DNA and creating favorable sites.

Among the many improvements of the technique that have been proposed for vertebrate cells, I will mention only those that can be applied, more or less successfully, to *Drosophila* material: (1) a brief posttransfection "glycerol boost" (Frost and Williams, *1978*); (2) the elimination of "carrier" DNA (Huttner *et al., 1981*); (3) the stimulation by UV preirradiation of the recipient cells (Nagata *et al., 1984*).

In their pioneering work on *Drosophila* cell transfection, Bourouis and Jarry (1983), Sinclair *et al.* (1983, and later papers), as well as Di Nocera and Dawid (1983) adopted a calcium procedure directly inspired by Wigler's protocol and, ever since, this standard Ca method, with only slight modifications, has been routinely used in most transformation experiments with *Drosophila* cells.

It is noteworthy that the buffered saline solutions for vertebrate cells, used at several stages of Wigler's original technique, have not even been adapted to the specific ionic requirement of cultured *Drosophila* cells, without any apparent damage to the cells or a drop in the method's efficiency (see note in Appendix 9.A).

The most important parameters of the operating procedure seem to be the following:

1. *Drosophila* cells have to be seeded 1 day prior to transfection at a concentration corresponding to a nonconfluent monolayer (for instance, 5×10^6 to 5×10^7 cells per 25-cm^2 culture flask).

2. 5 to 40 μg of plasmid DNA are used per culture flask.

3. One crucial point is the fineness of the DNA–calcium phosphate coprecipitate which forms when the DNA–CaCl$_2$ solution is added to an

equal volume of a buffered and phosphate-containing saline solution. It demands strict adjustment of pH and some gentle stirring during the mixing of both solutions.

4. When poured into culture flasks, the quality and apparent density of the snowy precipitate depend on differences in the compositions of the various culture media for *Drosophila* cells. The precipitate appears to be especially dense in Schneider's medium.

5. The optimal exposure time of the cell monolayer under the fine precipitate varies, according to the authors, from 4 h (the usual time for mammalian cells) to 24 h. It is a compromise between a prolonged DNA penetration and an obvious toxicity to the cells.

6. It is sometimes considered beneficial to complete the treatment with a posttransfection glycerol shock (15% in TBS, 2 min) (Martin and Bourouis, 1988). We have found it unnecessary.

When looking for stable transformants, it is advisable to allow transfected cells to recover in fresh standard culture medium for 24 to 48 h before subjecting them to selective conditions.

For more information, see detailed protocol in Appendix 9.A.

B. DEAE–Dextran-Mediated Transfection

Vaheri and Pagano (*1965*) were the first to observe an increase in the infectivity of poliovirus DNA in the presence of a polycationic substance, diethylaminoethyl–Dextran.

This DEAE–dextran rapidly became the most popular "facilitator" for the introduction of DNA into mammalian cells, at least until the development of the calcium method. Even today, this DEAE–dextran-mediated transfection technique remains an interesting alternative for cell lines that give a mediocre response to other treatments.

The transfection efficiency is proportional to exposure time. However, because of the toxicity of DEAE–dextran, an equilibrium must be found for each cell type between its concentration and the duration of the treatment. It may be advisable to use a 100 μg/ml concentration for several hours rather than 500 μg/ml for 30 min, as recommended in standard procedures.

In *Drosophila* cells, the DEAE–dextran-mediated transfection method was successfully used by Voellmy's group for the transfer of *hsp* genes (Lawson *et al.*, 1984, 1985; Amin *et al.*, 1985; Mestril *et al.*, 1986) and, similarly, by Riddihough and Pelham (1986) and Larocca (1986).

Furthermore, Gallagher *et al.* (1983) showed that DEAE–dextran could sensitize *Drosophila* cells to infection with RNA from black beetle virus (BBV) [see Chapter 11 and Selling and Kaesberg's protocol in Ashburner (1989) "*Drosophila*, A Laboratory Manual"].

The DEAE–dextran protocol for transferring DNA into *Drosophila* cells, as described by Lawson *et al.* (1985), is presented in Appendix 9.B.

C. Polybrene–Dimethyl Sulfoxide Method

Kawai and Nishizawa (*1984*) reported a new procedure for DNA transfection of chick embryo fibroblasts: the polycation Polybrene (1,5-dimethyl-1,5-diazaundecamethylene polymethobromide, molecular weight 6000) seems to mediate the adsorption of negatively charged DNA molecules to the cell surface, probably by the formation of cation bridges (Bond and Wold, *1987*); then, dimethyl sulfoxide (DMSO) might facilitate the uptake of adsorbed DNA by the cells.

Durbin and Fallon (*1985*) found this Polybrene method to be more effective than the calcium phosphate procedure for the introduction of purified DNA into cultured mosquito (*Aedes*) cells. It gave excellent reproducibility and required 10- to 20-fold less DNA. Although the DMSO posttreatment recommended by the Japanese authors in the original technique seemed to have little effect on transfection efficiency in mosquito cells, the authors kept a 10% DMSO shock in their protocol. Fallon (*1986*) defined the main parameters of this technique.

As for *Drosophila* cells, we have been able to establish (Echalier and Fourcade, 1994) that, at least with our experimental system, the Polybrene method gives better results than the calcium procedure with the *Kc* line, even though calcium-mediated transfection of *S2* cells gave an even higher yield (see Fig. 9.3). The efficiency of Polybrene mediation in several cell lines (*Kc-H, S2, S3, Dm1*) was also reported by Walker (1989).

See Appendix 9.C for a detailed protocol.

D. Poly(L-Ornithine) Method

During the 1960s, several other polycations were shown to augment the infectivity of viral nucleic acids for mammalian cells, one of them being the polyamino acid poly(L-ornithine) (Koch and Bishop, *1968*).

In early experiments of transformation (Wyss, 1981; Morganelli and Berger, 1984), *Drosophila* cells were successfully exposed to DNA com-

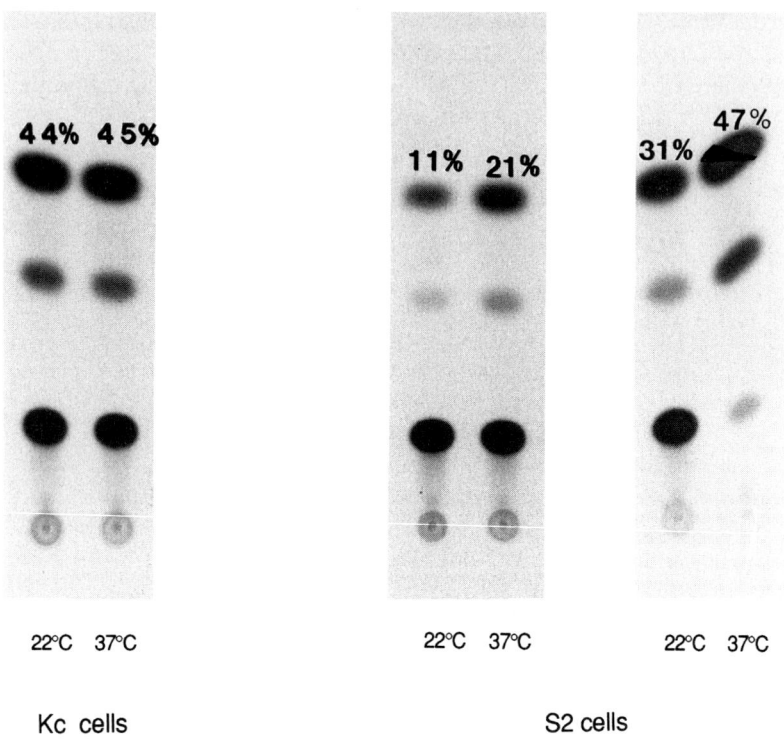

FIGURE 9.3   Comparison of calcium phosphate and Polybrene-mediated techniques of transfection: differences between *Drosophila* cell lines. *Kc* and *S2* cell lines were transfected with a "reporter" plasmid (pD64-CAT) in which the CAT coding sequence is driven by promoter sequences from the retrotransposon *1731* (see Chapter 10, and Appendix 9.C for technical details). From Echalier and Fourcade, 1994.

plexed with polyornithine (about 30 μg/ml final concentration) and results were comparable to those obtained with the DEAE–dextran procedure. This polyornithine technique should be limited to exceptional cases in which a cell line cannot be successfully transfected by any other method.

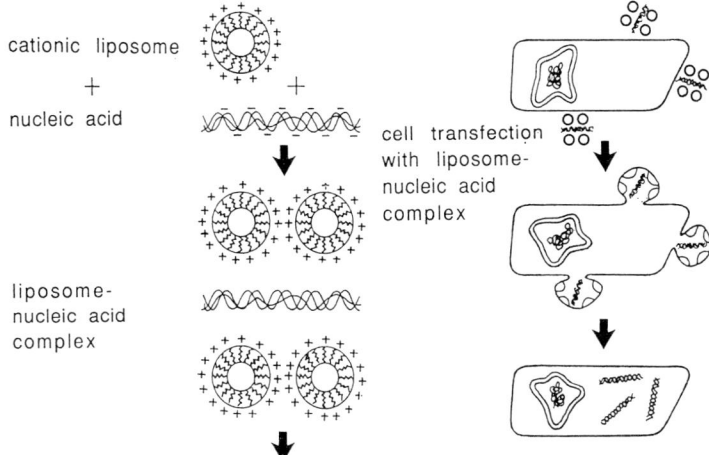

cationic liposome

$+$

nucleic acid

liposome-
nucleic acid
complex

cell transfection
with liposome-
nucleic acid
complex

FIGURE 9.4 Principle of liposome-mediated transfection. Redrawn from the Boehringer-Mannheim catalog by C. Montmory, Paris.

E. Cationic Liposome-Mediated Transfection

A novel cationic lipid molecule, DOTMA (*N*-[1-(2,3-dioleyloxy) propyl]-*N*,*N*,*N*-trimethylammonium chloride), was especially designed and synthesized to form vesicles with positive charges so that they would capture nucleic acids; such liposome/polynucleotide complexes (Fig. 9.4) should thus readily fuse with the plasma membrane and deliver functional DNA or RNA into cultured cells (Felgner *et al.*, *1987*).

A preparation of these DOTMA vesicles is sold commercially under the name of Lipofectin reagent (Gibco-BRL, Life Technologies, Inc., Gaithersburg, MD). General information on its use may be found in Felgner (*1991*) or in *Focus* (published by Life Technologies, Vol. 11, pp. 2 and 4).

This "lipofection"* proved particularly efficient (sometimes 100-fold more than the classical Ca method) in a variety of cell types, including *Drosophila* cell lines (*Kc*, *S2*) (Malone *et al.*, 1989).

Favored by the broad range of its applications (with DNA or RNA; in transient or stable expression), the simplicity of its use, as well as

* In connection with this liposome-mediated transfection procedure see Walter *et al.* (1986) for a method that allows a liposome-mediated delivery of antibodies to *Kc* cells.

its undeniable performances, cationic liposome-mediated transfection is becoming one of the favorite methods for eukaryotic gene delivery. A protocol for *Drosophila* cells, kindly sent to us by Dr. Sondergaard (University of Copenhagen), is given in Appendix 9.D.

Note: Mostly because of its significantly lower cost, but also because the absence of serum does not seem to be so strictly necessary, our laboratory (Fourcade-Peronnet, personal communication, 1995) prefers the DOTAP transfection reagent marketed by Boehringer-Mannheim. See also a recent report by Carlberg (1995).

F. Electroporation

This section is reproduced from Cherbas, L. *et al.* (1994) with the kind permission of the Cherbas.

Electroporation conditions have been optimized for transient expression in *Kc167* cells (S. Bodayla, M. Vaskova, and P. Cherbas, unpublished observations). The same protocol was used to transform stably the same subline. With minor modifications, this protocol can probably be applied to any other *Drosophila* cell line.

Following electroporation of *Kc167* cells with 20 μg DNA per plate, the levels of expression in transient assays and the frequencies of stable transformants are comparable to those obtained by calcium phosphate–DNA coprecipitation in the same line. The levels of expression and the frequency of stable transformants are approximately linear over at least a 10-fold range of input DNA concentrations, up to at least 20 mg per plate.

Judging by their appearance and rate of growth, the cells recover very rapidly from any ill effects of the procedure.

Supercoiled and linear DNAs could be successfully introduced into *Kc167* cells for both transient expression and stable transformation, but the structure of the plasmid DNA in stably transformed cells has not yet been characterized.

See the Cherbas' detailed protocol in Appendix 9.E.

V. SELECTION SYSTEMS

The successful integration of a transduced gene into the genome of individual cells remains a rare event, so that the isolation of these few transformants from huge populations of cultured cells demands the use of an efficient selection system.

In somatic genetics of mammalian cells, wide use has been made, for this purpose, of *auxotrophic* cells lines; that is, mutant lines unable to grow on a defined medium deprived of some specific metabolite but which may be rescued by acquisition of the corresponding "wild" allele. Unfortunately, very few recessive mutants of this kind are available in *Drosophila* cultured cell lines (see Chapter 3, Section III.D). Moreover, the relative complexity of all culture media for *Drosophila* cells does not easily permit the starvation for a single given nutrient.

There is only one brief report (Wyss, 1981) of a *APRT⁻ Kc* subline that could, in a proper TAM-selective medium, be saved by the integration of DNA from normal *Kc* cells (see Chapter 5, Section I.B.1). Incidentally, this work was the earliest known example of successful gene transfer in *Drosophila* cells.

In most cases, alternative methods for the selection of transfected. *Drosophila* cells have been exploited. They consist in transferring dominant selective markers, generally derived from bacterial plasmids, which confer on the cells either a new capacity for utilizing an unusual metabolite (e.g., *gpt* for xanthine) or a high level of resistance to a specific antibiotic (e.g., resistant DHFR, *neo,* or *hygro*). In these last systems, no special medium is needed for the selection, which is an obvious advantage, since the antibiotic is merely added to any standard culture medium.

A quite different selection method was offered by a cell line established from a thermosensitive mutant (*shibire^{ts}*) and which could be rescued at nonpermissive temperature by the allelic wild gene.

A. Methotrexate-Resistant Dihydrofolate Reductase

Folic acid, i.e., vitamin B_6, in its derivative form (tetrahydrofolate), is absolutely necessary for the generation of key precursor molecules in the synthesis of DNA and proteins. Its conversion is catalyzed by the enzyme dihydrofolate reductase (DHFR).

The drug methotrexate (mtx), as an analog of folic acid, binds tightly to the active site of the enzyme and, blocking it, will finally kill the cells.

There are, however, several types of DHFRs, usually encoded by bacterial plasmids, which prove more or less insensitive to the drug (for instance, the R67 plasmid of *E. coli,* identified by Fling and Elwell, *1980*). The transfer of their structural gene, via an expression vector, for instance, pHG of O'Hare *et al.* (*1981*), makes mammalian cells resistant to a high concentration of methotrexate (see also Chapter 5, Section I,B,2).

Bourouis and Jarry (1983) adapted this positive selection system to *Drosophila Kc* cells.

Their pHGCO vector is directly derived from O'Hare *et al.* (*1981*, see above). pHG, in which the SV40 early promoter was replaced by a *copia* promoter (see Section VI.A): the coding region of the DHFR gene from the prokaryotic R67 plasmid was located between a *copia* 5' LTR and the 3' end (i.e., splicing and polyadenylation sites) of the rabbit β-globin gene (Fig. 9.5A).

After transfection of a *Kc* subline and selection in a *D22* medium supplemented with 0.4 μg/ml methotrexate, resistant colonies appeared 3 weeks later and developed as stable transformants. Their growth rate remained the same with a methotrexate concentration of 4 μg/ml, whereas the rare spontaneous resistant cells observed in control cultures (frequency $<10^{-6}$) did not tolerate more than 2 μg/ml.

Southern analysis revealed the presence, in transformants, of a high number of plasmid molecules (1000 to 3000 per cell), very likely integrated into chromosomal DNA as large head-to-tail multimeric structures. Moreover, DHFR transcripts constituted a major band (of the expected size, 750 bp) among cytoplasmic RNAs from resistant cells.

Van der Straten *et al.* (1985, 1986) noticed, however, with the same pHGCO plasmid, but introduced into Schneider *S2* cells, that methotrexate selection (0.1 μg/ml *mtx* in *M3* medium) required a significantly longer period (5 to 6 weeks).

Bourouis and Jarry's vector and selection method have been successfully used in several experiments of cotransformation (Moss' thesis, 1985, and Moss *et al.*, 1985; Henikoff *et al.*, 1986). Moreover, for cointroducing a nonselectable marker, Martin and Jarry (1988) inserted the second transgene, driven by a distinct promoter, into a neutral region (i.e., in the middle of the pBR322 segment) of the selection vector itself.

B. *Ecogpt* and Purine Salvage Pathway via Xanthine

In eukaryotes, hypoxanthine–guanine phosphoribosyl transferase (HGPRT), i.e., one of the two enzymes of the purine nucleotide "salvage" pathway (see Chapter 5), catalyses the condensation of the free bases guanine and hypoxanthine (but not xanthine) with phosphoribosyl pyrophosphate to form guanylic and inosinic acids, respectively. In contrast, the analogous enzyme of *E. coli*, XGPRT (in short, *Ecogpt* or *gpt*) can also use xanthine for the formation of purine nucleotides.

So, when driven by a strong promoter (for instance the SV40 early promoter in the vector pSVgpt of Mulligan and Berg (*1981*), the bacterial *gpt* gene can confer on transfected mammalian cells the ability of using xanthine. It enables them to grow in a selective medium in which methotrexate (an analog of folic acid) blocks the endogenous purine synthesis, and xanthine is the sole available source of purines (see also Chapter 5, Section I.B.3).

Sinclair *et al.* (1983) adapted this "positive" selection system to the transformation of *Drosophila* cells. Their vectors were based on a cloned extrachromosomal *copia* circle (pBB5 of Flavell and Ish-Horowicz, 1981) (see Chapter 10, Section I.F) which contains two LTRs in tandem. The *Ecogpt* coding sequence and accompanying SV40 fragment (splicing and polyadenylation sites), both derived from Mulligan and Berg's plasmid, are positioned downstream from the second *copia* 5' LTR [for instance, at a distance of 850 bp in pCV2gpt (Fig. 9.5B) and only 11 bp in pCV31gpt].

The selective medium *M3X* ("X" for xanthine) is a modification of Shield and Sang *M3* medium for *Drosophila* cells (see Chapter 1). In order to exclude undesirable nucleotides, no yeast extract is used and the normal supplement of 10% fetal bovine serum is extensively extracted with Norit-GSX charcoal. To compensate for this, lactalbumin hydrolyzate and some salts and vitamins have to be added (see details in Appendix 9.F). This semidefined medium is made selective for expression of the *gpt* gene by the presence of 10^{-6} M methotrexate, xanthine (150 to 200 mg/liter), and thymidine (24 mg/liter). Originally, *M3X* contained, in addition, mycophenolic acid and adenine, but both were eliminated from a later version (Sinclair *et al.*, 1985) because *Drosophila* cells cannot salvage hypoxanthine (see Chapter 5).

In their first experiments on transformation of *D1* cells with vectors pCV2gpt or pCV102gpt (i.e., an identical construct, but with the *gpt* sequence oppositely orientated), Sinclair *et al.* (1983) observed the slow growth (doubling time: 3 days), after 4 to 6 weeks under selection, of resistant colonies at a frequency of about 10^{-4}. Apparently the plasmids remained extrachromosomal, with a maximum of two copies per cell, and they were rapidly lost in a nonselective medium. Curiously, the pCV102gpt construct proved as efficient as the correctly built pCV2gpt; but it seems very likely that it had been vastly rearranged: on Southern blots, the corresponding extrachromosomal DNA was indeed 3 kb smaller than the original 8.4-kb vector.

In contrast, in all further experiments of cotransfection carried out in the same laboratory (Burke *et al.*, 1984; Allday *et al.*, 1985; Sinclair *et al.*, 1985), and in which the same *copia–gpt* vectors were used for selection, stable double-transformant lines were obtained from *Drosophila melanogaster D1* cells and *Drosophila hydei DH33* cells (with frequencies

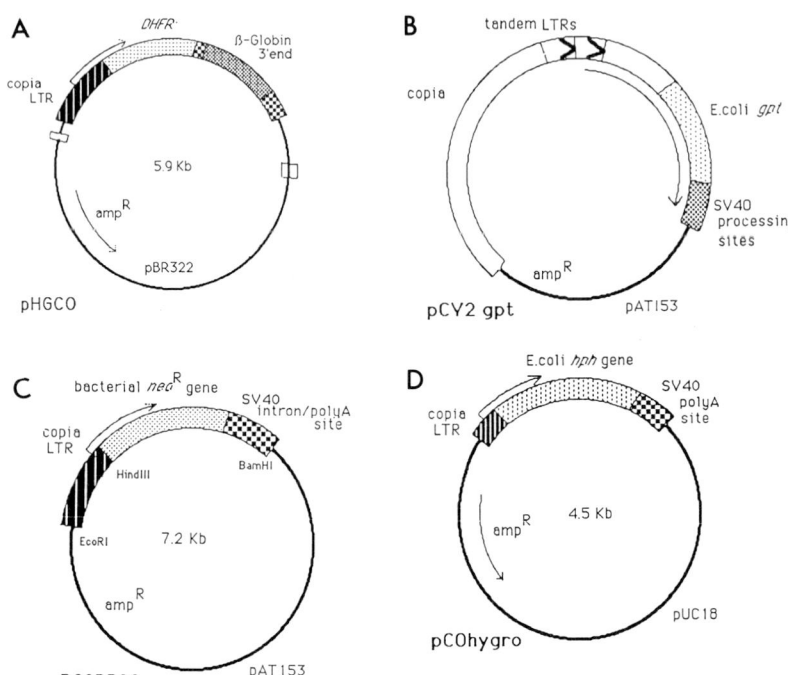

FIGURE 9.5   Selection plasmids (the relative lengths of the components) are not taken into account. (A) Methotrexate selection vector *pHGCO* (Bourouis and Jarry, 1983) is derived from O'Hare *et al.* (*1981*) plasmid pHG by simple substitution of a *copia* promoter: In a pBR322 (ampicillin resistance) the following are aligned: (1) *copia* LTR [353 bp* 5′ fragment from Dunsmuir *et al.* (1980) cloned *copia Dm* 5002] i.e., 5′ flanking 71 bp + whole 5′ LTR(276 bp) + 10 bp of the 5′ internal sequences. The arrow indicates the transcription initiation site; (2) *E. coli* R67 plasmid DHFR coding sequences (350 bp); (3) 3′ end of rabbit β-globin gene (i.e., second intervening sequence and polyadenylation sites). Exons are indicated by larger check patterns. As stigmata from O'Hare's construct, the plasmid keeps (open boxes) a short fragment of promoter and, downstream, a polyadenylation site of SV40 early gene. (B) *pCV2-gpt* (Sinclair *et al.,* *1983*). (1) The two tandem LTRs of a *copia* circle (from pBB5 of Flavell and Ish-Horowicz, *1981*); (2) *Ecogpt* coding sequence (located 850 bp downstream from the second LTR); (3) downstream, the small antigen splice sequences and poly(A) addition site from SV40. (*Note:* (2) and (3) derive from Mulligan and Berg (1980) pSV2-gpt.) (C) G418 selection vector *pcopneo* (Rio and Rubin, 1985) constructed by trimolecular ligation, in *pAT153* vector DNA (cleaved with

---

* The obvious discrepancy between the sum (353 bp) of the *copia* sequences and the number obtained by simple addition of the different ligated segments (357 bp) is due to the fact that the subcloned internal "71-bp" segment used by the authors comprised an extra *Bam*HI site (that is, 4 bp) which does not exist in the genuine *copia* sequence (personal communication from Bourouis to the author, 1995).

of transformation of about 1×10^{-4} and 5×10^{-4} to 10×10^{-4}, respectively). It appeared that multiple copies (from 20 to 200) of each transduced plasmid were integrated into the genome, primarily as head-to-tail concatemers.

Interestedly, this _Ecogpt_ gene can also be used as a "reporter" gene in transient expression systems (Burke _et al._, 1984; Sinclair _et al._, 1986). While control _Drosophila_ cells have little or no detectable HGPRT activity (see Chapter 5.I), the transduced _gpt_ allows the cells to incorporate [³H]guanine rapidly into acid-precipitable counts. This constitutes a valuable measure of its expression and, consequently, of the strength of the promoter used in the construct. It was verified that the levels of _gpt_ transcripts correlate well with the [³H]guanine incorporation assays.

In _Drosophila hydei DH33_ cells, where the expression of pCV2gpt was found to be maximal (see Section III), the incorporation of [³H]guanine is seen first at 48 h posttransfection, continuing up until 140 h, with a peak at 90 h. This transient incorporation of tritiated guanine was sufficient to permit an autoradiographic detection of the individual cells that, in culture populations, took up and expressed pCV2gpt (up to 20% of the population in _Drosophila hydei DH33_ cells) (Burke _et al._, 1984).

C. _neo_ Gene and Resistance to Antibiotic G418

G418 is an aminoglycoside antibiotic (Davies and Jiminez, _1980_) which, unlike related compounds (e.g., kanamycin, neomycin), can also block protein synthesis in eukaryotic cells.

Bacterial resistance to this group of antibiotics relies on the presence of transposons harboring genes coding for specific phosphotransferases: for instance, the gene _neo_ (1.5 kb) borne by the left end of _E. coli_ Tn5 (Jorgensen _et al._, _1979_), codes for APH(3')-II, i.e., aminoglycoside phosphotransferase type II.

EcoRI–_Bam_HI), of (1) (_Hind_III–_Apa_I but converted to _Eco_RI) _copia_ LTR-containing 1.6-kb DNA fragment (from cDM5002); (2) (_Hind_III–_Bam_HI) 2.5-kb fragment (from Southern and Berg pSV-neo) containing the neomycin resistance gene followed by SV40 messenger processing sequences. (D) Hygromycin B selection vector pCOhygro (Van der Straten _et al._, 1988, 1989). In _pUCI8_ (1) 5′ LTR of _copia_ (353-bp _Bam_HI fragment from Bourouis and Jarry, 1983, pHGCO); (2) _E. coli_ plasmid-borne hygromycin B phosphotransferase gene (_hph_ of Gritz and Davies, _1983_); 1026 bp (3) SV40 early poly(A) signal. [_Note:_ Another plasmid, pCOdhygro, differs by single removal of 25 bp upstream and 9 bp downstream of the LTR, including an ATG triplet (see text)].

Transduction of a hybrid vector carrying this bacterial gene driven by an efficient promoter can make mammalian cells highly resistant to the drug G418 (Colbère-Garapin *et al., 1981;* Southern and Berg, *1982).* Moreover, the neomycin transferase activity expressed by the cells can be assayed according to the method of Reiss *et al. (1984).*

G418 is also toxic for *Drosophila* cells (first reported by C. S. Parker, in a personal communication to Southern and Berg), although at much higher concentrations. So, the dominant selectable marker *neo* could be adapted to this material.

By ligating a *copia* LTR to the structural *neo* gene within the pAT153 vector, Rio and Rubin (1985) constructed a pcopneo plasmid (Fig. 9.5C) (derived from Southern and Berg's original plasmid and still followed by SV40 and mRNA processing sites).

Among transfected Schneider *S2* cells grown in the selective medium (G418 concentration as high as 1 mg/ml), resistant cells developed within 7 to 12 days with an estimated transformation frequency of about 10^{-4} to 10^{-5}.

Plasmid sequences appeared to be integrated in a single long array of 10–20 copies per cell, and they remained stable for many generations, even in the absence of selection.

With a very similar *copia–neo* vector and in the same selection medium (1 mg G418/ml), Van der Straten *et al.* (1988, 1989) also observed the growth of resistant populations of *S2* cells within *10* days.

Miyake *et al.* (1988), using a selection vector of the same type (i.e., *copia–neo*) in cotransformation, succeeded in "contaminating" the *copia*-free genome of a *Drosophila hydei* cell line with a *copia* element from *Drosophila simulans* and could thus demonstrate the causal relation with the production of virus-like particles (see Chapter 10, Section I.E).

Quite independently, Maisonhaute and Echalier (1986) used a derivative of the vector built by Colbère-Garapin *et al.* (1981, see above) in which the APH(3′)-II gene is under the control of the promoter region of the *herpes simplex I* thymidine kinase gene.

In a clonal *Kc* subline, the selective concentration of G418 had to be raised to 2 mg/ml. Transductants were obtained (with an estimated frequency of 10^{-5}) and could be grown in the same high concentration of the antibiotic for 18 months, with, nevertheless, a slightly slower multiplication rate.

Such a high level of resistance could be correlated with the characterization in transformed lines of specific neomycin phosphotransferase activity.

An APH(3′) coding sequence, but with hardly any surrounding plasmid sequence, was detected in the transformant genomic DNA (1 to 23 copies per cell). However, because of the profound rearrangements observed, one cannot be sure that the *neo* gene remained under the control of the heterologous HSVI TK promoter.

A practical limitation of this selection system, in all other respects very efficient, is the particularly high level of spontaneous resistance displayed by *Drosophila* cell lines. Selective concentrations of 1 to 2 mg/ml (or even more) G418 are needed to kill control cells (whereas only 100 to 400 μg/ml are sufficient for mammalian cells), and lethal concentrations have to be carefully defined for each *Drosophila* cell line. Furthermore, the drug is rather expensive.

Note: As was later suggested by Lycett (*1990*), this problem of high background resistance to G418 could perhaps be partly solved by using the Tn*903* (601) *neo* gene which has been shown to confer significantly higher levels of resistance than Tn*5* (Lang-Hinrichs *et al.*, *1989*).

D. *hph* Gene and Resistance to Hygromycin B

Hygromycin B (hygro, for short) is an aminocyclitol antibiotic produced by *Streptomyces hygroscopicus* (Pettinger *et al.*, *1953*). It inhibits protein synthesis in both prokaryotes and eukaryotes by interfering with translocation and causing mistranslation.

Hygromycin resistance in bacteria is due to plasmids carrying genes that code for various hygro-inactivating enzymes. For instance, the *hph* gene (1026 bp) cloned from *E. coli* pJR225 (Gritz and Davies, *1983*) corrresponds to a hygro-phosphotransferase.

The transfer of such bacterial hygromycin-resistance genes can confer on transformed mammalian cells the capacity for growth in the presence of 200 μg to 1 mg/ml hygromycin (Blochlinger and Diggelmann, *1984*).

Van der Straten *et al.* (1988, 1989; Johansen *et al.*, 1989) exploited this dominant selectable marker for stable cotransformation of *Drosophila* cells. In their selection vectors (pCOhygro and pCOdhygro), the hygromycin resistance gene *hph* is driven by a *copia* LTR (Fig. 9.5D).

The selection of resistant S2 cells in a medium with 300 μg/ml hygromycin B (Boehringer) was obtained within 2 to 3 weeks. Such transformation was stable, even in the absence of selective pressure, and *hygro* coding sequences were shown to be integrated as head-to-tail arrays at single or a few sites of the genome (40 to 120 copies per cell).

The authors concluded, from systematic comparisons, that (1) the hygromycin selection system is 2-fold more rapid than selection with methotrexate: cell death is readily apparent after 3 days in S2 control cultures treated with 300 to 600 μg/ml hygro; (2) it does not present the inconvenience of an important background of spontaneous resistance such as one observes with G418.

E. α-Amanitin Resistance

RNA polymerase II (or B) is a key enzyme in cell metabolism since it transcribes all usual messenger RNAs. It can be specifically inhibited by α-amanitin, a toxin from the mushroom *Amanita phalloides.*

Several *Drosophila* mutations of the enzyme, among which is the allele *C4*, were selected for their resistance to the drug (see also Chapter 5, Section V.A).

Jokerst *et al.* (*1989*), who cloned the gene encoding the large subunit of the enzyme, constructed a transformation plasmid (called pPC4) in which the coding sequence from the *C4* mutant gene was inserted into Rubin and Spradling's "Carnegie 4" vector. It could confer on transgenic flies a high level of resistance to α-amanitin. This selection system has today become one of the most extensively used systems for introducing a nonselectable gene into cultured cells by cotransfection and recovering the double-transformants. See Appendix 9.G for the protocol for α-amanitin selection.

F. Rescue of Heat-Sensitive *shibire*[ts] Mutant Cells

The principle of this selection system is completely different from that of preceding ones: it relies on the rescue, by wild-type DNA transfection, of heat-sensitive cell lines which would have died at elevated temperatures. Similar systems are available in mammalian cells (Kai *et al.*, *1983*).

Several recessive mutants of the *shibire* (*shi*[ts]) locus of D. *melanogaster* exhibit, in addition to a temperature-sensitive paralysis of adults and larvae, an embryonic cell lethality at nonpermissive temperatures (e.g., 29°C for *shi*[ts]) (Swanson and Poodry, *1981*) (see also Chapter 6, Section III.E).

A cell line was established from homozygous *shi*[ts] embryos (Sang, 1981) (see Chapter 3, Section III.D). It grows normally at 22°C but is killed after prolonged exposure to a temperature of 29°C. Sang *et al.* (1984) could rescue these temperature-sensitive *shi*[ts] cells by calcium phosphate-mediated transfection with wild-type DNA. Five days post-transfection, the cultures were shifted to 29°C and cells began to die after 1 week, except for a few transformed colonies which developed within additional 3 to 4 weeks (with a frequency of $\sim$120 per 2×10^6 transfected cells).

The authors showed that this rescue was sequence-dependent and that transformants remained temperature-resistant when propagated for 6 weeks without selective pressure (that is, at 22°C), which suggests a stable integration in the genome.

This selection system could, moreover, be efficiently used for cointroducing other recombinant plasmids.

Independently, Simcox *et al.* (1985) confirmed all these results by successfully cotransfecting an *hsp* gene into *shi^{ts}* cells. Moreover, they established *in vitro* some 30 distinct cell lines from different *Drosophila* heat-sensitive mutants, (see Chapter 3, Section III,D) which would all die within a few weeks at 30°C and, therefore, could be used in similar selection systems.

VI. EFFICIENT PROMOTERS

The correct expression of a eukaryotic gene requires the presence upstream from the coding sequence of a segment several hundred nucleotides long, called, *sensu lato,* its promoter. It consists of (1) a short region, close to the transcription start site and containing usually a typical TATA box, to which the RNA polymerase and its initiation complex will bind, and (2) more or less distally, one or a few regulatory regions displaying characteristic nucleotide motifs for the specific binding of trans-acting factors (which will interfere, positively or negatively, with the initiation complex assisting the polymerase).

Any transfected gene must, therefore, possess its own 5' control regions or, in order to obtain higher expression levels, a stronger foreign promoter has to be substituted for its upstream sequences. Furthermore, to ensure a correct termination of the messenger, the splicing and polyadenylation sites from another well-tested gene are ligated to the 3' extremity of the construct. Then, the chimeric gene has to be inserted into a bacterial plasmid (pBR322 or some derivative) for multiplication in *E. coli.* After DNA purification, such vector plasmids can be directly used without linearization for transfecting eukaryotic cells.

The utilization in *Drosophila* cells of plasmid vectors initially devised for mammalian cells rapidly showed that promoters that are very effective in vertebrate cells (for instance, viral promoters such as the SV40 early gene promoter or Rous sarcoma virus LTR) function only poorly in

Drosophila (Bourouis and Jarry, 1983; Sinclair *et al.*, 1983; etc.). The reverse, i.e., the weak efficiency of *Drosophila* gene promoters in vertebrate cells, proved to be true (McDonald *et al.*, 1988).

Investigators, therefore, have turned almost exclusively toward homologous *Drosophila* promoters. Nevertheless, in a search for putative exogenous promoters, Allday *et al.* (1985) observed that an Epstein-Barr virus nuclear antigen, under the control of its own regulatory sequences, was expressed as efficiently in *Drosophila* cells as in EBV-transformed mammalian cells; similarly, the human cytomegalovirus immediate early gene promoter/enhancer region proved a very strong promoter in *Drosophila* Kc cells (Sinclair, 1987).

The first candidates for the role of efficient homologous promoter in *Drosophila* cells were the LTR of the retrotransposon *copia* and the 5' region of *hsp70* genes, since transcripts of these two genomic sequences are particularly abundant in cultured *Drosophila* cell lines. Later on, a series of other promoters, from *Adh*, actin 5C, metallothionein genes, etc., were explored and the last two are, by the mid-1990s, among the most widely used promoters in chimeric constructs for transfecting *Drosophila* cells.

"Conditional" promoters, that is, regulatory regions of genes, the expression of which can be easily turned on by either a brief rise in temperature (such as *hsp* genes) or the simple addition of divalent cations (for instance, copper or cadmium for metallothionein genes), have a special advantage: the constitutive expression of any gene driven by such conditional promoters remains very low, or even nil, and thus does not interfere with the cell growth rate, whereas a rapid and high level of synthesis of the gene product can be easily triggered at any time.

These regulated systems, as has been clearly demonstrated with a metallothionein promoter in *Drosophila* cells (Bunch *et al.*, 1988; Johansen *et al.*, 1989), may be particularly useful for (1) various experimental protocols and for (2) potential mass production of some marketable protein by stable transformant cell lines (possibly equipped with a very high copy number of the transduced gene) (see Section IX,2 of this chapter).

A. Long Terminal Repeats of *copia* and Other Retrotransposons

The *copia* element is the prototype of the main group of mobile genetic elements present in the *D. melanogaster* genome (see Chapter 10). They are named retrotransposons because of their striking resemblance, in their structure as well as in their intracellular cycle (with its presumed retrotranscriptional step), to integrated forms of vertebrate retroviruses (=proviruses). They are framed at both extremities by two direct repetitive sequences of a few hundred nucleotides (long terminal repeats or LTRs). Let us recall that transcription of the element starts in its left (5′) LTR and ends within the right (3′) LTR. As in retroviruses, most of the transcriptional regulatory regions have been localized within LTR sequences or in close proximity (Fig. 9.6A).

Because of the particular abundance of *copia* transcripts in most cultured *D. melanogaster* cell lines (they may account for as much as 2 to 5% of total poly(A)$^+$ RNAs), the *copia* 5′ LTR was considered to be an efficient promoter and widely used in chimeric constructs.

It would seem worthwhile to pay some attention to the exact size of the *copia* 5′ LTR fragments that have been utilized by the different laboratories.

In the plasmid selection vector pHGCO (Fig. 9.5A) devised by Bourouis and Jarry (1983) and subsequently used in many works (Moss, 1985; Moss *et al.*, 1985; Henikoff *et al.*, 1986; Martin and Jarry, 1988; Van der Straaten *et al.*, 1988, 1989), the *copia* promoter which drives the bacterial DHFR sequence consists of a 353-bp sequence (*Bam*HI–*Sau*96I fragment from the *copia* clone cDm5002 of Dunsmuir *et al.*, 1980). It comprises 71 bp of left flanking sequence plus the complete 276 bp 5′ LTR plus 10 bp of 5′ internal sequences (see Fig. 9.5A).

Due to the presence of an ATG triplet in the short 10-bp downstream sequence of this *copia* promoter fragment and because their own laboratory had previously observed that such a methionine codon in the untranslated leader of any mRNA can reduce its level of translation, Van der Straaten *et al.* (1988, 1989) intentionally deleted in some constructions (for instance, pCOdhygro, Fig. 9.4D) 9 bp from the 3′ extremity of this promoter sequence, but without any obvious improvement.

Di Nocera and Dawid (1983), in order to construct their CAT expression vector (*copia–CAT1*), used a larger fragment from the same cloned *copia* (i.e., 1.6-kb *Eco*RI–*Apa*I segment from cDM5002); nevertheless, its 3′ end contains no more than 7 bp downstream from the LTR. Identical sequences were also utilized by Rio and Rubin (1985) (pcopneo, Fig. 9.5C), Thummel *et al.* (1986) (pC4copcat), and McDonald *et al.* (1989).

There are no details for the 3′ end of the *copia* fragment used by Kurata and Marunouchi (1985) in their *pDSV–ARS* constructs. It is merely stated that it is a 650-bp sequence derived from another *copia* clone of Dunsmuir *et al.* (1980).

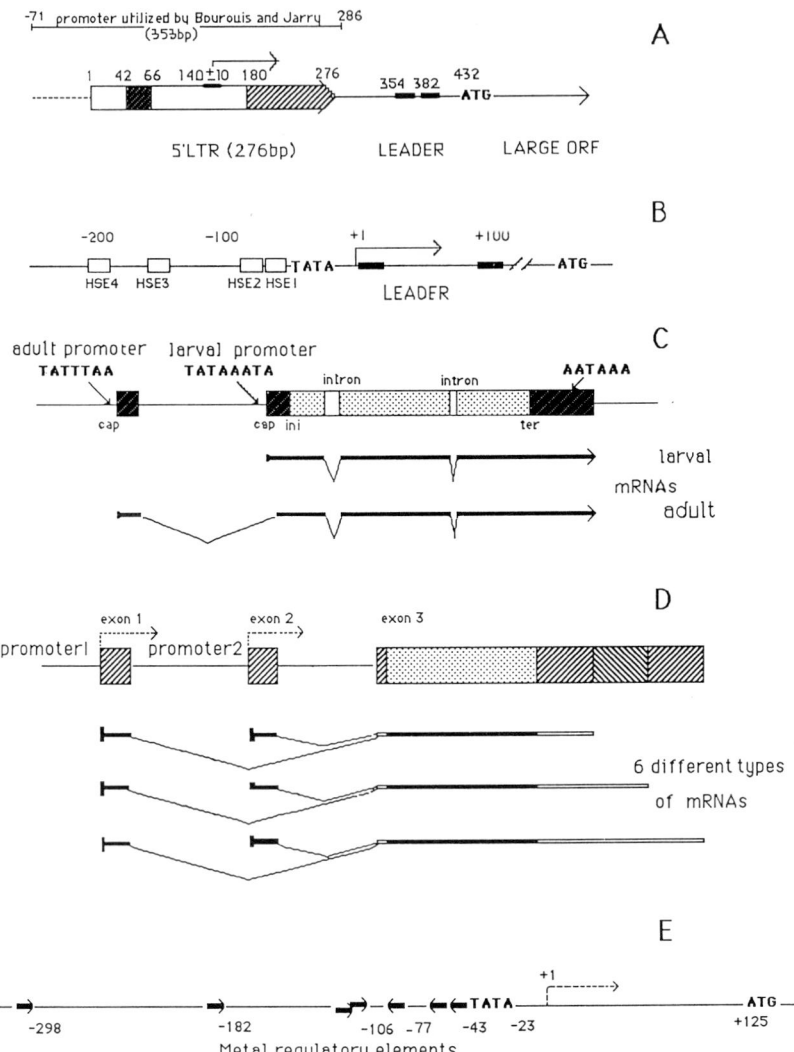

FIGURE 9.6   Structure of the promoters of several *Drosophila* genes. (A) Control regions of the 5' end of *copia* retrotransposon (from Mount and Rubin, *1985;* Sneddon and Flavell, *1989*): in the LTR and with respect to the multiple transcription start sites: (1) upstream sequence [(42 to 66 up to 112 (?)], and (2) downstream sequence (180 to 283); in the leader, two palindromic repeats resembling a motif of SV40 enhancer. (B) Regulatory regions of *Hsp70* gene (from Karch *et al.*, *1981;* McGarry and Lindquist, 1985; Bienz and Pelham, *1987*). HSE 1 to 4: Heat-shock consensus sequences (empty boxes) upstream of the TATA box. In the leader (242 bp), small solid boxes correspond to consensus sequences

Built in a different manner, the *copia–gpt* vectors of Sinclair *et al.* (1983) were based on a cloned extrachromosomal circular *copia*, with two LTRs one behind the other. The *E. coli gpt* coding sequence was inserted in the correct alignment downstream from the second 5' LTR; for instance, 820 bp downstream in pCV2gpt (Sinclair *et al.*, 1983; Burke *et al.*, 1985; Alldays *et al.*, 1985; Sinclair *et al.*, 1985) (Fig. 9.5B). They also used, apparently with the same success, another construct named pCV31gpt, with a shorter (11-bp) *copia* sequence 3' to the tandem LTRs (Burke *et al.*, 1984).

For a better delimitation of an optimal *copia* promoter, it is important to take into account the results of two "functional dissections" of *copia* LTRs.

When *Drosophila hydei DH33* cells are transiently transfected with the selection plasmid pCV2gpt (Fig. 9.5B and Section V.B), they become able to incorporate [^{3}H]guanine, which constitutes a direct assay of the promoter strength. By systematic deletions from the tandem LTRs and flanking sequences, Sinclair *et al.* (1986) concluded that (1) one *copia* LTR is sufficient for correct expression and that (2) the regulatory sequences reside entirely within the LTR and essentially downstream from the transcription start site. Removing this 3' end of the LTR (i.e., leaving only nucleotides 1–203 from the 276-bp LTR) reduces 10-fold the transient expression of *gpt*.

An analogous functional analysis, but carried out in *Drosophila melanogaster* cells transfected with a *copia*–CAT vector, led to contradictory results (Sneddon and Flavell, 1989). (1) There is an additional control region downstream from the 3' end of the LTR; it acts as an "enhancer,"

---

presumably involved in the translational control at high temperature. (C) *Adh* gene and its transcripts (from Benyajati *et al., 1983*). TATTTAA and TATAAATA represent the adult and larval putative promoters regions, respectively; *cap* indicates the RNA initiation sites; *ini* and *ter* indicate the translational initiation and termination codons; AATAAA refers to the putative polyadenylation signal. The larval-type primary transcript contains two introns, and the adult-type primary transcript contains an additional intron in the 5' untranslated region. (D) *Actin 5C* gene and its six possible transcripts (from Bond and Davidson, 1986). The dotted box represents protein-coding sequence, while the dark striped ones represent transcribed but untranslated sequences. Observe the two alternate leader exons and the three possible polyadenylation sites. (E) Regulatory regions of metallothionein gene *Mtn* (from Maroni *et al., 1986*). The seven small thick arrows mark 12-mers in which at least nine bases correspond to the consensus sequence identified as a metal regulatory element in mammals.

since it could be translocated several hundred nucleotides upstream from the LTR and in both orientations and still maintain its activity. It was not indicated whether this effect is correlated with the presence (already noted by Mount and Rubin, 1985) in this untranslated "leader" region of palindromic repeats which partially resemble a DNA-binding motif (specific of AP5 transcriptional factor) observed in the mammalian SV40 viral enhancer. In addition, it is not possible to establish from these data whether this *copia* sequence acts indeed at the transcriptional level or may be required for the stability of messengers. (2) As for the regulatory regions within the LTR itself, the authors observed striking differences when the same constructs were transduced into different *Drosophila* cell lines. In transfected *Drosophila melanogaster Kc* cells, two regions on either side of the major transcriptional start sites of *copia* 5' LTR (one upstream, from nucleotides 42 to 66, and another downstream, from 180 to 203) are required for a high level of *copia*–CAT expression. On the other hand, in transfected *D. hydei DH33* (in the genome of which *copia* elements do not normally exist), only the upstream region of the LTR is needed (for a possible interpretation, see Section III,5).

These contrasting results suggest that a *copia* 5' fragment containing not only the whole 5' LTR but also the adjacent leader sequence (or, at least, its enhancer-like motifs) (see Fig. 9.6.A) might prove a stronger promoter for chimeric constructs, at least in *D. melanogaster* host cells.

Note: If *copia* remains the prototype for *Drosophila* retrotransposons, and its LTR has been predominantly used as such, it should, however, not be ignored that the regulation of other retrotransposon families is being extensively studied. This might lead to a better control of the use of different putative promoters. For instance, current research carried out by our own laboratory (Ziarcyk *et al.*, 1989) on the regulatory regions of a *copia*-like element named *1731* and on its cellular transcriptional activators might result in the construction, by tinkering with its LTR, of a kind of "super-promoter" for *Drosophila* cells (see Chapter 10).

B. *Hsp* Promoters

A simple rise in temperature (from 25 to 37°C) has rapid and characteristic effecs on all *Drosophila* tissues, including cultured cell lines. Their normal program ceases, and the neosynthesis (or, at least, a dramatically increased synthesis) of a dozen specific proteins (i.e., heat-shock proteins, or HSPs) is immediately initiated; their mRNAs appear within a few minutes. The most abundant of those polypeptides has a molecular weight of 70,000 (HSP70) (See Chapter 7).

Extensive studies, requiring the cooperation of various approaches among which cell transfection systems have provided crucial information, have been devoted to an analysis of the 5' cis-regulatory sequences of *Drosophila hsp 70* genes and some other smaller *hsps*. They are, today, among the best-known eukaryotic promoters.

Transient expression and stable cultured cell transformation were used in such analyses. The general value of these methods was demonstrated by the fact that the patterns of expression of transduced whole *hsp* genes were usually found to be very similar in constitutive as well as in inducing conditions to those of endogenous genes (Morganelli and Berger, 1984; Sinclair *et al.*, 1985; McGarry and Lindquist, 1985; see, more especially, the comparative study by Larocca, 1986). Generally these *hsp* promoters presented a very low basal activity, which contrasted with a high efficiency after heat shock (and, for some of them, hormonal) induction.

As for *Drosophila hsp70* genes, Lawson *et al.* (1984, 1985) demonstrated in transfected *Drosophila* cells that the shortest promoter segment remaining fully active during heat-shock induction must retain some 90 bp upstream from the transcription start site. Their results confirmed the functional importance of the consensus heat-shock motif (HSE) previously identified by Pelham (*1982*) (see Chapter 7, Section IX,A).

Moreover, in a similar approach, McGarry and Lindquist (1985) established the crucial role of specific sequences present in the 5' untranslated leader region in the preferential translation of *hsp70* messengers at high temperature, which is one of the characteristics of *hsp* genes. Therefore, in any chimeric construction, the *hsp70* promoter segment should keep this leader sequence or, at least, its 5' half (see Fig. 9.6B).

Such intimate knowledge of the indispensable regulatory sequences of the 5' region of *hsp70* and of the optimal conditions for its induction, in addition to the fact that it has proved to be, at high temperature, one of the strongest promoters tested in *Drosophila* cells, has made the upstream regions of *hsp70* a model "conditional" promoter. It has been widely used in chimeric constructs.

Di Nocera and Dawid (1983) were the first to show that when fused to the CAT coding sequence in the proper orientation an *hsp70* promoter is able to drive the synthesis of functional CAT enzyme in transiently transfected *Drosophila* (*Drosophila melanogaster* and also *Drosophila immigrans*) cells. The enzymatic activity level rises substantially after temperature shift (20 to 90 min at 37°C), up to 30 times above the basal level, while the corresponding mRNAs become 60-fold more abundant. Note that in their hsp–cat1 plasmid (which was directly derived from Gorman's promoterless pSVOcat) the 1.2-kb *hsp* promoter comprised 1.1 kb of 5′ upstream sequences plus 65 bp downstream from the cap site.

Most people, in order to be sure of keeping the complete *hsp70* regulatory regions, prefer to use a large 5′ flanking region plus the leader sequence (Burke *et al.*, 1984; Rio and Rubin, 1985; Krasnow *et al.*, 1989). A model for these constructions is the "shuttle" vector pC4hspcat designed by Thummel *et al.* (1988) (see Section VIII); it contains a 1.6-kb segment [*Sal*I–*Pst*I fragment from one of the *hsp70* recombinant plasmids studied by Moran et al. (*1979*)], which includes 90 bp downstream from the start point of transcription.

It is worth pointing out the special use in various chimeric genes of what is called a "minimal" *hsp70* promoter. Reduced to 50 bp upstream from the transcription initiation site, it comprises only the TATA box, for potential anchoring of RNA polymerase and its complex, but without any heat-shock response element; so, this minimal promoter has to be completed by the insertion, in an upstream position, of any given heterogeneous regulatory sequence [for instance, the homeodomain binding sites in some "responder plasmids" of Jaynes and O'Farrell (1988), see Section II.6].

C. *Adh* Promoters

The genetics of *alcohol dehydrogenase* (*ADH;* EC 1.1.1.1) of D. *melanogaster* have been extensively studied. It is generally thought that its high activity level in the fly participates in the latter's tolerance of alcohol and may be important in an animal species which feeds and breeds in fermenting fruits.

The expression of this enzyme is very specific in stage and tissue distribution, but, although its messenger RNAs differ in larvae and imagos, *ADH* is coded by a single structural gene (cytogenetically located at band 35B2-3 in polytene chromosomes).

It was established (Benyajati *et al.*, *1983*) that the alternative activity of two distinct promoters, a proximal larval promoter and a more distal adult one, leads indeed to the synthesis of two *Adh* transcripts with different 5′ ends and different pathways for RNA splicing (Fig. 9.6C).[*]

---

[*] Heberlein *et al.* (1985) described accurate *in vitro* transcription from both distal and proximal promoters of *Adh* carried out by a nuclear extract from Kc cells.

Because of its high level of expression in many tissues, the *Adh* 5' regulating region may be considered a potentially efficient promoter in transfection systems. It is interesting to note, however, that the endogenous *Adh* gene remains inactive in most *Drosophila* cultured cell lines (Debec, 1974).

Benyajati and Dray (1984), after transient transfection into Schneider S2 cells of a plasmid containing the complete *ADH* coding sequence with its two promoters, could detect both specific mRNAs (about 0.01 to 0.02% of total cellular RNA, with two-thirds being of the adult type and one-third of the larval type). In close correlation with the level of these transcripts, the ADH protein could be immunologically characterized, and it represented 0.02% of total cytoplasmic proteins. It was observed that when the plasmid lacked the distal promoter only larval type RNAs were recovered, although their yield was not higher. Therefore, it is advisable to keep the two promoters in any hybrid construction that makes use of the *Adh* promoter.

As an example, Tijan's group used successfully such *ADH* tandem promoters (Courey and Tijan, 1988; Santoro *et al.*, 1988) to overexpress and thereby analyze, in this *Drosophila* heterologous cell system, the functional domains of two mammalian trans-activating factors, SP1 and CTF/NF-1 (which are lacking in *Drosophila*; see Chapter 5, Section V,C3).

D. *Actin 5C* Exon 1 Promoter

As in vertebrates, the various types of actin in *Drosophila melanogaster* are encoded by a dispersed multigene family, and their respective expressions are stage and tissue specific (Fyrberg *et al.*, 1980, 1983).

The *actin 5C* gene (named after its precise location on the polytene X chromosome) corresponds to a ubiquitous cytoskeletal form present in all nonmuscle cells. It is, therefore, normally expressed in *Drosophila* cultured cells and, moreover, was found to be positively modulated by ecdysterone (Berger *et al.*, 1978, 1979, 1981; Couderc *et al.*, 1980, 1982, 1983; see Chapter 8, Section II.D.1).

Note: With an *in vitro* transcription system derived from *Kc* nuclei extracts, Parker and Topol (1984) had previously demonstrated that sequences upstream from exon 1 are indeed a strong promoter.

Bond and Davidson (1986) established that the *actin 5C* gene has two alternative leader exons from which the transcription initiation can occur and three different polyadenylation sites (Fig. 9.6C). Using a CAT reporter system in transfected *Kc* cells, Bond-Matthews and Davidson (1988) confirmed that both leader exons are preceded by separate functional promoter regions and pointed out the presence of a "consensus" sequence with dyad symmetry [CC-(A-rich)$_6$-GG] which had also been observed in mammalian and chicken actin genes.

The *actin 5C* exon 1 promoter is currently used in the construction of many expression plasmid vectors.

The model is the "shuttle" vector pUChsneo-act (see Section VIII and Fig. 9.6B) designed by Thummel and Lipchitz (1988). Between the *act5C* exon1–promoter (2650 bp upstream plus 88 bp downstream from the transcription start site) and another *actin 5C* fragment including the three polyadenylation sites, there is a unique *Bam*HI site for the insertion of any given foreign ORF; this core was inserted into the pUChsneo P element vector, which makes either selection of stably transformed cells (by G418) or P element-mediated transformation of flies possible.

A simplified version, pP$_{ac}$, in which the same block (that is, the *actin 5C* promoter + *Bam*HI site + *act 5C* terminator sequences) was transferred into the simpler plasmid pUC18, has been widely used for building efficient "producer" vectors in several experimental systems involving transfected *Drosophila* cells (Jaynes and O'Farrell, 1988; Courey and Tijan, 1988; Driever and Nüsslein-Volhard, 1989; Winslow *et al.*, 1989; Krasnow *et al.*, 1989).

The hormonal regulation of this *actin 5C* promoter, the gene being induced by 20-hydroxyecdysone (see Chapter 8), was compared in conditions of transient expression or stable cell transformation (Martin and Jarry, 1988) (see Section IX.1).

E. Metallothionein Promoter(s)

Metallothioneins (MTs) are ubiquitous low-molecular-weight cysteine-rich proteins which bind heavy metals (i.e., Cu, Zn, Cd) and seem to protect organisms from their toxicity. Moreover, they are probably involved in zinc ion homeostasis.

In contrast with the close similarity between the multiple copies of MT genes in mammalian cells, the metallothionein system appears to be dual in *Drosophila melanogaster*. There are two different types of metallothioneins encoded by two distantly related genes. (1) *Mtn* (Maroni *et al., 1986*) whose cytogenetic locus is at 85E 10-15 in chromosome 3R and which might be preferentially copper-inducible (like *Neurospora* and *Saccharomyces* MTs). (2) *Mto* (Mokdad *et al., 1986*; see also Chapter

4, Section III.B.2) is localized at 92E in 3L and seems to be mostly cadmium inducible (like mammalian MTs).

In experimental transformation of animals or cultured cells, MT gene regulatory 5' regions have been shown to be very efficient promoters in chimeric gene constructs, and their main appeal is their easy inducibility by heavy metals ("conditional" promoter). This is also true for the 5' flanking sequences of *Drosophila* metallothionein genes. Otto *et al.* (1986) observed in a heterologous system (baby hamster kidney cells) that the sequence between -130 to -6 bp upstream from the transcription start site of *Drosophila Mtn* gene is sufficient to confer metal regulated expression on a transduced *Mtn*–thymidine kinase hybrid gene. It may be noted that such a segment comprises five out of the seven 12-bp motifs that are present in the 5' region of the *Mtn* gene and are very similar to the consensus mammalian metal regulatory elements (see Fig. 9.6E).

Most people, however, prefer to use a larger 5' segment of *Mtn*, which includes the whole set of putative metal regulatory motifs and leader sequences (*Eco*RI–*Stu*I fragment from Maroni's plasmid Dm13, i.e., -370 bp to $+54$bp, with respect to the transcription start site).

In a methodological analysis of such a system, Bunch *et al.* (1988) concluded that copper is a less toxic inducer than cadmium. At a concentration of 0.7 mM copper sulfate (which has no effect on the doubling rate of control S2 cells and does not induce a heat-shock response), transformed cells showed a dramatic increase in the levels of chimeric gene transcripts (30-fold induction from 24 to 48 h after treatment).

Likewise, when an *Mtn* promoter + *E. coli* GalK construction was stably introduced into *Drosophila* cells, Johansen *et al.* (1989) observed a strong induction with heavy metals (a 50-fold increase of both RNA and protein); but here the maximum induction was obtained with cadmium (10 μM CdCl$_2$), whereas a 20-fold-higher concentration of copper was required (200 μM CuSO$_4$). Such considerable GalK synthesis could be maintained for 14 days of continuous induction. In comparison, in quite similar constructs, the *Mtn* promoter was found to be 2- to 5-fold more efficient than *copia* 5' LTR (Van der Straten *et al.*, 1988).

Because the basal expression remained very low in the absence of inducer, even when several hundred copies of the transgene were stably integrated within the cell genome, and because, in contrast, high expression can be easily and rapidly induced, one must emphasize the considerable advantage of such an *Mtn* "conditional" promoter for the possible industrial production of foreign proteins by mass cultures of *Drosophila* cells (Johansen *et al.*, 1989). Thus, no constitutive expression of a potentially toxic product impedes cell proliferation until its overproduction is triggered (see Section IX.2).

Note: the promoter strength of the control regions of the other *Drosophila* metallothionein gene (*Mto*) is being investigated in various constructs (Wegnez *et al.*, personal communication, 1995).

F. BBV Late Promoter

See Chapter 11 for a discussion of the Black Beetle virus.

VII. REPORTER GENES

It is imperative that a clear distinction be made between the products of transduced genes, both transcripts and proteins, and those of their endogenous counterparts.

The simplest solution is to use various allelic forms differing in their coding (or transcribed) sequences. For instance, Morganelli and Berger (1984) took advantage of the existence of electrophoretic variants of the heat-shock protein HSP22. Similarly, Burke *et al.* (1984) and Sinclair *et al.* (1985) used a recombinant plasmid, pMH10A, containing a truncated form (called *hsp40*) of the 87A7 locus *hsp70* gene; it enodes a 1.8-kb messenger (instead of the normal 2.4-kb mRNA) and a 40-kDa protein.

With such systems being limited by the availability of mutant forms, McGarry and Lindquist (1985) deliberately created an in-frame deletion in the coding sequence of a cloned *hsp70* gene so that its product had a molecular weight of 44,000. Conversely, but for the same purpose, Larocca (1986) inserted an identifiable segment from the bacteriophage λ into the coding regions of the main *Drosophila hsp* genes.

Another possibility is offered by the transfer of a foreign gene which can be either the heterologous form of a gene normally expressed in *D. melanogaster* cells or even a gene which simply does not exist in this species.

Heterologous systems were developed by Tijan's laboratory for the *in vivo* functional analysis in *Drosophila* cells of mammalian transcription factors that had been shown to be absent from the *Drosophila* genome; for instance, *Sp1* (Courey and Tijan, 1988) and *CTF* (Santoro *et al.*, 1988).

The most convenient strategy for studying the regulatory regions of a given gene consists, however, in linking its upstream sequences, in the right orientation and correct phase, to the coding sequence of what is called a "reporter" gene because its product is easily identified and/or quantified. This is usually a bacterial enzyme whose quantitative assay is well standardized or which can, by its specific activity on a colorless substrate, give rise to cytochemical staining.

Indubitably, *E. coli* chloramphenicol acetyltransferase (*CAT*) is today the most extensively used reporter gene in *Drosophila* cells, as it is in mammalian cells.

E. coli β-galactosidase (*lacZ*), however, in addition to a reliable measurement of its activity, allows the visualization, by colored reactions with several chromogenic substrates, of the individual cells expressing the enzyme within a transfected population.

Moreover, whenever internal control (see Section II.2) is needed in an experimental system based, for instance, on a *CAT* assay, it is very convenient to cotransfect the cells with a control plasmid in which a standard promoter drives β-Gal; or reciprocally (for instance, in Driever and Nüsslein-Volhard, 1989).

Secondarily, several other enzymes were used as reporters in *Drosophila* cells: *E. coli gpt* (Burke *et al.*, 1984; see Section V.B); herpes simplex virus thymidine kinase (*hsv tk*) (Morganelli *et al.*, 1985); *E. coli* galactokinase (*galK*) (Van der Straten *et al.*, 1988, 1989).

A. Chloramphenicol Acetyltransferase

The sequence coding for the enzyme chloramphenicol acetyltransferase (CAT) is carried by a plasmid of *E. coli* named Tn9, which confers on bacteria a resistance to the antibiotic chloramphenicol (Kondo and Mitsuhashi, *1964*).

Because there is in vertebrate cells no endogenous enzymatic activity that can compete for the same substrate and because specific, rapid, and sensitive assays are available for CAT, Gorman *et al.* (*1982*) proposed the use of CAT as a "reporter" gene to measure promoter function in transiently transfected mammalian tissue culture cells.

Their original construct pSV2, comprised within a pBR322 vector a 773-bp fragment from the CAT gene (i.e., a 29-bp 5′ untranslated segment + the CAT coding sequence + 86 bp 3′ to the translation stop codon) driven by the simian virus 40 early promoter region and followed by SV40 splice and polyadenylation sites.

The CAT activity of cell extracts is monitored by the acetylation of [^{14}C]chloramphenicol and measured by silica gel thin-layer chomatography (Shaw and Brodsky, *1968*). Parent, mono-, and diacetylated forms of chloramphenical are easily separated (for instance, see Figs. 9.2C and 9.3) and the respective radioactive spots can be counted in a scintillation counter (see Appendices 9.H and 9.I).

Drosophila too does not contain any enzyme capable of acetylating chloramphenicol. Because it allows an accurate and relatively easy quantitation, the CAT assay is probably the most widely used system for monitoring promoter activity in transfected *Drosophila* cells.*

This method was first adopted by Di Nocera and Dawid (1983) who merely substituted *copia* or *hsp* promoters for the SV40 promoter in Gorman's original vector (*copia*–CAT and *HSP*–CAT).

The same CAT cartridge (i.e., the coding sequence of bacterial *CAT* + splice and polyadenylation sites of *SV40* t-antigene gene) was used in many constructs to test the putative promoter and transcriptional regulatory regions of a series of genes. We cite as examples: Glue protein Sgs3 (Martin, 1984; Martin and Jarry, 1988); Heat-shock *hsp70* (Di Nocera and Dawid, 1983; Saunders *et al.*, 1989; Thummel *et al.*, 1989); *hsp27* (Riddihough and Pelham, 1986); *hsp22* (Berger and Morganelli, 1989); primary ecdysone-responsive genes *Eips* (Cherbas and Andres, 1989); *copia* element (Di Nocera and Dawid, 1983; Martin, 1984; Sinclair *et al.*, 1986; Sinclair and Bryant, 1983; Martin and Jarry, 1988; Jaynes and O'Farrell, 1988; McDonald *et al.*, 1989; Thummel *et al.*, 1989; Saunders *et al.*, 1989; Sneddon and Flavell, 1989); *actin 5C* (Bond-Matthews and Davidson, 1988; Martin and Jarry, 1989); *60C β-tubulin* (Bruhat *et al.*, 1990); yolk polypeptides YP1 and 2 (Martin and Jarry, 1988); *74 E-F* early inducible gene (Thummel *et al.*, 1989); *rudimentary* (Saunders *et al.*, 1989); *1731* retrotransposon (Zyarczyk *et al.*, 1989); human cytomegalovirus major immediate early gene *HMCMV EI* (Sinclair, 1987); *gypsy* element inserted in *hsp82* intron (Dorsett *et al.*, 1989), etc . . .

In cotransfection experiments designed to analyze the putative action of a factor (over-expressed in the cells by a "producer" vector) on a target "responder" plasmid, this latter is usually based on a CAT construct (Santoro *et al.*, 1988; Courey *et al.*, 1988; Jaynes and O'Farrel, 1988; Winslow *et al.*, 1989; Krasnow *et al.*, 1989; Driever and Nüsslein-Volhard, 1989) (see Section II.6 and Fig. 9.2).

The success of this CAT system is demonstrated by the fact that there are several commercially available promoterless CAT vectors in which a

---

* It should be noted that, according to Cherbas, L. *et al.* (1994), CAT may be toxic at high levels of expression.

polylinker, in position 5′ to the CAT coding sequence, allows the easy insertion of any given promoter.

It may be noted, moreover, that Davies and Glover (1986), using an anti-CAT monoclonal antibody, were able to detect the individual cells expressing CAT within transfected cell populations using immunocyto-chemical methods.

B. *Escherichia coli* β-Galactosidase

The wide use of β-galactosidase as a marker gene is eukaryotic cells can be directly linked to methods of bacterial genetics. As a matter of fact, the enzyme β-galactosidase (β-D-galactoside galactohydrolase, EC 3.2.1.23) of *E. coli*, better known as *lacZ*, is the main component of the extensively studied *lac* operon (Jacob and Monod, *1961*).

Its activity can be accurately measured with chromogenic substrates, that is, colorless compounds which after hydrolysis yield colored products (see Appendix 9.J). Alternatively, the corresponding protein can be estimated from an ELISA assay using specific antibodies (Lawson *et al.,* 1985).

Moreover, there are efficient histochemical or immunocytochemical methods for staining β-galactosidase expressing cells (see Appendix 9.K).

It is mainly this last possibility that has been exploited in transgenic *Drosophila* flies. The deep-bluish color (developed, for instance, with the popular indigogenic substrate, X-Gal) allows a precise tissue localization of the enzyme expression pattern, even at the cellular level.

A problem was caused by the existence in wild type *Drosophila* of an endogenous β-galactosidase. Fortunately, it was shown (Best-Belpomme *et al.*, 1978; Fuerst *et al., 1987*) that *Drosophila* β-galactosidase, which is localized in the lysozomes, has a significantly lower pH optimum (~pH 5). So, if bacterial β-galactosidase is to be measured selectively, assays have to be performed at a pH slightly higher than neutral (~7.2).

In Drosophila cultured cells, at least in *Kc* cells and some other lines, Best-Belpomme *et al.* (1978) could not detect any basal β-galactosidase activity, but demonstrated that the endogenous enzyme is inducible by ecdysteroid hormones (see Chapter 8, Section II.C.2). This must be taken into account when one uses *lacZ* as a reporter gene.

Both possibilities offered by this marker gene (namely, quantitative estimation and cytochemical characterization) were utilized in transfection experiments with *Drosophila* cells.

Voellmy's group, in order to analyze the regulatory regions of heat-shock proteins, constructed a series of plasmids (Chapter 7) in which 5' fragments from *hsp70*, *hsp23*, or *hsp84* genes (comprising variously deleted upstream sequences, but also the complete leader and a few initial codons) were fused to a truncated (i.e., lacking the first seven codons) *E. coli lacZ* ORF (Lawson *et al.*, 1984; Amin *et al.*, 1985; Lawson *et al.*, 1985; Mestril *et al.*, 1986).

With similarly built vectors, Riddihough and Pelham (1986) studied the regulatory elements of the *hsp27* promoter. The *pC4βgal* vector, especially designed by Thummel *et al.* (1988) for easy insertion and testing of any putative promoter, is based on identical principles, but inserted into a "Carnegie 4" P element (Rubin and Spradling, 1983).

It must be pointed out that, in all the above mentioned works, a fusion protein was made, even though the lack of its N-terminal end has no effect on the specificity and activity of β-galactosidase; thus, the first 27 amino acid codons of *lacZ* may be removed without any apparent effect (Casadaban *et al.*, 1980). Nevertheless, in order to provide a constant and known *Drosophila* start codon for translation of *lacZ*, a *Drosophila Alcohol dehydogenase* gene N-terminus fragment (127 bp long and containing the *Adh* AUG start codon) was inserted by Thummel *et al.* (1988), in the correct orientation, in some plasmid vectors (for instance, pC4AUGβgal, Fig. 9.7A). This construct results, of course, in the synthesis of a fusion protein comprising 30 amino acids of *Adh* at its N terminus.

As mentioned above, histochemical staining, based on the specific effect of β-Gal on some indigogenic substrate (such as X-Gal), allows the identification of individual cells expressing the enzyme after transfection. This opportunity was exploited by several groups working with *Drosophila* cells (see in Appendix 9.K the simple staining protocol recommended by Sinclair and Bryant, 1987; see also in Thummel *et al.*, 1988).

VIII. SHUTTLE VECTORS

Results from cultured cell trasnfection systems must be systematically compared with and complemented by observations of gene transfer into whole animals. For this purpose, common chimeric constructs which permit an efficient gene introduction and possible genomic integration, either in cultured cells or in whole embryos, are particularly convenient. They are called shuttle vectors.

Thummel *et al.* (1988) built a series of such vectors, which are currently utilized by several laboratories. Fundamentally they are derived from the

Drosophila mobile P element (see Chapter 10) and, therefore, may be used for germline transformation of flies (according to the technique devised by Spradling and Rubin (*1982*); but in parallel they can also be used for cell culture transfection assays.

A first set of these plasmid vectors serves to test putative transcriptional regulatory sequences of a gene. For this purpose, they contain within a P-element vector from the Carnegie series (Rubin and Spradling, *1983*) a short polylinker for easy insertion of any studied foreign promoter, upstream from either a CAT or a *lacZ* reporter gene (see, for instance, pC4-AUG-βgal, Fig. 9.7A). Furthermore, a *white*[+] gene (capable of restoring eye pigmentation in *w*[−] flies) or a *rosy* gene may be added for easy identification of transformed flies.

Another group, on the other hand, is composed of expression vectors, (devised for efficient expression of any gene product). For instance, pUChsneo-act (Fig. 9.7B) consists of the original pUChsneo P-element vector of Steller and Pirotta (*1985*) into which both the strong and constitutive proximal promoter of exon 1 and the three polyadenylation sites of *Drosophila actin 5C* gene have been inserted (see Section VI.D). Between these two segments, a unique *Bam*HI site permits the insertion of any given foreign ORF. Moveover, as the *neo* gene confers resistance to the antibiotic G418 (see Section V.C), this vector can be used in cultured cells for the selection of stable transformants as well as for transient expression assays.

Similarly, Sass (1990) constructed P-transposable vectors containing an *hsp82–neo* fusion gene encoding a truncated heat-shock protein of *Drosophila pseudoobscura* and the bacterial neomycin phosphotransferase. They functioned both for selection of fly transformants (this fusion gene exhibits high levels of activity, even in the absence of heat shock) and in cell culture transfection assays.

Beside the fact that such shuttle vectors allow a direct and fruitful comparison between the two complementary approaches of gene transfer in *Drosophila*, they can also be extremely useful in pilot experiments for a rapid screening, by transient tests in cultured cells, of the validity of a given chimeric gene construct, before one embarks on onerous and time-consuming experiments of fly transformation.

IX. PROSPECTS

1. UTILIZATION IN DEVELOPMENTAL BIOLOGY

The DNA-mediated transfer of cloned genes, into either cultured cells or whole embryos, has rapidly become an indispensable tool in the molec-

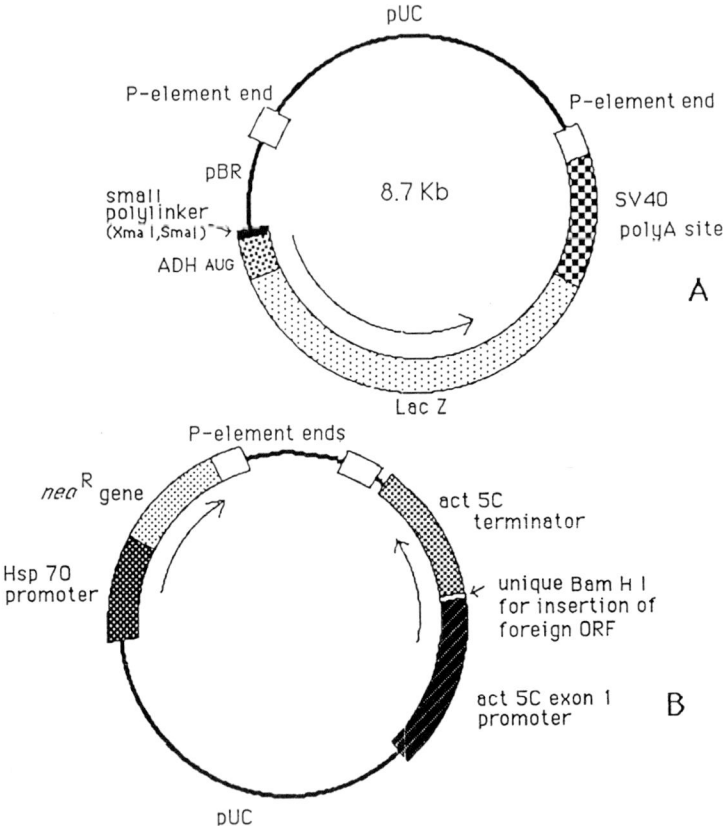

FIGURE 9.7   Shuttle vectors for transfection of both *Drosophila* organism and cultured cells (from Thummel *et al.*, 1988). (A) Reporter vector pC4-AUG-βgal. P-element ends (empty boxes) with pUC plasmid DNA derived from Carnegie 4. A short polylinker allows the insertion of any studied foreign promoter. The 127-bp *Sau*3A fragment from *Adh* gene (Benyajati *et al.*, 1983) contains the *Adh* start codon. *E. coli lacZ* ORF lacks the seven first codons and is followed by SV40 poly(A) addition signal. (B) Expression vector pUChsneo-act. Between the two ends of a P element, the *Hsp70* promoter (1100 bp of 5′ flanking DNA + 90 bp of transcribed "leader" sequence) drives the *neo* coding sequence; and the *act 5C* exon 1 promoter (2650 bp of 5′ flanking DNA and 88 bp downstream from the transcription start site) is separated by a unique *Bam*HI (or *Eco*RI) site (for the potential insertion of any given ORF to be expressed) from another *act 5C* fragment (1100 bp) including the three polyadenylation signals. Note that the orientation of *act 5C* transcription is opposite that of the *neo* gene (as indicated by arrows).

ular approach to the regulatory mechanisms that govern development and cell differentiation in higher organisms.

The leading role played by the *Drosophila* model in the current field of developmental biology (see Introduction of this book), as well as the value of the two experimental systems of gene induction developed in *Drosophila* cell lines (see Chapters 7 and 8), account for the successful expansion of this method of analysis.

The investigation may relate to (1) the function(s) and structure of any gene product; for instance, see studies of P-element transposase (Rio and Rubin, 1986), multienzyme Gart (Henikoff *et al.*, 1986), suppressor *su(Hw)* product (Spana *et al.*, 1988), or the *in vivo* assay of many cell adhesion molecules (Snow *et al.*, 1989; see Chapter 6); (2) a definition of upstream control regions of a given gene and analysis of their interactions with transcriptional regulators.

From this latter point of view, the cotransfection systems currently exploited by several groups of *Drosophila* embryologists appear especially promising: by systematically bringing together two by two (or sometimes even more) transfected genes into the same cultured cells, it should be possible to elucidate their molecular interactions and to determine their "pecking order" in the complex hierarchy of the myriad of genes which, as deduced from genetic data, control the formation of the embryonic pattern.

This type of cell culture approach is unhampered by the complexity of the organism context and permits the isolation *in vivo* of successive parts of the developmental genic circuitry. Thus, analysis of the combinatorial control of every key developmental gene by its mutliple regulators can be performed with ease and rapid quantification. Moreover, biochemical studies are greatly facilitated by the availability of large amounts of a single cell type. A vast new field is thereby opened to investigators, complementing the extensive analysis of the *Drosophila* developmental program already initiated by the genetic approach. This methodology is described at length in Chapter 5, Section V.C.2.

Such studies of gene regulation depend critically on the reliability of the cell culture system. It is crucial to ascertain that regulatory factors function in cultured cells in ways that correctly reflect their normal action as native target genes.

As stressed throughout this chapter, the expression and regulation of transfected genes were shown in most cases to be very similar to those of their endogenous counterparts.

One should be aware, however, that a few examples of ectopic expression (i.e., the expression in undifferentiated cell lines of some genes whose products were presumed to be restricted to a specific tissue or developmental stage) have at times been observed (see Section II.5). Likewise, even if the results of cotransfection assays of various regulatory homeotic genes seem generally consistent with those one would expect judging from genetic data, nevertheless some discrepancies with the situation in whole embyros have been occasionally noted (Jaynes and O'Farrell, 1988; Winslow *et al.,* 1989; Krasnow *et al.,* 1989). Such anomalies can be easily explained by the excessive levels of the effector protein or inappropriate ratios of the interacting regulators.

In close association with the important question of the validity of cell culture systems, the fine analysis carried out by Martin and Bourouis (1988) merits a special mention. They compared the constitutive expression and hormonal regulation of a series of 20-hydroxyecdysone-responsive genes when cells were either stably transformed or transiently transfected, and they pointed out some unexpected albeit significant differences.

These few "out of tune" observations do not, however, seriously question the general validity of cell culture systems nor lessen their real value as an analytic tool. They merely prompt a certain prudence in the interpretation of the results and emphasize the absolute necessity of comparing them with data obtained by other approaches; but is this not true of any experimental study?

2. POSSIBLE PRODUCTION OF PROTEINS OF INTEREST

Stably transformed *Drosophila* cell lines should be valuable not only for producing rare *Drosophila* proteins (for instance, regulatory factors or any tissue-restricted molecule) which are difficult to isolate from flies, but also for the industrial preparation of any marketable heterologous protein.

The production of a fully active protein requires, in higher organisms, correct processing (glycosylation, for instance) of the gene product. This may be a crucial reason for preferring eukaryotic host cells to any bacterial expression system for the *in vivo* synthesis of proteins of higher animals.

To this end, wide use is currently made of mammalian cell cultures. Insect cells, however, constitute a worthwhile alternative because of their (1) easy handling, (2) growth at room temperature with no requirement for a CO_2-controlled gaseous phase, (3) relatively inexpensive culture media (especially for *Drosophila Kc0* cells, which proliferate in serum-free media), (4) high cell densities in suspension cultures, and (5) lower risk of contamination by agents infectious to mammals.

Several studies have indeed demonstrated in insect cells and particularly in *Drosophila* cells that protein processing is fundamentally similar to that of mammalian cells. Fully active mammalian transcription factors can be overexpressed by transfected *Drosophila* cells and they appear to be compatible with the basal transcriptional apparatus of these cells (Santoro *et al.*, 1988; Courey and Tijan, 1988); furthermore, cellular localization signals which operate in mammalian cells are also recognized in *Drosophila* cells (Allday *et al.*, 1985). As examples of recent evaluations of the *Drosophila* system, we list (1) the proper expression of a locust tyramine receptor in transformed S2 cells (Vanden Broeck *et al.*, 1995), (2) a functional analysis of the linking interactions of human interleukin 5 and its receptor (Johanson *et al.*, 1995), and (3) the efficient formation and secretion of a human antibody (Kirkpatrick *et al.*, 1995) [via a BiP(*hsc72*)-mediated pathway; that is, in a manner which closely resembles the interaction of the homologous chaperone protein in mammalian cells].

Moreover, because any progress in the prion problem is of particular interest, the expression of a mammalian prion in *Drosophila* cells deserves a special mention. With the hope that *Drosophila* might become a fruitful model for the analysis of the neurodegenerative diseases caused by prions, Raeber *et al.* (1995) have, of late, initiated a series of transgenetic studies with the Syrian hamster prion protein (SHaPrP) ORF put under the control of a *Drosophila hsp70* promoter. In heat-shocked trangenic S2 cells, as in transformed flies, immunoblots revealed the synthesis of SHPrP (three major bands whose apparent molecular masses are slightly lower than that of the 33- to 35-kDa protein produced in hamster brain, due possibly to a lower level of glycosylation in insect cells). The prion protein is targeted to the cell surface and seems to be properly anchored by a glycolipid.

Nevertheless, one must admit that the recombinant baculovirus in lepidopteran cell lines has remained, so far, the most frequently used expression system for heterologous genes in insect cells (see Chapter 11, Section III.A). It has, however, the serious disadvantage of leading to a rapid lysis of the producing cells.

It is undeniable that stable transformant cell lines, with a high number of integrated gene copies and which, thanks to the use of certain conditional promoters (see Sections VI.B and VI.E), are able to oversynthesize the encoded protein at will, should constitute a much more convenient production system. *Drosophila* cell lines seem to be ideally suited for such a purpose. Let us, once again, point out their major assets: (1) an unrivaled knowledge of their genetic background, (2) the exceptionally high number of integrated gene copies (which is rather unusual, compared to mammalian cells), (3) the absence of endogenous viruses pathogenic for man or domestic mammals, and also the fact that the low temperature of insect cell cultures does not favor the multiplication of possible dangerous contaminants.

An interesting comparison has been recently made by Bernard *et al.* (1994) between the two systems of heterologous protein expression that are available in insect cells; namely, baculovirus/lepidopteran *Sf9* cells and stable recombinant *Drosophila S2* cells (in which expression is driven by the conditional promoter metallothionein). Both types of cells were grown in 15-liter bioreactors. The two test proteins were a secreted protein (the extracellular domain of human VCAM) and a transmembrane protein (the dopamine D4 receptor). In both systems, VCAM was correctly processed and secreted in the medium, and the D4 receptor was properly folded and integrated into cell membranes. In terms of yield, the *S2* system was more efficient in the production of the former protein, but the opposite was true for the latter. The advantage of the *Drosophila* system, however, lies essentially in the very high viability of expressing cells, whereas viral expression is lethal to lepidopteran cells.

Further technical advances, however, are still necesssary for the routine use of *Drosophila* cells. The efficiency of novel viral vectors is being explored and even if RNA engineering remains difficult RNA viruses should not be discarded *a priori;* for instance, Sindbis (Xiong *et al.,* 1989) or Black Beetle virus (Gallagher *et al.,* 1983; Dasmahapatra *et al.,* 1986; see Chapter 11). Moreover, it is likely that the strength of the promoters already available might be greatly increased after a fuller understanding of their specific regulation.

Be that as it may, the utilization of transformed and overexpressing *Drosophila* cell lines for the industrial production of heterologous proteins should have an important future.

APPENDIX 9

9.A. DNA–Calcium Phosphate Coprecipitation Method for Transfection of *Drosophila* Cells

The method presented here is slightly modified from Wigler *et al.* (*1979*).

1. SOLUTIONS

(1) 250 mM $CaCl_2$ solution (i.e., 3.7 g of $CaCl_2 \cdot 2H_2O$ in 100 ml distilled H_2O)
Filter sterilize and distribute into small tubes (1 ml).
Can be kept frozen ($-20°C$).

(2) 2X HEPES-buffered saline*
(per 100 ml distilled H_2O)

NaCl	280 mM	1.6 g
$Na_2HPO_4 \cdot 2H_2O$	1.5 mM	0.027 g
HEPES (Sigma)	50 mM	1.2 g

Adjust carefully to pH 7.08 with 1N NaOH. Filter sterilize and distribute in small tubes (1 ml) which may be stored in the refrigerator (4°C) for a few weeks.

2. OPERATING PROCEDURE

(1) One day prior to transfection Kc cells are seeded at 2×10^7 cells per plastic tissue culture flask (in 2 to 3 ml $D22$ medium supplemented with 5% FBS, as usual).

(2) Hypotonic shock**: for 2 h preceding the transfection, simply dilute the medium with an equal amount of sterile distilled water.

(3) Just before pouring the DNA–calcium precipitate, the medium is replaced with 5 ml of fresh medium + FBS.

(4) The preparation of DNA–calcium phosphate coprecipitate has to be made about 1 hr in advance. It consists of gently mixing a $CaCl_2$ solution containing plasmid DNA with an equal volume of twice-concentrated HEPES-buffered saline. Solutions have to be previously warmed to room temperature.

---

* Cherbas, L. *et al.* (1994) recommend using BBS (Chen and Okayama, *1987*) in place of the original HBS of Wigler *et al.*; this change of buffer might give roughly 10-fold higher expression in transient expression: 2 × BBS: 50 mM BES buffer (Calbiochem; pH 6.95), 280 mM NaCl and 1.5 mM Na_2HPO_4; filter sterilize and store at $-20°C$.

** We found it helpful to weaken the cell membrane with such a hypotonic pretreatment. In mammalian cell transfection, an osmotic shock was recommended by Alexander *et al.* (*1958*). As a matter of fact, when, in other protocols *Drosophila* cells are, prior to transfection, washed with any vertebrate saline solution, they are actually subjected to hypotonic conditions (300 mOsm instead of their 360 mOsm requirement).

The transfecting DNA (10 to 30 μg per flask) dissolved in TE is first added to the 250 mM CaCl$_2$ solution* and mixed.

(5) This DNA–calcium solution is then added dropwise to the same volume of 2X HEPES solution (and not the reverse). A sterile pipette with a cotton plug and a rubber tube is inserted into the mixing tube containing 2X HEPES and bubbles are introduced by blowing while the DNA–Ca solution is added.

The mixture displays progressively a characteristic opalescence. The coprecipitate will continue to form for 30 to 45 min at room temperature and without agitation.

(6) After gently pipetting this suspension, pour 0.5 ml of it into the 5 ml medium of each culture flask.

Let the coprecipitate sediment onto the cell monolayer and penetrate into the cells for 18 to 24 h at 25°C.

(7) Finally, replace the medium with 2 to 3 ml of fresh *D22* + FBS.

When looking for stable transformation, it is advisable to give the cell a chance to recover for 2 to 3 days before using the selective medium.

9.B. DEAE–Dextran-Mediated Transfection Method

This method is taken from Lawson *et al.* (1985); reproduced with the kind permission of Dr. Voellmy and Springer-Verlag, Inc.

1. *STOCK SOLUTIONS*

100X DEAE–dextran [Molecular weight 500,000 (Sigma)]
  10 mg/ml in Tris–saline (g/liter; NaCl, 7; glucose, 1;
  NaH$_2$PO$_4$ · H$_2$O, 0.07; Tris, 0.42; pH 7)
  Filter sterilize and keep in refrigerator
100X Chloroquine** Chloroquine diphosphate (Sigma)

---

* 0.5 ml of the final suspension will have to be added to each culture flask, which means that when for instance, four cell cultures have to be transfected, 1 ml of the DNA–calcium solution is added to 1 ml of 2X HEPES saline.

** Possible posttreatment with chloroquine. Luthman and Magnusson (*1983*) first reported that chloroquine could increase the proportion of polyoma DNA transfected into rodent cells up to 40%, probably by inhibiting degradation of the DNA adsorbed by the cells; so, this final step is currently adopted in most protocols of DEAE–dextran-mediated transfection, as follows: Just after the final wash, add 2 ml of complete medium +0.1 mM chloroquine for 4 h. Wash again and change the medium. The usefulness of such a chloroquine posttreatment has not been confirmed in the case of *Drosophila* cells (Echalier, unpublished).

10 mM, i.e., 5.16 mg/ml in distilled H_2O
Filter and keep in the refrigerator for up to 1 week

2. PROTOCOL

Nearly confluent *S2* cell cultures in 90-mm culture dishes are used
The cells are washed once carefully with Schneider's culture medium
Two milliliters of the transfection solution containing 100 μg/ml
DEAE–dextran and 2.5 to 5 μg/ml plasmid DNA in Schneider's
medium are added per dish
Incubation for 4 to 6 h at 25°C
After two washes with medium, the cell are fed with fresh medium
+ FBS

9.C. Polybrene-Mediated Transfection Protocol

This procedure is adapted to *Drosophila* cells, after Fallon (*1986*).

1. *(80X) POLYBRENE STOCK SOLUTION*

Polybrene (Sigma) 1 mg/ml distilled H_2O
Filter sterilize, dispense in aliquots, and store in refrigerator.

2. OPERATING PROCEDURE

On the preceding day, *Drosophila Kc* cells are seeded at a concentration
of 1 to 2 × 10^7 cells per culture flask.

The medium is replaced with 2 ml of serum-free medium containing
Polybrene (12.5 μg/ml) and plasmid DNA; the mixture is freshly prepared
by adding to 2 ml of *D22* medium, 25 μl of the Polybrene stock solution,
and, after stirring (important), 5 to 25 μg DNA.* A slight opalescence
is observed. Incubate for 6 to 18 h** at 25°C.

Two-tenths milliliters of pure DMSO (for a 10% final concentration)
is poured into the medium and mixed for a 3-min shock.

Replace with fresh medium supplemented with the usual percentage
of fetal bovine serum.

---

* According to Fallon, optimal amounts of DNA would be lower than with the calcium
method and expression efficiency might even be decreased with too much DNA.
** We obtained better results in *Kc* cells with an 18-h treatment.

9.D. Lipofectin-Mediated Transfection of *Drosophila* Cells

This procedure is kindly provided by Dr. Sondergaard (University of Copenhagen).

Transfection is performed in 25-cm^2 cell culture flasks containing 3.5 ml medium (Schneider's or Echalier's *D22*).

From a vigorously growing semiconfluent culture (5×10^6 cells/ml) seed a new bottle at 10^6 cells/ml (including 0.5 to 1 ml of spent medium) with fresh medium (no centrifugation).

Let cells grow for 24 hr (can be extended to 48 h), check under microscope that the culture has grown.

Gently remove the medium without disturbing the cells leaving 0.5 ml medium on the cells; add 3 ml fresh serum-free* medium, gently swirl the bottle, let sit for 10 min, and then repeat.

Add 50 μl DNA/Lipofectin mix prepared according to the Lipofectin protocol: 10 μg DNA** in 20 μl H$_2$O (or TE) is added to 30 μl Lipofectin (GIBCO-BRL, European Division, Paisley, Scotland) in a polystyrene*** tube, gently mixed and left for 15 min at room temperature before using.

After 24 h,**** FBS is added to obtain the necessary final concentration and the cells are allowed another day as expression period. Now the cells are ready for a transient assay or, alternately, the selective agent can be added in the case of permanently transformed cells.

This protocol worked well with *S2*, *Kc167*, and *Kc0* cells grown in Schneider's medium (*S2*) or *D22* (*Kc* lines).

9.E. Transfection of *Drosophila Kc* Cells by Electroporation

This procedure is reproduced from Cherbas, L. *et al.* (1994), with the kind permission of the authors. The protocol has been optimized for both transient transfection and stable transformation of *Kc167* cells, but with minor modifications it might probably be applied to any other *Drosophila* cell line.

---

* A transfection inhibitory factor occurs in serum.

** The optimal amount of DNA should be evaluated in a preliminary assay (depends on cell line, source of medium).

*** The mixture has a tendency to be adsorbed by polypropylene and glass.

**** The length of incubation period (before adding serum that arrests the transfection) should be optimized for each cell line.

1. SET-UP OF THE CELLS

Kc167, a subline from *Kc* cells which has been claimed to be more efficient for transfection (see Section III) are grown in M3 medium supplemented with Bacto-peptone (2.5 g/liter) and extra yeast extract (2 g/liter final concentration, instead of 1 g/liter in original M3).

Cells at the dense end of exponential growth should be diluted to $\sim 10^6$ cells/ml (in complete medium) 48 h before electroporation. Prepare one 10-ml plate for each plate of transfected cells that you wish to prepare. Incubate at 25°C. The cells will be at about 4×10^6/ml at the time of electroporation.

(The time between transfer and electroporation is important: a 24-h change in either direction leads to a 3- to 5-fold decrease in transfection efficiency).

2. ELECTROPORATION

Apparatus: Hoefer:PG200 Progenitor power supply with PG250 electroporation chamber and PG220C cuvette electrode (3.5-mm gap).

All steps are performed at room temperature.

Remove the cells from the plates, spin, and wash twice in 1/2 volume medium without serum. Resuspend at approximately 5×10^7 cells/ml medium without serum.

Use 0.8 ml (4×10^7 cells)/cuvette. Add DNA (up to 20 μg/plate) and let sit 5 min. (*Note:* Varying the concentration of cells in the electroporation cuvette does not seem to matter. The concentration specified here yields one plate of cells per transfection; to transfect more cells, use a higher concentration of cells in the cuvette and after electroporation dispense them among several plates, at 4×10^7 cells/plate. It is not clear that the 5-min wait matters. Removal of serum prior to electroporation increases reporter activity about 2-fold).

Shock cells at 440 V/cm; 1200 μF; 1 sec (i.e., complete decay) (*Note:* The optimum field strength is 400 to 500 V/cm, with a rapid fall-off of reporter activity outside this range. Their capacitance has a broad optimum between 800 and 1800 μF. Electroporation at room temperature gives about the same activity as electroporation on ice).

Allow the cells to recover for 10 min and then add the contents of the cuvette to 9.2 ml complete medium in a 100-mm plate.

9.F. MX3 Medium for *gpt* Selection

This formulation is based on Sinclair *et al.* (1983, 1985).

MX3 is based on Shield and Sang M3 medium (1977; see "Culture Media" in Chapter 1), but with the following modifications:

No yeastolate

In compensation, add, per liter:

   Lactalbumin hydrolyzate   200 mg

   Salts

$CuSO_4 \cdot 5\ H_2O$	10 μg
$FeSO_4 \cdot 7\ H_2O$	0.5 mg
$ZnSO_4 \cdot 7\ H_2O$	0.6 mg
$MgSO_4 \cdot 7\ H_2O$	0.3 mg

   Vitamins

      B_{12} 0.1 mg/l

   Grace's vitamin mix (see Chapter 1 on "Culture Media" and Appendix 1.B)

Supplementation with 10% heat-inactivated fetal bovine serum, but after extensive extraction with Norit GSX charcoal:

> Extraction of heat-inactivated serum with 5% Norit, for 3 h at room temperature, followed by an overnight extraction again with 5% Norit (4°C), followed by two 2-h extractions with 2% Norit (room temperature). Each extraction is done by gently stirring for the allocated time with a magnetic stirrer and between each extraction the charcoal serum is centrifuged in a bench centrifuge and then resuspended in fresh Norit for the next extraction. It is best to filter the serum through an 0.22 μm Millipore filter membrane after the last extraction (Sang and Sinclair, personal communication, 1984)

This medium is made selective for xanthine salvage with addition of:

Methotrexate	$10^{-6}\ M$
Xanthine	150 to 200 mg/liter
Thymidine	24 mg/liter
Adenine	13.4 mg/liter

9.G. *Drosophila* Cell Cotransformation with α-Amanitin Selection

This procedure has been provided by Dr. A. Bieber (Purdue University, West Lafayette, IN).

1. Introduction

The plasmid pPC4 encodes the α-amanitin-resistant RNA polymerase II gene from *Drosophila* (Jokerst *et al.*, 1989) and can be used to transform cultured *Drosophila* cells to α-amanitin resistance. Cotransformation with pPC4 and a nonselected plasmid yields transformed cells that contain both the selected (*pPC4*) and nonselected DNA sequences (see also page 474).

All culture procedures for amanitin selection can be carried out in commercially available complete Schneider's medium with 12.5% fetal bovine serum.

We have used Schneider's line 2 (*S2*) cells in our experiments, but these protocols could probably be adapted for use with any of the available *Drosophila* cell line.

2. TRANSFORMATION PROTOCOL

Plate 5 ml of cells at $1-2 \times 10^6$ cell/ml in 60-ml tissue culture plates. Allow the cells to sit overnight.

Ethanol precipitate 10 μg of *pPC4* with 10 μg of the nonselected plasmid* and resuspend in 0.5 ml of sterile water. To prepare a 1 ml calcium phosphate–DNA precipitate, add the 0.5 DNA solution to a sterile test tube and then add 65 μl of 2 M $CaCl_2$ (filter sterilized). Gently agitate the solution by bubbling air through a sterile cotton-plugged Pasteur pipette which is resting in the tube. While the aeration continues, add 0.5 of 2× HBS dropwise (1 to 2 drops/sec). Remove the pipette, cap the tube, and let it sit for about 30 min. After a few minutes the solution will take on a bluish tinge but no visible granular precipitate should form.

Add the 1 ml of precipitate dropwise over the surface of one plate of cells and allow the cells to sit for 15 to 18 h.**

Swirl the plates gently to dislodge any free precipitate and then decant the supernatant from each plate into separate 15 ml disposable centrifuge tubes. Add 4 ml of fresh media to the plate and swirl gently. Pellet any cells in the supernatant by a brief centrifugation (1 min at about 3/4 speed on a clinical centrifuge), discard the supernatant, and then resuspend the cells with the 4 ml of medium from the parental plate. Repeat this washing two other times and finally return the cells to the parental plate. Incubate the cells for 48 h to allow expression of the amanitin resistance.

3. α-AMANITIN SELECTION

After 48 h, dislodge the cells from the plate by blowing them off the plate with the medium. Pellet the cells and resuspend in 5 ml of fresh

---

* Bieber and colleagues have later switched to using 4 μg *pPC4* with 16 μg of the nonselected plasmid. This increases the copy number of the nonselected plasmid in the resulting transformants.

** The *S2* cell line does not do well if incubated under high calcium conditions for much longer than 15 to 18 hours. The extensive washing in step 3 is primarily designed to thoroughly wash out the calcium.

medium containing α-amanitin* at 5 μg/ml** (i.e., 25 μl of 1 mg/ml amanitin per 5 ml medium).

About 7 to 10 days after addition of the drug, many of the cells begin to lyse. The extent of cell lysis depends on the transformation efficiency. Poor transformations are followed by extensive cell lysis and it may take several weeks for the resistant cells to grow out. If the cell density gets too high or if the cellular debris on the plate gets too bad, wash the cells by pelleting and resuspend in fresh medium with α-amanitin at an appropriate density. (*Note:* As a control, do one transformation plate with 20 μg of the nonselected plasmid and no pPC4; and no resistant cells should grow out on this plate.)

As the resistant cells grow out, continue to passage the cells with 5 μg/ml amanitin, but after several passages, the amanitin treatment may be stopped.

At this point, the mixed population of transformed cells can be analyzed for the expression of the nonselected plasmid.

The cells can be cloned in soft agar to recover lines that express the nonselected plasmid at high levels.

9.H. CAT Assay

The procedure includes slight modifications from the original protocol of Gorman *et al.* (1982).

1. PREPARATION OF CELL EXTRACTS

24 to 48 h after transfection, *Drosophila* cells are harvested (by scraping the flasks with a curved glass pipette).

Wash twice with phosphate-buffered saline (137 mM NaCl, 27 mM KCl, 65 mM Na_2HPO_4, 15 mM KH_2PO_4, pH 6.8) and once with 0.25 M Tris buffer (each time 5 ml, then centrifuge at 2000 rpm at 4°C for 5 min).

Finally, the pellet is resuspended (dispersion by Vortex) in 100 μl 0.25 M Tris–HCl (pH 7.8).

Disruption by sonication (in ice-box) or by three freeze–thaw cycles (5 min in ethanol/solid CO_2, followed by 5 min at 37°C).

---

* α-Amanitin (Sigma) is dissolved at 1 mg/ml in 100% ethanol.

** At 5 μg/ml, Bieber sees no untransformed S2 cells that escape the selection. With *Kc* cells and *S1* cells, there are a few cells that escape the selection even at concentrations up to 10 μg/ml. He has not tried higher drug concentrations with those cell lines, but the number of cells escaping the selection is small and it is probably not a significant problem.

Extracts are then heated at 65°C for 5 min to destroy potential deacety-lase activities (not necessary for all cell types).

Cell debris and precipitated proteins are removed by centrifugation (10,000 rpm for 10 min at 4°C).

Total protein concentration is determined according to the method of Bradford (1976).

Aliquots of the cell extracts can be assayed immediately or stored for up to 2 weeks at −20°C (with little loss of activity).

2. ENZYME ASSAY

The assay mixture consists of cell extract (appropriate dilution; see footnote), 25 μl; [^{14}C]chloramphenicol (Amersham, 50 mCi/mmol; di-luted in water to 0.1 μCi/μl), 1 μl; 0.25 M Tris–HCl, up to 160 μl (pH 7.8).

After equilibration at 37°C (5 min), the reaction is initiated by adding Acetyl-Coenzyme A (lithium salt) fresh 4 mM solution in 0.25 M Tris–HCl, 20 μl.

Incubation at 37°C from 30 min to 3 h.*

The reaction is stopped with 1 ml of cold ethyl acetate, which is also used for extracting the chloramphenicol. Vortex well and after 5 min at room tem-perature the phases are separated by centrifugation (10,000 g; 5 min; 4°C).

The organic (superior) layer is collected, dried under vacuum [in a hood overnight or on a rotary evaporator (Speed-Vac)], then taken up in 20 μl ethyl acetate.

This is spotted onto silica gel thin-layer plates (Baker Phillipsburg, NJ Flex TLC silica gel 1B) and chromatographed in chloroform–methanol [95:5 (v/v), ascending] until solvent front reaches about three quarters of the plate height.

Visualization, by autoradiography [the dried plate is laid on X-ray film (Kodak), at room temperature, overnight] of chloramphen-icol and its acetylated forms: three or four spots are observed, i.e., in order of increasing mobility, nonacetylated chloramphenicol, 1-acetylchloramphenicol, 3-acetylchloramphenicol, and possibly 1,3-diacetylchloramphenicol.

For quantification, radioactive spots are cut out, eluted in scintillation fluid, and counted in a scintillation counter.

---

* It is important to point out that in order to keep the reaction within the linear range (i.e., from 5 to 50% conversion), standard conditions (i.e., optimal dilutions of the cell extract, and duration of the reaction) have to be carefully determined. The reaction must be limited to the monoacetylation step. Commercially available CAT may be used for positive control.

CAT activities are usually expressed as the percentage of conversion of [^{14}C]chloramphenicol into its acetylated forms, related to the protein content of the extract and the reaction time; but data may also be given as the pmol of [^{14}C]chloramphenicol converted to 1 + 3 monoacetylated forms per milligram of extract protein per minute. All values are corrected for background activity of mock transfected cells.

9.I. Direct Diffusion Assay for CAT Activity

This assay is from Neumann *et al.* (*1987*) (with permission of Eaton Publishing) and utilizes [^{3}H]acetyl-Coenzyme A as labeled substrate instead of the usual labeled chloramphenicol. It is a simple one-vial procedure that relies on the direct and selective diffusion of the labeled enzymatic reaction product, i.e., acetylchloramphenicol, into the water-immiscible liquid scintillation cocktail overlying the reaction mixture. Moreover, the scintillation vials, in which the assay is performed, are left to incubate at room temperature in the scintillation counter itself, so that a continuous data stream can be generated by direct monitoring of the in process reaction (instead of the single end-point determination given by standard methods).

1. PROTOCOL

Cell extract (adequate dilution has to be preliminarily defined for each experimental system) is added to a 7-ml glass miniscintillation vial containing sufficient 100 mM Tris–HCl (pH 7.8) to give a total volume of 50 μl.

If necessary (it depends on the cell type) and to destroy potential deacetylase activities, the solution is heated at 70°C for 15 min and then cooled to room temperature.

To the vial are added 200 μl of a freshly prepared reaction mix containing:

1 M Tris–HCl (pH 7.8)	25 μl
chloramphenicol (5 mM aqueous solution)	50 μl
[^{3}H]acetyl-CoA (NET-290L, 200 mCi/mmol from DuPont New England Nuclear Research Products)	0.1 μCi*

so that the final reaction volume is 250 μl.

---

* In the original procedure, 5 mCi [^{3}H]acetyl-CoA were used, which represented a prohibitive cost for repetitive CAT assays; but Eastman (*1987*) found that 0.1 μCi is sufficient and addition of unlabeled acetyl-CoA is not necessary.

Using a pipette, this reaction mixture is gently overlaid with 5 ml of the water-immiscible scintillation fluor (for instance, Econofluor from DuPont NEN Research Products).

The vials are left to incubate at room temperature in the scintillation counter. At selected time intervals (e.g., every 20 min), the individual vials are counted for 0.1 min.

Because only the acetylated chloramphenicol diffuses into the scintillation cocktail, the increase with time of the measurable radioactivity provides a kinetic analysis of the reaction.

9.J. β-Galactosidase Assay

This assay is taken from several sources (Miller, *1972*; Best-Belpomme *et al.*, 1978; Lawson *et al.*, 1985).

1. PRINCIPLE OF METHOD

o-Nitrophenyl-β-D-galactoside (ONPG) is a colorless substrate which, in the presence of β-galactosidase (β-Gal) is converted into galactose and o-nitrophenol. The yellow color of this latter substance can be measured by its absorption at 420 nm and in standard conditions the amount of o-nitrophenol produced is proportional to the amount of the enzyme and to the duration of the reaction.

2. MEASUREMENT PROTOCOL

Cells are scraped off the plates, collected by centrifugation and resuspended in 200 μl/dish of *Z buffer* [60 mM Na$_2$HPO$_4$; 40 mM NaH$_2$PO$_4$; 10 mM KCl; 1 mM MgSO$_4$; 50 mM β-mercaptoethanol (pH 7.0)] containing 0.5% Nonidet P-40.

Lysis by repeated pipetting or gentle agitation. The lysate is cleared by centrifugation (10,000 g; 5 min; 4°C).

Add to 200 μl of the supernatant 100 μl of 4 mg/ml ONPG (Sigma) in Z buffer.

The reaction lasts 1 to 2 h at 37°C and is stopped by the addition of 200 μl of 1 M Na$_2$CO$_3$.

After another centrifugation in order to remove insoluble debris, the absorbance through 1 cm is measured at 420 nm. The increase remains linear with time for at least 90 min; so the final measurement is generally made after 60 min.

To control against endogenous β-Gal activity and light scattering, subtract the absorbance of a nontransfected cell extract.

The relative activity will be given in arbitrary units/min/mg protein. *Note:* As an alternative, there is a Fluorometric assay of β-galactosidase, which was used in transfected *Drosophila* cell by Jaynes and O'Farrell (1988). According to Stuart *et al.* (1984) the assay buffer contains 100 μM 4-methylumbelliferyl-β-D-galactoside as a substrate for β-Gal. β-Gal activity was determined by removing 2 μl of the assay mixture, after 1 h, placing it in 1 ml of 0.5 mM NaOH and measuring the fluorescence at 445 nm (excitation at 365 nm).

9.K. Histochemical Staining for β-Galactosidase in *Drosophila* Cell Monolayers

This procedure is from Sinclair and Bryant (1987).

Analysis of β-Gal expression was carried out as follows.

Ninety-six-hour posttransfection culture plates were washed with phosphate-buffered saline (PBS) (pH 7.0).

One milliliter of solution C, made up from 2.275 ml of solution B (PBS containing 1.1 mM MgCl₂, 6.35 mM potassium ferricyanide, and 6.35 mM ferrocyanide) and 0.05 ml of solution A (10 mg/ml of X-Gal* in dimethyl formamide), was added to the cells and these were incubated at 37°C.

After 1 to 2 h, blue colonies are counted.

References

Alexander, H. E., Koch, G., Morgan Mountain, I., and Van Damme, O. (1958). Infectivity of RNA from Poliovirus in human cell monolayers. *J. Exp. Med.* **108**, 493–506.

Bacchetti, S., and Graham, F. L. (1977). Transfer of the gene for thymidine kinase to thymidine kinase-deficient human cells by purified herpes simplex viral DNA. *Proc. Natl. Acad. Sci. U.S.A.* **74**, 1590–1594.

Benyajati, C., Spoerel, N., Haymerle, H., and Ashburner, M. (1983). The messenger RNA for Alcohol Dehydogenase in *Drosophila melanogaster* differs in its 5′ end in different developmental stages. *Cell* **33**, 125–133.

Bienz, M., and Pelham, H. R. B. (1987). Mechanisms of heat-shock gene activation in higher eukaryotes. *Adv. Genet.* **24**, 31–72.

Blochlinger, K., and Diggelmann, H. (1984). Hygromycin B phosphotransferase as selectable marker for DNA transfer experiments with higher eucaryotic cells. *Mol. Cell. Biol.* **4**, 2929–2931.

Bond, B. J., and Wold, B. (1987). Poly-L-Ornithine-mediated transformation of mammalian cells. *Mol. Cell. Biol.* **7**, 2286–2293.

* X-Gal, 5-Bromo-4-chloro-3-indolyl-β-D-galactopyranoside (BRL).

Bradford, M. (1976). A rapid and sensitive method for the quantitation of microgram quantities of proteins, using the principle of protein-dye binding. *Anal. Biochem.* **72**, 248–254.

Casadaban, M., Chou, J., and Cohen, S. N. (1980). *In vitro* gene fusions that join an enzymatically active beta-galactosidase segment to amino-terminal fragments of exogeneous proteins: *E. coli* plasmid vectors for the detection and cloning of translational initiation signals. *J. Bacteriol.* **143**, 971–980.

Chen, C., and Okayama, H. (1987). High efficiency transformation of mammalian cells by plasmid DNA. *Mol. Cell. Biol.* **7**, 2745–2752.

Colbere-Garapin, F., Horodniceanu, F., Kourilsky, P., and Garapin, A. C. (1981). A new dominant hybrid selective marker for higher eukaryotic cells. *J. Mol. Biol.* **150**, 1–14.

Davies, J., and Jiminez, A. (1980). A new selective agent for eukaryotic cloning vectors. *Am. J. Trop. Med. Hyg.* **29**(5) (Suppl.), 1089–1092.

Dunsmuir, P., Brorien, W. J., Simon, M. A., and Rubin, G. M. (1980). Insertion of the *Drosophila* transposable element *copia* generates a 5 base pair duplication. *Cell* **21**, 576–579.

Durbin, J. E., and Fallon, A. M. (1985). Transient expression of the chloramphenicol acetyltransferase gene in cultured mosquito cells. *Gene* **36**, 173–178.

Eastman, A. (1987). An improvement to the novel rapid assay for chloramphenicol acetyltransferase gene expression. *Biotechniques* **5**, 730–732.

Fallon, A. M. (1986). Factors affecting polybrene-mediated transfection of cultured *Aedes albopictus* (Mosquito) cells. *Exp. Cell Res.* **166**, 535–542.

Felgner, P. L. (1991). Cationic liposome-mediated transfection with Lipofectin reagent. *In* "Methods in Molecular Biology," Vol. 7, pp. 81–89. Humana Press, Clifton, NJ.

Felgner, P. L., Gadek, T. R., Holm, M., Chan, H. W., Wenze, M., Northrop, J. P., Ringold, G. M., and Danielsen, M. (1987). Lipofection: A highly efficient lipid-mediated transfection procedure. *Proc. Natl. Acad. Sci. U.S.A.* **84**, 7413–7417.

Fling, M. E., and Elwell, L. P. (1980). Protein expression in *E. coli* minicells containing recombinant plasmids specifying Trimethoprim-resistant *J. Bacteriol.* **141**, 779–785.

Frost, E., and Williams, J. (1978). Mapping temperature-sensitive and host-range mutations of Adenovirus Type 5 by marker rescue. *Virology* **91**, 39–50.

Fuerst, T. R., Knipple, D. C., and McIntyre, R. J. (1987). Purification and characterization of beta-galactosidase-1 from *Drosophila melanogaster*. *Insect Biochem.* **17**, 1163.

Fyrberg, E. A., Mahaffrey, J. W., Bond, B. J., and Davidson, N. (1983). Transcripts of the 6 *Drosophila* Actin genes accumulate in a stage- and tissue-specific manner. *Cell* **33**, 115–123.

Fyrberg, E. A., Kindle, L., Davidson, N., and Sodja, A. (1980). The Actin genes of *Drosophila*: A dispersed multigene family. *Cell* **19**, 365–378.

Gorman, C. M., Moffat, L. F., and Howard, B. H. (1982). Recombinant genomes which express chloramphenicol acetyl transferase in mammalian cells. *Mol. Cell. Biol.* **2**, 1044–1051.

Graham, F. L., and Van der Eb, A. J. (1973). A new technique for the assay of infectivity of human adenovirus 5 DNA. *Virology* **52**, 456–467.

Gritz, L., and Davies, J. (1983). Plasmid-encoded hygromycin B resistance: The sequence of hygromycin B phosphotransferase gene and its expression in *Escherichia coli* and *Saccharomyces cerevisiae*. *Gene* **25**, 179–188.

Huttner, K. M., Barbarosa, J. A., Scangos, G. A., Pratcheva, D. D., and Ruddle, F. (1981). DNA-mediated gene transfer without carrier DNA. *J. Cell Biol.* **91**, 153–159.

Jacob, F., and Monod, J. (1961). Genetic regulatory mechanisms in the synthesis of proteins. *J. Mol. Biol.* **3**, 318–332.

Jokerst, R. S., Weeks, J. R., Zehring, W. A., and Greenleaf, A. L. (1989). Analysis of the gene encoding the largest subunit of RNA polymerase II in *Drosophila*. *Mol. Gen. Genet.* **215**, 266–275.

Jorgensen, R. A., Rothstein, S. F., and Reznikoff, W. S. (1979). A restriction enzyme cleavage map of Tn5 and location of a region encoding neomycin resistance. *Mol. Gen. Genet.* **177**, 65–72.

Kai, R., Sekiguchi, T., Yamashita, K., Sekiguchi, M., and Nishimoto, T. (1983). Transformation of temperature-sensitive growth mutants of BHK21 cell line to wild-type phenotype with Hamster and Mouse DNA. *Somat. Cell Genet.* **9**, 673–680.

Karch, F., Török, I., and Tissières, A. (1981). Extensive regions of homology in front of the two *hsp70* heat shock variant genes in *Drosophila melanogaster*. *J. Mol. Biol.* **148**, 219–230.

Kawai, S., and Nishizawa, M. (1984). New procedure for DNA transfection with polycation and dimethylsulfoxide. *Mol. Cell. Biol.* **4**, 1172–1174.

Koch, G., and Bishop, J. M. (1968). The effect of polycations on the interaction of viral RNA with mammalian cells: Studies on the infectivity of single- and double-stranded poliovirus RNA. *Virology* **35**, 9–17.

Kondo, E., and Mitsuhashi, S. (1964). Drug resistance of enteric bacteria. IV. Active transducing bacteriophage P1 CM produced by the combination of R factor with bacteriophage P1. *J. Bacteriol.* **88**, 1266–1276.

Lang-Hinrichs, C., Berndorff, D., Seefeldt, C., and Stahl, U. (1989). G418 resistance in the yeast *Saccharomyces cerevisiae*: Comparison of the neomycin resistance genes for Tn5 and Tn903. *Appl. Microbiol. Biotechnol.* **30**, 388–394.

Loyter, A., Scangos, G. A., and Ruddle, F. (1982). Mechanisms of DNA uptake by mammalian cells: Fate of exogenously added DNA monitored by the use of fluorescent dyes. *Proc. Natl. Acad. Sci. U.S.A.* **79**, 422–426.

Lycett, G. (1990). *R. Entomol. Soc. London Insect Mol. Genet. Newsletters* **4**, 1–3.

Luthman, H., and Magnusson, G. (1983). High efficiency polyoma DNA transfection of chloroquine treated cells. *Nucleic Acid Res.* **11**, 1295–1308.

Maroni, G., Otto, E., and Lastowski-Perry, D. (1986). Molecular and cytogenetic characterization of a metallothionein gene of *Drosophila*. *Genetics* **112**, 493–504.

Miller, J. H. (1972). *In* "Experiments in Molecular Genetics." Cold Spring Harbor Lab Press, Cold Spring Harbor.

Moran, L., Mirault, M. E., Tissieres, A., Lis, J., Schedl, P., Artavanis-Tsakonas, S., and Gehring, W. J. (1979). Physical map of two *Drosophila melanogaster* DNA segments containing sequences coding for the 70,000 Dalton heat-shock protein. *Cell* **17**, 1–8.

Moreno, S., and Nurse, P. (1990). Substrates for p34^{cdc2}: *In Vivo Veritas*. *Cell* **61**, 549–551.

Mount, S. M., and Rubin, G. M. (1985). Complete nucleotide sequence of the *Drosophila* transposable element *copia*: Homology between *copia* and retroviral proteins. *Mol. Cell. Biol.* **5**, 1630–1638.

Mulligan, R. C., and Berg, P. (1980). Expression of a bacterial gene in mammalian cells. *Science* **209**, 1422–1427.

Nagata, Y., Takagi, H., Morita, T., and Oishi, M. (1984). Stimulation of DNA-mediated transformation by UV irradiation of recipient (Mouse FM3A) cells. *J. Cell. Physiol.* **121**, 453–457.

Neumann, J. R., Moreney, C. A., and Russian, K. O. (1987). A novel rapid assay for chloramphenicol acetyltransferase gene expression. *Biotechniques* **5**, 444–447.

O'Hare, K., Benoist, C., and Breathnach, R. (1981). Transformation of mouse fibroblasts to methotrexate resistance by a recombinant plasmid expressing a prokaryotic dihydrofolate reductase. *Proc. Natl. Acad. Sci. U.S.A.* **78**, 1527–1531.

Otto, E., Young, J. E., and Maroni, G. (1986). Structure and expression of a tandem duplication of the *Drosophila* metallothionein gene. *Proc. Natl. Acad. Sci. U.S.A.* **83**, 6025–6029.

Pelham, H. R. B. (1982). A regulatory upstream promoter element in the *Drosophila hsp70* heat-shock gene. *Cell* **30**, 517–528.

Pettinger, R. C., Wolfe, R. N., Hoehn, M. M., Marks, P. N., Dailey, W. A., and McGuire, J. M. (1953). *Antibiot. Chemother.* **3**, 1268–1278.

Potter, S., Truett, M., Philipps, L. M., and Maher, A. (1980). Eucaryotic transposable genetic elements with inverted terminal repeats. *Cell* **20**, 639–647.

Reiss, B., Sprengel, R., Will, H., and Schaller, H. (1984). A new sensitive method for qualitative and quantitative assay of neomycin phosphotransferase in crude cell extracts. *Gene* **30**, 211–218.

Rubin, G. M., and Spradling, A. C. (1983). Vectors for P element-mediated gene transfer in *Drosophila. Nucleic Acids Res.* **11**, 6341–6351.

Sene, C., and Nicolau, C. (1981). *In* "Lysosome Drugs and Immunocompetent Cells" (C. Nicolau and A. Paraf, eds.), pp. 67–77. Academic Press, New York.

Shaw, W. V., and Brodsky, R. F. (1968). Characterization of chloramphenicol Acetyltransferase from chloramphenicol-resistant *Staphylococcus aureus. J. Bacteriol.* **95**, 28–36.

Southern, P. J., and Berg, P. (1982). Transformation of mammalian cells to antibiotic resistance with a bacterial under control of the SV40 early region promoter. *J. Mol. Appl. Genet.* **1**, 327–341.

Spradling, A. C., and Rubin, G. M. (1982). Transposition of cloned P elements into *Drosophila* germline chromosomes. *Science* **218**, 341–347.

Steller, H., and Pirotta, V. (1985). A transposable P vector that confers selectable G418 resistance to *Drosophila* larvae. *EMBO J.* **4**, 167–171.

Stuart, G. W., Searle, P. F., Chen, H. Y., Brinster, R. L., and Palmiter, R. D. (1984). A 12-base-pair DNA motif that is repeated several times in metallothionein gene promoters confers metal regulation to a heterologous gene. *Proc. Natl. Acad. Sci. U.S.A.* **81**, 7318–7322.

Swanson, M. M., and Poodry, C. A. (1981). The *shibire^ts* mutant of *Drosophila:* A probe for the study of embryonic development. *Dev. Biol.* **84**, 465–470.

Urieli-Shoval, S., Gruenbaum, Y., Sedat, J., and Razin, A. (1982). The absence of detectable methylated bases in *Drosophila melanogaster* DNA. *FEBS Lett.* **146**, 148–152.

Vaheri, A., and Pagano, J. S. (1965). Infectious poliovirus RNA: A sensitive method of assay. *Virology* **27**, 434–436.

Wigler, M., Pellicer, A., Silverstein, S., and Axel, R. (1978). Biochemical transfer of single-copy eucaryotic genes using total cellular DNA as donor. *Cell* **14**, 725–731.

Wigler, M., Pellicer, A., Silverstein, S., Axel, R., Urlaub, G., and Chasin, L. (1979). DNA mediated transfer of the adenine phosphoribosyltransferase locus into mammalian cells. *Proc. Natl. Acad. Sci. U.S.A.* **76**, 1373–1376.

Wigler, M., Silverstein, S., Lee, L. S., Pellicer, A., Cheng, Y. C., and Axel, R. (1977). Transfer of purified Herpes virus thymidine kinase gene to cultured mouse cells. *Cell* **11**, 223–232.

Yates, J. L., Warren, N., and Sugden, B. (1985). Stable replication of plasmids derived from Epstein-Barr virus in various mammalian cells. *Nature* **313**, 812–818.

10

Transposons

After Barbara McClintock (*1956*) had made the revolutionary discovery, in maize, that certain genomic sequences are able to move from one chromosome location to another, it was not by mere chance that *Drosophila* became the first higher animal in which such "jumping genes" could be characterized, because of the sophistication of its genetics.

The genome of *Drosophila melanogaster* contains a particular abundance and diversity of mobile genetic elements. These latter can constitute as much as 10% of its total nuclear DNA, and can be categorized in to several clear-cut classes, according to their different structures and putative mechanisms of transposition.

See general reviews on the different categories of transposons in "Mobile DNA" (Berg, D. E. and Howe, M. M., eds (*1989*). Reviews by Rubin (*1983*) and Georgiev (*1984*) still provide useful information.

Established cell lines were directly involved in the discovery of transposons in *Drosophila* and in their further study. When, during the late 1970s, technical progress enabled molecular biologists to tackle the functional organization of higher eukaryotic genomes, radiolabeled cDNA was prepared from the abundant poly(A) RNAs transcribed in *Drosophila* cultured cells, and they were used as probes for screening genomic libraries of recombinant plasmids. The elements *copia* (a name coined from a latin word meaning abundance) and *412* (Rubin *et al.*, 1976; Finnegan *et al.*, 1978), as well as several so-called *mdg* (for "mobile dispersed genes") (Georgiev *et al.*, 1977; Tchurikov *et al.*, 1978; Ilyin *et al.*, 1978, 1980a) were thereby rapidly identified.*

I. LONG TERMINAL REPEAT
RETROTRANSPOSONS: *copia*-LIKE ELEMENTS

This type of transposon is widely distributed throughout the genome of not only vertebrate and invertebrate animals, but also plants and fungi. They are the most important class of mobile genetic elements in *Drosophila melanogaster,* and the prototype is the *copia* element.

Their characteristic structure which resembles so strikingly that of the integrated "proviral" form of vertebrate retroviruses, and the fact that their transposition seems also to require the retrotranscription of a RNA intermediate, strongly suggests some evolutionary relationship (Temin, *1980;* Finnegan, *1983*).

---

* It is particularly noteworthy that the first transposons ever cloned in higher eukaryotes were elements from *Drosophila* (and not, as it might be thought, maize or yeast elements).

Framed, at both extremities, by long terminal repeats (LTRs) of a few hundred base pairs, with the typical U3-R-U5 organization, they comprise, in their central region, two coding sequences with significant homologies to the *gag* and *pol* polyproteins of retroviruses (i.e., corresponding to the structural proteins of the virion core and to the reverse transcriptase and its associated catalytic activities, respectively). A third open reading frame, which occurs in only a few families (for instance, *gypsy*), even though it shows no apparent sequence homologies, might perhaps, according to recent results (Syomin *et al.*, 1993; Kim *et al.*, 1994; Song *et al.*, 1994), encode a functional equivalent of retroviral *env* polypeptides (Fig. 10.1).

Some 15 to 20 different families have so far been identified in *Drosophila*, each of which is represented by 10 to 100 copies individually dispersed throughout the genome.

See reviews by Georgiev (*1984*), Finnegan and Fawcett (*1986*), Echalier (1989), and Finnegan (*1992*).

As will be specified below, the number of copies of these *copia*-like elements is generally increased in cell lines established *in vitro*. This amplification may vary considerably, in each cell line, from one family of retrotransposons to another. It is amusing to point out that, because of the widespread use of cultured cells in the molecular analysis of these mobile elements, the available cell lines determined, at least to a certain extent, which family of retrotransposons was studied by a given laboratory. For instance, *copia* or *412*, first identified by Anglo-Saxon investigators, are indeed particularly abundant in *Kc* or *S2* cells (i.e., two cell lines extensively used in laboratories of the "western" world), whereas Russian scientists discovered and primarily studied *mdg* elements which are largely amplified in their *67J25* cell lines.

A. Retrotransposons: Moderately Repeated, Dispersed, and Mobile

As has already been emphasized, the first *Drosophila* retrotransposons were identified *qua* genomic sequences corresponding to the most abundant transcripts in cultured cells.* The elective method for demonstrating the multiplicity of their copies, their dispersion throughout the genome and their mobility, by comparing one fly strain with another, was, of

---

* Retrospectively, it is evident that the most abundant class of poly(A)-containing RNAs that had been characterized by Spradling *et al.* (1975), during their pioneering analysis by *in situ* hybridization, corresponded to the usual chromosomic distribution of retrotransposons.

1 kb

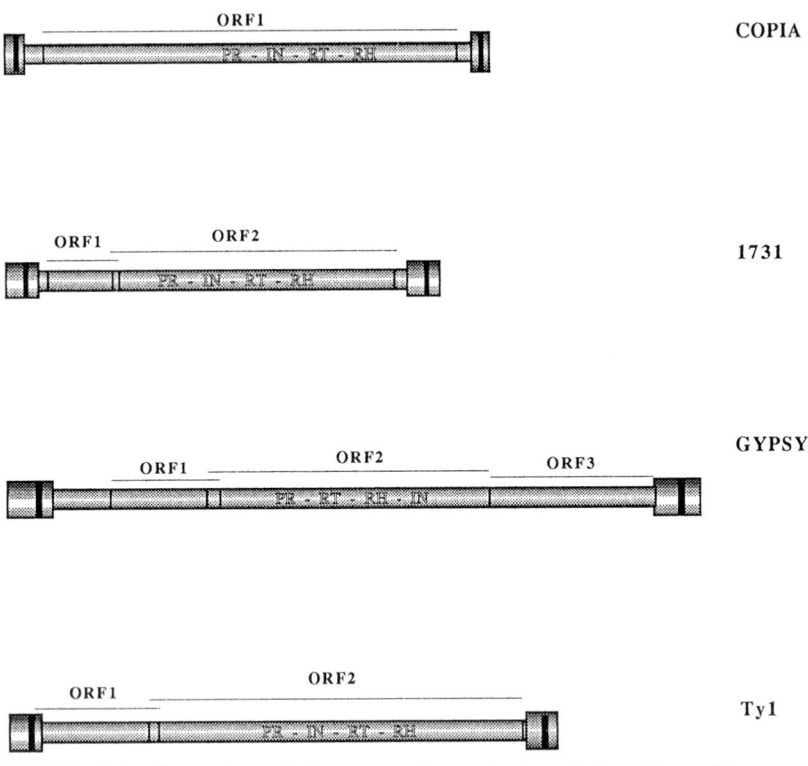

FIGURE 10.1   Comparison of the structural organization of three *Drosophila* retro-transposon families (*copia, 1731,* and *gypsy*) with a yeast retrotransposon (*Ty1*). All genetic elements and their different components (LTRs and ORFs) are schematically drawn on the same scale. The successive order, but not the relative size of the enzymatic domains of *Pol* is indicated. From Lacoste (1995); courtesy of Dr. Lacoste, Université P. et M. Curie, Paris.

course, *in situ* hybridization on polytene chromosomes. Yet, biochemical approaches with cultured cells provided useful confirmations.

1. The total number of copies of any given family of transposons can be estimated by two main methods: (a) ^{32}P-labeled nick-translated DNA

prepared from cultured cells is hybridized to a DNA probe made of the central region of the transposon and immobilized on HA nitrocellulose filters. In this manner, Tchurikov *et al.* (1978) found that about 0.5% of the total DNA from the cell line *67J25* forms hybrids with *mdg1* DNA, which corresponds to some 250 copies per haploid genome (note: the problem of amplification in cultured cells will be discussed in the following section). (b) As an alternative technique, a restriction fragment from any studied element is labeled to high specificity by nick-translation and allowed to reassociate with a large excess of nonradioactive DNA isolated from cultured cells. The number of copies can be deduced from reassociation kinetics. For instance, see the data of Potter *et al.* (1979) below.

2. The Southern technique confirmed the wide dispersion of transposons throughout the genome. If a restriction endonuclease capable of cutting the studied retrotransposon at one single site is used, the digestion of genomic DNA generates a large number of fragments which are recognizable by the probe and are heterogeneous in size. This heterogeneity is easily explained by the great diversity of the restriction sites in the regions flanking each copy. Very often, when the number of copies is particularly high, long smears are observed in addition to a few discrete bands.

3. Moreover, with the same method, a comparison of the autoradiographic patterns obtained from different cell lines should reveal transpositions, since new locations of the transposon result in different flanking sequences. Such differences in the genomic distribution of retrotransposons between various cell lines were first demonstrated by Tchurikov *et al.* (1978) for *mdg1* (between lines *67J25D* and *67J25G*), and by Potter *et al.* (1979) for *copia* and *412* (between *Kc* and *S2* cells).

These latter authors could even establish that one defined electrophoretic band, that they ascertained to be homologous in embryo and cultured cell DNAs (by the use of a probe corresponding to a unique flanking sequence), was smaller in the embryonic digested DNA than in its cultured cell counterpart, by the exact size (5 kb) of a *copia* element. This clearly means that the same genomic site, occupied in *S2* cells, was previously inhabited in the original embryos. Similar facts could be verified in four cases.

B. Amplification and Transposition in Cultured Cell Lines

The establishment *in vitro* of *Drosophila* cells is generally accompanied by a significant, and sometimes spectacular, increase in the number of copies of retrotransposons. The family concerned and the importance of this amplification depends on each cell line.

For instance, by comparison with their numbers in the genome of Oregon R embryos, a 3-fold increase was observed in S2 cells for *copia* and *297* elements, whereas the *412* family was unmodified. On the other hand, in *Kc* cells, *297* and also *412* elements were three times more abundant, whereas the *copia* number was only slightly modified (90 copies instead of 60) (Potter *et al.*, 1979; Rubin *et al.*, 1981). Similarly, a 7- to 8-fold amplification was reported for *mdg1* and as high as a 14-fold amplification for *mdg3* in the Russian *67J25* cell lines (Tchurikov *et al.*, 1978; Ilyin *et al.*, 1980a,c; Tchurikov and Ilyin, 1980; Tchurikov *et al.*, 1981; Georgiev, 1984).

This increased number of copies is not due to *in situ* tandem repetitions. Classical work by Potter *et al.* (1979) clearly established that not only the numbers, but also the genomic locations of three retrotransposons (*copia, 297, 412*) differed extensively in cultured cells from their initial distribution in the fly strains from which the lines derived. The additional copies were scattered to many alternative sites.

Moreover, there is considerable heterogeneity among individual cells in the same culture population. Probably impressed by such heterogeneity, the authors ventured the assumption that the rate of transposition in cell cultures was uniform. Combining the number of extra copies (i.e., 110 *copia* in S2 cells) and the 10 years during which the cell line had been grown *in vitro* at the time of their experiments, they estimated that this transposition rate might be as high as 10^{-3} to 10^{-4} transpositions per element per generation!

Russian investigators pointed out that amplification of retrotransposons in cell cultures might mainly affect certain specific variants, possibly endowed with a higher efficiency of transcription or reverse transcription. For instance, Bayev *et al.* (1984) showed that only copies of *mdg4* (*gypsy*) containing a specific restriction (*Hind*III) site, which is a rare form in the fly genome, were amplified in the *67J25D* line. Similarly, Ilyin *et al.* (1984) identified a minor *mdg3* variant (with a 1.3-kb deletion) in the fly genome. This deleted form was preferentially amplified in *Kc* cells, whereas in the *67J25D* line an enormous amplification (200 copies, i.e., a 13-fold amplification) concerned the full-size species. Lyubomirskaya *et al.* (1993), by devising a model system for the detection of new *gypsy* elements formed via reverse transcription, were able to compare the transposition abilities of two variants in stably transformed *Drosophila hydei* cells and found them to be indeed significantly different. Relevant to this problem of the possible occurrence of more active copies in a same retrotransposon family, Yoshioka *et al.* (1992) reported that the amplification of the *copia* element is much greater in cultured cells from the related species *Drosophila simulans* than in *Drosophila melanogaster* cell lines. Now, the reverse transcriptase (RT) activity associated with *Drosophila simulans* VLPs was found to be 25 times higher than that associated with *Drosophila melanogaster* VLPs and they differ, in their corresponding nucleotide sequences, by five substitutions.

The too rapidly and widely admitted assumption made by Potter and collaborators that retrotransposons continue to transpose steadily in culture cells deserved some reinvestigation in much more rigorous experimental conditions. Indeed, most continuous cell lines probably did not commence with a single cell (see Chapter 3) and, therefore, may be somewhat heterogeneous, from the start, with respect to the distribution of retrotransposons. Therefore, in order to distinguish between true transpositions and mere random drift of cell subpopulations, it is imperative that one works with cloned sublines.

A clonal subline from *Kc* cells was monitored for as long as 8 years (Echalier and Junakovic, 1988; Junakovic *et al.*, 1988). At regular intervals the line was subcloned in order to appraise the differences in transposon distribution among individual cells of the population. Each clone or subclone was partly frozen in liquid nitrogen, just after its isolation, and partly propagated under standard culture conditions, until the time of the analysis. All samples were then submitted to the Southern method, each blot being successively hybridized with internal fragments of 6 different retrotransposons (*copia, 1731, 412, 297, B104, mdg1*). Unexpectedly, only minor autoradiographic pattern modifications were observed among the successive cell samples, which implies a striking basic stability in the six different retrotransposons in *Drosophila Kc* cells throughout 8 years of culture.

A totally different picture emerged from the study of short-term cultures. The genomic distribution of the same six retrotransposons was analyzed in four newly established cell lines (less than 1 year old) and they were compared with not only the same *Drosophila* strain but also the laboratory fly stock from which they derived. In contrast with the preceding observations, a number of qualitative and quantitative differences could be detected, and they varied greatly from one sister cell line to another.

In order to eliminate the possible heterogeneity of the original fly stock, an additional cell line was established from an inbred *Drosophila* line (*inb-c*) which had just been subjected to 30 generations of brother–sister matings and in which the stability of the studied transposable elements had been verified. Again, in 3-month-old cultured cells, several changes in the restriction patterns were found, consistent with various types or rearrangements (such as amplification, transposition and excision) of the elements *copia, 1731, 412, 297,* and *mdg4* (*gypsy*); on the other hand, *B104, G,* and *Blood* elements appeared stable (Di Franco *et al.,* 1992).

It must be concluded that the amplification and considerable redistribution of retrotransposons, as previously described in many *Drosophila* cell lines seem to occur essentially during the initial period of the culture. They are perhaps induced by the stress(es) of the *in vitro* environment.

Furthermore, it is extremely tempting to correlate such a striking "burst" of transposition in young cultures with the "spontaneous" establishment of continuous cell lines (see Chapter 3). Any transposable element, with a strong promoter, might be randomly integrated in the vicinity of some protooncogene and, by enhancing the expression of this protooncogene, might promote an indefinite capacity for proliferation.

C. Retrotransposon Transcripts in Cultured Cells

Most retrotransposons are abundantly transcribed in *Drosophila* culture cell lines. For instance, the transcripts of *copia* may account for at least 3% of the poly(A)-containing RNA in *Kc0* cells, which led to the discovery of this transposable element (Finnegan *et al.,* 1978), and about 2% in *S2* cells (Falkenthal and Lengyel, 1980). Mossie *et al.* (1985) found, on the other hand, approximately twice as much poly(A)$^+$ *copia* RNA in *S2* cells (quantitated by dot-blot analysis) compared with *Kc0* cells, which might be in better agreement with the higher amplification of *copia* in Schneider's cells. Other retrotransposon families are also represented in this poly(A)$^+$ RNA, although in lesser proportions (0.5 to 1.5%). There are large variations between cell lines, probably correlated with the degree of amplification.

It is important to recall that retrotransposons are also normally transcribed in the organism *Drosophila,* according to specific developmental stage and tissular patterns, although their possible influence (or even function ?)* still remains questionable.

1. *copia*-SPECIFIC TRANSCRIPTS

Because *copia* is the prototype of this class of mobile elements, its transcription has been extensively studied and most of the molecular approaches have been carried out with cultured cells.

* See observation by Mozer and Benzer (1994).

a. Structure

Four discrete molecular weight species of *copia* poly(A)$^+$ RNAs have been identified: they are 5.2, 2.1, 1.3, and 0.8 long and are all transcribed in the same direction (Carlson and Brutlag, 1978; Falkenthal and Lengyel, 1980; Flavell *et al.*, 1980; Young and Schwartz, 1981; Schwartz *et al.*, 1982).

The two largest species are quantitatively dominant and their ratio is about equimolecular, at least in cultured cells.

The 5 kb RNA was shown to be complementary to all the sequences defining *copia* and, as expected from a full-length transcript from a provirus-like element (see Section I.D), carries the major portion (about 200 bp) of both LTRs. The 2 kb RNA is homologous to the 5' half of the *copia* sequence.

It was verified that all these *copia* RNAs, which were isolated by their binding to oligo(dT)-cellulose, do indeed possess a poly(A)$^+$ track and that the average length of this track is similar to that of all other *Drosophila* cytoplasmic poly(A)$^+$ RNAs. Moreover, Flavell *et al.* (1981) showed that the major 5' ends of both the 5 kb and 2 kb RNAs share an identical set of heterogeneous capped termini, predominantly composed of pyrimidine nucleotides. It was shown, later on, that the 2 kb RNA is generated throughout splicing. For instance, when a recombinant plasmid comprising a full-length *copia* sequence was introduced into a *Drosophila hydei* cell line (which does not possess any endogenous *copia*), both RNA species could be isolated, whereas, if there was a simple point mutation at the putative 3' slice site, only 5 kb RNA was observed (Yoshioka *et al.*, 1990).

b. Intracellular location and processing

Falkenthal and Lengyel (1980) calculated that, at steady state, there are about 1900 molecules of cytoplasmic *copia* RNAs (5 kb and 2 kb being in equimolecular amounts) per *Kc* cell, and they turn over in the cytoplasm with a half-life of 10 h (i.e., a much longer period than the average half-life of the intermediary decay class of cytoplasmic mRNAs).

copia-specific RNAs are also abundant in the nuclear compartment. At steady state, some 64% of the total labeled *copia* RNA was found in the nucleus of *Kc* cells (Schwartz *et al.*, 1982). Similarly, Falkenthal *et al.* (1982) estimated that, in terms of the number of molecules, 32% of the total [i.e., poly(A)$^+$ and poly(A)$^-$ RNAs of all sizes] are contained in the nucleus of *S2* cells, versus 68% in the cytoplasm.

Using continuous labeling, pulse-labeling and pulse-chase experiments, Falkenthal *et al.* (1982) carefully investigated the transcription, processing and size distribution of *copia* RNAs in cultured cells. They proposed the following processing scheme:

Transcription	Adenylation	Export	
$\longrightarrow$ Nuclear (A)$^-$	$\longrightarrow$ Nuclear (A)$^+$	$\longrightarrow$ Cytoplasmic (A)$^+$	
$\downarrow$ Decay	$\downarrow$ Decay	$\downarrow$ Decay	

Moreover, they established that only a very small fraction (<10%) of *copia* nuclear RNA migrates to the cytoplasm, when the remaining RNA turns over in the nucleus. The absolute rate of *copia* RNA synthesis was estimated to be approximately eighty five 5 kb molecules per nucleus per minute (which is consistent with the possibility that an unique copy is transcribed per haploid genome). Finally, only eight *copia* poly(A)$^+$ RNA molecules should enter the cytoplasm per cell per minute.

c. Are copia *RNAs functional messengers?*

Falkenthal and Lengyel (1980) observed their cosedimentation with polysomes in a sucrose gradient (with a peak for both 2 kb and 5 kb species in the fraction of tetrasomes) and their classical release after exposure to EDTA. On the other hand, Flavell *et al.* (1981) and Schwartz *et al.* (1982) asserted that *copia* RNAs are rarely, if ever, associated with ribosomes, even when they correspond to cytoplasmic preparations.

In vitro translation of *copia* RNA purified from cell cultures in a rabbit reticulocyte cell-free system led to the synthesis of several (6 or 7) polypeptides, ranging from 18,000 to 51,000 Da, the last one being the dominant product (Flavell *et al.*, 1980; Falkenthal and Lengyel, 1980). By using *copia* DNA as an inhibitor, the technique of "hybrid-arrested translation" proved the specificity of this synthesis.

The translational efficiency remained very low. Although *copia* RNA is by far the most abundant poly(A)$^+$ RNA in cultured cells, this quantitative dominance was not reflected in the translation products of total cytoplasmic polyadenylated RNA. It may be that a large number of these *copia* transcripts are not functional messengers. In this *in vitro* system at least, 5 kb was not efficiently translated and messenger activity seemed to be limited to 2 kb RNA.

It should be recalled that Mount and Rubin (1985) suggested that the production of the smaller messenger, which encodes only the 5' part of the large unique ORF, might be, for *copia* a mechanism whereby *gag* protein is produced in excess, whereas the complete 5 kb mRNA should correspond to a *gag–pol* fusion protein. This hypothesis was supported by later results from Brierley and Flavell (1990). Using β-Gal fusion constructs, they could demonstrate that the subgenomic 2 kb *copia* RNA is expressed much more efficiently (>10-fold) in transfected *DH33* than in the full genome length RNA.

Furthermore, by transferring into *Drosophila hydei* cells (remember that this species does not comprise any endogenous *copia*) a *copia* plasmid in which a 3 kb deletion of the internal region corresponded exactly to the normally spliced out sequence, Yoshioka *et al.* (1990) could prove that the 2 kb RNA contains sufficient information to make *copia* VLPs. By kinetic studies in an *in vitro* translation system, they showed, moreover, that this spliced RNA species encodes not only a *gag*-like protein but also a protease (just as deduced from the nucleotide sequence). A 50-kDa polyprotein precursor was cleaved into the 33-kDa major VLP protein and a smaller 23-kDa polypeptide corresponding to the protease domain. When a simple Asp to Ala mutation was substituted in the putative protease site, this autocatalytic process could be suppressed, with accumulation of the precursor.

2. TRANSCRIPTS OF OTHER RETROTRANSPOSON FAMILIES

The same general observations were made on the transcription, in cultured cells, of all studied retrotransposons: *412* (Rubin *et al.,* 1976; Finnegan *et al.,* 1978; Young and Schwartz, 1981; Schwartz *et al.,* 1982; Mossie *et al.,* 1985; Micard *et al.,* 1988); *mdg1, mdg2,* and *mdg3* (Tchurikov *et al.,* 1978; Ilyin *et al.,* 1978, 1980a,c,d; Georgiev *et al.,* 1981; Mossie *et al.,* 1985); *gypsy* (Ilyin *et al.,* 1980b; Mizrokhi and Mazo, 1991; see Section I.D.1.); *17.6* (Inouye *et al.,* 1986); *B104* (Scherer *et al.,* 1981, 1982); *1731* (Peronnet *et al.,* 1986; Fourcade-Peronnet, 1988; see Section I.D.3.).

Compared to the situation in the whole organism, retrotransposons are usually overexpressed in cultured cells and the specific transcripts of each family constitute a significant fraction (0.5 to 1.5%), albeit variable for each cell line, of polyadenylated RNAs. Yet, *B104,* although abundantly transcribed in embryos, did not give any appreciable hybridization with cultured cell RNA, at least from the *Kc* line (Scherer *et al.,* 1981, 1982).

Full-length and one or a few smaller transcripts are generally detected and their transcription runs from the 5′ LTR to the 3′ LTR, as is the rule in retroviruses. Putative initiation sites and polyadenylation signals could be identified in these repeated sequences (see the following section on transcriptional regulation).

Some peculiarities, however, have been reported. For instance, in *mdg1, mdg3,* and perhaps *mdg4 (gypsy),* both DNA strands might be transcribed (Ilyin *et al.,* 1980a,d; Chmeliauskaite and Ilyin, 1980; Georgiev *et al.,* 1981). Double-stranded RNA, homologous to these elements could be isolated. This was an artifact, the annealing being favored by the heating step of the preparation procedure, but it revealed, in any case, the presence in the cells of complementary RNAs, generated by symmetrical transcription. One DNA strand was however preferentially transcribed and the levels of its transcripts (corresponding to the main 5′ to 3′ direction) were about 20 times higher than those from the other strand.

D. Transcriptional Regulation of Retrotransposons

The regulation of a few retrotransposons only has been seriously investigated, and the work on the so-called *1731* element deserves special mention.

1. 1731: *A Steroid- and Heat Shock-responsive Retrotransposon*

This 4.6 kb transposable element was discovered during the cloning, by "differential screening," of *Drosophila* genomic sequences whose expression may be modulated by ecdysteroids in cultured cells (see Chapter 8). Several recombinant plasmids of Maniatis's genomic library were shown to hybridize more intensively with poly(A)$^+$ RNAs from β-ecdysone-treated *Kc* cells than with those from control cells. A few others, on the other hand, corresponded to decreased mRNA species (Couderc *et al.,* 1984). Two of these latter recombinant plasmids were soon found to contain the same novel retrotransposon (named *1731* from the plasmid denomination) (Peronnet *et al.,* 1986 a to c), whereas three others corresponded to the already-known *412* element (Micard *et al.,* 1988). Such retrotransposons with a "negative regulation" by steroids (Becker *et al.,* 1991) seemed to us an attractive model, particularly because *1731* was also shown to be modulated, but this time positively, by heat shock (see Chapter 7).

The complete sequencing of *1731* (4,648 bp) confirmed that its central region, with two main ORFs (one sequence *gag*-like and other *pol*-like), is flanked by two typical long terminal repeats (336 bp each, with 96% homology) (Fourcade-Peronnet *et al.,* 1988). A credible TATA box

(TATATAT, at −59) and two CAAT motifs (at −129 and −157) can be spotted on the coding strand; on the other hand, downstream there is a canonical polyadenylation signal (AATAAA, at +8) followed by a CA site of polyadenylation (at −118) (Fig. 10.2).

The identification, in *Kc* cells, of a major *1731* transcript (4.6 kb) corresponding to the full length of the element (Peronnet *et al.*, 1986), then the precise localization, by primer extension analysis, of the start site of initiation in the 5′ LTR (Ziarczyk *et al.*, 1989) demonstrated the functional division of both LTRs in three successive parts, in the order 5′-U3-R-U5-3′, which is characteristic of the LTR architecture of retroviruses and of their unidirectional promoter activity.

In addition, a series of putative regulatory motifs, very much like the consensus sequences of the specific binding sites of several well-known cellular transcriptional regulatory factors, were recognized in U3, upstream from the TATA box. Among these motifs, were (a) an imperfect 15 nucleotidic palindrome reminiscent of the binding site of a vertebrate

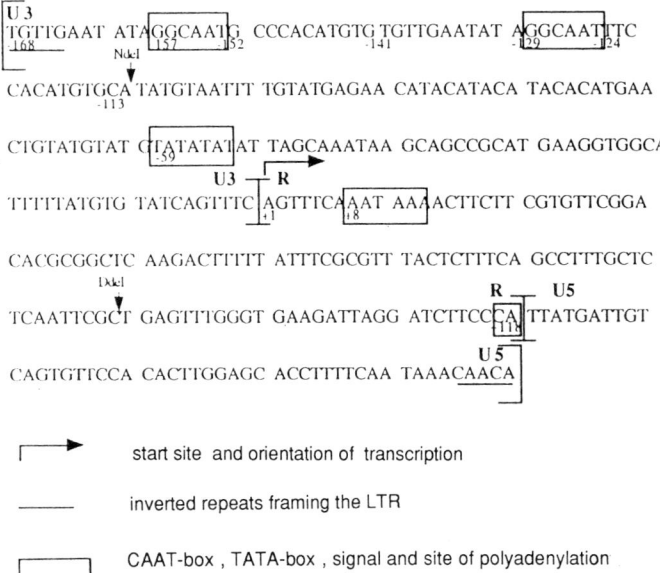

FIGURE 10.2 Nucleotidic sequence and definition of the U3, R, and U5 regions of the 5′ LTR of *Drosophila 1731* retrotransposon. From Peronnet *et al.* (1986); courtesy of Dr. Peronnet.

glucocorticoid receptor; (b) two sequences similar to the binding site of ecdysteroids (see Chapter 8); (c) a 14 nucleotide sequence with 72% homology to the heat shock-responsive element; and (d) 11 nucleotides resembling the binding site of NF-κB in HIV LTR and also three 10 nucleotide sequences very similar to the binding element of *dorsal* product.

Recombinant plasmids, in which the reporter gene *CAT* was put under the control of a complete LTR, or of various deleted segments of it, permitted a "functional dissection" of the LTR of *1731,* in transfected *Drosophila* S2 cells exposed or not to hormonal or heat-shock treatments (Ziarczyk *et al.,* 1989; Ziarczyk and Best-Belpomme, 1991; Ziarczyk, 1992). The main conclusions were (see Fig. 10.3) as follows:

Both the 5' and 3' LTRs constitute unidirectional promoters of identical strength.

The use of deleted constructs made it possible to define distinct functional domains. Beside a central "core promoter" (of about 200 bp), the last 76 nucleotides of U5 appeared as a negative regulatory sequence (or "silencer": its excision resulted in a 4-fold increase in the CAT activity), whereas the 5' end of U3 has, as a whole, a positive regulatory function ("enhancer"). As a matter of fact, this 5' extremity comprises a noticeable tandem of two direct and almost perfect repeats of 28 nucleotides and, when the proximal one is sufficient to multiply by 50-fold the CAT activity of the core promoter, the additional presence of the distal repeat results in a slight decrease.

The LTR, alone, is responsible for the hormonal negative modulation of *1731.* In S2 cells that had been transfected for 48 h with a *1731* LTR–CAT plasmid, the CAT activity dropped by at least a factor of two after 18 h of a treatment with 20-hydroxyecdysone (0.1 μM); as an internal control, it was verified that, concomitantly, the catalase activity was 5-fold increased (see Chapter 8). A decrease in the CAT activity was already detectable after 30 min of hormonal exposure.

Conversely, in similar experimental conditions, a 37°C heat shock promotes a 7-fold increase in the CAT activity after 4 h.

It could be shown that the same proximal repeat of 28 nucleotides that normally activates the core promoter is imperative for both negative regulation by ecdysteroids and positive regulation by heat shock, but this does not necessarily mean that other sequences are not also involved in these modulations.

A thorough analysis, using gel retardation assays of the many nuclear factors interacting with the regulatory regions of *1731* LTR is currently being conducted in our laboratory by Best-Belpomme, Fourcade-Peronnet

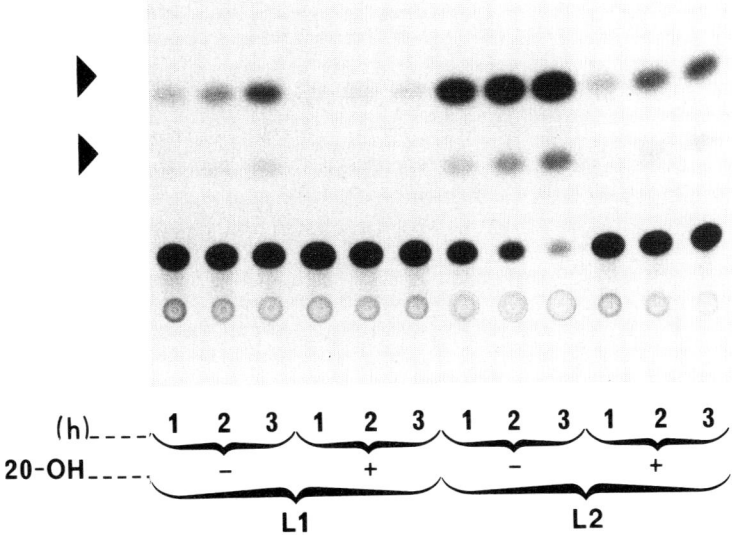

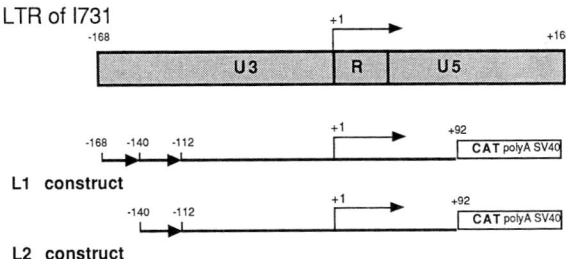

FIGURE 10.3 Functional dissection of the LTR of *1731* element. Influence of 20-hydroxyecdysone on the promoter activity of the LTR, as assayed after transfection with two different CAT constructs. In the vector L1, the reporter CAT sequence is driven by the segment-168 to +92 of the 5' LTR of *1731,* while in *L2* (segment −140 to +92) the more distal direct repeat had been deleted. Schneider *S2* cells were transfected with L1 or L2 vector DNA, then treated, or not, with 20-hydroxyecdysone (0.1 μM for 18 h) and the CAT activity was assayed (see Chapter 9, Appendix 9.H) 40 h after transfection (reaction times: 1 to 3 h). Autoradiography. Arrows indicate the acetylated forms deriving from radioactive chloramphenicol. From Ziarczyk (1992); courtesy of the author.

and collaborators. Provisional results were reported in Codani-Simonart's and Lacoste's doctoral theses (1993 and 1994, respectively), but most data are still unpublished (see Fig. 10.4).

A protein complex, binding to a synthetic oligonucleotide (called A) that copies the proximal repeat of the U3 region, seems to correspond to the 20-hydroxyecdysone receptor. It coelutes in the same 0.65 to 0.7 *M* KCl fraction from a heparin-agarose column and was characterized, in a retardation assay, by its specific attachment to an oligonucleotide

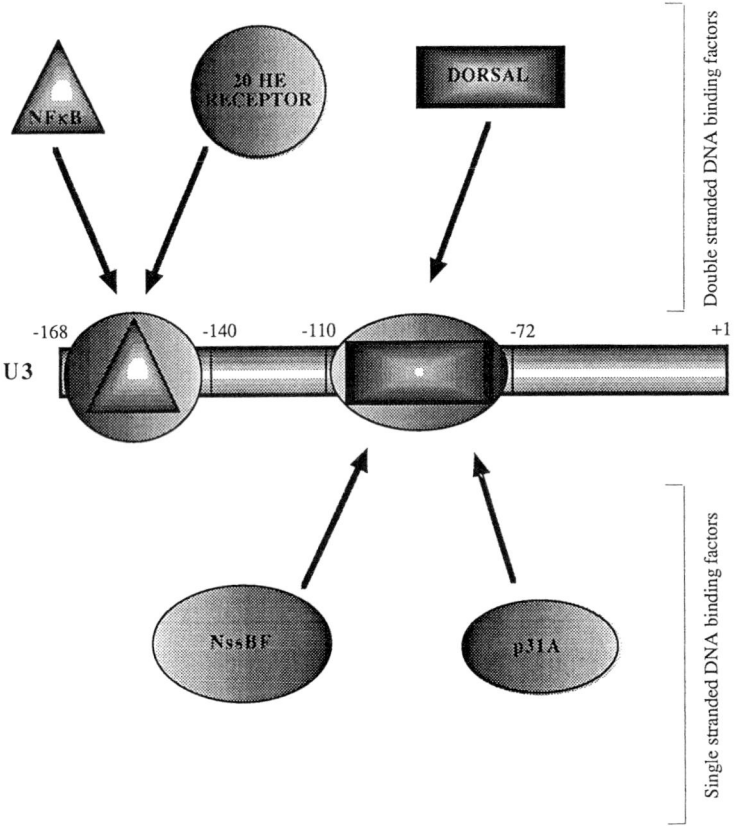

FIGURE 10.4 Putative transcriptional regulators that bind specifically to nucleotidic sequences of the regulatory regions (U3) of the LTR of *1731* retrotransposon. From Lacoste (1995); courtesy of Dr. Lacoste.

reproducing the ecdysone-responsive element of the *hsp27* gene (see Chapter 8). Confirmation was provided by its binding to tritiated Ponasterone and by its identification, on Western blots, with an antiecdysteroid receptor antibody (kindly sent by Professor Hogness). Now, the receptor/probe *hsp27* complex could be reduced, by 50%, by competition with 25 molar excess of the U3 oligonucleotide A. Symmetrically, the nuclear protein/oligo A complex was reduced by previous incubation with the *hsp27* probe.

On the other hand, the heat-shock factor (see Chapter 7), well characterized in the 0.35 M KCl fraction by its binding to a canonical heat-shock element, does not seem, so far, to attach itself to any segments of U3. Its binding might necessitate the cooperation of other factors, or the observed heat-shock effect might be indirect (other factors are known to be increased in stress conditions).

There is strong evidence that NF-κB-like factor(s) (most probably the product of *dorsal* or of the newly described *Dif*) binds to the *1731* LTR.

A novel nuclear factor (95 kDa) was named *NssBF* (nuclear single-strand DNA binding factor) because it specifically binds *in vitro* to the coding strand, i.e., the mRNA-like strand, of the *1731* LTR DNA, between position -99 and -74. There is no detectable binding to either the corresponding RNA, or to the complementary DNA strand (Fourcade-Peronnet *et al.*, 1992). The occurrence of similar "single-stranded DNA-binding proteins" has been reported in various cells, and they might be involved in replication or transcription. The recent discovery of another single-stranded DNA binding factor able to interact with the same region (and provisionally called *P31A*) was particularly intriguing. The sequencing of its cDNA confirmed that it was also a novel factor and its gene could be mapped to the 28EF region of polytene chromosomes. *In vitro* translation generated an 11 kDa protein which shows a 24% identity, especially in its helix–turn–helix, with a murine factor whose specificity and function are unknown (Lacoste, 1994).

It is noteworthy that DNA cleavage sites mediated *in vitro* by topoisomerase II could be identified in the LTR of *1731,* and one group of most frequent sites is, as it happens, located in the NssBF binding sequence (Nahon *et al., 1993*).

2. copia EXPRESSION AND INFLUENCE OF HOST CELL

As can be noted, the identified transcriptional modulators interacting with the regulatory regions of retrotransposons are cell-encoded factors, so that it seems reasonable to speculate that some of these LTR regulatory sequences were "borrowed" from the host genome during evolution (in the same way that oncogenic retroviruses "borrowed" cell protoonco-genes). In any case, the relative importance of the different regulatory elements of a given transposon should greatly depend on the array of transcriptional factors that may be present in a given cell type at a given time. This seems the best explanation of the somewhat contradictory results obtained by Sinclair *et al.* (1984) and Sneddon and Flavell (1989), after two independent "functional analyses" of the LTR of *copia*.

By systematic deletions from the two tandem LTRs and flanking se-quences of their selection plasmid *pCV2gpt* (see Chapters 5 and 9), Sin-clair and collaborators concluded, after transfection in *Drosophila hydei* cells, that one *copia* LTR is sufficient for correct expression and that the regulatory sequences reside entirely within the LTR and essentially downstream of the transcription start site(s). Removing this 3' end of the LTR reduced the transient expression of *gpt* by 10-fold.

A similar "functional dissection," but performed in *Drosophila mela-nogaster* cells transfected with a *copia*-CAT vector (Sneddon and Flavell, 1989), led to contradictory results (see Chapter 9, Section VI.A). They found an additional control region downstream of the 3' end of the LTR. This acts as an "enhancer," since it could maintain its activity when translocated several hundreds of nucleotides upstream of the LTR and in both orientations. It was not clarified whether this effect is correlated with the presence, in this untranslated "leader" sequence, of palindromic repeats (resembling an AP5-binding motif observed in the mammalian SV40 enhancer). Moreover, the data do not allow a distinction to be made between a transcriptional activation or an increase in the stability of messengers. Concerning the regulatory regions within the LTR itself, the authors observed striking differences when the same constructs were transfected into different *Drosophila* cell lines. In transfected *Drosophila melanogaster* cells, two regions on either side of the major transcriptional start sites of *copia* 5' LTR (one upstream, from nucleotide 42 to 66, and another downstream, from 180 to 203) are required for a high level of *copia*-CAT expression. Conversely, in transfected *Drosophila hydei* cells, only the upstream region is needed.

Recently, Cavarec and Heidmann (1993; Cavarec *et al.*, 1994) identi-fied, in the 5' untranslated region of *copia*, a sequence which is responsible for both positive and negative regulation by homeoproteins, and the results may, once more, vary with the host cell. The transcriptional activity was decreased 5- to 10-fold with *engrailed*, *even-skipped*, and *zerknüllt* products in cotransfected *Drosophila hydei* DH33 cells, whereas a 10-

to 30-fold increase was observed with *fushi tarazu* and the same *zerknüllt* in Schneider's *S2* cells.

The fact that the expression of a *copia* LTR-driven reporter gene was found to be much more efficient in all transfected *Drosophila hydei* cell lines than in several tested *Drosophila melanogaster* cell lines (Burke *et al.,* 1984) raises another point. Factors necessary for *copia* LTR expression might be limited in cultured *Drosophila melanogaster* cells, in which the number of *copia* elements is amplified, while this transposon is normally absent from the *Drosophila hydei* genome. An alternative explanation is the possibility that *copia* autoregulates its own expression. Bryant *et al.* (1991) established that cotransfection of a full genomic *copia* element, together with a CAT-expression vector under the control of a single *copia* LTR, resulted in a repression of the CAT activity. Cotransfection of the spliced 2 kb *copia* specific cDNA showed that this spliced product is, in itself, sufficient for a significant transregulation.

In a brief note, Braude-Zolotarjova and Schuppe (1987) reported that the *copia-CAT 1* plasmid constructed by DiNocera and Dawid (1983) is taken up more effectively and expressed more strongly in *Drosophila melanogaster 67j25D* cells than in *Drosophila virilis 79f7Dv3g* cells. This difference is specific (there is no such effect, for instance, on a Rous sarcoma virus LTR–CAT vector) and seems to be posttranscriptional, probably affecting the RNA stability. Thus, the cellular context can strongly influence the expression of a retroelement. On the other hand, the authors reported that a CAT vector driven by the promoter of a cellular gene (*larval serum protein 1*) was found to be positively modulated by *copia* cotransfection; such a transregulation of normal cellular genes by a retrotransposon might have far-reaching implications.

3. gypsy AND PRODUCT OF SUPPRESSOR OF HAIRY WING

The genomic insertion of the transposable element gypsy is responsible for many "spontaneous" mutations in *Drosophila melanogaster.* In addition, a "suppressor" is a gene whose mutations can reverse, at least partially, the mutant phenotypes induced by certain second-site mutations. For instance, *suppressor of Hairy wing* (*su(Hw)* corrects, to some extent, insertional mutations due to *gypsy*, such as the *yellow y^2* allele (characterized by the yellow color of fly wings and body, while bristles remain of a gray-black wild type). In this latter case, the transposon was located at -700 bp from the initiation start site of the *yellow* gene, separating the two tissue-specific enhancers governing wing and body coloration from the rest of the promoter.

Corces and collaborators (see review by Corces and Geyer, *1991*) could demonstrate that the suppressor acts via the transposon. The wild type *su(Hw)* protein possesses several of the characteristics of a transcriptional modulator and, with its multiple zinc finger motifs, binds to octomer repeats of *gypsy* and activates its transcription. It is the undue presence of this *su(Hw)* protein in the regulatory regions of the *yellow* gene (mediated by the *gypsy* insertion) and the resulting interferences with normal transcription factors which is, in fact, responsible for the disturbed expression of the y^2 gene. When the suppressor gene is mutated, affecting the DNA binding or transactivating activities of its product, such a deleterious influence may disappear.

1. A functional analysis, through transfection into *S2* cells of different *gypsy*-CAT constructs, revealed a pecularity in the promoter region of *gypsy* (Jarrell and Meselson, 1991). A 100-bp segment overlapping the initiation start site was sufficient for accurate and normal-level transcription. So, unlike most RNA polymerase II promoters, the promoter of *gypsy* needs downstream sequences for full expression. Furthermore, it was noted that the initiation site resembles a consensus sequence (TCAGTY) present in many other transposons or genes lacking a TATA box motif.

A similar approach, with various *gypsy*–CAT plasmids, had previously enabled Mazo *et al.* (1989) to establish that, in addition to a core promoter, there were two downstream and adjacent modulators, respectively negative and positive, in the untranslated sequences. A poly(A) block followed by a 23 nucleotide imperfect palindrome acts as a negative element, while the following region with 12 repeats of a 12 nucleotide sequence is a positive regulator. It is noteworthy that each of those repeats consists of a consensus octomeric motif which is homologous to that of several mammalian transcription enhancers. Moreover, using the method of retention of DNA-protein complexes on nitrocellulose filters, combined with proper oligonucleotide competitions, they showed that these two modulator elements bind to two different proteins from *S2* cell nuclear extracts. Then they could demonstrate, with extracts isolated from mutant pupae, that *suppressor of Hairy wing* mutations resulted in a decrease in the protein binding to the 12 repeat region (enhancer), while in *suppressor of fork* mutants the binding to the palindromic region (silencer) is reduced. Both sets of results are in perfect agreement with previous observations on the respective effects of these two suppressor mutations on the levels of *gypsy* RNA in flies.

As soon as the *su(Hw)* gene was cloned, Spana *et al.* (1988) confirmed the specific binding of its product to the enhancer region of *gypsy*. Whereas the bacterial fusion protein was incomplete, the full-length protein could be obtained and partially purified from nuclear extracts of *Drosophila* S2-M3 cells transformed with a metallothionein promoter-*su*(*Hw*) expression vector. The ability of this 110 kDa protein to interact with different regions of *gypsy* DNA was then tested, using filter binding and gel mobility shift assays. Competition experiments with various *gypsy* fragments and inhibition with anti-*su*(*Hw*) antibodies (prepared against the bacterial protein) showed the specific attachment, *in vitro*, on a 367 bp fragment of the transcribed untranslated region of *gypsy*. This sequence is precisely that which contains the 12 repeats of the octomeric motif. The footprinting method clarified the interaction of *su*(*Hw*) protein with these enhancer-like sequences.

Note: As mentioned above, when the transposable element is located within the regulatory regions of a given gene (for instance, in the *yellow* y^2 allele), this *su*(*Hw*) transcriptional activator can interfere with the normal functions of surrounding transactivating factors and, thereby, disturb the expression of the gene. As an alternative interpretation (Roseman *et al.*, 1993), the binding of *su*(*Hw*) protein is postulated to cause changes in the chromatin structure, which, in turn, inactivates the enhancer or silencer.

2. In many other cases, the mechanism of insertional mutation is different. The presence of the transposable element results in the transcription of abnormal interrupted transcripts. In order to analyze this mutational process, Dorsett *et al.* (1989) developed an interesting experimental model in transfected *Drosophila* culture cells. They chose the heat-shock *hsp82* gene, because it is well known to be relatively simple (a single intron) and easily inducible (see Chapter 7). The authors inserted a *gypsy* element, or specific fragments of it, into a defined site of the gene intron. When *gypsy* was in a parallel orientation, the amount of the spliced 1.8 kb transcript was greatly reduced, while a novel truncated 1.4 kb RNA was 40-fold more abundant. On the whole, however, the activity of the *hsp82* promoter did not seem to be affected. The length of this abnormal transcript suggested that its polyadenylation was directed by the polyadenylation signal AATAAA located at the nucleotide 257 of the *gypsy* 5' LTR. As a matter of fact, new constructs, with only short segments comprising this polyadenylation signal, gave rise to similar truncated transcripts; their amount could still be increased by the presence of an additional sequence, between 641 and 835 (We would like to point out that it is the same *gypsy* sequence, with direct repeats, that was, just above, shown to be an enhancer.)

Note: In quite a different context (see Chapter 7 on heat shock), in a recent paper by Lubomiskaya and Ilyin (1993) addressing the transcription of *gypsy* in heat-shocked cells, active degradation of preexisting *gypsy* transcripts was observed.

E. Virus-like Particles

Fortuitously discovered during the late 1960s in *in vivo* cultures of imaginal discs (propagated by successive transplantations into the abdomen of female flies), then in mutant tumoral imaginal discs and neoplastic blood cells, virus-like particles do, in fact occur in a variety of normal *Drosophila* tissues, of adult or larval origin (see reviews by Gateff, *1978*, 1988). They are particularly widespread in most *in vitro* cell lines from *Drosophila melanogaster* [Williamson and Kernaghan, 1972; Teninges and Plus, 1972 (see Fig. 10.5); Lengyel *et al.*, 1976]. In the *GM3* embryonic cell line, for instance, Saigo *et al.* (1980) estimated that the total number of VLPs per nucleus, in the log phase and in the stationary state was about 10^3 and 10^4, respectively.

1. MORPHOLOGY

In electron micrographs, virus-like particles may be observed singly or in clusters, sometimes in the form of quasi-crystalline inclusions. More frequently confined to the nucleus, they were also found in the cytoplasm. Their size is reported to vary from 40 to 60 nm. This difference might imply the existence of distinct populations of particles or might simply result from the different techniques of observation. Roundish, slightly elliptical or polyhedral, a VLP is typically composed of an electron-dense outer shell surrounding a less dense core.

2. ORIGIN

These intriguing particles did not resemble any known insect virus (see Chapter 11). Nevertheless, because their morphology is highly evocative of nonenveloped virions, they were at first interpreted as signs of some latent viral infection.

Interest in the problem was renewed when Heine *et al.* (1980) demonstrated that the fraction from cultured cells that contains VLPs is associated with a reverse transcriptase-like activity and the presence of endogenous templates. It became obvious that these torroidal-shaped bodies are very similar to the retrovirus-related A-type particles previously described

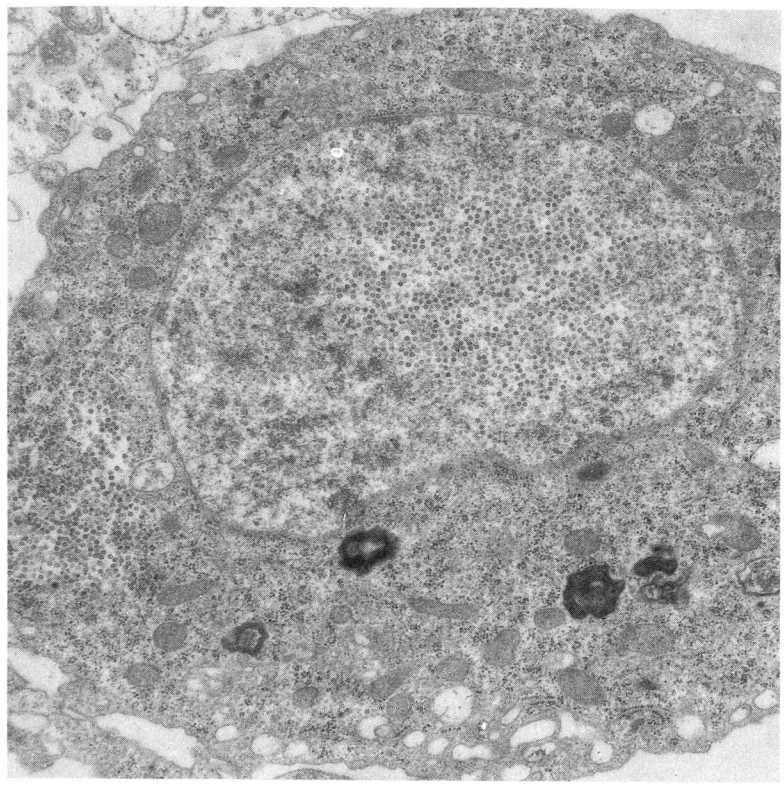

FIGURE 10.5 Virus-like particles in *Drosophila* cultured cells (*Kc* line). Although VLPs are primarily nuclear, a large group of similar cytoplasmic particles may be observed. Courtesy of Dr. D. Teninges, CNRS, Gif-sur-Yvette, France.

in vertebrate cells (such as the so-called mouse intracisternal A-type particles or IAPs).

Finally, Shiba and Saigo (1983) established that VLPs from a *Drosophila melanogaster* cell line (*GM2*) contain 5 kb RNA molecules homologous to the transposable element *copia*. It is fair to point out that 3 years before, Ilyin *et al.* (1980d; see also Georgiev *et al.*, 1981) had reported that, in *67J25D* cells, transcripts from *mdg1* and *mdg3* were concentrated predominantly in free RNP particles (sedimenting as an 80S peak) and they had already proposed a correct interpretation of these structures.

Definitive confirmation of the causal relationship between VLPs and retro-transposons was provided when Miyake *et al.* (1987, 1988) succeeded in transfecting a plasmid carrying a genomic *copia* into a *Drosophila hydei* cell line (i.e., a species which normally lacks *copia*-hybridizable sequences) and observed the production *de novo* of true *copia* VLPs.

These Japanese investigators have thus far been unable to detect homology between VLP RNA and DNA from transposon families other than *copia*. This might be due to the cell line they used, because it is well known (see above) that not all retrotransposons are amplified or transcribed at the same level in the different established cell lines. On the other hand, *mdg1* and *mdg3* probes recognized RNAs in VLPs from *67J25D* cells (Georgiev, 1984). More generally, it is thought that each retrotransposon family can generate its own VLPs, and that one is dealing with mixed populations of VLPs, which might account for observed differences in size and morphology. This opinion is strongly supported by an elegant, albeit unpublished, experiment by J.L. Becker (personal communication, 1988). When the four deoxynucleotide triphosphates (one of them being labeled with ^{32}P) were provided to a VLP fraction isolated from *Kc* cells, DNA probes could be recovered as a result of endogenous reverse transcription, and they hybridized to the genomic DNA of a series of tested retrotransposons (*copia*, but also *1731, 412,* and *297*).

3. COMPONENTS

On sucrose density gradients, VLPs can be isolated from sonicated *Drosophila* cultured cells (especially in the stationary state) as a subcellular fraction with a density of 1.22 g/ml (Heine *et al.*, 1980; Saigo *et al.*, 1980). They consist of proteins and RNAs (about 87% proteins and 13% RNA, according to Saigo and collaborators).

In electrophoretic analysis on polyacrylamide gels, the major protein component of *copia* VLPs seems to be a 31 kDa molecule (Shiba and Saigo, 1983). Its total amino acid composition was shown to be very close to the deduced residue sequence of the putative 5' *gag*-like part of the *copia* large ORF; therefore, it should be the main capsid protein. Moreover, the presence in VLPs of reverse transcriptase molecules was inferred from its detectable enzymatic activity (see Section I.F).

As we have just emphasized, the high molecular weight RNA (5 kb) included in VLPs (Heine, 1980; Shiba and Saigo, 1980; Emori *et al.*, 1985) was found to be homologous to genome-size transcripts of *copia* (Shiba and Saigo, 1983). With a reticulocyte lysate system, it could be translated *in vitro* into several polypeptides related to those contained in

VLPs, since they were precipitable by anti-VLP polyclonal sera. So, the large RNA appears to be the exact counterpart of the retroviral genome confined within the virion capsid. In addition, as is also the case in retroviral virions, there are variable amounts of host cell-specific RNA species in VLPs. Shiba *et al.* (1980) first characterized these small rRNAs and showed that the 5S and 4S fractions hybridized to *Drosophila* ribosomal and various tRNA genes, respectively; one of this latter category might correspond to the specific tRNA-like primer used in reverse transcription. Moreover, Gorelova *et al.* (1989) revealed, in VLPs isolated from *67J25D* cells, all the replication intermediates that might be expected from a retroviral model of reverse transcription (see Section I.F).

By studying the translation and fate of the *gag* protein encoded by the *1731* element, with the use of *1731-lacZ* recombinant plasmids (in which the reporter gene was inserted in frame with ORF1), Kim *et al.* (1993) demonstrated an efficient translation of the chimeric *gag* protein in transfected *Drosophila virilis* cells (where *1731* is normally absent) and its expected gathering in cytoplasmic VLPs.

4. FUNCTIONAL SIGNIFICANCE

It is currently admitted that VLPs are strictly intracellular structures, which is consistent with the lack of any envelope-like sequences in the genome of most retrotransposons. In any case, no budding process involving VLPs has ever been observed at the cell membrane.

It is important to emphasize that VLPs bear a striking resemblance to the cytoplasmic torroidal-shaped bodies (A-type particles) similarly produced by mouse proviral-like IAP genes. Virus-like particles have often been considered to be a type of defective retrovirus-like virions, but, in a more constructive interpretation, they might constitute organized foci of reverse transcription, i.e., an essential step in transposition (see Section I.F).

Yet, there have been a few recurrent reports of the isolation of extracellular VLPs from supernatants of *Drosophila* cell cultures (for instance, Georgiev, 1984; Syomin *et al.,* 1993), although it is possible that they simply derived from cell disruption. Such observations, especially the latter one that concerns *gypsy* and describes double-shelled particles, should arouse a renewal of interest, in the light of very recent assumptions (Kim *et al.,* 1994) that *gypsy* behaves like an infectious retrovirus; it should be recalled that this element is one of the four retrotransposons known to contain a large 3rd ORF, i.e., a putative *env* gene, downstream from the *pol*-like ORF.

F. Reverse Transcriptase and Intermediates of Reverse Transcription

Reverse transcriptase (RT) is the key enzyme in the biology of retroviruses and seems to play a similar role in the intracellular transposition cycle of retrotransposons. By inserting into the coding sequences of a yeast retrotransposon (*Ty*) a DNA segment framed by splice sites, Boeke *et al.* (*1985*) elegantly demonstrated that the transposition of this element proceeds through an RNA intermediate and its reverse transcription. Moreover, a correlation was shown to exist between the abundance of the transcripts and the frequency of transposition.

With the exception of *copia* and its unique *gag–pol* ORF, the genomes of retrotransposons contain a 2nd large ORF encoding a polyenzymatic precursor, named after its main domain (*pol*, i.e., RNA-dependent DNA polymerase). The typical organization of its successive enzymatic domains is 5′-protease–RNase H–reverse transcriptase–integrase-3′, as in retroviruses. In *copia* and *1731*, however, as in Yeast *Ty*, the arrangement is different, which can be explained by a mere relocation of the integrase into the junction of the protease and RT domains (Saigo, *1986*).

Strong evidence that transposition of *Drosophila* retrotransposons requires the reverse transcription of an RNA intermediate has been obtained from culture cells (and, concomitantly, from transgenic flies).

DNA–RNA complexes corresponding to every expected step of a reverse transcription process were detected in various cell lines (see below).

PCR analysis of the genomic DNA from *Kc* cells revealed the occurrence of a few intronless copies of *copia* in which a deleted segment coincided exactly with the intron that must be normally spliced in order to generate the small *copia* 2 kb RNA (Yoshioka *et al.*, 1991).

Plasmids carrying a 5′ end truncated *gypsy* were introduced into S2 and *Drosophila hydei DH14* cells. In the genome of stably transformed cells, new complete DNA copies could be detected by PCR and Southern blot analysis (Lyubomirskaya *et al.*, 1993). This reconstruction of a 5′ LTR is an unambiguous indication of reverse transcription.

Modeling their work on Boeke's demonstration for *Ty* transposition, Heidmann and collaborators engineered a *hsp70-lacZ*[TR] recombinant gene as an indicator of reverse transcription. A polyadenylation signal, framed by acceptor and donor splice-sites was inserted between the promoter and coding sequences of a reporter gene in order to halt expression of this reporter. Such an "indicator" gene was inserted into various transposable elements, for instance, I element (see Fig. 10.8) and the whole construct, via a *P* element vector, could be integrated into the genome of fly strains. If transposition occurs in transgenic flies through a reverse transcription (i.e., through an RNA intermediate liable to be spliced), the activity of the reporter gene should be recovered in the transposed copy and possibly detected, in the present case by a X-Gal enzymatic blue coloration. The system worked *in vivo* (Jensen and Heidmann, *1991*). In order to use it in cultured cell populations, Jensen (1993) recently established a few permanent cell lines from different transgenic fly strains. In two of them, according to preliminary observations (Jensen *et al.*, 1994), a few rare blue cells (with a frequency of 10^{-6} to 10^{-7}) were detected after X-Gal enzymatic staining.

1. REVERSE TRANSCRIPTASE

It was a breakthrough when Heine *et al.* (1980) discovered that VLPs isolated from *Drosophila* culture cells were associated with a RT-like activity. This was confirmed by several other groups (Shiba and Saigo, 1983; see more recent reports by Yoshika *et al.*, 1993; Lyubomirskaya *et al.*, 1993).

The rigorous demonstration of the specificity of this enzyme deserved, however, further study. It could be partially purified from a VLP fraction of *Kc0* cells by Becker *et al.* (1987), then by Lescault *et al.* (1988), in our own laboratory.

The differential diagnosis between RT and other DNA polymerases is primarily based on its preferential use of relatively specific ribopolymers as primer templates for initiation and elongation of DNA synthesis. For instance, the 1.2 g/ml density fraction prepared from *Kc* cells had the ability to utilize the synthetic polymethylcytidylate–oligodeoxyguanylate [poly(rCm)-oligo(dG)] which is considered to be a highly specific template primer for viral RT.

All polymerases need divalent ions as cofactors, and the activity of retroviral RT is greater in the presence of Mn^{2+} ions than in that of Mg^{2+} ions. Now, a similar preferential requirement for Mn^{2+} was noted with *Drosophila* RT, but the latter's activity was optimal at 25°C (which is the usual temperature for *Drosophila* cell growth). Moreover, this RT could be inhibited by HPA-23 (ammonium 21-tungstate 9-antimonate), an efficient inhibitor of the retrovirus RT.

Incidentally, Becker *et al.* (1988) observed that this RT activity increased in the hours that followed an ecdysteroid treatment, although it was not clarified whether this was correlated with a possible increase in the VLP population (as suggested by preliminary observations made by Debec and Marcaillou (personal communication, 1995).

Another step was the demonstration that a retrotransposon genome is indeed able to encode a functional RT. The best conserved parts of the putative reverse transcriptase-coding sequence of the *1731* element were subcloned and expressed in *Escherichia coli* (Champion *et al.*, 1992). This recombinant protein displayed a true RT activity. Furthermore, rabbit antisera directed against this bacterial protein recognized the VLP fraction from *Kc* cells and allowed the detection of an immunoreactive protein of about 110 kDa that was present only in *Drosophila melanogaster* cells, and not in *Drosophila virilis* or *Drosophila hydei* cell lines (two species whose genome does not contain the *1731* element).

2. Small Extrachromosomal DNA Circles

Stanfield and Helinski (1976) discovered a heterogeneous population of covalently closed circular DNAs in embryos and cultured cells of *Drosophila melanogaster*. The majority of them were located in the nucleus (3 to 40 per cell). More than 90% of the molecules had a mean contour length of 1.1 μm and their buoyant density was 1.703 g/ml. Such DNA circles were found to be complementary to middle repetitive sequences of the *Drosophila* genome, although not to ribosomal or to tRNA genes, and homologous to transcripts from different RNA fractions, so that it seemed therefore plausible that they might correspond to transposable elements (Stanfield and Lengyel, 1979, 1980).

In 1981, Flavell and Ish-Horowicz characterized circular copies of the *copia* element amid this heterogeneous population. In *Kc* cells, these circular *copia* molecules do not exceed one copy per 10 to 50 cells. Most of them are composed of a complete copy of the internal regions of the element, combined with either one (the dominant form, 4.7 kb long) or two LTRs (5 kb, four times less abundant). The striking analogy with the circular proviral forms of retroviruses was obvious. As observed among several *copia* circles with two LTRs (Flavell and Ish-Horowicz, 1983), the sequence heterogeneity of the junction between the two ligated LTRs seemed to argue against a creation by reverse transcription and favored rather the hypothesis of a derivation by imprecise excision of a genomic *copia*, then its propagation as a plasmid. After an analysis of the uptake of BUdR (a general method for discriminating nonconservative synthesis of DNA by reverse transcription from usual semiconservative replication of chromosomal DNA), Flavell (1984) had to admit that extrachromosomal circular *copia* are, in fact, generated by reverse transcription. It cannot, however, be excluded that a small part of them perhaps derive from excision through homologous (between the two LTRs or small flanking sequences) or nonhomologous recombination.

Similarly, free circular forms, corresponding to most of the retrotransposon families, were subsequently identified in *Drosophila* cultured cells (as in embryos): *412* (Shepherd and Finnegan, 1984; Junakovic and Ballario, 1984; Mossie *et al.*, 1985); *mdg1* and *mdg3* (Ilyin *et al.*, 1984; Junakovic and Ballario, 1984; Mossie *et al.*, 1985); *gypsy* (Mossie *et al.*, 1985); *1731* (Peronnet *et al.*, 1986).

The number of circular copies always remains extremely low: about one per 5 to 20 or sometimes hundreds of cells. Their relative abundance, in the different families of elements, varies (up to 20-fold) between the various cell lines and from cultured cells to embryos. For instance, taking advantage of the occurrence of a minor variant of *mdg3*, Ilyin *et al.* (1984) showed that this deleted form was selectively amplified and preferentially expressed in *Kc* cells, and the standard full-sized form in *67J25D* cells. Only the amplified species was found to be represented as extrachromosomal circles in the corresponding cell line. Thus, there seems to be some correlation between the number of elements integrated in the genome, the amount of specific poly(A)$^+$ RNA present in the cell and the abundance of the circular forms, which might be consistent with the hypothesis of a reverse transcription.

For a time, such circularization was considered to be a crucial step in the typical retroviral cycle before genomic integration takes place; but it has now been established that integration takes place through linear DNA copies (Brown *et al.*, *1989*) and DNA circles are instead interpreted as being a sort of functional "dead end."

Nevertheless, in transfected cell lines grown under selective conditions, the persistence for months of free circular *copia* copies in which the *Escherichia coli gpt* gene was inserted (thus conferring on the cells the ability to use exogenous purines; see Chapter 5) suggested that extrachromosomal circles might be perpetuated, as plasmids, by semiconservative replication (Sinclair *et al.*, 1983).

3. EXTRACHROMOSOMAL LINEAR DNA COPIES AND PUTATIVE RNA-DNA REPLICATIVE INTERMEDIATES

Flavell (1984) isolated from *Drosophila* cultured cells linear and complete extrachromosomal *copia* elements whose structure, with two LTRs, closely resembles the analogous preprovirus linears of retroviruses. They are not to be confused with linearized *copia* circles, since the majority of circles contain only one LTR. The analysis of BUdR uptake confirmed that the synthesis of these extrachromosomal molecules is not conservative, which is consistent with their presumed generation by reverse transcription.

The process of reverse transcription requires elaborate "molecular gymnastics" (Varmus) with defined RNA-DNA intermediates. Such DNA–RNA complexes are detectable in *Drosophila* cultured cells. For instance, Arkhipova *et al.* (1984) observed, in *67J25D* cells, perfect hybrid molecules between *mdg1* or *mdg3* DNAs and their corresponding poly(A)$^+$ RNAs. Similarly, in *Kc* cells, Becker *et al.* (1990) characterized DNA–RNA duplexes specific to *1731* and a few other retrotransposons.

A more thorough analysis (Arkhipova *et al.*, 1986) allowed the identification of (−) and (+) "strong-stop" DNAs corresponding to *mdg1, mdg3,* and *mdg4*. In the case of the (−) strong-stop DNA, it was shown that RNase treatment could cut off some 75 to 80 nucleotides from the sequence, which reflects the tRNA primer molecule that should be, typically, covalently attached to this (−) strong-stop complex.

It seems particularly noteworthy that such putative intermediates of reverse transcription could be located within the VLPs (Kikuchi *et al.,* 1986; Gorelova *et al.,* 1989). This strongly suggests that VLPs constitute organized foci of reverse transcription. The compartimentalization of RT might prevent the enzyme from inopportunely reverse transcribing any other cellular RNA.

In Fig. 10.6, a schematic representation of the presumed intracellular cycle of a retrotransposon is presented.

II. LINE-LIKE RETROPOSONS

Long interspersed nucleotidic elements, or "nonviral" retroposons, are another class of transposons that are also found in a wide range of eukaryotic organisms (from mammals to plants). The mechanism of their replicative transposition, although requiring, likewise, a reverse transcription from RNA intermediates, is probably rather different since they lack the terminal repeats of "viral" retrotransposons.

They have the typical structure of processed pseudogenes, with a 3′ poly(A) tail, often preceded by a polyadenylation signal (with the exception, however, of the *Drosophila* so-called *I* element that ends with 4 to 7 tandem repeats of the triplet TAA). The same "plus" strand contains two ORFs: the 5′ one encodes a possible nucleic acid-binding polypeptide

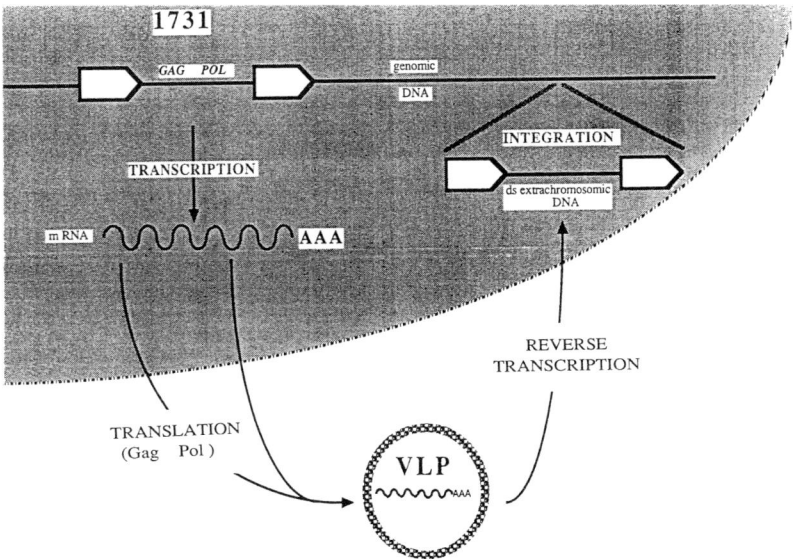

FIGURE 10.6   Presumed intracellular cycle of a *Drosophila* retrotransposon. Courtesy of Dr. Maisonhaute, CNRS–Université P. et M. Curie, Paris.

(with cysteine fingers like those of retroviral *gag*), and ORF2 codes for a polyprotein (*pol*) with reverse transcriptase and RNase H domains (but devoid of integrase homology) (see Fig. 10.7). Moreover, LINEs seem to possess an internal promoter near their 5' extremity, so that their frequently observed 5' truncated copies might be inactive.

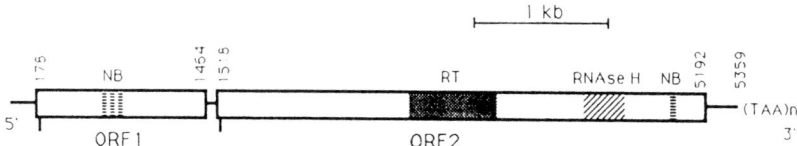

FIGURE 10.7   Structure of the *I* element. White boxes represent the two long open reading frames. Striped bars (NB) correspond to the nucleic acid-binding domains (three in ORF1, one in ORF2). The dark box (RT) represents the reverse transcriptase domain and the hatched box the RNase H domain. Short vertical lines below the white boxes indicate the position of the first methionine codon of each ORF. From Bucheton (1990); reproduced with the kind authorization of Dr. Bucheton and Elsevier Trends Journals.

Let us recall that *Drosophila* I factor controls the I-R (induced-reactive) "hybrid dysgenesis." In contrast with the almost nil effect of reciprocal crosses, when a male of a so-called "inducer" strain is crossed with a female of a "reactive" stock, the F_1 females (named SF) are sterile, because most of their eggs fail to hatch. Among the rare progeny, there is an unusually high rate of insertional mutation or chromosomal rearrangement. It was shown that only inducer fly strains contain 10 to 15 "active" I retroelements (5.37 1 bp long), whereas both types of strains contain incomplete "dead" copies in centromeric heterochromatin. The whole symptomatology of the dysgenesis is due to a very high frequency of I transposition in the germ line of SF females (see a short review by Bucheton, *1990*).

1. I-RETROELEMENT

In imitation of Boeke's demonstration of reverse transcription in the process of *Ty* transposition, recombinant *I* constructs (in which an intron-containing "indicator" gene was inserted) were used to demonstrate the RNA-mediate transposition of such tagged *I* elements, first in transgenic flies, then in cultured cells (Jensen and Heidmann, *1991;* Jensen, *1993;* Jensen *et al.,* *1994*):

The "indicator" gene with a spliceable block was a *lacZ* sequence driven by an *Hsp70* promoter (see preceding Section I.F); it was inserted into a large deletion of the ORF2. This marked *I* element was placed, moreover, downstream of a potent actin promoter. In transfected *Kc* cells, spliced DNA copies and specifically stained β-galactosidase-expressing cells (about 10^{-3} to 10^{-4} blue cell, after X-Gal staining) could not be observed before the cells had been cotransfected with an expression vector for the two normal ORFs of the *I* element. Therefore, over and above the fact that they constitute further evidence of reverse transcription during *I* transposition, these results showed that an *I* element with a defect in its reverse transcriptase coding domain can be complemented in trans for retrotranscription. The system offers a very suitable tool for analysis of the molecular mechanisms of *I* transposition (Fig. 10.8).

Because of the absence of LTRs (those complex assemblages which are rebuilt during the process of reverse transcription in retroviruses), the RNA template, in LINE transposition, must correspond to the full-length sequence. In other words, the promoter should lie entirely within the *I* element. Full-sized *I* transcripts could indeed be characterized in the ovaries of *SF* females, but the promoter activity of the 5' untranslated region of the *I* element was more easily assayed in cultured cells (McLean *et al.,* *1993*).

Transient transfection analysis in *S2* cells established that nucleotides +1 to +40 are sufficient for high promoter activity and accurate transcription initiation (starting at nucleotide +1). Only the initial GA(G/T)T motif, well conserved among other LINEs and several TATA-less gene promoters, seems to be absolutely necessary; but deletion to position +28 reduced CAT activity 2-fold, suggesting that some sequence between +28 and +40 also plays a positive role. Moreover, transcription initiation was sensitive to a low concentration (4 mg/ml) of α-amanitin, i.e., a specific inhibitor, at least at this concentration, of RNA polymerase II.

Bouhidel *et al.* (1994) have recently shown that the full-length transcript, which is the only transcript to have been observed from functional *I*-elements, may be used to translate both ORFs from their own initiation codons, at different rates.

2. JOCKEY

The involvement of RNA polymerase II in the transcription of LINE-like elements, and also the unusual internal position of their promoters and their peculiar structure had previously been demonstrated by Mizrokhi *et al.* (1988; see also Georgieva *et al.*, 1988) in the case of another nonviral retroposon, *Jockey*.

They identified, in poly(A)$^+$ RNA isolated from Schneider's cells, two transcripts corresponding, in length, to the full-sized (5 kb) and deleted (2.7 kb) copies which have been observed in *Jockey*; and it was shown that their initiation site coincides with the beginning of the element.

In the presence of α-amanitin (20 μg/ml) in a preparation of nuclei, the transcription of *Jockey* fell to 5% of its control value, which was identical to that of actin transcription and confirmed the participation of RNA polymerase II.

Then CAT constructs were used to transfect the cells, and the whole untranslated region of *Jockey* proved to be as efficient a promoter as that of the *gypsy* retrotransposon. The removal of the first 13 nucleotides completely abolished this activity (Fig. 10.9).

3. F ELEMENT

Similar transient transfection assays in *S2* cells led to the identification, in the 5' region of the mobile *F* element, of not only a promoter F_{in} (within the +1 to +30 interval) that transcribes in the inward direction and should control the formation of transposition RNA intermediates and ORF products, but also an additional and independent promoter F_{out} (within the interval +218 to +269) driving transcription in the opposite

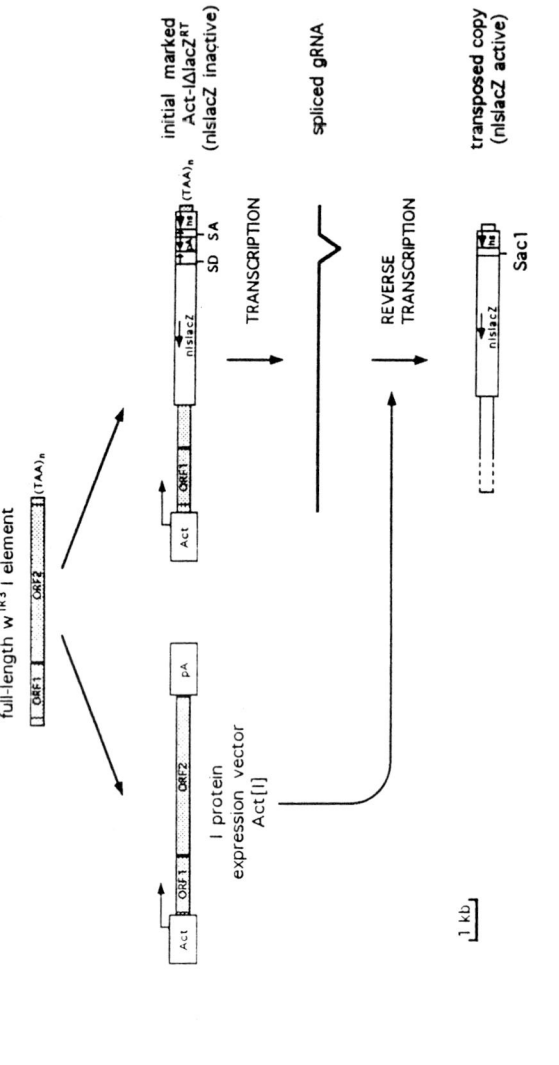

full-length w^{IR3} I element

I protein
expression vector
Act[I]

initial marked
Act-IΔlacZ^{RT}
(nlslacZ inactive)

TRANSCRIPTION

spliced gRNA

REVERSE
TRANSCRIPTION

transposed copy
(nlslacZ active)

Sac1

1 kb

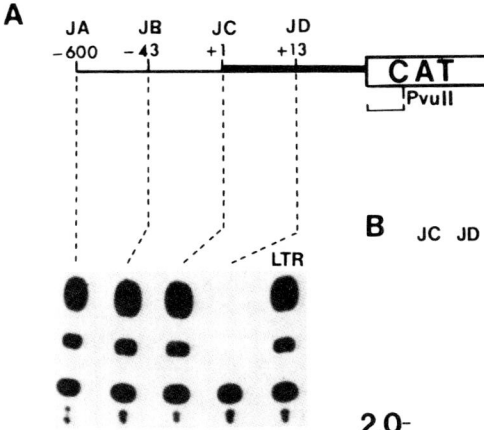

FIGURE 10.9 Schematic representation of *Jockey* CAT constructs and their expression in transfected *Drosophila* cells. (A) The CAT activities of constructs based on *Jockey* and *mdg4* LTR (the lane corresponding to this latter is simply indicated by LTR in the picture); (B) Northern blots with poly(A)⁺ RNA from culture cells transfected with pJC and pJD constructs. The hybridization probe is indicated by a bracket under the CAT coding sequence of (A). From Mizrokhi *et al.* (1988); reproduced with the kind authorization of Professor Ilyin and Cell Press.

FIGURE 10.8 Demonstration of reverse transcription in the process of transposition of the *Drosophila* I element.

The structures of the full-length w^{IR3} I element, with its two ORFs and 3' terminal TAA repeats, and of its two derived constructs are shown.

The expression vector for the I element coding sequences, *Act(I)*, contains the two ORFs under the control of the promoter (Act) and followed by the polyadenylation signal (pA) of the actin-5C gene.

The marked I element: The lacZ^{RT} indicator gene contains the β-galactosidase coding sequence fused to a nuclear location signal (nlslacZ), under the control of the *Drosophila* *hsp70* promoter (hs) (arrows indicate the orientation of each genetic element); in its initial configuration, nlslacZ is inactive because of a polyadenylation sequence (pA)inserted between the promoter and coding sequence and bracketed by splice donor and acceptor sites (SD and SA). This indicator gene is inserted into an I element (on a large deletion in ORF2 encompassing the reverse transcriptase domain). Moreover, the marked I element is also driven by the actin-5C promoter for high expression in cultured cells.

Transposed copy: Retrotransposition should result in a transposed copy in which the pA sequence have been removed by splicing of the RNA intermediate, thus resulting in an active nlslacZ gene (the SacI restriction site generated at the splice junction in a retrotransposed copy is indicated). See text. From Jensen (1993); reproduced by permission of Oxford University Press.

orientation (Minchiotti and Di Nocera, 1991; Contursi *et al.*, 1993). This is not artifactual, since "out" transcripts were also detectable in untransfected cells.

It should be noted, furthermore, that extrachromosomal DNA forms, hybridizing to the F element, have been reported in cultured cells (Mossie *et al.*, 1985).

III. BACTERIAL INSERTION-LIKE TRANSPOSONS: *P* ELEMENT

Bacterial insertion-like transposons belong to a completely distinct category of mobile elements which rather resemble those of prokaryotes, with, at their two extremities, inverted repeats of a few tens of base pairs.

The prototype is the extensively studied *P* element. "Complete" (or "autonomous") copies are 2.9 kb in length and their inverted repeats comprise a perfect 31 bp repeat at the termini and, at 125 bp from each end, another smaller (11 bp) inverted repeat. The internal region is a complex encoding sequence with four exons. Its differential splicing gives rise to either a transposase (the 87 kDa protein which is responsible for the mobilization of the element) or a putative 66 kDa repressor (at least, a serious candidate for a negative regulation of transposition) (see Fig. 10.10).

The transposition of *P* elements is highly regulated and causes, in *Drosophila* another syndrome of hybrid dysgenesis (named P–M), analogous to that (I–R) provoked by I elements. When males of a P (paternally contributing) strain are mated with females of an M (maternally contributing) fly stock, but not in the reciprocal cross, a series of genetic abnormalities appear in the progeny (atrophy of both male and female gonads in F_1; chromosomal aberrations; male recombinations).

So-called P fly strains (usually taken from natural populations) were found to carry 10 to 15 "complete" P elements and a heterogeneous population of 30 to 40 internally deleted ("defective") copies, whereas typical M strains (the majority of ancient laboratory stocks) have none.

It was shown that the complete splicing of the P transcripts, generating a functional transposase and, thereby, a mass mobilization of the P elements, is totally restricted to the germinal line of the F_1 progeny from a dysgenic cross (see review by Rio, *1991*).

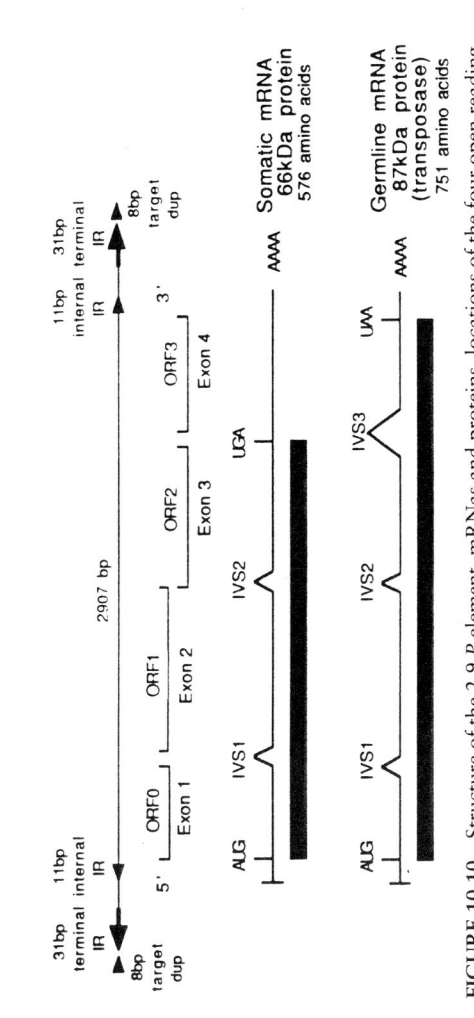

FIGURE 10.10 Structure of the 2.9 *P* element, mRNas and proteins, locations of the four open reading frames (ORFs) or exons are shown with three introns (IVS) joining them. The terminal 31 bp and internal 11 bp inverted repeats and the direct 8 bp target site replication are denoted by arrowheads. Translation initiation and termination codons are shown. Black thick lines indicate the P element-encoded polypeptides. From Rio (1991); reproduced with the authorization of Elsevier Trends Journals.

Since the famous paper by Spradling and Rio (*1982*), the *P* element because of its well-defined conditions of transposition, has become a universally adopted vector for gene transfer, and this technique is today an essential component of the toolkit of *Drosophila* geneticists.

See a complete review on the biology of the *P* element by Engels (*1989*).

Precisely because of the relative ease with which recombinant *P* elements can be introduced into fly strains, most studies have been carried out with transgenic flies. However, transient expression in cultured cells is currently used to test the efficiency of various constructs before launching time-consuming experiments of fly transformation (see Chapter 9, Section VIII). Moreover, important details or confirmation concerning the molecular mechanisms of transposition can be obtained from cultured cells.

1. 3RD INTRON SPLICING REPRESSION IN SOMATIC CELLS

When they were devising a general method for introducing foreign DNA into cultured cells and selecting stable transformants (see Chapter 9), Rio and Rubin (1985) inserted a fusion gene, in which the *P* element coding sequences are under the control of the strong *hsp70* promoter, into their *copia*–neomycin resistance plasmid. The genome of stably transformed *S2* cells contained a tandem array of 10 to 20 copies of this *pNHP* vector, and a prominent 2.8 kb mRNA, hybridizing to *P* probes, could be detected. After 1 h of 37°C heat shock, its amount was increased 20- to 30-fold, which represented as much as 0.5 to 1% of the total poly(A)⁺ RNA.

Using a similar transfection construct, but with a *P* element in which the (exon2-exon3) intervening sequence (or IVS3) had been previously removed, the same authors observed the synthesis of a 87 kDa protein which was shown, with specific antibodies, to derive from the four open reading frames of the *P* element (Rio *et al.*, 1986). It proved to be a functional transposase, since it could catalyze the transposition of nonautonomous *P* elements. On the other hand, when a normal *P* element was inserted into the vector, the cell expressed only a 66 kDa protein. These results, as a whole, were a conclusive demonstration (concomitant with that made by the same group using transgenic flies) that somatic cells cannot perform the complete splicing of *P* transcripts that generate the transposase.

Rio (1988) had developed an efficient *in vitro* splicing system from nuclear extracts of *Kc* cells. Thereby, he investigated the mechanism of specific repression, by *Drosophila* somatic tissues, of the splicing of the

P-element 3rd intron (Siebel and Rio, 1990). Although extracts from these culture cells were able to splice introns from various model genes (for instance *fushi tarazu*), as well as the two first introns of P, they did not detectably process substrates containing the P-element 3rd intron. On the other hand, a *P*-element IVS3 pre-mRNA was accurately spliced by heterologous human cell extracts. If increasing amounts of *Kc* nuclear extract were added to this HeLa cell system, a clear inhibitory effect was observed. This effect was specific, since it did not affect the splicing of the *ftz* intron, for instance, and was very likely caused by some protein(s) (a 5-min 100°C treatment destroyed the inhibitory effect). Potential inhibitors might bind preferentially to IVS3 RNA. Ultraviolet cross-linking experiments allowed the characterization of a 97 kDa protein, and its proper binding site could be localized within a short segment of the 5′ flanking exon, just upstream of the IVS3 5′ splice site.

It seemed that the block to splicing occurs during the early stages of spliceosome assembly. Further studies (Siebel *et al.,* 1992) with the same cell-free system highlighted the crucial importance of "pseudo"-5′ splice sites (i.e., 5′ splice site-like sequences) already noted in the vicinity of the true 5′ splice site. The attachment of an U1 small nuclear ribonucleoprotein particle (snRNP) and a multiprotein complex to two adjacent pseudo-5′ splice sites of the flanking exon probably prevents U1snRNP binding to the accurate 5′ splice site.

2. TRANSPOSASE AND TRANSPOSITION PROCESS

Transposase-overexpressing *S2* cells (*MT Δ2-3* line) were isolated by stable transformation with a plasmid in which the metal inducible metallothionein promoter was put in front of a *P* element sequence lacking the 3rd intron (Kaufman *et al.,* 1989). After copper induction, the transposase (with an apparent molecular mass of 92 kDa) could be purified exclusively from the nuclear fraction.

Because several prokaryotic transposases are sequence-specific DNA binding proteins that interact with the terminal sequence of the transposon, a DNase I protection analysis was performed on some 175 bp of both 5′ and 3′ ends of the P-element in the presence of purified transposase. This transposase showed a high affinity for a short nucleotidic sequence (with a 10 bp consensus motif) located internally and adjacent to the 31 bp inverted repeats at both 3′ and 5′ ends of the *P* element.

Furthermore, in the same paper, but with a *Kc* cell extract transcription system, the P promoter was defined from approximately 30 bp upstream to 15 bp downstream of the initiation site (previously located at nucleotide 87). Thus an overlap between this P promoter and the transposase binding site was shown to exist; a result which led to the possibility that the transcription of *P* is affected by the transposase or the smaller P-encoded 66-kDa protein (since they are colinear for 561 amino acids).

As a matter of fact, Kaufman and Rio (1991) demonstrated that purified transposase behaves like a transcriptional repressor in an *in vitro Kc* cell system, most probably by preventing the assembly of a RNA polymerase II complex onto the P promoter.

In order to study the molecular mechanism of P transposition, Kaufman and Rio (1992) developed an *in vitro* reaction system. A "donor plasmid" (containing a tetracycline resistance-marked P element with wild-type terminal sequences) and a "target plasmid" (carrying the ampicillin-resistance gene and the essential ColE1 plasmid replication origin) were incubated with an enriched protein fraction from transposase-producing S2 cells in the presence of nucleotide triphosphates and proper buffer components (especially Mg^{2+}). Putative transposition products were then introduced into *Escherichia coli* by electroporation. The efficiency of transposition could be scored by counting the number of bacterial colonies which grew in the presence of both ampicillin and tetracycline, relative to the total number of colonies that were solely resistant to ampicillin. Under optimal conditions, about 2×10^{-4} recombinant plasmids per target molecule were recovered.

Efficient transposition required partially purified transposase and a wild-type DNA sequence at the 3′ end of the P element. The structure of the transposition products revealed the creation of flanking 8 bp target site duplications (as is habitual after transposition *in vivo*). Moreover, all data supported the hypothesis of a "cut-and-paste" mechanism in the process of such *P* element transposition *in vitro*.

It is noteworthy that a transposase that is too highly purified was found to be inefficient, which suggests that other host-encoded proteins are necessary [perhaps the so-called "inverted repeat-binding protein" previously characterized by Rio and Rubin (1988) in *Kc* nuclear extract for its specific binding to the outer half of the 31 bp terminal inverted repeats of *P* element].

References

Berg, D. E., and Howe, M. M. (1989). "*Mobile DNA*," American Society of Microbiology, Washington DC, 972 pp.

Boeke, J. D., Garfinkel, D. J., Styles, C. A., and Fink, G. R. (1985). Ty elements transpose through an RNA intermediate. *Cell* **40,** 491–500.

Brown, P. O., Bowerman, B., Varmus, H. E., and Bishop, J. M. (1989). Retroviral integration: structure of the initial covalent product and its precursor, a role for the viral IN protein. *Proc. Natl. Acad. Sci.* **86,** 2525–2529.

Bucheton, A. (1990). I-transposable element and I-R hybrid dysgenesis in *Drosophila. Trends Genet.* **6,** 16–21.

Corces, V. G., and Geyer, P. K. (1991). Interactions of retrotransposons with the host genome: the case of the gypsy element of *Drosophila. Trends Genet.* **7,** 86–90.

Engels, W. R. (1989). P element in *Drosophila melanogaster In "Mobile DNA"* (D. E. Berg and M. M. Howe, eds.), pp. 437–484, American Society of Bacteriology, Washington DC.

Finnegan, D. J. (1983). Retroviruses and transposable elements—which came first? *Nature* **302,** 105–106.

Finnegan, D. J. (1992). Transposable elements. *In* "The genome of *Drosophila melanogaster*" (D. L. Lindsley, & G. G. Zimm, eds.), pp. 1096–1015. Academic Press, San Diego.

Finnegan, D. J., and Fawcett, D. H. (1986). Transposable elements in *Drosophila melanogaster. Oxford Surv. Eukaryotic Genome* **3,** 1–62.

Gateff, E. (1978). Malignant and benign "neoplasms" of *Drosophila melanogaster. In "The Genetics and Biology of Drosophila melanogaster"* (M. Ashburner and R. F. Wright, eds.) Vol. *2b,* pp. 181–275; Academic Press, San Diego.

Georgiev, G. P. (1984). Mobile genetic elements in animal cells and their biological significance. *Eur. J. Biochem.* **145,** 203–220.

Jensen, S., and Heidmann, T. (1991). An indicator gene for detection of germline retrotransposition in transgenic *Drosophila* demonstrates RNA-mediated transposition of the LINE 1 element. *EMBO J.* **10,** 1927–1937.

Kim, A., Tezrzian, C., Santamaria, P., Pelisson, A., Prud'homme, N., and Bucheton, A. (1994) Retroviruses in invertebrates: the *gypsy* retrotransposon is apparently an infectious retrovirus of *Drosophila melanogaster. Proc. Natl. Acad. Sci. U.S.A.* **91,** 1285–1289.

McClintock, B. (1956). Controlling elements and the gene. *Cold Spring Harbor Symp. Quant. Biol.* **21,** 197–216.

Mount, S. M., and Rubin, G. M. (1985). Complete nucleotide sequence of the *Drosophila* transposable element *copia:* homology between *copia* and retroviral proteins. *Mol. Cell. Biol.* **5,** 1630–1638.

Mozer, B. A. and Benzer, S. (1994). In growth by photoreceptor axons induces transcription of a retrotransposon in the developing *Drosophila* brain. *Dev.* **120,** 1049–1058.

Nahon, E., Best-Belpomme, M., and Saucier, J. M. (1993) Analysis of the DNA topoisomerase-II-mediated cleavage of the long terminal repeat of *Drosophila* 1731 retrotransposon. *Eur. J. Biochem.* **218,** 95–102.

Rio, D. C. (1991). Regulation of *Drosophila* P element transposition. *Trends Genet.* **7,** 282.

Roseman, R. R., Pirotta, V., and Geyer, P. K. (1993). The *su(Hw)* protein insulates expression of the *Drosophila melanogaster white* gene from chromosomal position effects. *EMBO J.* **12,** 435–442.

Rubin, G. M. (1983). Dispersed repetitive DNAs in *Drosophila. In "Mobile Genetic Elements"* (J. A., Shapiro, ed.) pp. 329–351, Academic Press, San Diego.

Saigo, K. (1986). Evolution and molecular architecture of *copia*-like transposons in *Drosophila*. *Adv. Biophys.* **21,** 79–88.

Song, S. U., Kurkulos, M., Gerasimova, T. Boeke, J. D., and Corces, V. G. (1994). An Env-like protein encoded by a *Drosophila* retroelement: evidence that *gypsy* is an infectious retrovirus. *Genes Dev.* **8,** 2046–2057.

Spradling, A. C., and Rubin, G. M. (1982). Transposition of cloned P elements into *Drosophila* germline chromosomes. *Science* **218,** 341–347.

Temin, H. (1980). Origin of Retroviruses from cellular moveable genetic elements. *Cell* **21,** 599–600.

11

Drosophila Viruses and Other Infections of Cultured Cells

B. Rickettsia

C. Malaria Parasite *Plasmodium*

Appendix 11

A. Preparation and Maintenance of Virus-free *Drosophila* Cell Lines

B. *Black Beetle Virus* Plaque Assay

C. Infection of Cells with *Black Beetle Virus* RNA

References

In the early 1960s when just *Escherichia coli* and a few viruses were considered adequate material for molecular biologists, animal cell monolayers were primarily developed as useful substrates for the multiplication of viruses. This applied to both invertebrate and vertebrate cell cultures.

Therefore, it was not by mere chance that A. Ohanessian, a member of L'Héritier's group who was studying the *Sigma* virus, the only known virus of *Drosophila*, at that time, joined our attempts to culture *Drosophila* cells. As soon as we had obtained primary cultures from embryonic cells, we observed the multiplication of this *Sigma* virus (from 4×10^3 infectious units to 6×10^5 units, after 2 months) in small subcultures grown in microdrops under paraffin oil (see Chapter 2) (Ohanessian and Echalier, 1967).

Moreover, even if *Drosophila* is not a natural vector for any of the vertebrate pathogenic viruses, it was established that the fly, as well as cell lines, could, after experimental inoculation, support the multiplication of various arthropod-borne viruses (such as *Sindbis* or vesicular stomatitis viruses) and develop the typical latent association that characterizes most "biological" transmissions through arthropod vectors. Thus, the investigation, in such a convenient experimental system, of the interactions between infectious agents and a host cell whose genome has been so extensively studied seemed particularly attractive.

I. *DROSOPHILA* HEREDITARY *SIGMA* VIRUS AND PERSISTENT INFECTIONS WITH VARIOUS ARBOVIRUSES

The *sigma* (σ) virus infection of *Drosophila*, of all the viral infections of higher organisms, is probably one that has been the most thoroughly studied from the genetic point of view (as far as virus and host genotypes are concerned).

The σ virus is normally transmitted exclusively by inheritance and, although, in all other respects, it does not seem to be pathogenic for its host, it confers on the flies what was initially assumed to be a pathognomonic symptom, namely, a peculiar sensitivity to carbon dioxide. Brief exposure to this gas, which is a simple method for anesthetizing *Drosophila* flies, results in rapid paralysis and death of certain (σ infested) fly strains. The phenomenon was discovered by L'Héritier and Teissier as early as *1937*.

Although noncontagious by simple contact, the agent responsible could also be transmitted by experimental inoculation in nonsensitive flies, and it was assumed to be a virus since it passed through membranes with a pore size as small as 180 nm. Yet, it was not until *1965* (Berkaloff *et al.*) that it could be observed by electron microscopy, and its typical bullet-like shape (Fig. 11.1) proved it to be very close to the rhabdoviruses *vesicular stomatitis viruses* (or VSVs).

The various associations between the virus *Sigma* and *Drosophila* are uncommonly complex and correspond to two main types of situation: (1) in the "stabilized condition," the vertical transmission of CO_2 sensitivity by females to all their offspring can be accounted for by a perpetuation of the virus in the germ cells; (2) in the "nonstabilized condition," i.e., either in inoculated flies or in the fraction of the offspring of males that inherits CO_2 sensitivity, the virus multiplies in somatic tissues and the invasion of the female germinal line (and resulting transmission to the progeny) remains infrequent and depends essentially on the viral genotype.

L'Héritier's laboratory (in Gif/Yvette, near Paris) has studied in depth a number of the genetic traits, both of the virus and of the fly, that are involved in the qualitative as well as quantitative control of their relationship. For instance, the so-called ρ and *ultra-ρ* flies (with immunity to superinfection, without CO_2 sensitivity) are maternal lines stabilized for defective viruses. On the other hand, at least five *Drosophila* genes, such as the *refractory (ref)* genes [for example, *ref(2) P*; see a very recent molecular study by Wyers *et al., 1995*], can affect the multiplication of *Sigma*.

A concise survey of this intricate hereditary association between host and virus was made by Brun and Plus (1980) in volume 2d of "The Genetics and Biology of Drosophila." See also a more recent review by Dezelee *et al.* (1987) which deals more extensively with rhabdoviruses in *Drosophila* cells.

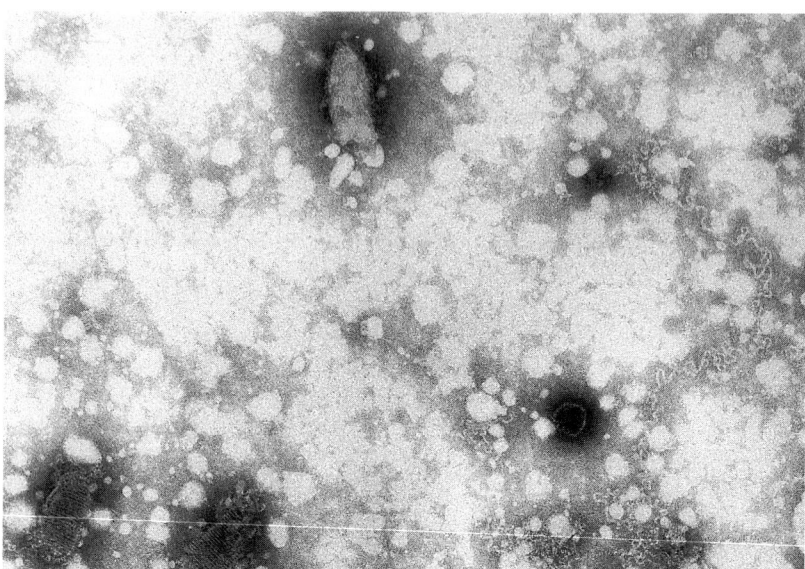

FIGURE 11.1   Bullet-shaped virions of the Rhabdovirus Sigma. Preparation of supernatant from infected *Drosophila Kc* cells. Negative staining. See at the surface of the complete virion (*top*) the typical spiculae of its envelope. In the other damaged particles (*left corner*), the spiraled nucleocapsid is clearly visible and it is unwound in the right part of the picture. Courtesy of Dr. D. Teninges, CNRS, Gif-sur-Yvette, France.

A. *Sigma* Virus

1. ESTABLISHMENT OF "CARRIER STATE" IN CULTURED CELLS

After the preliminary observation that *Sigma* virus could multiply, for several weeks, in subcultures from primary cultures of *Drosophila* embryonic cells (Ohanessian and Echalier, 1967, 1968), just as in whole disemboweled larvae maintained *in vitro* for a few days (Seecof, 1969), the availability of continuous cell lines made a more detailed and extensive analysis of the viral growth cycle possible.

In optimal conditions of infection [high inoculum concentration and presence of DEAE-dextran (50 μg/ml)] the maximum yield of *Sigma* virions is reached around the 40th h (0.7 Infectious Unit per 2 h per cell) (see Fig. 11.2). Thereafter, the virus release decreases, but will remain continuous and approximately constant for a period of more than 1 year (about 0.3 IU per cell per day). It must be emphasized that *Sigma* resistance

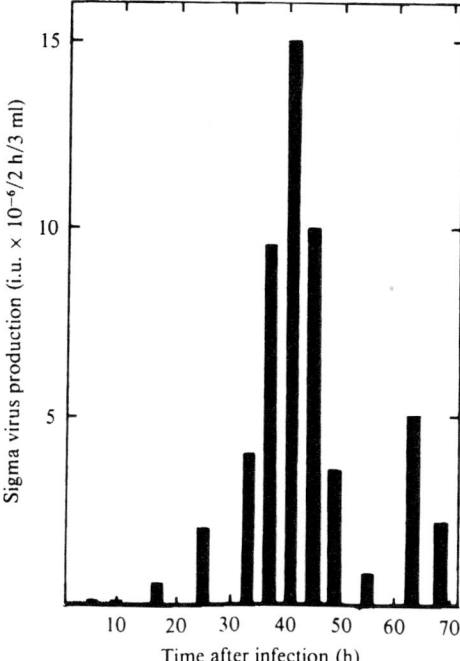

FIGURE 11.2 Sigma growth in *Drosophila* cells. Cells were infected with *Sigma* virus (2.7×10^9 IU/ml). The noncumulative virus production was determined at 2 h intervals. Two hours before harvest of virus-containing supernatant, the culture medium was removed, the cells were washed twice and 3 ml of fresh medium was added; after 2 h, the cell supernatant was collected and virus production during this interval of time was determined by injecting "standard" flies. Richard-Molard *et al.* (1984); reproduced with the authorization of the Society of General Microbiology.

to thermal inactivation in culture medium does not exceed 13 days. Moreover, no cytopathic effect was observed, although it is not known to what extent the parameters of the cell cycle are affected. This persistent infection therefore corresponds to a typical "carrier state" between *Sigma* virus and its host cell (and it is very similar to the so-called "stabilized condition" observed in whole flies) (Ohanessian, 1971; Richard-Molard *et al.*, 1984).

It was worthwhile comparing the infection efficiency of *Sigma* in *Drosophila* fly and cultured cells, respectively. Remember that the virus is titrated with a biological assay derived from the classical dilution end point titration method. One IU is defined as the amount of virus which, when inoculated into a fly, makes it sensitive to CO_2 with a probability of 0.63.

Instead of whole flies, small cell cultures, grown in individual wells of microtest plates, were inoculated with a serially diluted inoculum. Viral growth or no growth in each well was further assayed by injection of supernatants into sensitive flies so that the number of efficient particles could be deduced from the percentage of infected minicultures.

The ratio between the number of infectious units measured in *Drosophila* flies and that of efficient units as estimated in cell cultures was found to be 2.6 for a subline of *Kc* cells, which means that the results of both methods are notably similar. However, in another cell subline, *Sigma* virions were 50 times less efficient (which might, in fact, be due to an observed intercurrent infection of this cell subline by a quite distinct virus) (Ohanessian, 1973).

2. INTERACTIONS BETWEEN VARIANT STRAINS OF SIGMA VIRUS AND DIFFERENT MUTANT HOST CELLS

As already emphasized, the particular advantage of the *Sigma–Drosophila* model, in general virology, lies in the characterization of several variants, both in the virus strains and in the host flies, that affect their mutual relations.

Common strains of the *Sigma* virus are unable to multiply in certain *Drosophila* mutants, called, therefore, *refractory*. For instance, they do not develop in flies of a so-called "Paris" strain which are homozygous for the P^p allele of the gene *ref(2)P* (while the wild allele is named P^O). However, a few strains of *Sigma,* called P$^+$, could be adapted to growth in this *Drosophila* Paris strain (in contrast to the nonadapted P$^-$ *Sigma* strains).

In order to analyze these types of interactions, Richard-Molard (1975) succeeded in establishing three continuous cell lines homozygous for the P^p allele and one homozygous for the P^O (that is, wild-type) form.

The multiplication of P$^+$ and P$^-$ virus strains was studied in these different cell lines. Whereas cultured cells homozygous for P^O/P^O gave the same results with both viral strains, the *ref(2)P^p* lines did not allow the multiplication of the P$^-$ viral types. Thus, the *refractory* gene, with its restrictive effect on viral growth, is unquestionably expressed in cultured cell lines. Unfortunately, this work was interrupted by the contamination of all the cell lines with quite a different virus, probably of serum origin (see Section II.C).

A more complex virus–host association corresponds to the so-called ultra-ρ fly strains. Although they do not display the typical CO_2 sensitivity, they are immune to superinfection with the *Sigma* virus, and this seems to be due to the perpetuation of defective forms of the virus.

Richard-Molard (1973) showed that the immunity is maintained, in primary cultures, by embryonic cells from such ultra-ρ strains. Following the establishment of a continuous cell line, however, its viral yield, after infection, was found to be similar to that of control cell lines; the rapid rate of cell multiplication may, perhaps, have resulted in the loss of the defective virus information.

3. PROTEIN COMPONENTS AND GENOME ORGANIZATION OF SIGMA VIRUS

Infection of cell cultures provided clean preparations of *Sigma* virions for biochemical studies.

Five major proteins could be characterized (210, 68, 57, 44, and 25 kDa). The protein p68 was the only one to be glycosylated. Like the G protein of VSV, it could be solubilized with an anionic detergent, and electron microscopy showed that the treated virions had thus lost most of their spikes. The nucleocapsid seems to contain p44 and one or two minor proteins (p210 and, probably, p57).

The purified genome of the *Sigma* virus was, after ^{32}P-labeling, used as a probe to detect viral mRNAs in infected *Drosophila* cells. Then, the corresponding cDNA clones allowed the nucleotide sequencing of the different viral genes (Teninges *et al.*, 1987, 1993; Bras *et al.*, 1994; Landes *et al.*, 1995).

As is the rule in Rhabdoviridae, the *Sigma* genome is an unsegmented single-strand of RNA of negative sense (Mr: 42S). It consists of six genes with a gene overlap, in the order 3′-N-2-3-M-G-L-5′ (using the terminology of *VSV* structural components). More information can be found in the quoted papers. The predicted sequences of the gene products show various degrees of homology with their *VSV* counterparts, and the *Sigma* virus bridges the gap between the lyssa- and the vesiculovirus genera.

B. Vesicular Stomatitis Viruses

The striking resemblance between *Drosophila Sigma* virions and the Rhabdoviridae, and the wide host range of the latter virus family, probably correlated with its extreme genetic lability, led Printz (1970) to investigate the possible multiplication of the *vesicular stomatitis virus* (VSV) in *Drosophila*. It was indeed tempting to take advantage of both the considerable amount of information already collected on this prototype of rhabdovirus and the genetics of *Drosophila*. Serial inoculations of flies with a clone of VSV (Indiana serotype) resulted in the selection of a variant adapted to *Drosophila* (named VSV_D). The virus efficiency and its maximum yield increased rapidly, but remained uniform after the 10th passage. Moreover, a plaque size reduction was observed in chick fibroblasts. It is of particular note that VSV_D could also confer a typical CO_2 sensitivity on infected flies.

1. VSV GROWTH CYCLE IN DROSOPHILA CELLS

By infecting cultured cells (S2 line), Mudd *et al.* (1973) obtained very similar results to those of Printz in whole flies: a persistent noncytocidal infection was established. After many passages, viral mutants, better adapted to grow in *Drosophila* cells than the original VSV Indiana serotype (this latter called VSV_B, because it had been grown for many generations in murine BHK-21 cells), could be selected. In addition to an increased yield in the insect cells (see Fig. 11.3), these VSV_D variants usually displayed a temperature-sensitive phenotype and a small plaque character (in the plaque assays on BHK-21 cells).

The production kinetics of VSV (and also of another rhabdovirus, *Piry*) in *Drosophila* cells were, later, established by L'Héritier's group (Wyers *et al.*, 1983; Richard-Molard, 1984; see also a review by Dezelee *et al.*, 1987). After a 3 h lag period, viral production increased rapidly to a maximum attained 8 h after infection, then decreasing to reach a plateau at 15 h. This persistent infection is a true "carrier state" perpetuated through cell division, since it could be maintained for several months, in spite of the presence of specific anti-VSV antisera in the medium (Ohanessian, personal communication, 1984).

It had been established that fly strains, differing only in their chromosome III bearing two alleles of the *refractory* gene *ref(3)A*, were, respectively, permissive and restrictive toward a *Piry* virus mutant *ts*. Continuous embryonic cell lines were established. In the so-called VVP line (deriving from the restrictive strain), *ts* mutant multiplication was significantly impaired, compared to the wild-type virus *F*; however, the yield was only 2 to 3 times lower than in the permissive cell line VVV.

2. POSSIBLE MECHANISMS OF PERSISTENT INFECTION

The so-called "biological transmission" of a vertebrate (or plant) pathogenic virus via an arthropod, which implies some viral multiplica-

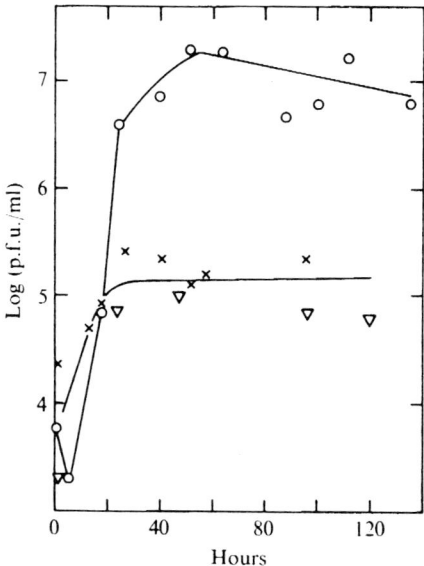

FIGURE 11.3 Adaptation of VSV to *Drosophila*. Single-step growth curves of VSV_B (i.e., previously grown on BHK 21 cells) compared with ninth passage VSV_D (i.e., grown in *Drosophila* cells) in cultured *Drosophila melanogaster* cells at 22°C.

About 1×10^7 *Drosophila* S2 cells in each glass plaque bottle were infected with 5 PFU/cell (plaque-forming units) of VSV_B or ninth passage VSV_D. After adsorption of virus for 1 h at room temperature, cells were washed three times with 2 ml CSM and then incubated at 22°C with 2 ml of CSM. Samples were removed, stored, and, finally, titrated at 28°C on BHK-21 cells. The curve for VSV_B is a composite of two separate experiments at 22 (v) and 28°C (x); VSV_D (o). Mudd *et al.* (1973); reproduced with the authorization of the Society of General Microbiology.

tion in the vector, albeit without any deleterious effect, should surely be correlated with the establishment of such a persistent and noncytopathic infection of insect cells (contrasting with the habitual destruction of mammalian host cells).

VSV infection of cultured *Drosophila* cells seemed an attractive, albeit experimental, model for the study of this "carrier state" process. For a general discussion of the model, see Dezelee *et al.* (1987b).

Wyers *et al.* (1980) showed that no inhibition or modification of the host cell protein synthesis occurred in VSV-infected *Drosophila* cells, neither during the first hours of active viral production nor when this production dropped to the carrier state plateau.

It is well known that, in mammalian cells, the VSV positive-strand leader RNA, the first product of viral transcription, is involved in the shut off of cell transcription (perhaps by interacting with polymerase complexes). *Drosophila* cells also synthesized this leader RNA, but in much lower amounts (some 70 molecules per cell, as opposed to 1000 to 2000 molecules per CER hamster cell) and, moreover, this leader RNA does not migrate, or only very slightly, into the nucleus. This might explain the absence of any blockage in the macromolecular synthesis of the insect host cell (Dezelee *et al.*, 1987a).

Yet, the restriction of viral production by simple competition with the active protein synthesis of the host cell is rather unlikely, and it is generally assumed that more specific cellular control exists. No interferon-like activities, however, could be characterized in insect cells, for instance, in *Sindbis*-infected *Kc* cells (Chany and Echalier, unpublished observations).

In any case, in actinomycin D pretreated *Drosophila* cells, VSV production was enhanced up to eight times (Wyers *et al.*, 1980).

All VSV structural proteins could be detected during the first hours of infection of *Drosophila* cells, but a relative deficiency in the synthesis of the G protein (i.e., the protein constitutive of the virion spikes) was noted, and it seemd to be due to a posttranscriptional step. Moreover, its molecular weight was found to be only 62,000, instead of the usual 64,000 value, which could perhaps be explained by differences in glycosylation and sialylation (Wyers *et al.*, 1980, 1989). The virions released by various *Drosophila* cell lines always contained four to five times less G protein than those produced by vertebrate cells, and the reduction of their spike number was confirmed by electron microscopy. This may account for their lower infectivity.

Furthermore, Blondel *et al.* (1983) established that the degree of phosphorylation of the so-called M protein was 3.5-fold higher in VSV grown in *Drosophila* cells. Since it is generally assumed that this matrix protein plays a crucial role, both in the regulation of VSV transcription and in the virion assembly at the cell surface, such an unusual state of phosphorylation might drastically affect virus multiplication in *Drosophila* cells.

RNA transcription of VSV, during the establishment of a latent infection in *Drosophila* cells, and more especially VSV replication [which requires the synthesis of full-length (42S) plus-strand antigenomes, and then, from these templates, the transcription of minus-strand genomes] were investigated by Blondel *et al.* (1988).

All viral messenger RNAs reached a maximum 6 h after infection, in amounts which, however, did not exceed a quarter of the values observed in vertebrate cells, then decreased to a plateau. Thereafter, as proved by pulse-chase experiments, this reduced and constant level of transcripts is mostly due to inhibition of their synthesis.

As transcription of messengers declined, there was a shift to replication: antigenomes and genomes, measured with adequate radiolabeled probes, accumulated until the 15th h. Later, a steady state was reached and, with a weak virion release, the antigenome and genome intracellular pool remained constant. So, several hundreds of copies per cell are amply sufficient for transmitting the viral information throughout the rapid cell divisions.

C. *Sindbis* Virus and Other Arboviruses

The term arboviruses (for "arthropod-borne" viruses) groups together several different families, in addition to most Rhabdoviridae.

The avian alphavirus *Sindbis* (normally transmitted by the bite of a mosquito) belongs to the Togaviridae (= "Group A" arboviruses). It could be experimentally better adapted to growth in *Drosophila*, through serial passages, in conditions quite similar to those observed with VSV, and has become, too, a convenient model for analyzing persistent viral infection in an Insect cell (Bras-Herreng, 1973, 1976).

Studying the multiplication dynamics of a "large plaque" clonal strain of *Sindbis* in *Kc* cells, Hannoun (1973) demonstrated the persistence of virus-carrying cells for 16 months, without any detectable cytopathic effect.

By plating on chick embryo cells a diluted cell suspension of the infected cultures, he estimated that some 12 to 26 cells per 1000 were able to induce "infectious centers." This low proportion of virus-containing cells was confirmed by immunofluorescence (with tagged anti-*Sindbis* antibodies). Did these cells correspond to a small "competent" subpopulation? The ratio between total virus titer of the culture and the number of infectious centers was found to be 6 to 36 PFU (plaque-forming units) per infected cell. This figure does not represent the virus yield per cell, but only corresponds to the number of intracellular mature virions present at a given time.

Concomitantly with experiments in whole flies, Bras-Herreng (1975, 1976) compared the growth curves, in *Kc* cells, of *Drosophila*-adapted clones of *Sindbis* and the wild-type virus (Fig. 11.4). Although the general profiles remained very similar, with a rapid increase after the 6th h, a maximum reached after 24 h, then a slightly decreasing plateau, the

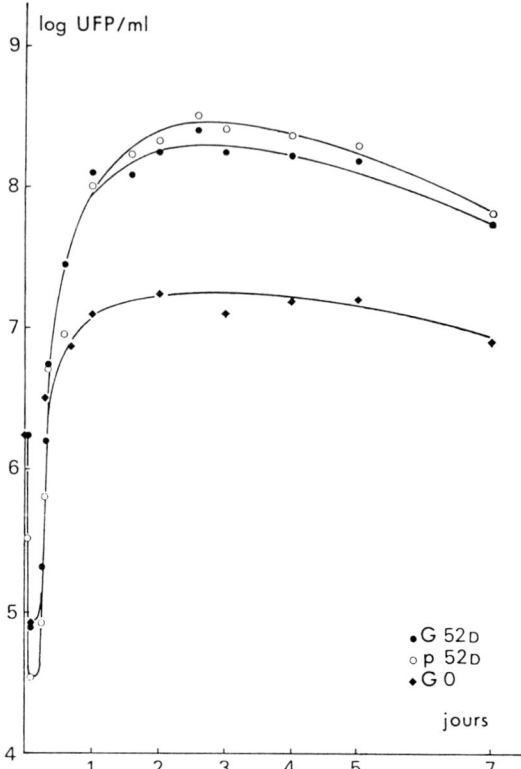

FIGURE 11.4   Adaptation of Sindbis virus to *Drosophila*. Growth kinetics, in cultured *Drosophila* cells, of virus-adapted clones (52nd passage on *Drosophila* cells: G 52D and p52D, large and small plaque variants) compared to wild virus (G0). UFP, plaque-forming units, as tested on vertebrate cells. Bras-Herreng (1976); reproduced with the authorization of the Institut Pasteur, Paris.

author observed significant differences, in both the infection efficiency (only 1 PFU is required per cell with adapted virus versus 6 PFU with the wild type inoculum) and viral production (a difference of 1 log unit). Anyhow, the yield per cell remained relatively small (at the best, 5 PFU/day).

Hannoun and Echalier (1971) methodically analyzed the susceptibility of *Kc* cells to a number of arboviruses belonging to the major groups (see Table 11.I): with Group A, as with *Sindbis,* multiplication was rapid and the titers of the culture fluids reached relatively high values. In Group

TABLE 11.I   Susceptibility of *Drosophila* Cell Line *Kc* to Arboviruses[a]

Viruses	Logarithm ID_{50} introduced per culture	Virus titers days after inoculations[b]			
		4th	7th	11th	14th
A Group					
Chikungunya	6.9	4.3	>5.2	>5.2	>5.2
Sindbis (G)	6.2	4.4	>5.2	>5.2	4.7
Sindbis (p)	6.1	3.4		3.3	3.3
B Group					
Dengue-1	2.5	N	1.7?	N	2.4
Dengue-2	2.5	N	2.2?	N	2.4
Dengue-3	2.7	N	N	N	1.9
Dengue-4	2.5	N	N	N	N
Yellow Fever FN	2.5	N	N	N	N
St Louis encephalitis	4.7	2.2	2.2	N	1.7
West Nile (Egypt 101)	5.0	>5.2	>5.2	>5.2	>5.2
West Nile (Halima)	4.5	4.9		4.7	>5.2
Japanese encephalitis	5.9	N	2.9	3.1	2.3
Central European encephalitis	4.5	2.3	2.8	2.2	2.2
Other groups					
Tahyna (Theobaldia)	5.5	2.7	4.5	4.2	3.7
Tahyna 488V	4.5				4.8
Tahyna 488N	4.5	3.2	4.8	4.5	3.4
Hesha	5.4	N	N	N	N
Sandfly Fever (Sic.)	3.5	N	N	N	2.2

[a] From Hannoun and Echalier (1971); reproduced with the permission of Springer-Verlag.

[b] Expressed as the logarithm of the ID_{50}/ml of culture fluid, as determined in BHK 21 cells. N means <1.2. The culture medium was changed at the end of the first day and at intervals of 8 to 10 days, thereafter.

B, although the *West Nile virus* multiplied exceptionally well, other viruses either grew to only low titers or did not grow at all. A "carrier state" was observed, in long-term experiments, with *dengue 2, Tahyna*, and *West Nile* viruses.

Last, it is noteworthy that Xiong *et al.* (1989) promoted the use of *Sindbis* as an efficient vector for gene expression in animal cells because of its particularly broad host range. A "reporter" gene (such as CAT) was substituted, in the single-stranded RNA genome, for the 3' sequences which encode the virion structural proteins. It must be recalled that, during the virus cycle, the corresponding subgenomic messenger RNA is transcribed (from a minus-strand antigenome) in particularly high amounts. Thus, CAT expression could be observed in a variety of transfected vertebrate and insect cells (among which was the *Drosophila S2* line). Moreover, when transfected cells were infected with normal "helper" *Sindbis* (supplying structural proteins), vector genomic RNAs were packaged into infectious particles, which may be convenient for production of large amounts of any given gene product by secondarily infected cultures.

II. ENDOGENOUS VIRUSES OF *DROSOPHILA* AND CELL LINES

More than one-third of the many natural populations and laboratory stocks of *Drosophila melanogaster* examined by Plus and collaborators (see the aforementioned review by Brun and Plus, 1980) were found to be endemically infected with one or more viruses.

Four different types of endogenous viruses have been identified in cultured cells: the very common picornaviruses (with three distinct serotypes, A, P, and C*), and a reovirus F. In addition, many cell lines are contaminated with a birnavirus, the so-called X virus, never found in fly populations and which might originate from the fetal bovine sera used for supplementing culture media (see Table 11.II).

Plus (1978, 1980) looked systematically for the presence of endogenous viruses in a variety of cultured cell lines of *Drosophila* from many different laboratories. As it can be noted in Table 11.III, the majority were found to be infected, although chiefly by the contaminant X virus (see below). The author, therefore, emphasized both the importance of starting with virus-free flies when establishing new cell lines and the necessity for screening each serum batch used for their maintenance. See practical advice in Appendix 11.A.

A. Picornaviridae

Picornaviridae are small nonenveloped icosahedral viruses (diameter: 25-30 nm) whose genome is a positive-sense single-strand RNA. Their localization and morphogenesis are cytoplasmic. Virions display a typical resistance to acid pH and organic solvents.

---

* This denomination corresponds simply to the initials of the country of origin of the fly population in which the virus was first characterized: for instance, *A* means Antilla.

TABLE 11.II   Differential Diagnosis of Endogenous or Contaminant Viruses[a] Observed in *Drosophila* Cell Lines

Location	Virion size (nm)	Shape and structure	Genome	Classification
Cytoplasm	25–30	icosahedral	positive single strand RNA	Picornaviruses (serotypes C, P, A)
	60	two-layered spherical capsid	10 segments of dsRNA	Reovirus (F = K?)
	60–70	icosahedral single shell	2 segments of dsRNA	Birnavirus X
Nucleus	36	icosahedral	1 segment	Duplornavirus

[a] VLPs are not considered here (see Chapter 10).

1. DROSOPHILA *C VIRUS*

Although quite similar in morphology and strictly cytoplasmic distribution, either as dispersed particles or in crystalline arrays, sometimes segregated in dense amorphous membrane-bound bodies) (see Fig. 11.5; see also Fig. 11.6A for size comparison with reoviruses), the three *Drosophila* picornaviruses can be distinguished by their serotypes, their pathogenicity, and their different tissue tropisms. *Drosophila* C virus (DCV) provokes, even at the first passage, a rapid death (within 3 days) of inoculated flies and is found, in abundance, in tracheal cells, whereas DAV and DPV reduce the life span by only one-half and cause female sterility. These latter can, moreover, be transovarially transmitted, and the organs preferentially infected, other than the ovaries, are the intestine and Malpighian tubules. (DPV: Plus and Duthoit, *1969;* Teninges and Plus, *1972;* DCV: Jousset *et al., 1972, 1977*).

Picornaviruses, and even the pathogenic DCV, do not seem to have any apparent deleterious effect on the growth of cultured cell populations. Is this related to the restricted spectrum of susceptible cell types which characterizes these viruses? Whatever the case, occasional viral invasion or rupture of a delicate equilibrium, under unknown culture conditions (for instance, change of the serum batch?), might perhaps explain certain periods of depression recorded by several investigators during the long-term subculturing of some of the cell lines.

The polypeptides induced in *S1* cells (more precisely a subline found to be virus free) by experimental infection with DCV were analyzed by

TABLE 11.III   Endogenous Viruses of Cell Lines of *Drosophila melanogaster*[a]

Cell lines[b]	Laboratory	Viruses detected
75B, 75F, 75M (*1*)	Gif-sur-Yvette (Ohanessian)	none
KB (*2*)	Gif-sur-Yvette (Ohanessian)	X virus
E₁, E₇, G, Ch (*3*)	Gif-sur-Yvette (Contamine)	none
E₇A, E₇B, OM₃, PE29 (*4*)	Gif-sur-Yvette (Simonet)	none
KC₀ (*2*)	Marseille (Jordan)	X virus, DPV, Picornavirus
KC₀ (*2*)	Basel (Gehring)	X virus
F6–19, F6–40, F6–47 (*2*)	Marseille (Jordan)	X virus, DPV Picornavirus
Schneider-1 (*5*)	Freiburg (Gateff)	DCV
Schneider-1 (*5*)	Auckland (Scotti)	none
Schneider-2 (*5*)	Freiburg (Gateff)	DCV
l(2)mbn, l(3)mbn (*6*)	Freiburg (Gateff)	DCV, F, virus
GM₁, GM₂, GM₃ (*7*)	Milan (Dolfini)	X virus, Picornavirus
GM₁ (*7*)	Auckland (Scotti)	none
1–35, 1–56, 1–59, 0–57, 3–38 (*8*)	Milan (Dolfini)	X virus, DPV, Picornavirus
Gehring-2 (*9*)	Basel (Gehring)	none
Gehring-2 (*9*)	Moscow (Kakpakov)	X virus
67j25D (*10*)	Moscow (Kakpakov)	X virus, DPV
75e7vg1, 75e7vg2, 75e7vg3, 75e7vg4 (*11*)	Moscow (Kakpakov)	X virus
75e7vg5 (*11*)	Moscow (Kakpakov)	X virus, Picornavirus
75e7vg7 (*11*)	Moscow (Kakpakov)	X virus, DAV
PE3, PE8, PE17, PE33, PE34/115 (*12*)	Moscow (Kakpakov)	X virus
E85, E42 (*13*)	Basel (Bernhard)	X virus

[a] From Brun and Plus (1980); reproduced with the kind permission of Dr. Brun (CNRS, Gif/Yvette, France and Academic Press).

[b] Key to references to cell lines (*1*) Ohanessian, personal communication; (*2*) Echalier and Ohanessian, 1970; (*3*) Contamine, personal communication; (*4*) Simonet, personal communication; (*5*) Schneider, 1972; (*6*) Gateff *et al.,* 1980; (*7*) Mosna and Dolfini, 1972; (*8*) Mosna and Barigozzi, 1976; (*9*) Kakpakov *et al.,* 1969; (*10*) Kakpakov *et al.,* 1977; (*11*) Kakpakov, personal communication; (*12*) Schneider and Blumenthal, 1978; (*13*) Bernhard, personal communication.

Moore *et al.* (1981b,c), and their data support the inclusion of this virus in the Picornaviridae group.

2. CRICKET PARALYSIS VIRUS

The serological relationship between DCV and another picornavirus isolated in a quite different Insect (*Cricket Paralysis Virus,* or CrPV), and

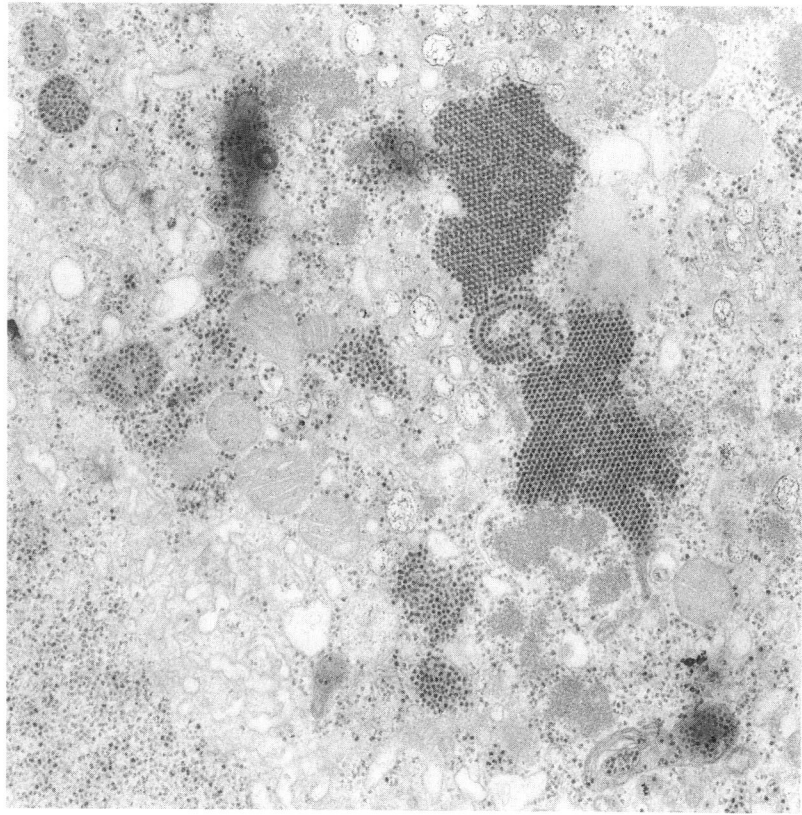

FIGURE 11.5   Picornaviruses in a *Drosophila* cell line. Small icosahedral virions (30 nm diameter), dispersed through the cytoplasm or arranged in crystalline arrays, are sometimes segregated in dense amorphous bodies. Courtesy of Dr. D. Teninges, CNRS, Gif-sur-Yvette, France.

also similarities in their host range, as tested by inoculation into a variety of insects, led Scotti (1976) to demonstrate the possible multiplication of this CrPV in a *Drosophila* cell line (Schneider's *S1*). Cytopathic effects made it possible to develop titration bioassays, i.e., end-point dilution or plaque assay methods (Scotti, 1977). Another cell line (*75B*, see Chapter 3), however, failed to support the multiplication of either CrPV or DCV, which demonstrates the differences in the susceptibilities of various cell lines (Plus *et al.*, 1978).

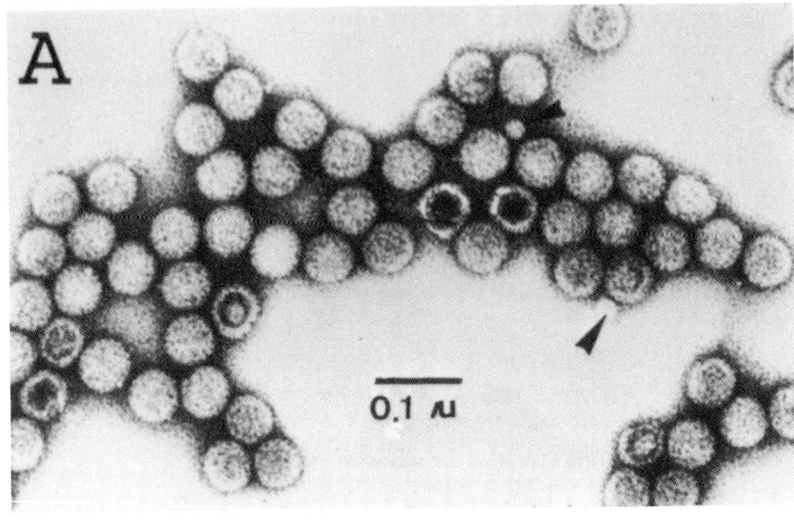

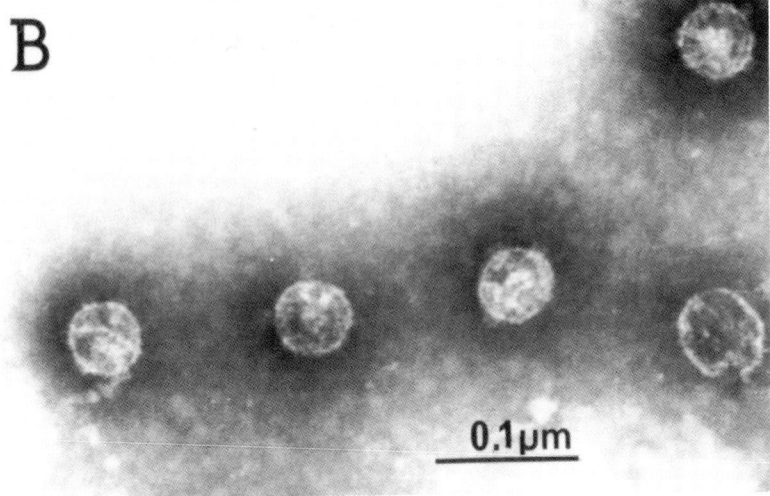

FIGURE 11.6 Reovirus type particles from *Drosophila* cell lines. (A) A group of purified F virus prepared from *l(2)bm* cells (from Gateff *et al.*, 1980; with permission from Elsevier Scientific Publishers). Note the double capsid seen on the few empty particles. The arrows point to C-picornaviruses which were constant contaminants of the preparations, due to the similar buoyant density of 1.34/ml; (B) Reovirus K particles from *Kc* cells (courtesy of Dr. D. Teninges). The typical structure of the inner capsid with its chimney-like protuberances is visible here, probably in relation to a different protocol of negative staining.

Using this convenient CrPV–*Drosophila S1* cell system to study the intracellular cycle of such an insect picornavirus, Moore and collaborators (Moore and Pullin, 1980; Moore *et al.,* 1980, 1981a) were able to characterize the major structural proteins of the virion and showed that they derived from the very complicated cleavage of high molecular (up to 205 kDa) precursor protein(s). Their results are essentially similar to those observed in DCV infection.

Moreover, the same authors (Moore *et al.,* 1981c) took advantage of the well-known heat-shock response of *Drosophila* cells (see Chapter 7) to investigate the "shut-off" mechanisms whereby picornaviruses (and this is true, too, for mammalian picornaviruses) can inhibit the host cell protein synthesis to different extents. CrPV infection of *S1* cells appeared to be as effective as actinomycin D in impeding the appearance, after temperature elevation, of the typical heat-shock proteins, although its action may not take place at the transcription level. Distinctively, in heat-shocked DCV-infected cells, HSPs were synthesized, albeit in reduced amounts.

3. CAN HUMAN PATHOGENIC ENTEROVIRUSES MULTIPLY IN DROSOPHILA CELLS?

Even before the resemblance between such *Drosophila* Picornaviruses and the genus Enterovirus was established, people wondered whether synanthropic flies, such as *Musca domestica* or even *Drosophila,* might not only harbor but mutliply human pathogens, more especially the poliovirus or coxsackievirus.

So, Klowden and Greenberg (1974) infected *Drosophila S2* cells (and concomitantly primary cultures from *Musca*) with coxsackievirus type B5 and poliovirus type 2. In both cases, the gradual decline of the virus titer, over the 2-week assay period, was essentially similar to that observed in cell-free medium. This *in vitro* failure does not, however, necessarily imply that these viruses are incapable of multiplying in whole insects.

B. Reovirus F (=K ?)

Teninges *et al.* (1979a) observed by electron microscopy a reovirus type particle in a persistently infected subline of *Kc* cells. It was named *Drosophila* K virus (DKV), according to its cell line origin, and it seems to correspond, although there is no conclusive evidence of identity, to the so-called *F* virus previously isolated from French laboratory fly stocks by Plus *et al.* (*1975*).

In her first description of the two cell lines established from malignant blood cells of lethal mutant larvae (*l(2)mbn* and *l(3)mbn;* see Chapter 3), Gateff *et al.* (1980) also reported the presence, in the cytoplasm, of a reo-type virus showing extensive similarities to F virus (see Fig. 11.6A). Viruses of the same type were detected in the Russian cell line *67j25D-G* by Alatort-sev *et al.* (1981).

Biochemical studies of purified virions (Haars *et al.*, 1980) confirmed the classification of *F* among the Reoviridae, according to the following criteria: (1) a nonenveloped, spherical capsid (diameter of 60 nm) composed of two layers (clearly visible on slightly dislocated particles) (see Fig. 11.6B); (2) a genome consisting of 10 segments of double-stranded RNA (according to the criteria of both E.M. observations and resistance to ribonuclease A and nuclease S1); (3) virus multiplication occurs in the cytoplasm; and (4) mRNA could be transcribed *in vitro* in a cell-free system under conditions similar to those described for mammalian reoviruses.

Note: Saigo *et al.* (1980) reported that crude extracts from various cell lines (*S2, Kc, Kc-H*) contained a dozen discrete double-stranded RNAs, the length distribution of which (from 1 to 4 kb) is very similar to the published lengths of the 10 dsRNA segments from human reoviruses.

C. Contaminant Birnavirus X

The so-called *X* virus (X standing for enigmatic) is the most widespread virus in *Drosophila* cultured cell line (see Table 11.III), whereas it has never been found in fly stocks. It seems to me, therefore, that the denomination DXV is incorrect. It was assumed to come from the fetal bovine serum routinely incorporated in culture media and could indeed be detected in a few serum batches [coming, precisely, from laboratories where most examined cell lines were found contaminated (Plus, 1980)]. Yet, this does not imply that it is a bovine virus. As a matter of fact, it does not resemble any of the latter known viruses and, moreover, if it was a bovine virus, one would expect it to grow even better in vertebrate cell cultures, which was not found to be the case in either primary cultures of calf kidney, or chick embryo cells, or several tested vertebrate cell lines (Teninges *et al.,* 1979a).

Its inoculation into flies causes an untimely death (within 12 days) with, just a few days before, a characteristic symptom of sensitivity to anoxia (lethal paralysis induced not only by CO_2 but also by pure nitrogen).

As described by Teninges *et al.* (1979a), the X viral particles, either scattered or in crystalline arrays, are exclusively cytoplasmic and generally observed at the periphery of a viroplasm. In ultrathin sections of infected cells, a dense core surrounded by an apparently single shell, hexagonal in shape (from side to side distance 62 nm and point to point 72 nm) is visible. In negatively stained preparations of virions, four protuberances, with tubular appearance, can be observed on each edge of the hexagon (Fig. 11.7). Extracellular particles are identical, and no budding was ever observed at the host cell membrane.

Six major polypeptides could be characterized from purified virions, and the genome was demonstrated to be a double-stranded DNA divided in two segments (Teninges, *1979*). Moreover, a RNA polymerase activity was found to be associated with the particles (Bernard and Petitjean, *1978*).

Infected cultured cells could be detected by immunofluorescence (with a rabbit antiserum prepared against an *X* virus preparation). Many tested cell lines presented 0.02% fluorescent cells. A virus-free subline from *Kc* cells enabled Teninges *et al.* (*1979a*) to study the kinetics of an experimental infection with *X* virus. The percentage of fluorescent cells reached 86% at 12 h and most cells underwent lysis after 26 h. However, a few cells escaped death and were able to occupy the whole flask within 10 days. Thereafter, a persistent infection was established and the percentage of fluorescent cells remained invariably close to 2% (with a viral yield estimated, from inoculated flies, to be 0.9×10^2 I.U. per fluorescent cell per day).

Plus *et al.* (1983) isolated from Debec's haploid cell line *1182* (see Chapter 3) a virus that was morphologically, biochemically and serologically similar to *X* virus, but which did not confer the typical anoxia sensitivity on inoculated flies. It may, therefore, be considered a nonpathogenic *X* mutant. The fact that the same "symptomless" *X* virus was also detected in the serum batch used for the maintenance of this cell line strongly supports the hypothesis that the *X* virus is indeed introduced into cell lines by the calf serum.

Because its genome consists of only two segments of double-stranded RNA, the *X* virus cannot be included in any of the genera of reoviridae. So, after careful biophysical and biochemical comparisons of five animal viruses (including the *X* virus and the infectious *Pancreatic necrosis virus* of trout), which all present a bisegmented dsRNA within 60 nm icosahedral particles, Dobos *et al.*(*1979*) proposed the creation of a new class (birnaviruses).

Note: Scott *et al.* (1980) discovered an apparently persistent infection of Schneider's *S2* line with another double-stranded RNA virus, albeit quite different in its nuclear localization, the size of its particles (36 nm), and the fact that its genome comprises a single segment of dsRNA. They considered it to belong to a distinct class ("duplornaviruses").

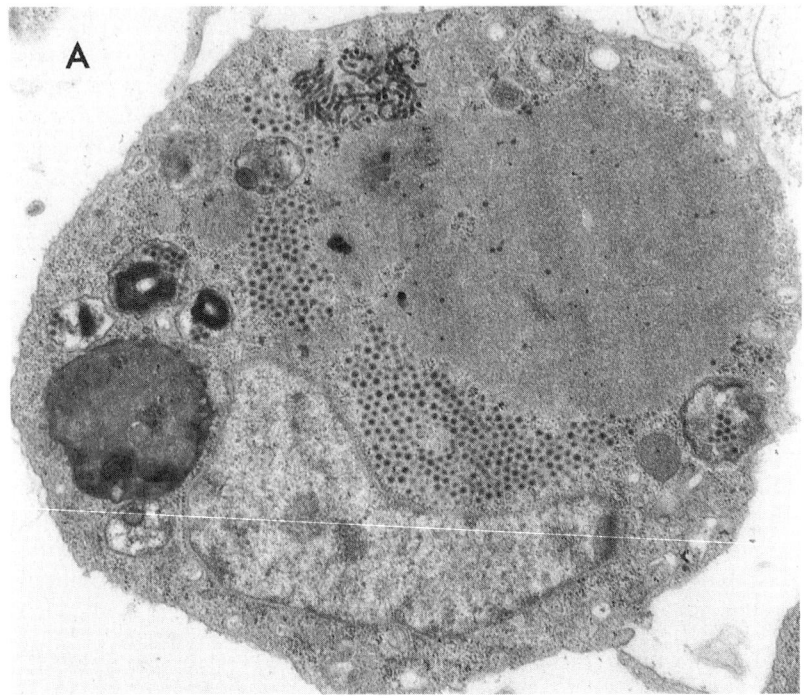

FIGURE 11.7   Birnavirus X, a contaminant of *Drosophila* cultured cells. (A) A cell from a persistently infested *Kc* subline. The virions are observed in relation to a large viroplasm; (B) Purified virions. Negative staining. Note the clearly polygonal shape (by comparison with the roundish outline of Reovirus in Fig. 11.6). Courtesy of Dr. D. Teninges, CNRS, Gif-sur-Yvette, France. *(continues)*

D. Virus-like Particles

See the discussion of Retrotransposons in Chapter 9.

III. LOOKING FOR EFFICIENT VIRAL PROMOTERS IN *DROSOPHILA* CELL LINES

Powerful viral promoters, such as the SV40 early promoter or Rous sarcoma virus LTR in mammalian cells, and baculovirus late promoter in lepidopteran cells, are extensively used in biotechnology today.

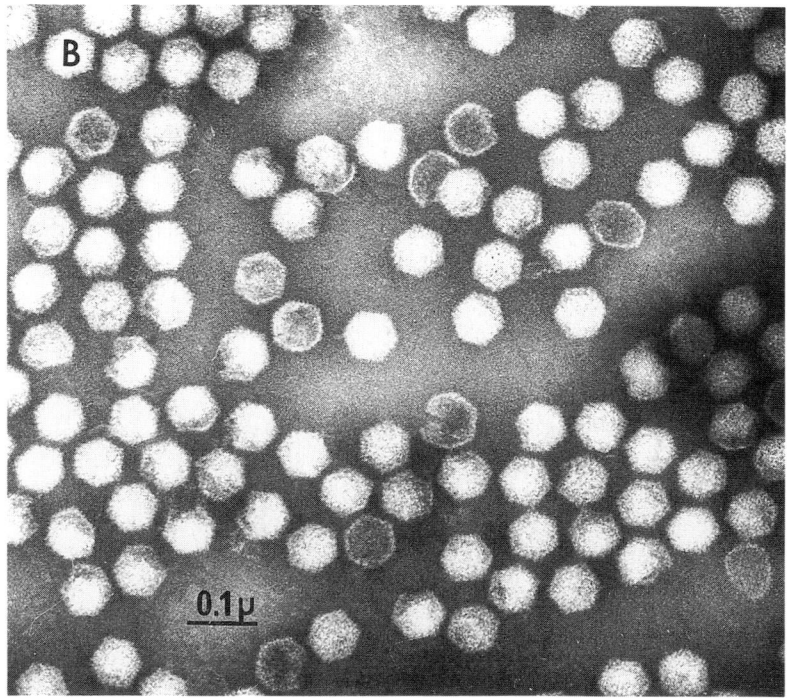

FIGURE 11.7   *(continued)*

A. *Drosophila* Cells are Nonpermissive for Baculoviruses

The recombinant Baculovirus/lepidopteran cell system has proved remarkably efficient for producing large amounts of foreign proteins of interest.

The most widely used baculovirus is the *Autographa californica* nuclear polyhedrosis virus (AcNPV) whose host range is restricted to lepidopteran species. A biphasic replication cycle yields two forms of infectious viruses. During the first 24 h, extracellular particles are released by budding and they spread the infection throughout the whole organism, whereas, from 18 h postinfection onward, nucleocapsids, within the nucleus of infected cells, are embedded in paracrystalline proteinaceous bodies (the typical polyhedra). Cell lysis liberates these stable capsules which, in the wild, are dissolved in the midgut of another caterpillar, thereby extending the sickness.

The polyhedron matrix is made of a single major protein, or *polyhedrin,* that may constitute more than 25% of the total protein mass in the cell. Its very late expression is driven by a particularly strong promoter, so that the standard procedure for producing a given protein from recombinant baculovirus is to place the corresponding coding sequence under the transcriptional control of this so-called "very late" promoter.

See two recent reviews by Maeda (*1994*) and by Lenz Goodman and McIntosh (*1994*). Important methodological advice on the baculovirus expression system will also be found in Summers and Smith (*1987*).

Because of the high host specificity of baculoviruses, it is not surprising that cultured *Drosophila* cells were found to be nonpermissive for AcNPV. In *Drosophila S1* cells infected with a baculovirus vector in which the *E. coli* β-galactosidase gene had been inserted downstream of the polyhedrin promoter, no β-Gal activity was detected (Pennock *et al.,* 1984). Likewise, in a Northern blot analysis, Rice and Miller (1986) observed a few early transcripts (i.e., mRNAs corresponding to the envelope glycoprotein of the budded virus phenotype), but did not reveal any late or very late transcripts (i.e., messengers for polyhedrin) in AcNPV-infected *Drosophila DL-1* cells.

In order to determine whether this failure was due to the fact that the virus had difficulty in entering the nucleus of nonpermissive cells or to a block in its complex expression cycle, Carbonell *et al.* (1985) employed a double recombinant baculovirus. It contained both a reporter gene β-galactosidase under the control of the polyhedrin promoter and a CAT gene driven by the LTR of rous sarcoma virus (this latter being known to work, albeit moderately, in *Drosophila* cells; see Section III.D). A mild cytopathic effect was noted after 1 or 2 days. CAT expression was observed 6 h after infection and it increased until 48 h persistant injection (PI), whereas there was, once more, no expression of β-Gal (Fig. 11.8). These results prove that (1) the baculovirus was adsorbed, then penetrated, and could be uncoated into an expressible form in *Drosophila* cells, and that (2) gene expression is clearly promoter dependent. CAT activity continued for 2 to 3 weeks, but it decreased with time, this probably being accounted for by mere cell multiplication. As a matter of fact, a direct quantification of viral DNA, with dot-blot analysis, indicated an only 4- to 5-fold increase, which represents approximately no more than two cycles of multiplication (*Note:* this observation is very relevant to the highly praised utilization of baculovirus as a biological pesticide with high host specificity and to its claimed innocuity for nontarget organisms.)

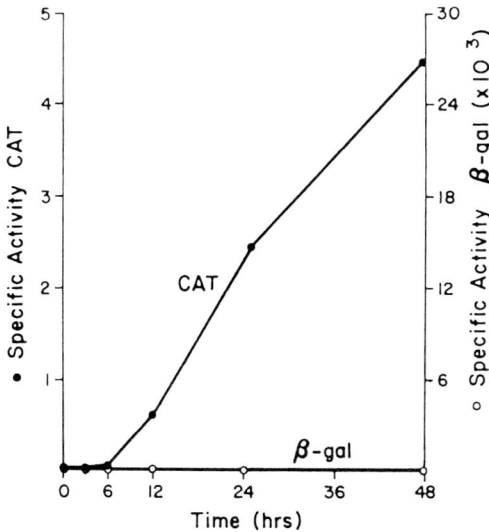

FIGURE 11.8 Gene expression in *Drosophila melanogaster* cells infected with a double recombinant reporter Baculovirus. Monolayers of *Drosophila melanogaster* cells were infected with *AcNPV L1LC-galcat* (see text) at a MOI of 50. The specific activities of CAT and β-galactosidase were determined at the indicated times postinfection. Carbonell *et al.* (1985); reproduced with the kind permission of Dr. Miller and the American Society of Microbiology Journals Department.

In further investigations, the reporter gene CAT was placed, in different recombinant AcNPV, under the transcriptional control of promoters which are normally active in the early, late, or very late stage of the viral cycle, respectively (Morris and Miller, 1992, 1993). In contrast to the usual temporal pattern and relative strengths of expression from these three types of promoters, as observed in permissive *Spodoptera* cells, the late promoter-driven expression, even though it is dependent on DNA replication, was surprisingly found to be higher in nonpermissive *Drosophila S1* cells than early promoter-driven expression. Moreover, limited expression driven by the very late polyhedrin promoter could also be detected (probably due to the use of more sensitive methods). At any rate, in analogous constructs, the *Drosophila hsp70* promoter proved stronger than any of the three viral promoters tested.

Because it is assumed that immediate early genes of baculovirus are transcribed by the host cell RNA polymerase II (in contrast to the later ones), Blissard and Rohrmann (1991) analyzed, by transient transfection, the early promoter of the major envelope glycoprotein (gp64) of another well-studied baculovirus. OpMPNV (from the lepidoptera *Orygia pseudotsugata*). p64 constructs were expressed, although at various levels, and transcripts could be concomitantly detected, not only in permissive lepidopteran cells but in *Drosophila S1* cells. This indicates that this early promoter requires no other viral gene products. The AcNPV product *pIE1*, however, which is known to be a transcriptional activator, was shown to be active after cotransfection with a *pIE1* expression vector. Note that this interest in early promoters must be correlated with current projects on insect biocontrol, the aim of which is to induce a gene coding for some insect-specific toxin via recombinant baculoviruses. If it could be expressed early, this might accelerate the lethal efficiency of the recombinant virus and, possibly, enlarge its host range.

B. Black Beetle Virus

The *Black Beetle Virus* (BBV), isolated from the New Zealand Coleoptera *Heteronychus arator* (Longworth and Archibald, 1975), belongs to a newly identified virus family named Nodaviridae (because their prototype is the *Nodamura* Virus). They are small riboviruses, resembling Picornaviruses in size (diameter 30 nm), but, inside a coat made of a single major protein (40,000), the genome consists of two different single-stranded messenger-sense RNAs, packaged in a common viral particle (see Hendry, 1991).

The particularly efficient multiplication of BBV in a *Drosophila* cell line allowed a thorough study, at the molecular level, of the distinct functions of its two genomic segments. Furthermore, the smallness of its genome (4500 bases) and the technical possibilities for engineering qualify BBV as a potentially attractive vector for the production of foreign protein in *Drosophila* cell cultures.

Friesen *et al.* (1980) established that BBV multiplies efficiently in Schneider's *Drosophila* line 1 cells (at least in a certain cell strain, named WR because it was sent directly by Schneider from the Walter Reed Army Institute of Research in Washington D.C.). Virus yields were found to be unusually high: 1 to 2 mg per 10^8 cells after 2 days of infection, which represents some 20% of the total cell proteins. Other Nodaviruses, such as *Flock house virus* (FHV) or Boolara virus (BOV) (but not *Nodamura* virus), can also be routinely grown in *Drosophila* cells (Dasgupta and Sgro, 1989; Zhong and Rueckert, 1993; Li and Ball, 1993; Schneemann *et al.*, 1994).

Infection with the wild-type BBV has not very cytolytic, but repeated passages through *Drosophila* cells resulted in the isolation of a vigorously replicating viral strain (called BBV-W17) which gives rise to substantial cytolysis (even at a multiplicity of 1 virion per cell). This enabled Selling and Rueckert (1984; see also Selling's thesis, 1986) to develop a reliable plaque assay, which was a decisive step for the genetic approach.

Practical information on the propagation of BBV in *S1* cells, transfection of genomic RNA and plaque assay method can be found in Ashburner's "*Drosophila*, a laboratory manual" (1989, protocols 50 to 53). These protocols are reproduced here, in Appendixes II.B and II.C, with the kind authorization of Professor Kaesberg.

Using this BBV-*S1* cell system, Rueckert, Kaesberg and collaborators carried out an extensive analysis of the viral organization and cycle. They systematically studied the following: virion and virus-induced polypeptides (Guarino *et al.*, 1981; Friesen and Rueckert, 1981; Crump and Moore, 1981) and, particularly, replicase (Guarino and Kaesberg, 1981; Saunders and Kaesberg, 1985); subgenomic RNAs and their expression (Friesen and Rueckert, 1982; Gallagher *et al.*, 1983; Guarino *et al.*, 1984; Dasgupta *et al.*, 1984; Dasmahapatra *et al.*, 1985); transcriptional controls (Friesen and Rueckert, 1984; Zhong and Rueckert, 1993).

Figure 11.9 sums up their main results which are also reviewed by Kaesberg (1987): the subgenomic RNA1 encodes an RNA-dependent RNA polymerase (or replicase) necessary for BBV replication and whose synthesis peaks early in infection (4 to 6 h p.i.), whereas RNA2 directs, in a late phase, the synthesis of the precursor of the coat protein. It was therefore suggested that the *raison d'être* of such a bipartite genome might well be to separate the genes involved in early (replicative) functions from those involved in late (packaging) functions.

A further step toward the utilization of BBV as a vector for foreign genes was the demonstration that the RNA transcripts that derived *in vitro* from cDNA copies of the two genome subunits of BBV are infectious to *Drosophila* cells. It is indeed technically difficult to engineer RNA molecules, but this new approach makes BBV readily modifiable by usual methods for DNA (Dasmahapatra *et al.*, 1986).

Recently, Dasgupta *et al.* (1994) studied the ability of Flock House Virus to establish and maintain a persistent infection in *Drosophila* cells: After the extensive initial lysis, about 1% of the cells survived and grew in colonies. Periodical plaque assays demonstrated that deriving cell lines remained persistently infected, albeit in a nonlytic fashion. They proved

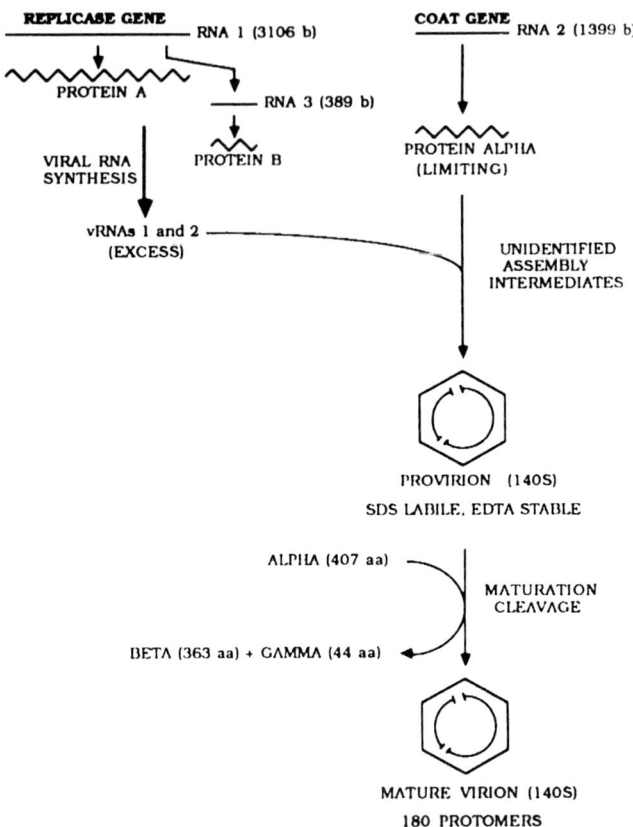

FIGURE 11.9   Model of nodavirus (*Black Beetle Virus*) morphogenesis. Coat protein precursor alpha is synthesized in limiting amounts relative to virions RNAs 1 and 2. Maturation cleavage begins after provirion assembly and generates stable mature virions. Protomer refers to protein alpha or its cleavage product (beta plus gamma). According to Gallagher and Rueckert (1988); reproduced with the kind permission of Dr. Rueckert and American Society of Microbiology Journals Department.

to be resistant to superinfection by FHV or other Nodaviruses. The establishment of PI is obviously caused by some cellular modification, since the virus showed, at first, no changes. In later passages, however, the size of the plaques decreased, which corresponded to a series of mutations affecting only RNA2, i.e., the gene for the coat protein.

C. Human Immunodeficiency Viruses

If HIV can be adsorbed at the surface of *Drosophila Kc0* cells, visualized by the use of fluorescein-labeled virions, and is perhaps even able to progress until the proviral insertion stage (as suggested by hybridation dots of high molecular weight DNA from infected cells with a radiolabeled HIV probe), completion of its replicative cycle is blocked in some unknown manner. No viral antigen could be detected by immunofluorescence and no reverse transcriptase activity was measurable in cell culture supernatants (a standard procedure for quantifying the release of viral particles) (Becker *et al.*, 1986).

McDonald *et al.* (1988, 1989) showed that this repression of HIV expression in *Drosophila* cells may be due to posttranscriptional processes. Comparing, by transient expression, the ability of *copia*, RSV, and HIV long terminal repeats to drive expression of the reporter gene CAT in *Drosophila S2* cells, they observed that, in spite of the synthesis of almost identical amounts of *cat* mRNAs, the level of CAT activity was more than 100-fold higher in *copia*-CAT transfected cells than in HIV-CAT transfected cells. In mammalian cells, with the same constructs, the exact opposite was the case. Among the many possible explanations, it could be that this repression is related to a sequestration of the transcripts within the nuclear compartment.

Nevertheless, a *Drosophila* cell expression system proved very efficient for producing and purifying significant amounts of the HIV-1 envelope glycoprotein gp120 (Ivey-Hoyle *et al.*, 1991; Culp *et al.*, 1991). *S2* cells were stably transformed with an expression vector in which the coding sequence was driven by the strong and "conditional" *Drosophila* metallothionein (Mt) promoter (see Chapter 9). Moreover, owing to the insertion of a tPA signal sequence (from the human plasminogen activator), the product was secreted. This *Drosophila* secreted product was recognized by gp120-specific monoclonal antibodies and bound to the CD4 receptor. It was highly glycosylated, although this glycosylation was primarily of the high mannose type in contrast to the complex glycans observed in mammalian systems.

On the other hand, in cells transformed with analogous gp160 constructs and after proper metal induction, no gp160 protein was immunodetectable (Ivey-Hoyle and Rosenberg, 1990). The authors reasoned that, as in mammalian cells, gp160 expression is likely to require the HIV regulatory Rev protein and this could be clearly demonstrated by cotransfection with a Rev-expression vector. Moreover, it was shown that the mechanisms of Rev action on *env* mRNA, that is, a facilitation of mRNA export to the cytoplasm, is well conserved in the insect cells. If the total synthesis of gp160 RNA (2.8 kb) was not affected by the presence of Rev, in contrast, the level of this mRNA in the cytoplasm was greatly increased in cotransfected cells.

D. Relative Strengths of Various Vertebrate Virus Promoters

It should be recalled that, in their pioneering experiments of gene transfer into *Drosophila* cells (see Chapter 9), DiNocera and Dawid (1983) and then Burke *et al.* (1984) noticed that certain promoters of vertebrate DNA or RNA tumor viruses, such as rous sarcoma virus LTR or SV40 early promoter, although quite efficient in vertebrate cells, function poorly in *Drosophila* cells. In any case, they were much less powerful than an LTR of the *Drosophila* transposon *copia*.

Sinclair (1987) reported that the major early gene promoter/enhancer region from human cytomegalovirus (or HCMV-IE) proved a very strong promoter in *Drosophila Kc* cells. Surprisingly, the same *pIEP1CAT* vector was not expressed in *Drosophila hydei* cells, even though this *DH33* cell line was capable of particularly high levels of expression with a *copia*-CAT vector.

Note: Kemble and Mocarki (1989) reported, incidentally, that a host cell protein, binding to a specific sequence of the cytomegalovirus regulatory region and present in a wide variety of human cell lines, could not be found in *Drosophila* Schneider's cells.

IV. EXPERIMENTAL STUDIES OF PARASITE–HOST CELL RELATIONS IN *DROSOPHILA* CELL LINES

A. Mycoplasmas

Mycoplasmas is the common name for a division of prokaryotes, the official taxonomic denomination of which, Mollicutes, refers to a total lack of cell wall; they are, therefore, extremely pleiomorphic.

They are parasites, commensals or saprophytes, and many of them are pathogens for humans, animals, or plants.

1. SPIROPLASMAS

Spiroplasmas are characterized by their helicoidal morphology, with rotatory and flexional motility, and have been isolated from both diseased and apparently healthy insects or plants. Their prototype, *Spiroplasma citri,* causes the symptoms of citrus "stubborn disease."

Although a few of them could be propagated in media based on insect tissue culture formulations (Jones *et al., 1977*), it was interesting to test their growth in insect cell culture systems in order to study possible deleterious effects on the host cells. Therefore, Steiner *et al.* (1982) infected systematically two different *Drosophila* cell lines (*S1* and milanese *I-XII* lines; see Chapter 3) with a variety of spiroplasmas isolated from plants or arthropods (see Table 11.IV).

TABLE 11.IV   Growth of Spiroplasmas in *Drosophila melanogaster* Cells[a]

Spiroplasma isolate	Growth	Culture death (days)	Titer (CFU/ml)	Adsorption
CSO	+	21	2.6×10^8	+
S. citri	+	14	1.0×10^9	+
Cactus	+	14	9.2×10^8	+
Lettuce	+	14	4.4×10^8	+
PP	+	7	2.5×10^9	+
BNR-1	+	5	3.3×10^9	+/−[b]
OBMG	+	5	3.0×10^9	+/−
HB	+	2	1.1×10^9	+
SMCA	+		1.0×10^{9c}	+
277F	+	5	5.5×10^9	+
SRO	−		ND[d]	+

[a] Approximately 10^5 to 10^7 spiroplasmal CFU or cells (SRO) were inoculated into *Dm-1* cell cultures. Culture death was determined by trypan blue stain; adsorption by dark-field microscopy and DNA staining. The titers were determined within 3 to 7 days, depending on the spiroplasma strain. CSO, Corn stunt organism; PP, powder puff isolate; BRN-1, isolated from tulip tree; OBMG, isolated from magnolia; HB, honey bee infection disease; SMCA, suckling mouse contaminant agent, isolated from ticks; 277F, isolated from ticks; SRO, sex-ratio organism. From Steiner *et al.* (1982); with the permission of *Infection and Immunity* and the American Society of Microbiology.

[b] +/− Questionable results.

[c] Color-changing units, checked by dark-field microscopy.

[d] Not done.

The peak titers were very similar to those observed in cell-free broth (although high concentrations of *Corn stunt organism* were clearly detected earlier in cell cultures). (See also several short papers by Kuroda *et al.,* 1988, 1989, 1992).

After fluorescent staining with Hoechst 33258, or in SEM, the parasites could be observed either swarming around or attached to the cell membrane.

The type and extent of the cytopathic effects varied according to the different spiroplasmas. For instance, HB (responsible of a rapidly lethal infection in honey bees) killed *Drosophila* cells in two days and CSO in 3 weeks, whereas SMCA produced only a chronic infection of the cultures.

SRO

The *sex ratio organism* (SRO) is a "spirochete-like" microorganism which is responsible for deviation from the normal 1 : 1 sex ratio in *Drosophila*. Discovered in *Drosophila willistoni* and *Drosophila nebulosa* by Malogolowkin and Poulson (*1957*), it could be transferred to *Drosophila melanogaster*. In infected strains, the agent is transovarially transmitted. The male progeny of infected mothers (at least all individuals with a single X chromosome) are killed more or less early during embryonic life, whereas embryos with two Xs survive. The primary target site for SRO was mapped to the ventral region of the blastoderm fate map.

When studying the effects of SRO on embryonic cells, Koana and Miyake (1983) grew primary cultures from single embryos (see Chapter 2) derived from a *Drosophila melanogaster* infested strain. In order to identify, *a posteriori,* the genotype of each culture, females homozygous for the G6PD-null allele (remember that this gene is X-linked) was crossed with wild-type (G6PD positive) males. After observation of the cultures had been completed, histochemical staining revealed whether cells contained two or only a single X [according to a method developed by Koana and Miyake (1982)]. More specifically in "male" cultures, neurons, hemocytes, and imaginal disc cells barely differentiate, while muscles and fat body cells were not so severely affected.

Later, Ueda *et al.* (1987) confirmed that SRO could transiently proliferate in primary cultures from infested embryos. In contrast, as seen in Table 11.IV, Steiner *et al.* (1982) could not grow the microorganism in the two established cell lines tested.

2. CONTAMINANT MYCOPLASMAS

It was assumed by the people in charge of the American Type Culture Collection (Hay *et al., 1989*) that some "5 to 35% of mammalian cell lines in current use are infected with mycoplasmas," and there is no reason why this should not also be true for many insect cell lines.

Contaminant mycoplasmas in cell cultures are of primarily bovine or human origin. Commercially available batches of bovine sera (to be added to most culture media) are certified as having been filtered and controlled for Mycoplasmas, but it is common knowledge that, because of their deformability, these microorganisms can be forced, even through 0.2 μm-pore filters. Several *Mycoplasma* species (especially *Mycoplasma orale*) are found in the oropharynx of many healthy people. Vertical laminar-flow biohazard hoods, currently used for cell culture, fortunately offer a better protection for and from human operators.

Even though heavy mycoplasma infections are not always too severely cytopathic, the culture growth rate and many of the biochemical and immunological properties of the cells can be altered by the presence of the microorganisms, so is essential to detect and eliminate such contaminations.

The best diagnosis resides in the microbiological cultivation procedure. Various mycoplasma agar or broth formulas are commercially available (for instance, in the Sigma catalog for cell culture ingredients). Owing to the tendency of mycoplasmas to penetrate deeply into solid media, colonies show a typical "fried egg" appearance, with an opaque center embedded in the agar and a translucent peripheral zone at its surface.

Another efficient detection method is the direct observation after DNA fluorescent staining with either fluorochrome Hoechst 33258 or the Dapi reagent. Mycoplasmas can be seen as numerous bright points clustered at the surface of cells or scattered in the background.

Biochemical tests, based mainly on enzymatic assays of various phosphorylases, are much less reliable. It is relevant that Steiner *et al.* (1982) reported that, if *Drosophila S2* cells had, as expected, no detectable uridine phosphorylase activity, the milanese *I-XII* line had significant enzyme levels (although it was proved, according to all other criteria, to be free from mycoplasmas).

It is advisable to get rid of any infected cell cultures before the infection spreads, through aerosols, to all other clean cultures in your laboratory. Mycoplasmas can, indeed, survive on laboratory benches, and even in laminar-flow hoods, for 2 to 3 days.

However, when it is imperative to keep an infected cell strain, it should be cured with proper antibiotics. Because they lack a cell wall, mycoplasmas are, of course, resistant to penicillin and other β-lactam antibiotics whose primary target is the biosynthesis of peptidoglycans, but they are usually sensitive to antibiotics that inhibit protein synthesis in prokaryotes (such as the tetracyclines). Bernhard *et al.* (1980) recommended the use of vibramycin (Pfizer, Zürich, 10 μg/ml), and Braude-Zolotarjova *et al.* (1986) could eradicate mycoplasmas from their *Drosophila virilis* cell lines with doxycycline (10 μg/ml, two passages). There are, on the market, several antibiotic mixtures and standardized protocols for eliminating mycoplasmas from mammalian cell cultures, and they should work with insect cells (do not forget to test first the particular sensitivity of your control cell lines).

Nevertheless, the treatment remains delicate and time consuming, and the best remedy is always careful prevention, with periodic controls (especially when one is confronted with a suspicious anomaly in the growth rate of the culture or individual cell appearance).

B. Rickettsia

Szollosi and Debec (1980) fortuitously discovered the presence, in the cytoplasm of *Drosophila* haploid cell lines *1182*, of many microorganisms that were identified as being Rickettsia.

In ultrathin sections, they appeared mainly as circular profiles of about 0.5 μm in diameter, more rarely as rods of 1 to 1.2 μm in length. A central area contained a network of filaments, representing probably the DNA genome (Fig. 11.10).

These organisms were clearly intravacuolar, so that three membranes could be distinguished: the plasma membrane of the host cell vacuole, then two tripartite membranes, the outer one (cell wall) separated from the inner one (plasma membrane) by a 6 nm space. The occurrence of 2 to 4 individuals within the same vacuole was evocative of recent binary fission.

Identical microorganisms were detected, albeit in lesser quantities, in embryos of the *Drosophila* mutant strain from which the cell lines were derived.

The strict intracellular location and morphological characteristics of this organism are very evocative of a rickettsia-type. Similar supposed symbionts have been periodically described in different tissues of *Drosophila*.

Infected cell lines, as well as corresponding fly strains, could be readily cured with antibiotics (e.g., tetracycline, 100 μg/ml).

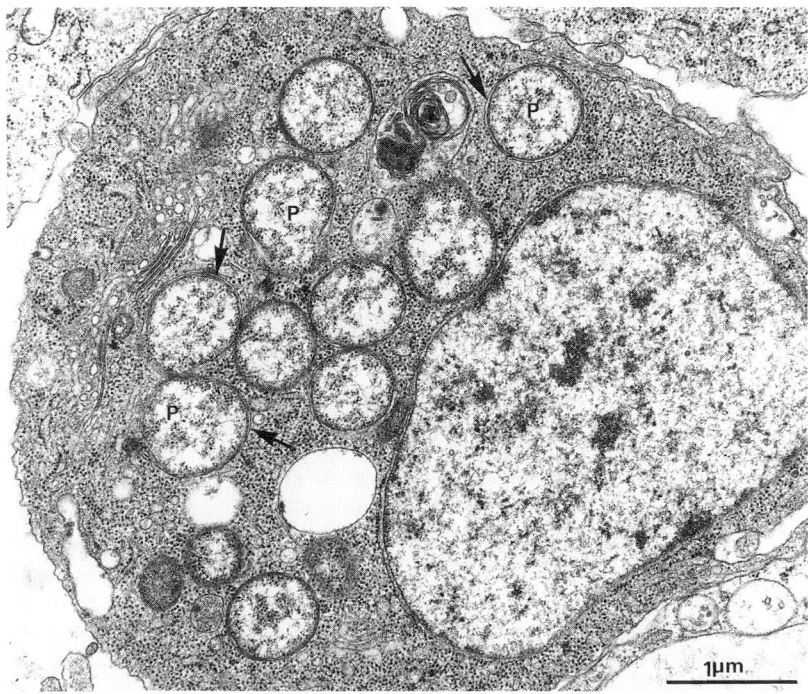

FIGURE 11.10   Rickettsiae in a *Drosophila* cell line *(1182)*. The prokaryotic organisms (P) are distributed in individual vacuoles, among the host cell organelles. The arrows point to the membranes of these vacuoles. Szollosi and Debec (1980); courtesy of Dr. Debec and with permission of Editions Scientifiques Elsevier, Paris.

C. Malaria Parasite *Plasmodium*

Although the different stages of the cycles of malaria agents that occur in the human host had been, for many years, obtained in cultures of proper cell types (i.e., in settled erythrocyte layers and in cultured human hepatocytes, respectively), their sporogonic cycle, which takes places in the mosquito vector, did not seem to be amenable to *in vitro* culture. In the past few years, however, Warburg and collaborators reported successful cocultures with *Drosophila S2* cells of the avian *Plasmodium gallinaceum* (Warburg and Miller, 1992), and of the human pathogen *Plasmodium falciparum* (Warburg and Schneider, 1993).

Because differentiating oocysts are characteristically located under the basement membrane of the midgut epithelium of the mosquito, Warburg and Miller (1992) suspected that extracellular matrix components may be important. So, they coated the bottom of culture flasks with Matrigel (Collaborative Research, Bedford, MA), a basement membrane-like product, derived from the murine Engelbreth–Holm–Swarm tumor and now available commercially (containing not only laminin and collagen type IV, but also heparan sulfate proteoglycan and several growth factors).

After production *in vitro* (see Warburg and Schneider, 1993, for details) of a high percentage of active ookinetes, they were seeded in the coated culture vessels, together with a suspension of *Drosophila S2* cells. They adhered to the Matrigel or were embedded in it, and they then transformed progressively into oocysts. Mature oocysts reached a diameter of 25 to 40 μm and, toward the end of the second week, some of them sporulated. Typically shaped sporozoites were observed (Fig. 11.11), and they expressed the characteristic circumsporozoite protein (as established by immunofluorescence with a monoclonal antibody).

The presence of *Drosophila* cells was absolutely necessary, although their role is unclear. It should be remembered that S2 cells, like several other *Drosophila* cell lines, are able to synthesize and secrete extracellular matrix components (see Chapter 6), and it was verified that *Drosophila* laminin and collagen IV adhere actively to *Plasmodium* oocysts, but the cells may also supply many other useful nutrients.

This work is an important achievement, since the entire life cycle of this very widespread human pathogen can now be studied *in vitro*.

APPENDIX 11

11.A. Preparation and Maintenance of Virus-free *Drosophila* Cell Lines

This procedure is taken from Plus (1978, 1980).

1. *VIRUS-FREE STOCKS OF FLIES*

They can be obtained by dechorionation of eggs from old mothers (20 days at 25°C) in a sodium hypochloride solution. The hereditary picornaviruses, P and A, are no longer transmitted by such old females and, besides, the virions, at the surface of the eggs, are destroyed by the hypochloride treatment.

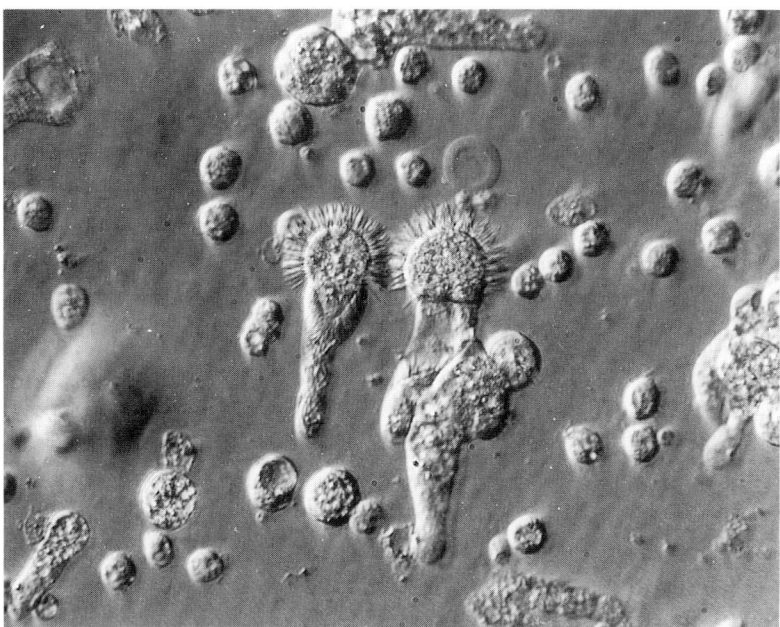

FIGURE 11.11 Sporogonic development of *Plasmodium gallinaceum* in coculture with *Drosophila* cells. Sporozoites still connected to the residual body show a flower-like appearance. All around are *Drosophila* S2 cells. 14th day *in vitro*. Bar: 30 μm. From Warburg and Miller (1992); courtesy of Dr. A. Warburg and reprinted with permission from American Association for the Advancement of Science, copyright (1992).

Virus-free flies have to be kept separately from the others and handled with sterile instruments.

Note: As for possible σ infection, the CO_2 sensibility test allows an easy diagnosis.

2. TEST FOR THE PRESENCE OF VIRUSES

(For screening cultured cell supernates and serum batch to be used.)

Aliquots of serum or of metabolized cell culture medium are diluted 2-fold with Ringer's saline solution and injected into about 50 flies from a previously verified virus-free stock.

Inoculated flies, reared at 25°C, are carefully observed for symptoms of disease.

Those exhibiting any abnormality are freshly crushed in a Ringer's solution. This extract is clarified by low centrifugation and the supernate is negatively stained with phosphotungstic acid, then carefully examined by electron microscopy for the presence of viral particles.

In the absence of any visible symptom, three further successive passages into control flies have to be performed in order to possibly promote viral multiplication, and the final fly extract is examined by electron microscopy.

3. FILTRATION OF CALF SERUM

Millipore filters of pore size of up to 45 nm can retain X virus.

11.B. *Black Beetle Virus* Plaque Assay

This assay is presented according to Selling and Kaesberg [from Ashburner (1989); courtesy of Professor Kaesberg, University of Wisconsin, Madison, WI, with the permission of Cold Spring Harbor Laboratory Press].

MATERIALS

Schneider's medium, 15% FBS (CGM = cell growth medium)
Plastic conical tube, 15 ml
Tissue culture dishes, 60 mm diameter
NHA buffer*
Low-melting temperature agarose, 1% in CGM (dissolve by heating to 65°C in a water bath for 10 min; cool to 37 to 40°C before adding to cells)
Isotonic buffer
MTT [3-(4,5-dimethylthiazol-2-yl)-2,5-diphenyltetrazolium bromide], 3 mg/ml in isotonic buffer

METHOD

1. Transfer *Drosophila* cells (at least 4×10^6 per assay) into a 15-ml conical tube. Determine volume and density and pellet (580 g for 5 min at room temperature).
2. Resuspend cells in CGM to a density of 4.2×10^7/ml. Dispense aliquots of 95 μl into 15-ml plastic conical tubes.

---

* Hepes (pH 7.0), 25 mM; NaCl, 125 mM; bovine serum albumin, 0.2% (w/v), trypan blue, 10 μg/ml.

3. Add 5 μl virus suspension (if diluted, do so in CGM). Vortex lightly and then shake tubes gently for 1 h at room temperature.
4. Add 5 ml of NHA buffer to virus/cell mixtures and then vortex. Pour into 60-mm diameter tissue culture dishes.
5. Allow 1 h for cells to attach to dishes. Remove buffer by aspiration and add 3.0 ml of a low-melting temperature agarose overlay. This will solidify in about 5 min. Then add 2 ml CGM. Incubate at 26°C.
6. After 50 h, develop plaques by pipetting, as several, well-scattered drops, 0.5 ml MTT solution onto overlay. Plaques appear as yellow circles against a blue-green background.

11.C. Infection of Cells with *Black Beetle Virus* RNA

This procedure is reproduced from Selling and Kaesberg [in Ashburner (1989); courtesy of Professor Kaesberg, University of Wisconsin, Madison, WI; with the permission of Cold Spring Harbor Laboratory Press].

MATERIALS

All reagents, buffers and vessels must be RNase-free
Schneider's medium + 15% FBS (CGM = cell growth medium)
Plastic centrifuge tubes
Plastic conical tubes, 15 ml
Isotonic buffer
IBDD buffer (isotonic buffer at pH 6.0 containing 320 μg/ml DEAE-Dextran
NHA buffer (see footnote on page 592)
Tissue-culture dishes, 60-mm diameter
Low-melting-temperature agarose, 1% i CGM (see Appendix 11.B)
MTT solution (see Appendix 11.B)

METHOD

1. Pellet an appropriate number of cells (at least 5 × 10⁶ per RNA sample) in an RNase-free plastic centrifuge tube (580 g for 5 min at room temperature).
2. Using a 10-ml pipette, resuspend cells in 10 to 15 ml of RNase-free isotonic buffer and pellet as in step 1.
3. Repeat step 2 twice more. During centrifugation, prepare inocula (total volume 10 μl) by pipetting water and RNA samples into conical plastic tubes.

4. Resuspend pellet to 5.6×10^7 cells in IBDD. Add 90 μl of cell suspension to each 10-μl RNA sample.
5. Incubate RNA and cell mixture for 20 min at room temperature. Vortex gently every 5 min.
   Note: At this point, the cells may be propagated and/or the proportion of infected cells assayed. For propagation, rinse cells with 5 ml CGM, pellet and resuspend to an appropriate density in CGM. Proceed as follows for assay.
6. Prepare indicator cells by diluting *Drosophila* cells (not those rinsed with isotonic buffer) to 8×10^5 cells/ml in NHA buffer. Dispense 5 ml aliquots into 15-ml plastic conical tubes.
7. Add 0.9 ml of CGM to transfected cells. Disperse by drawing through a pipette several times (a rubber bulb attached to a 1-ml pipette is convenient for this step) and then transfer aliquots (1 to 100 μl) to tubes containing indicator cells. Vortex and then pour into 60-mm diameter tissue-culture plates.
8. Allow cells to attach to surface (1 h). Remove buffer by aspiration and add 3 ml 1% low-melting temperature agarose. After agarose has solidified (5 min), overlay with 2 ml CGM. Incubate at 26°C.
9. Develop plates after 50 h by pipetting MTT solution onto overlay. Add MTT as several well-separated drops. Allow stain to develop for 2 h at 26°C. Plaques appear as yellow circles against a blue-green background.

References

Berkaloff, A., Bregliano, J. C., and Ohanessian, A. (1965). Mise en évidence de virions dans des drosophiles infectées par le virus héréditaire σ. *CR Acad. Sci. Paris* **260,** 5956.

Bernard, J. P., and Petitjean, A. M. (1978). In vitro synthesis of double stranded RNA by *Drosophila* X virus purified virions. *Biochem. Biophys. Res. Commun.* **83,** 763–770.

Bras-Herreng, F. (1973). Etude de l'évolution d'une population du virus Sindbis au cours de passages successifs chez la Drosophile. *Ann. Inst. Pasteur* **124A,** 507–533.

Dobos, P., Hill, B. J., Hallett, R., Kells, D. T. C., Becht, H., and Teninges, D. (1979). Biophysical and biochemical characterization of five animal viruses with bisegmented double-stranded RNA genomes. *J. Virol.* **32,** 593–605.

Hay, R. J., Macy, M. L., and Chen, T. R. (1989). Mycoplasma infection of cultured cells. *Nature* **339,** 487–488.

Hendry, D. A. (1991) Nodaviridae of invertebrates. *In "Viruses of Invertebrates"* (E. Kurstak, ed.), Dekker, New York, pp. 227–276.

Jones, A. L., Whitcomb, R. F., Williamson, D. L., and Coan, M. L. (1977). Comparative growth and primary isolation of Spiroplasmas in media based on Insect tissue culture formulations. *Phytopathology* **67,** 738–746.

Jousset, F. X., Bergoin, M., and Revet, B. (1977). Characterization of the *Drosophila* C virus. *J. Gen. Virol.* **34**, 269–285.

Jousset, F. X., Plus, N., Croizier, G., and Thomas, M. (1972). Existence chez Drosophile de deux groupes de Picornavirus de propriétés sérologiques et biologiques différentes. *CR Acad. Sci. Paris* **275**, 3043–3046.

Lenz Goodman, C., and McIntosh, A. H. (1994). Production of Baculoviruses for Insect cell control using cell culture. *In "Insect Cell Biotechnology"* (K. Maramorosch and A. H. McIntosh, eds.), pp. 33–56, CRC Press, Boca Raton, FL.

L'Heritier, P., and Teissier, G. (1937). Une anomalie physiologique héréditaire chez la Drosophile. *CR Acad. Sci. Paris* **205**, 1099–1101.

Longworth, J. F., and Archibald, R. D. (1975). A virus of black beetle, *Heteronychus arator* (F.) (Coleoptera: Scarabeidae) *N.Z. J. Zool.* **2**, 233–236.

Maeda, S. (1994). Expression of foreign genes in Insect cells using Baculovirus vectors. *In "Insect Cell Biotechnology"* (K. Maramorosch and A. H. McIntosh, eds.), pp. 1–31, CRC Press, Boca Raton, FL.

Malogolowkin, C., and Poulson, D. F. (1957). Infective transfer of maternally inherited abnormal sex-ratio in *Drosophila willistoni*. *Science* **126**, 32.

Plus, N., Croizier, G., Duthoit, J. L., David, J., Anxolabehere, D., and Periquet, G. (1975). Découverte, chez la Drosophile, de virus appartenant à trois nouveaux groupes. *CR Acad. Sci. Paris* **280**, 1501–1504.

Plus, N., and Duthoit, J. L. (1969). Un nouveau virus de *Drosophila melanogaster*, le virus P. *CR Acad. Sci. Paris* **268**, 2313–2315.

Printz, P. (1970). Adaptation du virus de la Stomatite Vésiculaire à *Drosophila melanogaster*. *Ann. Inst. Pasteur* **119**, 520–537.

Summers, M. D., and Smith, G. E. (1987). A manual of methods for baculovirus vectors and insect cell culture procedures. *Texas Agric. Exp. Sta.*, Bulletin 1555.

Teninges, D. (1979). Protein and RNA composition of the structural components of *Drosophila* X virus. *J. Gen. Virol.* **45**, 641–649.

Teninges, D., and Plus, N. (1972). P virus of *Drosophila melanogaster*, as a new picornavirus. *J. Gen. Virol.* **16**, 103–109.

Wyers, F., Petitjean, A. M., Dru, P., Gay, P., and Contamine, D. (1995). Localization of domains within the *Drosophila* Ref(2)P protein involved in the intracellular control of *Sigma* rhabdovirus multiplication. *J. Virol.* **69**, 4463–4470.

Bibliography

The references listed in this bibliography are strictly limited to research carried out on *Drosophila* cell or tissue cultures. All other references are listed at the end of the respective chapter. Note, also, that additional references added in proofs are listed in the Addendum on pages 689–690. The following key has been used to classify the topic presented in the reference: **B**, cell biology (growth factors, oncogenes, adhesion molecules); **C**, culture media and methods; **Cd**, imaginal discs; **Cp**, cell primary cultures; **Cl**, cell lines; **Ct**, tissue and organ cultures (except imaginal discs); **Cy**, cell cycle, karyology; **E**, ecdysone and other hormones; **H**, heat-shock; **M**, molecular biology; **R**, retrotransposons; **r**, review or handbook; **T**, transfection; and **V**, viruses or other infectious agents.

E   Abraham, E. G., Mounier, N., and Bosquet, G. (1993). Expression of a *Bombyx* cytoplasmic actin gene in cultured *Drosophila* cells: influence of 20-hydroxyecdysone and interference with expression of endogenous cytoplasmic actin genes. *Insect Biochem. Mol. Biol.* **23**, 905–912.

B   Abrams, J. M., Lux, A., Steller, H., and Krieger, M. (1992). Macrophages in *Drosophila* embryos and L2 cells exhibit scavenger receptor-mediated endocytosis. *Proc. Natl. Acad. Sci. U.S.A.* **89**, 10375–10379.

M   Ackerman, P., Glover, C. V., and Osheroff, N. (1988). Phosphorylation of DNA topoisomerase II *in vivo* and in total homogenates of *Drosophila* Kc cells. The role of casein kinase II. *J. Biol. Chem.* **263**, 12653–12660.

B   Alahiotis, S., and Berger, E. (1977). Isozyme and allozyme patterns in embryonic *Drosophila* cell culture lines. *Biochem. Genet.* **15**, 877–883.

597

V   Alatortsev, V. E., Ananiev, E. V., Gushchina, E. A., Grigoriev, V. B., and Gushchin, B. V. (1981). A virus of the reoviridae in established cell lines of *Drosophila melanogaster*. *J. Gen. Virol.* **54**, 23–31.

M   Albert, J. L., and Lingle, C. J. (1994). Activation of nicotinic acetylcholine receptors on cultured *Drosophila* and other insect neurons. *J. Physiol. (London)* **463**, 605–630.

TV Allday, M. J., Sinclair, J. H., MacGillivray, A. J., and Sang, J. H. (1985). Efficient expression of an Epstein-Barr nuclear antigen in *Drosophila* cells transfected with Epstein-Barr virus DNA. *EMBO J.* **4**, 2955–2959.

M   Alziari, S., Berthier, F., Touraille, S., Stepien, G., and Durand, R. (1985). Mitochondrial DNA expression in *Drosophila melanogaster*: neosynthesized polypeptides in isolated mitochondria. *Biochimie* **67**, 1023–1034.

E   Amero, S. A., Matunis, M. J., Hockensmith, J. W., Raychaudhuri, G., and Beyer, A. L. (1993). A unique ribonuclein complex assembles preferentially on Ecdysone-responsive sites in *Drosophila melanogaster*. *Mol. Cell. Biol.* **13**, 5323–5330.

TH Amin, J., Ananthan, J., and Voellmy, R. (1988). Key feature of heat shock regulatory elements. *Mol. Cell. Biol.* **8**, 3761–3769.

H   Amin, J., Fernandez, M., Ananthan, J., Lis, J. T., and Voellmy, R. (1994). Cooperative binding of heat shock transcription factor to the *Hsp70* promoter *in vivo* and *in vitro*. *J. Biol. Chem.* **269**, 4804–4811.

TH Amin, J., Mestril, R., Lawson, R., Klapper, H., and Voellmy, R. (1985). The heat-shock consensus sequence is not sufficient for *hsp70* gene expression in *Drosophila melanogaster*. *Mol. Cell. Biol.* **5**, 197–203.

H   Amin, J., Mestril, R., Schiller, P., Dreano, M., and Voellmy, R. (1987). Organization of the *Drosophila hsp70* heat-shock regulation unit. *Mol. Cell. Biol.* **7**, 1055–1062.

THE Amin, J., Mestril, R., and Voellmy, R. (1991). Genes for *Drosophila* small heat shock proteins are regulated differently by ecdysterone. *Mol. Cell. Biol.* **11**, 5937–5944.

CyM Ananiev, E. V., Polukarova, L. G., and Yurov, Y. B. (1977). Replication of chromosomal DNA in diploid *Drosophila melanogaster* cells cultured *in vitro*. *Chromosoma* **59**, 259–272.

CyM Ananiev, E. V., Yurov, Y. B., and Polukarova, L. G. (1977). Determination of the size of the replicons and rate of DNA replication in a culture of *Drosophila melanogaster* cells. *Mol. Biol.* **10**, 807–815.

TB Ananthan, J., Baler, R., Morrissey, D., Zuo, J., Lan, Y., Weir, M., and Voellmy, R. (1993). Synergistic activation of transcription is mediated by the N-terminal domain of *Drosophila fushi tarazu* homoprotein and can occur without DNA binding by the protein. *Mol. Cell. Biol.* **13**, 1599–1609.

E   Andres, A. J. (1990). An analysis of the temporal and spatial patterns of expression of the ecdysone-inducible genes *Eip28/29* and *Eip40* during development of *Drosophila melanogaster*. Ph.D. Thesis, Dept. of Biology, Indiana University.

E   Andres, A. J., and Cherbas, P. (1992). Tissue-specific ecdysone responses: regulation of the *Drosophila* genes *Eip 28/29* and *Eip 40* during larval development. *Development* **116**, 865–876.

E   Andres, A. J., and Cherbas, P. (1994). Tissue-specific regulation by ecdysone: Distinct patterns of *Eip28/29* expression are controlled by different ecdysone response elements. *Development* **15**, 320–331.

rE Andres, J. A., Fletcher, J. C., Karim, F. D., and Thummel, C. S. (1993). Molecular analysis of the initiation of insect metamorphosis: A comparative study of *Drosophila* ecdysteroid-regulated transcription. *Dev. Biol.* **160**, 388–404.

rE Andres, J. A., and Thummel, C. S. (1992). Hormones, puffs and flies: the molecular control of metamorphosis by ecdysone. *Trends Genet.* **8**, 132–138.

M Andrews, C. A. (1975). Surface properties and carbohydrate metabolism of cells of *Drosophila melanogaster* grown *in vitro*. Ph.D. Thesis, University of Michigan, Ann Arbor, MI.

M Andrews, C. A., and Rizki, T. M. (1978). Studies on lectin-induced agglutination of *Drosophila* embryonic cell lines. *J. Insect Physiol.* **24**, 9–12.

BCp Antley, R. M., and Fox, A. S. (1970). The relationship between RNA and protein synthesis and the aggregation of *Drosophila* embryonic cells. *Dev. Biol.* **22**, 282–297.

CdE Appel, L. F., Prout, M., Abu-Shumays, R., Hammonds, A., Garbe, J. C., Fristrom, D., and Fristrom, J. (1993). The *Drosophila stubble-stubbloid* gene encodes an apparent transmembrane serine protease required for epithelial morphogenesis. *Proc. Natl. Acad. Sci. U.S.A.* **90**, 4937–4941.

Cd Apple, R. T., and Fristrom, J. W. (1991). 20-hydroxyecdysone is required for, and negatively regulates, transcription of *Drosophila* pupal cuticle protein genes. *Dev. Biol.* **146**, 569–582.

R Arkhipova, I. R., Gorelova, T. V., Ilyin, Y. V., and Schuppe, N. G. (1984). Reverse transcription of *Drosophila* mobile dispersed genetic element RNAs: detection of intermediate forms. *Nucleic Acid Res.* **12**, 7533–7548.

R Arkhipova, I. R., and Ilyin, V. V. (1991). Properties of promoter regions of *mgd1 Drosophila* retrotransposon indicate that it belongs to a specific class of promoters. *EMBO J.* **10**, 1169–1177.

R Arkhipova, I. R., Mazo, A. M., Cherkasova, V. A., Gorelova, T. V., Schuppe, N. G., and Ilyin, Y. V. (1986). The steps of reverse transcription of *Drosophila* mobile dispersed genetic elements and U3-R-U5 structure of their LTRs. *Cell* **44**, 555–563.

M Arndt-Jovin, D. J., Udvardy, A., Garner, M. M., Ritter, S., and Jovin, T. M. (1993). Z-DNA binding and inhibition by GTP of *Drosophila* Topoisomerase II. *Biochemistry* **32**, 4862–4872.

H Arrigo, A. P. (1980a). Etudes des protéines induites par le choc thermique chez *Drosophila melanogaster*. Ph.D. Thesis, Université de Genève, Switzerland.

H Arrigo, A. P. (1980b). Investigation of the function of heat-shock proteins in *Drosophila melanogaster* tissue culture cells. *Mol. Gen. Genet.* **178**, 517–524.

H Arrigo, A. P. (1983). Acetylation and methylation of core histone as modified after heat or arsenite treatment of *Drosophila* tissue culture cells. *Nucleic Acids Res.* **11**, 1389–1404.

H Arrigo, A. P., and Ahmad-Zadeh, C. (1981). Localization of a small heat-shock protein (*hsp23*) in salivary glands of *Drosophila melanogaster*. *Mol. Gen. Genet.* **184**, 73–79.

H Arrigo, A. P., Darlix, J. L., Khandjian, E. W., Simon, M., and Spahr, P. F. (1985). Characterization of the prosome from *Drosophila* and its similarity to the cytoplasmic structures formed by the low molecular weight heat-shock proteins. *EMBO J.* **4**, 399–406.

H Arrigo, A. P., Fakan, S., and Tissieres, A. (1980). Localization of the heat shock-induced proteins in *Drosophila melanogaster* tissue culture cells. *Dev. Biol.* **78**, 86–103.

H Arrigo, A. P., and Pauli, D. (1988). Characterization of *HSP27* and three immunologically related polypeptides during *Drosophila* development. *Exp. Cell Res.* **175**, 169–183.

rB Artavanis-Tsakonas, S., and Simpson, P. (1991). Choosing a cell fate: a view to the *Notch* locus. *Trends Genet.* **7**, 403–408.

CtH Ashburner, M. (1970). Patterns of puffing activity in the salivary gland chromosomes of *Drosophila*. V. Responses to environmental treatments. *Chromosoma* **31**, 356–376.

CtE   Ashburner, M. (1971). Induction of puffs in polytene chromosomes of *in vitro* cultured salivary glands of *Drosophila melanogaster* by ecdysone and ecdysone analogues. *Nature New Biology* **230**, 222–224.

CtE   Ashburner, M. (1972). Patterns of puffing activity in the salivary gland chromosomes of *Drosophila*. VI. Induction by ecdysone in salivary glands of *Drosophila melanogaster* cultured *in vitro*. *Chromosoma* **38**, 255–281.

r   Ashburner, M. (1989). *Drosophila, Vol. 1. A laboratory handbook and Vol. 2. A laboratory manual,* Cold Spring Harbor Laboratory Press, Cold Spring Harbor.

CtE   Ashburner, M., Chihara, C., Meltzer, P., and Richards, G. (1974). Temporal control of puffing activity in polytene chromosomes. *Cold Spring Harbor Symp. Quant. Biol.* **38**, 655–662.

M   Attardi, L. D., and Tijan, R. (1993). *Drosophila* tissue-specific transcription factor NFT-1 contains a novel isoleucine-rich activation motif. *Genes Dev.* **7**, 1341–1353.

M   Attardi, L. D., Von Seggern, D., and Tijan, R. (1993). Ectopic expression of wild-type or dominant-negative mutant of transcription factor NTF-1 disrupts normal *Drosophila* development. *Proc. Natl. Acad. Sci. U.S.A.* **90**, 10563–10567.

B   Auld, V. J., Fetter, R. D., Broadie, K., and Goodman, C. S. (1995). *Gliotactin,* a novel transmembrane protein on peripheral glia, is required to form the blood-nerve barrier in *Drosophila. Cell* **81**, 757–767.

TM   Ayer, S., and Benyajati, C. (1990). Conserved Enhancer and Silencer elements responsible for differential *Adh* transcription in *Drosophila* cell lines. *Mol. Cell. Biol.* **10**, 3512–3523.

E   Ayer, S., and Benyajati, C. (1992). The binding site of a steroid hormone receptor-like protein within the adult *Adh* adult enhancer is required for high levels of tissue-specific alcohol dehydrogenase expression. *Mol. Cell. Biol.* **12**, 661–673.

E   Ayer, S., Walker, N., Mosammaparast, M., Nelson, J. P., Shilo, B. Z., and Benyajati, C. (1993). Activation and repression of *Drosophila* alcohol dehydrogenase distal transcription by two steroid receptor superfamily response element. *Nucleic Acids Res.* **21**, 1619–1627.

M   Azou, Y., and Laval, M. (1993). Sequence of the novel joints present in the amplified DNA of N-phosphonacetyl-L-aspartate resistant *Drosophila* cells: Implication on the mechanisms of amplification in these cells. *Biol. Cell* **77**, 155–164.

B   Baeze-Squiban, A., Best-Belpomme, M., and Marano, F. (1988. Effects of Deltamethrin on a *Drosophila* cell line *in vitro* cytotoxicity. Accumulation and metabolism. *In* "*Inv. Fish Tissue Culture*" (Y. Kuroda, E. Kurstak, and K. Maramorosch, eds.), pp. 58–59. Springer Verlag, New York.

B   Baeze-Squiban, A., Best-Belpomme, M., and Marano, F. (1989). Cytotoxicity, accumulation and metabolism of deltamethrin and pyrethroid insecticids in *Drosophila melanogaster* cells. *Pesticide Biochem. Physiol.* 1–21.

M   Baikenova, A. A., Braude-Zolotarjova, T. Y., Kakpakov, V. T., Ivanschenko, N. I., Korochkin, L. I., and Schuppe, N. G. (1986). (in Russian) Isoenzymic variations of esterase during the growth of cultured embryonic cells of *Drosophila virilis* Sturt. *Ontogenez* **17**, 535–537.

H   Ballinger, D. G., and Pardue, M. L. (1982). The subcellular compartmentalization of mRNAs in heat-shocked *Drosophila* cells. *In* "*Heat-shock from Bacteria to Man*" (M. J. Schlesinger, M. Ashburner, and A. Tissieres, eds.), pp. 183–190. Cold Spring Harbor Laboratory Press, Cold Spring Harbor.

H   Ballinger, D. G., and Pardue, M. L. (1983). The control of protein synthesis during heat-shock in *Drosophila* involves altered polypeptide elongation rates. *Cell* **33**, 103–114.

H   Ballinger, D., and Pardue, M. L. (1985). Mechanism of translational control in heat-shocked *Drosophila* cells. In *"Changes in Eukaryotic Gene Expression in Response to Environmental Stress"* (B. G. Atkinson and D. B. Walden, eds.), pp. 53–70. Academic Press, San Diego.

CyM   Bandara, L. R., Buck, V. M., Zamanian, M., Johnston, L. H., and La Thangue, N. B. (1993). Functional synergy between DP-1 and E2F-1 in cell cycle-regulating transcription factor DRTF1/E2F. *EMBO J.* **12**, 4317–4324.

rC   Barigozzi, C. (1971). *Drosophila* cell culture and its application for the study of genetics and virology. I. *Drosophila* cells *in vitro:* Behavior and utilization for genetic purpose. *Curr. Top. Microbiol. Immunol.* **55**, 209–218.

rC   Barigozzi, C. (1972). Invertebrate cell culture in Genetic Research. In *"Invertebrate Tissue Culture"* (C. Vago, ed.), Vol. II, pp. 163–180.

Cy   Barigozzi, C. (1973). Cariogamia e divisione di sincarion in cellule embrionale stabilizzate *in vitro* di *Drosophila melanogaster. Acta Embryologiae Experimentalis,* 225.

Cy   Barigozzi, C. (1982). Chromosomal phenomena in cell lines of *Drosophila*. In *"Invertebrate Cell Culture and Applications"* (K. Maramorosch and J. Mitsuashi, eds.) pp. 105–124. Academic Press, San Diego.

Cy   Barigozzi, C., Dolfini, S., Fraccaro, M., Halfer, C., Rezzonico-Raimondi, G., and Tiepolo, L. (1967). A first approach to the somatic cell genetics of *Drosophila melanogaster:* Studies on DNA replication. *Atti Ass. Genet. Ital.* **12**, 291–300.

Cy   Barigozzi, C., Dolfini, S., Fraccaro, M., Rezzonico Raimondi, G., and Tiepolo, L. (1966). *In vitro* study of the DNA replication patterns of somatic chromosomes of *Drosophila melanogaster. Exp. Cell Res.* **43**, 231–234.

Cy   Barigozzi, C., Fraccaro, M., Halfer, C., and Tiepolo, L. (1969). DNA replication of eu- and heterochromatin translocation of *Drosophila melanogaster. Dros. Inf. Serv.* **44**, 189.

Cy   Barigozzi, C., and Halfer, C. (1972). Constitutive heterochromatin in *Drosophila melanogaster. Chromosomes today,* 3, pp. 8–11. Wiley, New York.

M   Bautch, V. L., and Storti, R. V. (1983). Identification of a cytoplasmic tropomyosin gene linked to two muscle tropomyosin genes in *Drosophila. Proc. Natl. Acad. Sci. U.S.A.* **80**, 7123–7127.

R   Bayev, A. A., Lyubomirskaya, N. V., Dzhumagaliev, E. B., Ananiev, E. V., Amiantova, I. G., and Ilyin, Y. V. (1984). Structural organization of transposable element *mdg4* from *Drosophila melanogaster* and a nucleotide sequence of its long terminal repeat. *Nucleic Acids Res.* **12**, 3707–3723.

R   Beall, E. L., Admon, A., and Rio, D. C. (1994). A *Drosophila* protein homologous to the human p70 Ku autoimmune antigen interacts with the *P* transposable element inverted repeats. *Proc. Natl. Acad. Sci. U.S.A.* **91**, 12681–12685.

H   Beaulieu, J. F., Arrigo, A. P., and Tanguay, R. M. (1989). Interaction of *Drosophila* 27,000 M_r Heat-shock protein with the nucleus of heat-shocked and ecdysone-stimulated culture cells. *J. Cell Sci.* **92**, 29–36.

H   Beaulieu, J. F., and Tanguay, R. M. (1988). Members of the *Drosophila HSP70* family share ATP-binding properties. *Eur. J. Biochem.* **172**, 341–347.

M   Beck, L. A. (1982). Studies on a photoreactivating enzyme from *Drosophila* cultured cells. Ph.D. dissertation, University of California, Irvine, Irvine, CA.

M   Beck, L. A., and Sutherland, B. M. (1979). Purification of a photoreactivating enzyme from *Drosophila melanogaster. Am. Soc. Photobiol.* **7**, 130–131 [Abstract].

R   Becker, J., Becker, J. L., and Best-Belpomme, M. (1990). Characterization and purification of DNA-RNA complexes related with *1731* and *copia*-like transposons in a *Drosophila* cell line. *Cell. Mol. Biol.* **36**, 449–460.

H   Becker, J., Mezger, V., Courgeon, A. M., and Best-Belpomme, M. (1990). Hydrogen peroxide activates immediate binding of a *Drosophila* factor to DNA heat-shock regulatory element *in vivo* and *in vitro*. *Eur. J. Biochem.* **189**, 553–558.

H   Becker, J., Mezger, V., Courgeon, A. M., and Best-Belpomme, M. (1991). On the mechanism of action of H_2O_2 in the cellular stress. *Free Rad. Res. Comm.* **12–13**, 455–460.

ER Becker, J., Micard, D., Becker, J. L., Fourcade-Peronnet, F., Dastugue, B., and Best-Belpomme, M. (1991). Ecdysterone decreases the transcription level of the retrotransposons *1731* and *412* in a *Drosophila* cell line. *Cell. Mol. Biol.* **37**, 41–49.

Cy Becker, J. L. (1970). Fluorescence spécifique du chromosome Y, aprés coloration à la quinacrine, dans des cellules somatiques en culture de *Drosophila melanogaster*. *CR Acad. Sci. Paris* **271**, 2131–2133.

B   Becker, J. L. (1972). Fusions *in vitro* de cellules somatiques en culture de *Drosophila melanogaster*, induites par la Concanavaline A. *CR Acad. Sci. Paris* **275**, 2969–2972.

M   Becker, J. L. (1974a). Purine metabolism pathways in *Drosophila* cells grown *in vitro*: Phosphoribosyl transferase activities. *Biochimie* **56**, 779–781.

M   Becker, J. L. (1974b). Métabolisme des purines dans des cellules de *Drosophila melanogaster* en culture *in vitro*: Interconversion des purines. *Biochimie* **56**, 1249–1253.

M   Becker, J. L. (1975). Métabolisme des Purines dans des cellules de *Drosophila melanogaster* en culture *in vitro*: utilisation de l'adénosine. *CR Acad. Sci. Paris* **281**, 1171–1174.

M   Becker, J. L. (1976). Role of vitamin B12 in the reduction of ribonucleotides into deoxyribonucleotides in *Drosophila* cells grown *in vitro*. *Biochimie* **58**, 427–430.

M   Becker, J. L. (1978). Regulation of purine biosynthesis in cultured *Drosophila melanogaster* cells. I. Conditional activity of hypoxanthine-guanine-phosphoribosyltransferase and 5'-nucleotidase. *Biochimie* **60**, 619–625.

M   Becker, J. L. (1980). Regulation of purine biosynthesis in cultured *Drosophila melanogaster* cells. II. Relationships between hypoxanthine-guanine-phosphoribosyl-transferase and 5'-nucleotidase. *Biochimie* **62**, 665–670.

R   Becker, J. L., Barre-Sinoussi, F., Dormont, D., Best-Belpomme, M., and Chermann, J. C. (1987). Characterization of the purified RNA dependent DNA polymerase isolated from *Drosophila*. *Cell. Mol. Biol.* **33**, 225–235.

R   Becker, J. L., Barré-Sinoussi, F., Dormont, D., Best-Belpomme, M., and Chermann, J. C. (1988). RNA-dependent DNA polymerase is regulated by ecdysterone in a *Drosophila* cell line. *Cell. Mol. Biol.* **34**, 265–270.

V   Becker, J. L., Hazan, U., Nugeyre, M. T., Rey, F., Spire, B., Barré-Sinoussi, F., Georges, A., Teulières, L., and Chermann, J. C. (1986). Infection de cellules d'Insectes en culture par le virus HIV, agent du SIDA, et mise en évidence d'Insectes d'origine africaine contaminés par ce virus. *CR Acad. Sci. Paris* **303**, 303–305.

B   Becker, J. L., and Roussaux, J. (1981). 6-benzylaminopurine as a growth factor for *Drosophila melanogaster* cells grown *in vitro*. In "*Metabolism and Molecular Activities of Cytokinins*" (J. Guern and C. Péaux-Lenoel, eds.), pp. 319–328. Springer-Verlag, New York.

E   Beckers, C., Maroy, P., Dennis, R., O'Connor, J. D., and Emmerich, H. (1980). The uptake and release of (^{3}H)Ponasterone A by the *Kc* cell line of *Drosophila melanogaster*. *Mol. Cell. Endocrinol.* **17**, 51–59.

E   Beckers, C., Maroy, P., O'Connor, J. D., and Emmerich, H. (1980). Uptake and release of [³H]-Ponasterone A by the Kc cell line of *Drosophila melanogaster. In "Progress in Ecdysone Research"* (J. Hoffmann, ed.), pp. 335–347. Elsevier/North Holland Biomedical Press, Amsterdam.

C   Begg, M., and Cruickshank, W. J. (1963). A partial analysis of *Drosophila* larval haemolymph. *Proc. R. Soc. Edinb.* B, **68**, 215–236.

M   Belikov, S. V., Belgovsky, A. I., Partolina, M. P., Karpov, V. L., and Mirzabekov, A. D. (1993). Mapping and positioning DNA-binding proteins along genomic DNA. Structure of *Drosophila melanogaster* ribosomal "Alu-repeats" and 1.688 satellite chromatin. *Nucleic Acids Res.* **21**, 4796–4802.

Cy Belmont, A. S., Braunfeld, M. B., Sedat, J. W., and Agard, D. A. (1989). Large scale chromatin structural domains within mitotic and interphase chromosomes *in vivo* and *in vitro. Chromosoma* **98**, 129–143.

Cy Belmont, A. S., Sedat, J. W., and Agard, D. A. (1987). A 3-dimensional approach to mitotic chromosome structure: evidence for a complex hierarchical organization. *J. Cell Biol.* **105**, 77–92.

H   Bendena, W. G., Ayme-Southgate, A., Garbe, J. C., and Pardue, M. L. (1991). Expression of heat-shock locus *hsr-omega* in non-stressed cells during development of *Drosophila melanogaster. Dev. Biol.* **14**, 65–77.

H   Bendena, W. G., Fini, M. E., Garbe, J. C., Kidder, G. M., Lakhotias, S. C., and Pardue, M. L. (1989a). *Hsr-omega:* a different sort of heat-shock locus. *In "Stress-induced proteins"* (M. L. Pardue, J. R. Feramisco, and S. Lindquist, eds.), Vol. 96, pp. 3–14. ICN-UCLA Symp. Mol. Biol., New Ser., Alan Riss, New York.

H   Bendena, W. G., Garbe, J. C., Lahey-Traverse, K., Lakhotia, S. C., and Pardue, M. L. (1989b). Multiple inducers of the *Drosophila* heat-shock locus 93 D (*hsr-omega*): inducer-specific patterns of the three transcripts. *J. Cell Biol.* **108**, 217–228.

M   Benhamou, J. (1976). Etude structurale du RNA ribosomique 5S de *Drosophila melanogaster.* 98pp, *Thèse 3ème cycle Univ. d'Aix-Marseille II* (France).

M   Benhamou, J., and Jordan, B. R. (1976). Nucleotide sequence of *Drosophila melanogaster* 5S RNA: evidence for a general model. *FEBS Lett.* **62**, 146–149.

M   Benhamou, J., Jourdan, R., and Jordan, B. R. (1977). Sequence of *Drosophila* 5S RNA synthesized by cultured cells and by the insect at different developmental stages. Homogeneity of the product and homologies with other 5S RNAs at the level of primary and secondary structure. *J. Mol. Evol.* **9**, 279–298.

T   Benyajati, C., and Dray, J. F. (1984). Cloned *Drosophila* alcohol dehydrogenase genes are correctly expressed after transfection into *Drosophila* cells in culture. *Proc. Natl. Acad. Sci. U.S.A.* **81**, 1701–1705.

F   Benzakour, O. (1992). Contribution à l'étude du Transforming Growth Factor-beta (TGF-β). Mise en évidence, effets sur la croissance cellulaire et approche du mécanisme d'action. *Thèse de l' Université Paris XI,* Orsay, France.

F   Benzakour, O., Echalier, G., and Lawrence, D. (1990). *Drosophila* cell extracts contain a TGF-β like activity. *Biochem. Biophys. Res. Commun.* **169**, 1178–1184.

H   Berger, E. M. (1984). The regulation and functions of small *hsp* synthesis. *Dev. Genet.* **4**, 255–265.

EH Berger, E. (1988). Genetic programming of *Drosophila* cells by heat-shock and by ecdysterone. *Inv. Fish Tissue Culture,* (Y. Kuroda, E. Kurstak, and K. Maramorosch, eds.) pp. 123–126. Springer-Verlag, New York.

B   Berger, E., and Cox, G. (1979). A precursor of cytoplasmic actin in cultured *Drosophila* cells. *J. Cell Biol.* **81**, 680–683.

E   Berger, E. M., Cox, G., Ireland, R., and Weber, L. (1981). Actin content and synthesis in differentiating *Drosophila* cells in culture. *J. Insect Physiol.* **27**, 129–137.

B   Berger, E. M., Cox, G., Weber, L., and Kenney, J. S. (1981). Actin acetylation in *Drosophila* tissue culture cells. *Biochem. Genet.* **19**, 321–331.

E   Berger, E. M., Frank, M., and Abell, M. C. (1980). Ecdysone-induced changes in protein synthesis in embryonic *Drosophila* cells in culture. *In "Invertebrate Systems in Vitro"* (E. Kurstak, K. Maramorosch, and A. Dübendorfer, eds.), pp. 195–208. Elsevier/North Holland Biomedical Press, Amsterdam.

TEH   Berger, E. M., Goudie, K., Klieger, L., Berger, M., and DeCato, R. (1992). The juvenile hormone analogue methoprene inhibits ecdysterone induction of small heat-shock protein gene expression. *Dev. Biol.* **151**, 410–418.

E   Berger, E. M., Ireland, R., and Wyss, C. (1980). Patterns of peptide synthesis in *Drosophila* cells lines and their hybrids. *Somatic Cell Genet.* **6**, 719–729.

E   Berger, E. M., and Morganelli, C. M. (1984). *Drosophila* cells and ecdysterone: a model system for gene regulation. *In Vitro* **20**, 959–974.

EH   Berger, E. M., and Morganelli, C. M. (1990). Transcriptional regulation of *Drosophila* Hsp 22 gene. (Abst) IXth Ecdysone Workshop, Paris Sept. 1989, *Invert. Reprod. Dev.* **18**, 104.

E   Berger, E., Ringler, R., Alahiotis, S., and Frank, M. (1978). Ecdysone-induced changes in morphology and protein synthesis in *Drosophila* cell cultures. *Dev. Biol.* **62**, 498–511.

HE   Berger, E., and Rudolph, K. M. (1989). Heat-shock- and ecdysone-induced protein synthesis in *Drosophila* cells. *In "Invertebrate Cell System Applications"* (J. Mitsuhashi, ed.), Vol. I, pp. 155–168. CRC Press, Boca Raton, FL.

E   Berger, E. M., Sloboda, R. D., and Ireland, R. C. (1980). Tubulin content and synthesis in differentiating *Drosophila* cells in culture. *Cell Motil.* **1**, 113–129.

E   Berger, E. M., Vitek, M. P., and Morganelli, C. M. (1986). Genetic reprogramming of a *Drosophila* cell line by ecdysterone. *Techniques in Life Sciences, C2, In vitro Invertebrate Hormones and Genes*, C 220, 1–14, Elsevier Science Publishers, Ireland.

H   Berger, E. M., and Woodward, M. P. (1983). Small heat shock proteins in *Drosophila* may confer thermal tolerance. *Exp. Cell Res.* **147**, 437–442.

E   Berger, E., and Wyss, C. (1980). Acetylcholinesterase induction by β-ecdysone in *Drosophila* cell lines and their hybrids. *Somatic Cell Genet.* **6**, 631–640.

T   Bernard, A. R., Kost, T. A., Overton, L., Cavegn, C., Young, J., Bertrand, M., Yahia-Cherif, Z., Chabert, C., and Mills, A. (1994). Recombinant protein expression in a *Drosophila* cell line: comparison with the Baculovirus system. *Cytotechnology* **15**, 139–144.

TB   Berndorff, D., Gessner, R., Kreft, B., Schnoy, N., Lajous-Petter, A. M., Loch, N., Reutter, W., Hortsch, M., and Tauber, R. (1994). Liver-intestine Cadherin: molecular cloning and characterization of a novel Ca^{2+}-dependent cell adhesion molecule expressed in liver and intestine. *J. Cell Biol.* **125**, 1353–1369.

B   Bernhard, H. P. (1976). *Drosophila* cells: Fusion of somatic cells by polyethylene glycol. *Experientia* **32**, 786. [Abstract]

C   Bernhard, H. P., and Gehring, W. (1975). *Drosophila* cells: colony formation and cloning in agarose medium. *Experientia* **31**, 734. [Abstract]

Cl   Bernhard, H. P., Lienhard, S., and Regenass, U. (1980). Isolation and characterization of mutant *Drosophila* cell lines. *"In Invertebrate Systems in Vitro"* (E. Kurstak, K. Maramorosch, and A. Dübendorfer, eds.), pp. 14–26. Elsevier/North Holland Biomedical Press, Amsterdam.

Cp Bernstein, S. I. (1979). Ph.D. Thesis, Wesleyan University.

CpM Bernstein, S. I., and Donady, J. J. (1980). RNA synthesis and coding capacity of polyadenylated and non-polyadenylated mRNA from cultures of differentiating *Drosophila melanogaster* myoblasts. *Dev. Biol.* **79**, 388–398.

Cp Bernstein, S. I., Fyrberg, E. A., and Donady, J. J. (1978). Isolation and partial characterization of *Drosophila* myoblasts from primary cultures of embryonic cells. *J. Cell Biol.* **78**, 856–865.

M Bernstein, S. I., Glenn, A. W., Emerson, C. P. Jr., Minnie, M. E., and Donady, J. J. (1986). Purification of myosin heavy chain RNA from cultured *Drosophila* muscle cells and isolation of a myosin heavy chain gene. *Techn. Life Sci.* C2. *In vitro Invertebrate Hormones and Genes,* C219, 1–13. Elsevier Science Publishers, Ireland.

M Besser, von, H., Schnabel, P., Wieland, C., Fritz, E., Stanewsky, R., and Saumweber, H. (1990). The puff-specific *Drosophila* protein Bj6, encoded by the gene *no-on transient A,* shows homology to RNA-binding proteins. *Chromosoma* **100**, 37–47.

CyE Besson, M. T., Cordier, G., Quennedey, B., Quennedey, A., and Delachambre, J. (1987). Variability of ecdysteroid-induced cell cycle alterations in *Drosophila* Kc sublines. *Cell Tissue Kinet.* **20**, 413–425.

E Best-Belpomme, M. (1981). *Drosophila melanogaster* cells established *in vitro* are ecdysteroid target cells. In *"Regulations of Insect Development and Behaviour"* (F. Sehnal, and B. Cymborowsky, eds.), Vol. 22, pp. 715–722. Scient. Papers Inst. Organic Physical Chemistry, Wroclaw Tech. University, Poland.

E Best-Belpomme, M., and Courgeon, A. M. (1975). Présence ou absence de récepteurs saturables de l'ecdystérone dans des clones sensibles ou résistants de *Drosophila melanogaster* en culture *in vitro*. *CR Acad. Sci. Paris* **280**, 1397–1400.

E Best-Belpomme, M., and Courgeon, A. M. (1976a). Inductions protéiques par l'ecdystérone dans des clones cellulaires de *Drosophila melanogaster* cultivés *in vitro*. *CR Acad. Sci. Paris* **282**, 469–471.

E Best-Belpomme, M., and Courgeon, A. M. (1976b). Etude comparative des effets de l'α et de la β-ecdysone sur un clone de cellules diploïdes de *Drosophila melanogaster* en culture *in vitro*: induction protéique et modifications morphologiques. *CR Acad. Sci. Paris* **283**, 155–157.

E Best-Belpomme, M., and Courgeon, A. M. (1977). Ecdysterone and acetylcholinesterase activity in cultured *Drosophila* cells: inducible, non-inducible and constitutive clones or lines. *FEBS Lett.* **82**, 345–347.

E Best-Belpomme, M., and Courgeon, A. M. (1980). A critical period of ecdysterone action on sensitive clones of *Drosophila* cultured *in vitro*: the maturation of the cells. *Eur. J. Biochem.* **112**, 185–191.

E Best-Belpomme, M., Courgeon, A. M., and Echalier, G. (1980). Development of a model for the study of ecdysteroid action: *Drosophila melanogaster* cells established *in vitro*. In *"Progress in Ecdysone Research"* (J. Hoffmann, ed.), pp. 379–392. Elsevier/ North Holland Biomedical Press, Amsterdam.

E Best-Belpomme, M., Courgeon, A. M., and Rambach, A. (1978). β-Galactosidase is induced by hormone in *Drosophila melanogaster* cell cultures. *Proc. Natl. Acad. Sci. U.S.A.* **75**, 6102–6106.

E Best-Belpomme, M., and Ropp, M. (1982). Catalase is induced by ecdysterone and ethanol in *Drosophila* cells. *Eur. J. Biochem.* **121**, 349–355.

E Best-Belpomme, M., Sykiotis, M., and Courgeon, A. M. (1978). Antisera against ecdysteroid-induced proteins in an established line and a clone of *Drosophila melanogaster* cells. *FEBS Lett.* **89**, 86–88.

E   Bieber, A. J. (1986). Ecdysteroid inducible polypeptides in *Drosophila Kc* cells: Kinetics of mRNA induction and aspects of protein structure. Ph.D. Thesis, Harvard University, Cambridge, MA.

B   Bieber, A. J. (1994). Analysis of cellular adhesion in cultured cells. *In "Methods in Cell Biology"* Vol. 44, *Drosophila melanogaster*. Practical uses in cell and molecular Biology (L. S. B. Goldstein and E. A. Fyrberg, eds.), pp. 683–696. Academic Press, San Diego.

R   Bierley, C., and Flavell, A. J. (1990). The retrotransposon *copia* controls the relative levels of its gene products post-transcriptionally by differential expression from its two major mRNAs. *Nucleic Acids Res.* **18**, 2947–2951.

Cp   Bierly, L. (1985). Voltage-dependent currents in embryonic cultures of *Drosophila* neurons. *Soc. Neurosci. Abst.* **11**, 149.

Cp   Bierly, L. (1986). Potassium, sodium and calcium currents in embryonic cultures of *Drosophila* neurons. *Biophys. J.* **49**, 574a.

Cp   Bierly, L., and Leung, H. T. (1988). Ionic currents of *Drosophila* neurons in embryonic cultures. *J. Neurosci.* **8**, 4379–4393.

H   Biessmann, H., Falkner, F. G., Saumweber, H., and Walters, M. F. (1982). Disruption of the vimentin cytoskeleton may play a role in heat shock response. *In "Heat-shock from Bacteria to Man"* (M. Schlesinger, M. Ashburner, and A. Tissières, eds.), pp. 275–281. Cold Spring Harbor Laboratory, Cold Spring Harbor.

H   Biessmann, H., Levy, W. B., and McCarthy, B. J. (1978). *In vitro* transcription of heat shock-specific RNA from chromatin of *Drosophila melanogaster* cells. *Proc. Natl. Acad. Sci. U.S.A.* **75**, 759–763.

H   Biessmann, H., Wadsworth, S., Levy, W. B., and McCarthy, B. J. (1978). Correlation of structural changes in chromatin with transcription in the *Drosophila* heat shock response. *Cold Spring Harbor Symp. Quant. Biol.* **42**, 829–834.

MT   Biggin, M. D., Bickel, S., Benson, M., Pirotta, V., and Tijan, R. (1988). *Zeste* encodes a sequence-specific transcription factor that activates the *Ultrabithorax* promoter *in vitro*. *Cell* **53**, 713–722.

M   Biggin, M. D., and Tijan, R. (1988). Transcription factors that activate the *Ultrabithorax* promoter in developmentally staged extracts. *Cell* **53**, 699–711.

E   Biggs, W. H., III, and Zipursky, S. L. (1992). Primary structure, expression, and signal-dependent tyrosine phosphorylation of a *Drosophila* homolog of extracellular signal-regulated kinase. *Proc. Natl. Acad. Sci. U.S.A.* **89**, 6295–6299.

M   Birnboim, H. C., and Sederoff, R. (1975). Polypyrimidine segments in *Drosophila melanogaster* DNA. I. Detection of a cryptic satellite containing polypyrimidine/polypurine DNA. *Cell* **5**, 173–181.

M   Birnboim, H. C., Strauss, N. A., and Sederoff, R. R. (1975). Characterization of polypyrimidine in *Drosophila* and L-cell DNA. *Biochemistry* **14**, 1643–1647.

CdE   Birr, C. A., Fristrom, D., King, D. S., and Fristrom, J. W. (1990). Ecdysone-dependent proteolysis of an apical surface glycoprotein may play a role in imaginal disc morphogenesis in *Drosophila*. *Development* **110**, 239–248.

M   Bishop, J. O., Morton, J. G., Rosbash, M., and Richardson, M. (1974). Three abundance classes in Hela cell messenger RNA. *Nature* **250**, 199–203.

V   Blissard, G. W., and Rohrmann, G. F. (1991). Baculovirus *gp64* gene expression: analysis of sequences modulating early transcription and transactivation by *IE1*. *J. Virol.* **65**, 5820–5827.

V   Blondel, D., Dezelee, S., and Wyers, F. (1983). Vesicular Stomatitis Virus growth in *Drosophila melanogaster* cells. II. Modifications of viral protein phosphorylation. *J. Gen. Virol.* **64**, 1793–1799.

V   Blondel, D., Petitjean, A. M., Dezelee, S., and Wyers, F. (1988). *Vesicular Stomatitis Virus* in *Drosophila melanogaster* cells: regulation of viral transcription and replication. *J. Virol.* **62**, 277–284.

M   Blumberg, B., Mackrell, A. J., and Fessler, J. H. (1988). *Drosophila* basement membrane procollagen α*I(IV)*. II. Complete cDNA sequence, genomic structure and general implications for supermolecular assemblies. *J. Biol. Chem.* **263**, 18328–18337.

M   Blumberg, B., McKrell, A. J., Olson, P. F., Kurkinen, M., Monson, J. M., Natzle, J. E., and Fessler, J. H. (1987). Basement membrane procollagen IV and its specialized domain are conserved in *Drosophila*, mouse and man. *J. Biol. Chem.* **262**, 5947–5950.

CyM   Blumenthal, A. B. (1974). The mechanism and organization of DNA replication in *Drosophila* chromosomes. (Abst. 1st Inter. Cong. Radiation Res., Seattle, Washington, USA, July 1973) *Radiat. Res.* **59**, 254.

Cy   Blumenthal, A. B., Kriegstein, H. J., and Hogness, D. S. (1974). The units of DNA replication in *Drosophila melanogaster* chromosomes. *Cold Spring Harbor Symp. Quant. Biol.* **38**, 205–223.

Ct   Bohrmann, J. (1991). *In vitro* culture of *Drosophila* ovarian follicles: the influence of different media on development, RNA synthesis, protein synthesis and potassium uptake. *Roux's Arch. Dev. Biol.* **199**, 315–326.

M   Bond, B. J., and Davidson, N. (1986). The *Drosophila melanogaster* Actin 5C gene uses two transcription initiation sites and three polyadenylation sites to express multiple mRNA species. *Mol. Cell. Biol.* **6**, 2080–2088.

MT   Bond-Matthews, B., and Davidson, N. (1988). Transcription from each of the *Drosophila act5C* leader exons is driven by a separate functional promoter. *Gene* **62**, 289–300.

CtH   Bonner, J. J. (1981). Induction of *Drosophila* heat-shock puffs in isolated polytene nuclei. *Dev. Biol.* **86**, 409–418.

CtM   Bonner, J. J., Berninger, M., and Pardue, M. L. (1978). Transcription of polytene chromosomes and of the mitochondrial genome in *Drosophila*. *CSH Symp. Quant. Biol.* **42**, 803–814.

CtH   Bonner, J. J., and Kerby, R. L. (1982). RNA polymerase II transcribes all of the heat shock induced genes of *Drosophila melanogaster*. *Chromosoma* **85**, 93–108.

CdH   Bonner, J. J., and Pardue, M. L. (1976a). The effect of heat shock on RNA synthesis in *Drosophila* tissues. *Cell* **8**, 43–50.

CdE   Bonner, J. J., and Pardue, M. L. (1976b). Ecdysone-stimulated RNA synthesis in imaginal discs of *Drosophila melanogaster*. *Chromosoma* **58**, 87–99.

CtE   Bonner, J. J., and Pardue, M. L. (1977). Ecdysone-stimulated RNA synthesis in salivary glands of *Drosophila melanogaster*: Assay by *in situ* hybridization. *Cell* **12**, 219–225.

T   Bonner, J. J., Parks, C., Parker-Thornburg, J., Mortin, M. A., and Pelham, R. B. (1984). The use of promoter fusions in *Drosophila* genetics: isolation of mutations affecting the heat-shock response. *Cell* **37**, 979–991.

B   Bossie, C. A., and Sanders, M. M. (1993). A cDNA from *Drosophila melanogaster* encodes a lamin C-like intermediate filament protein. *J. Cell Sci.* **104**, 1263–1272.

R   Bouhidel, K., Terzian, C., and Pinon, H. (1994). The full-length transcript of the I factor, a LINE element of *Drosophila melanogaster,* is a potential bicistronic RNA messenger. *Nucleic Acids Res.* **22**, 2370–2374.

M   Bourbon, H. M., Martin-Blanco, E., Rosen, D., and Kornberg, T. B. (1995). Phosphorylation of the *Drosophila engrailed* protein at a site outside its homeodomain enhances DNA binding. *J. Biol. Chem.* **270**, 11130–11139.

CpM Bournias-Vardiabasis, N. (1985). *In "Alternative Methods in Toxicology"* (A. Goldberg, ed.), p. 317.

Cp Bournias-Vardiabasis, N. (1990). *Drosophila melanogaster* embryo cultures: an *in vitro* teratogen assay. *ATLA* **18**, 290–300.

CpH Bournias-Vardiabasis, N., and Buzin, C. H. (1986). Developmental effects of chemicals and the heat shock response in *Drosophila* cells: *Teratog. Carcinog. Mutagen.* **6**, 523–536.

CpH Bournias-Vardiabasis, N., and Buzin, C. H. (1987). Altered differentiation and induction of heat-shock proteins in *Drosophila* embryonic cells associated with teratogen treatment. *Banbury Rep., 26*. Developmental toxicology: Mechanisms and Risks, 3, Cold Spring Harbor Laboratory.

HCp Bournias-Vardiabasis, N., Buzin, C., and Flores, J. (1990). Differential expression of heat shock proteins in *Drosophila* embryonic cells following metal ion exposure. *Exp. Cell Res.* **189**, 177–182.

Cp Bournias-Vardiabasis, N., Buzin, C. H., and Reilly, J. G. (1983). The effect of 5-azacytidine and cytidine analogs on *Drosophila* cells in culture. *Roux's Arch. Dev. Biol.* **192**, 299–302.

Cp Bournias-Vardiabasis, N., and Flores, J. C. (1983). Drug metabolizing enzymes in *Drosophila melanogaster:* Teratogenicity of cyclophosphamide *in vitro. Teratog. Carcinog. Mutagen.* **3**, 255–262.

Cp Bournias-Vardiabasis, N., and Flores, J. C. (1986). Response of *Drosophila* embryonic cells to tumor promoters. *Toxicol. Appl. Pharmacol.* **85**, 196–206.

Cp Bournias-Vardiabasis, N., and Teplitz, R. L. (1982). Use of *Drosophila* embryo cell cultures as an *in vitro* teratogen assay. *Teratog. Carcinog. Mutagen.* **2**, 233–241.

CpM Bournias-Vardiabasis, N., Teplitz, R. L., Chernoff, G. F., and Seecof, R. L. (1983). Detection of teratogens in the *Drosophila* embryonic cell culture test: assay of 100 chemicals. *Teratology* **18**, 109–122.

TR Bourouis, M., and Jarry, B. (1983). Vectors containing a prokaryotic dihydrofolate reductase gene transform *Drosophila* cells to methotrexate-resistance. *EMBO J.* **2**, 1099–1104.

M Boyd, J. B., Golino, M. D., and Setlow, R. B. (1976). The *mei-9* mutant of *Drosophila melanogaster* increases mutagen sensitivity and decreases excision repair. *Genetics* **84**, 527–544.

M Boyd, J. B., and Harris, P. V. (1985). Isolation and characterization of a photorepair-deficient mutant in *Drosophila melanogaster. Genetics* **110** (Suppl.), 203.

M Boyd, J. B., and Harris, P. V. (1987). Isolation and characterization of a photorepair-deficient mutant in *Drosophila melanogaster. Genetics* **116**, 233–239.

M Boyd, J. B., Harris, P. V., Osgood, C. J., and Smith, K. E. (1980). Biochemical characterization of repair deficient mutants of *Drosophila. In "DNA Repair and Mutagenesis"* (W. M. Generoso, M. D. Shelby, and F. J. de Serres, eds.), pp. 209–221. Plenum Press, New York.

rM Boyd, J. B., Harris, P. V., Presley, J. M., and Narachi, M. (1983). *Drosophila melanogaster:* a model eukaryote for the study of DNA repair. *In "Cellular Responses to DNA Damage"* pp. 107–123, Alan Liss, New York.

rM Boyd, J. B., Harris, P. V., and Sakaguchi, K. (1988). Use of *Drosophila* to study DNA repair. *In "DNA Repair"* (E. C. Friedberg and P. C. Hanawalt, eds.), pp. 399–413. Marcel Dekker, New York.

M   Boyd, J. B., Sakaguchi, K., and Harris, P. V. (1990). *mus-308* mutants of *Drosophila* exhibit hypersensitivity to DNA cross-linking agents and are defective in a deoxyribonuclease. *Genetics* **125**, 813–819.

M   Boyd, J. B., and Setlow, R. B. (1976). Characterization of post replication repair in mutagen sensitive strains of *Drosophila melanogaster*. *Genetics* **84**, 507–526.

M   Boyd, J. B., and Shaw, K. E. S. (1982). Postreplication repair defects in mutants of *Drosophila melanogaster*. *Mol. Gen. Genet.* **186**, 289–294.

M   Boyd, J. B., Snyder, R. D., Harris, P. V., Presley, J. M., Boyd, S. F., and Smith, P. D. (1982). Identification of a second locus in *Drosophila melanogaster* required for excision repair. *Genetics* **100**, 239–257.

TM   Boyd, L., and Thummel, C. S. (1993). Selection of CUG and AUG initiator codons for *Drosophila E74A* translation depends on downstream sequences. *Proc. Natl. Acad. Sci. U.S.A.* **90**, 9164–9167.

Ct   Boyd, M., and Ashburner, M. (1977). The hormonal control of salivary gland secretion in *Drosophila melanogaster*: Studies *in vitro*. *J. Insect Physiol.* **23**, 517–523.

E   Bradbrook, D. A. (1990). A study of ecdysteroid entry into Insect cells. Ph.D. Thesis, University of Exeter, U.K.

Cp   Brainard, M. S., Zagotta, W. N., and Aldrich, R. W. (1987). Four distinct potassium channels in cultured *Drosophila* myotubes. *Soc. Neurosci. Abstr.* **13**, 578.

M   Brand, S. R., Worthington, J., McIntosh, D. P., and Bernstein, R. M. (1992). Antibody to a 63 kilodalton insect protein in ankylosing spondylitis. *Ann. Rheum. Dis.* **51**, 334–339.

V   Bras, F., Teninges, D., and Dezelee, S. (1994). Sequences of the N and M genes of the *Sigma* virus of *Drosophila* and evolutionary comparison. *Virology* **200**, 189–199.

V   Bras-Herreng, F. (1975). Multiplication du virus Sindbis dans des cellules de Drosophile cultivées *in vitro*. *Arch. Virol.* **48**, 121–129.

V   Bras-Herreng, F. (1976). Adaptation d'une population du virus Sindbis a *Drosophila melanogaster*. *Ann. Microbiol.* (*Inst. Pasteur*) **127B**, 541–565.

Cl   Braude-Zolotarjova, T. Ya., Kakpakov, V. T., and Nikoshkov, A. B. (1982). (in Russian) Obtention and description of a new line of *Drosophila virilis* embryonic cells. 4th Meeting of the Union of Societies of Geneticists and Selectionists (Vavilov Soc.), Vol. I, pp. 38–39, Kishinev, Shtiintza Press.

Cl   Braude-Zolotarjova, T. Ya., Kakpakov, V. T., and Schuppe, N. G. (1986). Male diploid cell line of *Drosophila virilis*. *In Vitro* **22**, 481–484.

T   Braude-Zolotarjova, T. Ya., and Schuppe, N. G. (1987). Transient expression of *hsp-CAT1* and *copia-CAT1* hybrid genes in *D. melanogaster* and *D. virilis* cells. *Dros. Inf. Serv.* **66**, 33.

H   Brevet, A., Plateau, P., Best-Belpomme, M., and Blanquet, S. (1985). Variations of *AP4A* and other dinucleosidephosphates in stressed *Drosophila* cells. *J. Biol. Chem.* **260**, 15567–15570.

M   Briata, P., D'Anna, F., Franzi, A. T., and Gherzi, R. (1993). AP-1 activity during normal human keratinocyte differentiation: Evidence for a cytosolic modulator of AP-1/DNA binding. *Exp. Cell. Res.* **204**, 136–146.

R   Brierley, C., and Flavell, A. J. (1990). The retrotransposon *copia* controls the relative levels of its gene products post-transcriptionally, by differential expression from its two major mRNAs. *Nucleic Acids Res.* **18**, 2947–2951.

Ct   Broadie, K., Shaker, H., and Bate, M. (1992). Whole-embryo culture of *Drosophila melanogaster*: development of embryonic tissues *in vitro*. *Roux's Arch. Dev. Biol.* **201**, 364–375.

H   Bronner-Becker, J. (1994). Le stress oxydatif dans l'activation des genes de choc thermique chez *Drosophila melanogaster*: une induction par le peroxyde d'hydrogène (H₂O₂).
Diplôme de l'Ecole Pratique des Hautes Etudes, Paris, 1–109.

rC   Brooks, M. A., and Kurti, T. J. (1981). Insect cell and tissue culture. *Adv. Rev. Entom.*
16, 27–52.

M   Brown, K., Havel, C. M., and Watson, J. A. (1983). Isoprene synthesis in isolated
embryonic *Drosophila* cells. II. Regulation of 3-hydroxy-3-methylglutaryl coenzyme A
reductase activity. *J. Biol. Chem.* 258, 8512–8518.

M   Brown, T. C., and Boyd, J. B. (1981). Abnormal recovery of DNA replication in UV-
irradiated cell cultures of *Drosophila melanogaster* which are defective in DNA repair.
Mol. Gen. Genet. 183, 363–368.

B   Brown, T. C., Harris, P. V., and Boyd, J. B. (1981). Effects of radiation on the survival of
excision-defective cells from *Drosophila melanogaster*. *Somatic Cell Genet.* 7, 631–644.

E   Bruhat, A., Tourmente, S., Chapel, S., Sobrier, M. L., Couderc, J. L., and Dastugue,
B. (1990). Regulatory elements in the first intron contribute to transcriptional regulation
of the β₃ tubuline gene by 20-hydroxyecdysone in *Drosophila Kc* cell. *Nucleic Acids Res.*
18, 2861–2867.

rV   Brun, G., and Plus, N. (1980). The viruses of *Drosophila. In "The Genetics and
Biology of Drosophila"* (M. Ashburner and T. R. F. Wright, eds.), Vol. 2d, pp. 625–702.
Academic Press, New York.

M   Brunet, C., Quan, T., and Craft, J. (1993). Comparison of the *Drosophila,* human
and murine SmB cDNAs: evolutionary conservation. *Gene* 124, 269–273.

M   Brutlag, D. L., and Peacock, W. J. (1975). Sequences of highly repeated DNA in
Drosophila melanogaster. In "The Eukaryotic Chromosome" Australian Nat. Univ.
Press, Canberra.

R   Bryant, L. A., Brierley, C., Flavell, A. J., and Sinclair, J. H. (1991). The Retrotransposon
copia regulates *Drosophila* gene expression, both positively and negatively. *Nucleic Acids
Res.* 19, 5533–5536.

CdE   Bullmore, D. (1977). The differential action of alpha and beta-ecdysone on the
division of imaginal disc cells of *Drosophila melanogaster in vitro. Thèse* de l'Institut
de Zoologie de i'Université de Fribourg, Suisse.

T   Bunch, T. A. (1988). Characterization and use of inducible promoters in *Drosophila*
cells. Ph.D. Thesis, Harvard University, Cambridge, MA.

B   Bunch, T. A., and Brower, D. L. (1992). *Drosophila* PS2 integrin mediates RGD-
dependent cell-matrix interactions. *Development* 116, 239–247.

rB   Bunch, T. A., and Bower, D. L. (1993). *Drosophila* Cell Adhesion Molecules. *Curr.
Top. Dev. Biol.* 28, 81–123.

TM   Bunch, T. A., and Goldstein, L. S. (1989). The conditional inhibition of gene expression in cultured *Drosophila* cells by antisense RNA. *Nucleic Acids Res.* 17, 9761–9782.

T   Bunch, T. A., Grinblat, Y., and Goldstein, L. S. B. (1988). Characterization and use
of the *Drosophila* metallothionein promoter in cultured *Drosophila melanogaster* cells.
Nucleic Acids Res. 16, 1043–1061.

M   Burckhardt, J., and Birnstiel, M. L. (1978). Analysis of histone messenger RNA of
Drosophila melanogaster by two-dimensional gel electrophoresis. *J. Mol. Biol.* 118,
61–79.

CdE   Burdette, W. J., Hanley, E. W., and Grosch, J. (1968). The effect of ecdysone on
the maintenance and development of ocular imaginal discs *in vitro. Texas Rep. Biol.
Med.* 26, 173–180.

M   Burgtorf, C., and Bünemann, H. (1994). Representative and efficient cloning of satellite DNAs based on PGFE pre-fractionation or restriction digests of genomic DNA. *J. Biochem. Biophys. Methods* **28**, 301–312.

TH   Burke, J. F., Pinchin, S. M., Ish-Horowicz, D., Sinclair, J. H., and Sang, J. H. (1984). Integration of *Drosophila* heat-shock genes transfected into cultured *Drosophila melanogaster* cells. *Somat. Cell. Mol. Gen.* **10**, 579–588.

T   Burke, J. F., Sinclair, J. H., Sang, J. H., and Ish-Horowicz, D. (1984). An assay for transient gene expression in transfected *Drosophila* cells, using (^{3}H) guanine incorporation *EMBO J.* **3**, 2549–2554.

E   Burtis, K. B. (1985). Isolation and characterization of an ecdysone inducible gene from *Drosophila melanogaster*. Ph.D. Thesis, Stanford University, Stanford, CA.

CtE   Butterworth, F. M., Tysell, B., and Waclawski, I. (1979). The effect of 20-hydroxyecdysone and protein on granule formation in the *in vitro* cultured fat body of *Drosophila*. *J. Insect Physiol.* **25**, 855–860.

H   Buzin, C. H. (1982). A comparison of the multiple *Drosophila* heat-shock proteins in cell lines and larval salivary glands by two-dimensional electrophoresis. *J. Mol. Biol.* **158**, 181–201.

CpH   Buzin, C. H., and Bournias-Vardiabasis, N. (1982). The induction of a subset of heat-shock proteins by drugs that inhibit differentiation in *Drosophila* embryonic cell cultures. *In "Heat-shock from Bacteria to Man"* (M. Schlesinger, M. Ashburner, and A. Tissières, eds.), pp. 387–394. Cold Spring Harbor Laboratory Press, Cold Spring Harbor.

CpH   Buzin, C. H., and Bournias-Vardiabasis, N. (1984). Teratogens induce a subset of small heat shock proteins in *Drosophila* primary embryonic cell cultures. *Proc. Natl. Acad. Sci. U.S.A.* **81**, 4075–4079.

Cp   Buzin, C. H., Dewhurst, S. A., and Seecof, R. L. (1978). Temperature sensitivity of muscle and neurone differentiation in embryonic cultures from the *Drosophila* mutant *shibire*[ts1]. *Dev. Biol.* **66**, 442–456.

H   Buzin, C. H., and Petersen, N. S. (1982). A comparison of the multiple heat-shock proteins in cell lines and larval salivary glands by two-dimensional gel electrophoresis. *J. Mol. Biol.* **158**, 181–201.

Cp   Buzin, C. H., and Seecof, R. L. (1981). Developmental modulation of protein synthesis in *Drosophila* primary embryonic cell cultures. *Dev. Genet.* **2**, 237–252.

Cp   Byerly, L., and Leung, H. T. (1988). Ionic currents of *Drosophila* neurons in embryonic cultures. *J. Neurosci.* **8**, 4379–4393.

E   Cade-Treyer, D., and Munsch, N. (1980). Immunoautographic study of the synthesis of an ecdysteroid amplified protein in a *Drosophila* cell line and a clone *in vitro*. *FEBS Lett.* **117**, 19–22.

B   Cagan, R. L., Krämer, H., Hart, A. C., and Zipursky, S. L. (1992). The *bride of sevenless* and *sevenless* interaction: Internalization of a transmembrane ligand. *Cell* **69**, 393–399.

H   Camato, R., and Tanguay, R. M. (1982). Changes in the methylation patterns of core histones during heat shock in *Drosophila* cells. *EMBO J.* **1**, 1529–1532.

B   Campbell, A. G., Fessler, L., Salo, T., and Fessler, J. H. (1987). *Papilin*: a *Drosophila* proteoglycan-like sulphate glycoprotein from basement membranes. *J. Biol. Chem.* **262**, 17605–17612.

V   Caproni, M. (1973). personal communication.

H   Carbajal, E. M., Beaulieu, J. F., Nicole, L. M., and Tanguay, R. M. (1993). Intramitochondriazl localization of the main 70kDa heat-shock cognate protein in *Drosophila* cells. *Exp. Cell Res.* **207**, 300–309.

H   Carbajal, E. M., Duband, J. L., Lettre, F., Valet, J. P., and Tanguay, R. M. (1986). Cellular localization of *Drosophila 83 kDa* heat shock protein in normal, heat shocked and recovering cultured cells with a specific antibody. *Biochem. Cell. Biol.* **64**, 816–825.

H   Carbajal, E. M., Valet, J. P., Charest, P. M., and Tanguay, R. M. (1990). Purification of *Drosophila* HSP 83 and immunoelectron microscopic localization. *Eur. J. Cell. Biol.* **52**, 147–156.

T   Carlberg, C. (1995). Highly reproducible transient transfection of *Drosophila SL-3* and Human *MCF-7* cell lines using DOTAP. *Biochemica* (Boehringer Mannheim) **2**, 19–20.

V   Carbonell, L. F., Klowden, M. J., and Miller, L. K. (1985). Baculovirus-mediated expression of bacterial genes in dipteran and mammalian cells. *J. Virol.* **56**, 153–160.

R   Carlson, M., and Brutlag, D. (1978). One of the *copia* genes is adjacent to satellite DNA in *Drosophila melanogaster*. *Cell* **15**, 733–742.

B   Carrow, G. M., Van Buskirk, R., and Wagner, J. A. (1988). Induction of process outgrowth in vertebrate and invertebrate cell lines by a 2-pyridinyl thiosemicarbazone. *Differentiation* **39**, 22–27.

H   Cartwright, I. L., and Elgin, S. C. R. (1986). Nucleosomal instability and induction of new upstream protein DNA associations accompany activation of four small-*hsp* genes in *Drosophila melanogaster*. *Mol. Cell. Biol.* **6**, 779–791.

Cp   Castiglioni, M. C., and Rezzonico-Raimondi, G. (1961a). First results of *in vitro* cultivation of *Drosophila melanogaster* tissues *Atti Ass. Genet. Ital.* **6**, 139–150.

Cp   Castiglioni, M. C. and Rezzonico Raimondi, G. (1961b) First results of tissue culture in *Drosophila*. *Experientia* **17**, 88–90.

Cp   Castiglioni, M. C., and Rezzonico-Raimondi, G. (1963). Differenze genotipiche tra ceppi diversi di *Drosophila melanogaster* relivate con la tecnica delle culture *in vitro*. *1rst Lombardo Rend. Sc.* **B97**, 117–134.

Cp   Castiglioni, M. C., and Rezzonico Raimondi, G. (1963). Genotypical differences between stocks of *D. melanogaster* revealed from culturing *in vitro*. *Experientia* **19**, 527–529.

R   Cavarec, L., and Heidmann, T. (1993). The *Drosophila copia* retrotransposon contains binding sites for transcriptional regulation by homeoproteins. *Nucleic Acids Res.* **21**, 5041–5049.

R   Cavarec, L., Jensen, S., and Heidmann, T. (1994). Identification of a strong transcriptional activator for the *copia* retrotransposon responsible for its differential expression in *D. hydei* and *D. melanogaster* cell lines. *Biochem. Biophys. Res. Commun.* **203**, 392–399.

M   Champlin, D. T., Frasch, M., Saumweber, H., and Lis, J. T. (1991). Characterization of a *Drosophila* protein associated with boundaries of transcriptionally active chromatin. *Genes Dev.* **5**, 1611–1621.

R   Champion, S., Maisonhaute, C., Kim, M. H., and Best-Belpomme, M. (1992). Characterization of the reverse transcriptase of *1731*, a *Drosophila melanogaster* retrotransposon. *Eur. J. Biochem.* **209**, 523–531.

E   Chang, E. S. (1985). Cellular juvenile hormone binding proteins. *Methods Enzymol.* **111**, 494–509.

E   Chang, E. S., Bruce, M. J., and Prestwich, G. D. (1985). Further characterization of the JH-binding protein from the cytosol of a *Drosophila* cell line: Use of a photo-affinity label. *Insect Biochem.* **15**, 197–204.

E   Chang, E. S., Coudron, T. A., Bruce, M. J., Sage, B. A., O'Connor, J. D., and Law, J. H. (1980). Juvenile hormone-binding protein from the cytosol of *Drosophila* Kc cells. *Proc. Natl. Acad. Sci. U.S.A.* **77**, 4657–4661.

E   Chang, E. S., Yudin, A. I., and Clark, W. H. (1982) Hormone action on a *Drosophila* cell line. *In Vitro* **18**, 297. [Abstract]

T   Chang, Y. L., King, B. O., O'Connor, M., Mazo, A., and Huang, D. H. (1995). Functional reconstruction of trans-regulation of the *Ultrabithorax* promoter by the products of two antagonistic genes, *trithorax* and *Polycomb. Mol. Cell. Biol.* 6601–6612.

M   Chao, Y., and Pellegrini, M. (1993). *In vitro* transcription of *Drosophila* rRNA genes shows stimulation by a Phorbol ester and serum. *Mol. Cell. Biol.* **13**, 934–941.

E   Chapel, S. (1985). Etude de la synthèse protéique sous contrôle hormonal dans une lignée de cellules de *Drosophila melanogaster* en culture. DEA Clermont-Ferrand II.

E   Chapel, S., Sobrier, M. L., Montpied, P., Micard, D., Bruhat, A., Couderc, J. L., and Dastugue, B. (1993). In *Drophila Kc* cells, 20-0HE induction of the 60C β3-tubulin gene expression is a primary transcriptional event. *Insect Mol. Biol.* **2**, 39–48.

T   Chavrier, P., Vesque, C., Galliot, B., Vigneron, M., Dolle, P., Duboule, D., and Charnay, P. (1990). The segment-specific *Krox-20* encodes a transcription factor with binding sites in the promoter region of the *Hox-1.4* gene. *EMBO J.* **9**, 1209–1218.

M   Chen, X., Farmer, G., Zhu, H., Prywes, R., and Prives, C. (1993). Cooperative DNA binding of P53 with TFIID (TBP): a possible mechanism for transcriptional activation. *Genes Dev.* **7**, 1837–1849.

CdH   Cheney, C. M., and Shearn, A. (1983). Developmental regulation of *Drosophila* imaginal discs: synthesis of *hsps* under non heat-shock conditions. *Dev. Biol.* **95**, 325–330.

E   Cherbas, L., Benes, H., Bourouis, M., Burtis, K., Chao, A., Cherbas, P., Crosby, M., Garfinkel, M., Guild, G., Hogness, D., Jami, J., Jones, C. W., Koehler, M., Lepesant, J. A., Martin, C., Maschat, F., Mathers, P., Meyerowitz, E., Moss, R., Pictet, R., Rebers, J., Richards, G., Roux, J., Schulz, R., Segraves, W., Thummel, C., and Vijayraghavan, K. (1986). Structural and functional analysis of some moulting hormone-responsive genes from *Drosophila. Insect Biochem.* **16**, 241–248.

E   Cherbas, L., and Cherbas, P. (1981). The effects of ecdysteroid hormones on *Drosophila melanogaster* cell lines. *In* "Adv. Cell Culture" (K. Maramoroosch, ed.), Vol. 1, pp. 91–124. Academic Press, New York.

E   Cherbas, L., Cherbas, P., Savakis, C., Demetri, G., Manteuffel-Cymborowska, M., Yonge, C. D., and Williams, C. M. (1980). Studies of ecdysteroid action on a *Drosophila* cell line. *In* "Invertebrate Systems in Vitro" (E. Kurstak, K. Maramorosch, and A. Dübendorfer, eds.), pp. 217–228. Academic Press, New York.

E   Cherbas, L., Fristrom, J. W., and O'Connor, J. D. (1984). The action of ecdysone in imaginal discs and *Kc* cells of *Drosophila melanogaster In* "*Biosynthesis, Metabolism and Mode of action of Invertebrate Hormones*" (J. Hoffmann, and M. Porchet, eds.). pp. 305–322. Springer Verlag, Berlin.

E   Cherbas, L., Koehler, M. M. D., and Cherbas, P. (1989). The effects of juvenile hormone on the ecdysone response of *Drosophila melanogaster Kc* cells. *Dev. Genet.* **10**, 177–188.

E   Cherbas, L., Lee, K., and Cherbas, P. (1990). The induction of Eip28/29 by ecdysone in *Drosophila* cell lines. IXth *Ecdysone Workshop,* Paris, Sept. 89. *Invert. Reprod. Dev.* **18**, 108. [Abstract]

E   Cherbas, L., Lee, K., and Cherbas, P. (1991). Identification of ecdysone response elements by analysis of the *Drosophila Eip28/29* gene. *Genes Dev.* **5**, 120–131.

rT   Cherbas, L., Moss, R., and Cherbas, P. (1994). Transformation techniques for *Drosophila* cell lines. *In* "*Methods in Cell Biology*" (L. Goldstein, and E. Fyrberg, eds.), Vol. 44, pp. 161–179. Academic Press, New York.

ET Cherbas, L., Schulz, R. A., Koehler, M. M. D., Savakis, C., and Cherbas, P. (1986). Structure of *Eip28/29* gene, an ecdysone-inducible gene from *Drosophila. J. Mol. Biol.* **189**, 617–631.

E Cherbas, L., Yonge, C. D., Cherbas, P., and Williams, C. M. (1980). The morphological response of *Kc-H* cells to ecdysteroids: hormonal specificity. *W. Roux's Arch. Dev. Biol.* **189**, 7–15.

rE Cherbas, P. (1993). Ecdysone-responsive genes (The IVth Karlson Lecture). *Insect Biochem. Mol. Biol.* **23**, 3–11.

E Cherbas, P., and Andres, A. (1990). The regulation of *Eip28/29* and *Eip40* by ecdysone in *Drosophila*. IXth Ecdysone Workshop, Paris, Sept. 89, *Invert. Reprod. Dev.* **18**, 108. [Abstract]

E Cherbas, P., Cherbas, L., Demetri, G., Manteuffel-Cymborowska, M., Savakis, C., Yonger, C. D., and Williams, C. M. (1980). Ecdysteroid Hormone effects on a *Drosophila* cell line. *In "Gene Regulation by Steroid Hormones"* (A. K. Roy, and J. H. Clark, eds.), pp. 278–305. Springer Verlag, New York.

E Cherbas, P., Cherbas, L., Lee, S. S., and Nakanishi, K. (1988). 26-(^{125}I)Iodoponasterone A is a potent ecdysone and a sensitive radioligand for ecdysone receptors. *Proc. Natl. Acad. Sci. U.S.A.* **85**, 2096–2100.

E Cherbas, P., Cherbas, L., Savakis, C., and Koehler, M. M. D. (1984). Ecdysteroid-responsive genes in a *Drosophila* cell line. *Am. Zool.* **21**, 743–750.

E Cherbas, P., Cherbas, L., and Williams, C. M. (1977). Induction of acetylcholinesterase activity by beta-ecdysone in a *Drosophila* cell line. *Science* **197**, 275–277.

E Cherbas, P., Trainor, D. A., Stonard, R. J., and Nakanishi, K. (1982). 14-Deoxymurister-one, a compound exhibiting exceptional moulting hormone activity. *Chem. Commun.* 1307–1308.

CdE Chihara, C. J., and Fristrom, J. W. (1973). Effects and interactions of juvenile hormone and beta-ecdysone on *Drosophila* imaginal discs cultured *in vitro. Dev. Biol.* **35**, 36–46.

Cd Chihara, C. J., and Fristrom, J. W. (1974). A juvenile hormone activity from extracts of *D. melanogaster. Dros. Inf. Serv.* **51**, 139–140.

CdE Chihara, C. J., Petri, W. H., Fristrom, J. W., and King, O. S. (1972). The assay of ecdysones and juvenile hormones on *Drosophila* imaginal discs *in vitro. J. Insect Physiol.* **18**, 1115–1123.

R Chmeliauskaite, V. C., and Ilyin, Y. V. (1980). (in Russian) Scattered reiterated genes of *Drosophila melanogaster* with varying locations. V. The nature of self-complementary RNA transcribed with chromosomal DNA. *Genetika* **16**, 1535–1550.

H Chomyn, A., and Mitchell, H. K. (1982). Synthesis of 84,000 dalton protein in normal and heat-shocked *Drosophila melanogaster* cells, as detected by specific antibodies. *Insect Biochem.* **12**, 105–114.

TM Chung, Y. T. (1990). Transcriptional regulation of the *Drosophila* cytoskeletal *Actin 5C* gene. Ph.D. Dissertation, Cornell University, Ithaca, New York.

TM Chung, Y. T., and Keller, E. (1990a). Regulatory elements mediating transcription from the *Drosophila melanogaster Actin 5C* proximal promoter. *Mol. Cell. Biol.* **10**, 206–216.

TM Chung, Y. T., and Keller, E. (1990b). Positive and negative regulatory elements mediating transcription from the *Drosophila melanogaster Actin 5C* distal promoter. *Mol. Cell. Biol.* **10**, 6172–6180.

TM   Chung, Y. T., and Keller, E. B. (1991). The TATA-dependent and TATA-independent promoters of the *Drosophila melanogaster* actin 5C-encoding gene. *Gene* **106**, 237–241.

E   Clement, C. Y. (1991). Responses of a *Drosophila* cell line to Insect steroid hormones. Ph.D. Thesis, University of Exeter, U.K.

E   Clement, C. Y., and Dinan, L. (1990). Morphological responses of a *Drosophila* cell line to 20-hydroxyecdysone. Abst IXth Ecdysone Workshop, Paris, Sept. 1989, *Invert. Reprod. Dev.* **18**, 109.

E   Clement, C. Y., Bradbrook, D. A., Lafont, R., and Dinan, L. (1993). Assessment of a microplate-based bioassay for the detection of ecdysteroid-like or antisteroid activities. *Insect Biochem. Mol. Biol.* **23**, 187–193.

H   Clos, J., Rabindran, S., Wisniewski, J., and Wu, C. (1993). Induction temperature of human heat-shock factor is reprogrammed in a *Drosophila* cell environment. *Nature* **364**, 252–255.

H   Clos, J., Westwood, J. T., Becker, P. B., Wilson, S., Lambert, K., and Wu, C. (1990). Molecular cloning and expression of hexameric *Drosophila* HSF subject to negative regulation. *Cell* **63**, 1085–1097.

M   Cock, de, J. G. R., van Hoffen, A., Wijnands, J., Molenaar, G., Lohman, P. H. M., and Eeken, J. C. J. (1992). Repairs of UV-induced (6-4) photoproducts measured in individual genes in the *Drosophila* embryonic *Kc* line. *Nucleic Acids Res.* **20**, 4789–4793.

M   Cock de, J. G. R., Klink, E. C., Ferro, W., Lohman, P. H. M., and Eeken, J. C. J. (1991). Repair of UV-induced cyclobutane pyrimidin dimers in the individual genes *Gart, Notch* and *white* from *Drosophila* cell lines. *Nucleic Acids Res.* **19**, 3289–3294.

M   Cock, de, J. G. R., Klink, E. C., Lohman, P. H. M., and Eeken, J. C. J. (1992a). Neither enhanced removal of cyclobutane pyrimidine dimers nor strand-specific repair is found after transcription induction of the β3-tubulin gene in a *Drosophila* embryonic cell line *Kc*. *Mutat. Res.* **293**, 11–20.

M   Cock, De, J. G. R., Klink, E. C., Lohman, P. H. M., and Eeken, J. C. J. (1992b). Absence of strand-specific repair of cyclobutane pyrimidin dimers in active genes from *Drosophila melanogaster* cell lines. *Mutat. Res.* **274**, 85–92.

R   Codani-Simonart, S. (1993). Approche des régulations de l'expression de *1731*, un rétrotransposon de Drosophile. *Thèse de Université Paris XI*, 1–85.

R   Codani-Simonart, S., Lacoste, J., Best-Belpomme, M., and Fourcade-Peronnet, F. (1993). Promoter activity of the *1731 Drosophila* retrotransposon in a human monocytic cell line. *FEBS Lett.* **325**, 177–182.

M   Colgan, J., and Manley, J. L. (1992). TFIID can be rate limiting *in vivo* for TATA-containing but not TATA-lacking RNA Polymerase II promoters. *Genes Dev.* **6**, 304–315.

M   Colgan, J., Wampler, S., and Manley, J. L. (1993). Interaction between a transcriptional activator and transcription factor IIB *in vivo*. Nature **362**, 549–552.

CtH   Compton, J. L., and Bonner, J. J. (1978). An *in vitro* assay for the specific induction and regression of puffs in isolated nuclei of *Drosophila melanogaster*. *Cold Spring Harbor Symp. Quant. Biol.* **42**, 835–838.

CtH   Compton, J. L., and McCarthy, B. J. (1978). Induction of the *Drosophila* heat-shock response in isolated polytene nuclei. *Cell* **14**, 191–201.

R   Contursi, C., Minchiotti, G., and Di Nocera, P. P. (1993). Functional dissection of two promoters that control sense and antisense transcription of *Drosophila melanogaster* F elements. *J. Mol. Biol.* **234**, 988–997.

M   Corbett, A. H., Guerry, P., Pflieger, P., and Osheroff, N. (1993). A pyrimidol (1,6-a) benzymidazole that enhances DNA cleavage mediated by eukaryotic topoisomerase II:

a novel class of topoisomerase II-targeted drugs with cytotoxic potential. *Antimicrob. Agents Chemother.* **37**, 2599–2605.

H   Corces, V., Holmgren, R., Freund, R., Morimoto, R., and Meselson, M. (1980). Four heat-shock proteins of *Drosophila melanogaster* coded within a 12 kilobase region in chromosome subdivision 67B. *Proc. Natl. Acad. Sci.* **77**, 5390–5393.

H   Corell, R. A., Riordan, J. A., and Gross, R. H. (1994). Chemical induction of stress proteins does not induce splicing thermotolerance under conditions producing survival thermotolerance. *Exp. Cell Res.* **211**, 189–196.

TM   Cornwell, M. M., and Smith, D. E. (1993). *Sp1* activates the MDR1 promoter through one of the two distinct G-rich regions that modulate promoter activity. *J. Biol. Chem.* **268**, 19505–19511.

H   Costlow, N., and Lis, J. T. (1984). High resolution mapping of DNase I-hypersensitive sites of *Drosophila* heat-shock genes in *Drosophila melanogaster* and *Saccharomyces cerevisiae. Mol. Cell. Biol.* **4**, 1853–1863.

E   Couderc, J. L., Becker, J. L., Sobrier, M. L., Dastugue, B., Best-Belpomme, M., Lepesant, J. A., and Pardue, M. L. (1984). Isolation and chromosomal localization of ecdysterone-responsive genes in a *Drosophila* cell line. *Chromosoma* **89**, 338–342.

E   Couderc, J. L., Cadic, A. L., Sobrier, M. L., and Dastugue, B. (1982). Ecdysterone induction of actin synthesis and polymerization in a *Drosophila melanogaster* cultured cell line. *Biochem. Biophys. Res. Commun.* **107**, 188–195.

E   Couderc, J. L., and Dastugue, B. (1980). Ecdysterone-induced modifications of protein synthesis in a *Drosophila melanogaster* cultured cell line *Biochem. Biophys. Res. Commun.* **97**, 173–181.

E   Couderc, J. L., Hilal, L., Sobrier, M. L., and Dastugue, B. (1987). 20-hydroxyecdysone regulates cytoplasmic actin gene expression in *Drosophila* cultured cells. *Nucleic Acids Res.* **15**, 2549–2561.

E   Couderc, J. L., Sobrier, M. L., Giraud, G., Becker, J. L., and Dastugue, B. (1983). Actin gene expression is modulated by ecdysterone in a *Drosophila* cell line. *J. Mol. Biol.* **164**, 419–430.

E   Couderc, J. L., Sobrier, M. L., Giraud, G., Micard, D., Dastugue, B., Ropp, M., Becker, J. L., Maisonhaute, C., Peronnet, F., Courgeon, A. M., and Best-Belpomme, M. (1984). Pleiotropic specific responses induced by ecdysterone in the cultured *Kc* cells of *Drosophila melanogaster. In "Biosynthesis, metabolism and mode of Action of Invertebrate Hormones"* (J. Hoffman and M. Porcher, eds.), pp. 293–297. Springer-Verlag, New York.

TM   Courey, A. J., Holtzman, D. A., Jackson, S. P., and Tijan, R. (1989). Synergistic activation by the glutamine-rich domains of human transcription factor *Sp1. Cell* **59**, 827–836.

TM   Courey, A. J., and Tijan, R. (1988). Analysis of Sp1 *in vivo* reveals multiple transcriptional domains, including a novel glutamine-rich activation motif. *Cell* **55**, 887–898.

E   Courgeon, A. M. (1972a). Effects of α- and β- ecdysone on *in vitro* diploid cell multiplication in *Drosophila melanogaster. Nature New Biol.* **238**, 250–251.

E   Courgeon, A. (1972b). Action of Insect hormone at the cellular level: Morphological changes of a diploid cell line of *Drosophila melanogaster* treated with ecdysone and several analogues *in vitro. Exp. Cell Res.* **74**, 327–336.

E   Courgeon, A. M. (1975a). Action of Insect hormones at the cellular level: II Differing sensitivity to β-ecdysone of several lines and clones of *Drosophila melanogaster* cells. *Exp. Cell Res.* **94**, 283–291.

E   Courgeon, A. M. (1975b). Action conjuguée de l'hormone juvénile et de l'ecdystérone sur des lignées cellulaires de Drosophile *in vitro*. *CR Acad. Sci. Paris* **280**, 2563–2566.

H   Courgeon, A. M., Becker, J., Maingourd, M., Maisonhaute, C., and Best-Belpomme, M. (1990). Early activation of heat-shock genes in hydrogen peroxide treated *Drosophila* cells. *Free Radical Res. Commun.* **9**, 147–155.

E   Courgeon, A. M., and Best-Belpomme, M. (1976). Action de l'ecdysterone sur des lignées et des clones cellulaires diploïdes de *Drosophila melanogaster* en culture *in vitro*. *Actualités sur les Hormones d'Invertébrés* (Colloques Intern. CNRS, N° 251), 383–392.

E   Courgeon, A. M., and Best-Belpomme, M. (1977). Modifications protéiques induites par l'ecdysone dans des lignées et des clones cellulaires de *Drosophila melanogaster*. *Bull. Soc. Zool. France* **102**, 289.

E   Courgeon, A. M., and Cailla, H. (1981). Cyclic AMP and cyclic GMP variations in several *Drosophila* embryonic cellular clones cultured *in vitro* with or without 20-hydroxyecdysone. *Exp. Cell Res.* **133**, 15–22.

E   Courgeon, A. M., and Cailla, H. L. (1984). Cyclic GMP production and excretion by *Drosophila* cells. Modulation by 20-hydroxyecdysone. *Insect Biochem.* **14**, 691–695.

H   Courgeon, A. M., Maisonhaute, C., and Best-Belpomme, M. (1984). Heat-shock proteins are induced by cadmium in *Drosophila* cells. *Exp. Cell Res.* **153**, 515–521.

H   Courgeon, A. M., Maingourd, M. Maisonhaute, C., Montmory, C., Rollet, E., Tanguay, R., and Best-Belpomme, M. (1993). Effects of Hydrogen peroxide on cytoskeletal proteins of *Drosophila* cells: comparison with heat-shock and other stresses. *Exp. Cell Res.* **204**, 30–37.

H   Courgeon, A. M., Rollet, E., Becker, J., Maisonhaute, C., and Best-Belpomme, M. (1988). Hydrogen peroxide (H_2O_2) induces actin and some heat-shock proteins in *Drosophila* cells. *Eur. J. Biochem.* **171**, 163–170.

H   Courgeon, A. M., Ropp, M., Rollet, E., Becker, J., Maisonhaute, C., and Best-Belpomme, M. (1988). (Abstr. P105, UCLA Symp. Mol. Cell. Biol.) *J. Mol. Biochem.* **12D** (Suppl.), 256.

E   Courgeon, A. M., Ropp, M., Rollet, E., Becker, J., Maisonhaute, Echalier, G., and Best-Belpomme, M. (1989). Relationships between ecdysterone-induced cellular differentiation and aerobiosis in an *in vitro* Drosophila cell system. *In "Invertebrate Cell System Applications"* (J. Mitsuhashi, ed.), pp. 93–97. CRC Press, Boca Raton, FL.

H   Craig, E. A., Ingolia, T. D., and Manseau, L. J. (1983). Expression of *Drosophila* heat-shock cognate genes during heat-shock and development. *Dev. Biol.* **99**, 418–426.

H   Craig, E. A., McCarthy, B. J., and Wadsworth, S. C. (1979). Sequence organization of two recombinant plasmids containing genes for the major heat-shock-induced proteins of *Drosophila melanogaster*. *Cell* **16**, 575–588.

H   Craine, B. L., and Kornberg, T. (1981a). Transcription of the major *Drosophila hsp* genes *in vitro*. *Biochemistry* **20**, 6584–6588.

H   Craine, B. L., and Kornberg, T. (1981b). Activation of the major *Drosophila* heat-shock genes *in vitro*. *Cell* **25**, 671–681.

Cp Cross, D. P. (1975). The culture *in vitro* of cells from lethal embryos of *Drosophila melanogaster*. Ph.D. Thesis, University of Sussex.

Cp Cross, D. P., and Sang, J. H. (1978a). Cell culture of individual *Drosophila* embryos. I. Development of wild-type cultures. *J. Embryol. Exp. Morphol.* **45**, 161–172.

Cp Cross, D. P., and Sang, J. H. (1978b). id. II. Culture of X-linked embryonic lethals. *J. Embryol Exp. Morphol.* **45**, 173–187.

V   Crump, W. A. L., and Moore, N. F. (1981). The polypeptides induced in *Drosophila* cells by a virus of *Heteronychus arator. J. Gen. Virol.* **52**, 173–176.

Cp Cl Cullen, C. F., and Milner, M. J. (1991). Parameters of growth in primary cultures and cell lines established from *Drosophila* imaginal discs. *Tissue Cell* **23**, 29–39.

VT Culp, J. F., Johansen, H., Helmig, B., Beck, J. Matthews, T. S., Delers, A., and Rosenberg, M. (1991). Regulated expression allows high level production and secretion of HIV-1*gp120* envelope glycoprotein in *Drosophila* Schneider cells. *Biotechnology* **9**, 173–177.

B   Cumberledge, S., and Krasnow, M. A. (1993). Intercellular signalling in *Drosophila* segment formation reconstructed *in vitro. Nature* **363**, 549–552.

Cp Cumberledge, S., and Krasnow, M. A. (1994). Preparation and analysis of pure cell populations from *Drosophila. In "Methods in Cell Biology"* (L. S. B. Goldstein and E. A. Fyrberg eds.), Vol. 44 p. 143. Academic Press, New York.

Cp Cunningham, I. (1961). Studies on the maintenance and growth of Insect tissues and cells *in vitro. M. Sci. Thesis,* University of Edinburgh.

Cp Currie, D. A., Milner, M. J., and Evans, C. W. (1988). The growth and differentiation *in vitro* of leg and wing imaginal disc cells from *Drosophila melanogaster. Development* **102**, 805–814.

M   Dan Garza, M., and Hartl, D. L. (1990). *Drosophila* nonsense suppressors functional analysis in *Saccharomyces cerevisiae, Drosophila* cultured cells and *Drosophila melanogaster. Genetics* **126**, 625–637.

R   Danilevskaya, O. N., Slot, F., Traverse, K. L., Hogan, N. C., and Pardue, M. L. (1994). *Drosophila* telomere transposon HeT-A produces a transcript with tightly bound protein. *Proc. Natl. Acad. Sci. U.S.A.* **91**, 6679–6682.

V   Dasgupta, R., Ghosh, A., Dasmahapatra, B., Guarino, L. A., and Kaesberg, P. (1984). Primary and secondary structure of Black Beetle Virus RNA2, the genomic messenger for BBV coat protein precursor. **12**, 7215–7223.

V   Dasgupta, R., Selling, B., and Rueckert, R. (1994). Flock house virus: a simple model for studying persistant infection in cultured *Drosophila* cells. *Arch. Virol.* **9** (Suppl.) 121–132.

V   Dasgupta, R., and Sgro, J. Y. (1989). Nucleotide sequences of three Nodavirus RNA2's: the messengers for their coat protein precursor. *Nucleic Acids Res.* **17**, 7525–7526.

V   Dasmahapatra, B., Dasgupta, R., Ghosh, A., and Kaesberg, P. (1985). Structure of the Black Beetle Virus genome and its functional implications. *J. Mol. Biol.* **182**, 183–189.

VT Dasmahapatra, B., Dasgupta, R., Saunders, K., Selling, B., Gallagher, T., and Kaesberg, P. (1986). Infectious RNA derived by transcription from cloned cDNA copies of the genomic RNA of an insect virus. *Proc. Natl. Acad. Sci. U.S.A.* **83**, 63–66.

TM Davies, J. A., Addison, C. F., Delaney, S. J., Sunkel, C., and Glover, D. M. (1986). Expression of the prokaryotic gene for Chloramphenicol aminotransferase in *Drosophila* under the control of larval serum protein I promoters. *J. Mol. Biol.* **189**, 13–24.

Cd Davis, K. T., and Shearn, A. (1977). *In vitro* growth of imaginal discs from *Drosophila melanogaster. Science* **196**, 438–440.

M   Dawid, I. B., and Botchan, P. (1977). Sequences homologous to ribosomal insertions occur in the *Drosophila* genome, outside the nucleolar organizer. *Proc. Natl. Acad. Sci. U.S.A.* **74**, 4233–4237.

M   Dawid, I. B., and Rebbert, M. L. (1986). Expression of ribosomal insertion in *Drosophila:* Sensitivity to intercalating drugs. *Nucleic Acids Res.* **14**, 1267–1277.

B   Dearolf, C. R., Topol, J., and Parker, C. S. (1989). The *caudal* gene is a direct activator of *fushi tarazu* transcription during *Drosophila* embryogenesis. *Nature* **341**, 340–343.

B Debec, A. (1974). Isozymic patterns and functional state of *in vitro* cultured cell lines of *Drosophila melanogaster*. I. *W. Roux's Arch.* **174**, 1–19.

B Debec, A. (1976). Isozymic patterns and functional states of cell lines of *Drosophila melanogaster* culured *in vitro* II. *W. Roux's Arch* **180**, 107–119.

Cl Debec, A. (1978). Haploid cell cultures of *Drosophila melanogaster*. *Nature* **274**, 255–256.

Cy Debec, A. (1984). Evolution of karyotype in haploid cell lines of *Drosophila melanogaster*. *Exp. Cell Res.* **151**, 236–246.

Cl Debec, A. (1986). Etude génétique de lignées cellulaires de *Drosophila melanogaster*. Thèse d'Etat, Univ. P.et M. Curie, Paris.

Cy Debec, A., and Abbadie, C. (1989). The acentriolar state of the *Drosophila* cell lines *1182*. *Biol. Cell* **67**, 307–311.

H Debec, A., Courgeon, A. M., Maingourd, M., and Maisonhaute, C. (1990). The response of the centrosome to heat shock and related stresses in a *Drosophila* cell line. *J. Cell Sci.* **96**, 403–412.

Cy Debec, A., Détraves, C., Montmory, C., Géraud, G., and Wright, M. (1995). Polar organization of gamma-tubulin in acentriolar mitotic spindles of *Drosophila melanogaster* cells. *J. Cell Sci.* **108**, 2645–2653.

B Debec, A., Mokdad, R., and Wegnez, M. (1985). Metallothioneins and resistance to cadmium poisoning in *Drosophila* cells. *Biochem. Biophys. Res. Commun.* **127**, 143–151.

B Debec, A., and Montmory, C. (1992). Cyclin B associated with centrosomes in *Drosophila* mitotic cells. *Biol. Cell* **75**, 121–126.

Cy Debec, A., Szöllösi, A. and Szöllösi, D. (1982) A *Drosophila melanogaster* cell line lacking centriole. *Biol. Cell* **44**, 133–138.

H Dellavalle, R. P., Petersen, R., and Lindquist, S. (1994). Preferential deadenylation of *Hsp70* mRNA plays a key role in regulating *Hsp70* expression in *Drosophila melanogaster*. *Mol. Cell. Biol.* **14**, 3646–3659.

Cd Demal, J. (1955). Differenciation d'ébauches imaginales de Diptères en culture *in vitro*. *Bull. Acad. R. Belg., Cl. Sci.* **5**, 1061–1071.

rCd Demal, J. (1961). Problèmes concernant la morphogenèse *in vitro* chez les Insectes. *Bull. Soc. Zool. France* **86**, 522–533.

Ct Demal, J., and Leloup, A. M. (1963). Essai de cultures *in vitro* d'organes d'Insectes. (Premier Collque Intern. Cult. Tissus Invert., Montpellier, Octobre 1962) *Ann. Epiphyties* **14**, 91–93.

B Dennis, R. D., and Haustein, D. (1981). Radio-iodination of cell-surface proteins in a *Drosophila* cell line. *Insect Biochem.* **11**, 699–705.

EB Dennis, R., and Haustein, D (1982). Ecdysteroid related changes in cell surface properties of a *Drosophila* cell line. *Insect Biochem.* **12**, 83–89.

H Desrosiers, R., and Tanguay, R. M. (1985). The modifications in the methylation patterns of *H2B* and *H3* after heat shock can be correlated with the inactivation of normal gene expression. *Biochem. Biophys. Res. Commun.* **133**, 823–829.

H Desrosiers, R., and Tanguay, R. M. (1986). Further characterization of the post-translational modifications of core histones in response to heat and arsenite stress in *Drosophila*. *Biochem. Cell Biol.* **64**, 750–757.

H Desrosiers, R., and Tanguay, R. (1988). Methylation of *Drosophila* histones at proline, lysine and arginine residues during heat shock. *J. Biol. Chem.* **263**, 4686–4692.

H Desrosiers, R., and Tanguay, R. M. (1989). Transcriptional inhibitors affecting topoisomerase II induce changes in histone methylation patterns similar to those induced by heat shock. *Biochem. Biophys. Res. Commun.* **162**, 1037–1043.

C   Dewurst S., and Sang, J. H. (1977). The development of a medium supplemented with egg extract for the maintenance of *Drosophila* cell lines. *In Vitro* **13**, 305–310.

BCp   Dewhurst, S. A., and Seecof, R. L. (1975). Development of acetylcholine metabolizing enzymes in *Drosophila* embryos and in cultures of embryonic *Drosophila* cells. *Comp. Biochem. Physiol.* **50C**, 53–58.

Cp   Dewhurst, J. A., and Seecof, R. L. (1979). Reactions programming enzyme production during *Drosophila* neurogenesis. *Insect Biochem.* **9**, 49–54.

V   Dezelée, S., Blondel, D., Wyers, F. and Petitjean, A. M. (1987a) Vesicular Stomatitis Virus in *Drosophila melanogaster* cells: lack of leader RNA transport into the nuclei and frequent abortion of the replication step. *J. Virol.* **61**, 1391–1397.

V   Dezélée, S., Wyers, F., Blondel, D., Petitjean, A. M., Contamine, D., Bras, F., and Teninges, D. (1987b). Rhabdoviruses in *Drosophila* cells. *In "Arboviruses in Arthropod Cells In Vitro"* (C. E. Yunker, ed.), pp. 111–133, CRC Press, Boca Raton, FL.

M   Dezzani, W., Harris, P. V., and Boyd, J. B. (1982). Repair of double-strand DNA breaks in *Drosophila*. *Mutat. Res.* **92**, 151–160.

H   DiDomenico, B. J., Bugaisky, G., and Lindquist, S. (1982a). Heat-shock and recovery are mediated by different translational mechanisms. *Proc. Natl. Acad. Sci.* **79**, 6181–6185.

H   DiDomenico, B. J., Bugaisky, G., and Lindquist, S. (1982b). The heat-shock response is self regulated at both the transcriptional and post-transcriptional levels. *Cell* **31**, 593–603.

B   Diederich, R. J., Matsuno, K., Hing, H., and Artavanis-Tsakonas, S. (1994). Cytosolic interaction between *deltex* and *Notch* ankyrin repeats implicates *deltex* in the *Notch* signaling pathway. *Development* **120**, 473–481.

R   Di Franco, C., Pisano, C., Fourcade-Peronnet, F., Echalier, G., and Junakovic, N. (1992). Evidence for the *de novo* rearrangements of *Drosophila* transposable elements induced by the passage to the cell culture. *Genetica* **87**, 65–73.

B   Dimarcq, J. L., Hoffmann, D., Meister, M., Bulet, P., Lanot, R., Reichhart, J. M., and Hoffmann, J. A. (1994). Characterization and transcriptional profiles of a *Drosophila* gene encoding an Insect defensin. A study in Insect immunity. *Eur. J. Biochem.* **221**, 201–209.

E   Dinan, L. (1985). Ecdysteroid receptors in a tumorous cell line of *Drosophila melanogaster*. *Arch. Insect Biochem. Physiol.* **2**, 295–317.

rE   Dinan, L., Spindler-Barth, M., and Spindler, K. D. (1990). Insect cell lines as tools for studying ecdysteroid action. (IXth Ecdysone Workshop, Paris, Sept. 1989) *Invert. Reprod. Dev.* **18**, 43–53.

E   Dinan, L., Whiting, P., Girault, J. P., and Lafon R. (1994). Novel ecdysteroid agonists and antagonists from plants. (Abst. in *XIth Ecdysone Workshop*, Ceske Budejovice, Tchec Republic, June 1994).

M   Dingermann, T., Sharp, S., Appel, B., DeFranco, D., Mounts, S., Hiermann, R., Pongs, O., and Söll, D. (1981). Transcription of cloned tRNA and 5S RNA genes in a *Drosophila* cell free extract. *Nucleic Acids Res.* **9**, 3907–3918.

T   Di Nocera, P. P. and Dawid, I. B. (1983) Transient expression of genes introduced into cultured cells of *Drosophila*. *Proc. Natl. Acad. Sci. U.S.A.* **80**, 7095–7098.

ET   Dobens, L., Rudolph, K. and Berger, E. M. (1991) Ecdysterone regulatory elements function as both transcriptional activators and repressors. *Mol. Cell. Biol.* **11**, 1846–1853.

F   Doctor, J. S., Hoffmann, F. M., and Olwin, B. B. (1991). Identification of a Fibroblast Growth Factor-binding protein in *Drosophila melanogaster*. *Mol. Cell. Biol.* **11**, 2318–2323.

F   Doctor, J. S., Jackson, P. D., Rashka, K. E., Visalli, M., and Hoffmann, F. M. (1992). Sequence, biochemical characterization and developmental expression of a new member of the TGF-β superfamily in *Drosophila melanogaster*. *Dev. Biol.* **151**, 491–505.

M   Doenecke, D., and McCarthy, B. J. (1975a). Protein content of chromatin fractions separated by sucrose gradient centrifugation. *Biochemistry* **14**, 1366–1372.

M   Doenecke, D., and McCarthy, B. J. (1975b). The nature of protein association with chromatin. *Biochemistry* **14**, 1373–1378.

Cp   Dolfini, S. (1969). Some comparative data on embryonic cells of *Drosophila melanogaster* cultured *in vitro*. *Experientia* **22**, 144.

rC   Dolfini, S. (1971). Cell culture in Diptera. In *"Invertebrate Tissue Culture"* (C. Vago, ed.), Vol. I, pp. 247–265. Academic Press, New York.

Cy   Dolfini, S. (1971). Karyotype polymorphism in a cell population of *Drosophila melanogaster* cultured *in vitro*. *Chromosoma* **33**, 196–208.

Cy   Dolfini, S. (1973). Further data on karyotype polymorphism of *Drosophila melanogaster* cells cultured *in vitro*. *3rd Intern. Coll. Invert. Tissue Cult.* (J. Rehacek, D. Blascovic, and W. F. Hink, eds.), pp. 143–158, Publish. House Slovak Acad. Sci.

Cy   Dolfini, S., Courgeon, A. M., and Tiepolo, L. (1970). The cell cycle of an established line of *Drosophila melanogaster* cells *in vitro*. *Experientia* **26**, 1020–1021.

Cy   Dolfini, S. (1976) Karyotype evolution in cell lines of *Drosophila melanogaster*. *Chromosoma* **58**, 73–86.

Cy   Dolfini, S., and Gottardi, A. (1966). Changes of chromosome number in cells of *Drosophila* cultured *in vitro*. *Experientia* **22**, 144–146.

Cy   Dolfini, S., Gottardi, A., and Rezzonico Raimondi, G. (1966) First results on changes of chromosome number in cells of *D. melanogaster* cultured *in vitro*. *Dros. Inf. Serv.* **41**, 110.

Cy   Dolfini, S., Gottardi, A., and Rezzonico Raimondi, G. (1955). Primi dati sulla variazione del numer cromosomico in cellule di *Drosophila melanogaster* coltivate *in vitro*. *Atti Ass. Genet. Ital.* **11**, 226–236.

Cy   Dolfini, S., and Tiepolo, L. (1968). The cell cycle of embryonic cells of *Drosophila melanogaster* cultured *in vitro*. *2nd Intern. Coll. Invert. Tissue Cult.*(sett. 1967,Tremezzo, Italy), 182–188.

CpM   Donady, J. J., Bernstein, S. I., Fyrberg, E. A., Murray, S. C., and Minnie, M. E. (1980). Changes in RNA and protein synthesis associated with myogenesis *in vitro*. In *"Invertebrate Systems In Vitro"* (E. Kurstak, K. Maramorosch, and A. Dubendorfer, eds.), pp. 291–301, Elsevier/North Holland Biomedical Press, Amsterdam.

Cp   Donady, J. J., and Fyrberg, E. (1975). A method for culturing *Drosophila* embryonic cells *in vitro* for nerve and muscle differentiation. *TCA Manual* **1**, 81–83.

Cp   Donady, J. J., and Fyrberg, E. (1977). Mass culturing of *Drosophila* embryonic cells *in vitro*. *TCA Manual* **3**, 685–687.

Cp   Donady, J. J., Murray, S. C., and Intres, R. C. (1986). Coordinate gene activity in *Drosophila* myogenesis *in vitro*. *Techn. Life Sci. C2. In vitro Invertebrate Hormones and Genes,* C216, I-15, Elsevier Science Publishers, Ireland.

Cp   Donady, J. J., and Seecof, R. L. (1971). The effect of lethal myospheroid on *Drosophila* muscle and nerve cell differentiation *in vitro*. *Genetics* **68**, (Meeting Genet. Soc. America), s15. [Abstract]

Cp   Donady J. J., and Seecof, R. L. (1972). Effect of the gene *lethal (1)myospheroid* on *Drosophila* embryonic cells *in vitro*. *In Vitro* **8**, 7–12.

Cp   Donady, J. J., Seecof, R. L., and Dewhurst, S. (1975). Actinomycin D sensitive periods in the differentiation of *Drosophila* neurons and muscle cells *in vitro*. *Differentiation* **4**, 9–14.

CpB   Donady, J. J., Seecof, R. L., and Fox, M. A. (1973). Differentiation of *Drosophila* cells lacking ribosomal DNA, *in vitro*. *Genetics* **73**, 429–434.

THR   Dorsett, D., Viglianti, G. A., Rutledge, B. J., and Meselson, M. (1989). Alteration of *hsp82* gene expression by the *gypsy* transposon and *suppressor* genes in *Drosophila melanogaster*. *Genes Dev.* **3**, 454–468.

MT   Driever, W., and Nüsslein-Volhard, C. (1989). The *bicoid* protein is a positive regulator of *hunchback* transcription in the early *Drosophila* embryo. *Nature* **337**, 138–143.

TM   Driever, W., Ma,J., Nusslein-Volhard, C., and Ptashne, M. (1989). Rescue of *bicoid* mutant *Drosophila* embryos by *bicoid* fusion proteins containing heterologous activating sequences. *Nature* **342**, 149–154.

H   Drummond, I. A. S., and Steinhardt, R. A. (1987). The role of oxidative stress in the induction of *Drosophila* heat shock proteins. *Exp. Cell Res.* **173**, 439–449.

H   Duband, J. L., Lettre, F., Arrigo, A. P., and Tanguay, R. M. (1986). Expression and localization of *hsp 23* in unstressed and heat-shocked *Drosophila* cultured cells. *Can. J. Genet. Cytol.* **28**, 1088–1092.

Cp   Dübendorfer, A. (1976). Metamorphosis of imaginal disc tissue grown *in vitro* from dissociated embryos of *Drosophila*. In *"Invertebrate Tissue Culture: Applications in Medicine, Biology and Agriculture"* (E. Kurstak, and K, Maramorosch, eds.) (Proc. 4th Intern., Conf. Invert. Tissue Cult., Mont Gabriel, 1975) pp. 151–159, Academic Press, New York.

E   Dübendorfer, A. (1986) Ecdysone C20-hydroxylation and conjugate formation by *Drosophila melanogaster* cell lines. *Insect Biochem* **16**, 645–651.

E   Dübendorfer, A. (1988). Feed-back-regulation of ecdysone C20-hydroxylation in primary cell cultures from *Drosophila* embryos (7th Intern. Conf. Invert. & Fish Tissue Cult., Ohito, Japan, 1987) in *Invert. & Fish Tissue Cult.* (Y. Kuroda, E. Kurstak, and K. Maramorosch, eds.), pp. 39–42, Jap. Sci. Soc. Press, Tokyo and Springer-Verlag, Berlin.

E   Dübendorfer, A. (1989). Ecdysteroid action in embryonic systems In *"Ecdysone, from Chemistry to Mode of Action"* (J. Koolman, ed.), pp. 421–425, G. Thieme Verlag, Stuttgart.

Cp   Dübendorfer, A., Blumer, A., and Deak, II. (1978). Differentiation *in vitro* of larval and adult muscles from embryonic cells of *Drosophila*. *W. Roux's Arch.* **184**, 233–249.

Cp   Dübendorfer, A., and Eichenberger, S. (1985). *In vitro* metamorphosis of insect cells and tissues: development and function of fat body cells in embryonic cell cultures of *Drosophila*. In *"Metamorphosis"* (M. Balls and M. Bownes, eds.) Oxford Univ. Press.

Cp   Dübendorfer, A., and Eichenberger-Glinz, S. (1980). Development and metamorphosis of larval and adult tissues of *Drosophila in vitro*. *Invert. Systems in Vitro* (5th Intern. Conf. Invert. Tissue Cult., Rigi-Kaltbad, Suitzerland, April 1979 (E. Kurstak, K. Maramorosch, and A. Dübendorfer, eds.), pp. 169–185. Elsevier/North Holland Biomedical Press, Amsterdam.

Cl   Dübendorfer, A., and Shields, G. (1972). Proliferation *in vitro* and *in vivo* of a cell line originally derived from imaginal disc cells. *Dros. Inf. Serv.* **49**, 43.

Cp   Dübendorfer, A., Shields, G., and Sang, J. H. (1974) Pattern formation by embryonic *Drosophila* imaginal cells cultivated *in vitro*. *Heredity* **33**, 138–139. [Abstract]

Cp   Dübendorfer, A., Shields, G., and Sang, J. H. (1975). Development and differentiation *in vitro* of *Drosophila* imaginal disc cells from dissociated early embryos. *J. Embryol. Exp. Morphol.* **33**, 487–498.

H   Duncan, R. F., Cavener, D. R., and Qu, S. (1995). Heat shock effects on phosphorylation of protein synthesis initiation factor proteins elF-4E and elF-2a in *Drosophila*. *Biochemistry* **34**, 2985–2997.

M Dusenbery, R. L., and Lee-Chen, S. F. (1988). Equivalence of UDS responses for established cell lines and primary culture cells derived from the *mei9a* and *mus201D1* excision repair-deficient strains of *Drosophila melanogaster*. *Mutat. Res.* **194**, 257–261.

M Dusenbery, R. L., McCormick, S. C., and Smith, P. D. (1983). *Drosophila* mutations at the *mei-9* and *mus(2)201* loci which block excision of thymidine dimers also block induction of unscheduled DNA synthesis by methyl methanesulfonate, ethyl methanesulfonate, N-methyl-N-nitrosourea, UV light and X-rays. *Mutat. Res.* **112**, 215–230.

M Dynlacht, B. D., Brook, A., Dembski, M., Yenush, L., and Dyson, N. (1994). DNA-binding and trans-activation properties of *Drosophila* E2F and DP proteins. *Proc. Natl. Acad. Sci. U.S.A.* **91**, 6359–6363.

B Echalier, G. (1971). Established diploid cell lines of *Drosophila melanogaster* as potential material for the study of genetics of somatic cells. *Curr Top Microb. Immun.* 55, 220–227.

rC Echalier, G. (1976). *In vitro* established lines of *Drosophila* cells and applications in physiological genetics. *Invert. Tissue Cult.: Applications in Medicine, Biology and Agriculture.* (4th Int. Conf. Invert. Tissue Cult., Mont Gabriel/Montréal, Can., 1975) (E. Kurstak, and K. Maramorosch, eds.), pp. 131–150, Academic Press, New York.

C Echalier, G. (1980). Necessity of radically new Insect cell culture methods. *In "Invertebrate Systems in Vitro"* (E. Kurstak, K. Maramorosch and A. Dubendorfer, eds.), pp. 589–592. Elsevier/North Holland Biomedical Press, Amsterdam.

rR Echalier, G. (1989). *Drosophila* retrotransposons: Interactions with genome. *Adv. Virus Res.* 36, 33–105.

Cl Echalier, G. (1989). A "citation classic" commentary: Daughters of Paris May Revolution, *Current Contents* **38**, 16.

T Echalier, G., and Fourcade-Peronnet, F. (1994). Differences in transfection efficiency between *Drosophila* cell lines. *In "Insect Cell Biotechnology"* (K. Maramorosch and A. H. McIntosh, eds.), pp. 141–156, Boca Raton, FL.

R Echalier, G., and Junakovic, N. (1988). Possible correlation between transposition of *copia*-like nomadic elements and establishment *in vitro* of continuous *Drosophila* cell lines. *Invert. and Fish Tissue Culture* (Y. Kuroda, E. Kurstak, and K. Maramorosch, eds.), pp. 111–114, Springer Verlag, New York.

Cl Echalier, G., and Ohanessian, A. (1968). Cultures *in vitro* de cellules de Drosophile. Obtention d'une souche à multiplication continue*. *2nd Intern. Coll. Invert. Tissue Cult.*(Tremezzo,Sett.1967), 174–181, Istituto Lombardo, Accad. Scienze & Lettere.

Cl Echalier, G., and Ohanessian, A. (1969). Isolement, en cultures *in vitro*, de lignées cellulaires diploides de *Drosophila melanogaster*. *CR Acad. Sci. Paris*, 268, 1771–1773.

Cl Echalier, G., and Ohanessian, A. (1970). *In vitro* culture of *Drosophila melanogaster* embryonic cells. *In Vitro* 6, 162–172.

Cp Echalier, G., Ohanessian, A., and Brun, G. (1965). Cultures "primaires" de cellules embryonnaires de *Drosophila melanogaster* (Insecte Diptère) *CR Acad. Sci. Paris* 261, 3211–3213.

* This paper should not figure in this bibliography, because it is likely that our primary cultures were contaminated with lepidopteran cell of a Grace's line, which were being grown in the laboratory.

B   Echalier, G., and Proust, J. (1973). Analysis, with the technique of *in vivo* transplantation of the capacities of differentiation of *Drosophila* cells from an established line grown *in vitro;* preliminary results. *3rd Coll. Invert. Tissue Culture,* Smolenice (Tchecoslovaquia, June 71) (J. Rehacek, D. Blaskovic, W. F. Hink, eds.) Publ. House of Slovak Acad. Sci.

Cd   Edwards, J. S., Milner, M. J., and Chen, S. W. (1978). Integument and sensory nerve differentiation of *Drosophila* leg and wing imaginal discs *in vitro.* W. *Roux's Arch* **185**, 59–77.

Cl   Eide, P. E. (1975). Establishment of a cell line from long-term primary embryonic House fly cell cultures. *J. Insect Physiol.* **21**, 1431–1438.

Cp   Eide, P. E., and Chang, T. H. (1969). Cell cultures from dispersed embryonic House fly tissues. *Exp. Cell Res.* **54**, 302–308.

M   Eissenberg, J. C., Ge, Y. W., and Hartnett, T. (1994). Increased phosphorylation of HP1, a heterochromatin-associated protein of *Drosophila,* is correlated with heterochromatin assembly. *J. Biol. Chem.* **269**, 21315–21321.

M   Elgin, S. C. R., and Hood, L. E. (1973). Chromosomal proteins of *Drosophila* embryos *Biochemistry* **12**, 4984–4991.

M   Elgin, S. C. R., and Miller, D. W. (1978). Preparation of nuclei, chromatin and chromosomal proteins of *Drosophila* embryos. *In* "Genetics and Biology of Drosophila" (M. Ashburner and T. R. F. Wright, eds), Vol. 2A, pp. 145–150. Academic Press, New York.

BT   Elkins, T., Hortsch, M., Bieber, A. J., Snow, P. M., and Goodman, CS.J. (1990b). *Drosophila* fasciclin I is a novel homophilic adhesion molecule that along with fasciclin III can mediate cell sorting. *J. Cell. Biol.* **110**, 1825–1832.

B   Elkins, T., Zinn, K., McAllister, L., Hoffmann, F. M., and Goodman, C. S. (1990a). Genetic analysis of a *Drosophila* neural adhesion molecule: Interaction of Fasciclin I and Abelson Tyrosine kinase mutations. *Cell* **60**, 565–575.

V   Emeny, J., and Lewis, M. J. (1984). *Sigma* virus of *Drosophila* as a vector model. *In* "Vectors in Virus Biology" (M. A. Mayo, and K. A. Harrap, eds.), pp. 93–112.

R   Emori, Y., Shiba, T., Kanaya, S., Inoue, S., Yuki, S., and Saigo, K. (1985) The nucleotide sequence of *copia* and *copia*-related RNA in *Drosophila* virus-like particles. *Nature* **315**, 773–776.

HB   Engel, M., and Cornelius, G. (1995). Involvement of protein kinase C in activation of *Drosophila fos* and *hsp70. Cell Physiol. Biochem.* **5**, 313–317.

B   Engström, Y., Kadalayil, L., Sun, S. C., Samakovlis, C., Hultmark, D., and Faye, I. (1993). κB-like motifs regulate the induction of immune genes in *Drosophila. J. Mol. Biol.* **232**, 327–333.

C   Ephrussi, B., and Beadle, G. (1936). A technique of transplantation for *Drosophila. Am. Nature* **70**, 218–225.

Cd   Eugene, O. M., Yund, M. A., and Fristrom, J. W. (1979). Preparative isolation and short-term organ culture of imaginal discs of *Drosophila melanogaster. Tissue Cult. Assn Manual* **5**, 1055–1062.

Cy   Faccio Dolfini, S. (1974a). The distribution of repetitive DNA in the chromosomes of cultured cells of *Drosophila melanogaster. Chromosoma* **44**, 383–391.

Cy   Faccio Dolfini, S. (1974b). Spontaneous chromosome rearrangements in an established cell line of *Drosophila melanogaster. Chromosoma* **47**, 253–261.

Cy   Faccio Dolfini, S. (1976a). Variazioni del cariotipo in popolazioni di cellule di Drosofila coltivate *in vitro. Atti Convegni Lincei* **14**, 375–396.

Cy   Faccio Dolfini, S. (1976b). Karyotype evolution in cell lines of *Drosophila melanogaster. Chromosoma* **58**, 73–86.

Cy   Faccio Dolfini, S. (1977). Karyotype evolution in two cell lines of *Drosophila melanogaster* maintained in different culture conditions. *Dros. Inf. Serv.* **52**, 46.

Cy   Faccio Dolfini, S. (1978). Sister chromatid exchanges in *Drosophila melanogaster* cell lines *in vitro*. *Chromosoma* **69**, 339–347.

CtCy   Faccio Dolfini, S. (1987). The effect of distamycin A on heterochromatin condensation of *Drosophila* chromosomes. *Chromosoma* **95**, 57–62.

Cy   Faccio Dolfini, S. Bonifazio Razzini, A. (1983). High resolution of heterochromatin of *Drosophila melanogaster* by distamycin A. *Experientia* **39**, 1492–1404.

Cy   Faccio Dolfini, S., and Halfer, C. (1976). Variazioni del cariotipo in popolazioni di cellule di *Drosophila* coltivate *in vitro*. *Colloquio Genetica di Populazioni, Accad. Naz. Lincei,* Roma, **14**, 375–396.

Cy   Faccio Dolfini, S., and Halfer, C. (1978). Spontaneous chromosomal changes in established cell populations of *Drosophila melanogaster*. In *"Origin and Natural History of Cell Lines"* (Int. Symp. Accad. Lincei Roma, C. Barigozzi ed.), 125–147, Alan Riss, New York.

Cd   Fain, M. J. and Schneiderman, H. A. (1979) Wound healing and regenerative response of fragments of the *Drosophila* wing imaginal disc cultured *in vitro*. *J. Insect Physiol.* **25**, 913–924.

M   Fairman, R., and Brutlag, D. L. (1988). Expression of the *Drosophila* type II topoisomerase developmentally regulated. *Biochemistry* **27**, 560–565.

H   Falkenburg, P. E., Haass, C., Kloetzel, P. M., Niedel, B. Kopp, F., Kuehn, L., and Dahlmann, B. (1988). *Drosophila* small cytoplasmic 19S ribonucleoprotein is homologous to the Rat multicatalytic proteinase. *Nature* **331**, 190–192.

R   Falkenthal, S. V. (1980) *Copia* gene expression in *Drosophila* cultured cells. Ph.D. Thesis, University of California, Los Angeles, Los Angeles.

R   Falkenthal, S., Graham, M. L., Korn, E. L., and Lengyel, J. A. (1982). Transcription, Processing and Turnover of RNA from the *Drosophila* mobile genetic element *copia*. *Dev. Biol.* **92**, 294–305.

R   Falkenthal, S., and Lengyel, J. A. (1980). Structure, translation and metabolism of the cytoplasmic *copia* ribonucleic acid of *Drosophila melanogaster*. *Biochemistry* **19**, 5842–5850.

H   Falkner, F. G., and Biessmann, H. (1980). Nuclear proteins in *Drosophila melanogaster* cells after heat-shock and their binding to homologous DNA. *Nucleic Acids Res.* **8**, 943–955.

M   Falkner, F. G., Saumweber, H., and Biessmann, H. (1981). Two *Drosophila melanogaster* proteins related to intermediate filament proteins of Vertebrate cells. *J. Cell. Biol.* **91**, 175–183.

Cdp   Fausto-Sterling, A., and Hsieh, L. (1983). The behavior during the initial phase of *in vitro* aggregation of dissociated imaginal disc cells from *Drosophila melanogaster*. *Dev. Biol.* **100**, 339–349.

Cdp   Fausto-Sterling, A., and Hsieh, L. (1987). *In vitro* culture of *Drosophila* imaginal disc cells: aggregation, sorting out and differentiative abilities. *Dev. Biol.* **120**, 284–293.

B   Fausto-Sterling, A., Muckenthaler, F. A., Hsieh, L., and Rosenblatt, P. L. (1985). Some determinants of cellular adhesiveness in an embryonic cell line from *Drosophila melanogaster*. *J. Exp. Zool.* **234**, 47–55.

HT   Feder, J. H., Rossi, J. M., Solomon, J., Solomon, N., and Lindquist, S. (1992). The consequences of expressing *hsp70* in *Drosophila* cells at normal temperatures. *Genes Dev.* **6**, 1402–1413.

F   Fehon, R. G., and Johansen, K. M. (1988). Evidence for processing of the *Notch* protein in *Drosophila* tissue culture cells and the developing nervous system (Abst. Crete Meeting, Summer 1988).

TB Fehon, R. G., Kooh, P. J., Rebay, I., Regan, C., Xu, T., Muskavitch, M. A., and Artavanis-Tsakonas, S. (1990). Molecular interactions between the protein products of the neurogenic loci *Notch* and *Delta*, two EGF-homologous genes in *Drosophila*. *Cell* 61, 523–534.

V   Fedorova, G. I., Podchernjaeva, R. Y., Amchenkova, A. M., Nikita, N. I., Blinova, V. K., Veselovkaya, O. V., and Kakpakov, V. T. (1974). (in Russian) Study of interactions of vertebrate viruses and Insect polyhedrosis virus with cultured diploid embryonic cells of *Drosophila*. *Cytologia Genet.* 8, 396–399.

E   Fernandez-Almonacid, R. (1992). The *Drosophila melanogaster* Insulin Receptor homolog: Biochemical characterization and molecular cloning. Ph.D. Thesis, Cornell University.

E   Fernandez-Almonacid, R., and Rosen, O. M. (1987). Structure and ligand specificity of the *Drosophila melanogaster* insulin receptor. *Mol. Cell. Biol.* 7, 2718–2727.

B   Fernandez, R., Tabarini, D., Azpiazu, N., Frasch, M., and Schlessinger, J. (1995). The *Drosophila* insulin receptor homolog: a gene essential for embryonic development encodes two receptor isoforms with different signaling potential. *EMBO J.* 14, 3373–3384.

M   Ferro, W. (1985). Studies on mutagen-sensitive strains of *Drosophila melanogaster*: V Biochemical characterization of a strain (*ebony*) that is UV- and X-ray sensitive and deficient in photorepair. *Mutat. Res.* 149, 399–408.

M   Fessler, L. I., Campbell, A. G., Duncan, K. G., and Fessler, J. H. (1987). *Drosophila* laminin; characterization and localization. *J. Cell Biol.* 105, 2383–2391.

rB Fessler, J. H., and Fessler, L. I. (1989). *Drosophila* extra-cellular matrix. *Ann. Rev. Cell Biol.* 5, 309–339.

B   Fessler, J. H., Lunstrum, G. P., Duncan, K. G., Campbell, A. G., Sterne, R., Bachinger, H. P., and Fessler, L. J. (1984). Evolutionary constancy of basement membrane components. *In* "The role of Extracellular Matrix in Development" (R. Trelstad, ed.), pp. 89–121, Alan Riss, New York.

rB Fessler, L. I., Nelson, R. E., and Fessler, J. H. (1994). Extracellular matrix of *Drosophila*. *Methods Enzymol.* 245, 271–294.

H   Findly, R. C., and Pederson, T. (1981). Regulated transcription of the genes for actin and heat-shock proteins in cultured *Drosophila* cells. *J. Cell Biol.* 88, 323–328.

H   Fini, M. E., and Pardue, M. L. (1988). Novel behaviors of the transcripts of *Hsr-93D* of *Drosophila melanogaster*. (Abstr. P202, UCLA Symp. Mol. Cell Biol.), *J. Mol. Biochem.* 12D (Suppl) 268.

H   Fini, M. E., Bendena, W. G., and Pardue, M. L. (1989). Unusual behavior of the cytoplasmic transcript of *hsr-omega*: an abundant, stress-inducible RNA that is translated but yields no detectable protein product. *J. Cell Biol.* 108, 2045–2057.

rR Finnegan, D. J., and Fawcett, D. H. (1986). Transposable elements in *Drosophila melanogaster*. *Oxford Surveys Eukaryotic Genes* 3, 1–62.

R   Finnegan, D. J., Rubin, G. M., Young, M. W., and Hogness, D. S. (1978). Repeated gene families in *Drosophila melanogaster*. *Cold Spring Harbor Symp. Quant. Biol.* 42, 1053–1063.

Ct Fischer, I., and Gottschewski, G. (1939). GewebeKultur bei *Drosophila*. *Naturwiss.* 27, 391–392.

R   Flavell, A. J. (1984). Role of reverse transcription in the generation of extrachromosomal *copia* mobile genetic elements. *Nature* 310, 514–516.

R Flavell, A. J., and Ish-Horowicz, D. (1981). Extrachromosomal circular copies of the eukaryotic transposable element *copia* in cultured *Drosophila* cells. *Nature* **292**, 591–595.

R Flavell, A. J., and Ish-Horowicz, D. (1983). The origin of extrachromosomal circular *copia* elements. *Cell* **34**, 415–419.

R Flavell, A. J., Levis, R., Simon, M. A., and Rubin, G. M. (1981). The 5′ termini of RNAs encoded by the transposable element *copia*. *Nucleic Acids Res.* **9**, 6279–6291.

R Flavell, A. J., Ruby, S. W., Toole, J. J., Roberts, B. E., and Rubin, G. M. (1980). Translation and developmental regulation of RNA encoded by the eukaryotic transposable element *copia*. *Proc. Natl. Acad. Sci. U.S.A.* **77**, 7107–7111.

B Fogerty, F. J., Fessler, L. I., Bunch, T. A., Yaron, Y., Parker, C. G., Nelson, R. E., Brower, D. L., Gullberg, D., and Fessler, J. H. (1994). Tiggrin, a novel *Drosophila* extracellular matrix protein that functions as a ligand for *Drosophila* $\alpha_{PS2}\beta_{PS}$ integrins. *Development* **120**, 1747–1758.

RE Fourcade-Peronnet, F. (1988). Analyse moléculaire d'un rétrotransposon dont l'expression est modulée par une hormone stéroide. *Thèse Doctorat, Université Paris VI*, 1–125.

R Fourcade-Peronnet, F., d'Auriol, L., Becker, J., Galibert, F., and Best-Belpomme, M. (1988). Primary structure and functional organization of *Drosophila 1731* retrotransposon. *Nucleic Acids Res.* **16**, 6113–6125.

R Fourcade-Peronnet, F., Codani-Simonart, S., and Best-Belpomme, M. (1992). A nuclear single-stranded-DNA-binding Factor interacts with the Long Terminal Repeat of the *1731 Drosophila* retrotransposon. *J. Virol.* **66**, 1682–1687.

REH Fourcade-Peronnet, F., Ziarczyk, P., Simonart, S., Maisonhaute, C., and Best-Belpomme, M. (1990). Regulation of the *1731* retrotransposon expression. Abst. IXth Ecdysone Workshop, Paris Sept. 1989, *Invert. Reprod. Dev.* **18**, 113.

Ct Fowler, G. (1973). *In vitro* cell differentiation in the testes of *Drosophila hydei*. *Cell Differ.* **2**, 33–42.

Ct Fowler, G., and Johannisson, R. (1976). Single-cyst *in vitro* spermatogenesis in *Drosophila hydei*. (4th Intern. Invert. Tissue Cult. 1975) *Invert. Tissue Cult., Applications in Medicine, Biology and Agriculture* (E. Kurstak and K. Maramorosch, eds.), pp. 161–172 Academic Press, New York.

Ct Fowler, G. L., and Uhlmann, J. (1974). Single-cyst *in vitro* spermatogenesis in *Drosophila hydei*. *Dros. Inf. Serv.* **51**, 81.

*Cp Fox, A. S., Horikawa, M., and Ling, L-N. L. (1968). The use of *Drosophila* cell cultures in studies of differentiation. *In Vitro* **3**, 65–84.

M Franke, A., DeCamillis, M., Zink, D., Cheng, N., Brock, H. W., and Paro, R. (1992). *Polycomb* and *Polyhomeotic* are constituents of a multimeric protein complex in chromatin of *Drosophila melanogaster*. *EMBO J.* **11**, 2941–2950.

M Frasch, M. (1985). Charakterisierung chromatinassoziierter Kernprotein von *D. melanogaster* mit Hilfe monoklonar Antikorper. Ph.D. Thesis, Eberhard Karls Universität, Tübingen, Germany.

M Frasch, M., Paddy, M., and Saumweber, H. (1988). Developmental and mitotic behaviour of two novel groups of nuclear envelope antigens of *Drosophila melanogaster*. *J. Cell Sci.* **90**, 247–263.

M Frasch, M., and Saumweber, H. (1989). Two proteins from *Drosophila* nuclei are bound to chromatin and are detected in a series of puffs on polytene chromosomes. *Chromosoma* **97**, 272–281.

* See footnote on page xxii of the Introduction to this book.

M Fredieu, J. R., and Mahowald, A. P. (1993). Characterization of a putative *Drosophila* GTP-binding protein. J. Cell Sci. **105,** 81–91.

Ct Fredieu, J. R., and Mahowald, A. P. (1994). Glycoconjugate expression during *Drosophila* embryogenesis. *Acta Anat. Basel* **149,** 89–99.

B Freeman, M. (1994). Misexpression of the *Drosophila argos* gene, a secreted regulator of cell determination. *Development* **120,** 2297–2304.

V Friesen, P. D., and Rueckert, R. R. (1981). Synthesis of Black Beetle Virus proteins in cultured *Drosophila* cells: differential expression of RNAs 1 and 2. *J. Virol.* **37,** 876–886.

V Friesen, P. D., and Rueckert, R. R. (1982). Black Beetle Virus: messenger for protein B is a subgenomic viral RNA. *J. Virol.* **42,** 986–995.

V Friesen, P. D., and Rueckert, R. R. (1984). Early and late functions in a bipartite RNA virus: evidence for translational control by competition between viral mRNAs. *J. Virol.* **49,** 116–124.

V Friesen, P., Scotti, P., Longworth, J., and Rueckert, R. (1980). Black Beetle Virus: propagation in *Drosophila* Line 1 cells and an infection-resistant subline carrying endogenous Black Beetle Virus-related particles. *J. Virol.* **35,** 741–747.

Cd Fristrom, J. W. (1968). Hexosamine metabolism in imaginal discs of *Drosophila melanogaster. J. Insect Physiol.* **14,** 729–740.

Cd Fristrom, J. W. (1972). The biochemistry of imaginal disc development. *In* "The Biology of Imaginal Disks: Results and Problems in Cell Differentiation" (H. Ursprung, and R. Nothiger eds.) Vol. 5, pp. 109–154, Springer-Verlag, Berlin.

rCd Fristrom, J. W. (1981). *Drosophila* imaginal discs as a model for the study of metamorphosis. *In* "Metamorphosis: A problem in Developmental Biology" (L. I. Gilbert and E. Frieden, eds.), 2nd ed., pp. 217–240. Springer-Verlag, Berlin.

CdE Fristrom, J. W., Chihara, C. J., Kelly, L., and Nishiura, J. T. (1976). The effects of juvenile hormone on imaginal discs of *Drosophila in vitro:* The role of the inhibition of protein synthesis. *In* "The Juvenile Hormones" (L. I. Gilbert, ed.), pp. 432–448, Plenum Press, New York.

CdE Fristrom, J. W., Doctor, J., Fristrom, D. K., Logan, W. R., and Silvert, D. J. (1982). The formation of the pupal cuticle by *Drosophila* imaginal discs in vitro. *Dev. Biol.* **91,** 337–350.

Cd Fristrom, J. W., Fristrom, D. K., Fekete, E., and Kuniyuki, A. H. (1977). The mechanism of evagination of imaginal discs of *Drosophila melanogaster. Am. Zool.* **17,** 671–684.

CdE Fristrom, J. W., Gregg, L., and Siegel, J. (1974). The effects of β-ecdysone on protein synthesis in imaginal discs of *Drosophila melanogaster* cultured in vitro. I. The effect on total protein synthesis. *Dev. Biol.* **41,** 301–313.

CdE Fristrom, J. W., and Kelly, L. (1976). Effects of β-ecdysone and juvenile hormone on the Na/K dependent ATPase activity in imaginal discs of *Drosophila melanogaster. J. Insect Physiol.* **22,** 1697–1707.

Cd Fristrom, J. W., and Knowles, B. B. (1967). Studies on protein synthesis in imaginal discs of *Drosophila melanogaster. Exp. Cell. Res.* **47,** 97–107.

CdE Fristrom, J. W., Logan, W. R., and Murphy, C. (1973). The synthetic and minimal culture requirements for evagination of imaginal discs of *Drosophila melanogaster in vitro. Dev. Biol.* **33,** 441–456.

CdE Fristrom, J. W., and Yund, M. A. (1976). Characteristics of the action of ecdysones on *Drosophila* imaginal discs cultured in vitro. *In* "Invertebrate Tissue Culture, Research and Applications" (K. Maramorosch, ed.), pp. 161–178, Academic Press, New York.

CdE Fristrom, J. W., and Yund, M. A. (1980). A comparative analysis of ecdysteroid action in larval and imaginal tissues of *Drosophila melanogaster. In* "Progress in Ecdysone

Research" (J. Hoffmann, ed.), pp. 349–362, Elsevier/North Holland Biomedical Press, Amsterdam.

Cy   Fuchs, J. P., Giloh, H., Kuo, C. H., Sauweber, H., and Sedat, J. (1983). Nuclear structure: determination of the fate of the nuclear envelope in *Drosophila* during mitosis, using monoclonal antibodies. *J. Cell Sci.* **64**, 331–349.

Cd   Fujio, Y. (1960). Studies on the development of eye-antennal discs of *Drosophila melanogaster* in tissue culture. I. Effects of the facet-increasing substances upon the growth and differentiation of eye-antennal discs. *Jpn. J. Genet.* **35**, 361–370.

Cd   Fujio, Y. (1962). *idem* II. Effects of substances secreted from cephalic complexes upon eye-antennal discs of eye-mutant strains. *Jpn. J. Genet.* **37**, 110–117.

Cp   Furst, A., and Mahowald, A. P. (1984). Rapid immunofluorescent screening procedure using primary cell cultures or tissue sections. *J. Immunol. Methods* **70**, 101–109.

Cp   Furst, A., and Mahowald, A. P. (1985a). Differentiation of primary embryonic neuroblasts in purified neural cell cultures from *Drosophila*. *Dev. Biol.* **109**, 184–192.

CpCy   Furst, A., and Mahowald, P. (1985b). Cell division cycle of cultured neural precursor cells from *Drosophila*. *Dev. Biol.* **112**, 467–476.

CpM   Fyrberg, E. A. (1978). *In vitro* differentiation of embryonic *Drosophila malanogaster* myoblasts. Ph.D. Thesis, Wesleyan University, Middletown, CT.

Cp   Fyrberg, E. A., Bernstein, S. I., and Vijay Raghavan, K. (1994). Basic methods for *Drosophila* muscle Biology. *In Methods in Cell Biology, Drosophila melanogaster.* Practical uses in Cell and Mol. Biol. (L. S. B. Goldstein and E. A. Fyrberg, eds.), Vol. 44, 237, Academic Press, New York.

CpM   Fyrberg, E. A., and Donady, J. J. (1978). Protein synthesis during the differentiation of embryonic *Drosophila melanogaster* myoblasts. *J. Cell Biol.* **79**, 33a.

CpM   Fyrberg, E. A., and Donady, J. J. (1979). Actin heterogeneity in primary embryonic culture cells from *Drosophila melanogaster*. *Dev. Biol.* **68**, 487–502.

Cp   Fyberg, E., Donady, J. J., and Bernstein, S. (1977). Isolation of myoblasts from primary mass cultures of embryonic *Drosophila* cell. *Tissue Cult. Assn. Manual* **3**, 689–690.

E   Galewsky, S., Hope, J. K., and Rickoll, W. L. (1988). The effects of monensin on 20-hydroxyecdysone-induced glycoprotein secretion and aggregation in *Drosophila S3* cells. *J. Insect Physiol.* **34**, 661–668.

E   Galewsky, S., and Rickoll, W. L. (1989). 20-hydroxyecdysone induced aggregation of *Drosophila* S3 cells is inhibited by antibodies to a hormone-dependent extracellular glycoprotein. *Roux's Arch. Dev. Biol.* **198**, 14–18.

V   Gallagher, T. M. (1987). Ph.D Thesis, University of Wisconsin, Madison, WI.

VT   Gallagher, T. M., Friesen, P. D., and Rueckert, R. R. (1983). Autonomous replication and expression of RNA I from Black Beetle Virus. *J. Virol.* **46**, 481–489.

V   Gallagher, T. M., and Rueckert, R. R. (1988). Assembly-dependent maturation cleavage in provirions of a small icosahedral Insect ribovirus. *J. Virol.* **62**, 3399–3406.

H   Garbe, J. C., Bendena, W. G., Alfano, M., and Pardue, M. L. (1986). A *Drosophila* heat-shock locus with a rapidly diverging sequence but a conserved structure. *J. Biol. Chem.* **261**, 16889.

H   Garbe, J. C., and Pardue, M. L. (1986). Heat-shock locus 93D of *Drosophila melanogaster*: a spliced RNA most strongly conserved in the intron sequence. *Proc. Natl. Acad. Sci. U.S.A.* **83**, 1812–1816.

M   Garber, M., Panchanathan, S., Fan, R. S., and Johnson, D. L. (1991). The phorbol ester 12-O-tetradecanoylphorbol-13-acetate induces specific transcription by RNA polymerase III in *Drosophila* Schneider cells. *J. Biol. Chem.* **266**, 20598–20601.

M   Garber, M. E., Vilalta, A., and Johnson, D. L. (1994). Induction of *Drosophila* DNA polymerase III gene expression by the phorbol ester 12-O-tetradecanoylphorbol-13-acetate (TPA) is mediated by transcription factor IIIB. *Mol. Cell. Biol.* 14, 339–347.

E   Garcia, J. V., Fenton, B. W., and Rosner, M. R. (1988). Isolation and characterization of an Insulin-degrading enzyme from *Drosophila melanogaster*. *Biochemistry* 27, 4237–4244.

F   Garcia, J. V., Gehm, B. D., and Rosner, M. R. (1989). An evolutionary conserved enzyme degrades transforming factor-alpha as well as insulin. *J. Cell Biol.* 109, 1301–1307.

F   Garcia, J. V., Stoppelli, M. P., Decker, S., and Rosner, M. R. (1989). An insulin EGF-binding protein from *Drosophila* has insulin-degrading activity. *J. Cell Biol.* 108, 177–182.

F   Garcia, J. V., Stoppelli, M. P., Thompson, K. L., Decker, S. J., and Rosner, M. R. (1987). Characterization of a *Drosophila* protein that binds both EGF and Insulin-related Growth Factor. *J. Cell Biol.* 105, 449–456.

M   Garza, D., Medhora, M. M., and Hartl, D. L. (1990). *Drosophila* nonsense suppressors: functional analysis in *S. cerevisiae*, *Drosophila* tissue culture and *Drosophila melanogaster*. *Genetics* 126, 625–637.

CI   Gateff, E. (1978). The genetics and epigenetics of neoplasms in *Drosophila*. *Biol. Rev.* 53, 123–168.

R   Gateff, E. (1988). Retrovirus-like particles, Reoviruses and *c-sarc* expression in malignant tumors of *Drosophila melanogaster*. "*Eukaryotic Transposable Elements as mutagenic agents.* Banbury Rep., 30, 299–307, Cold Spring Harbor.

CIV   Gateff, E., Gissmann, L., Shrestha, R., Plus, N., Pfister, H., Schröder, J. and Zur Hausen, H. (1980). Characterization of two tumorous blood cell lines of *Drosophila melanogaster* and the viruses they contain. *In "Invertebrate Systems In Vitro"* (E. Kurstak, K. Maramorosch, and A. Dübendorfer, eds.), pp. 517–533. Elsevier/North Holland Biomedical Press, Amsterdam.

B   Gauger, A., Fehon, R. G., and Schubiger, G. (1985). Preferential binding of imaginal disk cells to embryonic segments of *Drosophila*. *Nature* 313, 395–397.

T   Gavis, E. R., and Hogness, D. S. (1991). Phosphorylation, expression and function of the *ultrabithorax* protein family in *Drosophila melanogaster*. *Development* 112, 1077–1093.

TB   Gay, N. J., Poole, S., and Kornberg, T. (1988). Association of the *Drosophila melanogaster engrailed* protein with specific soluble nuclear protein complexes. *EMBO J.* 7, 4291–4297.

F   Geiser, A. B., Busam, K. J., Kim, S. J., Lafyatis, R., O'Reilly, M. A., Webbing, R., Roberts, A. G., and Sporn, M. B. (1993). Regulation of the transforming growth factor-β1 and β3 promoters by transcription factor SP1. *Gene* 129, 223–228.

B   Georgel, P., Kappler, C., Langley, E., Gross, I., Nicolas, E., Reichhart, J. M., and Hoffmann, J. A. (1995). *Drosophila* immunity. A sequence homologous to mammalian interferon consensus response element enhances the activity of the diptericin promoter. *Nucleic Acids Res.* 23, 1140–1145.

B   Georgel, P., Meister, M., Kappler, C., Lemaitre, B., Reichhart, J. M., and Hoffmann, J. A. (1993). Insect immunity: The diptericin promoter contains multiple functional regulatory sequences homologous to mammalian acute-phase response elements. *Biochem. Biophys. Res. Commun.* 197, 508–517.

rR   Georgiev, G. P. (1984). Mobile genetic elements in animal cells and their biological significance. *Eur. J. Biochem.* 145, 203–220.

R   Georgiev, G. P., Ilyin, Y. V., Chmeliauskaite, V. G., Ryskov, A. P., Kramerov, D. A., Skryabin, K. G., Krayev, A. S., Lukanidin, E. M., and Grigoryan, M. S. (1981). Mobile dispersed genetic elements and other middle repetitive DNA sequences in the genomes of *Drosophila* and Mouse: Transcription and biological significance. *Cold Spring Harbor Symp. Quant. Biol.* **45**, 641–654.

rR   Georgiev, G. P., Ilyin, Y. V., Ryskov, A. P., and Gerasimova, T. I. (1986). Mobile DNA sequences and their possible role in evolution. *In DNA Systematics"* (S. K. Dutta, ed.), Vol. 1, pp. 19–46, CRC Press, Boca Raton, FL.

MR   Georgiev, G. P., Ilyin, Y. V., Ryskov, A. P., Tchurikov, N. A., and Yenikolopov, G. N. (1977). Isolation of eukaryotic DNA fragments containing structural genes and the adjacent sequences. *Science* **195**, 394–397.

R   Georgieva, S. G., Mizrokhi, L., Krichevskaia, A. A., and Ilyin, Y. V. (1988). (in Russian) *Drosophila* mobile element *jockey* is similar to LINEs and is transcribed by RNA polymerase II from the internal promoter. *Genetika* **24**, 1353–1363.

B   Gerke, V. (1989). Consensus peptide antibodies reveal a widespread occurence of Ca^{+}/ lipid-binding proteins of the *annexin* family. *FEBS Lett.* **258**, 259–262.

Cp   Gerson, I., Seecof, R. L., and Teplitz, R. L. (1976). Ultrastructural differentiation during embryonic *Drosophila* myogenesis *in vitro*. *In Vitro* **12**, 612–622.

ME   Gertler, F. B., Chiu, C. Y., Richter-Mann, L., and Chin, D. J. (1988). Developmental and metabolic regulation of the *Drosophila melanogaster* 3-Hydroxy-3-Methylglutaryl Coenzyme A Reductase. *Mol. Cell. Biol.* **8**, 2713–2721.

F   Gertler, F. B., Hill, K. K., Clark, M. J., and Hoffmann, F. M. (1993). Dosage-sensitive modifiers of *Drosophila melanogaster* *abl* tyrosine kinase function: *prospero*, a regulator of axonal growth and *disabled*, a novel tyrosine kinase substrate. *Genes Dev.* **7**, 441–453.

H   Giardina, C., and Lis, J. T. (1993). Polymerase processivity and termination on *Drosophila* heat-shock genes. *J. Biol. Chem.* **268**, 23806–23811.

M   Giardina, C., Perez-Riba, M., and Lis, J. T. (1992). Promoter melting and TFIID complexes on *Drosophila* genes *in vivo*. *Genes Dev.* **6**, 2190–2200.

TM   Gibson, K. R., Vanek, P. G., Kaloss, W. D., Collier, G. B., Connaughton, J. F., Angelichio, M., Livi, G. P., and Fleming, P. J. (1993). Expression of Dopamine-β-Hydroxylase in *Drosophila* Schneider 2 cells. Evidence for a mechanism of a membrane binding other than uncleaved signal peptide. *J. Biol. Chem.* **268**, 9490–9495.

M   Gill, G., Pascal, E., Tseng, Z. H., and Tijan, R. (1994). A glutamin-rich hydrophobic patch in transcription factor Sp1 contacts the dTAFII110 component of the TFHHD complex and mediates transcriptional activation. *Proc. Natl. Acad. Sci. U.S.A.* **91**, 192–196.

H   Gilmour, D. S., and Elgin, S. C. R. (1987). Localization of specific Topoisomerase I interactions with the transcribed region of active heat-shock genes by using the inhibitor Camptothecin. *Mol. Cell. Biol.* **7**, 141–148.

H   Gilmour, D. S., and Lis, J. T. (1985). *In vivo* interactions of RNA polymerase II with genes of *Drosophila melanogaster*. *Mol. Cell. Biol.* **5**, 2009–2018.

H   Gilmour, D. S., and Lis, J. (1986). RNA polymerase II interacts with the promoter region of the non induced *hsp70* gene. *Mol. Cell. Biol.* **6**, 3984–3989.

H   Gilmour, D. S., Pfugfelder, G., Wang, J. C., and Lis, J. T. (1986). Topoisomerase I interacts with transcribed regions in *Drosophila* cells. *Cell* **44**, 401–407.

M   Giorgi, D., Laval, D., and Pardo, D. (1983). Amplification of the *rudimentary* gene in a PALA-resistant *Drosophila* cell line. *FEBS Lett.* **162**, 374–378.

?E   Giron, M. D., Havel, C. M., and Watson, J. A. (1993). Isopentenoid synthesis in eukayotic cells. An initiating role for post-translational control of 3-hydroxy-3-methylglutaryl coenzyme A. *Arch. Biochem. Biophys.* **302**, 265–271.

M Glover, D. M. (1977). Cloned segment of *Drosophila melanogaster* rDNA containing new types of sequence insertion. *Proc. Natl. Acad. Sci. U.S.A.* **74**, 4932–4936.

H Glover, C. V. C. (1982a). Heat-shock induces dephosphorylation of a ribosomal protein in *Drosophila*. *Proc. Natl. Acad. Sci.* **79**, 1781–1785.

H Glover, C. V. C. (1982b). Heat shock effects on protein phosphorylation in *Drosophila*. *In "Heat shock from Bacteria to Man"* (M. J. Schlesinger, M. Ashburner, and A. Tissières, eds.), pp. 227–234, Cold Spring Harbor Laboratory Press, Cold Spring Harbor.

M Glover, D. M., and Hogness, D. S. (1977). A novel arrangement of the 18S and 28S sequences in a repeating unit of *Drosophila melanogaster* rDNA. *Cell,* **10**, 167–176.

M Glover, D. M., White, R. L., Finnegan, D. J., and Hogness, D. S. (1975). Characterization of six cloned DNAs from *Drosophila melanogaster,* including one that contains the genes for rRNA. *Cell* **5**, 149–157.

M Goldring, E. S., Brutlag, D. L., and Peacock, W. J. (1975). Arrangement of highly repeated DNA in *Drosophila melanogaster*. *In "The Eukaryote Chromosome"* Australian Nat. Univ. Press, Canberra.

E Gonzales-Pacanowska, D., Arison, B., Havel, C. M., and Watson, J. A. (1988). Isopentenoid synthesis in isolated embryonic *Drosophila* cells. Farnesol catabolism and omega-oxidation. *J. Biol. Chem.* **263**, 1301–1306.

rE Goodman, W. G., and Chang, E. S. (1985). Juvenile hormone cellular and hemolymph binding proteins. *In "Comprehensive Insect Physiology, Biochemistry and Pharmacology"* (G. A. Kerkut, and L. I. Gilbert, eds.), Vol. 7, pp. 491–510. Pergamon Press, Elmsford, New York.

H Gordon, E. D., Mora, R., Meredith, S. C., and Lindquist, S. L. (1987). Hypusine formation in eukaryotic Initiation Factor 4D is not reversed when rates or specificity of protein synthesis is altered. *J. Biol. Chem.* **262**, 16590–16595.

R Gorelova, T. V., Resnick, N. L., and Schuppe, N. G. (1989). Retrotransposon transposition intermediates are encapsidated into virus-like particles. *FEBS Lett.* **244**, 307–310.

M Gossen, M., Pak, D., Acharya, J., and Botchan, M. (1995). Identification of a putative origin recognition complex in *Drosophila* embryo extracts. *Abst. Meeting on Eukaryotic DNA replication,* Cold Spring Harbor Laboratory, Sept. 1995, 3.

Cy Gottardi, A. O. (1968). Karyotype selection in cultures with different intitial cell concentrations (*Drosophila melanogaster*). *2nd Intern. Coll. Invert. Tissue Cult.* (Sett. 1987, Tremezzo, Italy), 189–200.

Cd Gottschewski, G. (1958). Uber das Wachstum von *Drosophila* Augenimaginal-scheiben *in vitro*. *Naturwiss.* **45**, 400.

Cd Gottschewski, G. (1960). Morphogenetische Untersuchungen an *in vitro* Wachsen den Augenanlagen von *Drosophila melanogaster*. *W. Roux' Arch. Entw.* **152**, 204–229.

Cd Gottschewski, G., and Fischer, I. (1939). Uber das Pigmentausbildungs vermögen von *Drosophila melanogaster* Augenanlagen *in vitro*. *Naturwiss.* **27**, 584.

Cd Gottschewski, G., and Querner, W. (1961). Beobachtung an explantierten frühen Entwicklungsstaden der Augenlage von *Drosophila melanogaster*. *W. Roux Arch. EntwMech. Org.* **153**, 168–175.

Ct Gould-Somero, M., and Holland, L. (1974). The timing of RNA synthesis for spermiogenesis in organ culture of *Drosophila* testis. *W. Roux Arch.* **174**, 133–148.

B Gow, C. H., Chang, H. Y., Lih, C. J., Chang, T. W., and Hui, C. F. (1993). Analysis of the *Drosophila* gene for the laminin B1 chain. *DNA Cell Biol.* **12**, 573–587.

B Gratecos, D., Krejci, E., and Semeriva, M. (1990). Calcium-dependent adhesion of *Drosophila* embryonic cells. *Roux's Arch. Dev. Biol.* **198**, 411–419.

B Gratecos, D., Naidet, D., Astier, M., Thierry, J. P., and Semeriva, M. (1983). *Drosophila* fibronectin: a protein that shares properties similar to those of its mammalian homologue. *EMBO J.* 7, 215–223.

Ct Greenberg, J. R. (1968). Ribosomal RNA synthesis in cultured *Drosophila* salivary glands. *J. Cell Biol.* 39, 55a.

B Grenningloh, G., Bieber, A. J., Rehm, E. J., Snow, P. M., Traquina, Z. R., Hortsch, M., Patel N. H., and Goodman, C. S. (1990). Molecular genetics of neuronal recognition in *Drosophila*: Evolution and function of immunoglobin superfamily cell adhesion molecules. *Cold Spring Harb. Symp. Quant. Biol.* 45, 327–340.

MT Grimaldi, G., and Di Nocera, P. P. (1986). Transient expression of *Drosophila melanogaster* rDNA promoter into cultured *Drosophila* cells. *Nucleic Acids Res.* 14, 6417–6432.

M Grimaldi, G., and Di Nocera, P. P. (1988). Multiple repeated units in *Drosophila melanogaster* ribosomal DNA spacer stimulate rRNA precursor transcription. *Proc. Natl. Acad. Sci. U.S.A.* 85, 5502–5506.

TM Grimaldi, G., Fiorentini, P., and Di Nocera, P. P. (1990). Spacer promoters are orientation-dependent activators of pre-rRNA transcription in *Drosophila melanogaster*. *Mol. Cell. Biol.* 10, 4667–4677.

E Gronemeyer, H., Harry, P., and Alberga, A. (1983). A reappraisal of ecdysteroid binding in *Drosophila*. *Mol. Cell. Endocrinol.* 32, 171–178.

V Guarino, L. A., Ghosh, A., Dashmahapatra, B., Dasgupta, R., and Kaesberg, P. (1984). Sequence of the Black Beetle Virus subgenomic RNA and its location in the viral genome. *Virology* 139, 199–203.

V Guarino, L. A., Hruby, D. E., Ball, L. A., and Kaesberg, G. P. (1981). Translation of Black Beetle virus RNA and heterologous viral RNA in cell-free lysate derived from *Drosophila melanogaster*. *J. Virol.* 37, 500–505.

V Guarino, L. A., and Kaesberg, P. (1981). Isolation and characterization of an RNA-dependant RNA polymerase from Black Beetle Virus-infected *Drosophila melanogaster* cells. *J. Virol.* 40, 379–386.

Cd Guillermet, C. et Mandaron, P. (1978) L'endovagination: un type particulier de développement des disques imaginaux de drosophile observé en culture *in vitro*. *CR Acad. Sci. Paris,* 287, 483–486.

CdE Guillermet, C., and Mandaron, P. (1980). *In vitro* imaginal disc development and moulting hormone. *J. Embryol. Exp. Morphol.* 57, 107–118.

BCp Gulberg, D., Fessler, L. I., and Fessler, J. H. (1994). Differentiation, extracellular matrix synthesis, and integrin assembly by *Drosophila* embryo cells cultured on vitronectin and laminin substrates. *Dev. Dyn.* 199, 116–128.

T Gumucio, D. L., Rood, K. L., Blanchard-McQuate, K. L., Gray, T. A., Saulino, A., and Collins, F. S. (1991). Interaction of *Sp1* with the human gamma-globin promoter: binding and transactivation of normal and mutant promoters. *Blood* 78, 1853–1863.

M Guo, M., Lo, P. C. H., and Mount, S. M. (1993). Species-specific signals for the splicing of a short *Drosophila* intron *in vitro*. *Mol. Cell. Biol.* 13, 1104–1118.

B Gvozdev, V. A., Birstein, V. Y., Kakpakov, V. T., and Polukarova, L. G. (1971). (in Russian) Activity of sex-linked gene in cultivated *in vitro* sublines of *Drosophila melanogaster* embryonic cell. *Ontogenez* 2, 304–310 (in English translation: Activity of a sex-linked gene in D. *melanogaster* embryonic cell sublines. 1972, *Soviet J. Dev. Biol.* 2, 243–248.)

B Gvozdev, V. A., Birstein, V. J., Polukarova, L. G., and Kakpakov, V. T. (1971). Expression of the sex-linked genes in the established aneuploid sublines of *Drosophila melanogaster*. *Dros. Inf. Serv.* 46, 68.

Cp Gvozdev, V. A. and Kakpakov, V. T. (1968a). (in Russian) Culture of embryonic cells of *Drosophila melanogaster in vitro*. *Genetika* **4**, 129–142. [in English translation (Plenum Publ. Corp., NY): *idem.*, 1972, *Soviet Genet.* **4**, 226–235].

C Gvozdev, V. A., and Kakpakov, V. T. (1968b). The medium for cell cultivation of *Drosophila melanogaster in vitro*. *Dros. Inf. Serv.* **43**, 200.

CI Gvozdev, V. A., and Kakpakov, V. T. (1970). Establishment of female embryonic cell sublines of *D. melanogaster in vitro*. *Dros. Inf. Serv.* **45**, 110.

E Gvozdev, V. A., Kakpakov, V. T., and Mukhovatova, L. M. (1973). Effect of beta-ecdysone on cell growth and synthesis of macromolecules in the established embryonic cell lines of *D. melanogaster*. *Dros. Inf. Serv.* **50**, 121.

E Gvozdev, V. A., Kakpakov, V. T., Mukhovatova, L. M., Polukarova, L. G., and Tarantul, V. Z. (1974). (in Russian) Effect of ecdysterone on cell growth and synthesis of macromolecules in the established embryonic cell lines of *Drosophila melanogaster*. *Ontogenez* **5**, 33–42. (in English translation: Influence of ecdysterone on the growth of cells and synthesis of macromolecules in established cell lines of *Drosophila melanogaster*. 1975, *Soviet J. Dev. Biol.* **5**, 29–36.)

C Gvozdev, V. A., Polukarova, L. G., and Kakpakov, V. T. (1981). (in Russian) Culture of organs, tissues, and cells *in vitro*. In *"Biochemistry and Genetics of Drosophila"* pp. 126–156, Novossibirsk Sci. Press, Novossibirsk, Sovietic Union.

V Haars, R., Zentgraf, H., Gateff, E., and Bautz, F. A. (1980). Evidence for endogenous Reovirus-like particles in a tissue culture cell line from *Drosophila melanogaster*. *Virology* **101**, 124–130.

H Haass, C., Klein, V., and Kloetzel, P. M. (1990). Developmental expression of *Drosophila melanogaster* small heat-shock proteins. *J. Cell Sci.* **96**, 413–418.

Cy Halfer, C. (1978). Karyotype evolution in an originally XY cell line of *D. melanogaster*. A case of heterochromatic increase *in vitro*. *Chromosoma* **68**, 149–163.

Cy Halfer, C., and Barigozzi, C. (1972). Prophasis synapsis in mitoses of *Drosophila* somatic cells. *Genetica Iberica* **24**, 211–232.

Cy Halfer, C., and Barigozzi, C. (1973). Prophase synapsis in somatic cells of *Drosophila melanogaster*. In *"Chromosomes Today"* (J. Wahrman and K. R. Lewis, eds.), Vol. 4, pp. 181–186, Wiley, New York.

Cy Halfer, C., and Barigozzi, C. (1977). Behaviour of the Y chromosomes during prophase synopsis in tetraploid cells of *Drosophila melanogaster* grown *in vitro*. *Caryologia* **30**, 231–235.

M Halfer, C., and Petrella, L. (1976). Cell fusion induced by lysolecithin and concanavalin A in *Drosophila melanogaster* somatic cells cultured *in vitro*. *Exp. Cell Res.* **100**, 399–404.

Cy Halfer, C., Tiepolo, L., Barigozzi, C., and Fraccaro, M. (1969). Timing of DNA replication of translocated Y chromosome sections in somatic cells of *Drosophila melanogaster*. *Chromosoma* **27**, 395–408.

Cl Cy Halfer, C., Privitera, E., and Barigozzi, C. (1980). A study of spontaneous chromosome variations in seven cell lines derived from *Drosophila* stocks marked by translocations. *Chromosoma* **76**, 201–218.

E Hall, L. M. C., Mason, P. J., and Spierer, P. (1983). Transcripts, genes and bands in 316,000 base pairs of *Drosophila* DNA. *J. Mol. Biol.* **169**, 83–96.

M Hamer, D. H., and Thomas, C. A., Jr. (1975). The cleavage of *Drosophila melanogaster* DNA by restriction endonuclease. *Chromosoma* **49**, 243–267.

T Han, K., Levine, M. S., and Manley, J. L. (1989). Synergistic activation and repression of transcription by *Drosophila* homeobox proteins. *Cell* **56**, 573–583.

T Han, K., and Manley, J. L. (1993). Transcriptional repression by the *Drosophila Even-skipped* protein: definition of a minimal repression domain. *Genes Dev.* 7, 491–503.

H Han, S., Udvardy, A., and Schedl, P. (1985). Novobiocin blocks the *Drosophila* heat-shock response. *J. Mol. Biol.* 183, 13–29.

Cd Hanly, E. W. (1961). Tissue culture of *Drosophila*. Pteridine and eye pigmentation in *Drosophila melanogaster*. *Diss. Abstr.*, 22, 980.

Cd Hanly, E. W. C., Fuller, C. W., and Stanley, M. S. (1967). The morphology and development of *Drosophila* eye I. *In vivo* and *in vitro* pigment deposition. 2. *In vitro* development of ommatidial bristles. *J. Embryol. Exp. Morphol.* 17, 491–501.

Cd Hanly, E. W., and Hemmert, W. H. (1967). Morphology and development of the *Drosophila* eye. 2. *In vitro* development of ommatidial bristles. *J. Embryol. Exp. Morphol.* 17, 501.

V Hannoun, C. (1973). Quantitative aspects of Arbovirus multiplication in a diploid cell line of *Drosophila melanogaster*. *Proc. 3rd Intern. Coll. Invertebrate Tissue Cult.*, Bratislava, Czechoslovakia (J. Rehacek, D. Blaskovic, and W. F., Hink, eds.), pp. 413–421, Publ. House Slovak Acad. Sci., Bratislava.

V Hannoun, C., and Echalier, G. (1971). Arbovirus multiplication in an established diploid cell line of *Drosophila melanogaster*. *In "Arthropod Cell Cultures and their Applications to the Study of Viruses"* (E. Weiss, ed.), pp. 227–230, Springer-Verlag, New York.

Cy Hanson, C. V. (1978). Mass isolation of metaphase chromosomes from *Drosophila melanogaster*. *In "The Genetics and Biology of Drosophila"* (M. Ashburner and T. R. F. Wright, eds.), Vol. 2a, pp. 140–145. Academic Press, New York.

BCy Hanson, C. V., and Hearst, J. E. (1974). Bulk isolation of metaphase chromosomes from an *in vitro* cell line of *Drosophila melanogaster*. *Cold Spring Harbor Symp. Quant. Biol.* 38, 341–345.

M Harris, P. V., and Boyd, J. B. (1980). Excision repair in *Drosophila*: Analysis of strand breaks appearing in DNA of *mei-9* mutants following mutagen treatment. *Biochim. Biophys. Acta* 610, 116–129.

M Harris, P. V., and Boyd, J. B. (1993). Re-evaluation of excision repair in the *mus 304, mus 306*, and *mus 308* mutants of *Drosophila*. *Mutat. Res.* 301, 51–55.

B Harrison, D. A., Binari, R., Stines Nahreini, T., Gilman, M., and Perrimon, N. (1995). Activation of a *Drosophila* Janus kinase (JAK) causes hematopoietic neoplasia and developmental defects. *EMBO J.* 14, 2857–2865.

M Hart, A. C., Krämer, H., and Zipursky, S. L. (1993). Extracellular domain of the *boss* membrane ligand acts as an antagonist of the *sev* receptor. *Nature* 361, 731–736.

E Havel, C. M., Silberklang, M., Friend, D. S., McCarthy, B. J., and Watson J. A. (1980). Sterol free cultured eukaryotic cells *Fed. Proc.* 39, 2098. [Abstract 2602]

Cp Hayashi, I., and Perez-Magallanes, M. (1994). Establishment of pure neuronal and muscle precursor cell cultures from *Drosophila* early gastrula stage embryos. *In Vitro Cell. Dev. Biol.* 30A, 202–208.

rB Hayashi, S., and Scott, M. P. (1990). What determines the specificity of action of *Drosophila* homeodomain proteins? *Cell* 63, 883–894.

TM Hayward, D. C., and Glover, D. M. (1988). Analysis of the *Drosophila* rDNA promoter by transient expression. *Nucleic Acids Res.* 16, 4253–4268.

M Heberlein, U., England, B., and Tijan, R. (1985). Characterization of *Drosophila* transcription factors that activate the tandem promoters of the alcohol dehydrogenase gene. *Cell* 41, 965–967.

R   Heine, C. W., Kelly, D. C., Avery, R. J. (1980). The detection of intracellular retrovirus-like entities in *Drosophila melanogaster* cell cultures. *J. Gen. Virol.* **49**, 385–395.

CtM   Helmsing, P. G. (1970). Protein synthesis of polytene nuclei *in vitro. Biochim. Biophys. Acta* **224**, 579.

TM   Henikoff, S., Keene, M. A., Sloan, J. S., Bleskan, J., Hards, R., and Patterson, D. (1986). Multiple purine pathway enzyme activities are encoded at a single genetic locus. *Proc Natl. Acad. Sci. U.S.A.* **83**, 720–724.

H   Henikoff, S., and Meselson, M. (1977). Transcription of two heat shock loci in *Drosophila. Cell* **12**, 441–451.

H   Hess, M. A., and Duncan, R. F. (1994). RNA/protein interactions in the 5′-untranslated leader of *hsp70* mRNA in *Drosophila* lysates. Lack of evidence for specific protein binding. *J. Biol. Chem.* **269**, 10913–10922.

TB   Heuven, van den, M., Harryman-Samos, C., Klingensmith, J., Perrimon, N., and Nusse, R. (1993). Mutations in the segment polarity genes *wingless* and *porcupine* impair secretion of the *wingless* protein. *EMBO J.* **12**, 5293–5302.

rCl   Hink, W. F (1976). Compilation of Invertebrate cell lines and culture media. *In* "Invertebrate Tissue Culture, Research, and Applications" (K. Maramorosch, ed.), p. 319, Academic Press, New York.

rCl   Hink, W. F. (1980). The 1979 compilation of Invertebrate cell lines and culture media. *In "Invertebrate Systems in Vitro"* (E. Kurstak, K. Maramorosch, and A. Dübendorfer, eds.), pp. 553–578, Elsevier, Amsterdam.

rCl   Hink, W. F., and Bezanson, D. R. (1985). Invertebrate cell culture media and cell lines. *In "Techniques in the Life Sciences. Techniques in Setting Up and Maintenance of Tissue and Cell Cultures"* (E. Kurstak, ed.), Vol. C1, pp. 1–30, Elsevier Scientific Publishers, Ireland.

B   Hirano, S., Ui, K., Miyake T., Uemura, T., and Takeichi, M. (1991). *Drosophila* PS Integrins recognize vertebrate *vitronectin* and function as cell-substratum adhesion receptors *in vitro. Development* **113**, 1007–1016.

Cy   Hirose, F., Yamaguchi, M., Handa, H., Inomata, Y., and Matsukage, A. (1993). Novel 8-base pair sequence (*Drosophila* DNA replication-related Element) and specific binding factor involved in the expression of *Drosophila* genes for Polymerase α and Proliferating Cell Nuclear Antigen. *J. Biol. Chem.* **268**, 2092–2099.

M   Hirose, F., Yamaguchi, M., and Matsukage, A. (1994). Repression of regulatory factor for *Drosophila* DNA replication-related gene promoters by *zerknullt* homeodomain protein. *J. Biol. Chem.* **269**, 2937–2942.

M   Hirose, F., Yamaguchi, M., Nishida, Y., Masutani, M., Miyazawa, H., Hanaoka, F., and Matsukage, A. (1991). Structure and expression during development of *Drosophila melanogaster* of the gene for DNA polymerase α. *Nucleic Acids Res.* **19**, 4991–4998.

Cl   Hirschi, M., and Boyd, J. B. (1981). Characterization of established cell lines derived from mutagen-sensitive *Drosophila* strains. *In Vitro* **17**, 796–804.

V   Hirumi, H. (1976). Viral, microbial, and extrinsic cell contamination of insect cell cultures. *In "Invertebrate Tissue Culture: Research Applications"* (K. Maramorosch, ed.), pp. 233–268, Academic Press, New York.

M   Hodges, D., and Bernstein, S. I. (1992). Suboptimal 5′ and 3′ splice sites regulate alternative splicing of *Drosophila melanogaster* myosin heavy chain transcripts *in vitro. Mech. Dev.* **37**, 127–140.

M   Hodges, D., and Bernstein, S. I. (1994). Preparation and use of nuclear extracts from *Drosophila* cells for *in vitro* splicing. *In "Methods in Mol. Genetics"* (K. W. Adolph, ed.), Vol. **54**, part C, 91–108. Academic Press, New York.

M   Hoey, T., Dynlacht, B. D., Peterson, M. G., Pugh, B. F., and Tijan, R. (1990). Isolation and characterization of the *Drosophila* gene encoding the TATA box binding protein, TFIID. *Cell* **61**, 1179–1186.

M   Hoey, T., Weinzierl, R. O. J., Gill, G., Chen, J. L., Dynlacht, B. D., and Tijan, R. (1993). Molecular cloning and functional analysis of *Drosophila TAF110* reveals properties expected of coactivators. *Cell* **72**, 247–260.

HE   Hoffman, E., and Corces, V. (1986). Sequences involved in temperature and ecdysterone-induced transcription are located in separate regions of a *Drosophila* heat-shock gene. *Mol. Cell. Biol.* **6**, 663–673.

H   Hogan, N. C., Traverse, K. L., Sullivan, D. E., and Pardue, M. L. (1994). The nucleus-limited *Hsr-omega-n* transcript is a polyadenylated RNA with a regulated intranuclear turnover. *J. Cell Biol.* **125**, 21–30.

TH   Holdridge, Ch., and Dorsett, D. (1991). Repression of *hsp70* heat shock gene transcription by the *suppressor of Hairy wing* protein of *Drosophila melanogaster*. *Mol. Cell. Biol.* **11**, 1894–1900.

M   Holmes, J., Jr., Clark, S., and Modrich, P. (1990). Strand-specific mismatch correction in nuclear extracts of human and *Drosophila melanogaster* cell lines. *Proc. Natl. Acad. Sci. U.S.A.* **87**, 5837–5841.

H   Holmgren, R., Corces, V., Morimoto, R., Blackman, R., and Meselson, M. (1981). Sequence homologies in the 5′ regions of four *Drosophila* heat-shock genes. *Proc. Natl. Acad. Sci. U.S.A.* **78**, 3775–3778.

Cd   Horikawa, M. (1956a). Tryptophan metabolism of the eye discs of *Drosophila melanogaster* in tissue culture (in Japanese) *Jpn. J. Gen.* **31**, 295.

Cd   Horikawa, M. (1956b). Tryptophan metabolism in the eye discs of *Drosophila melanogaster* in tissue culture. *Dros. Inf. Serv.* **30**, 122–123.

Cd   Horikawa, M. (1957). Growth, differentiation, and tryptophan metabolism in eye discs of *Drosophila melanogaster* in tissue culture. *Dros. Inf. Serv.* **31**, 124.

Cd   Horikawa, M. (1958a). Tissue culture analysis of delayed lethal irradiation effects in *Drosophila melanogaster*. *Dros. Inf. Serv.* **32**, 126–127.

Cd   Horikawa, M. (1958b). Developmental-genetic studies of tissue cultured eye discs of *D. melanogaster*. I. Growth, differentiation and tryptophan metabolism. *Cytologia* **23**, 468–477.

Cd   Horikawa, M. (1960). Developmental-genetic studies of tissue-cultured eye-antennal discs of *Drosophila melanogaster*. II. Effects of the metamorphic hormone (cephalic complex) upon growth and differentiation of eye-antennal discs, and strain differences in relation to metamorphic hormone. *Jpn. J. Genet.* **35**, 76–83.

*Cp   Horikawa, M., and Fox, A. S. (1964). *In vitro* cultivation of embryonic cells of *Drosophila melanogaster*. *Genetics* **50**, 256.

*Cp   Horikawa, M., and Fox, A. S. (1964). Culture of embryonic cells of *Drosophila melanogaster in vitro*. *Science* **145**, 1437–1439.

Cp   Horikawa, M., and Kuroda, Y. (1959). The *in vitro* cultivation of blood cells of *Drosophila melanogaster*. *Dros. Inf. Serv.* **33**, 139–140.

Cp   Horikawa, M., and Kuroda, Y. (1959). *In vitro* cultivation of blood cells of *D. melanogaster* in a synthetic medium. *Nature* **159**, 2017–2018.

*Cp   Horikawa, M., Ling, L. N., and Fox, A. S. (1966). Long-term culture of embryonic cells of *Drosophila melanogaster*. *Nature* **210**, 183–185.

---

* See footnote on page xxii of the Introduction to this book.

Cp Horikawa, M., Ling, L. N., and Fox, A. S. (1967). Effects of substrates on gene-controlled enzyme activities in cultured embryonic cells of *Drosophila*. *Genetics* **55**, 569–583.

Ct Horikawa, M., and Sugahara, T. (1960a). Studies on the effect of radiation on living cells in tissue culture I. Radiosensitivity of various imaginal discs and organs in larvae. *Radiation Res.* **12**, 266–275.

Ct Horikawa, M., and Sugahara, T. (1960b). Studies on the effect of radiation on living cells in tissue culture 2. Radiosensibility of cells isolated from various imaginal discs and organs of larvae. *Radiation Res.* **13**, 825–831.

B Hortsch, M., and Bieber, A. J. (1991). Sticky molecules in not-so-sticky cells. *Trends Biochem. Sci.* **16**, 283–287.

rB Hortsch, M., and Goodman, C. S. (1991). Adhesion molecules in *Drosophila*. *Annu. Rev. Cell Biol.* **7**, 505.

B Hortsch, M., Wang, Y. E., Mazrikar, Y., and Bieber, A. J. (1995). The cytoplasmic domain of the *Drosophila* cell adhesion molecule *Neuroglian* is not essential for its homophilic adhesive properties in *S2* cells. *J. Biol. Chem.* **270**, 18809–18817.

T Hoshijima, K., Inoue, K., Higuchi, I., Sakamoto, H., and Shimura, Y. (1991). Control of *doublesex* alternative splicing by *transformer* and *transformer-2* in *Drosophila*. *Science* **252**, 833–836.

Cp Huff, R., Furst, A., and Mahowald, A. P. (1989). *Drosophila* embryonic neuroblasts in culture: autonomous differentiation of specific neurotransmitters. *Dev. Biol.* **134**, 146–157.

B Illmensee, K. (1976). Nuclear and cytoplasmic transplantation in *Drosophila*. In "*Insect Development*" (P. A. Lawrence, ed.), Blackwell Scientific Publications, Oxford.

B Illmensee, K. (1978). *Drosophila* chimeras and the problem of determination. In "*Genetic Mosaic and Cell Differentiation. Results and Problems in Cell Differentiation*" (W. J. Gehring, ed.), Vol. 9, pp. 51–59, Springer-Verlag, Berlin.

R Ilyin, Y. V., Chmeliauskaite, V. G., Ananiev, E. V., and Georgiev, G. P. (1980a). Isolation and characterisation of mobile dispersed genetic elements, *mdg3*, in *Drosophila melanogaster*. *Chromosoma* **81**, 27–53.

R Ilyin, Y. V., Chmeliauskaite, V. G., Ananiev, E. V., Lyubomirskaya, N. V., Kulguskin, V., Bayev, A. A., and Georgiev. G. P. (1980c). Mobile dispersed genetic element MDG1 of *Drosophila melanogaster*. *Nucleic Acids Res.* **8**, 5333–5346.

R Ilyin, Y. V., Chmeliauskaite, V. G., and Georgiev, G. P. (1980b). Double-stranded sequences of *Drosophila melanogaster*: relation to mobile dispersed genes. *Nucleic Acids Res.* **8**, 3439–3457.

R Ilyin, Y. V., Chmeliauskaite, V. G., Kulguskin, V. V., and Georgiev, G. P. (1980d). Mobile dispersed genetic element MDG1 of *Drosophila melanogaster*: transcription pattern. *Nucleic Acids Res.* **8**, 5347–5361.

R Ilyin, Y. V., Karavanov, A. A., Mazo, A. M., Mizrokhi, L. J., Priimagi, A. F., Sedkov, Y. A., and Cherkasova, V. A. (1986). (in Russian) Observation of sequences in mobile elements of *Drosophila* specifically binding with nuclear proteins. *Dokl. Akad. Nauk SSSR* **290**, 720–723.

R Ilyin, Y. V., Schuppe, N. G., Lyubomirskaya, N. V., Gorelova, T. V., and Arkhipova, I. R. (1984). Circular copies of mobile dispersed genetic elements in cultured *Drosophila melanogaster* cells. *Nucleic Acids Res.* **12**, 7517–7531.

R Ilyin, Y. V., Tchurikov, N. A., Ananiev, E. V., Ryskov, A. P., Yenikolopov, G. N., Limborska, S. A., Maleeva, N. E., Gvozdev, V. A., and Georgiev, G. P. (1978). Studies on the DNA fragments of Mammals and *Drosophila* containing structural genes and adjacent sequences. *Cold Spring Harbor Symp. Quant. Biol.* **42**, 959–969.

R Ilyin, Y. V., Tchurikov, N. A., Lyubomirskaya, N. V., and Georgiev, G. P. (1979). Scattered reiterated genes of *Drosophila melanogaster* with varying location. *Genetika* **15**, 775–784.

ET Imhof, M. O., Rusconi, S., and Lezzi, M. (1993). Cloning of a *Chironomus tentans* cDNA encoding a protein (cEcRH) homologous to the *Drosophila melanogaster* ecdysteroid receptor (dEcR). *Insect Biochem. Mol. Biol.* **23**, 115–124.

MT Inoue, H., Baetge, E. E., and Hersh, L. B. (1993). Enhancer containing unusual GC box-like sequences on the human choline acetyltransferase gene. *Brain Res. Mol. Brain Res.* **20**, 299–304.

M Inoue, K., Hoshijima, K., Higuchi, I., Sakamoto, H., and Shimura, Y. (1992). Binding of the *Drosophila transformer* and *transformer-2* proteins to the regulatory elements of *doublesex* primary transcripts for sex-specific RNA processing. *Proc. Natl. Acad. Sci. U.S.A.* **89**, 8092–8096.

TM Inoue, K., Hoshijima, K., Sakamoto, H., and Shimura, Y. (1990). Binding of the *Drosophila sex-lethal* gene product to the alternative splice site of *transformer* primary transcript. *Nature* **344**, 461–463.

R Inouye, S., Hattori, K., Yuki, S., and Saigo, K. (1986). Structural variations in the *Drosophila* retrotransposons *17.6*. *Nucleic Acids Res.* **14**, 4765–4778.

B Ip, Y. T., Reach, M., Engström, Y., Kadalayil, L., Cail, H., Gonzàles-Crespo, S., Tatei, K., and Levine, M. (1993). *Dif*, a *dorsal*-related gene that mediates an immune response in *Drosophila*. *Cell* **75**, 753–763.

EH Ireland, R. C., and Berger, E. M. (1982). Synthesis of low molecular weight heat shock peptides stimulated by ecdysterone in a cultured *Drosophila* cell line. *Proc. Natl. Acad. Sci. U.S.A.* **79**, 855–859.

EH Ireland, R. C., Berger, E. M., Sirotkin, K., Yund, M. A., Osterbur, D., and Fristrom, J. (1982). Ecdysterone induces the transcription of four heat-shock genes in *Drosophila* S3 cells and imaginal discs. *Dev. Biol.* **93**, 498–507.

M Ireland, L., Szyszko, J., and Krause, M. (1982). Small nuclear RNAs from *Drosophila* Kc-H cells: characterization and comparison with mammalian RNAs. *Mol. Biol. Rep.* (The Hague) **8**, 97–101.

M Ishimi, Y., and Kikuchi, A. (1991). Identification and molecular cloning of yeast homolog of nucleosome assembly protein 1 which facilitates nucleosome assembly *in vitro*. *J. Biol. Chem.* **266**, 7025–7029.

V Ivey-Hoyle, M., Culp, J. S., Chaikin, M. A., Hellmig, B. D., Matthews, T. J., and Sweet, R. W. (1991). Envelope glycoproteins from biologically diverse isolates of Immunodeficiency viruses have widely different affinities for CD4. *Proc. Natl. Acad. Sci. U.S.A.* **88**, 512–516.

TV Ivey-Hoyle, M., and Rosenberg, M. (1990). *Rev*-dependent expression of HIV type 1 gp160 in *Drosophila melanogaster* cells. *Mol. Cell. Biol.* **10**, 6152–6159.

H Jack, R., Gehring, W., and Brack, C. (1981). Protein component from *Drosophila* larval nuclei showing sequence specificity for a short region near a major heat shock protein gene. *Cell* **24**, 321–331.

TM Jackson, M. R., Song, E. S., Yang, Y., and Petterson, P. A. (1992). Empty and peptide-containing conformers of class I major histocompatibility complex molecules expressed in *Drosophila melanogaster* cells. *Proc. Natl. Acad. Sci. U.S.A.* **89**, 12117–12121.

B Jackson, M. R., Cohen-Doyle, M. F., Peterson, P. A., and Williams, D. B. (1994). Regulation of MHC class I transport by the molecular chaperone, *Calnexin* (p88, 1P90). *Science* **263**, 384–387.

H Jackson, R. J. (1982). The cytoplasmic control of protein synthesis. *In "Protein Biosynthesis in Eukaryotes"* (R. Perez-Bercoff, ed.), pp. 363–418, Plenum Press, New York.

H Jackson, R. J. (1986). The heat-shock response in *Drosophila Kc161* cells: mRNA competition is the main explanation for reduction of normal biosynthesis. *Eur. J. Biochem.* **158**, 623–634.

Ct Jacob, J., and Sirlin, J. L. (1964). Synthesis of RNA *in vitro* stimulated in *Drosophila* salivary glands by 1,1,3-tricyano-2 amino-1 propene. *Science* **144**, 1011–1012.

V Jaronski, S. T. (1984). Microsporidia in cell culture. *Adv. Cell Culture* **3**, 183–229.

R Jarrell, K. A., and Meselson, M. (1991). *Drosophila* retrotransposon promoter includes an essential sequence at the initiation site and requires a downstream sequence for full activity. *Proc. Natl. Acad. Sci.* **88**, 102–104.

M Jarry, B. (1976). Isolation of a multifunctional complex containing the first three enzymes of pyrimidine biosynthesis in *Drosophila melanogaster*. *FEBS Lett.* **70**, 71–75.

T Jaynes, J. B., and O'Farrell, P. H. (1988). Activation and repression of transcription by homeodomain-containing proteins that bind a common site. *Nature* **336**, 744–749.

T Jaynes, J. B., and O'Farrell, P. H. (1991). Active repression of transcription by the *Engrailed* homeodomain protein. *EMBO J.* **10**, 1427–1433.

R Jensen, S. (1993). Mécanismes moléculaires et régulation de la transposition d'éléments génétiques mobiles de Drosophile marqués par des gènes indicateurs. *Thèse de Doctorat*, Université Paris 6, 1–291.

R Jensen, S., Cavarec, L., Dhellin, O., and Heidmann, T. (1994). Retrotransposition of a marked *Drosophila* Line-like I element in cells in culture. *Nucleic Acids Res.* **22**, 1484–1488.

RT Jensen, S., and Heidmann, T. (1991). An indicator gene for detection of germline retrotransposition in transgenic *Drosophila* demonstrates RNA-mediated transposition of the *LINE 1* element. *EMBO J.* **10**, 1927–1937.

T Johansen, H., Straten van der, A., Sweet, R., Otto, E., Maroni, G., and Rosenberg, M. (1989). Regulated expression at high copy number allows production of a growth-inhibitory oncogene product in *Drosophila* Schneider cells. *Genes Dev.* **3**, 882–889.

T Johanson, K., Appelbaum, E., Doyle, M., Hensley, P., Zhao, B., Abdel-Meguid, S. S., Young, P., Cook, R., Carr, S., Matico, R., Cusimano, D., Dul, E., Angelicio, M., Brooks, I., Winborne, E., McDonnell, P., Morton, T., Bennett, D., Sokoloski, T., McNulty, D., Rosenberg, M., and Chaiken, I. (1995). Binding interactions of human Interleukin 5 with its receptor alpha subunit. *J. Biol. Chem.* **270**, 9459–9471.

BT John, A., Smith, S. T., and Jaynes, J. B. (1995). Inserting the *Ftz* homeodomain into *Engrailed* creates a dominant transcriptional repressor that specifically turns off *Ftz* target genes *in vivo*. *Development* **121**, 1801–1813.

M Johnson, F. B., and Krasnow, M. A. (1990). Stimulation of transcription by an *Ultrabithorax* protein *in vitro*. *Genes Dev.* **4**, 1044–1052.

E Johnson, T. K., Brown, L. A., and Denell, R. E. (1983). Changes in cell surface proteins of cultured *Drosophila* cells exposed to 20-hydroxyecdysone *Roux's Arch. Dev. Biol.* **192**, 103–107.

CpB Joliot, A., Le Roux, I., Volovitch, M., Bloch-Gallego, E., and Prochiantz, A. (1993). (Neurotrophic activity of homeopeptide) *CR Soc. Biol.* **187**, 24–27.

M   Jordan, B. R. (1974). 2S RNA, a new ribosomal RNA component in cultured *Drosophila* cells. *FEBS Lett.* **44**, 39–42.

M   Jordan, B. R. (1975). Demonstration of intact 26 S ribosomal RNA molecules in *Drosophila* cells. *J. Mol. Biol.* **98**, 277–280.

M   Jordan, B. R., and Glover, D. M. (1977). 5.8S and 2S rDNA is located in the "transcribed spacer" region between the 18S and 26S rRNA genes in *Drosophila melanogaster. FEBS Lett.* **78**, 271–274.

M   Jordan, B. R., Jourdan, R., and Jacq, B. (1976). Late steps in the maturation of *Drosophila* 26 S ribosomal RNA; Generation of 5.8 S and 2S RNAs by cleavages occurring in the cytoplasm. *J. Mol. Biol.* **101**, 85–105.

R   Junakovic, N., and Ballario, P. (1984). Circular extrachromosomal *copia*-like transposable elements in *Drosophila* tissue culture cells. *Plasmid* **11**, 109–115.

R   Junakovic, N., Di Franco, C., Best-Belpomme, M., and Echalier, G. (1988). On the transposition of *copia*-like nomadic elements in *Drosophila* cultured cells. *Chromosoma* **97**, 212–218.

rV   Kaesberg, P. (1987). Organization of Bipartite Insect Viruses genomes: the genome of Black Beetle Virus. *In "The Molecular Biology of the Positive Strand RNA Viruses"* (D. J. Rowland, M. A. Mayo, and B. W. J. Mahy, eds.), pp. 207–218, Academic Press, New York.

r   Kakpakov, V. T. (1988). Cultivation of cells and tissues of invertebrates. *In "Methods of Cell Culture"* pp. 241–250, Leningrad Sci. Press.

rCpCl   Kakpakov, V. T. (1989). (in Russian) Problems and characterization of cultures of somatic cells of *Drosophila*. Ph.D. Thesis, Moscow, 1–47.

C   Kakpakov, V. T., and Gvozdev, V. A. (1968). Maintenance of diploid embryonic cells of *Drosophila melanogaster in vitro. Dros. Inf. Serv.* **43**, 142.

C   Kakpakov, V. T., and Gvodzev, V. A. (1974). (in Russian) Culture of cells and tissues of Insects. *In "Methods of Developmental Biology"* pp. 190–195, Moscow Sci. Press.

Cl   Kakpakov, V. T., Gvodzev, V. A., Platova, T. P., and Polukarova, L. G. (1969). (in Russian) *In vitro* establishment of embryonic cell lines of *Drosophila melanogaster. Genetika* **5**, 67–75 (in English translation: Establishment *in vitro* of embryonic cell lines of *Drosophila melanogaster.* 1972 *Soviet Genetics* **5**, 1647–1655. Plenum Publishing, New York).

Cl   Kakpakov, V. T., Gvozdev, V. A., Polukarova, L. G., Birstein, V. Ja., and Platova, T. P. (1969). (in Russian) Culture of continuous lines of *Drosophila melanogaster* embryonic cells *in vitro.* Growth pattern, karyotype and function of sex-linked gene. *2nd Conf. Dept. Radiobiol. of Kurchatov's Institute of Atomic Energy* (June 1969), **1**, 61–76.

Cy   Kakpakov, V. T., and Kakpakova, E. S. (1993). Stability and variability of the genome of embryonic cell lines during 26 years uninterrupted cultivation *in vitro. In "Cell Cultures in Biotechnology and Veterinary"* Moscow.

C   Kakpakov, V. T., and Polukarova, L. G. (1975). (in Russian) Fusion and polyploidization of Insect cells by action of Concanavalin A. *Dokl. Akad. Nauk SSR (ser. Biol.)* **223**, 209–212.

B   Kakpakov, V. T., and Polukarova, L. G. (1977). Effect of concanavalin A on established cultures of *Drosophila* and Mosquito diploid embryonic cells. *Dros. Inf. Serv.* **52**, 26.

Cl   Kakpakov, V. T., Polukarova, L. G., and Cherdanzeva, E. M. (1977). Some new embryonic cell lines in *Drosophila melanogaster. Dros. Inf. Serv.* **52**, 110.

Cy   Kakpakov, V. T., Polukarova, L. G., and Gvozdev, V. A. (1971). (in Russian) Stability and variability of karyotypes in sublines of *Drosophila melanogaster* embryonic cells cultivated *in vitro* for a long time. *Ontogenez* **2**, 295–302. (in English translation: Stability and variability of karyotype in continually cultivated sublines of *D. melanogaster* embryonic cells. 1972, *Soviet J. Dev. Biol.* **2**, 236–242. Plenum Publishing, New York).

E   Kakpakov, V. T., Shamshuzin, A. A., Spektor, V. I., and Muntjan, G. V. (1974). (in Russian) Repression of long cultivated embryonic cell lines by action of Insect hormones and cyclic AMP. *Proc. Acad. Sci. Moldav* **3**, 76–81.

ECp   Kambysellis, M. P., and Schneider, I. (1975). *In vitro* development of Insect cells. III. Effects of ecdysone on neonatal larval cells. *Dev. Biol.* **44**, 198–203.

M   Kanaar, R., Roche, S. E., Beall, E. L., Green, M. R., and Rio, D. C. (1993). The conserved pre-mRNA splicing factor U2AF from *Drosophila*: requirement for viability. *Science* **262**, 569–573.

B   Kania, A., Han, P. L., Kim, Y. T., and Bellen, H. (1993). *Neuromusculin*, a *Drosophila* gene, expressed in peripheral neuronal precursors and muscles, encodes a Cell Adhesion Molecule. *Neuron* **11**, 673–687.

B   Kappler, C., Meister, M., Lagueux, M., Gateff, E., Hoffmann, J., and Reichhart, J. M. (1993). Insect immunity: Two 17 bp repeats nesting a κB-related sequence confer inducibility to the diptericin gene and bind a polypeptide in Bacteria-challenged *Drosophila*. *EMBO J.* **12**, 1561–1568.

Cy   Kar, A., and Mukherjee, A. S. (1993). Induction and characterization of premature chromosome condensation in *Drosophila* synkaryons and implications to dosage compensation. *Indian J. Exp. Biol.* **31**, 210–214.

H   Karpov, V. L., Preobrazhenskaya, O. V., and Mirzabekov, A. D. (1984). Chromatin structure of *hsp70* genes activated by heat-shock: selective removal of histone from the coding region and their absence from the 5′ region. *Cell* **36**, 423–431.

M   Käs, E., and Laemmli, U. K. (1992). *In vivo* topoisomerase II cleavage of the *Drosophila* histone and satellite III repeats: DNA sequence and structural characteristics. *EMBO J.* **11**, 705–716.

M   Käs, E., Poljak, L., Adachi, Y., and Laemmli, U. K. (1993). A model for chromatin opening: stimulation of topoisomerase II and restriction enzyme cleavage of chromatin by distamycin. *EMBO J.* **12**, 115–126.

RT   Kaufman, P. D., Doll, R. F., and Rio, D. C. (1989). *Drosophila* P element transposase recognizes internal P element DNA sequences. *Cell* **59**, 359–371.

R   Kaufman, P. D., and Rio, D. C. (1991). *Drosophila* P element transposase is a transcriptional repressor *in vitro*. *Proc. Natl. Acad. Sci. U.S.A.* **88**, 2613–2617.

R   Kaufman, P. D., and Rio, D. C. (1992). P-element transposition *in vitro* proceeds by a cut-and-paste mechanism and uses GTP as cofactor. *Cell* **69**, 27–39.

M   Kavenoff, R., Klotz, L. C., and Zimm, B. H. (1974). On the nature of chromosome-sized DNA molecules. *Cold Spring Harbor Symp. Quant. Biol.* **38**, 1–8.

M   Kavenoff, R., and Zimm, B. H. (1973). Chromosome-sized DNA molecules from *Drosophila*. *Chromosoma* **42**, 1–27.

H   Kawata, Y., Fujiwara, H., and Ishikawa, H. (1988). Low molecular RNA of *Drosophila* cells which is induced by heat shock. I. Synthesis and its effect on protein synthesis. *Comp. Biochem. Physiol.* **91B**, 145–153.

H   Kawata, Y., Fujiwara, H., Shiba, T., Miyake, T., and Ishikawa, H. (1988). Low molecular weight RNA of *Drosophila* cells which is induced by heat shock. II. Structural properties. *Comp. Biochem. Physiol.* **91B**, 155–157.

B   Keith, F. J., and Gay, N. J. (1990). The *Drosophila* membrane receptor *Toll* can function to promote cellular adhesion. *EMBO J.* **9**, 4299–4306.

B   Kellum, R., Raff, J. W., and Alberts, B. (1995). Heterochromatin protein 1 distribution during development and during cell cycle in *Drosophila* embryos. *J. Cell Sci.* **108**, 1407–1418.

HE Kelly, S. E., and Cartwright, I. L. (1989). Perturbation of chromatin architecture on ecdysone induction of *Drosophila melanogaster* small heat shock protein genes. *Mol. Cell. Biol.* **9**, 332–335.

V   Kemble, G. W., and Mocarski, E. S. (1989). A host cell protein binds to a highly conserved sequence element (pac-2) within the cytomegalovirus a sequence. *J. Virol.* **63**, 4715–4728.

M   Kephart, D. D., Marshall, N. F., and Price, D. H. (1992). Stability of *Drosophila* RNA polymerase II elongation complexes *in vitro*. *Mol. Cell. Biol.* **12**, 2067–2077.

M   Kephart, D. D., Peele Price, M., Burton, Z. F., Finkelstein, A., Greenblatt, J., and Price, D. H. (1993). Cloning of a *Drosophila* cDNA with sequence similarity to human transcription factor RAP74. *Nucleic Acids Res.* **21**, 1319.

M   Kephart, D. D., Wang, B. Q., Burton, Z. F., and Price, D. H. (1994). Functional analysis of *Drosophila* factor 5 (TFIIF), a general transcription factor. *J. Biol. Chem.* **269**, 13536–13543.

M   Kerjean, P., Cerini, C., Semeriva, M., and Mirande, M. (1994). The multienzyme complex containing nine aminoacyl-tRNA synthetases is ubiquitous from *Drosophila* to mammals. *Biochem. Biophys. Acta* **1199**, 293–297.

M   Kerr, L. D., Ransone, L. J., Wamsley, P., Schmitt, M. J., Boyer, T. S., Zhou, Q., Berk, A. J., and Verma, I. M. (1993). Association between proto-oncoprotein *Rel* and TATA-binding protein mediates transcriptional activation by NF-kb. *Nature* **365**, 412–419.

M   Kiehart, D. P., and Feghali, R. (1986). Cytoplasmic myosin from *Drosophila melanogaster*. *J. Cell Biol.* **103**, 1517–1525.

R   Kikuchi, Y., Ando, Y., and Shiba, T. (1986). Unusual priming mechanism of RNA-directed DNA synthesis in *copia* retrovirus-like particles of *Drosophila*. *Nature* **323**, 824–826.

R   Kim, M. H., Coulondre, C., Champion, S., Lacoste, J., Best-Belpomme, M., and Maison-haute, C. (1993). Translation and fate of the *gag* protein of 1731, a *Drosophila melanogaster* retrotransposon. *FEBS Lett.* **328**, 183–188.

M   Kim, S. T., Malhotra, K., Smith, C. A., Taylor, J. S., and Sancar, A. (1994). Characterization of (6-4) photoproduct DNA photolyase. *J. Biol. Chem.* **269**, 8535–8540.

Cl Kim, Y. T., and Wu, C. F. (1991). Distinctions in growth cone morphology and motility between monopolar and multipolar neurons in *Drosophila* CNS cultures. *J. Neurobiol.* **22**, 263–275.

Cl Kim, Y. T., and Wu, C. F. (1987). Reversible blockage of neurite development and growth cone formation in neuronal cultures of a temperature-sensitive mutant of *Drosophila*. *J. Neurosci.* **7**, 3245–3255.

F   Kimchie, Z., Segev, O., and Lev, Z. (1989). Maternal and embryonic transcripts of *Drosophila* proto-oncogenes are expressed in Schneider 2 culture cells but not in *l(2)gl* transformed neuroblasts. *Cell Differ. Dev.* **26**, 79–86.

T   Kirkpatrick, R. B., Ganguly, S., Angelichio, M., Grigo, S., Shatzman, A., Silverman, C., and Rosenberg, M. (1995). Heavy chain dimers as well as complete antibodies are efficiently formed and secreted from *Drosophila* via a BIP-mediated pathway. *J. Biol. Chem.* **270**, 19800–19805.

B   Kirkpatrick, R. B., Matico, R. E., McNulty, D. E., Strickler, J. E., and Rosenberg, M. (1995). An abundantly secreted glycoprotein from *Drosophila melanogaster* is related to mammalian secretory proteins produced in rheumatoid tissues and by activated macrophages. *Gene* **153**, 147–154.

CtE Kiss, I., and Molnar, I. (1980). Metamorphic changes of wild type and mutant *Drosophila* tissues induced by 20-hydroxy ecdysone *in vitro*. *J. Insect Physiol.* **26**, 391–401.

H Kloetzel, P. M., and Bautz, K. F. (1983). Heat-shock proteins are associated with hnRNA in *Drosophila melanogaster* tissue culture cells. *EMBO J.* **2**, 705–710.

H Kloetzel, P. M., and Haass, P. C. (1988). The small HSPs of *Drosophila melanogaster* form globular cytoplasmic 16S RNP-particles. (Abstr. P O23, UCLA Symp. Mol. Cell. Biol.) *J. Cell. Biochem.* **12D** (Suppl.), 251.

V Klowden, M., and Greenberg, B. (1974). House fly and *Drosophila* cell cultures as hosts for human enteroviruses. *J. Med. Entom.* **11**, 428–432.

M Klukas, C. K., and Dawid, I. B. (1976). Characterization and mapping of mitochondrial ribosomal RNA and mitochondrial DNA in *Drosophila melanogaster*. *Cell* **9**, 615–625.

E Knibiehler, B., Bernadac, A., Mirre, C., and Rosset, R. (1982). Pattern of RNA synthesis and morphological changes induced at the nucleolar level by ecdysone treatment in a *Drosophila Kc* cell line derivative. *Biol. Cell* **46**, 239–248.

B Knibiehler, B., Mirre, C., and Rosset, R. (1982). Nucleolar organizers structure and activity in a nucleolus without fibrillar centres: The nucleolus in a *Drosophila* established cell line. *J. Cell Sci.* **57**, 351–364.

Cy Koana, T., and Miyake, T. (1982a). A histochemical method to identify the genotype of single embryo cultures of *Drosophila melanogaster*. *Jpn. J. Genet.* **57**, 79–87.

Cy Koana, T., and Miyake, T. (1982b). Histochemical determination of the genotype of primary cultures from single *Drosophila* embryos. *Dros. Inf. Serv.* **58**, 158–159.

Cp Koana, T., and Miyake, T. (1983). Effects of the sex-ratio organism on *in vitro* differentiation of *Drosophila* embryonic cells. *Genetics* **104**, 113–122.

TE Koelle, M. R., Talbot, W. S., Segraves, W. A., Bender, M. T., Cherbas, P., and Hogness, D. S. (1991). The *Drosophila EcR* gene encodes an Ecdysone Receptor, a newly member of the Steroid Receptor superfamily. *Cell* **67**, 59–77.

M Kohorn, B. D., and Rae, P. M. M. (1982a). Accurate transcription of truncated ribosomal DNA templates in *Drosophila* cell-free system. *Proc. Natl. Acad. Sci. U.S.A.* **79**, 1501–1505.

M Kohorn, B. D., and Rae, P. M. M. (1982b). Non-transcribed spacer sequences promote *in vitro* transcription of *Drosophila* ribosomal DNA. *Nucleic Acids Res.* **10**, 6879–6886.

M Koken, M., Reynolds, P., Bootsma, D., Hoeijmakers, J., Prakash, S., and Prakash, L. (1991). *Dhr 6*, a *Drosophila* homolog of the yeast DNA-repair gene *RAD6*. *Proc. Natl. Acad. Sci. U.S.A.* **88**, 3832–3836.

Cy Koval, T. M. (1983). Radiosensitivity of cultured insect cells. *Radiation Res.* **96**, 127–134.

T Krämer, H., Cagan, R. L., and Zipursky, S. L. (1991). Interaction of *bride of sevenless* membrane-bound ligand and the *sevenless* tyrosine-kinase receptor. *Nature* **352**, 207–212.

CdE Kramerov, A. A., Metakoskii, E. V., Mukha, D. V., and Gvozdev, V. A. (1981). Induction of the synthesis of specific glycoprotein in the process of metamorphosis of the imaginal discs of *Drosophila melanogaster* induced by ecdysterone. *Dokl. Akad. Nauk SSSR* **259**, 226–229.

EB Kramerov, A. A., Metakovsky, E. V., Polukarova, L. G., and Gvozdev, V. A. (1983). Glycoprotein patterns in different established cell lines of *Drosophila melanogaster* responding to 20-hydroxyecdysone. *Insect Biochem.* **13**, 655–663.

B Kramerov, A. A., Muxha, D. V., Metakovsky, E. V., and Gvozdev, V. A. (1986). Glycoproteins containing sulfated chitin-like carbohydrate moiety are synthesized in an established *Drosophila melanogaster* cell line. *Insect Biochem.* **16**, 417–432.

B Kramerov, A. A., Rozovsky, Y. M., Baikova, N. A., and Gvozdev, V. A. (1990). Cognate chitinoproteins are detected during *Drosophila melanogaster* development and in cell cultures from different Insect species. *Insect Biochem.* **20**, 769–775.

E Kraminsky, G. P., Clark, W. C., Estelle, M. A., Gietz, R. D., Sage, B. A., O'Connor, J. D., and Hodgetts, R. B. (1980). Induction of translatable mRNA for dopa-carboxylase in *Drosophila*: an early response to ecdysterone. *Proc. Natl. Acad. Sci. U.S.A.* **77**, 4175–4179.

TM Krantz, D. E., and Zipursky, S. L. (1990). *Drosophila chaoptin*, a member of the leucin-rich repeat family, is a photoreceptor cell-specific adhesion molecule. *EMBO J.* **9**, 1969–1977.

Cp Krasnow, M. A., Cumberle, D. G. E., Manning, G., Herzenberg, L. A., and Nolan, G. P. (1991). Whole animal cell sorting of *Drosophila* embryos. *Science* **251**, 81–85.

T Krasnow, M., Saffman, E., Kornfeld, K., and Hogness, D. (1989). Transcriptional activation and repression by *Ultrabithorax* proteins in cultured *Drosophila* cells. *Cell* **57**, 1031–1043.

HM Kruger, C., and Benecke, B. J. (1981). *In vitro* translation of *Drosophila* heat-shock and non heat-shock mRNAs in heterologous and homologous cell-free systems. *Cell* **23**, 595–603.

H Kruger, C., and Benecke, B. J. (1982). Translation and turnover of *Drosophila* heat-shock and non heat-shock mRNAs. In *"Heat shock from Bacteria to Man"* (M. J. Schlesinger, M. Ashburner, and A. Tissière, eds.), pp. 191–197, Cold Spring Harbor Laboratory, Cold Spring Harbor.

M Kubaneishvili, M. S., Kakpakov, V. T., and Schuppe, N. G. (1981). (in Russian) Regulation of the number of ribosomal genes in repeatedly plated cell cultures. *Dokl. Akad. Nauk SSR* **257**, 714–716.

M Kubaneishvili, M. Sh., Kakpakov, V. T., and Schuppe, N. G. (1983). (in Russian) Regulation of rRNA synthesis in *Drosophila melanogaster* embryonic established cells with various sex chromosome content. *Genetika* **19**, 1005–1012. (in English translation, *Soviets Genetics,* **19**, 786–792, Plenum Publishing, New York.)

B Kubota, K., and Gay, N. J. (1995). The *dorsal* protein enhances the biosynthesis and stability of the *Drosophila IkB* homologue cactus. *Nucleic Acids Res.* **23**, 3111–3118.

B Kubota, K., Keith, F. J., and Gay, N. J. (1993). Relocalization of *Drosophila dorsal* protein can be induced by a rise in cytoplasmic calcium concentration and the expression of constitutively active but not wild-type *Toll* receptors. *Biochem. J.* **296**, 497–503.

B Kubota, K., Keith, F. J., and Gay, N. J. (1995). Wild type and constitutively activated forms of the *Drosophila Toll* receptor have different patterns of N-linked glycosylation. *FEBS Lett.* **365**, 83–86.

CdE Kuniyuki, A. H., and Fristrom, J. W. (1977). The effect of β-ecdysone on the rate of peptide chain elongation in imaginal disks of *Drosophila melanogaster*. *Insect Biochem.* **7**, 169–174.

M Kuo, C. H., Gilon, H., Blumenthal, A. B., and Sedat, J. W. (1982). A library of monoclonal antibodies to nuclear proteins from *Drosophila melanogaster* embryos. Characterization by a cultured cell assay. *Exp. Cell Res.* **142**, 141–154.

T Kurata, N., and Marunouchi, T. (1988). Retention of autonomous replicating plasmids in cultured *Drosophila* cells. *Mol. Gen. Genet.* **213**, 359–363.

Cd   Kuroda, Y. (1954a). The culture of the eye discs of *Drosophila*. *Zool. Mag.* **63**, 75.

Cd   Kuroda, Y. (1954b). The culture of eye-discs of *Bar* mutants of *Drosophila*. *Zool. Mag.* **63**, 463.

Cd   Kuroda, Y. (1954c). The culture of wing discs of *Drosophila melanogaster*. *Jpn. J. Genet.* **29**, 163.

Cd   Kuroda, Y. (1954d). Tissue culture of wing discs of *D. melanogaster*. *Dros. Inf. Serv.* **28**, 127–128.

Ct   Kuroda, Y. (1956). Synthetic medium for the tissue culture of *Drosophila*. *Dros. Inf. Serv.* **30**, 161 162.

Cd   Kuroda, Y. (1958). Comparative study of the wing discs of *vestigial* series of *Drosophila melanogaster* in tissue culture. *Dros. Inf. Serv.* **32**, 134–135.

Cd   Kuroda, Y. (1959). (in Japanese) The tissue culture of the wing disc in the *vestigial* series of *Drosophila melanogaster*. *Med. J. Osaka Univ.* **10**, 1–16.

Cp   Kuroda, Y. (1963). *In vitro* cultivation of single cells from the embryonic blastoderm of *Drosophila melanogaster*. *Dros. Inf. Serv.* **38**, 89–90.

Cd   Kuroda, Y. (1966). *In vitro* cultivation of single cells from *Drosophila melanogaster* larvae. *Ann. Rep., Nat. Inst. Genet., Japan* **17**, 37–38.

Cd   Kuroda, Y. (1967). Histogenetic aggregation of dissociated imaginal disc cells of *Drosophila melanogaster* larvae in rotation culture. *Ann. Rep. Nat. Inst. Genet., Japan* **18**, 27–28.

Cd   Kuroda, Y. (1968a). Differentiation of ommatidium-forming cells of *Drosophila melanogaster* in culture. *Ann. Rep. Nat. Inst. Genet., Japan* **19**, 21–22.

CdE   Kuroda, Y. (1968b). Effects of ecdysone analogues on differentiation of eye-antennal discs of *Drosophila melanogaster* in culture. *Ann. Rep., Nat. Inst. Genet., Japan* **19**, 22–23.

Cp   Kuroda, Y. (1968c). Growth and differentiation of embryonic cells of *Drosophila melanogaster in vitro*. *Proc. XIIth Int. Congr. Genetics* **1**, 231.

CdE   Kuroda, Y. (1969a). The effects of ecdysone analogues on the differentiation of eye-antennal discs cultured in a chemically defined medium. *Dros. Inf. Serv.* **44**, 99–100.

Cd   Kuroda, Y. (1969b). Characteristic aggregation pattern of dissociated imaginal disc cells of *Drosophila melanogaster* larvae in rotation culture. *Dros. Inf. Serv.* **44**, 109–110.

Cp   Kuroda, Y. (1969c). (in Japanese) Growth and differentiation of embryonic cells of *Drosophila melanogaster in vitro*. *Jpn. J. Genet.* **44** (Suppl. 1), 42–50.

Ct   Kuroda, Y. (1969d). An attempt to obtain long-term culture cells from *Drosophila melanogaster*. *Ann. Rep. Nat. Inst. Genet., Japan* **20**, 28–29.

Cd   Kuroda, Y. (1970a). Differentiation of ommatidium-forming cells of *Drosophila melanogaster* in organ culture. *Exp. Cell Res.* **59**, 429–439.

Cd   Kuroda, Y. (1970b). Effects of X-irradiation on the differentiation of eye-antennal discs of *D. melanogaster* in organ culture. *Dros. Inf. Serv.* **45**, 172.

Cd   Kuroda, Y. (1970c). Effects of BUdR, actinomycin D and puromycin on the differentiation of eye-antennal discs of *D. melanogaster* in organ culture. *Dros. Inf. Serv.* **45**, 173.

Cp   Kuroda, Y. (1970d). Growth-stimulating effect of peptone on *Drosophila* ovarian cells in culture. *Ann. Rep. Nat. Inst. Genet., Japan* **21**, 42.

Ct   Kuroda, Y. (1971a). Effects of various sera and Insect blood on the growth of embryonic tissues from *Drosophila melanogaster* in culture. *Dros. Inf. Serv.* **46**, 82.

Ct   Kuroda, Y. (1971b). Effects of substances with ecdysone and juvenile hormone activity on the growth of embryonic tissues from *Drosophila melanogaster* in culture. *Dros. Inf. Serv.* **46**, 104.

Cp Kuroda, Y. (1971c). Fibroblastic cells derived from pupal ovary of *Drosophila melanogaster* in culture. *Dros. Inf. Serv.* **47**, 55.

Ct Kuroda, Y. (1971d). *In vitro* studies on spermatogenesis of *Drosophila melanogaster*. *Ann. Rep. Nat. Inst. Genet., Japan* **22**, 27.

Cp Kuroda, Y. (1971e). Growth and tissue-specificity of Drosophila embryonic cells in culture. *Ann. Rep. Nat. Inst. Genet., Japan* **22**, 28–29.

Ct Kuroda, Y. (1972). Differentiation of pupal testis in culture. *Dros. Inf. Serv.* **48**, 33.

Cp Kuroda, Y. (1973a). Macromolecular requirements of embryonic *Drosophila* cells in culture. *Proc. 3rd Coll. Invert. Tissue Culture* (J. Rehacek, D. Blaskovic and W. F. Hink, eds.) Publ. House Slovak Acad. Sci., 187–194.

Cp Kuroda, Y. (1973b). *In vitro* cultivation of developmentally defect cells from *Deep Orange* embryos in *Drosophila melanogaster*. *Ann. Rep. Nat. Inst. Genet., Japan* **24**, 19.

Cp Kuroda, Y. (1974a). Studies on *Drosophila* embryonic cells *in vitro*. I. Characteristics of cell types in culture. *Dev. Growth Different.* **16**, 55–66.

Cp Kuroda, Y. (1974b). *In vitro* activity of cells from genetically lethal embryos of *Drosophila*. *Nature* **252**, 40–41.

Ct Kuroda, Y. (1974c). Spermatogenesis in pharate adult testes of *Drosophila* in tissue cultures without ecdysone. *J. Insect Physiol.* **20**, 637–640.

Ct Kuroda, Y. (1974d). Ovarian cells from pharate adults of *Drosophila* in tissue culture. *Zool. Mag.* (Tokyo) **83**, 203–206.

Cp Kuroda, Y. (1974e). Prolonged survival of cells from lethal embryos homozygous or hemizygous for *dor* in *Drosophila melanogaster* in tissue culture. *Dros. Inf. Serv.* **51**, 30.

Cp Kuroda, Y. (1974f). Effective substances having a repair action of *dor* embryonic defects in *Drosophila melanogaster*. *Ann. Rep. Nat. Inst. Genet., Japan* **25**, 23.

Cp Kuroda, Y. (1974g). Analysis of time specificity of *dor$^+$* gene action in embryonic development of *Drosophila melanogaster*. *Ann. Rep. Nat. Inst. Genet., Japan* **25**, 23.

Cp Kuroda, Y. (1976). A tissue-specific defect in developmentally lethal cells from *fused* embryos in *D. melanogaster* in culture. *Ann. Rep. Nat. Inst. Genet. Japan* **26**, 28.

Cp Kuroda, Y. (1977a). Studies on *Drosophila* embryonic cells *in vitro*. II. Tissue- and time-specificity of a lethal gene, *deep orange*. *Dev. Growth Differ.* **19**, 57–66.

Cp Kuroda, Y. (1977b). *In vitro* analysis of tissue-specific defects in *rudimentary* embryonic cells of *Drosophila melanogaster*. *Ann. Rep. Nat. Inst. Genet. Japan* **27**, 38.

Cp Kuroda, Y. (1978). Pyrimidine requirements of cultured *rudimentary* embryonic cells of *Drosophila melanogaster*. *Ann. Rep. Nat. Inst. Genet. Japan* **28**, 43–44.

Cp Kuroda, Y. (1979). Comparison of effective lethal phases of embryos from three complementation groups of the *rudimentary* of *Drosophila melanogaster*. *Ann. Rep. Nat. Inst. Genet., Japan* **29**, 35.

Ct Kuroda, Y. (1980). *In vitro* cultivation of gonads from a female sterile mutant *fs231* of *Drosophila melanogaster*. *Ann. Rep. Nat. Inst. Genet., Japan* **30**, 45.

Ct Kuroda, Y. (1981a). The growth and differentiation of germ cells of *Drosophila melanogaster* in culture. *Ann. Rep. Nat. Inst. Genet., Japan* **31**, 43.

Cp Kuroda, Y. (1981b). *In vitro* differentiation of adult structures from embryonic cells of *Drosophila melanogaster*. *Ann. Rep. Nat. Inst. Genet., Japan* **32**, 49.

rC Kuroda, Y. (1982a). *Drosophila* tissue culture: Retrospect and prospect. *In "Invertebrate Cell Culture Applications"* (K. Maramorosch and J. Mitsuhashi, eds.), pp. 53–104, Academic Press, New York.

Cp Kuroda, Y. (1982b). Differentiation of adult structures from *Drosophila* embryonic cells in culture. *In "The Ultra-structure and Functioning of Insect Cells"* (H. Akai, R. C. King, and S. Morohoshi, eds.), p. 91, Soc. for Insect Cells, Japan.

Cp   Kuroda, Y. (1982c). *In vitro* formation of adult structures from cells of lethal embryos of *Drosophila melanogaster. Ann. Rep. Nat. Inst. Genet., Japan* 33, 29.

Cp   Kuroda, Y. (1984). *In vitro* formation of adult structures from lethal embryonic cells of *Drosophila melanogaster. Ann. Rep. Nat. Inst. Genet., Japan* 35, 35.

Ct   Kuroda, Y. (1986a). *In vitro* studies on the spermatogenesis of *Drosophila melanogaster. In "Techniques in the Life Sciences C2. In vitro Invertebrate Hormones and Genes"* C214, 1–6, Elsevier Science Publishers, Ireland.

Cp   Kuroda, Y. (1986b). Differentiation of adult structures in cultures of embryonic tissues from *Drosophila melanogaster. In "Techniques in the Life Sciences"* C215, 1 6, Elsevier Science Publishers, Ireland.

Cp   Kuroda, Y. (1988). Artificial expression of tissue-specific characteristics in embryonic lethal cells of *Drosophila melanogaster. Ann. Rep. Nat. Inst. Genet., Japan* 39, 63.

rC   Kuroda, Y. (1994). Japanese invertebrate cell culture pioneers. *In "Arthropod Cell Culture Systems"* (K. Maramorosch and A. H. McIntosh, eds.) pp. 193–215, CRC Press, Boca Raton, FL.

CdE Kuroda, Y., and Minato, K. (1967). (in Japanese) Differentiation of imaginal discs of *Drosophila melanogaster* cultured in synthetic medium and effects of some hormonal substances. *Ann. Rep. Nat. Inst. Genet., Japan* 18, 28–29.

V   Kuroda, Y., Oichi, K., and Shimada, Y. (1989). Detection of SR spiroplasms in embryonic tissues of *Drosophila melanogaster* by an electron microscope. *Ann. Rep. Nat. Inst. Genet., Japan* 40, 59.

Cp   Kuroda, Y., Sakaguchi, B., and Oishi, K. (1988). Studies on sex differentiation of embryonic cells of *Drosophila melanogaster. Ann. Rep. Nat. Inst. Genet., Japan* 38, 65.

V   Kuroda, Y., Sakagushi, B., Oishi, K., and Shimada, Y. (1988). Selective infection of SR spiroplasms in cultured embryonic cells of *Drosophila melanogaster. Ann. Rep. Nat. Inst. Genet., Japan* 39, 64.

Cp   Kuroda, Y., and Shimada, Y. (1983). Electron microscopic studies on the ultrafine structures of *in vitro* differentiated cells of *Drosophila melanogaster. Ann. Rep. Nat. Inst. Genet., Japan* 34, 37.

Cp   Kuroda, Y., and Shimada, Y. (1988). Differentiation of embryonic cells of *Drosophila* studied with electron microscope. *In "Invertebrate and Fish Tissue Cultures"* (Y. Kuroda, E. Kurstak, and K. Maramorosch, eds.), pp. 95–99, Springer Verlag, New York.

Cp   Kuroda, Y., and Shimada, Y. (1989). Electron microscopic studies on *in vitro* differentiated cells from *Drosophila* embryos. *In "Invertebrate Cell System Applications"* (J. Mitsuhashi, ed.), Vol. 1, pp. 77–89, CRC Press, Boca Raton, FL.

V   Kuroda, Y., Shimada, Y., Sakaguchi, B., and Oichi, K. (1992). Effects of sex-ratio (SR)–spiroplasma infection on *Drosophila* primary embryonic cultured cells and on embryogenesis. *Zool. Sci.* (Zool. Soc. Japan) 9, 283–291.

CtE Kuroda, Y., and Tamura, S. (1955a). The tissue culture of tumors in *Drosophila melanogaster.* I. The nature of the melanotic tumors and the effect of metamorphic hormone upon the melanotic growth of tumors. *Zool. Mag.* (Tokyo) 64, 380–384.

CtE Kuroda, Y., and Tamura, S. (1955b). Effect of metamorphic hormone on the growth of melanotic tumors in *Drosophila melanogaster in vitro. Dros. Inf. Serv.* 29, 133.

Ct   Kuroda, Y., and Tamura, S. (1956a). A technique for the tissue culture of melanotic tumors of *Drosophila melanogaster* in the synthetic medium. *Med. J. Osaka Univ.* 7, 137–144.

Ct   Kuroda, Y., and Tamura, S. (1956b). The tissue culture of tumors in *Drosophila melanogaster.* II. The effect of phenylthiocarbamide upon the melanotic growth of tumors. *Zool. Mag.* (Tokyo) 65, 219–222.

Ct Kuroda, Y., and Tamura, S. (1956c). Effect of DDC (diethyldithiocarbamate) on the melanotic growth of tumors in *Drosophila melanogaster* in tissue culture. *Dros. Inf. Serv.* **30**, 126.

Cp Kuroda, Y., and Tamura, S. (1956d). The tissue culture of tumors in *D. melanogaster*. III. The effects of Cu-ion upon the melanotic growth of tumors. *Zool. Mag.* (Tokyo) **65**, 301–305.

Cp Kuroda, Y., and Tamura, S. (1956e). Effects of Fe^{+++} on the melanotic growth of tumors in *D. melanogaster* in tissue culture. *Dros. Inf. Serv.* **30**, 125–127.

Cp Kuroda, Y., and Tamura, S. (1957). The tissue culture of tumors in *D. melanogaster*. IV. The effects of Fe^{+++} upon the melanotic growth of tumors. *Zool. Mag.* (Tokyo) **66**, 6–10.

Ct Kuroda, Y., and Tamura, S. (1958). Resistance to parathion of various organs in *D. melanogaster* in tissue culture. *Dros. Inf. Serv.* **32**, 135.

Cd Kuroda, Y., and Yamaguchi, K. (1955). Effects of the cephalic complex on the eye discs of *D. melanogaster in vitro*. *Dros. Inf. Serv.* **29**, 133–134.

CdE Kuroda, Y., and Yamaguchi, K. (1956). The effects of the cephalic complex upon the eye discs of *Drosophila melanogaster*. *Jpn. J. Genet.* **31**, 98–103.

F Kussick, S. J., and Cooper, J. A. (1992), Overexpressed *Drosophila src64B* is phosphorylated at its carboxy-terminal tyrosine, but is not catalytically repressed, in cultured *Drosophila* cells. *Oncogene* **7**, 2461–2470.

FT Kutoh, E., Margot, J. B., and Schwander, J. (1993). Genomic structure and regulation of the promoter of the Rat Insulin-like Growth Factor binding protein-2 gene. *Mol. Endocrinol.* **7**, 1205–1216.

M Kutoh, E., and Schwander, J. (1993). Sp1 interacts with the consensus sequence for *Egr-1* gene product with a cellular factor(s) and activates the transcription through this element. *Biochem. Biophys. Res. Commun.* **194**, 1475–1482.

R Lacoste, J. (1995). De la régulation du rétrotransposon 1731 de *Drosophila melanogaster*. Rôle de l'élément NssBF; caractérisation d'un transrépresseur interagissant spécifiquement avec cet élément. *Thèse Doctorat*, Univ. Paris VI (France).

R Lacoste, J., Codani-Simonart, S., Best-Belpomme, M., and Peronnet, F. (1995). Characterization and cloning of *p11*, a transrepressor of *Drosophila melanogaster* retrotransposon *1731*. *Nucleic Acids Res.* **23**, 5073–5079.

M Laird, C. D., Chooi, W. Y., Cohen, E. H., Dickson, E., Hutchinson, N., and Turner, S. H. (1974). Organization and transcription of DNA in chromosomes and mitochondria of *Drosophila*. *Cold Spring Harb. Symp. Quant. Biol.* **38**, 311–327.

Cy Lajoie-Mazenc, I., Tollon, Y., Destrave, C., Julian, A., Moisand, C., Gueth-Hallonet, A., Debec, A., Salles-Passador, I., Puget, A., Mazarguil, H., Raynaud-Messina, B., and Wright, M. (1994). Recruitment of antigenic gamma-tubulin during mitosis in animal cells: presence of gamma-tubulin in the mitotic spindle. *J. Cell Sci.* **107**, 2824–2837.

Ct Lakhotia, S. C., and Mukherjee, A. S. (1970). Activation of a specific puff by benzamide in *Drosophila melanogaster*. *Dros. Inf. Serv.* **45**, 108.

E Landon, T. M., Sage, B. A., Seeler, B. J., and O'Connor, J. D. (1988). Characterization and partial purification of the *Drosophila Kc* cell ecdysteroid receptor. *J. Biol. Chem.* **263**, 4693–4697.

HET Larocca, D. (1986). Ecdysterone and heat shock induction of transfecting and endogenous heat shock genes in cultured *Drosophila* cells. *J. Mol. Biol.* **191**, 563–567.

M Laval, D., Azou, M., Giorgi, D., and Rosset, R. (1986). Overproduction of the first three enzymes of Pyrimidine Nucleotide Biosynthesis in *Drosophila* cells resistant to N-phosphonacetyl-L-aspartate. *Exp. Cell Res.* **163**, 381–395.

EHT   Lawson, R., Mestril, R., Luo, Y., and Voellmy, R. (1985). Ecdysterone selectively stimulates the expression of a 23,000-Da heat-shock protein-galactosidase hybrid gene in cultured *Drosophila* cells. *Dev. Biol.* **110**, 321–330.

HT   Lawson, R., Mestril, R., Schiller, P., and Voellmy, R. (1984). Expression of heat shock-β-galactosidase hybrid genes in cultured *Drosophila* cells. *Mol. Gen. Genet.* **198**, 116–124.

TM   Le, H. B., Vaisanen, P. A., Johnson, J. L., Raney, A. K., and McLachlan, A. (1994). Regulation of transcription from the human muscle phosphofructokinase P2 promoter by the Sp1 transcription factor. *DNA Cell Biol.* **13**, 173–185.

E   Lee, K. (1990). The identification and characterization of ecdysone response elements. Ph.D. Thesis, Indiana University.

TB   Leeuwen, van, F. *et al.* (see Van Leeuwen)

H   Lefrère, V., and Duncan, R. F. (1994). Heat shock-induced repression of proteolysis: poly(A)-binding protein degradation patterns can illusorily suggest its specific loss during heat shock. *Nucleic Acids Res.* **22**, 1640–1642.

H   Leicht, B., Biessmann, H., and Bonner, J. J. (1985). (Abst.) The small heat shock proteins of *Drosophila* associate with the cytoskeleton. *In "Abstr. 1985 Meeting on Heat shock"* (E. Craig *et al.*, eds.), p. 60, Cold Spring Harbor Laboratory, Cold Spring Harbor.

H   Leicht, B. G., Biessmann, H., Palter, K. B., and Bonner, J. J. (1986). Small heat shock proteins of *Drosophila melanogaster* associate with the cytoskeleton. *Proc. Natl. Acad. Sci. U.S.A.* **83**, 90–94.

CdE   Lemaire, M. F., and Thummel, C. S. (1990). Splicing precedes polyadenylation during *Drosophila* E74A transcription. *Mol. Cell. Biol.* **10**, 6059–6063.

H   Lengyel, J. A., and Pardue, M. L. (1975). Analysis of hnRNA made during heat-shock in *Drosophila melanogaster* cultured cells. *J. Cell Biol.* **67**, 240a.

M   Lengyel, J. A., and Penman, S. (1975). hnRNA size and processing as related to different DNA content in two Dipterans: *Drosophila* and *Aedes*. *Cell* **5**, 281–290.

M   Lengyel, J. A., and Penman, S. (1977). Differential stability of cytoplasmic RNA in a *Drosophila* cell line. *Dev. Biol.* **57**, 243–253.

H   Lengyel, J. A., Ransom, L. J., Graham, M. L., and Pardue, M. L. (1980). Transcription and metabolism of RNA from the *Drosophila melanogaster* heat-shock puff site 93D. *Chromosoma* **80**, 237–252.

CMR   Lengyel, J. A., Spradling, A., and Penman, S. (1975). Methods with insect cells in suspension culture. II. *Drosophila melanogaster*. *In "Methods in Cell Biology"* (D. M. Prescott, ed.), Vol. 10, pp. 195–208, Academic Press, New York.

B   Leptin, M., Aebersold, R., and Wilcox, M. (1987). *Drosophila* position-specific antigens resemble the Vertebrate fibronectin-receptor family. *EMBO J.* **6**, 1037–1043.

M   Le Roux, I., Joliot, A. H., Bloch-Gallego, E., Prochiantz, A., and Volovitch, M. (1993). Neurotrophic activity of the *Antennapedia* homeodomain depends on its specific DNA-binding properties. *Proc. Natl. Acad. Sci. U.S.A.* **90**, 9120–9124.

R   Lescault, A., Becker, J. L., Barré-Sinoussi, F., Chermann, J. C., Best-Belpomme, M., and Ono, K. (1989). Characterization of a reverse transcriptase activity associated with retrovirus-like particles in a *Drosophila* cell line. *Cell. Mol. Biol.* **35**, 163–171.

Cp   Lesseps, R. J. (1965). Culture of dissociated *Drosophila* embryos: Aggregated cells differentiate and sort out. *Science* **148**, 502–503.

Cp   Leung, H. T., and Byerly, L. (1987). Calcium currents in embryonic cultures of *Drosophila* neurons. *Soc. Neurosci. Abst.* **13**, 101.

H Levinger, L., and Varshavsky, A. (1981). Heat-shock proteins of *Drosophila* are associated with nuclease-resistant, high-salt-resistant nuclear structures. *J. Cell Biol.* **90**, 793–796.

H Levinger, L., and Varshavsky, A. (1982a). Selective arrangement of variant nucleosomes within the *Drosophila melanogaster* genome and the heat-shock response. In *"Heatshock from Bacteria to Man"* (M. J. Schlesinger, M. Ashburner, and A. Tissière, eds.), pp. 115–120, Cold Spring Harbor Laboratory, Cold Spring Harbor.

H Levinger, L., and Varshavsky, A. (1982b). Selective arrangement of ubiquinated and D1 protein-containing nucleosomes within the *Drosophila* genome. *Cell* **28**, 375–385.

M Levinger, L., Vasisht, V., Greene, V., and Arjun, I. (1992). The effect of stem I and loop A on the processing of 5S rRNA from *Drosophila melanogaster*. *J. Biol. Chem.* **267**, 23683–23687.

M Levinger, L., Vasisht, V., Greene, V., Bourne, R., Birk, A., and Kolla, S. (1995). Sequence and structure requirements for *Drosophila* tRNA5′- and 3′-end processing. *J. Biol. Chem.* **270**, 18903–18909.

M Levis, R., and Penman, S. (1977). The metabolism of poly(A)$^+$ and poly(A)$^-$ hnRNA in cultured *Drosophila* cells studied with a rapid uridine pulse-chase. *Cell* **11**, 105–113.

M Levis, R., and Penman, S. (1978a). 5′-terminal structure of poly(A)$^+$ cytoplasmic messenger RNA and of poly(A)$^+$ and poly(A)$^-$ heterogeneous nuclear RNA of cells of the Dipteran *Drosophila melanogaster*. *J. Mol. Biol.* **120**, 105–113.

M Levis, R., and Penman, S. (1978b). Processing steps and methylation in the formation of the ribosomal RNA of cultured *Drosophila* cells. *J. Mol. Biol.* **121**, 219–238.

HM Levy, A., and Noll, M. (1981). Chromatin fine structure of actin and repressed genes. *Nature* **289**, 198–203.

H Levy, A., and Noll, M. (1982). Chromatin structure of *hsp70* genes of *Drosophila*. In *"Heat-shock from Bacteria to Man"* (J. Schlesinger, M. Ashburner, and A. Tissières, eds.), pp. 99–107, Cold Spring Harbor Laboratory Press, Cold Spring Harbor.

M Levy, W. B., and McCarthy, B. J. (1975). Messenger RNA complexity in *Drosophila melanogaster*. *Biochemistry* **14**, 2440–2446.

rH Lewin, B. (1975). Specific response of *Drosophila* cells to heat shock. *Nature* **255**, 276–277.

CtH Lewis, M., Helmsing, P. J., and Ashburner, M. (1975). Parallel changes in puffing activity and patterns of protein synthesis in salivary glands of *Drosophila*. *Proc. Natl. Acad. Sci. U.S.A.* **72**, 3604–3608.

E Lezzi, M., and Wyss, C. (1976). The antagonism between juvenile hormone and ecdysone. In *"The Juvenile Hormones"* (L. I. Gilbert, ed.), pp. 252–269, Plenum Press, New York.

H Li, D., and Duncan, R. F. (1995). Transient acquired thermotolerance in *Drosophila*, correlated with rapid degradation of *Hsp 70* during recovery. *Eur. J. Biochem.* **231**, 454–465.

V Li, Y., and Ball, L. A. (1993). Nonhomologous RNA recombination during negative-strand synthesis of FlockHouse Virus RNA. *J. Virol.* **67**, 3854–3860.

M Lieber, T., Kidd, S., Alcamo, E., Corbin, V., and Young, M. W. (1993). Antineurogenic phenotypes induced by truncated *Notch* proteins indicate a role in signal transduction and may point to a novel function of *Notch* in nuclei. *Genes Dev.* **7**, 1949–1965.

M Lieber, T., Wesley, C. S., Alcamo, E., Hassel, B., Krane, J. F., Campos-Ortega, J. A., and Young, M. W. (1992). Single amino acid substitutions in EGF-like elements of *Notch* and *Delta* modify *Drosophila* development and affect cell adhesion *in vitro*. *Neuron* **9**, 847–859.

Ct   Leibrich, W. (1981). *In vitro* differentiation of single cysts of spermatocytes of *Drosophila hydei*. *Dros. Inf. Serv.* **56**, 82–84.

Ct   Liebrich, W. (1981). *In vitro* spermatogenesis in *Drosophila*. I. Development of isolated spermatocyte cysts from wild-type *D. hydei*. *Cell Tissue Res.* **220**, 251–262.

Ct   Liebrich, W., and Kociok, N. (1988). The possible role of the Y chromosome during male germ cell differentiation in *Drosophila*. *In "Invert. Fish Tissue Culture"* (Y. Kuroda, E. Kurstak, and K. Maramorosch, eds.) pp. 104–106, Springer Verlag, New York.

MT   Lienhard-Schmitz, M., and Beauerle, P. A. (1991). The p65 subunit is responsible for the strong transcription activating potential of NF-κB. *EMBO J.* **10**, 3805–3817.

H   Lindquist McKenzie, S. (1977). Translation control of protein synthesis in *Drosophila* (Abst. Ann. Meeting Amer. Soc. Cell Biol., RT346) *J. Cell Biol.* **75**.

H   Lindquist, S. (1980a). Varying patterns of protein synthesis in *Drosophila* during heat shock: implications for regulation. *Dev. Biol.* **77**, 463–479.

H   Lindquist, S. (1980b). Translation efficiency of heat-induced messages in *Drosophila melanogaster* cells. *J. Mol. Biol.* **137**, 151–158.

H   Lindquist, S. (1981). Regulation of protein synthesis during heat shock. *Nature* **293**, 311–314.

rH   Lindquist, S. (1986). The heat shock response. *Annu. Rev. Biochem.* **55**, 1151–1191.

rH   Lindquist, S., and Craig, E. A. (1988). The heat-shock proteins. *Annu. Rev. Genet.* **22**, 631–677.

H   Lindquist, S., and Didomenico, B. (1985). Coordinate and noncoordinate gene expression during heat-shock: a model for regulation. *In "Changes in Eukaryotic Gene Expression in Response to Environmental Stress"* (B. G. Atkinson and D. B. Walden, eds.), pp. 71–90, Academic Press, New York.

rH   Lindquist, S., Didomenico, B., Bugaisky, G., Kurtz, S., Petko, L., and Sonoda, S. (1982). Regulation of the heat-shock response in *Drosophila* and Yeast. *In "Heat Shock from Bacteria to Man"* (M. J. Schlesinger, M. Ashburner, and A. Tissière, eds.), pp. 167–175, Cold Spring Harbor Laboratory, Cold Spring Harbor.

H   Lindquist, S., McGarry, T. J., and Golic, M. (1988). Use of antisense RNA in studies of the heat-shock response. *Curr. Commun. Mol. Biol., CSH Lab.* 71–77.

H   Lindquist-McKenzie, S., Henikoff, S., and Meselson, M. (1975). Localization of RNA from heat-induced polysomes at puff sites in *Drosophila melanogaster*. *Proc. Natl. Acad. Sci. U.S.A.* **72**, 1117–1121.

H   Lindquist-McKenzie, S., and Meselson, M. (1977). Translation *in vitro* of *Drosophila* heat-shock messages. *J. Mol. Biol.* **117**, 279–283.

C   Lindquist, S. L., Sonoda, S., Cox, T., and Slusser, K. (1982). Instant medium for *Drosophila* tissue culture cells. *Dros. Inf. Serv.* **58**, 163–164.

Cp   Ling, L. N. L., Horikawa, M., and Fox, A. S. (1966). Aggregation of dissociated *Drosophila* embryonic cells. *Dros. Inf. Serv.* **41**, 108.

Cp   Ling, L. N., Horikawa, M., and Fox, A. S. (1970). Aggregation of dissociated cells from *Drosophila* embryos. *Dev. Biol.* **22**, 264–281.

H   Lis, J. T., Neckameyer, W., Dubensky, R., and Costlow, N. (1981). Cloning and characterization of nine heat-shock-induced mRNAs of *Drosophila melanogaster*. *Gene* **15**, 67–80.

H   Lis, J. T., Prestidge, L., and Hogness, D. S. (1978). A novel arrangement of tandemly repeated genes at a major heat-shock site in *Drosophila melanogaster*. *Cell* **14**, 901–919.

H   Livak, K. J., Freund, R., Schweber, M., Wensink, P. C., and Meselson, M. (1978). Sequence organization and transcription at two heat-shock loci in *Drosophila. Proc. Natl. Acad. Sci. U.S.A.* **75**, 5613–5617.

CdE   Logan, W., Fristrom, D., and Fristrom, J. W. (1975). Effects of ecdysone and juvenile hormone on DNA metabolism of imaginal discs of *Drosophila melanogaster. J. Insect Physiol.* **21**, 1343–1354.

B   Londershausen, M., Kamman, V., Spindler-Barth, M., Spindler, K. D., and Thomas, H. (1988). Chitin synthesis in Insect cell lines. *Insect Biochem.* **18**, 631–636.

B   Lopez, J. M., Song, K., Hirshfeld, A. B., Lin, H., and Wolfner, M. F. (1994). The *Drosophila fs(1)Ya* protein, which is needed for the first mitotic division, is in the nuclear lamina and in the envelopes of cleavage nuclei, pronuclei and nonmitotic nuclei. *Dev. Biol.* **163**, 202–211.

H   Love, J. D., and Minton, K. W. (1985). Screening of lambda library for differentially expressed genes using *in vitro* transcripts. *Analyt. Biochem.* **150**, 429–441.

H   Love, J. D., Vivino, A. A., and Minton, K. W. (1985). Detection of low level gene induction, using *in vitro* transcription: heat-shock genes. *Gene Anal. Technol.* **2**, 100–107.

H   Love, J. D., Vivino, A. A., and Minton, K. W. (1986). Hydrogen peroxide toxicity may be enhanced by heat shock gene induction in *Drosophila. J. Cell. Physiol.* **126**, 60–68.

H   Lubsen, N. H., Sondermeijer, P. J. A., Pages, M., and Alonso, C. (1978). *In situ* hybridization of nuclear and cytoplasmic RNAs to locus 2-48BC in *Drosophila hydei. Chromosoma* **65**, 199–212.

M   Luchnik, A. N., Hisamutdinov, T. A., and Georgiev, G. P. (1988). Inhibition of transcription in eukaryotic cells by X-irradiation: relation to the loss of topological constraint in closed DNA loops. *Nucleic Acids Res.* **16**, 5175–5190.

Cp   Lüer, K., and Technau, G. M. (1992). Primary culture of single ectodermal precursors of *Drosophila* reveals a dorso-ventral pre-pattern of intrinsic neurogenic and epidermogenic capabilities of the early gastrula stage. *Development* **116**, 377–385.

M   Lunstrum, G. P., Bachinger, H. P., Fessler, L. I., Duncan, K. G., Nelson, R. E., and Fessler, J. H. (1988). *Drosophila* basement membrane procollagen IV. I. Protein characterization and distribution. *J. Biol. Chem.* **263**, 18318–18327.

M   Lunstrum, G. P., and Fessler, J. H. (1980). Cell adhesion in Invertebrates: a potential basement membrane collagen of *Drosophila melanogaster*. Abst. 484, 9th Ann. ICN-UCLA Symp. Mol. Cell. Biol., *J. SupraMol. Struct.* **4** (Suppl.), 183.

TEH   Luo, Y., Amin, J., and Voellmy, R. (1991). Ecdysterone receptor is a sequence-specific transcription factor involved in the developmental regulation of heat-shock genes. *Mol. Cell. Biol.* **11**, 3660–3675.

R   Lyubomirskaya, N. V., Arkhipova, I. R., and Ilyin, Y. V. (1993). Transcription of *Drosophila* mobile element *Gypsy* (*mdg4*) in heat-shocked cells. *FEBS Lett.* **325**, 233–236.

R   Lyubomirskaya, N. V., Avedisov, S. N., Surkov, S. A., and Ilyin, Y. V. (1993). Two *Drosophila* retrotransposon *gypsy* subfamilies differ in ability to produce new DNA copies via reverse transcription in *Drosophila* cultured cells. *Nucleic Acids Res.* **21**, 3265–3268.

M   McCarthy, B. J., Nishiura, J. T., Doenecke, D., Nasser, D. S., and Johnson, C. B. (1974). Transcription and chromatin structure. *Cold Spring Harbor Symp. Quant. Biol.* **38**, 763–771.

M   McCullough, A. J., and Schuler, M. A. (1993). AU-rich intronic elements affect pre-mRNA 5' splice site selection in *Drosophila melanogaster. Mol. Cell. Biol.* **13**, 7689–7697.

V   McDonald, J. F., Josephs, S. F., Wong-Staal, F., and Strand, D. J. (1989). HIV-1 expression is post-transcriptionally repressed in *Drosophila* cells. *AIDS Res. Human Retroviruses* **5**, 79–85.

TR   McDonald, J. F., Strand, D. J., Brown, M. R., Paskewitz, S. M., Csink, A. M., and Voss, S. H. (1988). Evidence of host-mediated regulation of retroviral element expression at the post-transcriptional level. In *"Eukaryotic Transposable Elements as Mutagenic Agents"* (M. E. Lambert, J. F. McDonald, and I. B. Weinstein, eds.), pp. 219–234, Banbury Report 30, Cold Spring Harbor Laboratory.

TH   McGarry, T. J., and Lindquist, S. (1985). The preferential translation of *Drosophila hsp* 70 mRNA requires sequences in the untranslated leader. *Cell* **42**, 903–911.

TH   McGarry, T. J., and Lindquist, S. (1986). Inhibition of heat-shock protein synthesis by heat-inducible antisense RNA. *Proc. Natl. Acad. Sci. U.S.A.* **83**, 399–403.

H   McKenzie, S. L. (1976). Protein and RNA synthesis induced by heat treatment in *Drosophila melanogaster* tissue culture cells. Ph.D. Thesis, Harvard University, Cambridge, MA.

HM   McKenzie, S. L., and Meselson, M. (1975). Translation *in vitro* of *Drosophila* heat-shock messages. *J. Mol. Biol.* **117**, 279–283.

R   McLean, C., Bucheton, A., and Finnegan, D. J. (1993). The 5′ untranslated region of the I factor, a long interspersed nuclear element-like retrotransposon of *Drosophila melanogaster* contains an internal promoter and sequences that regulate expression. *Mol. Cell. Biol.* **13**, 1042–1050.

F   Madhavan, K., Bilodeau-Wentworth, D., and Wadsworth, S. C. (1985). Family of developmentally regulated, maternally expressed *Drosophila* RNA species detected by a *v-myc* probe. *Mol. Cell. Biol.* **5**, 7–16.

C   Mahowald, A. P. (1994). Mass isolation of fly tissues. *Methods Cell Biol.* **44**, 129–142.

T   Maisonhaute, C., and Echalier, G. (1986). Stable transformation of *Drosophila* Kc cells to antibiotic resistance with the bacterial neomycin resistance gene. *FEBS Lett.* **197**, 45–49.

T   Malone, R. W., Felgner, P. L., and Verma, I. M. (1989). Cationic liposome-mediated RNA transfection. *Proc. Natl. Acad. Sci. U.S.A.* **86**, 6077–6081.

Cd   Mandaron, P. (1970). Développement *in vitro* des disques imaginaux de la Drosophile. Aspects morphologiques et histologiques. *Dev. Biol.* **22**, 298–320.

Cd   Mandaron, P. (1971). Sur le mécanisme de l'évagination des disques imaginaux chez la Drosophile. *Dev. Biol.* **25**, 581–605.

Cd   Mandaron, P. (1973a). Effects of alpha-ecdysone, beta-ecdysone and inokosterone on the *in vitro* evagination of *Drosophila* leg discs and the subsequent differentiation of imaginal integumentary structures. *Dev. Biol.* **31**, 101–113.

Cd   Mandaron, P. (1973b). Rôle de la membrane cellulaire au cours de l'évagination *in vitro* des disques imaginaux de la Drosophile. *CR Acad. Sci. Paris* **276**, 3167–3170.

Cd   Mandaron, P. (1974). Sur le mécanisme de l'évagination des disgues imaginaux de Drosophile cultivés *in vitro*: Effets de diverses substances affectant la membranes cellulaire. *W. Roux' Arch.* **175**, 49–63.

Cd   Mandaron, P. (1976a). Synthèse d'ARN et de protéines au cours du développement des disques imaginaux de patte de Drosophile. *In "Actualités sur les Hormones d'Invertébrés, Coll. Intern. CNRS"* **251**, 475–481, Editionsd du CNRS, Paris.

Cd   Mandaron, P. (1976b). Synthèse d'ARN et de protéines dans des disques de patte de Drosophile cultivés *in vitro*. *W. Roux's Arch.* **179**, 169–183.

Cd   Mandaron, P. (1976c). Ultrastructure des disques de pattes de Drosophile cultivés *in vitro.* Evagination, sécrétion de la cuticule nymphale et apolysis. *W. Roux's Arch.* **179,** 185–196.

Cd   Mandaron, P. (1980). *Drosophila* imaginal disc development *in vitro. In "Invertebrate Systems in Vitro"* (E. Kurstak, K. Maramorosch, and A. Dübendorfer, eds.), pp. 149–165, Elsevier, Amsterdam.

Cd   Mandaron, P., and Guillermet, C. (1978). Analyse microcinématographique de l'évagination des disques d'aile et de patte de Drosophile cultivés *in vitro. CR Acad. Sci. Paris* **287,** 257–260.

Cd   Mandaron, P., Guillermet, C., and Sengel, P. (1977). *In vitro* development of *Drosophila* imaginal discs: hormonal control and mechanism of evagination. *Am. Zool.* **17,** 661–670.

Cd   Mandaron, P., and Sengel, P. (1973). Effect of Cytochalasine B on the evagination *in vitro* of leg imaginal discs. *Dev. Biol.* **32,** 201–207.

M   Manteuil, S., Hamer, D. H., and Thomas, C. A. Jr. (1975). Regular arrangement of restriction sites in *Drosophila* DNA. *Cell* **5,** 413–422.

Cy   Marcaillou, C., Debec, A., Lauverjat, S., and Saihi, A. (1993). The effect of heat shock response on ultrastructure of the centrosome of *Drosophila* cultured cells in interphase: possible relation with changes in the chemical state of calcium. *Biochem. Cell. Biol.* **71,** 507–517.

rC   Marks, E. P. (1980). Insect tissue culture: an overview 1971–1978. *Annu. Rev. Entomol.* **25,** 73–101.

E   Maroy, P., Dennis, R., Beckers, C., Sage, B. A., and O'Connor, J. D. (1978). Demonstration of an ecdysteroid receptor in a cultured cell line of *Drosophila melanogaster. Proc. Natl. Acad. Sci. U.S.A.* **75,** 6035–6038.

TE   Martin, M. (1984). Transformation de cellules de Drosophile en culture; contribution à l'étude de gènes régulés par l'ecdysone. DEA Univ. Clermont-Ferrand II, France.

TER   Martin, M., and Jarry, B. (1988). Expression of transfected genes in a *Drosophila* cell line in transient assay and stable transformation. *J. Insect Physiol.* **34,** 691–699.

Cdr   Martin, P., and Schneider, I. (1978). *Drosophila* organ culture. *In "The Genetics and Biology of Drosophila"* (M. Ashburner and T. R. F. Wright, eds.), Vol. 2A, pp. 219–264, Academic Press, New York.

CdE   Martin, P., and Shearn, A. (1980). Development of *Drosophila* imaginal discs *in vitro*: Effects of ecdysone concentration and insulin. *J. Exp. Zool.* **211,** 291–301.

H   Martins de Sa, C., Rollet, E., Grossi de Sa, M. F., Tanguay, R., Best-Belpomme, M., and Scherrer, K. (1989). Prosomes and heat-shock complexes in *Drosophila melanogaster* cells. *Mol. Cell. Biol.* **9,** 2672–2681.

C   Marunouchi, T., and Miyake, T. (1978). Substitution of inosine for yeastolate in the culture medium for *Drosophila* cells. *In Vitro* **14,** 1010–1014.

B   Matsuno, K., Diederich, R. J., Go, M. J., Blaumueller, C. M., and Artavanis-Tsakonas, S. (1995). *Deltex* acts as a positive regulator of *Notch* signaling through interactions with the *Notch* ankyrin repeats. *Development* **121,** 2633–2644.

B   Maus, N., Stuurman, N., and Fisher, P. A. (1995). Diassembly of the *Drosophila* nuclear lamina in a homologous cell-free system. *J. Cell Sci.* **108,** 2027–2035.

H   Mayrand, S., and Pederson, T. (1983). Heat shock alters nuclear ribonucleoprotein assembly in *Drosophila* cells. *Mol. Cell. Biol.* **3,** 161–171.

M   Mazabraud, A., and Garel, J. P. (1979). Analysis of tRNA population from *Drosophila melanogaster* by means of polyacrylamide gel mapping. *FEBS Lett.* **105,** 70–76.

TR   Mazo, A. M., Mizrokhi, L. J., Karavanov, A. A., Sedkov, Y. A., Krichevskaja, A. A., and Ilyin, Y. V. (1989). Suppression in *Drosophila: su(Hw)* and *su(f)* gene products interacts with a region of *gypsy (mdg4)* regulating its transcriptional activity. *EMBO J.* **8**, 903–911.

BT   Meadows, L. A., Gell, D., Broadie, K., Gould, A. P., and White, R. A. H. (1994). The cell adhesion molecule, *connectin,* and the development of the *Drosophila* neuromuscular system. *J. Cell Sci.* **107**, 321–328.

M   Mermod, N., O'Neill, E. A., Kelly, T. J., and Tijan, R. (1989). The proline-rich transcriptional activator of CTF/NF-1 is distinct from the replication and DNA-binding domain. *Cell* **58**, 741–753.

BT   Messmer, S., Franke, A., and Paro, R. (1992). Analysis of the functional role of the *Polycomb* chromo domain in *Drosophila melanogaster. Genes Dev.* **6**, 1241–1254.

H   Mestril, R., Rungger, D., Schiller, P., and Voellmy, R. (1985). Identification of a sequence element in the promoter of the *Drosophila hsp23* gene that is required for its heat shock activation. *EMBO J.* **4**, 2971–2976.

HET Mestril, R., Schiller, P., Amin, J., Klapper, H., Ananthan, J., and Voellmy, R. (1986). Heat shock and ecdysterone activation of the *Drosophila melanogaster hsp23* gene; a sequence element implied in developmental regulation. *EMBO J.* **5**, 1667–1673.

E   Metakovsky, E. V., Cherdantseva, E. M., and Gvozdev, V. A. (1977). Action of ecdysterone on surface membrane glycoproteins of *Drosophila* cells in culture. *Molekulyarnaya Biol.* **11**, 158–170.

E   Metakovsky, E. V., and Gvozdev, V. A. (1978). Changes of cell surface glycoprotein pattern in the established cell line of *Drosophila melanogaster* as a result of ecdysterone action. *Dros. Inf. Serv.* **53**, 174.

E   Metakovsky, E. V., Kakpakov, V. T., and Gvozdev, V. A. (1975). (in Russian) Effect of ecdysterone on subcultured cells of *Drosophila melanogaster:* Stimulation of high molecular weight polypeptide synthesis and changes in cell surface properties. *Dokl. Akad. Nauk. SSSR* **221**, 960–963.

RE   Micard, D., Couderc, J. L., Sobrier, M. L., Giraud, G., and Dastugue, B. (1988). Molecular study of the retrovirus-like transposable element *412,* a 20-hydroxyecdysone responsive repetitive sequence in *Drosophila* cultured cells. *Nucleic Acids Res.* **16**, 455–470.

rC   Millam Stanley, M. S. (1972). Cultivation of Arthropod cells. In *"Growth, Nutrition and Metabolism of Cells in Culture"* (G. H. Rothblat and V. J. Cristofalo, Eds.), Vol. II, pp. 327–370, Academic Press, New York.

Cd   Milner, M. J. (1975). The development *in vitro* of the imaginal discs of *Drosophila melanogaster.* Ph.D Thesis, University of Sussex.

CdE Milner, M. J. (1977). The eversion and differentiation of *Drosophila melanogaster* leg and wing imaginal discs cultured *in vitro* with an optimal concentration of beta-ecdysone. *J. Embryol. Exp. Morphol.* **37**, 105–117.

CdE Milner, M. J. (1977). The time during which β-ecdysone is required for the differentiation *in vitro* and *in situ* of wing imaginal discs of *Drosophila melanogaster. Dev. Biol.* **56**, 206–212.

Cd   Milner, M. J. (1980). Epithelial and pattern integration in *Drosophila* eye-antennal discs cultured *in vitro. In "Invertebrate Systems in Vitro"* (E. Kurstak, K. Maramorosch, and A. Dübendorfer, eds.), pp. 135–148, Elsevier/North Holland, Amsterdam.

CdE Milner, M. J. (1985). Culture medium parameters for the eversion and differentiation of *Drosophila melanogaster* imaginal discs *in vitro. Dros. Inf. Serv.* **61**, 194–195.

E Milner, M. J., and Dübendorfer, A. (1982). Tissue-specific effects of the Juvenile Hormone analogue *ZR 515* during metamorphosis in *Drosophila* cell cultures. *J. Insect Physiol.* **28**, 661–668.

Cd Milner, M., and Haynie, J. L. (1979). Fusion of *Drosophila* eye-antennal imaginal discs during differentiation *in vitro*. *W. Roux's Arch.* **185**, 363–370.

CdE Milner, M. J., and Muir, J. (1987). The cell biology of *Drosophila* wing metamorphosis *in vitro*. *W. Roux's Arch. Dev. Biol.* **196**, 191–201.

CdE Milner, M. J., and Sang, J. H. (1974). Relative activities of alpha and beta-ecdysone for the differentiation *in vitro* of *Drosophila melanogaster* imaginal discs. *Cell* **3**, 141–143.

CdE Milner, M. J., and Sang, J. H. (1977). Active ion transport and beta-ecdysone induced differentiation of *Drosophila melanogaster* imaginal discs cultured *in vitro*. *J. Embryol. Exp. Morphol.* **37**, 118–131.

RT Minchiotti, G., and Di Nocera, P. P. (1991). Convergent transcription initiates from oppositely oriented promoters within the 5′ end regions of *Drosophila melanogaster* F elements. *Mol. Cell. Biol.* **11**, 5171–5180.

H Mirault, M. E., Goldschmidt-Clermont, M., Artavanis-Tsakonas, S., and Schedl, P. (1979). Organization of the multigenes for the 70 000-dalton protein in *Drosophila melanogaster*. *Proc. Natl. Acad. Sci.* **76**, 5254–5258.

H Mirault, M. E., Goldschmidt-Clermont, M., Moran, L., Arrigo, A. P., and Tissière, A. (1978). The effect of heat shock on gene expression in *Drosophila melanogaster*. *Cold Spring Harbor Symp. Quant. Biol.* **42** (2), 819–827.

Cl Miyake, T. (1984a). (in Japanese) *In vitro* culture of *Drosophila* cells. Establishment and maintenance of cell lines. *Tissue Culture* **10**, 232–236.

Cl Miyake, T. (1984b) (in Japanese). Establishment of cell lines from *Drosophila* embryos. *In "Biotechnology Manual: Biology and Molecular Biology of Drosophila"* (T. Miyake, K. Saigo, T. Shiba, and R. Ueda, eds.), pp. 41–45, Kodansha Science, Tokyo, Japan.

Cl Miyake, T. (1985). (in Japanese) Establishment and characterization of cell lines from DNA repair mutants of *Drosophila melanogaster*. *Saibo* **17**, 11–14.

Cl Miyake, T. (1986a). (in Japanese) Establishment and characterization of imaginal disc cell lines of *Drosophila melanogaster*. *Tissue Cult. Res. Commun.* **5**, 137–139.

R Miyake, T. (1986b). (in Japanese) Causal relation between movable genetic elements and virus-like particles proven by DNA transfection into cultured cells of *Drosophila*. *Tissue Cult. Res. Commun.* **5**, 140–142.

R Miyake, T. (1987). Transfection of *Drosophila melanogaster* transposable elements into the *Drosophila hydei* cell line. *In "Biotechnology Invertebrate Pathology Cell Cultures"* (K. Maramorosch, ed.), chap. 16, pp. 251–263, Academic Press, New York.

rC Miyake, T. (1988). (in Japanese) *In vitro* cell culture in a Diptera, *Drosophila melanogaster*. *Rep. New Technology*, 88–107 (Res. Assoc. for Biotechnol. Agric. Chemicals).

rC Miyake, T. (1989). (in Japanese) Cell manipulation. *In "Developmental Genetics of Drosophila melanogaster"* (Y. Hotta and M. Okada, eds.), chap. 3, pp. 58–70, Maruzen, Japan.

rC Miyake, T. (1991a). (in Japanese) Cell culture of *Drosophila* and its applications. *Byotai Seiri* **10**, 294–303.

rC Miyake, T. (1991b). (in Japanese) *Drosophila* cell culture and its application to Molecularbiology. *Tissue Cult. Res. Commun.* **10**, 51.

Cl Miyake, T. (1991c). (in Japanese) *In vitro* culture of imaginal disc cells of *Drosophila melanogaster* and establishment of cell lines. *Brain Technonews* **25**, 11–14.

rC Miyake, T., and Koana, T. (1992). (in Japanese) The method for cell culture. *Collection of articles on experiments in Biochemistry*, 3rd series, **14**, (Differentiation, Development, Aging), 74–84, (The Japanese Biochemical Soc., ed.) Tokyo Kagaku Dozin Co. Ltd.

RT Miyake, T., Mae, N., Shiba, T., and Kondo, S. (1987). Production of VLPs by the transposable genetic element *copia* of *Drosophila melanogaster*. *Mol. Gen. Genet.* **207**, 29–37.

R Miyake, T., Mae, N., Shiba, T., Kondo, S., Saigo, K., and Hattori, K. (1988). Production of retrovirus-like particles by transposons in genome of *Drosophila simulans. In "Invertebrates and Fish Tissue Culture"* (Y. Kuroda, E. Kurstak, and K. Maramorosch, eds.), pp. 115–118, Jap. Sci. Soc. Press, Springer Verlag, New York.

C Miyake, T., Saigo, K., Marunouchi, T., and Shiba, T. (1977). Suspension culture of *Drosophila* cells employing a gyratory shaker. *In Vitro* **13**, 245–251.

Cl Miyake, T., and Ueda, R. (1984a). (in Japanese) Experimental procedures for establishing cultured cell lines from *Drosophila melanogaster. In "Handbook of Somatic Genetics Experiment Procedures"* (H. Koyama, D. Ayusawa, and T. Seno, eds.), pp. 314–318, Tanpakushitsu, Kakusawa, Kosqo, Kyoritsu, Tokyo.

Cl Miyake, T., and Ueda, R. (1984b). (in Japanese) Establishment of cell lines of *Drosophila melanogaster. Protein Nucleic Acid Enzyme* **2** (Suppl.), 314–318.

RT Mizrokhi, L. J., Georgieva, S. G., and Ilyin, Y. V. (1988). *Jockey*, a mobile *Drosophila* element similar to mammalian LINEs, is transcribed from the internal promoter by RNA polymerase II. *Cell* **54**, 685–691.

R Mizrokhi, L. J., and Mazo, A. M. (1991). Cloning and analysis of the mobile element *gypsy* from *D. virilis. Nucleic Acids Res.* **19**, 913–916.

BE Moir, A. (1974). An immunological characterization of *Drosophila* cell lines and the effect of hormones on cell lines. Ph.D. Thesis, Oxford University, U.K.

B Moir, A., and Roberts, D. B. (1976). Distribution of antigens in established cell lines of *Drosophila melanogaster. J. Insect Physiol.* **22**, 299–307.

B Moiseenko, E. V., and Kakpakov, V. T. (1974). The absence of hypoxanthine-guanine phosphoribosyltransferase in extracts of *Drosophila melanogaster* flies and established embryonic diploid cell line. *Dros. Inf. Serv.* **51**, 44.

B Moiseenko, E. V., and Kakpakov, V. T. (1975). (in Russian) About the resistance of *Drosophila* cells to 6-mercaptopurine and 8-azaguanine. *Genetika* **11**, 160–162.

B Mokdad, R., Debec, A., and Wegnez, M. (1987). Metallothionein genes in *Drosophila melanogaster* constitute a dual system. *Proc. Natl. Acad. Sci. U.S.A.* **84**, 2658–2662.

F Monkovic, D. D., VanDusen, W. J., Petroski, C. J., Garsky, V. M., Sardana, M. K., Zavodszky, P., Stern, A. M., and Friedman, P. A. (1992). Invertebrate aspartyl/asparaginyl beta-hydroxylase: potential modification of endogenous Epidermal Growth Factor-like modules. *Biochem. Biophys. Res. Commun.* **189**, 233–241.

E Montpied, P. (1986). Régulation de l'expression de la famille multigénique beta-tubuline par la 20-hydroxyecdysone dans les cellules de Drosophile en culture. *Thèse de Doctorat*, Université Clermont-Ferrand II (France).

E Montpied, P., Sobrier, M. L., Chapel, S., Couderc, J. L., and Dastugue, B. (1988). 20-hydroxyecdysone induces the expression of one β-tubulin gene in *Drosophila Kc* cells. *Biochim. Biophys. Acta* **949**, 79–86.

V Moore, N. F., Kearns, A., and Pullin, J. S. K. (1980). Chacterization of Cricket Paralysis Virus-induced polypeptides in *Drosophila* cells. *J. Virol.* **33**, 1–9.

V Moore, N. F., and Pullin, J. S. K. (1980). The use of Invertebrate tissue culture to characterize picorna-like viruses of Insects: intra-cellular proteins induced by Cricket

Paralysis Virus in infected *Drosophila* cells. *In "Invertebrate Systems in vitro"* (E. Kurstak, K. Maramorosch, and A. Dübendorfer, eds.), pp. 403–410, Elsevier/North Holland, Amsterdam.

VH Moore, N. F., Pullin, J. S. K., and Reavy, B. (1981). Inhibition of heat-shock proteins in *Drosophila melanogaster* cells infected with insect picornavirus. *FEBS Lett.* **128**, 93–96.

V Moore, N. F., Reavy, B., and Pullin, J. S. K. (1981). Processing of Cricket Paralysis Virus induced polypeptides in *Drosophila* cells: production of high-molecular weight polypeptides by treatment with iodoacetamide. *Arch. Virol.* **68**, 1–8.

V Moore, N. F., Reavy, B., Pullin, J. S. K., and Plus, N. (1981). The polypeptides induced in *Drosophila* cells by *Drosophila* C virus (strain Ouarzazate). *Virology* **112**, 411–416.

H Moran, L., Mirault, M. E., Arrigo, A. P., Goldschmidt-Clermont, M., and Tissières, A. (1978). Heat-shock of *Drosophila melanogaster* induces the synthesis of new messenger RNAs and proteins. *Phil. Trans. R. Soc. London B* **283**, 391–406.

TH Morganelli, C. M., and Berger, E. M. (1984). Transient expression of homologous genes in *Drosophila* cells. *Science* **224**, 1004–1006.

TE Morganelli, C. M., and Berger, E. M. (1986). Effects of 20-OH-ecdysone on *Drosophila* cells: regulation of endogenous and tranfected genes. *Insect Biochem.* **16**, 233–240.

HET Morganelli, C. M., Berger, E. M., and Pelham, H. R. B. (1985). Transcription of small *hsp-tk* hybrid genes is induced by heat shock and by ecdysterone in transfected *Drosophila* cells. *Proc. Natl. Acad. Sci. U.S.A.* **82**, 5865–5869.

E Moritz, T. H., Edstrom, J. E., and Pongs, S. (1984). Cloning of a gene localized and expressed at the ecdysteroid regulated puff 74 EF in salivary gland of *Drosophila* larvae. *EMBO J.* **3**, 289–295.

VT Morris, T. D., and Miller, L. K. (1992). Promoter influence on baculovirus-mediated gene expression in permissive and non-permissive Insect cell lines. *J. Virol.* **66**, 7397–7405.

VT Morris, T. D., and Miller, L. K. (1993). Characterization of productive and nonproductive ACMNPV infection in selected Insect cell lines. *Virology* **197**, 339–348.

C Mosna, G. (1972). Medium for *Drosophila* cells *in vitro* without serum. *Dros. Inf. Serv.* **49**, 60.

C Mosna, G. (1973). Obtaining a nearly defined medium for *Drosophila* cells. *Accad. Naz. Lincei* **54**, 811–812.

Cy Mosna, G. (1979). The chromosomes in *Drosophila* cell lines since their start. *Istituto Lombardo (Rend. Sc) B113* 17–27.

C Mosna, G. (1983). Colony formation of *Drosophila* cells in semisolid medium containing agarose. *Experientia* **39**, 774–775.

C Mosna, G., and Barigozzi, C. (1976). Stimulation of growth by insulin in *Drosophila* embryonic cells *in vitro*. *Experientia*, **32**, 855–856.

Cl Mosna, G., and Dolfini, S. (1972a). New continuous cell lines of *Drosophila melanogaster*. Morphological characteristics and karyotypes. *Dros. Inf. Serv.* **48**, 144–145.

ClCy Mosna, G., and Dolfini, S. (1972b). Morphological and chromosomal characterization of three new continuous cell lines of *Drosophila melanogaster*. *Chromosoma* **38**, 1–9.

M Mosna, G., Pulcini, M., and Ghidoni, A. (1984). Chromosomal changes in methotrexate-resistant cell lines of *Drosophila melanogaster*. *Genetica* **65**, 199–203.

T Moss, R. (1985). Analysis of a transformation system for *Drosophila* tissue culture cells. Ph.D. Thesis, Harvard University, Cambridge, MA.

T Moss, R., Cherbas, L., Koehler, M., Bunch, T., and Cherbas, P. (1985). Transformation of *Drosophila* cells in culture: plasmid recombination and expression of non selectable markers (Abst. Ann. Meeting, 124). *In Vitro* **21**, 3.

R   Mossie, K. G., Young, M. W., and Varmus, H. E. (1985). Extrachromosomal DNA forms of *copia*-like transposable elements, F elements and middle repetitive DNA sequences in *Drosophila melanogaster*. *J. Mol. Biol.* **182**, 31–43.

B   Muckenthaler, F. A., and Fausto-Sterling, A. (1981). Normal and lectin-mediated aggregation in a *Drosophila* cell line. *W. Roux's Arch. Dev. Biol.* **190**, 118–122.

V   Mudd, J. A., Leavitt, R. W., Kingsbury, D. T., and Holland, J. J. (1973). Natural selection of mutants of vesicular stomatitis virus by cultured cells of *Drosophila melanogaster*. *J. Gen. Virol.* **20**, 341–351.

Cd   Mukha, D. V., Kramerov, A. A., and Gvozdev, V. A. (1987). Nascent glycoprotein with chitin-like carbohydrate component in *Drosophila* imaginal discs. *Insect Biochem.* **17**, 919–927.

CdE   Mukhovatova, L. M., and Kakpakov, V. T. (1974). Influence of α- and β-ecdysone on the differentiation of the imaginal discs of *Drosophila melanogaster,* cultured *in vitro*. *Ontogenez* **6**, 80–87.

M   Munks, R. J. L., Moore, J., O'Neill, L., and Turner, B. M. (1991). Histone H4 acetylation in *Drosophila*: Frequency of acetylation at different sites defined by immunolabelling with site-specific antibodies. *FEBS Lett.* **284**, 245–248.

E   Munsch, N., and Cade-Treyer, D. (1982). Effect of an ecdysteroid on DNA synthesis, DNA polymerase *a* and thymidine kinase activities of *Drosophila melanogaster* cells *in vitro*. *Experientia* **38**, 1197–1199.

M   Murtif, V. L., and Rae, P. M. M. (1985). *In vivo* transcription of rDNA spacers in *Drosophila*. *Nucleic Acids Res.* **13**, 3221–3239.

C   Nakajima, S., and Miyake, T. (1976). Effective colony formation in *Drosophila* cell lines using conditioned medium. (4th Intern. Conf. Invert. Tissue Cult., Mont Gabriel, 1975). In *"Invert. Tissue Culture. Applications in Medicine, Biology and Agriculture"* (E. Kurstak and K. Maramorosch, eds.) pp. 279–287, Academic Press, New York.

B   Nakajima, S., and Miyake, T. (1978). Cell fusion between temperature-sensitive mutants of a *Drosophila* cell line. *Somat. Cell Genet.* **4**, 131–141.

M   Narachi, M. A., and Boyd, J. B. (1987). The *giant* (*gt*) mutants of *Drosophila melanogaster* alter DNA metabolism. *Mol. Gen. Genet.* **199**, 500–506.

Cd   Nardi, J. B., and Willis, J. H. (1979). Control of cuticle formation by wing imaginal discs *in vitro*. *Dev. Biol.* **68**, 381–395.

M   Natori, S., Worton, R., Boshes, R. A., and Ristow, H. (1973). An RNA polymerase from *Drosophila*. *Insect Biochem.* **3**, 91–102.

CdE   Natzle, J. E. (1993). Temporal regulation of *Drosophila* imaginal disc morphogenesis: a hierarchy of primary and secondary 20-hydroxy-ecdysone-responsive loci. *Dev. Biol.* **155**, 516–532.

Cd   Natzle, J. E., Fristrom, D. K., and Fristrom, J. W. (1988). Genes expressed during imaginal disc morphogenesis: IMP-E1, a gene associated with epithelial cell rearrangement. *Dev. Biol.* **129**, 428–438.

Cd   Natzle, J. E., Hammonds, A. S., and Fristrom, J. W. (1986). Isolation of genes active during hormone induced morphogenesis in *Drosophila* imaginal discs. *J. Biol. Chem.* **261**, 5575–5583.

M   Nguyen, I. D., and Boyd, J. B. (1977). The *meiotic-9* (*mei-9*) mutants of *Drosophila melanogaster* are deficient in repair replication of DNA. *Mol. Gen. Genet.* **158**, 141–147.

BCp   Nguyen, P., Bournias-Vardiabasis, N., Haggren, W., Adey, W. R., and Phillips, J. L. (1995). Exposure of *Drosophila melanogaster* embryonic cell cultures to 60-Hz sinusoidal magnetic fields: Assessment of potential teratogenic effects. *Teratology* **51**, 273.

Cy Nichols, W. W., Bradt, C., Dwight, S., and Bowne, W. (1972). Somatic pairing in Dipteran cells in culture. *Cytogenetics* **11**, 46–52.

H Nicole, L., and Tanguay, R. M. (1987). On the specificity of antisense RNA to arrest *in vitro* translation of mRNA coding for *Drosophila* HSP 23. *Biosci. Rep.* **7**, 239–246.

B Nikoshkov, A. B., and Kakpakov, V. T. (1981). Dosage compensation of sex-linked genes in established cell lines of *D. melanogaster*. *Dros. Inf. Serv.* **56**, 103.

CdE Nishiura, J. T., and Fristrom, J. W. (1975). Effect of Insect hormones on RNA polymerases of mass-isolated imaginal discs of *Drosophila melanogaster* cultured *in vitro*. *Proc. Natl. Acad. Sci. U.S.A.* **72**, 2984–2988.

H Nolan, N. L., and Kidwell, W. R. (1982). Effects of heat shock on poly (ADP ribose) synthetase on DNA repair in *Drosophila* cells. *Rad. Res.* **90**, 187–203.

TB Norris, J. L., and Manley, J. L. (1992). Selective nuclear transport of the *Drosophila* morphogen *dorsal* can be established by a signaling pathway involving the transmembrane protein *Toll* and Protein Kinase A. *Genes Dev.* **6**, 1654–1667.

TB Norris, J. L., and Manley, J. L. (1995). Regulation of *dorsal* in cultured cells by *Toll* and *tube: tube* function involves a novel mechanism. *Genes Dev.* **9**, 358–369.

MT Nose, A., Mahajan, V. B., and Goodman, C. S. (1992). *Connectin*: a homophilic Cell adhesion molecule expressed on a subset of muscles and the motoneurons that innerve them in *Drosophila*. *Cell* **70**, 553–567.

rC Oberlander, H. (1980). Tissue culture methods. *In "Cuticle Techniques in Arthropods"* (T. A. Miller, ed.), pp. 253–272, Springer Verlag, New York.

HM O'Brien, T., Hardin, S., Greenleaf, A., and Lis, J. T. (1994). Phosphorylation of RNA polymerase II C-terminal domain and transcriptional elongation. *Nature* **370**, 75–77.

H O'Brien, T., and Lis, J. T. (1991). RNA Polymerase II pauses at the 5' end of the transcriptionally induced *Drosophila hsp70* gene. *Mol. Cell. Biol.* **11**, 5285–5290.

H O'Brien, T., and Lis, J. T. (1993). Rapid changes in *Drosophila* transcription after an instantaneous heat shock. *Mol. Cell. Biol.* **13**, 3456–3463.

H O'Brien, T., Wilkins, R. C., Giardina, C., and Lis, J. T. (1995). Distribution of GAGA protein on *Drosophila* genes *in vivo*. *Genes Dev.* **9**, 1098–1110.

E O'Connor, J. D. (1985). Ecdysteroid action at the molecular level. *In "Comprehensive Insect Physiology, Biochemistry and Pharmacology"* (G. A. Kerkut and L. I. Gilbert, eds.), Vol. 8, pp. 85–98, Pergamon Press, Oxford.

E O'Connor, J. D., and Chang, E. S. (1981). Cell lines as a model for the study of metamorphosis. *In "Metamorphosis: a problem in Developmental Biology"* (L. I. Gilbert and E. Frieden, eds.), 2nd ed., pp. 241–261, Plenum Press, New York.

H O'Connor, D., and Lis, J. T. (1981). Two closely linked transcription units within the 63B heat-shock puff locus of *Drosophila melanogaster* display strikingly different regulation. *Nucleic Acids Res.* **9**, 5075–5092.

E O'Connor, J. D., Maroy, P., Beckers, C., Dennis, R., Alvarez, C. M., and Sage, B. A. (1980). Ecdysteroid receptors in cultured *Drosophila* cells. *In "Gene Regulation by Steroid Hormones"* (A. K. Roy and J. H. Clarke, eds.), pp. 261–277, Springer-Verlag, New York.

B Oda, H., Uemura, T., Shiomi, K., Nagafuchi, A., Tsukita, S., and Tadeichi, M. (1993). Identification of a *Drosophila* homologue of α-catenin and its association with the *armadillo* protein. *J. Cell. Biol.* **1231**, 1133–1140.

Cp O'Dowd, D. K., Gee, J. R., and Smith, M. A. (1995). Sodium current density correlates with expression of specific alternatively spliced sodium channel mRNAs in single neurons. *J. Neurosci.* **15**, 4005–4012.

T   Oh, S. K., Scott, M. P., and Sarnow, P. (1992). Homeotic gene *Antennapedia* mRNA contain 5'-non coding sequences that confer translational initiation by internal ribosome binding. *Genes Dev.* **6**, 1643–1653.

V   Ohanessian, A. (1971). *Sigma* virus multiplication in *Drosophila* cell lines of different genotypes. In *"Arthropod Cell Cultures and their Applications to the Study of Viruses"* (E. Weiss, ed.), *Curr. Top. Microb. Immunol.* **55**, 230–233, Springer-Verlag, New York.

V   Ohanessian, A. (1973). Comparative study of *Sigma* virus infectivity in whole *Drosophila* and in two *Drosophila* cell lines. *Proc. 3rd Intern. Colloq. Invertebrate Tissue Cult.* 405–411, (Bratislava, Czechoslovakia).

V   Ohanessian, A., and Echalier, G. (1967). Multiplication of *Drosophila* hereditary virus (*Sigma* virus) in *Drosophila* embryonic cells cultivated *in vitro*. *Nature* **213**, 1049–1050.

V   Ohanessian, A., and Echalier, G. (1968). Multiplication du virus héréditaire *Sigma* de la Drosophile dans des cellules de Drosophile cultivées *in vitro*. *2nd Coll. Intern. Culture Tissus Invertébrés* (Tremezzo, Ital., Sept. 1967), 227–232, Istituto Lombardo Accad. Sc. Lett.

TM   Ohkuma, Y., Horikoshi, M., Roeder, R. G., and Desplan, C. (1990). Binding site-dependent direct activation and repression of *in vitro* transcription by *Drosophila* homeodomains. *Cell* **61**, 475–484.

M   Ohtani, K., and Nevins, J. R. (1994). Functional properties of a *Drosophila* homolog of the E2F1 gene. *Mol. Cell. Biol.* **14**, 1603–1612.

M   Olson, P. F., Fessler, L. I., Nelson, R. E., Sterne, R. E., Campbell, A. G., and Fessler, J. H. (1990). *Glutactin*, a novel *Drosophila* basement membrane-related glycoprotein with sequence similarity to serine esterases. *EMBO J.* **9**, 1219–1227.

B   Olson, P. F., Sterne, R., Fessler, L. I., and Fessler, J. H. (1987). *Entactin*, a sulfated glycoprotein of *Drosophila* basement membrane. *J. Cell Biochem.* **11C** (Suppl.), 26.

B   O'Neill, E. M., Ellis, M. C., Rubin, G. M., and Tijan, R. (1995). Functional domain analysis of *glass*, a zinc-finger-containing transcription factor in *Drosophila*. *Proc. Natl. Acad. Sci. U.S.A.* **92**, 6557–6561.

M   Orlando, V., and Paro, R. (1993). Mapping *Polycomb*-repressed domains in the *Bithorax* complex using *in vivo* formaldehyde cross-linked chromatin. *Cell* **75**, 1187–1198.

H   Ornelles, D. A., and Penman, S. (1990). Prompt heat-shock and heat-shifted proteins associated with the nuclear matrix-intermediate filament scaffold in *Drosophila melanogaster* cells. *J. Cell Sci.* **95**, 393–404.

M   Osgood, C., and Boyd, J. B. (1979). *Drosophila* DNA repair endonuclease specificity for apurinic/apyrimidinic sites in DNA. *Genetics* **91** (Suppl.), 91. [Abstract]

M   Osgood, C., and Boyd, J. B. (1982). Apurinic endonuclease from *Drosophila melanogaster*: reduced enzymatic activity in excision-deficient mutants of the *mei-9* and *mus(2)201* loci. *Mol. Gen. Genet.* **186**, 235–239.

Cy   Ottaviano Gottardi, A. (1968). Karyotype selection in cultures with different initial concentrations (*D. melanogaster*) *2nd Intern. Coll. Intern. Invert. Tissue Cult.*, Tremezzo, Sept. 67, Istituto Lombardo, Accad. Sc. Lett., 1968, 189–200.

Cy   Ottaviano Gottardi, A. (1969). Genetic analysis of growth pattern in cell populations *in vitro*. *Atti Accad. Naz. Lincei* R. classe Sc. fis., mat. natur. VIII, **47**, 70–78.

Cd   Paine-Saunders, S., Fristrom, M. D., and Fristrom, J. W. (1990). The *Drosophila* IMP-E2 gene encodes an apically secreted protein expressed during imaginal disc morphogenesis. *Dev. Biol.* **140**, 337–351.

BM   Paddy, M. R., Belmont, A. S., Saumweber, H., Agard, D. A., and Sedat, J. W. (1990). Interphase nuclear envelope lamins form a discontinuous network that interacts with only a fraction of the chromatin in the nuclear periphery. *Cell* **62**, 89–106.

HCp   Palter, K. B., Watanabe, M., Stinson, L., Mahowald, A. P., and Craig, E. A. (1986). Expression and localization of *Drosophila melanogaster hsp 70 cognate* proteins. *Mol. Cell. Biol.* **6**, 1187–1203.

F   Panganiban, G. F., Rashka, K. E., Neitzel, M. D., and Hoffmann, F. M. (1990). Biochemical characterization of the *Drosophila decapentaplegic* protein, a member of the *TGF-β* family of growth factors. *Mol. Cell. Biol.* **10**, 2669–2677.

F   Papageorge, A. G., Defeo-Jones, D., Robinson, P., Temeles, G., and Scolnick, E. M. (1984). *Saccharomyces cerevisiae* synthesises proteins related to the p21 gene product of *ras* genes found in Mammals. *Mol. Cell. Biol.* **4**, 23–29.

Cl   Paradi, E. (1972). Establishment of a new cell line from *Drosophila* mutant embryo. *Dros. Inf. Serv.* **49**, 53.

Cl   Paradi, E. (1973). (in Hungarian) Establishment of *in vitro* cultivated cell line of *Drosophila* embryo. *Biologiai Közl.* **21**, 11–14.

H   Pardue, M. L., Ballinger, D. G., and Scott, M. P. (1981). The expression of heat-shock genes in *Drosophila melanogaster*. In *"Developmental Biology Using Purified Genes"* (D. D. Brown, ed.) *ICN-UCLA Symp. Mol. Cell Biol.,* **28**, 415–427.

rH   Pardue, M. L., Bendena, W. G., Fini, M. E., Garbe, J. C., Hogan, N. C., and Traverse, K. L. (1990). *Hsr-omega*, a novel gene encoded by a *Drosophila* heat-shock puff. *Biol. Bull.* **179**, 77–86.

H   Pardue, M. L., Bendena, W. G., and Garbe, J. C. (1987). Heat-shock puffs and response to environmental stress: structure and function of eukaryotic chromosomes. *Results Probl. Cell Different.* **14**, 121–131.

M   Park, W. J., Liu, J., and Adler, P. N. (1994). The *frizzled* gene of *Drosophila* encodes a membrane protein with an odd number of transmembrane domains. *Mech. Dev.* **45**, 127–137.

B   Parker, C. G., Fessler, L. I., Nelson, R. E., and Fessler, J. H. (1995). *Drosophila* UDP-glucose: glycoprotein glucosyltransferase: sequence and characterization of an enzyme that distinguishes between denatured and native proteins. *EMBO J.* **14**, 1294–1303.

M   Parker, C. S., and Topol, J. (1984a). A *Drosophila* RNA polymerase II transcription factor contains a promoter region specific DNA binding activity. *Cell* **36**, 357–369.

HM   Parker, C. S., and Topol, J. (1984b). A *Drosophila* RNA polymerase II transcription factor binds to the regulatory site of an *hsp 70* gene. *Cell* **37**, 253–262.

M   Parker, G. F., Williams, P. J., Butters, T. D., and Roberts, D. B. (1991). Detection of the lipid-linked precursor oligosaccharide of N-linked protein glycosylation in *Drosophila melanogaster*. *FEBS Lett.* **290**, 58–60.

rH   Parsell, D. A., and Lindquist, S. (1993). The function of heat-shock proteins in stress tolerance: degradation and reactivation of damaged proteins. *Annu. Rev. Genet.* **27**, 437–496.

H   Parsell, D. A., Taulien, J., and Lindquist, S. (1993). The role of heat shock proteins in thermotolerance. *Phil. Trans. R. Soc. London Biol.* **339**, 279–285.

M   Pascal, E., and Tijan, R. (1991). Different activation domains of Sp1 govern formation of multimers and mediate transcriptional synergism. *Genes Dev.* **5**, 1646–1656.

MCp   Paterson, B. M., Shirakata, M., Nakamura, S., Dechene, C., Walldorf, U., Elridge, J., Dübendorfer, A., Frasch, M., and Gehring, W. J. (1992). Isolation and functional comparison of *Dmyd*, the *Drosophila* homologue of the Vertebrate myogenic determination genes with CMD1. *Symp. Soc. Exp. Biol.* **46**, 89–109.

Cp   Paterson, B. M., Walldorf, V., Eldridge, J., Dübendorfer, A., Frasch, M., and Gehring, W. J. (1991). The *Drosophila* homologue of vertebrate myogenic-determination gene encodes a transiently expressed nuclear protein marking primary myogenic cells. *Proc. Natl. Acad. Sci. U.S.A.* **88**, 3782–3786.

M   Paulsen, R. E., Weaver, C. A., Fahrner, T. J., and Milbrandt, J. (1992). Domains regulating transcriptional activity of the inducible orphan receptor NGF1-B. *J. Biol. Chem.* **267**, 16491–16496.

B   Pearson, A., Lux, A., and Krieger, M. (1995). Expression cloning of dSR-C1, a class C macrophage-specific scavenger receptor from *Drosophila melanogaster. Proc. Natl. Acad. Sci. U.S.A.* **92**, 4056–4060.

E   Peel, D. J. (1991). Cell interactions and the response to ecdysteroids of *Drosophila* imaginal disc cell lines. Ph.D. Thesis, University of St Andrews, U.K.

B   Peel, D. J., Johnson, J. A., and Milner, M. J. (1990). The ultrastructure of imaginal disc cells in primary cultures and during cell aggregation in continuous line cells. *Tissue Cell* **22**, 749–758.

Cl   Peel, D. J., and Milner, M. J. (1990). The diversity of cell morphology in cloned cell lines derived from *Drosophila* imaginal discs. *Dev. Biol.* **198**, 479–483.

E   Peel, D. J., and Milner, M. J. (1992a). The response of *Drosophila* imaginal disc cell lines to ecdysteroids. *Roux's Arch. Dev. Biol.* **202**, 23–35.

B   Peel, D. J., and Milner, M. J. (1992b). The expression of PS integrins in *Drosophila melanogaster* imaginal disc cell lines. *Roux's Arch. Dev. Biol.* **201**, 120–123.

V   Pennock, G. D., Shoemaker, C., and Miller, L. K. (1984). Strong and regulated expression of *E. coli* β-galactosidase in Insect cells with a Baculovirus vector. *Mol. Cell. Biol.* **4**, 399–406.

H   Perisic, O., Xiao, H., and Lis, J. T. (1989). Stable binding of *Drosophila* heat-shock factor to head-to-head and tail-to-tail repeats of a conserved 5bp recognition unit. *Cell* **59**, 797–806.

M   Perkins, K. K., Dailey, G. A., and Tijan, R. (1988). Novel *Jun-* and *Fos*-related proteins in *Drosophila* are functionally analogous to enhancer factor AP-1. *EMBO J.* **13**, 4265–4273.

RE   Peronnet, F., Becker, J. L., Becker, J., d'Auriol, L., Galibert, F., and Best-Belpomme, M. (1986a). *1731*, a new retrotransposon with hormone modulated expression. *Nucleic Acids Res.* **14**, 9017–9033.

RE   Peronnet, F., Becker, J. L., Becker, J., Lescault, A., D'Auriol, L., Galibert, F., and Best-Belpomme, M. (1988). *1731*, a new retrotransposon with hormone modulated expression. *Inv. Fish Tissue Cult.* (Y. Kuroda, E. Kurstak, and K. Maramorosch, eds.), 43–45, Springer Verlag, New York.

ER   Peronnet, F., Rollet, E., Maisonhaute, C., Becker, J. L., Becker, J., Courgeon, A. M., Echalier, G., and Best-Belpomme, M. (1986c). From transcript modulations to protein phosphorylation. A short survey of some effects of ecdysteroid. *Insect Biochem.* **16**, 199–201.

E   Peronnet, F., Ropp, M., Rollet, E., Becker, J. L., Becker, J., Maisonhaute, C., Pernodet, J. L., Vuillaume, M., Courgeon, A. M., Echalier, G., and Best-Belpomme, M. (1986b). *Drosophila* cells in culture as a model for the study of ecdysteroid action. *In* "*Techniques in the Life Sciences*" C2. *Techniques in In vitro Invertebrate Hormones and Genes.* C206, pp. 1–20, Elsevier Science Publishers, New York.

E   Peronnet, F., Ziarczyk, P., Rollet, E., Courgeon, A. M., Becker, J., Maisonhaute, C., Echalier, G., and Best-Belpomme, M. (1989). *Drosophila* cell lines as a model for studying the mechanisms of ecdysteroid action. *In* "*Ecdysone*" (J. Koolman, ed.), pp. 378–383, G. Thieme Verlag, Stuttgart.

Cl Petersen, N. S., Riggs, A. D., and Seecof, R. L. (1977). A method for establishing cell lines from *Drosophila melanogaster* embryos. *In Vitro* **13**, 36–40.

H Petersen, R., and Lindquist, S. (1988). The *Drosophila hsp 70* message is rapidly degraded at normal temperature and stabilized by heat shock. *Gene* **72**, 161–169.

H Petersen, R. B., and Lindquist, S. (1989). Regulation of HSP synthesis by messenger RNA degradation. *Cell Regul.* **1**, 135–149.

B Petersen, U. M., Björklund, G., Ip, Y. T., and Engström, Y. (1995). The *dorsal*-related immunity factor, *Dif*, is a sequence-specific transactivator of *Drosophila Cecropin* gene expression. *EMBO J.* **14**, 3146–3158.

Cd Petri, W. H., Fristrom, J. W., Stewart, D. J., and Hanly, E. W. (1971). The *in vitro* synthesis and characteristics of ribosomal RNA in imaginal discs of *Drosophila melanogaster*. *Mol. Gen. Genet.* **110**, 245–262.

Ct Petri, W. H., Mindrinos, M. N., Lombard, M. F., and Margaritis, L. H. (1979). *In vitro* development of the *Drosophila* chorion in a chemically defined organ culture medium. *W. Roux's Arch.* **186**, 351–362.

Cd Pino-Heiss, S., and Schubiger, G. (1989). Extra-cellular protease production by *Drosophila* imaginal discs. *Dev. Biol.* **132**, 282–291.

M Planques, Y. (1988). *Thèse Docteur-Ingénieur,* Univ. Clermont-Ferrand, France.

rV Plus, N. (1978). Endogenous viruses of *Drosophila melanogaster* cell lines: their frequency, identification and origin. *In Vitro* **14**, 1015–1021.

V Plus, N. (1980). Further studies on the origin of the endogenous viruses of *Drosophila melanogaster* cell lines. *In "Invertebrate Systems In Vitro"* (E. Kurstak, K. Maramorosch, and A. Dübendorfer, eds.), pp. 435–439, Elsevier, Amsterdam.

V Plus, N., Croizier, G., Renganum, C., and Scotti, P. D. (1978). Cricket Paralysis Virus and *Drosophila C* Virus: Serological analysis and comparison of capsid polypeptides and host range. *J. Invert. Pathol.* **31**, 296–302.

V Plus, N., Veyrunes, J. C., Croizier, L., and Debec, A. (1983). A symptomless *Drosophila* X virus from haploid *Drosophila* cell lines and from foetal calf serum: A further indication of the exogenous origin of this virus. *Ann. Virol. (Inst. Pasteur)* **134**, 293–300.

Ct Poluektova, E. V., Kakpakov, V. T., and Mitrofanov, V. G. (1985). (in Russian) Puffing differences in salivary gland chromosomes of *Drosophila virilis* in various media. *Ontogenez* **16**, 375–381.

CtE Poluektova, E. V., Mitrofanov, V. G., and Kakpakov, V. T. (1980). (in Russian) The action of Insect hormones on the puffing of salivary gland chromosomes of *Drosophila virilis,* cultivated *in vitro*. I. Puffing changes after cultivation in the medium S-50. *Ontogenez* **11**, 175–180.

Cy Polukarova, L. G., Kakpakov, V. T., and Gvozdev, V. A. (1975). Chromosomal variability in the transplantable cultures of cells of *Drosophila melanogaster*. *Genetika* **11**, 46. (english translation: *Soviet Genetics,* 1975, **11**, 575–579).

CdE Postlethwait, J. H., and Schneiderman, H. A. (1968). Effect of an ecdysone on growth and cuticle formation of *Drosophila* imaginal discs cultured *in vitro*. *Biol. Bull.* **135**, 431–432.

R Potter, S. S., Brorein, W. J., Dunsmuir, P., and Rubin, G. M. (1979). Transposition of elements of the *412, copia* and *297* dispersed repeated gene families in *Drosophila*. *Cell* **17**, 415–427.

M Preiser, P. R. (1990). *In vitro* processing of *Drosophila melanogaster* 5S RNA. *Ph.D. Dissertation,* University of Delaware, Newark, DE, U.S.A.

M   Preiser, P. R., and Levinger, L. F. (1991a). *In vitro* processing of *Drosophila melanogaster* 5S RNA: 3′ end effects and requirements for internal domains of mature 5S RNA. *J. Biol. Chem.* **266**, 7509–7516.

M   Preiser, P. R., and Levinger, L. F. (1991b). *Drosophila* 5S RNA processing requires the 1-118 base pair and additional sequence proximal to the processing site. *J. Biol. Chem.* **266**, 23602–23605.

M   Preiser, P., Vasisht, V., Birk, A., and Levinger, L. (1993). Poly(U)-binding protein inhibits *Drosophila* pre-5S RNA 3′Exonuclease digestion. *J. Biol. Chem.* **268**, 11553–11557.

M   Price, B. D., and Laughton, A. (1993). The isolation and characterization of a *Drosophila* gene encoding a putative NAD-dependent methylenetetrahydrofolate. *Biochim. Biophys. Acta* **1173**, 94–98.

M   Price, D. H., Sluder, A. E., and Greenleaf, A. L. (1987). Fractionation of transcription factors for RNA polymerase II from *Drosophila Kc* cell nuclear extracts. *J. Biol. Chem.* **262**, 3244–3255.

M   Price, D. H., Sluder, A. E., and Greenleaf, A. L. (1989). Dynamic interaction between a *Drosophila* transcription factor and RNA polymerase II. *Mol. Cell. Biol.* **9**, 1465–1475.

Cy Privitera, E. (1980). *Drosophila* cell line tested for the presence of active NORs by silver staining. *Chromosoma* **81**, 431–437.

M   Pugh, B. F., and Tijan, R. (1991). Transcription from a TATA-less promoter requires a multisubunit TFIID complex. *Genes Dev.* **5**, 1935–1945.

M   Pulido, D., Campuzano, S., Koda, T., Modolell, J., and Barbacid, M. (1992). *Drtk, a Drosophila* gene related to the *trk* family of neurotrophin receptors, encodes a novel class of neural cell adhesion molecules. *EMBO J.* **11**, 391–404.

EH Rabilloud, T. (1985). Etude des variations des protéines nucléaires des cellules en culture de *Drosophila melanogaster* dans différentes conditions d'expression du génome. *Thèse 3ème Cycle*, Univ. Paris VII.

H   Rabilloud, T., Vincens, P., Hubert, M., Pennetier, J. L., and Tarroux, P. (1985). Changes in nuclear proteins induced by heat-shock in *Drosophila* cultured cells. *FEBS Lett.* **184**, 278–284.

H   Rabindran, S. K., Wisniewski, J., Li, L., Li, G. C., and Wu, C. (1994). Interaction between heat shock factor and HSP70 is insufficient to suppress induction of DNA-binding activity *in vivo*. *Mol. Cell. Biol.* **14**, 6552–6560.

HT Rabindran, S. K., Haroun, R. I., Clos, J. Wisniewski, J., and Wu, C. (1993). Regulation of Heat-shock Factor trimer formation: role of a conserved leucine zipper. *Science* **259**, 230–234.

M   Raeber, A. J., Muramoto, T., Kornberg, T. B., and Prusiner, S. B. (1995). Expression and targeting of Syrian hamster prion protein induced by heat shock in transgenic *Drosophila melanogaster*. *Mech. Dev.* **51**, 317–327.

Cd Raikow, R., and Fristrom, J. W. (1971). Effects of beta-ecdysone on RNA metabolism of imaginal disks of *Drosophila melanogaster*. *J. Insect Physiol.* **17**, 1599–1614.

TV Raney, A. K., Le, H. B., and McLachlan, A. (1992). Regulation of transcription from the Hepatitis B virus major surface antigen promoter by the *Sp1* transcription factor. *J. Virol.* **66**, 6912–6921.

M   Rao, J. P., and Sodja, A. (1992). Further analysis of a transcript nested within the actin 5C gene of *Drosophila melanogaster*. *Biochim. Biophys. Res. Commun.* **184**, 400–407.

H   Rasmussen, E. B., and Lis, J. T. (1993). *In vivo* transcriptional pausing and cap formation on three *Drosophila* heat shock genes. *Proc. Natl. Acad. Sci. U.S.A.* **90**, 7923–7927.

TB Rebay, I., Fleming, R. J., Fehon, R. G., Cherbas, L., Cherbas, P., and Artavanis-Tsakonas, S. (1991). Specific EGF repeats of *Notch* mediate interactions with *Delta* and *Serrate:* Implications for *Notch* as a multifunctional receptor. *Cell* **67**, 687–699.

E Rebers, J. (1984). Structure and expression of an ecdysone-inducible gene. *Ph.D. Thesis,* Harvard University, Cambridge, MA.

CI Regenass, U., and Bernhard, H. P. (1979). *Rudimentary* mutants of *Drosophila melanogaster:* Isolation and characterization of pyrimidine auxotrophic cell lines. *W. Roux's Arch.* **187**, 167–177.

B Reichhart, J. M., Georgel, P., Meister, M., Lemaitre, B., Kappler, C., and Hoffmann, J. (1993). Expression and nuclear translocation of the *rel/NFκB*-related morphogen *dorsal* during the immune response of *Drosophila*. *CR Acad. Sci. Paris (Life Sci.)* **316**, 1218–1224.

M Renault, S., Degroote, F., and Picard, G. (1993). Despite its high representation in extrachromosomal circular DNAs from *Drosophila* embryos, the dodecasatellite does not allow autonomous replication in cultured cells. *Biol. Cell* **79**, 51–54.

CtE Rensing, L. (1969). Circadian Rhythmik von *Drosophila* Speichel Drüsen *in vivo, in vitro* und nach Ecdysonzugabe. *J. Insect Physiol.* **15**, 2285–2303.

Cp Rey, B. M. (1969). The culture of *Drosophila* embryonic cells in *H5* medium. *Dros. Inf. Serv.* **44**, 69.

Ct Rezzonico Raimondi, G., and Ghini, C. (1963). Cultures de tissus de la Drosophile. *Ann. Epiphyties* **14**, 153–159.

Ct Rezzonico Raimondi, G., Ghini, C., and Dolfini, S. (1964). Comparative observations on the behaviour *in vitro* of cephalic ganglia of wild homozygous and heterozygous *Drosophilae*. *Experientia* **20**, 440.

Cp Rezzonico Raimondi, G., and Gottardi, A. (1967). Genotypically controlled behaviour of embryonic cells of *Drosophila melanogaster* cultured *in vitro*. *J. Insect Physiol.* **13**, 523–529.

V Rice, W. C., and Miller, L. K. (1986/87). Baculovirus transcription in the presence of inhibitors and in nonpermissive *Drosophila* cells. *Virus Res.* **6**, 155–172.

CtE Richard, D. S., Applebaum, S. W., Sliter, T. J., Baker, F. C., Schooley, D. A., Reuter, C. C., Henrich, U. C., and Gilbert, L. I. (1989). Juvenile hormone bisepoxide biosynthesis *in vitro* by the ring gland of *Drosophila melanogaster:* a putative JH in the higher Diptera. *Proc. Natl. Acad. Sci. U.S.A.* **86**, 1421–1425.

V Richard-Molard, C. (1973). Etude de la multiplication du virus *Sigma* dans plusieurs cultures primaires et dans une lignée continue de cellules de Drosophile issues d'embryons perpétuant un virus *Sigma* défectif. *CR Acad. Sci. Paris,* **277**, 121–124.

CIV Richard-Molard, C. (1975). Isolement de lignées cellulaires de *Drosophila melanogaster* de différents génotypes et étude de la multiplication de deux variants du rhabdovirus *Sigma* dans ces lignées. *Arch. Virol.* **47**, 139–146.

V Richard-Molard, C. (1984). Etude du cycle de développement de Rhabdovirus dans les cellules d'Insectes, en utilisant des cultures de cellules de Drosophile. *Thèse Doctorat d'Etat,* Univ. Paris-Orsay, France.

V Richard-Molard, C., Blondel, D., Wyers, F., and Dezelee, S. (1984). Sigma virus: Growth in *Drosophila* cell culture; Purification; Protein composition and Localization. *J. Gen. Virol.* **65**, 91–99.

C Richard-Molard, C., and Ohanessian, A. (1977). Methode de clonage des cellules de Drosophile. Sensibilité aux rayons X de plusieurs lignées cellulaires. *W. Roux's Arch.* **181**, 135–149.

CdE   Rickoll, W. L., and Fristrom, J. W. (1983). The effects of 20-hydroxyecdysone on the metabolic labeling of membrane proteins in *Drosophila* imaginal discs. *Dev. Biol.* **95**, 275–287.

CdE   Rickoll, W. L., and Galewsky, S. (1987). Antibodies recognizing 20-hydroxyecdysone-dependent cell surface antigens during morphogenesis in *Drosophila*. *W. Roux's Arch.* **196**, 434–444.

E   Rickoll, W. L., and Galewsky, S. (1988). 20-hydroxyecdysone increases the metabolic labeling of extra cellular glycoprotein in *Drosophila* S3 cells. *Insect Biochem.* **18**, 337–345.

E   Rickoll, W. L., Stachowiak, J. A., Galewsky, S., Junio, M. A., and Hayes, E. S. (1986). Differential effects of 20-hydroxyecdysone on cell interactions and surface proteins in *Drosophila* cell lines. *Insect Biochem.* **16**, 211–224.

HET   Riddihough, G., and Pelham, H. R. B. (1986). Activation of the *Drosophila* hsp27 promoter by heat shock and by ecdysone involves independant and remote regulatory sequences. *EMBO J.* **5**, 1653–1658.

HET   Riddihough, G., and Pelham, H. R. B. (1987). An ecdysone response element in the *Drosophila* hsp27 promoter. *EMBO J.* **6**, 3729–3734.

B   Riemer, D., and Weber, K. (1994). The organization of *Drosophila* lamin C: limited homology with vertebrate genes and lack of homology versus the *Drosophila* lamin *Dmo* gene. *Eur. J. Cell Biol.* **63**, 299–306.

M   Rio, D. C. (1988). Accurate and efficient pre-mRNA splicing in *Drosophila* cell-free extracts. *Proc. Natl. Acad. Sci. U.S.A.* **85**, 2904–2908.

rRT   Rio, D. C. (1991). Regulation of *Drosophila* P-element transposition. *Trends Genet.* **7**, 282–287.

PT   Rio, D. C., Laski, F. A., and Rubin, G. (1986). Identification and Immunochemical analysis of biologically active *Drosophila* P element transposase. *Cell* **44**, 21–32.

T   Rio, D. C., and Rubin, G. M. (1985). Transformation of cultured *Drosophila melanogaster* cells with a dominant selectable marker. *Mol. Cell. Biol.* **5**, 1833–1838.

R   Rio, D. C., and Rubin, G. M. (1988). Identification and purification of a *Drosophila* protein that binds to the terminal 31 bp inverted repeats of the P transposable element. *Proc. Natl. Acad. Sci. U.S.A.* **85**, 8929–8933.

M   Risau, W., Symmons, P., Saumweber, H., and Frasch, M. (1983). Non packaging and packaging proteins of hnRNA in *Drosophila melanogaster*. *Cell* **33**, 529–541.

CpV   Rizki, R. M., and Rizki, T. M. (1990). Parasitoid virus-like particles destroy *Drosophila* cellular immunity. *Proc. Natl. Acad. Sci. U.S.A.* **87**, 8388–8392.

B   Rizki, R. M., and Rizki, T. M. (1991). Effect of Lamellolysin from a parasitoid wasp on *Drosophila* blood cells *in vitro*. *J. Exp. Zool.* **257**, 236–244.

B   Rizki, R. M., Rizki, T. M., and Andrews, C. A. (1975). *Drosophila* cell fusion induced by wheat germ agglutinin. *J. Cell Sci.* **18**, 113–121.

V   Rizki, T. M., and Rizki, R. M. (1994). Parasitoid induced cellular immunodeficiency in *Drosophila*. *Ann. NY Acad. Sci.* **712**, 178–184.

B   Rizki, T. M., Rizki, R., and Andrews, C. A. (1977). The surface features of *Drosophila* embryonic cell lines. *Dev. Growth Differ.* **19**, 345–356.

Cy   Rizzino, A., and Blumenthal, A. B. (1978). Synchronization of *Drosophila* cells in culture. *In Vitro* **14**, 437–442.

Cd   Robb, J. A. (1969). Maintenance of imaginal discs of *Drosophila melanogaster* in chemically defined media. *J. Cell Biol.* **41**, 876–884.

B Roberts, D. B. (1975). *Drosophila* antigens: their spatial and temporal distribution, their function and control. *Curr. Top. Dev. Biol.* **9**, 167–189.

M Robinson, M. J., Martin, B. A., Gootz, T. D., McGuirk, P. R., Moynihan, M., Sutcliffe, J. A., and Osheroff, N. (1991). Effects of quinolone derivatives on eukaryotic Topoisomerase II. A novel mechanism for enhancement of enzymatic-mediated DNA cleavage. *J. Biol. Chem.* **266**, 14585–14592.

EH Rollet, E. (1988). Etude de régulations post-transcriptionnelles dans des cellules de Drosophile sous l'effet de l'ecdysterone et en réponse au stress cellulaire. *Thèse Doctorat d'Etat*, Univ, Paris VI, France.

HE Rollet, E., and Best-Belpomme, M. (1986). *Hsp26* and *27* are phosphorylated in response to heat-shock and ecdysterone in *Drosophila melanogaster* cells. *Biochem. Biophys. Res. commun.* **141**, 426–433.

E Ropp, M., Calvayrac, R., and Best-Belpomme, M. (1986). Ecdysterone enhances cellular oxygen consumption and superoxide dismutase activity in *Drosophila* cells. In *"Superoxide and superoxide dismutase in Chemistry, Biology and Medicine"* (G. Rotillo, ed.), pp. 313–315. Elsevier, New York.

H Ropp, M., Courgeon, A. M., Calvayrac, R., and Best-Belpomme, M. (1983). The possible role of the superoxide ion in the induction of heat-shock and specific proteins in aerobic *Drosophila* cells during return to normoxia after a period of anaerobiosis. *Can. J. Biochem. Cell Biol.* **61**, 456–461.

B Rosetto, M., Engström, Y., Baldari, C. T., Telford, J. L., and Hultmark, D. (1995). Signals from the IL-1 receptor homolog, *Toll,* can activate an immune response in a *Drosophila* hemocyte cell line. *Biochem. Biophys. Res. Commun.* **209**, 111–116.

E Rosset, R. (1978). Effects of ecdysone on a *Drosophila* cell line. *Exp. Cell Res.* **111**, 31–36.

H Rougvie, A. E., and Lis, J. T. (1988). The RNA polymerase II molecule at the 5′ end of the uninduced *hsp70* gene of *Drosophila melanogaster* is transcriptionally engaged. *Cell* **54**, 795–804.

H Rougvie, A. E., and Lis, J. T. (1990). Post initiation transcriptional control in *Drosophila melanogaster.* *Mol. Cell. Biol.* **10**, 6041–6045.

H Rowe, T. C., Wang, J. C., and Liu, L. F. (1986). *In vivo* localization of DNA topoisomerase II cleavage sites on *Drosophila* heat-shock chromatin. *Mol. Cell. Biol.* **6**, 985–992.

rR Rubin, G. M. (1983). Dispersed repetitive DNA in *Drosophila.* In *"Mobile Genetic Elements"* Chap. 8, pp. 329–361. Academic Press, New York.

R Rubin, G. M., Brorein, W. J., Dunsmuir, P., Flavell, A. J., Strobel, J. J., Toole, J. J., and Young, E. (1981). *Copia*-like transposable elements in the *Drosophila* genome. *Cold Spring Harbor Symp. Quant. Biol.* **45**, 619–628.

rR Rubin, G. M., Finnegan, D. J., and Hogness, D. S. (1976). The chromosomal arrangement of coding sequences in a family of repeated genes. In *"Progress in Nucleic Acid Research"* (W. Cohn and E. Volkin, eds.), Vol. 19, pp. 221–226.

H Rubin, G. M., and Hogness, D. S. (1975). Effect of heat-shock on the synthesis of low molecular weight RNAs in *Drosophila:* Accumulation of a novel form of 5S RNA. *Cell* **6**, 207–213.

TM Rushlow, C. A., Han, K., Manley, J. L., and Levine, M. (1989). The graded distribution of the *dorsal* morphogen is initiated by selective nuclear transport in *Drosophila.* *Cell* **59**, 1165–1177.

MT Russo, M. W., Matheny, C., and Milbrandt, J. (1993). Transcriptional activity of the zing finger protein NGFI-A is influenced by its interaction with a cellular factor. *Mol. Cell. Biol.* **13**, 6858–6865.

M   Ryffel, G. U., Doenecke, D., Nasser, D. S., Levy, W. B., Siegel, J. G., and McCarthy, B. J. (1975). The distribution of histones and specific DNA sequences in fractionated chromatin. *Proc. ICN-UCLA Winter Conf. Dev. Biol.,* Squaw Valley, CA.

F   Sadowski, H. B., and Gilman, M. Z. (1993). Cell-free activation of a DNA-binding protein by EGF. *Nature* **362**, 79–83.

E   Sage, B. A., Horn, D. H. S., Landon, T. M., and O'Connor, J. D. (1986). Alternative ligands for measurement and purification of ecdysteroid receptors in *Drosophila Kc* cells. *Arch. Insect Biochem. Physiol.* suppl. I, 25–33.

E   Sage, B. A., and O'Connor, J. D. (1985). Measurement and characterization of ecdysteroid receptors. *Meth. Enzymol.* **111**, 458–468.

E   Sage, B. A., Tanis, M. A., and O'Connor, J. D. (1982). Characterization of ecdysteroid receptors in cytosol and naive nuclear preparations of *Drosophila Kc* cells. *J. Biol. Chem.* **257**, 6373–6379.

M   Saigo, K., Millstein, L., and Thomas, C. A. (1981). The organization of *Drosophila melanogaster* Histone genes. *Cold Spring Harbor Symp. Quant. Biol.* **45**, 815–827.

V   Saigo, K., Reilly, J. G., and Thomas, C. A., Jr. (1980). Double-stranded RNA in *Drosophila melanogaster* cultured cells. *Biochem. Biophys. Acta* **607**, 530–535.

R   Saigo, K., Shiba, T., and Miyake, T. (1980). Virus-like particles of *Drosophila melanogaster* containing t-RNA and 5S ribosomal RNA 1. Isolation and purification from cultured cells and detection of low molecular weight RNAs in the particles. *In Invertebrate Systems in Vitro* (E. Kurstak, K. Maramorosch, and A. Dubendorfer, eds.), pp. 411–424. Elsevier North-Holland Biomedical Press, Amsterdam.

M   Saigo, K., Ueda, R., and Miyake, T. (1983). Polymorphism and stability of histone gene clusters in *Drosophila melanogaster* cultured cells. *Biochem. Biophys. Acta* **740**, 390–401.

M   Saigo, K., Ueda, R., and Miyake, T. (1986). Structural stability of histone gene clusters in *Drosophila melanogaster* cultured cells. Techniques in the Life Sciences, C2, *In Vitro Invertebrate Hormones and Genes,* C217 (Kurstak, E., ed.), 1–7, Elsevier Sci. Publ. Ireland.

Cp Saito, M., and Wu, C. F. (1993). Ionic channels in cultured *Drosophila* neurons. *In Comparative and Molecular Neurobiology* (Y. Pichon, ed.), pp. 366–389, Birkauser, Basel.

Cp Salvaterra, P. M., Bournias-Vardiabasis, N., Nair, T., Hou, G., and Lieu, C. (1987). *In vitro* neuronal differentiation of *Drosophila* embryonic cells. *J. Neurosci.* **7**, 10–22.

B   Samakovlis, C., Asling, B., Boman, H. G., Gateff, E., and Hultmark, D. (1992). *In vitro* induction of *cecropin*-genes, an immune response in a *Drosophila* blood cell line. *Biochim. Biophys. Res. Commun.* **188**, 1169–1175.

B   Samakovlis, C., Kimbrell, D. A., Kylsten, P., Engström, A., and Hultmark, D. (1990). The immune response in *Drosophila*: pattern of *cecropin* expression and biological activity. *EMBO J.* **9**, 2969–2976.

F   Sampath, T., Rashka, K. E., Doctor, J. S., Tucker, R. F., and Hoffmann, F. M. (1993). *Drosophila* Transforming Growth Factor β superfamily proteins induces endochondral bone formation in mammals. *Proc. Natl. Acad. Sci. U.S.A.* **90**, 6004–6008.

M   Sanders, M. M. (1981). Identification of H2b as a heat shock protein in *Drosophila* (Abst. 21st Ann. Meeting Am. Soc. Cell Biol., 3070) *J. Cell Biol.* **91**(2).

H   Sanders, M. M. (1981). Identification of histone *H2b* as a heat-shock protein in *Drosophila*. *J. Cell Biol.* **91**, 579–583.

H   Sanders, M. M., John-Alder, K., and Sherwood, A. C. (1988). Anion transport is linked to heat shock induction. (Abstr. P004, UCLA Symp. Mol. Cell. Biol.) *J. Cell. Biochem.* **12D**(Suppl.) 241.

HM   Sanders, M. M., Feeney-Triemer, D., Olsen, A. S., and Farrell-Towt, J. (1982). Changes in protein phosphorylation and Histone H2b disposition in heat-shock in *Drosophila*. In "*Heat-shock from Bacteria to Man*" (M. J. Schlesinger, M. Ashburner, and A. Tissières, eds.), pp. 235–242, Cold Spring Harbor Laboratory, Cold Spring Harbor.

H   Sanders, M. M., Triemer, D. F., and Brief, B. (1982). Heat-shock changes the cytoskeleton in *Drosophila*. *J. Cell Biol.* **95**, (Abst. 11044), 237a.

H   Sanders, M. M., Triemer, D. F., and Olsen, A. (1985) The regulation of translation in heat shock in *Drosophila*. In "1985 meeting on Heat shock" (E. Craig, *et al.*, eds.), Cold Spring Harbor Laboratory Press, Cold Spring Harbor, 76. [Abstract]

H   Sanders, M. M., Triemer, D. F., and Olsen, A. S. (1986). Regulation of protein synthesis in heat-shocked *Drosophila* cells. *J. Biol. Chem.* **261**, 2189–2196.

rCl   Sang, J. H. (1980). *Drosophila* cell lines. (5th Intern. Conf. Invert. Tissue Cult., Rigi-Kaltbad, Apr. 79) *Invert. Systems in Vitro* (E. Kurstak, K. Maramorosch, and A. Dübendorfer, eds.), pp. 3–11. Elsevier/North Holland Biomedical Press, Amsterdam.

rCpCl   Sang, J. H. (1981). *Drosophila* cells and cell lines. In *Adv. Cell Cult.* (K. Maramorosch, ed.), Vol. 1, pp. 125–182. Academic Press, New York.

T   Sang, J. H., and Sinclair, J. H. (1985). The use of transfected tissue culture cells to study regulation of *Drosophila* yolk protein genes. *In Vitro* **21**, 42A (Ann. meeting Abst. N° 127).

T   Sang, J. H., Sinclair, J. H., Burke, J. F., and Ish-Horowicz, D. (1984). Rescue of a *Drosophila* temperature-sensitive mutant cell line by DNA transfection. *Somatic Cell Mol. Gen.* **10**, 573–577.

M   Santaren, J. F., Van Damme, J., Puype, M., Vandekerckhove, J., and Garcia-Bellido, A. (1993). Identification of *Drosophila* wing imaginal disc proteins by two-dimensional gel analysis and microsequencing. *Exp. Cell Res.* **206**, 220–226.

TM   Santoro, C., Mermod, N., Andrews, P. C., and Tijan, R. (1988). A family of human CCAAT-box binding proteins active in transcription and DNA replication: cloning and expression of multiple cDNAs. *Nature* **334**, 218–224.

B   Sap, J., Jiang, Y. P., Friedlander, D., Grumet, M., and Schlessinger, J. (1994). Receptor tyrosine phosphatase *R-PTP-k* mediates homophilic binding. *Mol. Cell. Biol.* **14**, 1–9.

T   Sartorelli, V., Webster, K. A., and Kedes, L. (1990). Muscle-specific expression of the cardiac alpha-actin gene requires MyoD1, CArG-box binding factor and Sp1. *Genes Dev.* **4**, 1811–1822.

T   Sass, H. (1990). P-transposable vectors expressing a constitutive and thermoinducible *hsp82-neo* fusion gene for *Drosophila* germline transformation and tissue-culture transfection. *Gene* **89**, 179–186.

ECl   Sater, A. K., Woods, D. F., and Poodry, C. A. (1984). Cell surface proteins of *Drosophila*. II. A comparison of embryonic and ecdysone-induced proteins. *Dev. Biol.* **104**, 1–8.

Ct   Sato, J. D., and Roberts, D. B. (1983). Synthesis of *Larval Serum Protein 1* and 2 of *Drosophila melanogaster* by third instar fat body. *Insect Biochem.* **13**, 1–5.

T   Sauer, F., and Jäckle, H. (1991). Concentration-dependent transcriptional activation or repression by *Krüppel* from a single binding site. *Nature* **353**, 563–566.

MT   Sauer, F., and Jäckle, H. (1993). Dimerization and the control of transcription by *krüppel*. *Nature* **364**, 454–457.

M   Saumweber, H., Symmons, P., Kabisch, R., Will, H., and Bonhoeffer, F. (1980). Monoclonal antibodies against chromosomal proteins of *Drosophila melanogaster*. Establishing of antibody producing lines and partial characterization of corresponding antigens. *Chromosoma* **80**, 253–275.

V Saunders, K., and Kaesberg, P. (1985). Template-dependent RNA polymerase from Black Beetle Virus-infected *Drosophila melanogaster* cells. *Virology* **147**, 373–381.

T Saunders, S. E., Rawls, J. M., Wardle, C. J., and Burke, J. F. (1989). High efficiency expression of transfected genes in *Drosophila melanogaster* haploid (*1182*) cell line. *Nucleic Acids Res.* **17**, 6205–6216 + Corrigendum 6775–6776.

E Savakis, C., Demetri, G., and Cherbas, P. (1980). Ecdysteroid-inducible polypeptides in a *Drosophila* cell line. *Cell* **22**, 665–674.

E Savakis, C., Koehler, M. M. D., and Cherbas, P. (1984). cDNA clones for the ecdysone-inducible polypeptide (EIP) mRNAs of *Drosophila Kc* cells. *EMBO J.* **3**, 235–243.

E Savre-Train, I. (1990). Effet de l'hormone juvénile sur les activités mitochondriales des cellules KcO% de Drosophile. *Thèse Univ. Blaise Pascal*, Clermont-Ferrand, France.

E Schaltmann, K., and Pongs, O. (1982). Identification and characterization of the ecdysone receptor in *Drosophila melanogaster* by photoaffinity labeling. *Proc. Natl. Acad. Sci. U.S.A.* **79**, 6–10.

H Schedl, P., Artavanis-Tsakonas, S., Steward, R., Gehring, W. J., Mirault, M. E., Goldschmidt-Clermont, M., Moran, L., and Tissières, A. (1978). Two hybrid plasmids with *D. melanogaster* sequences complementary to mRNA coding for the major heat shock protein. *Cell* **14**, 921–929.

R Scherer, G., Tschudi, C., Perera, J., Dekius, H., and Pirotta, V. (1982). *B104*, a new dispersed repeated gene family in *Drosophila melanogaster* and its analogies with retroviruses. *J. Mol. Biol.* **157**, 435–452.

T Schmitz, M. L., and Bauerle, P. A. (1991). The p65 subunit is responsible for the strong transcription activating potential of NF-kB. *EMBO J.* **10**, 3805–3817.

V Schneemann, A., Gallagher, T. M., and Rueckert, R. R. (1994). Reconstitution of Flock House provirions: a model system for studying structure and assembly. *J. Virol.* **68**, 4547–4556.

Ct Schneider, I. (1963). *In vitro* culture of *Drosophila* organs and tissues. (Abstr. 908, Meet. Genet. Soc. America) *Genetics* **48**, 908.

Cd Schneider, I. (1964). Differentiation of larval *Drosophila* eye-antennal discs *in vitro*. *J. Exp. Zool.* **156**, 91–104.

Cd Schneider, I. (1966). Histology of larval eye-antennal disks and cephalic ganglia of *Drosophila* cultured *in vitro*. *J. Embryol. Exp. Morphol.* **15**, 271–279.

rCt Schneider, I. (1967). Insect tissue culture. *In* "*Methods in Developmental Biology*" (F. H. Wilt, and N. K. Wessells, eds.), Crowel-Collier, New York.

rC Schneider, I. (1971). Cultivation of dipteran cells *in vitro*. *In Arthropod Cell Cultures and their Applications to the Study of Viruses* (E. Weiss, ed.), pp. 1–12, Springer-Verlag, New York.

Cl Schneider, I. (1971). Embryonic cell lines of *Drosophila melanogaster*. *Dros. Inf. Serv.* **46**, 111.

Cl Schneider, I. (1972). Cell lines derived from late embryonic stages of *Drosophila melanogaster*. *J. Embryol. Exp. Morphol.* **27**, 353–365.

C Schneider, I. (1988). Citation Classic. *Current Contents* 7, 17.

rC Schneider, I., and Blumenthal, A. B. (1978). *Drosophila* cell and tissue culture. *In* "*The Genetics and Biology of Drosophila*" (M. Ashburner and T. R. F. Wright, eds.), Vol. 2a, pp. 265–315. Academic Press, New York.

E   Schräder, M., Wyss, A., Sturzenbecker, L. J., Grippo, J. F., Lemotte, P., and Carlberg, C. (1993). RXR-dependent and RXR-independent transactivations by retinoic receptors. *Nucleic Acids Res.* **21**, 1231–1237.

H   Schuldt, C., and Kloetzel, P. M. (1985). Analysis of cytoplasmic 19S ring-type particles in *Drosophila* which contain *hsp23* at normal growth temperature. *Dev. Biol.* **110**, 65–74.

H   Schuldt, C., Kloetzel, P. M., and Bautz, E. K. (1989). Molecular organization of RNP complexes containing *P11* antigen in heat-shocked and non-heat-shocked *Drosophila* cells. *Eur. J. Biochem.* **181**, 135–142.

E   Schulz, R. A., Cherbas, L., and Cherbas, P. (1986). Alternative splicing generates two distinct *EIP28/29* gene transcripts in *Drosophila* Kc cells. *Proc. Natl. Acad. Sci. U.S.A.* **83**, 9428–9432.

E   Schulz, R. A., Shlomchik, W., Cherbas, L., and Cherbas, P. (1989). The *Drosophila Eip28/29* gene and its upstream neighbors: Structural overlap and diverse expression. *Dev. Biol.* **131**, 515–523.

R   Schwartz, H. E., Lockett, T. J., and Young, M. W. (1982). Analysis of Transcripts from two families of nomadic DNA. *J. Mol. Biol.* **157**, 49–58.

B   Schweitzer, R., Shaharabany, M., Seger, R., and Shilo, B. Z. (1995). Secreted *Spitz* triggers the DER signaling pathway and its limiting component in embryonic ventral ectoderm determination. *Genes Dev.* **9**, 1518–1529.

HV   Scott, M. P., Fostel, J. M., and Pardue, M. L. (1980). A new type of virus from cultured *Drosophila* cells: Characterization and use in studies of the heat-shock response. *Cell* **22**, 929–941.

M   Scott, M. P., and Pardue, M. L. (1981). Translational control in lysates of *Drosophila melanogaster* cells. *Proc. Natl. Acad. Sci. U.S.A.* **78**, 3353–3357.

M   Scott, M. P., Storti, R. V., Pardue, M. L., and Rich, A. (1979). Cell-free protein synthesis in lysates of *Drosophila melanogaster* cells. *Biochemistry* **18**, 1588–1594.

V   Scotti, P. D. (1976). Cricket paralysis virus replicates in cultured *Drosophila* cells. *Intervirology* **6**, 333–342.

V   Scotti, P. D. (1977). End-point dilution and Plaque Assay methods for titration of Cricket Paralysis Virus in cultured *Drosophila* cells. *J. Gen. Virol.* **35**, 393–396.

C   Sederoff, R., and Clynes, R. (1974). A modified medium for culture of *Drosophila* cells. *Dros. Inf. Serv.* **51**, 153.

M   Sederoff, R., Lowenstein, L., and Birnboim, H. C. (1975). Polypyrimidine segments in *Drosophila melanogaster* DNA. II. Chromosome location and nucleotide sequences. *Cell* **5**, 183–194.

V   Seecof, R. L. (1969). Sigma Virus multiplication in whole-animal culture of *Drosophila*. *Virology* **38**, 134–139.

C   Seecof, R. L. (1971). Phosphate-buffered saline for *Drosophila*. *Dros. Inf. Serv.* **46**, 113.

Cp   Seecof, R. L. (1980). Differentiation of primary embryonic neuroblasts in purified neural cell cultures in *Drosophila*. *TCA Manual* **5**, 1019–1022.

Cp   Seecof, R. L., and Alléaume, N. (1968). Differentiation of cells in culture made from single *Drosophila melanogaster* embryos. *Genetics* **60**, 224. [Abstract]

Cp   Seecof, R. L., Alléaume, N., Teplitz, R. L., and Gerson, I. (1971). Differentiation of neurons and myocytes in cell cultures made from *Drosophila* gastrulae. *Exp. Cell Res.* **69**, 161–173.

CpE Seecof, R. L., and Dewhurst, S. (1974). Insulin is a *Drosophila* hormone and acts to enhance the differentiation of embryonic *Drosophila* cells. *Cell Different.* **3**, 63–70.

Cp   Seecof, R. L., and Dewhurst, S. (1976). A 5-bromodeoxyuridine-sensitive interval during *Drosophila* myogenesis. *Differentiation* **6**, 27–32.

Cp   Seecof, R. L., and Donady, J. J. (1971). *Drosophila* myoblast differentiation *in vitro*. (Meeting Genet. Soc. America) *Genetics*, **68**. [Abstract s60].

Cp   Seecof, R. L., and Donady, J. J. (1972). Factors affecting *Drosophila* neuron and myocyte differentiation. *Mech. Ageing Dev.* **1**, 165–174.

Cp   Seecof, R. L., Donady, J. J., and Fiorio, P. (1973). Formation of axon to myocyte contacts in *Drosophila* cell cultures. *Am. Zool.* **13**, 331–336.

Cp   Seecof, R. L., Donady, J. J., and Teplitz, R. L. (1973). Differentiation of *Drosophila* neuroblasts to form ganglion-like clusters of neurons *in vitro*. *Cell Different.* **2**, 143–149.

Cp   Seecof, R. L., Gerson, I., Donady, J. J., and Teplitz, R. L. (1973). *Drosophila* myogenesis *in vitro*: The genesis of "small" myocytes and myotubes. *Dev. Biol.* **35**, 250–261.

Cp   Seecof, R. L., and Teplitz, R. L. M. (1971). *Drosophila* neuron differentiation *in vitro*. *Curr. Top. Microb. Immunol. In "Arthropod Cell Cultures and their Applications to the Study of Viruses"* (E. Weiss, ed.), **55**, 71–75.

Cp   Seecof, R. L., Teplitz, R. L., Gerson, I., Ikeda, K., and Donady, J. J. (1972). Differentiation of neuromuscular junctions in culture of embryonic *Drosophila* cells. *Proc. Natl. Acad. Sci. U.S.A.* **69**, 566–570.

Cp   Seecof, R. L., and Unanue, R. L. (1968). Differentiation of embryonic *Drosophila* cells *in vitro*. *Exp. Cell Res.* **50**, 654–660.

V   Selling, B. (1986). Infectivity of Black Beetle Virus in cultured *Drosophila* cells. Ph.D. Thesis, University of Wisconsin, Madison, WI.

V   Selling, B. H., and Rueckert, R. R. (1984). Plaque Assay for Black Beetle Virus. *J. Virol.* **51**, 251–253.

Cd   Sengel, P., and Mandaron, P. (1969). Aspects morphologiques du développement *in vitro* des disques imaginaux de la Drosophile. *CR Acad. Sci, Paris* **268**, 405–407.

M   Seto, E., Lewis, B., and Shenk, T. (1993). Interaction between transcription factors Sp1 and YY1. *Nature* **365**, 462–464.

B   Shalev, A., Pla, M., Ginsburger-Vogel, T., Echalier, G., Logdberg, L., Bjorck, L., Colombani, J., and Segal, S. (1983). Evidence for alpha-microglobulin-like and H2-like antigenic determinants in *Drosophila*. *J. Immunol.* **130**, 297–302.

M   Sharkov, I. V., Filippova, M. A., Strots, O. V., Borisevich, I. V., Bogachev, S. S., Baricheva, E. M., and Shilov, A. G. (1993). (in Russian) *Drosophila* β-heterochromatin: molecular organization and function. Characterization of the DNA sequences from proximal β-heterochromatin, associated with the nuclear envelope of *Drosophila melanogaster*. *Genetika* **29**, 393–402.

M   Sharp, S., DeFranco, D., Silberklang, M., Hosbach, H. A., Schmidt, T., Kubli, E., Gergen, J. P., Wensink, P. C., and Söll, D. (1981). The initiator tRNA genes of *Drosophila melanogaster*: evidence for a tRNA pseudogene. *Nucleic Acids Res.* **9**, 5867–5882.

Cd   Shearn, A., Davis, K. T., and Hersperger, E. (1978). Transdetermination of *Drosophila* imaginal discs cultured *in vitro*. *Dev. Biol.* **65**, 536–540.

Cd   Shearn, A., Davis, K. T., and Hersperger, E. (1980). Transdetermination of *Drosophila* imaginal discs cultured *in vitro*. *In "Invertebrate Systems in Vitro"* (E. Kurstak, K. Maramorosch, and A. Dübendorfer, eds.), pp. 125–133, Elsevier/North Holland Biomedical Press, Amsterdam.

CtE   Shemshedini, L., Lanoue, M., and Wilson, T. G. (1990). Evidence for a JH receptor involved in protein synthesis in *Drosophila melanogaster*. *J. Biol. Chem.* **265**, 1913–1918.

R   Shepherd, B. M., and Finnegan, D. J. (1984). Structure of circular copies of the *412* transposable element present in *Drosophila melanogaster* tissue culture cells, and isolation of a free *412* Long Terminal Repeat. *J. Mol. Biol.* **180**, 21–40.

BT   Sherwood, A. C., John-Alder, K., Biessmann, H., and Sanders, M. M. (1989). Overexpression of a 123-kDa anion transport inhibitor binding protein and two cytoskeleton proteins in *Drosophila Kc* cell variants resistant to disulfonic stilbenes. *J. Biol. Chem.* **264**, 1829–1836.

H   Sherwood, A. C., and Sanders, M. M. (1984). Ion movements occur during the heat-shock response in *Drosophila Kc* cells. *J. Cell Biol.* **99**, Abst. 1673 Am. Soc. Cell Biol., 452a.

R   Shiba, T., and Saigo, K. (1983). Retrovirus-like particles containing RNA homologous to the transposable element *copia* in *Drosophila melanogaster*. *Nature* **302**, 119–124.

R   Shiba, T., Saigo, K., and Miyake, T. (1980). VLPs of *Drosophila melanogaster* containing t-RNA and 5S ribosomal RNA II Isolation and characterization of low molecular weight RNAs. *In "Invertebrate Systems in Vitro"* (K. Maramorosch and A. Dübendorfer, eds.), pp. 425–433. Elsevier/North-Holland Medical Press, Amsterdam.

R   Shiba, T., Suzuki, Y., Miyake, T., and Saigo, K. (1986). Relation between a retrovirus and movable genetic element in *Drosophila melanogaster*. *In "Techniques in In Vitro Invertebrate Hormones and Genes"* (E. Kurstak, ed.), pp. 1–15.

Cp   Shields, G., Dübendorfer, A., and Sang, J. H. (1975). Differentiation *in vitro* of larval cell types from early embryonic cells of *Drosophila melanogaster*. *J. Embryol. Exp. Morphol.* **33**, 159–175.

Cp   Shields, G., and Sang, J. H. (1970). Characteristics of five cell types appearing during *in vitro* culture of embryonic material from *Drosophila melanogaster*. *J. Embryol. Exp. Morphol.* **23**, 53–69.

C   Shields, G., and Sang, J. H. (1977). Improved medium for culture of *Drosophila* embryonic cells. *Dros. Inf. Serv.* **52**, 161.

F   Shilo, B. Z., and Weinberg, R. A. (1981). DNA sequences homologous to Vertebrate oncogenes are conserved in *Drosophila melanogaster*. *Proc. Natl. Acad. Sci. U.S.A.* **78**, 6789–6792.

Cp   Shimada, Y., and Kuroda, Y. (1985). Differentiation of embryonic cells of *Drosophila melanogaster* studied with the electron microscope. *Ann. Rep. Nat. Inst. Genet., Japan* **36**, 44.

Cp   Shimada, Y., and Kuroda, Y. (1986). The fine structure of embryonic cells of *Drosophila melanogaster* cultured in the presence of ecdysterone. *Ann. Rep. Nat. Inst. Genet., Japan* **37**, 47.

M   Shinomiya, T., and Ina, S. (1993). DNA replication of Histone gene repeats in *Drosophila melanogaster* tissue culture cells: multiple initiation sites and replication pause sites. *Mol. Cell. Biol.* **13**, 4098–4106.

H   Shuey, D. J., and Parker, C. S. (1986a). Bending of promoter DNA on binding of heat shock transcription factor. *Nature* **323**, 459–461.

H   Shuey, D. J., and Parker, C. S. (1986b). Binding of *Drosophila* heat shock gene transcription factor to the *hsp70* promoter. *J. Biol. Chem.* **261**, 7934–7940.

R   Siebel, C. W., Fresco, L. D., and Rio, D. C. (1992). The mechanism of somatic inhibition of *Drosophila* P-element pre-mRNA splicing: multiprotein complexes at an exon pseudo-5' splice site control U1 snRNP binding. *Genes Dev.* **6**, 1386–1401.

R   Siebel, C. W., and Rio, D. C. (1990). Regulated splicing of the *Drosophila P* transposable element 3rd intron *in vitro*: Somatic repression. *Science* **24**, 1200–1208.

CdE Siegel, J., and Fristrom, J. W. (1974). The effect of β-ecdysone on protein synthesis in imaginal discs of *Drosophila melanogaster* cultured *in vitro*. II. Effects on synthesis in specific cell fractions. *Dev. Biol.* **41**, 314–330.

rCd Siegel, J., and Fristrom, J. W. (1978). The biochemistry of imaginal disc development. *In* "The Genetics and Biology of Drosophila" (M. Ashburner and T. R. F. Wright, eds.), Vol. 2a, pp. 317–394, Academic Press, New York.

H Sierra, J. M., and Zapata, J. M. (1994). Translational regulation of the heatshock response. *Mol. Biol. Rep.* **19**, 211–220.

M Silberkang, M., Havel, C., Friend, D. S., McCarthy, B. J., and Watson, J. A. (1983). Isoprene synthesis in isolated embryonic *Drosophila* cells. I. Sterol-deficient eukaryotic cells. *J. Biol. Chem.* **258**, 8503–8511.

rCd Silvert, D. J., and Fristrom, J. W. (1980). Biochemistry of imaginal discs: retrospect and prospect. *Insect Biochem.* **10**, 341–355.

HT Simcox, A. M., Cheney, C. M., and Shearn, A. (1985). A deletion of the 3' end of the *Drosophila melanogaster* hsp70 gene increases stability of mutant mRNA during recovery from heat shock. *Mol. Cell. Biol.* **5**, 3397–3402.

Cl Simcox, A. A., Sobeih, M. M., and Shearn, A. (1985). Establishment and characterization of continuous cell lines derived from temperature-sensitive mutants of *Drosophila melanogaster*. *Somat. Cell Mol. Genet.* **11**, 63–70.

T Simon, M. A., Bowtell, D. D., and Rubin, G. M. (1989). Structure and activity of the *sevenless* protein: a protein tyrosine kinase receptor required for photoreceptor development in *Drosophila*. *Proc. Natl. Acad. Sci. U.S.A.* **86**, 8333–8337.

F Simon, M. A., Kornberg, T. B., and Bishop, J. M. (1983). Three loci related to the *src* oncogene and tyrosine-specific protein kinase activity in *Drosophila*. *Nature* **302**, 837.

TV Sinclair, J. H. (1987). The human cytomegalovirus immediate early gene promoter is a strong promoter in cultured *Drosophila melanogaster* cells. *Nucleic Acids Res.* **15**, 2392.

TE Sinclair, J. H., and Bryant, L. A. (1987). 20-hydroxyecdysone increases levels of transient gene expression in transfected *Drosophila* cells. *Nucleic Acids Res.* **15**, 9255–9261.

TR Sinclair, J. H., Burke, J. F., Ish-Horowicz, D., and Sang, J. H. (1986). Functional analysis of the transcriptional control regions of the *copia* transposable element. *EMBO J.* **5**, 2349–2354.

TR Sinclair, J. H., Sang, J. H., Burke, J. F., and Ish-Horowicz, D. (1983). Extra-chromosomal replication of *copia*-based vectors in cultured *Drosophila* cells. *Nature* **306**, 198–200.

TH Sinclair, J. H., Saunders, S. E., Burke, J. F., and Sang, J. H. (1985). Regulated expression of a *Drosophila melanogaster* Heat shock locus after stable integration in a *Drosophila hydei* cell line. *Mol. Cell Biol.* **5**, 3208–3213.

CtC Singleton, K., and Woodruff, R. I. (1994). The osmolarity of adult *Drosophila* hemolymph and its effect on oocyte-nurse cell electrical polarity. *Dev. Biol.* **161**, 154–167.

H Sinibaldi, R., and Morris, P. W. (1981). Putative function of *Drosophila melanogaster* heat shock proteins in the nucleoskeleton. *J. Biol. Chem.* **256**, 10735–10738.

T Sitzlers, S., Oldendurg, I., Petersen, G., and Bautz, E. K. F. (1991). Analysis of the promoter region of the house keeping gene *Dm140* by sequence comparison of *D. melanogaster* and *D. virilis*. *Gene* **100**, 155–162.

H Small, D., Nelkin, B., and Vogelstein, B. (1985). The association of transcribed genes with the nuclear matrix of *Drosophila* cells during heat-shock. *Nucleic Acids Res.* **13**, 2413–2431.

Cy Smith, D. E., and Fisher, P. A. (1989). Interconversion of *Drosophila* nuclear lamin isoforms during oogenesis, early embryogenesis, and upon entry of cultured cells into mitosis. *J. Cell Biol.* **108**, 255–265.

CyH Smith, D. E., Gruenbaum, Y., Berrios, M., and Fisher, P. A. (1987). Biosynthesis and interconversion of *Drosophila* nuclear lamin isoforms during normal growth and in response to heat shock. *J. Cell Biol.* **105**, 771–790.

T Smith, J., and Calos, M. P. (1995). Autonomous replication in *Drosophila melanogaster* tissue culture cells. *Chromosoma* **103**, 597–605.

M Smoller, D., Friedel, C., Schmid, A., Bettler, D., Lam, L., and Yedvobnick, B. (1990). The *Drosophila* neurogenic locus *mastermind* encodes a nuclear protein unusually rich in amino-acid homopolymer. *Genes Dev.* **4**, 1688–1700.

Cp Smouse, D., and Perrimon, N. (1990). Genetic dissection of a complex neurological mutant, *polyhc   _otic*, in *Drosophila*. *Dev. Biol.* **139**, 169–185.

RT Sneddon, A., and Flavell, A. J. (1989). The transcriptional control region of the *copia* retrotransposon. *Nucleic Acids Res.* **17**, 4025–4035.

MT Snow, P. M., Bieber, A. J., and Goodman, C. S. (1989). Fasciclin III: a novel homophilic adhesion molecule in *Drosophila*. *Cell* **59**, 313–323.

E Sobrier, M. L. (1983). Regulation par l'ecdystérone de l'expression génique dans les cellules de Drosophile en culture. *Thèse Doctorat*, Univ. Clermont-Ferrand, France.

EM Sobrier, M. L., Chapel, S., Couderc, J. L., Micard, D., Lecher, P., Somme-Martin, G., and Dastugue, B. (1989). 20-OH-ecdysone regulates 60C β-tubulin gene expression in *Kc* cells and during *Drosophila* development. *Exp. Cell Res.* **184**, 241–249.

EM Sobrier, M. L., Couderc, J. L., Chapel, S., and Dastugue, B. (1986). Expression of a new β-tubulin subunit is induced by 20-hydroxyecdysone in *Drosophila* cultured cells. *Biochem. Biophys. Res. Commun.* **134**, 191–200.

M Soeller, W. C., Poole, S. J., and Kornberg, T. (1988). *In vitro* transcription of the *Drosophila engrailed* gene. *Genes Dev.* **2**, 68–81.

H Solomon, J. M., Rossi, J. M., Golic, K., McGarry, T., and Lindquist, S. (1991). Changes in *hsp70* after thermotolerance and heat shock regulation in *Drosophila*. *New Biol.* **3**, 1106–1120.

Ct Sommé-Martin, G., Colardo, J., Beydon, P., Blais, C., Lepesant, J. A., and Lafont, R. (1990). P1 gene expression in *Drosophila* larval fat body. Induction by various ecdysteroids. *Arch. Insect Biochem. Physiol.* **15**, 43–56.

M Sommer, U., and Spindler, K. D. (1991a). Demonstration of β-N-acetyl-D-glucosaminidase and β-N-acetyl-D-hexosaminidase in *Drosophila Kc* cells. *Arch. Insect Biochem. Physiol.* **17**, 3–13.

M Sommer, U., and Spindler, K. D. (1991b). Physical properties of beta-N-acetyl-D-glucosaminidase and beta-N-acetyl-D-hexosaminidase from *Drosophila Kc* cells. *Arch. Insect Biochem. Physiol.* **18**, 45–53.

T Sondergaard, L. (1991). *Drosophila* cell culture as an expression system for foreign genes. *Abst. 12th EDRC*, Mainz, Sept. 1991.

Cl Sondergaard, L. (1994). Morphology of *Drosophila* S2 cells in different culture conditions (letter). *In Vitro Cell Dev. Biol. Anim.* **1**, 18–20.

Cl Sondermeijer, P. J. A., Derksen, J. W. M., and Lubsen, N. H. (1980). Established cell lines of *Drosophila hydei*. *In Vitro* **16**, 913–914.

H Sondermeijer, P. J. A., and Lubsen, N. H. (1978). Heat-shock peptides in *Drosophila hydei* and their synthesis *in vitro*. *Eur. J. Biochem.* **88**, 331–339.

BT   Song, E. S., Yang, Y., Jackson, M. R., and Peterson, P. A. (1994). *In vivo* regulation of the assembly and intracellular transport of Classes I Major Histocompatibility Complex molecules. *J. Biol. Chem.* **269**, 7024–7029.

H   Southgate, R., Mirault, M. E., Ayme, A., and Tissieres, A. (1985). Organization, sequences and induction of heat-shock genes. In *"Changes in Eukaryotic Gene Expression in Response to Environmental Stress"* (B. G. Atkinson and D. B. Walden, eds.), pp. 3–30, Academic Press, New York.

RT   Spana, C., Harrison, D. A., and Corces, V. G. (1988). The *Drosophila melanogaster suppressor of Hairy-wing* protein binds to specific sequences of the *gypsy* retrotransposon. *Genes Dev.* **2**, 1414–1423.

E   Spencer, C. A., Stevens, B., O'Connor, J. D., and Hodgetts, R. B. (1983). A novel form of DOPA decarboxylase produced in *Drosophila* cells in response to 20-hydroxyecdysone. *Can. J. Biochem. Cell Biol.* **61**, 818–825.

E   Spindler-Barth, M., and Spindler, K. D. (1987). Anti-ecdysteroids and receptors. In *"Receptor-mediated Antisteroid Action"* (Agarwal, ed.), Walter de Gruyten, Berlin.

H   Spradling, A., Pardue, M. L., and Penman, S. (1977). Messenger RNA in heat-shocked *Drosophila* cells. *J. Mol. Biol.* **109**, 559–587.

H   Spradling, A., Penman, S., and Pardue, M. L. (1975). Analysis of *Drosophila* mRNA by *in situ* hybridization: Sequences transcribed in normal and heat shocked cultured cells. *Cell* **4**, 395–404.

B   Spragg, J. H., Bebbington, C. R., and Roberts, D. B. (1982). Monoclonal antibodies recognizing cell surface antigens in *Drosophila melanogaster*. *Dev. Biol.* **89**, 339–352.

Cy   Spray, D. C., Cherbas, L., Cherbas, P., Morales, E. A., and Carrow, G. M. (1989). Ionic coupling and mitotic synchrony of siblings in a *Drosophila* cell line. *Exp. Cell Res.* **184**, 509–517.

B   Spring, J., Paine-Saunders, S. E., Hynes, R. O., and Bernfield, M. (1994). *Drosophila syndecan:* conservation of a cell-surface heparan sulfate proteoglycan. *Proc. Natl. Acad. Sci. U.S.A.* **91**, 3334–3338.

R   Stanfield, S., and Helinski, D. R. (1976). Small circular DNA in *Drosophila melanogaster*. *Cell* **9**, 333–345.

R   Stanfield, S. W., and Lengyel, J. A. (1979). Small circular DNA of *Drosophila melanogaster:* chromosomal homology and kinetic complexity. *Proc. Natl. Acad. Sci. U.S.A.* **76**, 6142–6146.

R   Stanfield, S. W., and Lengyel, J. A. (1980). Small circular DNA of *Drosophila melanogaster:* homologous transcripts in the nucleus and cytoplasm. *Biochemistry* **19**, 3873–3878.

M   Staufenbiel, M., and Deppert, W. (1982). Intermediate filament systems are collapsed onto the nuclear surface after isolation of nuclei from tissue culture cells. *Exp. Cell Res.* **138**, 207–214.

V   Steiner, T., McGarrity, G. J., and Phillips, D. M. (1982). Cultivation and partial characterization of Spiroplasmas in cell culture. *Infect. Immunol.* **35**, 296–304.

E   Stepien, G. (1988). Contrôle de l'expression du génome mitochondrial par l'hormone juvénile dans les cellules de Drosophile en culture. *Thèse Doctorat,* Univ. Clermont-Ferrand, France.

E   Stepien, G., Renaud, M., Savre, I., and Durand, R. (1988). Juvenile hormone increases mitochondrial activities in *Drosophila* cells. *Insect Biochem.* **18**, 313–321.

Ct   Stern, C. (1940). Growth *in vitro* of the testis of *Drosophila*. *Growth* **4**, 377–382.

E   Stevens, B. (1981). Regulation of the *Drosophila melanogaster* cell cycle by ecdysteroids. Ph.D. Thesis, University of California, Los Angeles, Los Angeles, CA.

ECy Stevens, B., Alvarez, C. M., Bohman, R., and O'Connor, J. D. (1980). An ecdysteroid-induced alteration in the cell-cycle of cultured *Drosophila* cells. *Cell* **22**, 675–682.

E Stevens, B., and O'Connor, J. D. (1982). The acquisition of resistance to ecdysteroids in cultured *Drosophila* cells. *Dev. Biol.* **94**, 176–182.

Cy Stevens, B., and O'Connor, J. D. (1983). The effect of morphogenetic hormones on the cell cycle of cultured *Drosophila* cells. In *"Genetic Expression in the Cell Cycle"* (G. M. Padilla and K. S. McCarthy, eds.), pp. 269–287, Academic Press, New York.

C Stewart, B. A., Atwood, H. L., Renger, J. J., Wang, J., and Wu, C. F. (1994). Improved stability of *Drosophila* larval neuromuscular preparations in haemolymph-like physiological solutions. *J. Comp. Physiol.* **175**, 179–191.

F Stoppelli, M. P., Garcia, J. V., Decker, S. J., and Rosner, M. R. (1988). Developmental regulation of an insulin-degrading enzyme from *Drosophila melanogaster*. *Proc. Natl. Acad. Sci. U.S.A.* **85**, 3469–3473.

Cp Storti, R. V., Horovitch, S. J., Scott, M. P., Rich, A., and Pardue, M. L. (1978). Myogenesis in primary cell cultures from *Drosophila melanogaster*: protein synthesis and actin heterogeneity during development. *Cell* **13**, 589–598.

H Storti, R. V., Scott, M. P., Rich, A., and Pardue, M. L. (1980). Translational control of protein synthesis in response to heat shock in *D. melanogaster* cells. *Cell* **22**, 825–834.

R Strand, D., Lambert, D., Arthur, W. L., and McDonald, J. F. (1985). The *Drosophila* retroviral-like element *copia* is transcriptionally and transpositionally responsive to heat-shock. (Abstr at the "Heat-shock" Meeting, Cold Spring Harbor Laboratory).

T Straten van der, A. (see Van der Straten)

M Sugino, A., and Nakayama, K. (1980). DNA polymerase alpha mutants from a *Drosophila* cell line. *Proc. Natl. Acad. Sci. U.S.A.* **77**, 7049–7053.

T Sun, X. H., Lis, J. T., and Wu, R. (1988). The positive and negative transcriptional regulation of the *Drosophila Gapdh-2* gene. *Genes Dev.* **2**, 743–753.

H Sutherland, J. D., Kozlova, T., Tzertzinis, G., and Kafatos, F. C. (1995). *Drosophila* hormone receptor 38: A second partner for *Drosophila* USP suggests an unexpected role for nuclear receptors of the nerve-growth factor-induced protein B type. *Proc. Natl. Acad. Sci. U.S.A.* **92**, 7966–7970.

Cp Suzuki, N., and Wu, C. F. (1984). Fusion of dissociated neurons in culture. *Neurosci. Res.* **1**, 437–442.

M Sweitzer, S. M., Calvo, S., Kraus, M. H., Finbloom, D. S., and Larner, A. C. (1995). Characterization of a Stat-like DNA binding activity in *Drosophila melanogaster*. *J. Biol. Chem.* **270**, 16510–16513.

E Swiderski, R. E., and O'Connor, J. D. (1986). Modulation of novel-length DOPA decarboxylase transcripts by 20-hydroxyecdysone in a *Drosophila melanogaster* Kc cell subline. *Mol. Cell Biol.* **6**, 4433–4439.

R Syomin, B. V., Kandror, K. V., Semakin, A. B., Tsuprun, U. L., and Stepanov, A. S. (1993). Presence of the *gypsy* (MDG4) retrotransposon in extra-cellular virus particles. *FEBS Lett.* **323**, 285–288.

CtM Szabo, P., Edler, R., Steffensen, D. M., and Uhlenbeck, O. C. (1977). Quantitative *in situ* hybridization of ribosomal species to polytene chromosomes of *Drosophila melanogaster*. *J. Mol. Biol.* **115**, 539–563.

V Szöllösi, A., and Debec, A. (1980). Presence of Rickettsias in haploid *Drosophila melanogaster* cell lines. *Biol. Cellulaire* **38**, 129–134.

Cy Szöllözi, A., Ris, H., Szöllözi, D., and Debec, A. (1986). A centriole-free *Drosophila* cell line. A high voltage EM study. *Eur. J. Cell Biol.* **40**, 100–104.

M   Talerico, M., and Berget, S. M. (1994). Intron definition in splicing of small *Drosophila* introns. *Mol. Cell. Biol.* **14**, 3434–3445.

rH   Tanguay R. M. (1983). Genetic regulation during heat shock and function of heat shock proteins: a review. *Can. J. Biochem.* **61**, 387–394.

H   Tanguay, R. M. (1985). Intracellular localization and possible functions of heat-shock proteins. *In "Changes in Eukaryotic Gene Expression in Response to Environmental Stress"* (B. G. Atkinson and D. B. Walden, eds.), pp. 91–113, Academic Press, New York.

rH   Tanguay, R. (1988). Transcriptional activation of heat shock genes in eukaryotes. *Biochem. Cell Biol.* **66**, 584–593.

H   Tanguay, R. M., Camato, R., Lettre, F., and Vincent, M. (1983). Expression of histone genes during heat shock and in arsenite treated *Drosophila Kc* cells. *Can. J. Biochem.* **61**, 414–420.

H   Tanguay, R., and Desrosiers, R. (1988a). Methylation of core Histones and regulation of transcription during heat shock. (Abstr. P132, UCLA Symp. Mol. Cell. Biol.) *J. Mol. Biochem.* **12D** (Suppl.), 264.

H   Tanguay, R., and Desrosiers, R. (1988b). Histone methylation and modulation of gene expression in response to heat shock and chemical stress in *Drosophila*. *In "Advances in Post-translation Modifications of Proteins and Ageing"* (V. Zappia, ed.), pp. 353–362, Plenum Press, New York.

rH   Tanguay, R. M., and Desrosiers, R. (1990). Post-transcriptional methylation of histones and heat shock proteins in response to heat shock and chemical stresses. *In "Protein Methylation"* (W. K. Paik and S. Kim, eds.), pp. 139–153, CRC Press, Boca Raton, FL.

H   Tanguay, R. M., Duband, J. L., Lettre, F., Valet, J. P., Arrigo, A. P., and Nicole, L. (1985). Biochemical and immunocytochemical localization of *hsps* in *Drosophila* cultured cells. *Ann. N.Y. Acad. Sci.* **455**, 712–714.

H   Tanguay, R. M., and Vincent, M. (1982). Intracellular translocation of cellular and heat-shock induced proteins upon heat-shock in *Drosophila Kc* cells. *Can. J. Biochem.* **59**, 67–73.

M   Tarantul, V. Z., Kakpakov, V. T., and Gvozdev, V. A. (1971). Protein, RNA and DNA synthesis in the established lines of diploid cells of *Drosophila melanogaster in vitro*. *Dros. Inf. Serv.* **47**, 76.

M   Tautz, D., and Dover, G. A. (1986). Transcription of the tandem array of ribosomal DNA in *Drosophila melanogaster* does not terminate at any fixed point. *EMBO J.* **5**, 1267–1273.

R   Tchurikov, N. A., and Ilyin, Y. V. (1980). (in Russian) Multiple dispersed *Drosophila melanogaster* genes with varying location. IV. The properties of gene*Dm 225*. *Genetika* **16**, 391–400.

R   Tchurikov, N. A., Ilyin, Y. V., Ananiev, E. V., and Georgiev, G. P. (1978). The properties of gene Dm 225, a representative of dispersed repetitive genes in *Drosophila melanogaster*. *Nucleic Acids Res.* **5**, 2169–2187.

R   Tchurikov, N. A., Ilyin, Y. V., Skyrabin, K. G., Ananiev, E. V., Bayev, A. A., Krayev, A. S., Zelentsova, E. S., Kulguskin, V. V., Lyubomirskaya, N. V., and Georgiev, G. P. (1981) General properties of mobile dispersed genetic elements in *Drosophila melanogaster*. *Cold Spring Harb. Symp. Quant. Biol.* **45**, 655–665.

R   Tchurikov, N. A., Zelentsova, E. S., and Georgiev, G. P. (1980). Clusters containing different mobile dispersed genes in the genome of *Drosophila melanogaster*. *Nucleic Acids Res.* **8**, 1243–1258.

E   Tempel, K., Emmerich, H., and Gateff, E. (1981). Morphological and physiological reactions of two tumorous blood cell lines of *Drosophila melanogaster*, upon application of different ecdysteroids. *Verh. Dtsch. Zool. Ges.* 277, Gustav Fisher Verlag, Stuttgart.

T   TenHarmsel, A., Austin, R. J., Savenelli, N., and Biggin, M. D. (1993). Cooperative binding at a distance by *even-skipped* protein correlates with repression and suggests a mechanism of silencing. *Mol. Cell. Biol.* **13**, 2742–2752.

V   Teninges, D., Bras, F., and Dezelee, S. (1993). Genome organization of the *Sigma* Rhabdovirus: six genes and a gene overlap. *Virology* **193**, 1018–1023.

V   Teninges, D., and Bras-Herreng, F. (1987). Rhabdovirus *Sigma*, the hereditary CO_2 sensitivity agent of *Drosophila:* nucleotide sequences of a cDNA clone encoding the glycoprotein. *J. Gen. Virol.* **68**, 2625–2638.

V   Teninges, D., Ohanessian, A., Richard-Molard, C., and Contamine, D. (1979a). Isolation and biological properties of *Drosophila* X Virus. *J. Gen. Virol.* **42**, 241–254.

V   Teninges, D., Ohanessian, A., Richard-Molard, C., and Contamine, D. (1979b). Contamination and persistent infection of *Drosophila* cell lines by reovirus-type particles. *In Vitro* **15**, 425–428.

B   Theopold, U., Dal Zotto, L., and Hultmark, D. (1995). FKBP 39, a *Drosophila* member of proteins that bind the immunosuppressive drug FK506. *Gene* **156**, 247–251.

B   Theopold, U., Pinter, M., Daffre, S., Tryselius, Y., Friedrich, P., Nässel, D. R., and Hultmark, D. (1995). CalpA, a *Drosophila* calpain homolog specifically expressed in a small set of nerve, midgut and blood cells. *Mol. Cell. Biol.* **15**, 824–834.

Cy   Thibodeau, A., and Vincent, M. (1991). Monoclonal antibody CC-3 recognizes phosphoproteins in interphase and mitotic cells. *Exp. Cell Res.* **195**, 145–153.

T   Thisse, C., Perrin-Schmidt, F., Stoetzek, C., and Thisse, B. (1991). Sequence-specific transactivation of the *Drosophila twist* gene by the *dorsal* gene product. *Cell* **65**, 1191–1201.

T   Thomas, G. H., and Elgin, S. C. R. (1988). The use of the gene encoding the α-amanitin-resistant subunit of RNA polymerase II as a selectable marker in cell transformation. *Dros. Inf. Serv.* **67**, 85.

E   Thomas, H. E., Stunnenberg, H. G., and Stewart, A. F. (1993). Heterodimerization of the *Drosophila* ecdysone receptor with retinoid X receptor and *ultraspiracle*. *Nature* **362**, 471–475.

CtE   Thomasson, W. A., and Mitchell, H. K. (1972). Hormonal control of protein granule accumulation in fat bodies of *Drosophila melanogaster* larvae. *J. Insect Physiol.* **18**, 1885–1899.

F   Thompson, K., Decker, S., and Rosner, M. R. (1984). Identification of an EGF receptor homolog in *Drosophila melanogaster;* (Abst *24th Ann. Meeting Am. Soc. Cell Biol.*, *n°1529) J. Cell Biol.* **99**, 414a.

F   Thompson, K. L., Decker, S. J., and Rosner, M. R. (1985). Identification of a novel receptor in *Drosophila* for both Epidermal Growth Factor and Insulin. *Proc. Natl. Acad. Sci. U.S.A.* **82**, 8443–8447.

E   Thummel, C. S. (1989). The *Drosophila* E74 promoter contains essential sequences downstream from the start site of transcription. *Genes Dev.* **3**, 782–792.

T   Thummel, C. S., Boulet, A. M., and Lipshitz, H. D. (1988). Vectors for *Drosophila* P element-mediated transformation and tissue culture transfection. *Gene* **74**, 445–456.

CdE   Thummel, C. S., Burtis, K. C., and Hogness, D. S. (1990). Spatial and temporal patterns of E74 transcription during *Drosophila* development. *Cell* **61**, 101–111.

CIB Todo, T., Ueda, R., Miyake, T., and Kondo, S. (1985). Hypersensitivity to ultraviolet light and chemical mutagens of a cell line established from excisionless *Drosophila* strain *mus 201*. *Mutat. Res.* **145**, 165–170.

HM Topol, J., Ruden, D. M., and Parker, C. S. (1985). Sequences required for *in vitro* transcriptional activation of a *Drosophila hsp70* gene. *Cell* **42**, 527–537.

H Török, I., and Karch, F. (1980). Nucleotides sequences of heat shock activated genes in *Drosophila melanogaster*. I. Sequences in the regions of the 5' and 3' ends of the *hsp70* gene in the hybrid plasmid 56h8. *Nucleic Acids Res.* **8**, 3105–3123.

M Toung, Y. P., Hsieh, T. S., and Tu, C. P. (1990). *Drosophila* glutathione S-transferase 1-1 shares a region of sequence homology with the Maize glutathione S-transferase III. *Proc. Natl. Acad. Sci. U.S.A.* **87**, 31–35.

E Tourmente, S., Chapel, S., Dreau, D., Drake, M. E., Bruhat, A., Couderc, J. L., and Dastugue, B. (1993). Enhancer and silencer elements within the first intron mediate the transcriptional regulation of the β3-tubulin gene by 20-hydroxyecdysone in *Drosophila* Kc cells. *Insect Biochem. Mol. Biol.* **23**, 137–143.

M Trosko, J. E., and Wilder, K. (1973). Repair of UV-induced pyrimidine dimers in *Drosophila melanogaster* cells *in vitro*. *Genetics* **73**, 297–302.

Ct Tulchin, N., Mateyko, G. M., and Kopac, M. J. (1967). *Drosophila* salivary glands *in vitro*. *J. Cell Biol.* **34**, 891–897.

VCp Ueda, R., Koana, T., and Miyake, T. (1987). Transient proliferation of Sex ratio organisms of *Drosophila* in a primary culture from infected embryos. *Jpn J. Genet.* **62**, 85–93.

VCp Ueda, R., Koana, T., and Miyake, T. (1989). Transient proliferation of Sex ratio Organism of *Drosophila* in a primary cell culture from infected embryos. *In* "*Invertebrate Cell System Applications*" (J. Mitsuhashi, ed.), pp. 77–84, CRC Press, Boca Raton, FL.

Cp Ueda, R., and Miyake, T. (1982). (in Japanese) *In vitro* culture of imaginal discs of *Drosophila*. *Jpn J. Genet.* **36**, 46–51.

rC Ueda, R., and Miyake, T. (1985). *In vitro* culture of *Drosophila* cells. *Biol. Sci. News* **164**, 17–18.

Cl Ui, K., Nishihara, S., Sakuma, M., Togashi, S., Ueda, R., Miyata, Y., and Miyake, T. (1994). Newly established cell lines from *Drosophila* larval CNS express neural specific characteristics. *In Vitro Cell Dev. Biol.* **30A**, 209–216.

Cl Ui, K., Ueda, R., and Miyake, T. (1987). Cell lines from imaginal discs of *Drosophila melanogaster*. *In Vitro Cell. Dev. Biol.* **23**, 707–711.

Cl Ui, K., Ueda, R., and Miyake, T. (1988a). Continuous cell lines from imaginal discs of *Drosophila melanogaster*. *In Invertebrate Fish Tissue Culture* (Y. Kuroda, E. Kurstak, and K. Maramorosch, eds.), pp. 251–254, Jpn Sci. Soc. Press, Tokyo and Springer-Verlag, Berlin.

Cl Ui, K., Ueda, R., and Miyake, T. (1988b). *In vitro* cultures of cells from different kinds of imaginal discs of *Drosophila melanogaster*. *Jpn J. Genet.* **63**, 33–41.

Cl Ui, K., Ueda, R., and Miyake, T. (1989). *In vitro* culture of cells from dissociated imaginal discs of *Drosophila melanogaster*. *In* "*Invertebrate Cell System Applications*" (J. Mitsuhashi, ed.), pp. 221–231, CRC Press Boca Raton, FL.

Cl Ui, K., Sakuma, M., Nishiara, S., *et al.* (1989). Neurotransmitter analysis in cell lines from larval CNS of *Drosophila melanogaster*. *Jpn. J. Genet.* **64**, 492.

Cl Ui, K., Togashi, S., and Ueda, R., *et al.* (1988). Establishment of cell lines from larval central nervous system of *Drosophila melanogaster*. *Jpn. J. Genet.* **63**, 606.

Cl Ui-Tei, K., Nishihara, S., Sakuma, K., Matsuda, K., Miyake, T., and Miyata, Y. (1994). Chemical analysis of neurotransmitter candidates in clonal cell lines from *Drosophila* central nervous system. *Neurosci. Lett.* **174**, 85–88.

M Vallett, S. M., Brudnak, M., Pellegrini, M., and Weber, H. W. (1993). *In vivo* regulation of rRNA transcription occurs rapidly in non-dividing *Drosophila* cells in response to a Phorbol Ester and serum. *Mol. Cell. Biol.* **13**, 928–933.

TB Vanden Broeck, J., Vulsteke, V., Huybrechts, R., and De Loof, A. (1995). Characterization of a cloned locust tyramine receptor cDNA by functional expression in permanently transformed *Drosophila S2* cells. *J. Neurochem.* **64**, 2387–2395.

T Van der Straten, A., Johansen, H., Rosenberg, M., and Sweet, R. W. (1989). Introduction and constitutive expression of gene products in cultured *Drosophila* cells using Hygromycin B selection. *In* "*Methods Mol. Cell. Biol.*" Vol. I, pp. 1–8, Wiley, New York.

T Van der Straten, A., Johansen, H., Sweet, R., and Rosenberg, M. (1988). Efficient expression of foreign genes in cultured *Drosophila melanogaster* cells using hygromycin B selection. *In* "*Invertebrate and Fish Tissue Culture*" (Y. Kuroda, E. Kurstak, and K. Maramorosch, Eds.), pp. 131–134, Springer Verlag, New York.

T Van der Straten, A., Johansen, H., Sweet, R., and Rosenberg, M. (1989). Efficient expression of foreign genes in cultured *Drosophila melanogaster* cells using Hygromycin B selection. *In* "*Invertebrate Cell System Applications*" (J. Mitsuhashi, ed.), Vol. I, pp. 183–195, CRC Press, Boca Raton, FL.

TB Van Leeuwen, F., Harryman-Samos, C., and Nusse, R. (1994). Biological activity of soluble *wingless* protein in cultured imaginal disc cells. *Nature* **368**, 342–344.

rC Vaughn, J. L., and Goodwin, R. H. (1976). Preparation of several media for the culture of tissues and cells from invertebrates. *Tissue Cult. Assoc. Manual* **3**, 527–537.

H Vazquez, J., Pauli, D., and Tissieres, A. (1993). Transcriptional regulation in *Drosophila* during heat shock: a nuclear run-on analysis. *Chromosoma* **102**, 233–248.

H Velazquez, J. M., DiDomenico, B., and Lindquist, S. (1980). Intracellular localization of heat-shock proteins in *Drosophila*. *Cell* **20**, 679–689.

H Velazquez, J. M., and Lindquist, S. (1984). *hsp70*: nuclear concentration during stress and cytoplasmic storage during recovery. *Cell* **36**, 655–662.

H Velazquez, J. M., Sonoda, S., Bugaisky, G., and Lindquist, S. (1983). Is the major *Drosophila* heat shock protein present in cells that have not been heat-shocked? *J. Cell. Biol.* **96**, 286–290.

H Vincent, M., and Tanguay, R. M. (1982). Different intracellular distribution of heat-shock and arsenite-induced proteins in *Drosophila* Kc cells. Possible relation with the phosphorylation and translocation of a major cytoskeletal protein *J. Mol. Biol.* **162**, 365–378.

EH Vitek, M. P., and Berger, E. M. (1984). Steroid and high-temperature induction of the small heat-shock protein genes in *Drosophila*. *J. Mol. Biol.* **178**, 173–189.

E Vitek, M. P., Kreissman, S. G., and Gross, R. H. (1981). The isolation of ecdysterone-inducible genes by hybridization substraction chromatography. *Nucleic Acids Res.* **9**, 1191–1202.

E Vitek, M. P., Morganelli, C. M., and Berger, E. M. (1984). Stimulation of cytoplasmic actin gene transcription and translation in cultured *Drosophila* cells by ecdysterone. *J. Biol. Chem.* **259**, 1738–1743.

M Voelker, R. A., Gibson, W., Graves, J. P., Sterling, J. F., and Eisenberg, M. T. (1991). The *Drosophila suppressor of Sable* gene encodes a polypeptide with regions similar to those of RNA-binding proteins. *Mol. Cell. Biol.* **11**, 894–905.

H   Voellmy, R., Goldschmidt-Clermont, M., Southgate, R., and Tissieres, A. (1981). A DNA segment isolated from chromosomal site 67B in *Drosophila melanogaster* contains four closely linked heat-shock genes. *Cell* **23**, 261–270.

CpM   Volk, T. (1992). A new member of the *spectrin* superfamily may participate in the formation of embryonic muscle attachments in *Drosophila*. *Development* **116**, 721–738.

CpM   Volk, T., Fessler, L. I., and Fessler, J. H. (1990). A role for *Integrin* in the formation of sarcomeric architecture. *Cell* **63**, 525–536.

H   Wadsworth, S. C. (1982). A family of related proteins is encoded by the major *Drosophila* heat-shock gene family. *Mol. Cell. Biol.* **2**, 286–292.

rT   Walker, V. (1989). Gene transfer in insects. *Adv. Cell Cult.* **7**, 87–124.

T   Walker, V. K., Schreiber, M. L., Purvis, C., George, J., Wyatt, G. R., and Bendena, W. G. (1991). Yolk polypeptide gene expression in cultured *Drosophila* cells. *In Vitro, Cell. Dev. Biol.* **A27**, 121–127.

BT   Wallny, H. J., Sollami, G., and Karjalainen, K. (1995). Soluble mouse histocompatibility complex class II molecules produced in *Drosophila* cells. *Eur. J. Immunol.* **25**, 1262–1266.

B   Walter, M. F., Uster, P. S., and Deamer, D. W. (1986). Liposome-mediated delivery of antibody to a *Drosophila* cell line. *Eur. J. Cell Biol.* **40**, 195–202.

M   Walters, M. F., and Biessmann, H. (1984). Intermediate-sized filaments in *Drosophila* tissue culture cells. *J. Cell. Biol.* **99**, 1468–1477.

M   Wang, M. Y., and Wang, C. (1993). Characterization of glucose transport system in *Drosophila Kc* cells. *FEBS Lett.* **317**, 241–244.

B   Wang, W. C., Zinn, K., and Bjorkman, P. J. (1993). Expression and structural studies of *Fasciclin I*, an Insect adhesion molecule. *J. Biol. Chem.* **268**, 1448–1455.

E   Wang, X., Chang, E. S., and O'Connor, J. B. (1989). Purification of the *Drosophila Kc* cell juvenile hormone binding protein. *Insect Biochem.* **19**, 327–335.

V   Warburg, A., and Miller, L. H. (1992). Paragonic development of a Malaria parasite *in vitro*. *Science* **255**, 448–450.

V   Warburg, A., and Schneider, I. (1993). Sporogonic development of *Plasmodium falciparum in vitro*. *Exp. Parasitol.* **76**, 121–126.

EM   Watson, J. A., Havel, C. M., Lobos, D. V., Baker, F. C., and Morrow, C. J. (1985). Isoprenoid synthesis in isolated embryonic *Drosophila* cells: Sterol-independent regulatory signal molecule is distal to isopentenyl 1-pyrophosphates. *J. Biol. Chem.* **260**, 14083–14091.

E   Watson, J. A., Havel, C. M., Morrow, C., and Baker, F. C. (1984). Modulation of *Kc* cell 3-hydroxy-3-methylglutaryl Coenzyme A reductase by mevalonate and its analogues. *Fed. Proc.* **43**, 1732 [Abstract]

Cd   Wehmann, H. J. (1969). Fine structure of *Drosophila* wing imaginal discs during early stages of metamorphose. *W. Roux'Arch.* **163**, 375–390.

RCd   Wehmann, H. J., and Brager, M. (1971). VLPs in *Drosophila*: Constant appearance of VLPs in imaginal discs *in vitro*. *J. Inv. Pathol.* **18**, 127–130.

FT   Werner, H., Bach, M. A., Stannard, B., Roberts, C. T. Jr., and LeRoith, D. (1992). Structural and functional analysis of the insulin-like growth factor I receptor gene promoter. *Mol. Endocrinol.* **6**, 1545–1558.

H   Westwood, J. T., Clos, J., and Wu, C. (1991). Stress-induced oligomerization and chromosomal relocalization of heat-shock factor. *Nature* **353**, 822–827.

H   Westwood, J. T., and Steinhardt, R. (1988). The effect of heat shock inducers on protein degradation. (Abstr. P219, UCLA Symp. Mol. Cell. Biol.) *J. Mol. Biochem.* **12D** (Suppl.), 273.

H Westwood, J. T., and Wu, C. (1993). Activation of heat shock factor: conformational change associated with a monomer-to-trimer transition. *Mol. Cell. Biol.* **13**, 3481–3486.

M White, R., and Hogness, D. (1977). R loop mapping of the 18S and 28S sequences in the long and short repeating units of *Drosophila melanogaster* rDNA. *Cell* **10**, 177–192.

F Wides, R. J., Zak, N. B., and Shilo, B. Z. (1990). Enhancement of tyrosine kinase activity of the *Drosophila* Epidermal Growth Factor homolog by alterations of the transmembrane domain. *Eur. J. Biochem.* **189**, 637–645.

H Wieben, E. D., and Pederson, T. (1982). Small nuclear riboproteins of *Drosophila*; evolutionary conservation of U1 RNA-associated proteins and their behavior during heat-shock. *Mol. Cell. Biol.* **2**, 914–920.

H Wiederrecht, G., Shuey, D. J., Kibbe, W. A., and Parker, C. S. (1987). The *Saccharomyces and Drosophila* heat shock transcription factors are identical in size and DNA binding properties. *Cell* **48**, 507–515.

B Wilcox, M., Brown, N., Piovant, M., Smith, R. J., and White, R. A. H. (1984). The *Drosophila* position-specific antigens are a family of cell surface glycoprotein complexes. *EMBO J.* **3**, 2307–2313.

T Williams, T., Admon, A., Luscher, B., and Tijan, R. (1988). Cloning and expression of AP-2, a cell-type-specific transcription factor that activates inducible enhancer elements. *Genes Dev.* **2**, 1557–1569.

R Williamson, D. C., and Kernaghan, R. P. (1972). VLPs in Schneider's *Drosophila* cell lines. *Dros. Inf. Serv.* **48**, 58–59.

E Wing, K. D. (1988). RH 5849, a nonsteroidal ecdysone agonist: effects on a *Drosophila* cell line. *Science* **241**, 467–469.

E Wing, K. D. (1990). Recent findings on the nonsteroidal ecdysone mimic RH5849. *Int. J. Invert. Reprod.* **18**, 133.

T Winslow, G., Hayashi, S., Krasnow, M., Hogness, D., and Scott, M. (1989). Transcriptional activation by the *Antennapedia* and *fushi tarazu* proteins in cultured *Drosophila* cells. *Cell* **57**, 1017–1030.

B Wisniewski, J. R., and Schulze, E. (1992). Insect proteins homologous to mammalian high mobility group protein 1. Characterization and DNA-binding properties. *J. Biol. Chem.* **267**, 17170–17177.

E Woods, D. F., and Poodry, C. A. (1983). Cell surface proteins of *Drosophila*. I. Changes induced by 20-hydroxyecdysone. *Dev. Biol.* **96**, 23–31.

Cd Woods, D. F., Rickoll, W. L., Birr, C., Poodry, C. A., and Fristom, J. W. (1987). Alterations in the cell surface proteins of *Drosophila* during morphogenesis. *W. Roux's Arch. Dev. Biol.* **196**, 339–346.

H Wu, C. (1980). The 5' ends of *Drosophila* heat-shock genes in chromatin are hypersensitive to DNase I. *Nature* **286**, 854–860.

H Wu, C. (1982). Chromatin structure of heat-shock genes. In *"Heat-shock, from Bacteria to Man"* (M. J. Schlesinger, M. Ashburner, and A. Tissières, eds.), pp. 91–97, Cold Spring Harbor Laboratory Press, Cold Spring Harbor.

H Wu, C. (1984a). Two protein-binding sites in chromatin implicated in the activation of heat-shock genes. *Nature (London)* **309**, 229–234.

H Wu, C. (1984b). Activating protein factor binds *in vitro* to upstream control sequences in heat-shock gene chromatin. *Nature* **311**, 81–84.

Cp Wu, C. F. (1988). Neurogenetic studies of *Drosophila* central nervous system neurons in culture. In *"Cell Culture Approaches to Invertebrate Neuroscience"* (D. J. Beadle, G. Lees, and S. B. Kater, eds.), pp. 149–187, Academic Press, New York.

H   Wu, C., Hansell, A., Lambert, K., Walker, B., Wilson, S., and Zimarino, V. (1988). Regulation of heat shock gene regulation. (Abstr. P 019 UCLA Symp. Mol. Cell. Biol.) *J. Cell. Biochem.* **12D** (Suppl.), 249.

Cp Wu, C. F., Suzuki, N., and Poo, M. M. (1981). Cell culture and chemically induced fusion of dissociated neurons of *Drosophila*. *Soc. Neurosci.* **7**, 598. [Abstract]

Cp Wu, C. F., Suzuki, N., and Poo, M. M. (1983). Dissociated neurons from normal and mutant *Drosophila* larval central nervous system in cell culture. *J. Neurosci* **3**, 1888–1899.

H   Wu, C., Wilson, S., Walker, B., David, I., Paisley, T., Zimarino, V., and Ueda, H. (1987). Purification and properties of *Drosophila* heat shock activator protein. *Science* **238**, 1247–1253.

H   Wu, C., Wong, Y. C., and Elgin, S. C. R. (1979). The chromatin structure of specific gene. II. Disruption of chromatin structure during activity. *Cell* **16**, 807–814.

V   Wyers, F., Blondel, D., Petitjean, A. M., and Dezelee, S. (1989). Restricted expression of viral glycoprotein in Vesicular Stomatitis Virus-infected *Drosophila melanogaster* cells. *J. Gen. Virol.* **70**, 213–218.

V   Wyers, F., Richard-Molard, C., Blondel, D., and Dezelee, S. (1980). Vesicular Stomatitis Virus growth in *Drosophila melanogaster* cells: G protein deficiency. *J. Virol.* **33**, 411–422.

E   Wyss, C. (1976). Juvenile hormone analogue counteracts growth stimulation and inhibition by ecdysone in clonal *Drosophila* cell line. *Experientia* **32**, 1272–1274.

M   Wyss, C. (1977). Purine and pyrimidine salvage in a clonal *Drosophila* cell line. *J. Insect Physiol.* **23**, 739–747.

B   Wyss, C. (1979a). Cloning of *Drosophila* cells: Effects of vitamins and Yeast extract components. *Somatic Cell Genet.* **5**, 23–28.

B   Wyss, C. (1979b). TAM selection of *Drosophila* somatic cell hybrids. *Somatic Cell Genet.* **5**, 29–37.

EB Wyss, C. (1980a). Cell hybrid analysis of ecdysone sensitivity and resistance in *Drosophila* cell lines. In *"Invertebrate Systems in Vitro"* (E. Kurstak, K. Maramorosch, and A. Dübendorfer, eds.), pp. 279–289, Elsevier/North Holland Biomedical Press, Amsterdam.

EB Wyss, C. (1980b). Loss of ecdysterone sensitivity of a *Drosophila* cell line after hybridization with embryonic cells. *Exp. Cell Res.* **125**, 121–126.

E   Wyss, C. (1981a). *Drosophila* cells respond specifically to insulin. *Experientia* **37**, 664. [Abstract]

T   Wyss, C. (1981b). Transformation of a mutant *Drosophila* cell line *in vitro*. *Experientia* **37**, 665. [Abstract]

CE Wyss, C. (1982a). Ecdysterone, insulin and fly extract needed for the proliferation of normal *Drosophila* cells in defined medium. *Exp. Cell Res.* **139**, 297–307.

C   Wyss, C. (1982b). CalGF, a cationic low molecular weight growth factor from *Drosophila melanogaster* and the nutritional requirements of *KcHP* cells. *Insect Biochem.* **12**, 515–522.

C   Wyss, C. (1988). Hormone and growth factor effects on the proliferation of Dipteran cell lines in a defined medium. In *"Invertebrate and Fish Tissue Culture"* (E. Kuroda, E. Kurstak, and K. Maramorosch, eds.), pp. 19–22, Jpn. Sci. Soc. Press and Springer Verlag.

C   Wyss, C., and Bachmann, G. (1976). Influence of amino acids, mammalian serum, and osmotic pressure on the proliferation of *Drosophila* cell lines. *J. Insect Physiol.* **22**, 1581–1586.

E   Wyss, C., and Eppenberger, H. (1978). Morphological and proliferative response of Schneider's *Drosophila* cell line 3 to ecdysterone. *Experientia* **34**, 961–962. [Abstract]

H   Xiao, H., and Lis, J. T. (1989). Heat-shock and developmental regulation of the *Drosophila melanogaster hsp83* gene. *Mol. Cell. Biol.* **8**, 1746–1753.

H   Xiao, H., Perisic, O., and Lis, J. T. (1991). Cooperative binding of *Drosophila* heat-shock factors to arrays of a conserved 5bp unit. *Cell* **64**, 585–593.

VT   Xiong, Ch., Levis, R., Shen, P., Schlesinger, S., Rice, Ch. M., Huang, H. V. (1989). *Sindbis* Virus: An efficient, broad host range vector for gene expression in animal cells. *Science* **243**, 1188–1191.

TF   Xu, J., Thompson, K. L., Shephard, L. B., Hudson, L. G. and Gill, G. N. (1993). T_3 receptor suppression of Sp1-dependent transcription from the Epidermal Growth Factor receptor promoter via overlapping DNA-binding sites. *J. Biol. Chem.* **268**, 16065–16073.

BT   Yamaguchi, M., Hayashi, Y., Nishimoto, Y., Hirose, F., and Matsukage, A. (1995). A nucleotide sequence essential for the function of DRE, a common promoter element for *Drosophila* DNA replication-related genes. *J. Biol. Chem.* **270**, 15808–15814.

T   Yamaguchi, M., Hirose, F., Nishida, Y., and Matsukage, A. (1991). Repression of the *Drosophila* proliferating-cell nuclear antigen gene promoter by *zeknüllt* protein. *Mol. Cell. Biol.* **11**, 4909–4917.

B   Yamaguchi, M., Nishida, Y., Moriuchi, T., Hirose, F., Hui, C. C., Suzuki, Y., and Matsukage, A. (1990). *Drosophila* proliferating cell nuclear antigen (cyclin) gene: Structure, expression during development, and species binding of homeodomain proteins to its 5′-flanking region. *Mol. Cell. Biol.* **10**, 872–879.

TB   Yanagawa, S. I., van Leeuwen, F., Wodarz, A., Klingensmith, J., and Nusse, R. (1995). The *Dishevelled* protein is modified by *Wingless* signaling in *Drosophila*. *Genes Dev.* **9**, 1087–1097.

M   Yokomori, K., Admon, A., Goodrich, J. A., Chen, J. L., and Tijan, R. (1993). *Drosophila* TFIIA-L is processed into two subunits that are associated with the TBF/TAF complex. *Genes Dev.* **7**, 2235–2245.

E   Yoshinaga, S. K., and Yamamoto, K. R. (1991). Signaling and regulation by a mammalian glucocorticoid receptor in *Drosophila* cells. *Mol. Endocrinol.* **5**, 844–853.

R   Yoshioka, K., Honma, H., Zushi, M., Kondo, S., Togashi, S., Miyake, T., and Shiba, T. (1990). Virus-like particle formation of *Drosophila copia* through autocatalytic processing. *EMBO J.* **9**, 535–541.

R   Yoshioka, K., Kanda, H., Akiba, H., Enoki, M., and Shiba, T (1991). Identification of an unusual structure in the *Drosophila* transposable element *copia*: evidence for *copia* transposition through an RNA intermediate. *Gene* **103**, 179–184.

R   Yoshioka, K., Kanda, H., Takamatsu, N., Togashi, S., Kondo, S., Miyake, T., Sakaki, Y., and Shiba, T. (1992). Efficient amplification of *Drosophila simulans copia* directed by high-level reverse transcription activity associated with virus-like particles. *Gene* **120**, 191–196.

H   Yost, H. J., and Lindquist, S. (1986). RNA splicing is interrupted by heat shock and is rescued by heat shock protein synthesis. *Cell* **45**, 185–193.

H   Yost, H. J., Petersen, R. B., and Lindquist, S. (1990). Posttranscriptional regulation of Heat shock protein synthesis in *Drosophila*. In *"Stress Proteins in Biology and Medicine"* (R. I. Morimoto, A. Tissières, and C. Georgopoulos, eds.), 379–409, Cold Spring Harbor Laboratory Press, Cold Spring Harbor, New York.

MR   Young, M. W. (1981). Repeated DNA sequences in *Drosophila*. In *"Genetic Engineering: Principles and Methods"* (J. K. Sertlow and A. Hollaender, eds.), Vol. 3, pp. 109–128, Plenum Press, New York.

rR   Young, M. W., and Schwartz, H. E. (1981). Nomadic gene families in *Drosophila*. *Cold Spring Harbor Symp. Quant. Biol.* **45** (2), 629–640.

E   Yudin, A. I., Clark, W. H., and Chang, E. S. (1982). Ecdysteroid induction of cell surface contacts in a *Drosophila* cell line. *J. Exp. Zool.* **219**, 399–403.

CdE   Yund, M. A. (1979). Specific binding of 20-hydroxyecdysone to nuclei of imaginal discs of *Drosophila melanogaster*. *Mol. Cell. Endocrinol.* **14**, 19–35.

CdE   Yund, M. A. (1980). Imaginal disc of *Drosophila*: an *in vitro* system for the study of ecdysteroid action. In *"Invertebrate Systems In Vitro"* (E. Kurstak, K. Maramorosch, and A. Dübendorfer, eds.), pp. 229–237, Elsevier/North-Holland Biomedical Press, Amsterdam.

CdE   Yund, M. A. (1989). Imaginal discs as a model for studying Ecdysteroid action. In *"Ecdysone, from Chemistry to Mode of Action"* (J. Koolman, ed.), pp. 384–392, G. Thieme Verlag, Stuttgart.

CdE   Yund, M. A., and Fristrom, J. W. (1975). Uptake and binding of β-ecdysone in imaginal discs of *Drosophila melanogaster*. *Dev. Biol.* **43**, 287–298.

CdE   Yund, M. A., King, D. S., and Fristrom, J. W. (1978). Ecdysteroid receptors in imaginal discs of *Drosophila melanogaster*. *Proc. Natl. Acad. Sci. U.S.A.* **75**, 6039–6043.

E   Yund, M. A., and Osterbur, D. L. (1985). Ecdysteroid receptors and binding proteins. In *"Comparative Insect Physiol., Biochem., Pharmacology"* (G. A. Kerkut and L. I. Gilbert, eds.), pp. 473–490, Pergamon Press, New York.

F   Zak, N. B., and Shilo, B. Z. (1990). Biochemical properties of the *Drosophila* EGF receptor homolog (DER) protein. *Oncogene* **5**, 1589–1593.

V   Zhang, P., Raney, A. K., and McLachlan, A. (1993). Characterization of functional Sp1 transcription factor binding sites in the Hepatitis B nucleocapsid promoter. *J. Virol.* **67**, 1472–1481.

Cp   Zhao, M. L., Sable, E. O., Saito, M., Iverson, L. E., and Wu, C. F. (1993). Host-dependent *Shaker* cDNA expression in cultured *Drosophila* "giant" neurons by germline transformation. *Soc. Neurosci.* **18**, 78. [Abstract]

Cp   Zhao, M. L., Sable, E. O., Iverson, L. E., and Wu, C. F. (1995). Functional expression of *Shaker* K$^+$ channels in *Drosophila* "giant" neurons derived from *Sh* cDNA transformants: distinct properties, distribution and turnover. *J. Neurosci.* **15**, 1406–1418.

B   Zheng, Y., Jung, M. K., and Oakley, B. (1991). gamma-Tubulin is present in *Drosophila melanogaster* and *Homo sapiens* and is associated with the centrosome. *Cell* **65**, 817–823.

V   Zhong, W., and Rueckert, R. R. (1993). Flock House Virus: Down-regulation of subgenomic RNA 3 synthesis does not involve coat protein and is targeted to synthesis of its positive strand. *J. Virol.* **67**, 2716–2722.

RHE   Ziarczyk, P. (1992). Fonction promotrice et régulation des LTRs de *1731*, un rétro-transposon de *Drosophila melanogaster*. *Thèse Doct. Univ. Paris XII*, 1–239.

RTHE   Ziarczyk, P., and Best-Belpomme, M. (1991). A short 5' region of the Long Terminal Repeat is required for regulation by hormone and heat shock of *Drosophila* retrotransposon *1731*. *Nucleic Acids Res.* **19**, 5689–5693.

RTE   Ziarczyk, P., Fourcade-Peronnet, F., Simonart, S., Maisonhaute, C., and Best-Belpomme, M. (1989). Functional analysis of the long terminal repeats of *Drosophila* *1731* retrotransposon: promoter function and steroid regulation. *Nucleic Acids Res.* **17**, 8631–8643.

H   Zimarino, V., Tsai, Ch., and Wu, C. (1990). Complex modes of heat shock factor activation. *Mol. Cell. Biol.* **10**, 752–759.

H   Zimarino, V., Wilson, S., and Wu, C. (1990). Antibody-mediated activation of *Drosophila* heat-shock factor *in vitro*. *Science* **249**, 546–549.

H   Zimarino, V., and Wu, C. (1987). Induction of sequence-specific binding of *Drosophila* heat shock activation protein without protein synthesis. *Nature* **327**, 727–730.

Cy   Zuffardi, O., Tiepolo, L., Dolfini, S., Barigozzi, C., and Fraccaro, M. (1971). Changes in the fluorescence patterns of translocated Y chromosome fragments in *Drosophila melanogaster*. *Chromosoma* **34**, 274–280.

TM   Zuo, P., Stanojevic, D., Colgan, J., Han, K., Levine, M., and Manley, J. L. (1991). Activation and repression of transcription by the *gap* proteins *hunchback* and *Krüppel* in cultured *Drosophila* cells. *Genes Dev.* **5**, 254–264.

ADDENDUM

rH   Bienz, M., and Pelham, H. R. B. (1987). Mechanisms of heat shock gene activation in higher eukaryotes. *Adv. Genet.* **24**, 31–72.

Ct   Bienz-Tadmor, B., Smith, H. S., and Gerbi, S. A. (1991). The promoter of DNA puff gene II/9–1 of *Sciara coprophila* is inducible by ecdysone in late prepupal salivary glands of *Drosophila melanogaster*. *Cell Reg.* **2**, 875–888.

E   Chang, E. S., and O'Connor, J. D. (1981). Effect of morphogenetic hormones on cultured cells. *In "Metamorphosis"* (L. I. Gilbert, ed.), pp. 241–261, Plenum Press, New York.

E   Clement, C. Y., Bradbrook, D. A., Lafont, R., and Dinan, L. (1993). Assessment of a microplate-based bioassay for the detection of ecdysteroid-like or antiecdysteroid activities. *Insect Bioch. Mol. Biol.* **23**, 187–193.

E   Dinan, L. (1995). A strategy for the identification of ecdysteroid receptor, agonists and antagonists from plants. *Eur. J. Entomol.* **92**, 271–283.

E   Dinan, L., Whiting, P., Girault, J. P., and Lafon, R. (1994). Novel ecdysteroid agonists and antagonists from plants. [Abst.] *XIth Intern. Ecdysone Workshop, Ceské-Budejovice,* Czeck Rep., July 1994, p. 53.

CdE   Fristrom, J. W., Raikow, R., Petri, W. H., and Stewart, D. (1970). *In vitro* evagination and RNA synthesis in imaginal discs of *Drosophila melanogaster*. *In "Problems in Biology: RNA in Development,"* Park City Intern. Symposium, 1969 (E. W. Hanly, ed.), pp. 381–401, Univ. Utah Press, Salt Lake City, UT.

BT   Gross, I., Georgel, P., Kappler, C., Reichhart, J. M., and Hoffmann, J. A. (1996). *Nucleic Acids Res.* **24**, 1238–1245.

R   Haoudi, A., Kim, M. H., Champion, S., Best-Belpomme, M., and Maisonhaute, C. (1995). The gag polypeptides of the *Drosophila* 1731 retrotransposon are associated to virus-like particles and to nuclei. *FEBS Lett.* **377**, 67–72.

Cp   Hayashi, I., Perez-Magallanes, M., and Rossi, J. M. (1992). Neurotrophic factor-like activity in *Drosophila*. *Biochem. Biophys. Res. Comm.* **184**, 73–77.

C   Hirumi, H. (1963). An improved device for cultivating cells *in vitro* and for observation under high power phase magnification. *Contr. Boyce Thompson Inst.* **22**, 113–115.

Cp   LaBonne, S. G., and Furst, A. (1989). Differentiation *in vitro* of neural precursor cells from normal and *pecanex* mutant *Drosophila* embryos. *J. Neurogenet.* **50C**, 53.

Cy   Lehner, C. F., and O'Farrell, P. H. (1989). Expression and function of *Drosophila* Cyclin A during embryonic cell cycle progression. *Cell* **56**, 957–968.

CP Li, C., and Meinertzhagen, I. A. (1992). Neurite outgrowth in primary culture of Drosophila photoreceptor and optic lobe cells. Soc. Neurosci. Abstr. 18, 1468.

Cp Masur, S. K., Kim, Y. T., and Wu, C. F. (1990). Reversible inhibition of endocytosis in cultured neurons from the Drosophila temperature-sensitive mutant shibire[ts1]. J. Neurogenet. 6, 191.

Cp O'Dowd, D. K., and Aldrich, R. W. (1988). Voltage-clamp analysis of sodium channels in wild-type and mutant Drosophila neurons. J. Neurosci. 8, 3633.

Cp O'Dowd, D. K., Germeraad, S. E., and Aldrich, R. W. (1989). Alterations in the expression and gating of Drosophila sodium channels by mutations in the para gene. Neuron 2, 1301.

Ct Petersen, N., and Mitchell, H. K. (1981). Recovery of protein synthesis after heat shock: prior heat shock treatment affects the ability of cells to translate mRNA. Proc. Natl. Acad. Sci. U.S.A., 78, 1708–1711.

M Ramain, P., Bourouis, M., Dretzen, G., Richards, G., Sobkowiak, A., and Bellard, M. (1986). Changes in the chromatin structure of Drosophila glue genes accompany developmental cessation of transcription in wild type and transformed strains. Cell 45, 545–553.

Cp Rezzonico-Raimondi, G., and Ghini, C. M. (1962). New observations on cells of Drosophila melanogaster cultivated in vitro. Dros. Inf. Serv., 37, 122.

Ct Ritossa, F. (1962). A new puffing pattern induced by temperature shock and DNP in Drosophila. Experientia 18, 571–573.

B Rosay, P., Colas, J. F., and Maroteaux, L. (1995). Dual organization of the Drosophila neuropeptide receptor NKD gene promoter. Mech. Dev. 51, 329–339.

F Schweitzer, R., Howes, R., Smith, R., Shilo, B. Z., and Freeman, M. (1995). Inhibition of Drosophila EGF receptor activation by the secreted protein Argos. Nature (London), 376, 699.

Cp Solc, C. K., and Aldrich, R. W. (1988). Voltage-gated potassium channels in larval CNS neurons of Drosophila. J. Neurosci. 8, 2556.

Cp Solc, C. K., and Aldrich, R. W. (1990). Gating of single non-Shaker A-type potassium channels in larval Drosophila neurons. J. Gen. Physiol. 96, 135–165.

Cp Solc, C. K., Zagotta, W. N., and Aldrich, R. W. (1987). Single-channel and genetic analyses reveal two distinct A-type potassium channels in Drosophila. Science 236, 1094–1098.

CP Suzuki, N., and Wu, C. F. (1984). Altered sensitivity to sodium channel-specific neurotoxins in cultured neurons from temperature-sensitive paralytic mutants of Drosophila. J. Neurogenet. 1, 225.

Cp Wu, C. F., Sakai, K., Saito, M., and Hotta, Y. (1989). Giant Drosophila neurons differentiated from cytokinesis-arrested embryonic neuroblasts. J. Neurobiol. 21, 499.

Cp Yamamoto, D., and Suzuki, N. (1989a). Characterization of single nonactivating potassium channels in primary neuronal cultures of Drosophila. J. Exp. Biol. 145, 173.

Cp Yamamoto, D., and Suzuki, N. (1989b). Properties of single chloride channels in primary neuronal cultures of Drosophila. Biochim. Biophys. Acta 986, 187.

Index